Dislocations in Solids

Volume 11

Dislocations in Solids

Volume 11

Edited by

F. R. N. NABARRO

School of Physics
University of the Witwatersrand
Johannesburg, South Africa

and

M. S. DUESBERY†

Fairfax Materials Research Inc.
Springfield, VA, USA
(deceased)

N·H

2002
ELSEVIER
Amsterdam • Boston • London • New York • Oxford • Paris
San Diego • San Francisco • Singapore • Sidney • Tokyo

ELSEVIER SCIENCE B.V.
Sara Burgerhartstraat 25
P.O. Box 211, 1000 AE Amsterdam, The Netherlands

First edition 2002

Library of Congress Cataloging in Publication Data
A catalog record from the Library of Congress has been applied for.

British Library Cataloguing in Publication Data
A catalogue record from the British Library has been applied for.

ISBN: 0-444-50966-6 (volume)

ISBN: 0-444-85269-7 (set)

⊗ The paper used in this publication meets the requirements of ANSI/NISO Z39.48-1992 (Permanence of Paper).

Printed in The Netherlands

Preface

As Cottrell argues in this volume, work hardening is perhaps the most intractable of the remaining problems of modern physics. It is also one of great economic importance. No wonder, then, that theorists who have a special understanding of one of the mechanisms involved have tended to underestimate the importance (perhaps under different conditions) of other mechanisms, leading to what Friedel calls "repeated Homeric disputes at international meetings", and the situation in which the topics involved "lost some of their importance, owing to the mere emphasis of the contestants".

When M.S. (Mike) Duesbery joined me as editor of this series, we realized that the powerful new experimental techniques of diffraction and the modern possibility of massive computations could well lead to a resolution of some of the old disputes, and we decided to devote a volume solely to this topic. We thought it would be useful to invite three of the pioneers of the field, Cottrell, Friedel and Hirsch, to comment on the contributions, and to give the authors an opportunity to reply briefly to their comments. The process turned out to be time-consuming, and in the end not all the contributors had the opportunity to reply to Hirsch's comments.

Sadly, during this period, Mike Duesbery fell ill, and he passed away on 21 June 1999.

Born in England, on 20 May 1942, Duesbery studied at Downing College, Cambridge, and completed his Ph.D. under Hirsch. He then moved with Hirsch to Oxford, and after two years joined Basinski at the National Research Council of Canada in Ottawa. After a time the direction of research in that laboratory changed, and in 1987 Duesbery moved to Washington D.C. Initially he belonged to the U.S. Naval Research Laboratory, but in 1990 he and Norman Louat founded the independent consultancy of Fairfax Materials Research Inc.

Duesbery's largely computational research was guided by his strong experimental background, and was principally concerned with the way in which the detailed atomistic structure of the core of a dislocation governs its mobility. With Basinski and Taylor, he analyzed the three-fold splitting of the core of the screw dislocation in a b.c.c. metal, which controls the differences between the plastic behaviours of b.c.c. and f.c.c. metals. Then, with Basinski, Vitek and Bowen, he examined the influence of components of the applied stress other than the Schmid driving stress. These were pioneering studies. Probably his most remarkable achievement was the three-dimensional atomistic analysis of the formation of kinks on screw dislocations in b.c.c., a work well ahead of its time. It was followed by another fundamental analysis, the mechanism of cross slip in f.c.c. To these were added many other contributions to the theory of crystal plasticity.

Personally, Duesbery was a genial and cooperative colleague; he and his wife Wendy offered their friends a most hospitable home.

F.R.N. Nabarro

Commentary. A brief view of work hardening

A.H. Cottrell

Department of Materials Science and Metallurgy
University of Cambridge
United Kingdom

1. A hard problem

It is sometimes said that the turbulent flow of fluids is the most difficult remaining problem in classical physics. Not so. Work hardening is worse. Admittedly, the great complicating feature of fluid flow – the presence of non-linear inertial energy terms in the basic equations – is ordinarily absent from work hardening. Indeed, a crucial assumption in all standard theories of plastic deformation is that dislocations move 'non-relativistically', implying that when accelerating between separate obstacles they do not accumulate and store enough kinetic energy to help them significantly in overcoming these obstacles. In other words the entire theory is quasi-static. If there were a significant dynamic effect of fast dislocation motion, crystal plasticity would show many features very different from those actually observed.

That is about the only simplifying advantage work hardening has. Whereas fluid dynamics can be treated by continuum methods, so that everything can be reduced to the purely mathematical problem of solving standard differential equations, there is no similar escape in work hardening, for the discrete structures of dislocations render the theory intrinsically atomistic, even though in their lengthwise dimension dislocations are macroscopic objects, governed mainly by a classical physics which is unusual in being not reducible to continuum theory. Moreover, very little of the mechanical work done remains stored in the dislocated structure. Of course, as Kubin, Fressengeas and Ananthakrishna (this volume) have emphasized, work hardening will eventually have to be expressed in terms of partial differential equations since they are the only language of field theories. Another unusual and extremely complicating feature is of course that dislocations are lines, not the familiar point particles of mainstream physics – and flexible lines at that – so that the standard methods of particle theory are inapplicable. The only other area of classical physics with even remotely similar features is that of soap froths and this is of little help, being so weakly developed. Furthermore, neither of the two main strategies of theoretical many-body physics – the statistical mechanical approach; and the reduction of the many-body problem to that of the behaviour of a single element of the assembly – is available to work hardening. The first fails because the behaviour of the whole system is governed by that of weakest links, not the average, and is thermodynamically irreversible. The second fails because dislocations are flexible lines, interlinked and entangled, so that the entire system behaves more like a single object of extreme structural complexity and deformability, that Nabarro and I once compared to a bird's nest, rather than as a set of separate small and simpler elementary bodies. Of course, the properties of single dislocations have long been well-established: their Burgers vectors; edge, screw and jog orientations; elastic fields and energies; glide, climb, cross-slip and sessile kinematic modes; reactions, combinations and dissociations; Frank–Read sources; etc. These properties provide the alphabet in which the story of work hardening must be written. But only the alphabet, no more than that.

In the light of the above, people have not surprisingly sought other general principles to provide a unifying framework for a work hardening theory. The formation of cell structures (see below) has encouraged comparison with other cellular formations, such as convection cells in fluids, heated from below. This has been criticized on the grounds that these self-organized structures are far from equilibrium, whereas work-hardened dislocation structures are not. It is evident that in a heavily dislocated crystal, deforming at moderate rates, most dislocations are at rest for most of the time and so are in a temporary and local state of mechanical equilibrium. In fact, the evolution of a plastically deforming crystal could be described by the biological term 'punctuated equilibrium', long rest periods of dislocations being interrupted occasionally by brief punctuations of energetic activity.

There is merit in the comparison with self-organized structures, however. What triggers the convection cells is mechanical force – just as it triggers plastic yielding – in this case the buoyancy force. The large heat flows are merely the means for producing the density gradient which generates this force. There are other examples, more directly similar to plastic failure, such as the buckling failure of a compressed strut. The self-organization which emerges from such failures occurs because it is the long-wave modes of the system which give way first to the triggering forces and these naturally produce those large-scale heterogeneities of structure which we regard as 'organized'.

The other general principle, introduced by Kuhlmann-Wilsdorf (cf. this volume), is the LEDS (low-energy dislocation structures) principle. In essence, this recognizes that dislocations in a multiply-dislocated crystal will adjust their relative positions so as to screen out one another's long-range stress fields, which would otherwise be a source of high elastic energy. Thus, a single dislocation will attract another of opposite sign, to form a close pair. If screws they may even annihilate by cross slip. If edges they will form a closely bound dipole. The LEDS principle applies similarly to large dislocation groupings, such as pile-ups and braids. The general rule is that, if a Burgers circuit round such a group reveals an uncompensated net Burgers vector, the material is expected to adjust the positions of nearby dislocations and even make new ones, for example by activating Frank–Read sources, so as to screen out the long-range field of the group. Similarly within the group, where small-scale adjustments produce local screening. Again, tilt and twist boundaries obey the LEDS principle. Moreover, they do not terminate within the crystal since a large circuit round them would then show them to be superdislocations. They must either run through to a free surface or form closed cells with no uncompensated net Burgers vector. These LEDS conclusions represent an ideal, of course, which a real crystal approaches, imperfectly. In terms of punctuated equilibrium, these two general principles appear as complementary, not contradictory. The self-organizing instabilities are responsible for the punctuations, while the LEDS principle takes care of the equilibrium in between them.

2. *The classical evidence*

Even with such guiding principles the complexities of work hardening defy pure theoretical reasoning. The only practical way forward has been through experiment, to identify the key features involved. The alphabet theory can then be applied, in a much more modest way, to try to make sense of them. Because of this, the subject divides into two stages. The earlier 'classical' one covered all the experimental evidence gathered before the application of transmission electron microscopy (TEM). The 'modern' period is dominated by what the electron microscope, aided by related methods such as etch-pitting, has told us.

A promising start to the theory of work hardening was made many years ago, inspired by observed stress–strain curves on single crystals of soft pure metals particularly f.c.c. ones such as copper and aluminium. Oriented for single glide these show, after an initial short period of 'easy glide' – i.e. Stage I in which the hardening is very slight, at a rate of only about $10^{-4}\mu$, where μ is the shear modulus – a smooth entry into the main Stage II where the flow stress rises linearly with strain at a rate of about $\mu/300$. Thereafter, at some higher stress, which is reduced at higher temperatures, the hardening rate progressively diminishes, giving an approximately parabolic curve in this Stage III. Later Stages IV and V have also been identified.

A strong clue to the basis of Stage II was given by the observation that this stage was entered immediately, with no Stage I, in crystals oriented for double or multiple glide. Equally significantly, in crystals of c.p.h. structure such as magnesium, zinc and cadmium, in which single glide only is possible, on the basal planes, there was only Stage I and no strong work hardening. All this meant obviously that Stage II hardening was due to the mutual interference and obstruction of dislocations moving on intersecting glide systems, with different Burgers vectors, a conclusion that was supported by the realization that such dislocations could combine to form sessile obstacles. The 'overshooting' effect of latent hardening also showed convincingly that dislocations in one family of slip planes offer strong resistance to dislocations attempting to cut through them on an intersecting family of planes. A very simple picture of Stage II hardening was thus possible, analogous to road traffic when the movements of vehicles are halted in traffic jams by an accident between differently moving ones at a cross-roads.

But it left numerous questions unanswered. What triggers the movement of dislocations on the secondary systems which then block those on the primary glide system? How do the obstacles, if so formed, stop the progress of an entire slip band, presumably all the way back to its source? What determines the values of the work hardening coefficients? How does thermal energy play its part, leaving the work hardening coefficients and the Stage I–II transition apparently unchanged, but promoting an earlier Stage II–III transition? How is the Cottrell–Stokes law (that the temperature dependence of the flow stress for a given state of work hardening is, over a wide range of temperatures and such states, independent of the extent of the work hardening) explained?

There are also the problems of work softening, notably the sharp yield drop which occurs at the beginning of plastic flow at an upper temperature after work hardening first at a lower temperature, showing that the low-temperature work hardened state is unstable under flow at the higher temperature. Furthermore, and even more striking, continued plastic deformation at this higher temperature can raise the flow stress eventually above that at which the yield drop began, which shows that a state of work hardening cannot be adequately represented by a single numerical parameter such as the density of dislocations. The flow stress of a work hardened material must therefore be a sophisticated function, at least two-parametered, of the dislocation pattern belonging to this state. The obvious deduction is that the hardening depends on the distribution of the dislocations, i.e. on a heterogeneity of the structure, as well as on the dislocation density.

The study of the thermal effects has provided other insights into work hardening. It has long been known, of course, that crystal plasticity is much the same at near-zero temperatures as at, for example, room temperature. Nevertheless, there are significant differences. A crystal of, for example, aluminium, if strained into Stage II at room temperature, then unloaded approximately to 'freeze' its dislocated state, thereafter cooled to near-zero and finally reloaded, will recommence its plastic deformation at a flow stress about 50% higher than before. The significance of this is that it is known, from alphabet properties, that a dislocation segment can overcome a localized obstacle with a reasonable frequency (e.g. 1/s) only if the activation energy for this is no more than about 30 kT, which at room temperature is about 0.75 eV. But dislocation energies only fall to as low as this when they relate to atomically-sized events, not the mesoscopic ones characteristic of long-range interactions. It thus follows that the observed temperature dependence of the flow stress indicates, importantly, that major obstacles contributing to work hardening are highly localized, on an atomic scale.

This, together with the fact that strong hardening occurs when dislocation lines on intersecting glide systems meet and interact, leads directly to the forest theory in which work hardening is due to the glide dislocations having to cut through the cores of forest dislocations, whose lines thread the glide plane like tree trunks, obstacles of only a few atomic diameters across.

Observations show that the flow stress, measured as a fraction of the shear modulus, becomes independent of temperature above about room temperature (until much higher temperatures, where new processes appear), at a value which in f.c.c. metals remains a large fraction of the zero-point flow stress. This has led to the view that the total work hardening has two components, a short-range T-dependent one and a long range T-independent one. The Cottrell–Stokes behaviour would then require these to maintain the same ratio, for all work hardening at any given temperature, which is puzzling. A simpler view is that both are produced by the same obstacles, which Nabarro [1] has shown can be explained by assuming a reasonable shape for the energy 'hill' which represents an obstacle. The activation energy then falls within the grasp of thermal fluctuations only so long as the applied stress reaches up towards the zero-point flow stress. At lower stresses the hill presents an obstacle much too large for 30 kT.

This further strengthens the case for forest hardening as the basis of Stage II. As regards the rate of work hardening in this stage it follows from alphabet properties that the increase in flow stress, $\Delta\sigma$, should go as

$$\Delta\sigma = \alpha\mu b\sqrt{\rho}, \quad (1)$$

where b is the Burgers vector length, ρ is the density of forest dislocations and $\alpha \simeq 0.3$, which is in excellent agreement with observation over a wide range of densities (Saada and Veyssière, this volume).

The weak hardening of Stage I raises the important question of the origin of the obstacles which eventually lead into Stage II. At the beginning of plastic deformation, dislocations move out from their sources under an applied stress marginally greater than the 'friction stress' which, in soft f.c.c. metals, is contributed mainly by impurity atoms but also slightly by vacancies and the occasional grown-in dislocation. Statistical variations in the distribution of these defects produce locally hard regions where gliding dislocations may be brought temporarily to rest. The stress field of such a one, sometimes reinforced by that of a follower from the same source, may then trigger a nearby source into activity, especially if this source is of the same slip system and lies immediately in front of it, where the dislocation stress acts with the applied stress. Sources on other, inclined, slip systems are less likely to be activated since they carry smaller resolved components of the applied stress in a crystal oriented for single slip. We thus expect a forward source to become active and so feed a dislocation of opposite sign back to the obstructed one, while also sending the forward section of its created loop still further ahead. The back-going dislocation then usually forms a dipole with the obstructed one, in which each screens the long-range field of the other.

This dipole is not fixed to the lattice and so can be pushed along by a third dislocation, for example coming from the first source. However, the friction stress now acts on all three, whereas there is only one uncompensated dislocation to push them along. A higher applied stress is needed or, more probably, further uncompensated dislocations from that source. But as these pile into this embryonic obstacle their internal stresses are expected to trigger more dislocations from the nearby sources, with back-going ones then forming more dipoles, so that the obstacle grows in size, complexity and resistance. The ensuing pile-up stresses, which also then intensify, eventually reach the level at which secondary sources, on other slip systems, are triggered to feed their dislocations into the obstacle. The obstacle now becomes fully immobilized by the interactions between these intersecting dislocations which have mutually incompatible glide directions. At this stage such obstacles have grown so strong that large increases of applied stress are required to keep the plastic flow going. Stage I is ended and Stage II begun. A lesson from this is that, once an obstacle begins to form, however weak initially, the efforts of the material to screen its stresses lead inevitably to its enlargement and strengthening. Obstacle strengthening is thus autocatalytic. This might be regarded as a third general principle of work hardening, the autocatalytic principle, alongside the self-organization and LEDS principles.

This picture of Stage I is basically similar to that of Seeger (cf. Zaiser and Seeger, this volume).

3. *Electron microscopic evidence*

When the electron microscope began to show dislocations it seemed that the work hardening problem would soon be solved simply by directly seeing the dislocation structures of work hardened states. All that theory would then have to do would be to link these structures to the basic properties of dislocations and to estimate the flow stresses of the structures. But it has not gone at all like that. During plastic straining dislocations are produced in such vast numbers that it is difficult to observe the genesis and movement of individuals. More importantly, the observed patterns, which have now been documented in detail, are not at all what might have been expected from the classical approach, which failed completely to predict them. The problem has thus grown larger and become one of explaining both the mechanical and microscopic (including also etch-pit studies) observations and of reconciling them.

Prominent in the observed structures are the *braids*, which begin in Stage I and lead into Stage II. They consist of edge dislocation dipole bundles, spread out raggedly over several layers of the primary glide plane. The opposite components of the dipoles evidently come from adjoining slip bands which meet, approximately 'head-on', in the region where a braid is formed. In Stage II the braids include dislocations of secondary glide systems. As the hardening goes into Stage III the braids develop into closed *cells*, the spacing of which then decreases inversely with the increasing flow stress, which is Kuhlmann-Wilsdorf's principle of similitude. At still higher stresses (Stage IV) the ragged dipole structure of the cell walls gradually disappears, leaving these walls then more sharply defined by geometrically necessary dislocations. Consistent with the braid structure, the slip bands are not confined to single lattice planes, but resemble thin ellipsoids, spread out with their major axes along a set of adjoining lattice planes and with a minor–major axis ratio of about 1: 50, as emphasized by Brown (this volume). This spread allows the orientation of the slip bands to deviate slightly from that of the crystallographic glide plane, by the order of 1 degree of angle. The interiors of the bands are remarkably free of dislocations, apart from a few long straggling ones, but the faces of the bands are carpeted with thin networks dislocations in which sessile Lomer–Cottrell locks are abundant.

4. *Heterogeneity*

The general feature exposed by the electron microscopic observations is that work hardened dislocation structures are heterogeneous on a mesoscopic scale. This set two new problems for the theory: to deduce the flow stress from the starting point of a given observed heterogeneous structure; and to explain the formation of these structures and their evolution with plastic strain. The first of these has proved to be the easier and more fully developed.

Of the several theories of the flow stress, as recounted in this volume, that of Mughrabi (cf. Mughrabi and Ungár, this volume) stands out for its simplicity and firm experimental foundation. Its key feature is that there are, basically, two different kinds of region in the heterogeneous structure: the interiors of braids and cell walls, where the local dislocation density is very high: and the larger regions between them, where observation shows that the density is remarkably low. It follows, from eq. (1), that the 'local' flow stress is much higher in the first regions than in the second. Mughrabi then adopts an analogue with composite materials that are made up of hard and soft components, which is to take the overall flow stress as a simple average of the two component flow stresses, weighted according to the volume fractions. This successfully gives the overall flow stress with remarkably little theoretical assumption. Admittedly, eq. (1) was focused on the forest hardening theory, but even this particularization is unnecessary because the general form of this relation is common to many possible work hardening mechanisms, being derivable by purely dimensional reasoning.

Thus, structural heterogeneity leads to mechanical heterogeneity. It then follows that there must exist in the material a set of internal stresses which are balanced, positive against negative, on the same mesoscopic scale as the braid/cell structure. This is because the spatially constant applied stress lies below the local flow stress in the hard regions and above it in the soft ones; and so must be accompanied by an additional stress, the internal stress, which acts with it in the hard regions and against it in the soft ones, in both cases bringing the local driving stress to the local flow stress. The elastic strains corresponding to these internal stresses have been identified and measured by X-ray diffraction, giving important evidence in support of this mechanical heterogeneity. The internal stress in the hard regions is necessarily more intense than in the soft ones.

This stress is of course produced by obstructed dislocations. For example, one which is released in the soft region to the left of a wall glides freely until it reaches the left side of the wall where it is halted by the obstacles there. Its stress field then provides a back stress on its parent soft region, opposing the applied stress there, and a forward stress on the wall and region to the right of it. A compensating dislocation may then come in from the latter region and be halted on the right face of the wall, but the stress from it on the region to the left, although aiding the applied stress there, is weaker since it is further away, by the thickness of the wall. This picture of internal stress formation is quite similar to Brown's concept of stress formation in slip bands (cf. Brown, this volume).

Mughrabi's theory deals with the mesoscopic behaviour of obstructed dislocations and can be adapted to most elementary processes of obstruction such as forest hardening. In place of the assumed homogeneity of earlier theories we now have the same forest processes taking place in a heterogeneous dislocation distribution. There is an important new feature, however. Because the flow stress goes as the square root of the dislocation density, a crystal with a fixed average dislocation density becomes softer as these dislocations rearrange themselves into increasingly heterogeneous patterns. This follows from the assumption that the overall flow stress is a weighted average of the local flow stresses in the high-ρ and low-ρ regions; and from a generalization of the simple fact that the value of $\frac{1}{2}[(1+x)^{\frac{1}{2}}+(1-x)^{\frac{1}{2}}]$ falls further below its maximum, 1, as x increases from 0 (i.e. homogeneity) towards 1. The overall flow stress is thus not simply a function only of the average density, but also varies with heterogeneity, as the work softening experiments indicated. In terms of eq. (1), heterogeneity reduces the value of α.

At high temperatures, where rapid climb can annihilate edge dislocation dipoles, the heterogeneity softening is eventually overwhelmed by softening due to the reduction in dislocation density in the walls, which simplify down into small-angle cell boundaries. The overall flow stress then becomes only marginally greater that that of the soft regions.

5. *Stage III*

Stage III softening requires both stress and temperature. Stress, obviously, since it does not begin until the applied stress has climbed some way up the Stage II curve, although the gradualness of the transition may mean that it begins, imperceptibly, much earlier.

Temperature is involved since the transition occurs at markedly lower stresses at high temperatures. There appear to be two low-T softening processes; cross-slip of screw dislocations; and forest-cutting. The position on the annihilation of screw dipoles by cross slip is unclear. The traditional view is that the onset of Stage III is marked by the start of extensive cross slipping, but Brown (this volume) argues that such cross slip is already completed by the start of this stage and Saada and Veyssière (this volume) give evidence that there is some cross slip even in Stage I. These latter authors also show that cross slip appears to occur more easily and readily than expected from calculations of the activation energy required to make the constriction and transfer into the cross-slip plane. They conclude that jogs may be needed on the screw dislocation to facilitate its cross slip. Presumably these would be formed by forest cutting.

A glide dislocation at or in a wall feels three stresses driving it against the wall's obstacles. First, the applied stress itself. Second, the applied stress again, in the magnified form of the forward-acting Mughrabi stress. Third is another internal stress, this time an 'interior' stress due to the attractions of the corresponding glide dislocations of opposite sign on the other side of the wall, which is thus independent of the applied stress. Throughout Stage II these stresses help the obstructed dislocations to overcome their obstacles, aided in this by thermal energy fluctuations ($\simeq 30\ kT$), so enabling them to get closer to their opposite numbers. These two sets of approaching dislocations can thus be regarded as dipoles, very open at first but gradually closing up. As they do so, their long-range stresses diminish, through mutual screening, so that the backward-acting Mughrabi stress on the hinterland is also reduced, thus allowing more slip to start up there and throw yet more glide dislocations against the wall.

Through Stage II these processes go forward together, the driving stress and the average obstacle stress (which increases as the forest spacing decreases, for a constant resisting force from an obstacle) both rising in a running balance. But as the plastic deformation accumulates, throwing increasingly large numbers of glide dislocations at and into the walls, so the interior stress, pulling these dislocations into dipole configurations, increases and becomes a more dominating component of the driving stress. The running balance between driving and obstacle stresses can now be met without the need to increase the applied stress so much. The work hardening slope thus begins to fall below that of Stage II. Stage III has begun. At higher temperatures these same processes occur, but the greater thermal energy allows the fluctuations, about $30\ kT$, to carry a larger share of the burden of overcoming the obstacles and so Stage III begins earlier. At sufficiently higher temperatures where dislocation climb becomes rapid, the loss of dislocations by annihilation softens the wall material. The incentive for the outer dislocations to pull together towards the centre is thereby weakened but the ensuing diminution of the interior stress has a small effect on the flow stress, compared with that of this softening, so that Stage III weakening becomes more intense.

Yield-drop work softening fits in with this behaviour [2]. After a preliminary work hardening straining at a low temperature and then unloading, the material, when reloaded at a substantially higher temperature, starts off in the same mechanical state as at the end of the prestraining. But because $30\ kT$ is now much greater, obstacles which were impassable at the lower temperature, so preventing the dislocations coming together, can now be overcome. There is a consequential closing together of dislocation dipoles in the walls and ensuing reduction of their back stresses in the hinterland. Vigorous plastic flow can thus resume there, giving the yield drop. After then working through and beyond the Lüders strain, the density of dislocations in the walls builds up to a higher level, so eventually raising the flow stress at this temperature above that of the yield drop, the material now no longer being in the mechanically unstable state of the yield drop, but in the normal work hardened state characteristic of this temperature and amount of plastic deformation.

6. *Dislocation patterning*

How are the dislocation structures, as shown by electron microscopy and etch pitting, formed? And what has happened to Frank–Read sources and pile-ups, those stalwarts of the classical theories? A tenuous link between the classical and modern phases can perhaps be seen in Stage I deformation. The steps formed where slip lines emerge at a free surface continue to provide strong evidence for Frank–Read sources, at least before secondary slip begins and forests form. And the dipoles which begin in Stage I can be understood in terms of the interactions of dislocations approaching on parallel planes, thereby eliminating uncompensated pile-ups with their long-range stress fields, in conformity with the LEDS principle. But after Stage I "experimentally it is difficult and largely impossible to connect the two, slip line and dislocation structures" (Kuhlmann-Wilsdorf, this volume). Again, "after dislocations have begun to interact... among each other... at the end of Stage I... multiply acting dislocation sources" (i.e. Frank–Read sources) "are not expected, nor are they ever observed". Presumably, the network of entangled dislocations then provides singly-acting sources, in the form of sections of dislocation lines on many potential slip planes. The heavy jogging observed on some dislocations in Stage II could also help provide such sources. This source structure is consistent with the deduction, as emphasized by Brown (this volume), that slip bands develop into thin, elongated ellipsoids of finite thickness, in which slip seems to be distributed homogeneously over all the slip planes. The slight inclination of the major axes of these ellipses to the crystallographic slip plane remains a problem, however, which has been discussed by Brown.

Given that, in Stage II, slip can spread from plane to plane of the same family, by means such as the above, braids can develop into walls. A braid in, say, a horizontal primary glide system serves as an obstacle, not merely to the dislocations of its own slip planes, but also to those of neighbouring parallel planes just above and below it. Primary glide dislocations in these latter then also stop at the top and bottom regions of the braid, so extending it vertically, eventually into a wall.

If there is some rotation between the two hinterlands before and behind a wall, so that the wall also serves as a tilt or twist boundary, then this will favour the creation of other, elastically accommodating, boundaries, which convert the wall into a rotated cell without long-range stresses (Kuhlmann-Wilsdorf, this volume). At high temperatures where only the geometrically necessary dislocations survive the annealing processes, such cells become the familiar domain structure.

The final problem is to understand similitude; the fact that the scale of the mesoscopic heterogeneous structures, including the spacing of obstacles and size of dislocation cells, varies inversely with the flow stress. This has proved to be a difficult challenge. There are three major theories. In that of Brown (this volume) the rising forces on the dislocations in the walls, as work hardening increases, force some of these to become new sources of slip. The slip bands they then create push the sides of the walls apart, so multiplying their number. The theory of Kuhlmann-Wilsdorf (this volume) has an attractive simplicity. The 'turbulent' deformation of work hardened material inevitably produces local elastic bending on a mesoscopic scale. Where this becomes too severe it pays, in energy economy, to replace the bent lattice by a small-angle dislocation boundary. This is a kind of localized polygonisation. These boundaries are then obstacles around which new walls of dipoles, secondary slip, dislocation interactions and reactions, develop. The third theory, that of Nabarro [3], is a development from the Mughrabi theory. It starts with a homogeneous distribution of glide dislocations over the whole region of a potential cell, at a certain density. A perturbation is then considered which increases the density in one half of the cell and equally decreases it in the other half. The change in energy due to this heterogenisation is then calculated, in terms of the change in strain-field energy of the dislocations and the work done by the applied stress through their rearrangement. From this a critical cell size is deduced, above which the energy favours the change. The theory gives this size as varying inversely with the flow stress and also about 13 times larger than the spacing of the dislocations in their homogeneous distribution, both in agreement with experimental observations.

7. Conclusion

This has been a personal view of the work hardening problem, with no attempt to cover all the theories and opinions that festoon this subject. The chapters in this book give an excellent guide to these. It has also been a very limited view, with nothing on b.c.c. metals, or alloys, intermetallic compounds or non-metals. Nevertheless, it has focused on what is generally taken to be the central battleground of the subject, Stages I, II, and III of soft f.c.c. metal crystals. That a personal view is still possible, almost 70 years after the first dislocation theory of work hardening, is perhaps surprising, but the diversity to be seen in the chapters of this book make it inevitable. The theory is mainly still at the stage of merely being interpretative, not predictive. And it is very far from being a standard theory of classical physics, founded on an agreed set of axioms which lead to the master equations from the solutions of which all the basic features of the subject unfold.

It may never develop into such a theory. Work hardening is quite unique, unlike any other branch of physical science. While thoroughly physical in its elements and obviously belonging to the 'hard science' end of the inorganic world, its characteristics – huge populations of interacting individuals, competition and cooperation, organization and evolution of family groups – are more like those of biology. Ecology in particular. It may be that the theory of work hardening will always be more Darwin than Newton.

Acknowledgements

I am grateful to Professor Fray for making available the facilities of the Department of Materials Science and Metallurgy, University of Cambridge, during the course of this work.

References

[1] F.R.N. Nabarro, Acta Metal. Mater. 38 (1990) 161.
[2] A.H. Cottrell, Phil. Mag. Let. 81 (2001) 23.
[3] F.R.N. Nabarro, Phil. Mag. A80 (2000) 759.

Commentary

J. Friedel

2, rue Jean-François Gerbillon
75006, Paris
France

1. Introduction

This new volume in a well known series is devoted to problems of *plasticity of crystals*, and especially on work hardening processes in metals, alloys and some metallic compounds.

Plasticity is the context in which dislocations were first invoked in crystals in the 1920's. If the principle of their essential role is well established since the middle of last century, it is typically one of those fields where it is hard to go from the microscopic level (of individual dislocation glide, cross slip and climb) to the macroscopic one (of work hardening, recovery, creep and fatigue). It is therefore not surprising that this is still a very active field of research.

This volume covers the topic effectively, mostly from the fundamental point of view, but with some applications to superalloys.

It is clear however that the texts concentrate on *work hardening* in crystals, and especially *f.c.c. metals*, where Stage II provides, at least at low temperature, a strikingly large and constant work-hardening rate, responsible for both the characteristic plasticity and resilience of these materials. This topic in f.c.c. metals is treated in most chapters and is obviously a background of comparison for other crystalline materials. This might have been stressed in the title of the book; but the matter is obviously of paramount importance ever since the bronze age and not really settled, at least in detail.

This opportunity might also have been taken to review the role of short-range order in alloys, that of high-temperature self diffusion in metals and more generally what one understands of plasticity in non-metallic crystals (ionocovalent or molecular) or in nanocrystalline solids (amorphous, quasicrystalline, mesomorphic solids or sand heaps...). All these topics require more than the rapid allusions one finds here. The comparison with the main subject of this volume would be of interest, both as far as microscopic properties of dislocations or their equivalents are concerned and for mesoscopic instabilities such as slip lines, a general feature met in most cases.

2. Two fundamental questions

To come back to work hardening in f.c.c. metals, there seems to me to be two different fundamental questions.

2.1. The problem of hardening

It seems now fairly generally accepted that, at least in Stages II and III of single crystals under tension, the stress necessary to produce a given strain is related to some average dislocation density in the 'forest' that pierces the slip planes of the mobile dislocation responsible for the strain. As explained in detail by G. Saada in his thesis, the temperature independent relative stress σ/μ necessary for a mobile dislocation to cut across this forest in f.c.c. metals is essentially due to attractive trees, while repulsive ones merely provide a temperature dependent relative stress due to jog creation, which disappears at high temperatures. Crossing attractive trees also provides an effective mechanism for producing the many vacancies and interstitials created in work hardening [1].

This background, partly recalled in this volume, was made more obscure in the 1970's by repeated Homeric disputes at international meetings between A. Seeger and P.B. Hirsch. Looking back, the two topics involved have lost some of their importance, owing to the mere emphasis of the contestants!

2.1.1. Long-range stresses

The forest, as first imagined, was due to a three-dimensional dislocation network with various Burgers vectors; these dislocation loops were intuitively assumed to compensate their long-range stresses at distance large compared with the mesh of the dislocation network. It was in fact a direct extension of F.C. Frank's description of the mosaic structure. An increasing number of evidences showed however that in Stages II and III of tensile tests as in fatigue, f.c.c. single crystals showed a systematic bias in the dislocation networks developed. This had been well known for a long time in the Bauschinger effect, where, after plastic strain in a given direction, the elastic limit in the opposite direction is much smaller, in fact often very small. This could be formally described by the development of a biased structure with a long-range internal stress of the same order of magnitude as the direct plastic limit. This conclusion was later confirmed by shifts of X ray lines and, most directly, by Essmann's studies of bent dislocation loops observed under the electron microscope.

These long-range stresses, which are thus very real, were first interpreted as due to *dislocation pile-ups*, at the end of the slip lines which develop in increasing numbers during Stages II and III and in fatigue. This was in line with the early observations of dislocations in such pile-ups by etch pits in α brass by P.A. Jacquet, then by transmission electron microscopy in austenitic 18-8 stainless steel by P.B. Hirsch and his group. These pictures gave a simple description of slip lines in f.c.c. metals as loops produced by Frank–Read sources and piled up against obstacles such as Cottrell–Lomer locks. However, when I showed in 1953 P.A. Jacquet's pictures to N.F. Mott and F.C. Frank, they immediately retorted that, in pure three-dimensional metals, the long-range stresses of such pile ups should be effectively *relaxed* by the activation of secondary sources due to the mosaic structure. Indeed later work by P.B. Hirsch and many others showed that no clear pile ups could be related, in such ductile metals, to the ends of slip lines. Similar stress relaxations are met in all ductile crystals, at the tips of twin or martensitic lamellae or of cracks, as amply demonstrated e.g. in the many fracture studies following the accidents of the Comets.

In ductile f.c.c. metals, long-range stresses are then more likely related to *inhomogeneities* in the local density of dislocations, as observed by electron microscopy in most of Stages II and III of tensile tests and in fatigue. This is well explained in this volume by the major discoverer of this effect, H. Mughrabi. In the cellular structure thus developed, a polarization of the edges of the 'walls', with high dislocation density, produces a long-range stress in the volume of the 'cells', with low dislocation density, which compensates for their lower frictional stress. In fact, a simple model shows that the effective frictional force thus developed is not much below that expected from the average square root of the dislocation density of the structure: although present and explaining the bias responsible for the Bauschinger effect, long-range stresses do *not* affect, in such a case, the direct plastic limit, which remains essentially related to the average dislocation density.

The origin of the cellular structures developed in Stage III and in fatigue of f.c.c. metals is now clearly attributed to *cross slipping* of the screw part of the mobile dislocations.

This shows in the inspection of slip lines, but follows also from any detailed analysis of the cross slip. As shown by the work of B. Escaig and others, this is a process made easier by suitable applied stresses, both in the initial and cross slipping planes, and by the presence of jogs. Splitting of the core of dislocations makes the process more difficult, but less than initially computed by G. Schoeck and A. Seeger, who had assumed total recombination of the partials over a finite length of dislocation. The direct anticorrelation of the extension of Stage III at low temperatures with the dislocation splitting provides the best indication of cross slipping in that stage, whatever the details of its analysis.

One can understand that cross slipping in Stage III destabilizes the three-dimensional dislocation network and helps the edge parts to rearrange into low-energy walls of high density. The question is less clear for pure Stage II, i.e. temperatures which in pure metals such as Al, Ni or even Cu, are well below room temperature. Is there still a cellular structure? If so, would it be due to a destabilizing of the three-dimensional dislocations network by a general process of work softening to which we will come back in section 2.2? Clearly, more experiments at low temperatures are needed.

2.1.2. Dragging of jogs

This was, for a time, thought to be possibly an important factor in the frictional force opposed to moving dislocations. However, at least in f.c.c. metals, jogs seem to glide easily, with no great Peierls–Nabarro force. There should then be only a small and measurable dragging of the jog when the moving dislocation is near to screw orientation. The dragging force due to the climb of the moving jog produces a cusp on such dislocation that does not necessarily contain the Burgers vector of the jog, then inducing a climb of the jog. At low temperatures or high speeds, when the point defects created are not appreciably moving, this is equivalent to a trailing of a nearly edge dislocation dipole of atomic height from each tree of the forest with a screw component. By cutting across the forest, a mobile dislocation thus multiplies its dragging jogs, in much the same way as a tip of a crack produces steps of atomic height by crossing the 'forest'. In both cases, the dragging force however saturates: the jogs, created at random on the moving dislocation should induce pairs of neighbouring jogs closer than their average distance to move closer by an interaction of the cusps created on the mobile dislocation. They should then annihilate or double their height by coalescent climb. This is the equivalent of rivers of progressively fewer and higher steps created in the case of a propagating crack. This process is probably at the origin of the many 'debris' and trailing dipoles observed in plastically deformed ionic solids as well as metals. When high and long enough, such essentially edge dipoles could act as Frank–Read sources. The broadening of a slip line could also be explained, perhaps more easily than by double cross slip of a screw dislocation [2].

2.2. The problem of work-hardening rates

From what precedes, one expects the work-hardening rate to be related to changes with strain of the average dislocation density.

This is not straightforward to analyze, as the example of Stage II of f.c.c. metals shows. It is well known that, in that case, the macroscopic strain can be related to the *slip lines* observed to develop at the surface of the sample. Qualitatively, in the easy glide of Stage I, slip seems to concentrate on macroscopic slip lines, along the active slip system, while in Stage II the new slip lines developed have a length which shortens with increasing strain; in Stage III, neighbouring slip lines connect by cross slip. It is also remarkable that, at least in Stage II, the total macroscopic slip can be accounted for by the sum of the slip on each new slip line, as measured by its extent and its strain-independent height.

One can argue that the stress relaxation around the tips of slip lines in ductile metals adds no measurable macroscopic strain. One could also argue that the lengths of active slip lines should decrease in Stage II, when each slip line meets an increasing density of forest due to other slip systems. One can even understand that the work-hardening rate of Stage II in f.c.c. metals is higher than in b.c.c. ones, because of the many possibilities of creating Cottrell–Lomer locks in the f.c.c. structure. However the fundamental questions concerning slip lines remain to date unanswered:

1. What are the active sources?
2. Why do they rapidly produce slip lines of many atomic distances height?
3. Which exact process stops the progress of their tips?
4. What increase of forest density is produced by the stress relaxations at the tip of the slip lines?

It is tempting to try and apply to this problem the concepts evolved in cases of developing plastic instabilities such as the Portevin–Lechatelier effect in alloys recalled by L. Kubin et al.

Indeed the concentration of plastic deformation on some particular slip planes is certainly helped in many cases by the fact that the first slip over an atomic length is often more difficult than further slip along the same plane. This is true for crystals with dislocations blocked by impurities, but also for crystalline alloys with short-range order, quasicrystals, amorphous structures or sand heaps. In pure and ductile metals, one can wonder whether a similar effect could come from a rearrangement of the three-dimensional network of dislocations, when a mobile dislocation – or a whole piled up group – crosses it. Nobody has properly considered this question, either experimentally or theoretically. But one might well think of a *work softening process* due to such rearrangements.

It is indeed not very reasonable to treat the 'forest' as a rigid body, if one takes into account the rearrangements due to the cutting of attractive trees by a mobile dislocation. These rearrangements could then lead to a thinning down of the forest in the neighbourhood of the slip line; or more likely the instability produced on the three-dimensional network under plastic strain might lead to three-dimensional cells of lower dislocation density. Thus the cellular structure reported in Stage II might not be related here to cross slipping processes. This is a field of investigation in metals with well developed Stage II, such as Cu well below room temperature.

It is also clear that the development and evolution of such cellular structures in Stage II might provide a better understanding of the Frank–Read sources responsible for the nucleation of new slip lines. Here too, there is, in well developed Stage II, a total lack of investigations so far.

3. Conclusion

In conclusion, this volume presents, as it should, a very balanced view of the advances and remaining problems in work hardening of metals. May its reading stimulate the interest in this essential part of materials science!

References

[1] Most references in this paper can be found in the chapters of this volume or in J. Friedel, Dislocations (Pergamon Press, London, 1964).

[2] As pointed out by Nabarro to the author, E.J.H. Wessels and F.R.N. Nabarro (Acta Metallurgica 19 (1971) 915) have considered another work softening process in crystals with widely split dislocations, where the necessary constriction to produce an elementary jog is larger in energy than for lengthening that jog into a multiple one.

Commentary

P.B. Hirsch

Department of Materials
University of Oxford
Parks Road, Oxford, OX1 3PH
UK

1. Introduction

On reading through this excellent set of review articles, one is impressed on the one hand by the wealth of new knowledge, ideas and approaches, by the extent of the clarification which has occurred over the years, particularly by the development of the 2-phase model, and on the other by the fact that detailed dislocation modelling of work hardening generally still falls short of the requirement of fully validated models which can be incorporated in constitutive descriptions of plasticity. My comments will be confined to the contributions on linear work hardening in f.c.c. single crystals, and on work hardening in single crystals of $L1_2$ alloys.

2. Work-hardening rate in Stage II of face-centred cubic crystals

2.1. Introduction

Models to explain the rate of linear work hardening in f.c.c. crystals are discussed in the chapters by Zaiser and Seeger (chapter 56, [1]), Brown (chapter 58, [2]), and Kuhlmann-Wilsdorf (chapter 59, [3]). Mughrabi and Ungar's article (chapter 60, [4]) focuses on the flow stress on the composite model, rather than on the work-hardening rates. In Kuhlmann-Wilsdorf's model of linear work-hardening, the slip line length, a key parameter in any work-hardening theory, is introduced from experiment. The following remarks will therefore be confined to the models presented by Zaiser and Seeger and Brown, in which all the parameters are estimated from physical principles. These two models represent good examples of the so-called "long-range stress" and "forest" theories of work hardening, which were the subject of vigorous debate in the 1960s, as related in chapters 56 and 59. Some comments will also be made about the relationship between these two models and an earlier attempt in 1964 to treat the screening of long-range stresses from terminating slip bands [5,6].

So, what exactly is the difference between the Zaiser and Seeger [1] and Brown [2] models which leads to such rather different explanations of Stage II work hardening? The answer lies in the assumptions made about the nature of the slip bands. Zaiser and Seeger

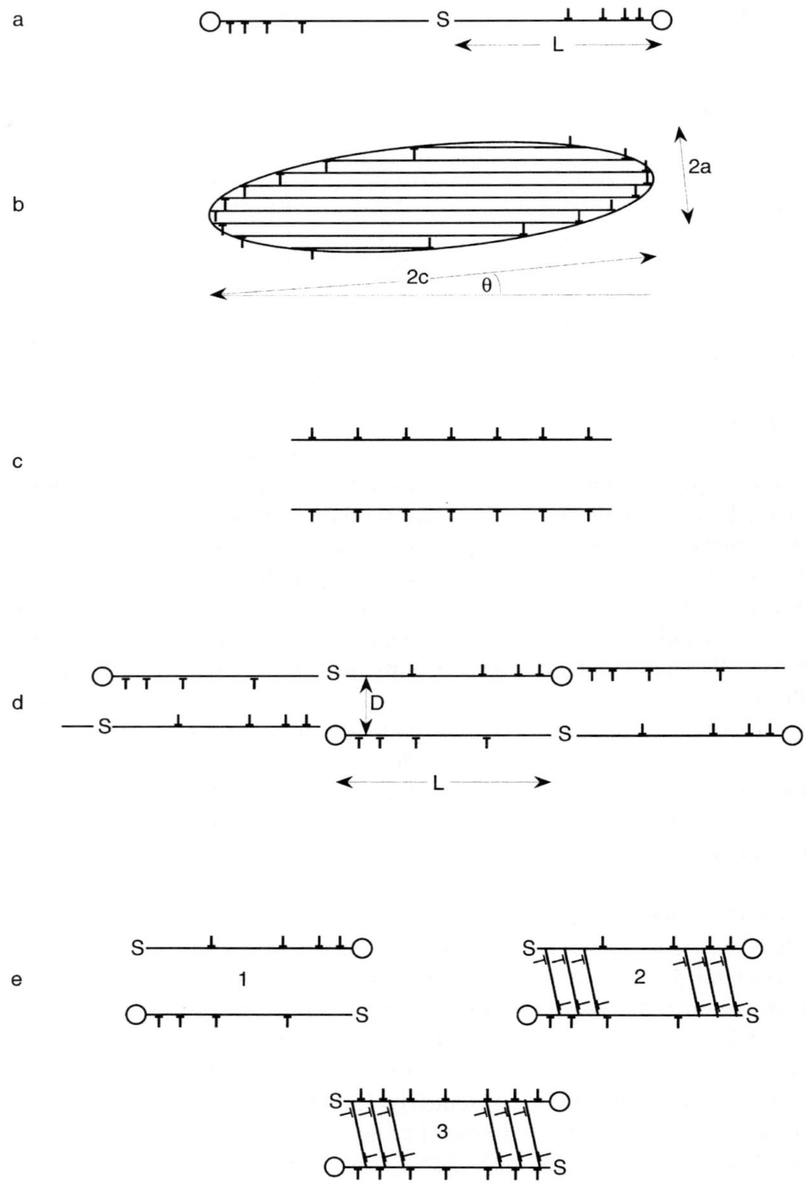

Fig. 1. (a) Slip band on the classical Seeger model, consisting of groups of dislocations emitted by a source piled-up at obstacles, generating long-range stresses. (b) Elongated ellipsoidal slip band model of Brown [2] which is characterised by low elastic energy and Eshelby accommodation factor. (c) Dislocation multipole. (d) Cooperative operation of dislocation sources on parallel planes, producing dislocations of different sign on top of one another. (e) Transformation of overlapping piled-up groups of dislocations of opposite sign into approximate multipole structures by the action of forest dislocations generated to reduce the internal stress. The steps are in the order 1, 2, 3.

assume that the individual slip events consist of the collective motion of n dislocations of opposite sign moving away from each other, resulting in groups of n dislocations of the same sign terminating in the crystal (fig. 1(a)). The n dislocations do not have to be confined to one plane as assumed in the original Seeger model [7]. The crucial assumption is made that the individual slip events are statistically independent. This results in the superposition of the long-range tails of the stress fields of groups of n dislocations within a radius of the order of the slip line length. According to Zaiser and Seeger, such long-range stresses can be relieved only by secondary slip on the same scale, which is contrary to experimental evidence. These long-range stresses then control a substantial part of the flow stress in Stage II.

On the other hand, in Brown's model [2] the independent slip events (the slip bands) are assumed to consist of collective motions of equal numbers of dislocations of opposite sign, arranged in uniformly sheared ellipsoids elongated along the slip direction, $\boldsymbol{b}$ but tilted by 1–2° relative to this direction, in a plane normal to the primary glide plane and parallel to $\boldsymbol{b}$. The ellipsoid arrangement ensures uniform stress within the slip band [8]. Because of the elongated shape the Eshelby accommodation factor γ [8] is very small, and the shear in the slip band is accommodated easily in the matrix. The total elastic energy in the matrix and the slip band, the elastic shear stress in the slip band, and the long-range stresses in the matrix are all very small. The slight tilt results in either tensile or compressive stresses within the slip bands which do not affect slip on the primary system, which are then largely relieved by forest slip across the band. The microstructure of the deformed crystal is then assumed to consist of randomly close packed arrangements of equal numbers of ellipsoidal slip bands of both tilts such that any remaining internal stresses within the bands cancel out. Figs. 1(a), (b) show schematically the geometries of individual slip bands assumed in the two models. In Brown's model the remaining stress concentrations at the tips of the bands extend only over a distance of the order of the minor axis of the ellipse, $2a$, i.e. the distance between neighbouring planes of high dislocation density. In Seeger's theory these planes of high dislocation density would be neighbouring pile-ups.

Because of the slight angular tilt, the dislocation arrangement in Brown's slip band is similar to a multipolar group on closely spaced slip planes (spacing $2a$), over a major fraction of the length of the slip band. The average shear stress parallel to the primary plane in a multipole would be zero. Outside the multipole the stresses decrease rapidly with distance beyond $\sim 2a$, because the contributions from the dislocations of opposite sign cancel out. In Brown's model it is then assumed that the flow stress is controlled by the forest dislocations at the external interfaces of the slip bands.

At the present stage of the theoretical development of work-hardening theories, it is still necessary to assume a particular structure which fits best the observed dislocation distributions. It is not possible to predict generally from first principles which structure would evolve during the deformation. It is therefore necessary to make a "reality check" on the proposed models, i.e. an assessment of which model approximates most closely to experimental observations.

One of the striking characteristics of the observed dislocation microstructure on the meso-scale in Stage II is the existence of carpets approximately parallel to the primary slip planes, stacked parallel to one another somewhat irregularly, and separating relatively dislocation-free bands, whose misorientations relative to each other alternate in sign.

There are many examples in the literature, and references to the pioneering work of Steeds, Essmann, and Basinski and others are given in the various chapters. The observed structures are also described in previous reviews [9,10]. An example for a single crystal deformed into Stage II is given in chapter 59, fig. 35, which refers however to the b.c.c. crystal of niobium deformed into Stage II taken from the work of Foxall, Duesbery and Hirsch [11]. Such structures are likely to originate initially from correlated slip events of dislocations of opposite sign, similar to those which might be expected from models of the type suggested by Brown and Hirsch and Mitchell (fig. 13 of [6]); i.e. from simultaneous or successive slip events which are correlated to produce dislocation structures in which the dislocations of opposite sign are arranged in adjacent arrays approximately parallel to the primary slip planes. Because such multipole type arrays can be converted by short-range forest slip on the conjugate CI and critical systems A VI into the observed structures with alternating tilts, we believe that these correlated slip events involving positive and negative dislocations in adjacent arrays of a Brown type model capture more realistically the essence of the deformation process than the model of Zaiser and Seeger. Brown's model will now be discussed in the next section.

2.2. Brown's model for linear work hardening

The tensile and compressive fibre stresses arising from the tilts of the ellipsoidal bands relative to the primary plane are reduced by forest slip. After relaxation the remaining fibre stress and the forest dislocation spacing depend on the spacing of the primary dislocations; the local flow stress of the primary dislocations in the soft parts of the band is in turn controlled by the forest spacing at the ellipsoid interfaces. The calculation presented in the paper assumes that the stress relaxation is driven entirely by the internal stress, and that the effect of the applied stress on the forest systems is negligible.

In Brown's model the work-hardening rate of an individual slip band is proportional to the angle θ made by the ellipsoid major axis with the primary slip plane. This angle plays a crucial role in the model. θ is assumed to be constant during Stage II, and from considerations of the packing of tilted ellipsoids is put equal to c/a (c and a are the major and minor semi axes of the ellipsoid). The ratio c/a is determined by a condition that the stress concentration at the ends of the slip band must be equal to the local forest stress, σ_{loc} at the obstacle, i.e. according to eq. (4.2) (chapter 58)

$$\sigma\sqrt{\frac{c}{a}} = \sigma_{\text{loc}} = \sigma/f, \tag{1}$$

where f is the volume fraction of the obstacles, σ the applied stress. The following points arise:

(i) The model is essentially 2-dimensional, i.e. the ellipsoid in fig. 1(b) is an elliptical cylinder, whose axis is parallel to the edge dislocations. But in practice the slip bands are finite along that direction. The fate of the screw dislocations terminating the slip band is not considered.

(ii) As Professor Brown has pointed out, no ellipsoids of the type envisaged in the model have been observed. Brown's ellipsoids would correspond to pairs of adjacent lines of high dislocation densities observed in the etch pit patterns of copper single crystals in single slip orientations on the cross-slip plane by Livingston [12] and Basinski and Basinski [13], after deformation at room temperature and 4.2 K respectively. While a given line of etch pits deviates from strict alignment along the trace of a single (111) primary plane by a tilt about the edge dislocation direction, as might be expected on Brown's model, indicating slip on several closely spaced planes, it seems to the writer not possible to pair-up adjacent lines of slip in the etch pattern after deformation at either temperature. The main difference between the results of these experimental studies is that after room temperature deformation the structure seems much less regular than after deformation at 4.2 K, showing greater deviations of the arrays from the trace of the (111) planes. Although no comparisons of the mean angles of deviation have been made, it would appear from the pictures that the mean angle θ is larger at R.T. than at 4.2 K. On the other hand the work-hardening rate θ_{II}/μ, where μ is the shear modulus, is effectively independent of temperature [see e.g. fig. 5.22 in [14]]. On Brown's model this would appear to require θ to be nearly independent of temperature.

A similar trend in angular deviations is also apparent in TEM sections viewed along the primary Burgers vector direction (compare the results of Steeds [15] and Basinski [16] for copper). It should be noted however that tilts observed in these cases are about the screw dislocation direction. The temperature dependence of this effect suggests that it is due to cross slip. The same mechanism occurring during expansion of loops from a source can lead to the deviations of edge dislocations from their original glide planes.

(iii) Brown's treatment of the relief of the internal stress, identified as the fibre stress, in terms of the spacings of forest and primary dislocations is an important advance on the early attempt at this problem by Hirsch and Mitchell [5,6]. Brown assumes that the relaxation of the fibre stress is driven entirely by the internal stress. From eq. (3.1) it turns out that after relaxation the shear stress arising from the fibre stress resolved on the conjugate or critical system is

$$\sigma_s = \alpha\mu e_p \theta\sqrt{3/2}, \tag{2}$$

where α is the forest hardening constant, e_p the plastic engineering strain. Using equation 3.3 for the flow stress σ_p on the primary system, it follows that

$$\sigma_s/\sigma_p = \sqrt{3}/2\beta, \tag{3}$$

β (the fraction of the fibre stress which is relaxed) is found to be $\sim 3/4$, so that $\sigma_s \sim \sigma_p$. But for a crystal orientation in the centre of the triangle, the Schmid factor ratio for the Lomer–Cottrell forming conjugate and critical systems (CI and A VI) relative to the primary system is about 0.5. Thus the amount of relaxation will be affected somewhat by the applied stress, and depending on the sign of the fibre stress, the degree of relaxation will be enhanced or reduced. This effect may also lead to a small orientation dependence of θ_{II}. It should be noted, however, that before relaxation begins the shear stress on the relevant forest systems from the fibre stress is about four times larger than the component from the applied stress.

(iv) Equation (1) assumes that the ends of the tilted ellipsoids act as stress concentrators similar to those of mode II shear cracks of length c. But the centre region of the tilted ellipsoid has a dislocation arrangement similar to that of a multipole, and the stress concentration in eq. (1) may be overestimated.

(v) As Professor Brown points out himself [17], the relaxation analysis leads to an estimate of $n \sim 90$, which is considerably larger than the estimate from slip line observations, ~ 20 [18].

(vi) The length of the slip bands decreases with increasing plastic strain. Some average needs to be used to determine the volume fraction of obstacles at any particular strain in the model.

(vii) The existence of concentrated shear stresses of the same sign at both ends of the tilted ellipsoid implies the existence of shear stresses of the opposite sign within at least the end parts of the soft slip band. This needs to be taken into account. If the strain compatibility condition in the soft and hard (obstacle) parts of the microstructure are met, then the flow stress depends on the volume average of the square root of the forest dislocation density of the two phases [4]. This still remains to be addressed together with the comparison of the predictions of the 2-phase model with the experimental relation between forest density and flow stress [13].

(viii) The geometric argument to determine the volume fraction of the obstacles is not convincing. The ellipsoids can still be packed if the walls are thicker and the ellipsoids smaller.

2.3. Stress relaxation in multipole and dipole type pile-up arrays

In Brown's model the stress relaxation is calculated for the fibre stress in the multipole part of the assumed ellipsoidal slip band configuration. This suggests that it is worth while to consider other configurations which either are multipoles or have similarities with them. De Lange, Jackson and Nathanson [19] have calculated the fibre stress (using the continuous dislocation distribution method) between long sheets parallel to the glide planes of regularly spaced screw and edge dislocations of opposite sign (an edge dislocation multipole is shown in fig. 1(c)). They have shown that for both types of arrays the shear stresses between the sheets on the primary system are very small, while for the edge dislocations, as expected, the fibre stress exerts large shear stresses on the Lomer–Cottrell forming conjugate and critical systems, similar to the ellipsoidal model. For the screw arrays there is a large shear stress on the cross-slip plane. It is suggested that arrays approximating to multipolar structures can be formed as follows:

Consider the structure produced by cooperative operation of sources as envisaged in [5, 6], the dislocations being stopped at obstacles. Figure 1(d) is an adaptation of part of fig. 13 in [6]. In the region between two operating sources dislocations of opposite sign are generated on two closely spaced slip planes. In the central part of the array, the configuration approximates to a multipole while at the ends near the obstacles the configuration has pile-up character, i.e. somewhat similar to the arrangement in Brown's tilted ellipsoids.

Figure 8 of [6] shows contours of the shear stress on the primary system, consisting of the internal stress plus applied stress, due to two overlapping pile-ups of edge dislocations of opposite sign. It is clear that the stress field is very non-uniform and that there is a substantial back stress in the central region. However it follows from the work of Mitchell [20], and Basinski and Mitchell [21] (see also [6]) that there are substantial shear stresses on a number of systems, including the conjugate CI and critical A VI systems, sufficient to cause local slip on these systems, forming Lomer–Cottrell locks near the piled-up regions. The forest dislocations of opposite sign move away towards the tails of the adjacent pile-ups of opposite sign. This results in further emissions from the source, increasing the density of the primary dislocations in the tail regions, reacting with the CI and A VI dislocations to form locks. These steps are shown schematically in fig. 1(e) in the sequence 1, 2, 3. The net result is a more even distribution of primary dislocations on the slip planes, thus tending to transform the original piled-up groups into near multipolar configurations over a major part of the length of the array. A similar mechanism should occur for screw pile-ups, where slip on the cross-slip plane would play an important role. The back shear stress on the primary system in the central region between the two arrays of dislocations should then reduce considerably, and the residual stress between the edge dislocation arrays should approximate to the fibre stress. The long-range stresses outside the near multipole array cancel out at distances larger than their separation. It should then become reasonable to estimate the equilibrium relaxation state by applying Brown's methodology.

On this basis, following his analysis, the plastic engineering strain e_{p} produced by the two sources is

$$e_{\mathrm{p}} \sim nb/D, \tag{4}$$

where n is the number of edge dislocations of one sign in each sheet, D the separation of the sheets. Then the fibre stress is approximately

$$\sigma_{\mathrm{F}} = E_{\mathrm{y}} nb/L = E_{\mathrm{y}} e_{\mathrm{p}}(D/L), \tag{5}$$

where L is the length of the array (equal to half the slip-line length). The average primary dislocation spacing L_{p} is

$$L_{\mathrm{p}} = bL/e_{\mathrm{p}}D. \tag{6}$$

Comparing (4), (5), (6) with the equivalent expressions in chapter 58, it is clear that the equations are identical, with θ replaced by (D/L). The extent of fibre stress relaxation is identical, and the flow stress in the soft interior of the multipole, σ_{p} is given by

$$\sigma_{\mathrm{p}} \sim \alpha\beta\mu e_{\mathrm{p}}(D/L). \tag{7}$$

Provided (D/L) remain constant is Stage II, this gives linear hardening. In fact, since Brown equates θ approximately to $a/c = D/L$, the treatments are identical as far as the relaxation part of the model is concerned, and all the consequences which follow, which are discussed in Brown's earlier paper [17]. As for Brown's model, the scale of the structure, i.e. (D/L), may be obtained from the condition that the concentrated stress at the obstacle must be sufficient to permit a new slip band to penetrate it; i.e. to be equal to the local flow stress in the obstacle. However it is not clear what the volume fraction of obstacle should be in this case. I have already commented on this condition for Brown's model in section 2.2. For the transformed pile-up into multipole model discussed in the preceding paragraphs, a convincing argument for the scale (D/L) has yet to be formulated.

There appears to be some measure of support for the cooperative source operation model considered here, from TEM images of sections normal to the Burgers vector. A consequence of this model is that the black and white stripes should alternate at the sources (see fig. 1(d)), with the boundaries of the bands, i.e. the sheets of high dislocation density, continuing at least approximately on the same primary (111) plane on both sides of the source. Figure 2 shows examples in a micrograph copied from fig. 12 of [16], of networks in Cu single crystals deformed at 4.2 K, viewed along the Burgers vector. (There is another in fig. 13 of [10].) Possible source positions have been indicated at S. The structures are obviously more complex than the ideal envisaged in fig. 1(d), and there are clearly many dislocations within the bands, which would be consistent with "fine slip" of the type expected in the model of Brown.

2.4. Conclusions

The prominent dislocation structures observed by TEM in Stage II of the stress-strain curve of single crystals of pure metals in single slip orientations are sheets of positive and negative dislocations roughly parallel to the primary (111) planes, stacked so that edge dislocations in adjacent sheets alternate in sign. The aspect ratio of these arrays, i.e. the separation to length is small, typically $\sim 1/30$. Their essential feature is that they approximate to multipole structures, which accommodate the plastic strain easily, and as shown by Jackson and Siedersleben [22] largely cancel out the long-range shear stresses on the primary system. Furthermore, as pointed out by the same authors and by Brown [17], the resultant fibre stress has large shear stress components on secondary systems, which stabilise these structures. It is this observed systematic pairing of dislocations of opposite sign above one another which distinguishes Brown's model and its variants from that due to Zaiser and Seeger, and leads to different conclusions about the contribution of long-range shear stresses to the flow stress. Because of this pairing of primary dislocations no long-range secondary slip is required to relieve the shear stress. On the other hand the relaxation of the fibre stress between the sheets comprising the approximate multipole configuration is accomplished by short-range slip on secondary systems. These secondary dislocations interact with those on the primary system and largely control the flow stress in the soft regions.

Why and how does this pairing occur? In Brown's model the structures arise as a result of positive and negative primary dislocations from many sources forming ellipsoidal slip bands whose strain is easily accommodated in the matrix. On the model involving cooperative operation of sources, a source close to the head of a pile-up which is forming would be favoured to operate, leading to the pairing (fig. 1(d)). This leads to a relay race.

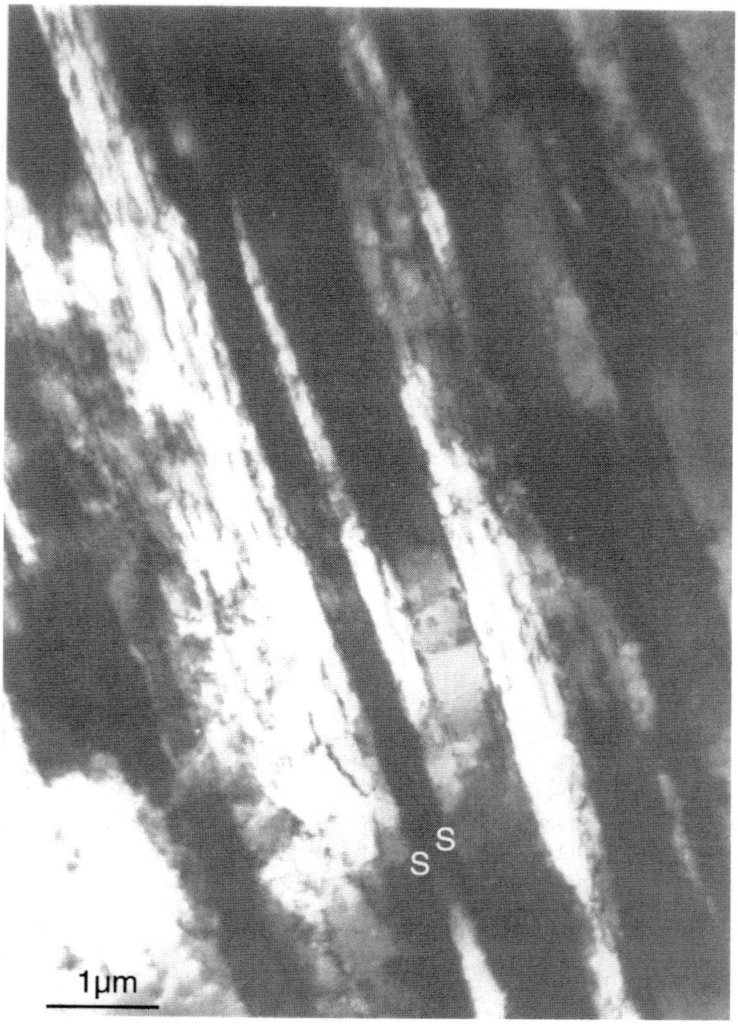

Fig. 2. TEM micrograph of single crystal of Cu deformed at 4.2 K in Stage II, showing bands of alternating contrast approximately parallel to the primary slip plane. Beam direction is the primary Burgers vector [101]; foil plane is (101). Possible source positions are marked at S where the contrast changes from black to white. (Taken from fig. 12 of Basinski [16].)

Zaiser and Seeger argue that this sort of shielding simply reduces the back stress at the source, resulting in further emission of dislocations. However, this ignores the fact that the mechanism of the internal stress relaxation generates secondary dislocations, and the flow stress increases, the secondaries now becoming the controlling contribution to the flow stress. This limits any emission to additional primary dislocations at the tails of the arrays, which interact with the secondary dislocations generated at the neighbouring tips of the pile-ups (see section 2.3). The effect on source emission of the back stress from glissile dislocations in a single slip plane model is replaced by that of forest hardening in the near multipole structure.

Much remains to be done. Some of the points are noted in section 2.2. Other outstanding questions include the orientation dependence of the work-hardening rate, and the fact that linear hardening is observed in alloys with a high friction stress due to solution hardening and with low stacking-fault energy, in which the microstructure in Stage II differs substantially from that of the pure metals [23]. However, in my view Brown's approach following his work with Pederson and Stobbs [24] together with Jackson's contributions [25] and those due to Mughrabi [26] represent a major advance in understanding and modelling, by identifying the importance of pairing of the dislocations in multipole arrays implicit in the ellipsoidal model, and of the fibre stress, as the agent responsible for creating the forest. This realisation then enables the relaxation and the flow stress to be treated approximately analytically in a relatively simple way. The models are consistent with the observed structures, which have low energy (small accommodation factor) as envisaged by Kuhlmann-Wilsdorf [3], and at the same time are stabilised by forest. However, as Brown points out, even at this stage it is possible only to demonstrate the consistency of the model, rather than predicting it from first principle.

3. *Work hardening in L1$_2$ alloys*

3.1. Introduction

Sun and Hazzledine in their review article in volume 10 of this series [27] concluded that "Quantification of a work-hardening model for L1$_2$ alloys remains an outstanding challenge in dislocation mechanics". Chapter 62 by Viguier, Martin and Bonneville [28] leaves little doubt that this challenge still remains. As pointed out by these authors, one of the reasons is that most studies have concentrated on elucidating the origin of the yield stress anomaly, and that experimental results on work hardening are rather limited.

In order to develop realistic models of work hardening, it is essential first to establish the mechanism controlling the flow stress. In f.c.c. metals this has been achieved using two types of experiments: (i) change of temperature or strain-rate experiments to establish the temperature/strain-rate dependent component of the flow-stress, pioneered (for temperature changes) by Cottrell and Stokes [29]; (ii) correlations between flow stress and different types of dislocations to determine the type of interaction which is controlling. Such experiments have been reviewed extensively in [10], and are also discussed by Saada and Veyssière in chapter 61 [30] (see e.g. fig. 6 in [30]). These experiments demonstrated the importance of interactions with forest dislocations for f.c.c. metals. On the other hand the work-hardening rate at any particular point on the stress-strain curve is controlled by the dislocation storage rate. For primary dislocations this is given by $\mathrm{d}\rho_\mathrm{p}/\mathrm{d}\epsilon_\mathrm{p} = (bL)^{-1}$, where $\mathrm{d}\rho_\mathrm{p}$ is the increment in primary dislocation density, L the average distance travelled by a dislocation in a plastic strain interval $\mathrm{d}\epsilon_\mathrm{p}$, and b is the Burgers vector. To determine the work-hardening rate it is then necessary to establish the relation between ρ_p and the flow-stress determining dislocations.

3.2. Strain-rate jump tests in single crystals of $L1_2$ alloys

Extensive strain-rate jump tests as a function strain at different temperatures were carried out by Ezz and Hirsch [31] for single crystals of $Ni_3(Al,Hf)B$ and by Ezz for single crystals of Ni_3Ga [32]. These experiments have been mentioned in chapter 62, but not discussed, except for comments that the interpretation is not obvious and that there is some controversy about the reversibility of the flow stress. Both these statements are true. In particular the nature of the transients requires further detailed examination and interpretation. They are complicated by the fact that slip occurs not only on octahedral planes, but also on cube planes, the latter being strongly temperature dependent. However, any proposed model of work hardening has to explain the results of the strain-rate jump tests. For that reason it is worth while to summarise them briefly here, and to consider their possible implications, particularly as they represent an attempt to use the same approach as was used successfully for f.c.c. metals to elucidate the nature of the mechanism controlling the flow stress.

Figure 3 shows curves of yield stress (RSS) as a function of temperature for single crystals of $Ni_3(Al,Hf)B$ (composition 75% Ni, 22.7% Al, 1.51% Hf, 0.2% B) deformed in compression at strains of 0.01% and 0.2% respectively, as well as work-hardening rates (WHR) measured at strains of 0.06%, 0.15% and 1.5% strain. The figure also shows the nature of the transients when the strain rate is increased in sudden jumps of 20 times. The transients in domains I and II in fig. 3 are accompanied by yield drops as for f.c.c. crystals in Stage II (see [10]); in region I the transients are smooth, in region II serrated flow is observed. In region III the yield drop is replaced by work hardening which decreases with increasing strain (reminiscent of Stage III for f.c.c. crystals), and beyond the peak a much larger smooth transient of a different nature is observed, which is characteristic of cube slip. Changes in the nature of the transient with temperature have also been recorded by other workers (see chapter 62, section 2.1.1.2). Details of the measurements of the stress change $\delta\tau$ are given in [31].

Figure 4 shows the strain-rate sensitivity $\beta = \delta\tau/\delta(\ln \dot{e}_p)$ plotted for different temperatures as β/T against $(\tau - \tau_y)$, where τ_y is the yield stress measured at 0.01% strain (see fig. 3). Up to about 700 K, the point of inflexion on the yield stress versus temperature curve, accurate measurements of $\delta\tau$ can be made, particularly at high strains (stresses). No measurements are possible for strains less than about 0.5%. The following points may be noted: (i) Cottrell–Stokes like behaviour is observed, but unlike the case for pure f.c.c. metals the lines of β (or β/T) extrapolated to $\beta = 0$ do not pass through zero stress, but through points on the stress axis corresponding approximately to the yield stress at about 0.01% strain. (ii) β is independent of the strain-rate jump ratio as expected from the thermodynamic analysis, assuming a constant mobile dislocation density during the strain jump. (iii) Typical values of $\delta\tau/(\tau - \tau_y)$ for a jump ratio of 20 times are about 1%, similar to those for f.c.c. metals. (iv) $\beta/T = k/v^x$, where v^x is the apparent activation volume, k the Boltzmann constant (see chapter 62, section 3.1.2.2). v^x decreases with increasing stress $(\tau - \tau_y)$, and $v^x(\tau - \tau_y) = \text{constant}$.

What is the interpretation of these results? If we write $v^x = blw$ where l is the spacing of the obstacles, and w the obstacle width, and if $(\tau - \tau_y) = \alpha\mu b/l$, as might be expected from forest or jog hardening, then the observed Cottrell–Stokes like behaviour follows,

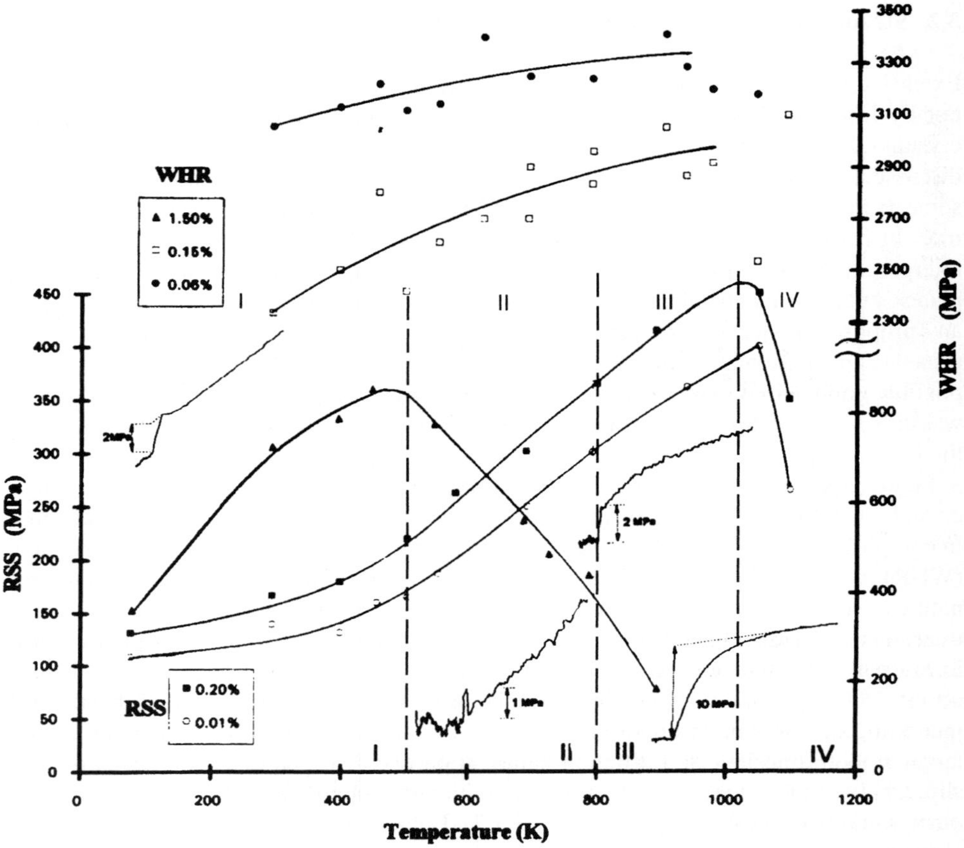

Fig. 3. Resolved shear stress measured at strains of 0.01% and 0.2% as a function of temperature for single crystals of Ni_3(Al,Hf)B, divided into four regions, each characterised by a distinct stress transient when the strain rate is increased by a factor of 20; crystal orientation 1 (see fig. 4). The figure also shows work-hardening rates measured at strains of 0.06%, 0.15% and 1.5% (taken from [42]).

i.e. $v^x(\tau - \tau_y) = \alpha\mu b^2 w =$ constant, provided α, w are independent of stress/strain. For a discussion about the basis of the Cottrell–Stokes law see Nabarro [33] and also chapter 61. The similarity between the magnitudes of $\delta\tau/(\tau - \tau_y)$ for the $L1_2$ alloy and f.c.c. metals, suggests that the rate-controlling obstacles may be forest dislocations.

The flow stress τ can then be written

$$\tau = \tau_y(T) + \tau_f(e_p, \dot{e}_p, T), \tag{8}$$

where $\tau_y(T)$ is dependent on temperature only, and $\tau_f(e_p, \dot{e}_p, T)$ is the work-hardening stress, which depends on strain, strain-rate and temperature, and obeys a Cottrell–Stokes law. $\tau_y(T)$ has the characteristic of an internal stress which is independent of strain and strain-rate.

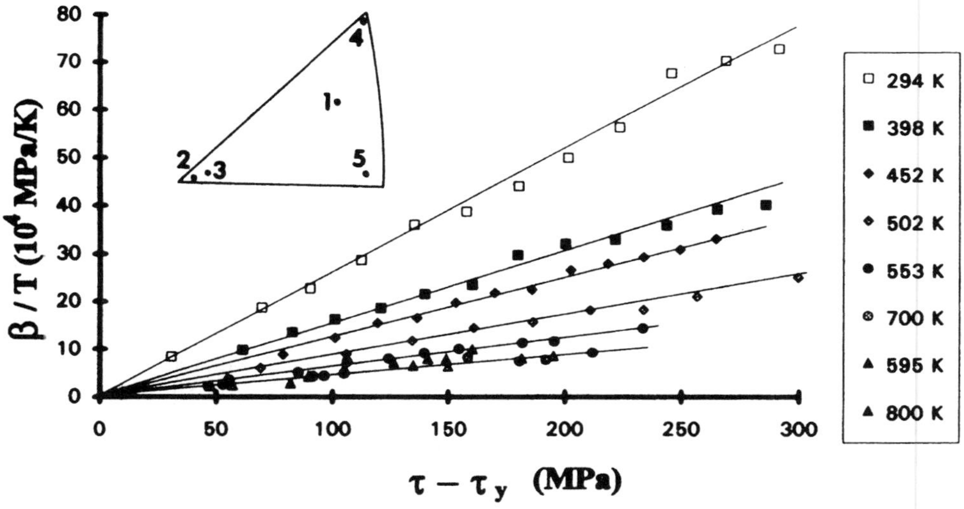

Fig. 4. The strain-rate sensitivity parameter β/T at different temperatures as a function of the work-hardening stress for a crystal of $Ni_3(Al,Hf)B$, orientation 1 (adapted from [42]).

It should be emphasized that eq. (8) is only valid for the range of conditions under which measurements could be made, i.e. typically for strains $> 0.5\%$, and for temperatures up to about 700 K. In reality for smaller strains at any temperature the curve of β against τ is likely to bend around and approach $\beta = 0$ at $\tau = 0$ asymptotically. This is sketched in fig. 8 of [32]. It is clear then that the average strain-rate dependence $\delta\tau/\tau$ for small strains, effectively in the microstrain region, is considerably less that the slope of $\delta\tau/\tau_f$ for larger strains in the macroplasticity region.

Why does the flow stress have this form? Macroscopic slip is generated by the expansion of dislocation loops; both edge and screw dislocations have to advance. Since the classic work of Thornton, Davies and Johnston [34] is it known that edge dislocations are responsible for plastic flow in the microstrain region. The implication of the low strain-rate sensitivity $\delta\tau/\tau$ in this domain is that the edge dislocations move effectively athermally in relatively obstacle-free regions and that hardening takes place by an athermal process as occurs when the edge dislocations move and pull out screw dipoles. These consist of Kear–Wilsdorf locks joined by edge-character superkinks [35,36]. Eventually the edge dislocations become trapped (exhaustion). As the stress increases, shorter lengths of edge dislocations move, the dipoles become narrower [37] and the superkinks tend to increase in length, until the stress is sufficient for them to become unstable, bow out, annihilate the screws to which they are connected or form APB tubes [31], and jogs on the cube cross-slip plane, and generate new edge and screw dislocations. At this critical stress the screws effectively begin to propagate by the lateral motion of superkinks of critical length, which depends only on temperature. This screw propagation stress corresponds to the onset of loop expansion, source operation and macroscopic yielding. The rate of storage of screw dislocations with increasing strain should also then decrease, since their average slip distance increases.

When sources begin to operate, groups of dislocations may trap one another, and forest dislocations are then likely to be generated to reduce the local internal stresses, just as occurs in f.c.c. crystals in Stage II linear hardening. τ_y in eq. (8) can be considered as the stress below which hardening occurs by the athermal generation of screw dipoles and by exhaustion of the mobile edge dislocations, associated with a very low strain-rate dependence, and above which forest interactions and jogs resulting from the screw annihilation process become important. The flow stress then becomes more strain-rate dependent, becoming controlled by forest interactions and jogs. But τ_y is also expected to be of the order of the macroscopic yield stress, i.e. when sources can operate, which is how τ_y has been identified in previous publications on the subject [31,37]. In practice, fig. 3 shows that τ_y at 0.01% strain is not much smaller than the conventional flow stress defined at 0.2% strain, and follows the same anomalous variations with temperature.

With increasing plastic strain more forest dislocations and jogs on (010) are generated. But there will always be an athermal component of the flow stress reflecting the need for the superkinks of critical size to bow out unstably, to ensure that screws can propagate by the superkink mechanism; this is τ_y in eq. (8). The physical picture is that for macroscopic plastic flow the superkinks have to bow out *and* overcome the forest and jog obstacles. Or, alternatively, if new loops are nucleated from superkinks in edge dislocations pulling out screw dipoles, the critical step may be for the edge dislocations to generate the dipole *and* overcome a forest dislocation (or jog), as proposed in [37]. This then is the suggested physical interpretation of τ_y in eq. (8). It should also be noted that if we decompose the forest hardening stress τ_f into an athermal component τ_{fa}, and a temperature and strain-rate dependent component τ_{fth}, as in the original Seeger formulation [38] (see also chapter 61, section 3.2.3), the "effective stress" in the thermally activated process is $(\tau - \tau_y - \tau_{fa})$. Without this decomposition of τ_f, the effective stress is $(\tau - \tau_y)$.

This discussion may be compared with that concerning Ni_3Al in section 6.1 of chapter 62. The interpretation of the nature of the hardening in the microstrain region suggested above agrees with that in chapter 62, with the added qualification that the strain-rate sensitivity is low. But the suggestion made there that the anomalous yield stress is associated with the effective part of the stress necessary to overcome obstacles is at variance with the interpretation offered here; namely that the yield stress is a strain-rate independent stress, and that the strain-rate dependence is entirely due to work hardening and in particular to forest or jog hardening, which is consistent with the Cottrell–Stokes type behaviour. It should be noted that at room temperature the orientation dependence of β as a function of strain correlates with work-hardening rates [31]. The temperature dependence of the yield stress arises from the temperature dependence of the mean free path of screw dislocations.

With reference to existing models of work-hardening rates due to Caillard [37] (see chapter 62, section 6.2.2.3), and Devincre, Veyssière and Saada [38], it may be mentioned that neither accounts for the observed variation of the strain-rate sensitivity with stress.

Finally it should be noted that the above model applies only over the range of temperatures over which eq. (8) has been established, i.e. roughly up to the point of inflexion on the yield stress versus temperature curve. It would correspond to the low-temperature regime in which computer simulations suggest that the screws propagate by the lateral sliding of superkinks [40].

Table 1
Flow stress and dislocation densities

Temp. (K)	e_p (%)	τ_f (MPa)	ρ_p ([$\bar{1}$01] screws) (cm^{-2})	ρ_{cc} ([$\bar{1}$01] (010)) (cm^{-2})	Comments	τ_f (calc) (MPa) ($\alpha = 1/3$)
294	2	57	1.1×10^9	negligible	Cross-slip on (1$\bar{1}$1); densities unknown, high density of APB tubes	–
600	2	95	$\sim 4 \times 10^9$	3.6×10^9	Bowing on (010); some secondary octahedral slip	70
720	2.4	85	$\sim 3 \times 10^9$	1.9×10^9	Fewer primary screws, extensive bowing on (010); inhomogenous slip on secondary systems and on [$\bar{1}$10] (001)	51
870	1.5	92.5		3.3×10^9		67

3.3. Work-hardening stress, forest dislocations, and jogs on (010)

In chapter 62, section 4, Viguier, Martin and Bonneville [28] discuss the correlation of total dislocation density with flow stress. As the authors point out, there are relatively few data in the literature, and there are even fewer which distinguish between dislocations of different types. The latter data are however more useful for elucidating the mechanism controlling the flow stress. Table 1 lists results from TEM studies of dislocation distributions in single crystals of Ni_3(Al, Hf)B, (orientation (1) see fig. 4), deformed at different temperatures [42]. It lists τ_f, the densities of primary screw dislocations (ρ_p) and primary dislocations on the cube cross-slip plane (010) (ρ_{cc}), values of τ_f calculated assuming ρ_{cc} constitutes the forest density, and comments on activity on other slip systems. Taking α the forest hardening parameter as $1/3$, at 600, 720 and 870 K about $2/3$ of the flow stress can be accounted for by interactions between primaries on octahedral planes and threading primary dislocations resulting from cube cross-slip. In addition there is evidence for inhomogenous secondary slip on other systems, with quite high local densities [42]. Although these results suggest that forest dislocations may play an important hardening role, the results are too few to reach any definite conclusions, and the observation at room temperature does not fit the model. It has been suggested that work-hardening is caused at least in part by jogs on (010), which are subject to lattice friction, and which at low and moderate temperatures trail APB tubes formed by the transformation of KW locks [43], the APB tubes themselves also causing hardening [53]. The increase in work-hardening rates at low temperatures is attributed to an increase in density of KW locks and therefore of jogs on (010), while at high temperatures the decrease in work-hardening rates is due to an increased mobility of jogs on (010), which also prevents the formation of APB tubes [43]. (This differs from the suggestion made in Chapter 62, section 6.2.2.2 [28] that the disappearance of APB tubes is due to diffusion.) Shi [54] noted that for the Ni_3(Al,Hf) alloy used in her study of APB tubes, the drop in work-hardening coincides with the disappearance of APB tubes at the same temperature (about 860 K). Further work is needed to determine the relative contributions of this mechanism and of forest interactions to work-hardening. However both mechanisms are qualitatively compatible with the observed Cottrell–Stokes type behaviour of the strainrate dependence of the flow stress [31].

3.4. Reversibility of flow stress in γ-TiAl compounds

One of the remarkable features of the plastic properties of Ni_3Al and related $L1_2$ compounds is that the yield stress in the anomalous region is partially reversible with temperature, i.e. a specimen slightly predeformed at T_1, and then redeformed at a lower temperature T_2, yields at a stress lower than that at T_1 and only slightly above the yield stress of a virgin specimen deformed at T_2. The increase in stress reflects part of the workhardening at T_1. This phenomenon, which is associated with slip by superdislocations in these compounds, is described in some detail by Viguier, Martin and Bonneville in chapter 62, section 5.2. It is fair to say that while some suggestions have been made in the literature [37,44,45], the precise mechanism responsible for this remarkable effect is not understood.

Similar tests have been performed on single crystals of γ-TiAl; these are described in chapter 62, section 5.1. For γ-TiAl the situation is more complicated in that different slip systems, i.e. superdislocation slip of the type $\langle 011]\{111\}$, ordinary dislocation slip $\frac{1}{2}\langle 110]\{111\}$, and $\frac{1}{2}\langle 112]\{111\}$ slip can operate for different crystal orientations, and at different temperatures for a given crystal orientation. There are four experimental studies described in chapter 62. Three of these by Stücke, Dimiduk and Hazzledine [46], Inui, Matsumuro, Wu and Yamaguchi [47] and Jiao, Bird, Hirsch and Taylor [48] consisted of mechanical tests as well as TEM observations of the dislocations operating at the predeformation temperature T_1, and for virgin (unpredeformed) crystals at T_2. Both T_1 and T_2 lay within the anomalous region. The fourth study by Mahapatra, Chan and Pope [49] was limited to mechanical tests.

The first three studies included tests for different crystal orientations: one for which ordinary slip was expected to operate at T_1 and T_2, Jiao et al. [48], and three for which superdislocation slip was expected [46–48]. For the first of these, Jiao et al. [48] found that for a crystal with nominal composition Ti54.5at%Al axial orientation $[\bar{1}36]$, predeformed at 973 K and restrained at RT, the yield stress is not reversible, i.e. the yield stress at T_2 is equal to the flow stress at T_1 of the prestrained crystal. TEM showed that ordinary slip occurs at T_1 and for a virgin crystal at T_2 (RT), although there is also a small contribution from superdislocations at RT (see figs 3 and 4 of [50]).

For the superdislocation orientations, Stücke et al. [46] found that for a crystal with nominal composition Ti56at%Al, with axial orientation near [010], predeformed at 773 K and redeformed at RT, the yield stress is again irreversible. TEM showed superdislocations for virgin crystals deformed at 873 K and at RT [51], and it seems likely that superdislocation slip operated also at 773 K.

The experiments of Mahapatra et al. [49] were also carried out for crystal orientations [001] and [011], for which superdislocation slip is expected to operate, although this was not confirmed by TEM. For both orientations of crystals with nominal composition Ti56at%Al, predeformed at 973 K and redeformed at RT, the yield stress was found to be irreversible in agreement with the results of Stücke et al. [46].

[9] and stress-balance arguments due to Brown and Stobbs (1971) [10]. To use Cottrell's phrase, the theory is one component of the 'alphabet' necessary to write the story of work hardening. It is a part of the alphabet absent from all books on dislocations, except for Mura's (1987) [11]. To go from alphabet to story, it is necessary to decide on an actual value for the volume fraction of obstacles. If Kuhlmann-Wilsdorf's principle of similitude is accepted, both the volume fraction and the average geometrical disposition of obstacles must be constant throughout stage II hardening. I tentatively suggest that the volume fraction is equal to the fraction of space between random close-packed spheres, about 0.2.

3.2. Deformable obstacles

There are two types of geometrically necessary dislocations. The first type comprises 'Orowan loops', equivalent to the edge dislocations of opposite sign pressing against the walls, as in Mughrabi and Ungár's fig. 5. Such dislocations produce kinematical hardening, manifested by permanent softening in a Bauschinger experiment. The forward work-hardened stress is increased, but the backward stress is reduced. The reason for this is that the glide dislocations sample only one component of the stress-balanced composite, not both of them. The second type of geometrically necessary dislocations is produced by plastic relaxation of the large local stresses caused by the Orowan loops. Such dislocations can form a bewildering variety of patterns, prismatic loops caused by cross-slip or rotations by secondary slip, etc. The forest of these produces isotropic hardening, equally effective both backwards and forwards. Because of slack in the position of glide dislocations as they move from one obstacle to another, stress-reversal produces a rounded approach of the stress-strain curve to its asymptotic shape, but no permanent softening. The main result of Pedersen et al. is that by comparison with composites containing only tiny amounts of non-deformable obstacles, the permanent softening in pure copper single crystals is negligible. The inescapable conclusion is that the obstacles in work-hardened copper are deformable: they are themselves just on the point of deformation by the applied stress. If this is the case, Pedersen et al. demonstrate that the flow stress is governed by the average forest stress, or rule of mixtures, exactly as Mughrabi says. The Eshelby factor denoted Γ by Mughrabi and Ungár, and γ by Pedersen et al., does not come in. The internal stresses, though they may be large, do not contribute a term to the flow stress. One way of seeing this is to recall that the average stress along any straight line is equal to the applied stress, because the distribution of internal stress is balanced and cannot contribute. Where the dislocation is inside the obstacle, the forward stress there is exactly balanced by the local flow stress, so the dislocation is straight. Where the dislocation is in the 'matrix', the region outside the obstacles, the applied stress less the back stress is again exactly balanced by any local forest stress, so the dislocation is straight. The dislocation is thus entirely straight as it moves through the work-hardened material encountering obstacles, and a virtual work argument shows that the stress necessary to move it forward is just the average friction stress, given by the rule of mixtures.

It is an extraordinary fact that deformable obstacles, just themselves on the point of deformation, but embedded in a dislocation-free matrix, cause a flow stress equally well regarded as due to the back stress in the matrix, as to the frictional forest stress. However, since the magnitude of the back stress in the matrix results from the strength of the forest, the formula for the work-hardening must depend only upon the forest, not upon any feature of the back stress, such as the shape of the obstacle, etc. I believe this to be the true resolution of the famous arguments between the 'forest hardeners' and the 'back-stress hardeners'.

3.3. Which is the continuous phase?

The question arises: which is the continuous phase, obstacle or matrix? From the point of view of Eshelby's theory, one must decide which of these, if either, is to be modelled as the 'ellipsoid'. Mughrabi, and Pedersen et al., think of the matrix as continuous, and the obstacles as discrete ellipsoids. However, if the obstacles form a continuous phase of carpets and braids, derived from the excluded volume in an assembly of random close-packed spheres, then the obstacles are the continuous phase. My assumption now is that it is the slip-bands which are discrete ellipsoids. If Eshelby's famous theorem about the uniformity of stress in an elastic inclusion can be inverted, one could say that because the dislocation-free, hence very soft, slip-band has a uniform stress of zero, then it must be ellipsoidal. Here, then, is a first-class puzzle for the mathematicians: is it possible to prove the inverse of Eshelby's theorem? or to disprove it?

3.4. Persistent slip bands

We turn briefly to the vexed question of the internal stress in persistent slip bands (PSBs). Mughrabi and Ungár present new irrefutable evidence (their fig. 7) that in the unloaded state the walls of dipoles (the 'rungs' in the ladder structure of the PSBs) cause internal shear stress in the space between them [12]. The stress opposes the one most recently applied and has a magnitude of about one-quarter the saturation stress. It is a classic Mughrabi stress, not to be confused with the fibre stress, although it is about the same magnitude.[2] We conclude that although the walls are deformable enough to allow edge dislocations into them to refine the dipoles of which they are composed, they are strong enough to force screw dislocations to bow between them. They are strong but deformable obstacles. According to recent descriptions by Pedersen (2000) [13], in the dipole distribution of the walls, the largest dipoles are on the point of separation under the action of the maximum applied stress, and smaller dipoles unfavourably placed at bends in the walls will also break up. If the walls are deformable, the argument (see section 3.2) above suggests that the internal stress caused by the walls does not contribute to the saturation stress.

4. *How many state parameters?*

Finally, we come to Cottrell's statement that "...the flow stress of a work-hardened material must... be a sophisticated function, at least two-parametered, of the dislocation pattern

[2] I am indebted to Prof. Mughrabi for correspondence on these points.

belonging to this state." I certainly support this view, and believe that there are three essential parameters, corresponding to the confinement of slip bands in three dimensions. Firstly, there are screw dipoles, which can be annihilated by cross-slip without matter flow; secondly, there are the edge dipoles, mostly of vacancy type, which can be turned into necklaces of circular prismatic loops by pipe diffusion and fully annihilated by long-range diffusion; and thirdly, there are the carpets and tilt walls of small-angle boundaries, enclosing misoriented cells caused by secondary slip. These are much more resistant to annealing, and may be removed by subgrain coarsening or recrystallisation. While work hardening is progressing, all three types of obstacle maintain equal strength, but changes of temperature or strain-rate can strengthen or weaken one of them relative to the others. In the restoration of equilibrium caused by flow under the new conditions, anomalous hardening or softening will be observed. The most spectacular of these effects is the work softening observed by Cottrell and Stokes (1955) [14].

References

[1] R.P. Feynman, in: The Feynman Lectures, Vol. II, eds Richard P. Feynman, Robert B. Leighton and Matthew Sand (Addison-Wesley Publishing Co. Inc., Reading, Mass., Palo Alto, London, 1964) Ch. 30, §9.

[2] A.H. Cottrell, The Mechanical Properties of Matter (John Wiley and Sons, N.Y., London, Sydney, 1964). This is now frustratingly out of print.

[3] L.L. Lisiecki and O.B. Pedersen, Acta Met. Mater. 39 (1991) 1449–1456.

[4] U. Holzwarth and U. Essmann, Appl. Phys. A58 (1994) 197–220.

[5] L.M. Brown, in: Dislocation Modelling of Physical Systems, eds. M.F. Ashby, R. Bullough, C.S. Hartley and J.P. Hirth (Pergamon Press, Oxford, 1980) pp. 51–68.

[6] L.M. Brown, Materials Science and Engineering A285 (2000) 35–42.

[7] O.B. Pedersen, L.M. Brown and W.M. Stobbs, Acta Metall. 29 (1981) 1843–1850.

[8] J.D. Eshelby, Proc. R. Soc. A241 (1957) 376–396.

[9] K. Tanaka and T. Mori, Acta Metall. 18 (1970) 931–939.

[10] L.M. Brown and W.M. Stobbs, Phil. Mag. 23 (1971) 1185–1199.

[11] T. Mura, Micromechanics of Defects in Solids, 2nd edn (Martinus Nijhoff, Dordrecht, 1987).

[12] H. Mughrabi and T. Ungár, this volume.

[13] O.B. Pedersen, in: Multiscale Phenomena in Plasticity, eds J. Lépinoux et al. (Kluwer Academic Publishers, Amsterdam, 2000) pp. 83–97.

[14] A.H. Cottrell and R.J. Stokes, Proc. Roy. Soc. A233 (1955) 17–34.

Contents

Volume 11

CONTENTS OF VOLUMES 1–10

VOLUME 1. The Elastic Theory
1979, 1st repr. 1980; ISBN 0-7204-0756-7

VOLUME 2. Dislocations in Crystals
1979, 1st repr. 1982; ISBN 0-444-85004-x

VOLUME 3. Moving Dislocations
1980; 2nd printing 1983; ISBN 0-444-85015-5

VOLUME 4. Dislocations in Metallurgy
1979; 2nd printing 1983; ISBN 0-444-85025-2

VOLUME 5. Other Effects of Dislocations: Disclinations
1980; 2nd printing 1983; ISBN 0-444-85050-3

VOLUME 6. Applications and Recent Advances
1983; ISBN 0-444-86490-3

VOLUME 7
1986; ISBN 0-444-87011-3

VOLUME 8. Basic Problems and Applications
1989; ISBN 0-444-70515-5

VOLUME 9. Dislocations and Disclinations
1992; ISBN 0-444-89560-4

VOLUME 10. Dislocations in Solids

1996; ISBN 0-444-82370-0

CHAPTER 56

Long-range Internal Stresses, Dislocation Patterning and Work-hardening in Crystal Plasticity

MICHAEL ZAISER* and ALFRED SEEGER

Max-Planck-Institut für Metallforschung
Institut für Physik
Heisenbergstr. l, D-70569 Stuttgart
Germany

*Present address: Center for Materials Science and Engineering, The University of Edinburgh, Sanderson Building, The King's Buildings, Edinburgh EH1 1DT, United Kingdom

Dislocations in Solids
Edited by F. R. N. Nabarro and M. S. Duesbery

Contents

1. Introduction

The investigations of the plastic deformation of crystals at the end of the 19th and in the first third of the 20th century revealed several insights that are still basic to our present understanding of the field.

(i) The detection of 'slip lines' on rock salt and various metals [1] led to the conclusion that permanent deformation was achieved by crystal blocks sliding over each other along crystallographic planes without substantial loss of cohesion. From his observations of slip lines on 'natural' metal single crystals, Mügge was able to establish [110] (111) glide in Cu, Ag, and Au as early as 1899, well before through the work of Laue, Friedrich, and Knipping [2–5] X-rays had become available as a tool for studying crystal structures, crystallographic orientations, and the perfection of crystals.

(ii) In 1906, Hort [6] showed that most of the mechanical work W expended in the plastic deformation of α-Fe was liberated as heat and that the fraction W_{st}/W stored as 'latent heat' was of the order of magnitude 0.1 only. At the time, this fraction was tentatively interpreted as the energy required to transform the original crystalline state into another structure. Later, Taylor [7] tried to associate the stored energy with 'micro-cracks' formed in regions of stress concentration. Detailed measurements by Farren and Taylor [8], Satō [9], Rosenhain and Stott [10], Taylor and Quinney [11], and Quinney and Taylor [12] on α-Fe, several steels, Cu, Ag, Ni, Al as well as on various alloys gave the same order of magnitude of W_{st}/W. Satō [9] noted that W_{st}/W tended to decrease with increasing deformation, an observation that was subsequently confirmed by many workers (cf., e.g., [13,14]).

(iii) X-ray diffraction studies of deformed crystals confirmed not only the conclusions on the crystallography of the glide processes drawn earlier from surface observations but demonstrated convincingly that plastic deformation preserved the crystal structure as well as the specific volume within the experimental accuracy achievable at the time. The 'asterism' shown by the X-ray diffractograms of deformed crystals [7,15–17] was correctly interpreted as being caused by local rotations of the crystal structure around an axis lying in the glide plane perpendicular to the glide direction [7].

(iv) Schmid's law, formulated in 1924 [18] as an empirical law describing quantitatively how in uniaxial tests on Cd single crystals the onset of slip on the basal plane depended on the crystallographic orientation of the stress axis, demonstrated that in single crystals oriented for slip in one slip system β (so-called single slip) the extensive quantity that governs the slip process is the (external) shear stress, σ_{ext}^{β}, resolved in this slip system. The corresponding intensive quantity is the resolved shear strain in the slip system, ε^{β}. Hence in this case the mechanical work per unit volume may be written as

$$W = \int \sigma_{ext}^{\beta}(\varepsilon^{\beta})\,d\varepsilon^{\beta}, \qquad (1)$$

where the integral is to be taken over a given strain path. (Note that this expression ceases to be valid when in the course of deformation other slip systems become active and contribute significantly to the total plastic strain.)

(v) In some crystal structures (notably f.c.c. metals and h.c.p. metals with basal glide, which happened to be those that were first investigated in detail) crystal plasticity is an essentially *athermal* phenomenon. A landmark was the demonstration by Meissner, Polanyi, and Schmid [19] that suitably orientated Cd and Zn single crystals could be deformed by slip at 1.2 K. The critical resolved shear stress required for the initiation of slip in the basal slip system in high-purity Cd crystals at 1.2 K was shown to be less than 2.3 MPa. For basal slip in a hexagonal crystals the theoretical shear strength, defined as the resolved shear stress at which slip sets in a perfect crystal, may be estimated as [20]

$$\sigma_{\text{th}} \sim (c_{11} - c_{12})/60, \tag{2}$$

where c_{ij} denote Voigt's elastic constants. Inserting the low-temperature numerical value $c_{11} - c_{12} = 38 \times 10^3$ MPa of Cd gives us $\sigma_{\text{th}} \approx 600$ MPa. Hence the low-temperature critical shear stress was less than $1/100$ of the theoretical shear strength.

(vi) The changes in the *physical* properties introduced by cold work (but not the change in the specimen shape!) can be partially or totally reversed by annealing well below the melting temperature. Two basically different repair mechanisms have to be distinguished, viz. recovery (not involving formation of new grains) and recrystallization (formation of new grains without previous melting). The fact that shape changes cannot be recovered by annealing shows that the plastic strain cannot even approximately be used as a 'state variable'.

After forerunners by Prandtl [21], Dehlinger [22], and Yamaguchi [17], the concept of dislocations in crystals appeared in the literature almost simultaneously in papers by Orowan [23], Polanyi [24], and Taylor [25] in 1934. The developments during the second third of the twentieth century demonstrated convincingly that dislocation theory was capable of accounting for most of the above-mentioned observations (and many more) not only qualitatively but – at least in some instances – even quantitatively and that it provided the appropriate framework for planning future experimental work. The main approach during this period may be characterized as 'bottom-to-top' theory or as 'mechanistic'.

The *mechanistic* approach to dislocation theory starts from the properties of individual dislocations and certain refinements such as the concepts of extended dislocations spanning stacking-fault ribbons between partial dislocations, of kinks, of jogs, and of constrictions. In going beyond the 'one-dislocation picture' it considers, on the one hand, the long-range interaction between dislocations through their stress fields as given by the linearized theory of elasticity with emphasis on the rôle of particular dislocation arrangements such as pile-ups or small-angle grain boundaries and, on the other hand, short-range interactions between dislocations leading, e.g., to the *annihilation* of dislocations of opposite sign of the same slip system and to *reactions* between dislocations of different glide systems. These reactions, which may result in sessile dislocation configurations, dislocation nodes, and dislocation intersections, are essential for the defining feature of plastic deformation, viz. the permanency of the deformation when the load on the specimen is removed.

The approach to dislocation theory outlined in the preceding paragraph proved exceedingly successful in accounting, often quantitatively, for physical phenomena that are governed primarily by the properties of individual dislocations. Examples are the anelasticity due to dislocation movement or the Peierls-barrier-controlled flow stress of body-centred cubic metals at low and intermediate temperatures. To some extent the mechanistic approach was assisted by the possibility to observe individual dislocations by transmission electron microscopy (TEM) and thus to test many of its predictions.

Agreement between the expectations of dislocation theory and the observations on isolated or few dislocations does not necessarily mean that the observed configurations are particularly relevant in other circumstances, e.g., when many dislocations are involved (cf. sections 3–5). In fact, it must be admitted that in quantitatively describing situations involving high dislocation densities, the mechanistic approach had only limited success. As an alternative that is more appropriate for describing situations and processes involving many dislocations, the present contribution emphasizes the *holistic* viewpoint. It will be presented as a 'top-to-bottom' approach that starts out from general notions and basic principles (section 2), and that subsequently introduces specific mechanisms and notions step by step.

The holistic approach is most powerful in situations in which the long-range interactions between many dislocations are dominant (this is often the case if the dislocation density is high) and/or if the dislocation mobilities are high. In such situations various types of collective phenomena may arise. Theoretical treatments of such phenomena require the development of adequate mathematical tools for the quantitative characterization of many-dislocation systems and their stress fields (section 3). Section 4 deals with experimental methods that are capable of giving us quantitative statistical information on dislocation patterns and the internal stresses produced by them. Sections 5 and 6 are devoted to the calculation of statistical signatures of dislocation patterns and internal-stress fields within the framework of a stochastic model of dislocation dynamics. Cellular dislocation patterning, the evolution of local lattice curvatures and misorientations, and work-hardening are treated within a common framework which links the evolution of inhomogeneous dislocation microstructures to the fundamental processes of energy dissipation by dislocation motion and dislocation reactions.

In order to keep the treatment reasonably concise, we confine ourselves to plastic deformation by slip (as opposed to twinning and large-scale dislocation climb) and focus on situations in which collective phenomena dominate. Since low dislocation mobilities counteract collective phenomena we say little on b.c.c. metals and take our examples almost entirely from f.c.c. metals, although the general ideas have much wider applicability.

2. *General aspects of crystal plasticity and work-hardening*

In accordance with the greek origin of 'theory',[1] the current section outlines the common conceptual framework of theories of plastic deformation from both the thermodynamic and mechanistic ('modelling') viewpoint. It concentrates on the plastic deformation of crystalline materials in circumstances where dislocation–dislocation interactions dominate. Furthermore, a number of notations that are essential for structuring the discussion are introduced and used throughout the chapter.

[1] $\theta\varepsilon\omega\rho\iota\alpha =$ 'a looking at, viewing, contemplation'.

2.1. Length and time scales in plastic deformation

Even with the restriction indicated above, crystal plasticity presents a multiscale problem in time and in space of considerable complexity. On the one hand, it involves interconnected processes on length scales that extend from the atomistic scale, on which the arrangement of single atoms is considered, up to the macroscopic scale given by the specimen size. Similarly, the time scales range from the atomic vibration periods (10^{-13}–10^{-12} s) to the times required for performing the experiments and observing their outcome, which may be several hours or even days.

Appropriate *length scales* may be defined as follows:

- The *atomistic scale* deals with the arrangement of and the interactions between individual atoms. These interactions govern the dislocation core structures and, therefore, influence dislocation mobilities and short-range interactions between dislocations. Examples of processes which have to be considered on this scale are (i) the motion of dislocations with extended cores and the overcoming of the Peierls barriers of the first and second kind[2] [26,27], (ii) the cross-slip of extended screw dislocations [28] and the elimination of narrow dislocation dipoles of predominant screw character, (iii) the collapse of narrow dislocation dipoles of non-screw character, resulting in the formation of intrinsic atomic defects (vacancies or self-interstitials) and/or their agglomerates [29], (iv) the formation of sessile (i.e. virtually immobile) dislocation core configurations such as Lomer–Cottrell [30] or Kear–Wilsdorf [31] locks and their break-up under the influence of large stresses, (v) the formation and motion of jogs in dislocations.
- On the *microscopic* scale the elementary 'units' of plastic deformation are dislocation segments or – if, for simplicity, we consider processes in two dimensions only, cf. sections 3 and 5 – isolated dislocation lines. On this scale, dislocations are treated as line singularities in an elastic continuum. The appropriate tool for calculating their stress fields and the elastic interactions mediated by these fields is continuum mechanics. In many cases the dislocation motion can be described by force–velocity relationships into which the atomistic dislocation properties enter through parameters such as electron or phonon drag coefficients or the stress-dependent free enthalpies of kink-pair formation or cross-slip. The length scale on which this description is appropriate is the mean dislocation spacing $\rho^{-1/2}$, where ρ is a suitable measure of the dislocation density. Owing to the gradual accumulation of dislocations, the scale usually *decreases* during the plastic deformation.
- The *mesoscopic scale* is the spatial scale on which the evolution of the dislocation system may be described in terms of dislocation densities and dislocation correlation functions (cf. section 3). On this scale, it is expedient to consider separately the external stress $\boldsymbol{\sigma}_{\text{ext}}$ and the internal stress $\boldsymbol{\sigma}^{(\text{me})}$ arising from the superposition of the stress fields of a large number of dislocations and spatially varying on the length scale λ of the variations of the dislocation density. This scale is typically larger than $\rho^{-1/2}$ by one to two powers of ten [32]. Apart from rare exceptions, on the mesoscopic scale the external stress $\boldsymbol{\sigma}_{\text{ext}}$ may be considered as spatially constant.

[2] Peierls barriers of the first kind are the energy barriers between equivalent positions of a straight dislocation; these are overcome by kink-pair formation. Peierls barriers of the second kind are the barriers which have to be overcome by a kink moving *along* an isolated dislocation line.

Under certain circumstances it may be appropriate to introduce a hierarchy of mesoscopic scales, e.g. in polycrystals whose grain size is small compared with the macroscopic dimensions. In this example two mesoscopic scales arise naturally. Mesoscopic scale I deals with the dislocation patterns (cells, subgrains, etc.) within the grains, whereas mesoscopic scale II comprises many grains. On the mesoscopic scale II the distributions of the orientations (known as texture), of the sizes, and of the shapes of the crystallites are considered. Since the present contribution focuses on the dislocation dynamics in single crystals and on the subgrain scale of polycrystals, the term 'mesoscopic' refers always to the mesoscopic scale I unless specified otherwise.

- On the *macroscopic scale*, the specimen may be considered as being composed of macroscopic volume elements whose extensions are large compared to the mesoscopic length scales of the microstructure. On this scale, it is in many cases possible to describe the plastic response by deterministic constitutive laws which result from averaging over the dynamics of the dislocation system on microscopic and mesoscopic scales.

In dislocation dynamics, three different *time scales* may be distinguished:

(i) The shortest time scale is characterized by the time required for the elementary 'steps' of dislocation motion. Depending on circumstances, it may be determined by the scattering of phonons or electrons by dislocations, or by the much longer time required for dislocations to overcome Peierls barriers or localized obstacles with the help of thermal activation.

(ii) The dislocation motion on the mesoscopic scale I often occurs in 'avalanches' leading to the formation of slip lines or slip bands which involve many dislocations or dislocation segments. The characteristic duration of the spatio-temporal fluctuations associated with the avalanches defines the second (intermediate) time scale.

(iii) The dislocation patterns in deformed crystals exhibit features such as cell patterns, kink walls etc. that persist much longer than the time scale of dislocation avalanches. We denote such long-lived features by the term *dislocation microstructure*; their lifetime defines the third (longest) time scale. More generally, 'microstructure' refers to long-lived features in the arrangement of many elementary defects; hence it pertains to the mesoscopic length scale. The relationships between length scales, defect structure, and internal stresses are schematically illustrated in fig. 1.

If the motion of dislocations is controlled by the atomistic configuration of their cores, with dislocation–dislocation interactions being of secondary importance, we may draw conclusions on macroscopic crystal plasticity directly from the elementary mechanisms which govern the motion of dislocations on the atomistic level. A classical example is the plasticity of body-centred cubic (b.c.c.) metals below the knee temperature T_K, i.e., at temperatures at which the flow stress is controlled by the thermally activated nucleation of kink pairs on screw dislocations [33,34]. Other examples are the flow-stress anomaly of ordered intermetallic compounds, which has been discussed in terms of non-planar core splitting of superdislocations leading to the formation of immobile dislocation locks [31], or the plasticity of quasicrystals where the motion of dislocations is necessarily

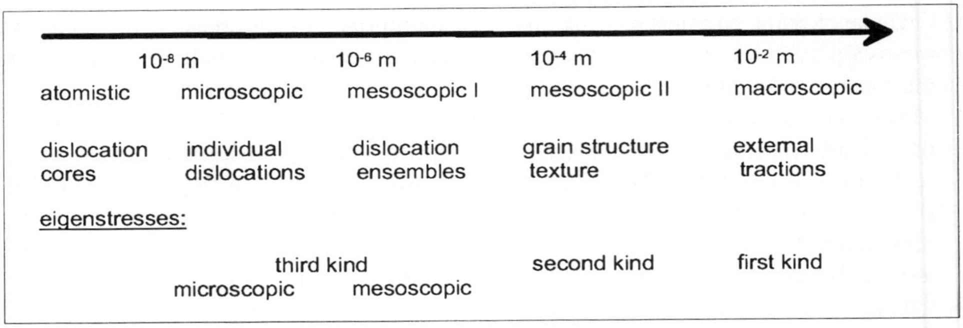

Fig. 1. Length scales associated with dislocation systems, microstructure, and internal stresses.

accompanied, and hindered, by the formation of structural defects, the so-called phasons [35].[3]

2.2. Eigenstresses and further notations

The preceding consideration on length scales bears a close relationship to the concept of eigenstresses that emerged at the beginning of the last century [36]. Following Masing's terminology [37], we denote as *eigenstresses* $\boldsymbol{\sigma}_{(\mathrm{I})}$ *of the first kind* internal stresses which arise from the non-uniformity of macroscopic deformation. *Eigenstresses* $\boldsymbol{\sigma}_{(\mathrm{II})}$ *of the second kind* are stresses which ensure compatible deformation of a polycrystal and which, therefore, may vary strongly from grain to grain. They belong to the mesoscopic scale II and will be referred to as intergranular stresses. *Eigenstresses of the third kind* are associated with dislocations; they belong to the microscopic scale but may also appear on the mesoscopic scale I. Thus, a further distinction is appropriate: *Long-range stresses* which vary on the characteristic scale λ of dislocation-density variations will be called *mesoscopic* eigenstresses of the third kind or, for short, mesoscopic stresses, while the internal-stress variations in the vicinity of single dislocations will be called *microscopic* eigenstresses of the third kind or simply *microscopic stresses.* The mesoscopic stresses are denoted by $\boldsymbol{\sigma}^{(\mathrm{me})}_{(\mathrm{III})}$, while the microscopic stress fluctuations are denoted by $\delta\boldsymbol{\sigma}_{(\mathrm{III})}$. (For a formal definition of these quantities, see section 3.2.) The tensor of the local stress acting in a microscopic volume element,

$$\boldsymbol{\sigma} = \boldsymbol{\sigma}_{\mathrm{ext}} + \boldsymbol{\sigma}_{(\mathrm{I})} + \boldsymbol{\sigma}_{(\mathrm{II})} + \boldsymbol{\sigma}^{(\mathrm{me})}_{(\mathrm{III})} + \delta\boldsymbol{\sigma}_{(\mathrm{III})}, \tag{3}$$

[3]Quasicrystals can be described in terms of projections of six-dimensional periodic hyperlattices on three-dimensional space. The analogue to a complete dislocation in a crystal is characterized by a 6D hyperlattice vector. Its projection on 3D space is not a lattice vector. Hence, dislocations in a 3D quasicrystal are necessarily partial dislocations and must trail stacking faults (the 'phason wall') behind them. The same conclusion may be arrived at by considering dislocations in the sequence of approximants of a quasicrystal. Clearly, in the limit of very large approximant unit cells any Burgers vectors of finite length belong to a partial dislocation.

is the sum of the external stress, the eigenstresses of the first and second kind, and the mesoscopic and microscopic eigenstresses of the third kind. The volume average of the microscopic stresses vanishes over a mesoscopic volume $\Delta V \approx \lambda^3$, the volume average of the mesoscopic stresses vanishes over a grain (or, more generally, over any volume $\gg \lambda^3$), and the volume average of the eigenstresses of the second kind vanishes if taken over a volume containing many grains. The volume average of the eigenstresses of the first kind vanishes over the macroscopic volume V_0 of the deformed body. The external stress fulfills the relationship

$$\{\boldsymbol{\sigma}_{\text{ext}}\} := \frac{1}{V_0}\int_{V_0} \boldsymbol{\sigma}_{\text{ext}}\, \mathrm{d}^3 r = \frac{1}{V_0}\int_{V_0} \boldsymbol{\sigma}(\boldsymbol{r})\, \mathrm{d}^3 r = \frac{1}{V_0}\int_{\partial V_0} \boldsymbol{r} \otimes \boldsymbol{\tau}\, \mathrm{d}^2 r, \tag{4}$$

where ∂V_0 is the surface of the body, $\boldsymbol{\tau}$ are the applied surface tractions, and $\otimes$ denotes the dyadic product. Here and in the following, macroscopic volume averages are denoted by curly brackets $\{\cdots\}$.

In single crystals or on the subgrain scale of polycrystals, only the (mesoscopic and microscopic) eigenstresses of the third kind are position dependent. Since we shall exclusively consider processes on this scale, we shall drop the superscripts (I)–(III) and denote as ‘internal stresses’ $\boldsymbol{\sigma}_{\text{int}} = \boldsymbol{\sigma}^{(\text{me})} + \delta\boldsymbol{\sigma}$ the eigenstresses of the third kind. The remainder, namely the sum $\boldsymbol{\sigma}_{\text{ext}} + \boldsymbol{\sigma}_{(\text{I})} + \boldsymbol{\sigma}_{(\text{II})}$, is treated as a constant ‘external stress’. As a further simplification (in the present context only rarely a serious one), we assume this ‘external stress’ to be uniaxial.

Unless otherwise stated, the dislocations are assumed to slip on crystallographic planes. Slip plane (with normal $\boldsymbol{n}^\beta$) and slip vector $\boldsymbol{e}_{\text{e}}^\beta$ define the slip system β. The slip vector is a unit vector in the slip direction of edge dislocations, which coincides with the direction of the Burgers vector $\boldsymbol{b}^\beta = b\boldsymbol{e}_{\text{e}}^\beta$. Together with a third unit vector $\boldsymbol{e}_{\text{s}}^\beta = \boldsymbol{e}_{\text{e}}^\beta \times \boldsymbol{n}^\beta$, which denotes the direction of motion of screw dislocations on the slip plane with normal $\boldsymbol{n}^\beta$, these vectors span a Cartesian coordinate system which we use for characterizing dislocation configurations in the slip system β. The components of the tensor $\dot{\boldsymbol{\varepsilon}}^{\text{pl}} = \{\dot{\varepsilon}_{kl}^{\text{pl}}\}$ of the plastic deformation rate are obtained from the shear-strain rates ε^β in the different slip systems according to

$$\dot{\varepsilon}_{kl}^{\text{pl}} = \sum_\beta M_{kl}^\beta \dot{\varepsilon}^\beta. \tag{5}$$

The projection tensors M_{kl}^β are given by $M_{kl}^\beta = (b_k^\beta n_l^\beta + n_k^\beta b_l^\beta)/(2b)$, where b_k^β and n_l^β denote the components of the Burgers vector $\boldsymbol{b}^\beta$ and of the slip plane normal $\boldsymbol{n}^\beta$, respectively. Slip systems with the same slip plane are called coplanar. Stresses resolved in a slip system are denoted by the superscript β. In single crystals, the slip system with the highest Schmid factor (ratio between resolved shear stress $\sigma_{\text{ext}}^\beta$ and the uniaxial stress) is called the primary system, while all other systems are denoted as secondary systems. In f.c.c. crystals under uniaxial stress with stress axis in $\langle 100\rangle$ and $\langle 111\rangle$ directions several $\langle 110\rangle\{111\}$ slip systems (8 for $\langle 100\rangle$, 6 for $\langle 111\rangle$) have a common non-zero Schmid factor while the Schmid factor of the remaining systems is zero. The former are denoted as active and the latter as inactive systems.

2.3. Classification of dislocation motions

Plastic deformation may be defined as deformation of a solid which persists, on the time scale of observation, after the load causing the deformation has been removed. For crystalline solids deforming by dislocation motion, this implies that dislocations during unloading do not return to their original positions – in other words: the unloading path of the dislocation ensemble (envisaged in a high-dimensional space spanned by the positions, tangent and Burgers vectors of the dislocation line segments) is different from the loading path and leads, in general, to a final configuration that differs from the initial one. It has been proposed to denote dislocation motions exhibiting this property of *kinematic* irreversibility as *non-inversive* [38]. Non-inversive behaviour implies that in general neither neighbourship relations between dislocation segments nor the initial dislocation length are recovered during unloading.

The notion of non-inversivity includes thermodynamic irreversibility, but inversive dislocation motions (i.e., motions in which the loading and the unloading paths coincide) are not necessarily thermodynamically reversible. Consider a dislocation segment pinned between two fixed anchoring points. When a stress is applied below a critical level, the segment will bow out between the anchoring points but returns to its original configuration when the stress is gradually removed. This motion is inversive but in general not thermodynamically reversible: Thermodynamic reversibility requires that the loading be done in a quasistatic manner. If loading is done at finite rate, electron or phonon drag effects will lead to friction and energy dissipation. The stress–strain diagram characterizing the behaviour of an assembly of such segments during a loading–unloading path is a closed loop with finite area, corresponding to the dissipated energy per unit volume. Assume now that the stress exceeds the critical level for the segment to bow out and act as a Frank–Read source [39]. Then a sequence of loops will be emitted. As long as these loops move through an obstacle-free crystal, their motion is still inversive as the loops will run back and disappear when the stress is removed. The process becomes non-inversive, however, as soon as parts of these loops reach the crystal surface, react with other dislocations, cross-slip or climb, or get trapped due to interactions with other imperfections acting as dislocation obstacles. Irrespective of the loading mode (quasi-static or not), deformation in this case proceeds in a thermodynamically irreversible manner.

When the initial shape of a plastically deformed specimen is restored by reverse straining, the initial internal state of the material is in general not restored. There are, however, apparent exceptions to this rule. A very important one may occur in strain-controlled cyclic deformation. Here, a fixed plastic strain amplitude is imposed and reversed during each deformation cycle. After many cycles, the specimen may reach a state of cyclic saturation in which adding a further cycle leaves the statistical properties of the microstructure virtually unchanged in spite of the fact that non-inversive dislocation motions have taken place both during forward and reverse straining. This is due to the fact that the mesoscopic dislocation microstructure reaches a (quasi-)stationary state of *dynamic equilibrium* even though the microscopic arrangement of the individual dislocations changes continually.[4]

[4]This dynamic equilibrium state is quasi-stationary, since on a much larger time scale the microstructure does change, e.g., by the nucleation and propagation of fatigue cracks.

Table 1
Dichotomies in plastic deformation.

Thermodynamics	*Reversible:* Mechanical work expended during loading is completely recovered during unloading.	*Irreversible:* Mechanical work is partly transformed into heat.
Dislocation motion	*Inversive:* Dislocation ensemble follows the same path in configurational space during loading and unloading.	*Non-inversive:* Dislocation ensemble follows different paths in configurational space during loading and unloading.
Microstructure	*Stationary:* Microstructure remains statistically invariant during deformation along a given strain path.	*Non-stationary:* Statistical properties of dislocation microstructure change in the course of deformation.

The concept of *dynamic equilibrium* in plastic deformation is defined only with respect to a fixed straining path. For instance, dislocation microstructures which are in dynamic equilibrium in strain-controlled cyclic deformation will lose this property when subjected to subsequent unidirectional straining. Dynamic equilibrium may also be reached in large-strain unidirectional deformation; again, the quasi-stationary subgrain patterns which evolve in this case may be unstable under subsequent deformation along a different straining path, e.g. under cyclic deformation.

Table 1 summarizes the preceding discussion in terms of three dichotomies that must be clearly distinguished because they belong to different levels of description: reversibility vs. irreversibility (thermodynamic), inversivity vs. non-inversivity (kinematic) and stationarity vs. non-stationarity (statistical).

2.4. Thermodynamics of plastic deformation

From the viewpoint of thermodynamics, a specimen undergoing plastic deformation under a load (which may even be its own gravity) is an open, strongly dissipative system. During the deformation the specimen is subjected to an energy flux with an 'input' of mechanical work from and an 'output' of heat to the experimental set-up. The mechanical work expended per unit volume during a strain increment $d\varepsilon$ is $dW = \boldsymbol{\sigma}_{ext}(\boldsymbol{\varepsilon})\,d\varepsilon$. The strongly dissipative nature of plastic deformation is borne out by the fact that, as mentioned in section 1, only a small fraction of this work, dW_{st}, increases the internal energy of the crystal, while the larger part is dissipated as heat. The non-dissipated energy, W_{st}, is stored mainly in the elastic stress fields of dislocations. The 'energy storage rate' $\eta_{st} = (dW_{st}/dW)$ is typically less than 10% and decreases with increasing deformation (fig. 2). Under quasi-stationary conditions, such as in high-cycle fatigue or in high-strain unidirectional deformation, it may become unmeasurably small.

On the microscopic scale, several processes contribute to the dissipation of mechanical energy and the concomitant entropy production: (i) all processes that lead to a viscous damping of the dislocation motion, including the thermally activated overcoming of Peierls barriers or of localized obstacles, the dragging of foreign atoms and their re-orientation, the scattering of phonons and electrons by moving dislocations; (ii) the generation of intrinsic atomic defects (vacant lattice sites and self-interstitials) by non-conservative motion of jogs and the subsequent annihilation of these defects at internal sinks; (iii) dislocation reactions

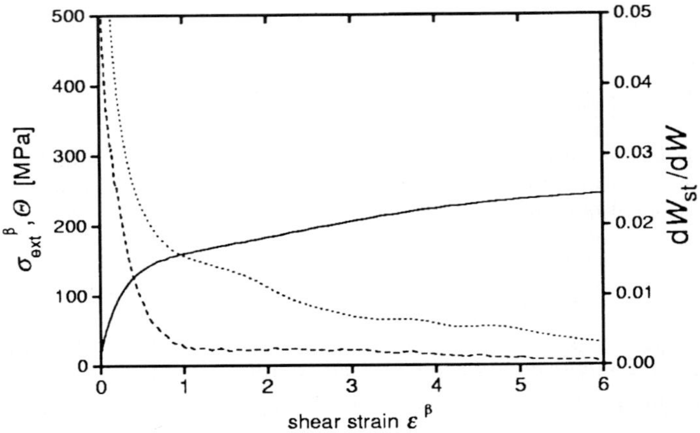

Fig. 2. Strain hardening and energy storage in a Cu polycrystal deformed in torsion at room temperature, after Kaps [40]. Full line: stress–strain curve (shear stress σ_{ext}^{β}) vs. shear strain ε^{β}), dashed line: hardening coefficient $\Theta = \partial\sigma_{ext}^{\beta}/\partial\varepsilon^{\beta}$, dotted line: stored-energy ratio $\eta_{st} = dW_{st}/dW$.

and dislocation annihilation. In low-speed deformation, the dissipation due to processes (i) and (ii) may be strong enough for the motion of dislocations to proceed in an overdamped manner. In this case, the dislocation motion is unaffected by the inertia of the dislocations, and the rules of quasi-static stress equilibrium apply in spite of a non-vanishing plastic deformation rate.

An important consequence of the strongly dissipative character of dislocation motions and reactions is that the ensuing dislocation patterns cannot be deduced from general principles of energy minimization. This general statement drawn from the thermodynamics of open systems [41,42] requires some comments when applied to dislocation systems in deforming crystals. In a high-dimensional configurational space, for such systems we may define an energy functional which includes the energy of the externally applied elastic field as well as the internal elastic fields and the core energies of the dislocations. In low-temperature, low-speed deformation, the dislocation system follows approximately a path of steepest descent with respect to this energy functional. When the external load is removed, the system relaxes into some local minimum of the internal energy. In this sense, it is trivial that any dislocation pattern observed after unloading corresponds to a minimum-energy structure. On the other hand, it is also trivial that the configuration of lowest energy is the perfect crystal with all dislocations removed. In sufficiently pure crystals this 'ideal' may be approximately achieved in high-temperature deformation or by appropriate annealing processes.

From the preceding it follows that the problem is not whether after unloading dislocation arrangements corresponding to local energy minima can be observed (they can!) but how out of the uncountable multitude of possible metastable configurations those emerging under given experimental conditions are selected. In view of its complexity, the problem has not yet been solved. On the other hand, the main issues have become clear, as will be discussed in the remainder of this subsection.

In materials with high dislocation mobility such as the f.c.c. metals, dislocation motion proceeds in a strongly intermittent manner in space and in time (cf. section 2.5). This means that, at a fixed location, the plastic strain rate is close to zero most of the time and that the plastic deformation takes place in isolated 'bursts' where the instantaneous strain rate may exceed the average (imposed) strain rate by several orders of magnitude [43].[5] While the imposed strain rate is carried by a small number of dislocations which move rapidly, at any given moment the large majority of the dislocations in the crystal is virtually at rest [43]. This is the reason why many properties of the dislocation arrangements can be understood in terms of static stress equilibrium or, equivalently, of arguments based on energy minimization [44]. While intermittence of dislocation motion allows us to account for *local* features of the dislocation arrangements in terms of energy minimization, it makes it virtually impossible to understand the *mesoscopic evolution* of the dislocation system within the framework of equilibrium or linear irreversible thermodynamics. At any given moment, the processes responsible for the evolution of the microstructure are essentially confined to a very small fraction of the crystal volume. Virtually all mechanical work done by the external tractions is dissipated into heat in these 'active slip volumes'. Hence, locally the rates of entropy production may be extremely high. As expected for far-from-equilibrium processes, the motions of dislocations within the active slip volume proceed in a collective and self-organized manner that can *not* be described in terms of local and uncorrelated relaxations of individual dislocation segments towards configurations of lower energy. The motion of one dislocation segment may trigger the motions of others, and the collective dynamics arising from this process can be understood in energetic terms only if the global energy functional of the entire dislocation system is considered. In tests with prescribed strain rate, however, the external tractions are continually adjusted so that the system can never reach a minimum of this functional.

Analogies of solids undergoing plastic deformation with biological systems have been discussed elsewhere [45,46]. An equally instructive but more transparent analogy is that of a sandpile to which grains are added at a small rate. Most of the time, practically all grains in the sandpile are at rest, and force equilibrium (or energy minimization) arguments can tell us a lot about local configurations and grain arrangements. When grain inertia is neglected, the evolution of the sandpile can be understood in terms of relaxation processes with a relaxation rate that depends on the downward slope of an energy functional which is slowly changed by the addition of grains. The dynamics of the relaxation is characterized by collective motions where moving grains trigger the motion of others. When the external driving is slow, the energy added from outside is dissipated through an intermittent sequence of avalanches which exhibit a power-law size distribution [47]. This type of complex dynamics has served as a paradigmatic example for the notion of self-organized criticality [48].

Returning to the plastic deformation of f.c.c. metals by slip, we may state that it exhibits several analogies with this type of critical behaviour [49,50]: (i) the evolution of the dislocation system takes place in discrete 'slip events' involving the collective motion of many dislocations; (ii) the external driving is slow, since the external stress can be

[5] On the surface, these bursts manifest themselves by the rapid formation of slip lines and slip bands.

considered constant during a slip event;[6] (iii) the size distribution of 'slip events' as monitored by acoustic emission has been reported to follow a power law [51,52] (cf. section 5.1). Hence, we may state that plastic deformation of f.c.c. metals resembles much more the intermittent dynamics of critical systems such as sandpiles [48] or driven interfaces [53,54] than the flow of a viscous fluid, even if the deformation is macroscopically 'smooth'. It is important to realize that collective dislocation motions are not confined to plastic instabilities [55,56] but constitute a generic feature of the microscopic and mesoscopic dynamics of dislocation systems in materials with high dislocation mobility.

2.5. Intermittency of plastic flow

2.5.1. Dissipation and fluctuations: quantitative estimates

As a measure of the 'jerkiness' of dislocation motion one may consider the relative fluctuations of the dislocation velocities. Hähner [57,58] has proposed a line of reasoning which allows the magnitude of these fluctuations to be estimated as follows: Consider an ensemble of dislocations moving in a fluctuating internal-stress field.[7] We define the effective resolved shear stress acting on a dislocation segment moving in slip system β as the sum of the external and internal stresses at the position of the segment, $\sigma^\beta_{\text{eff}} := \sigma^\beta_{\text{ext}} + \sigma^\beta_{\text{int}}(\boldsymbol{r})$. When dislocation inertia can be neglected, the energy dissipation rate due to the motion of this segment is proportional to the product $\sigma^\beta_{\text{ext}} v^\beta$, where v^β is the velocity of the segment. The energy per unit volume and time dissipated by all dislocations of slip system β is $\rho^\beta b \langle \sigma^\beta_{\text{eff}} v^\beta \rangle = \rho^\beta b (\sigma^\beta_{\text{ext}} \langle v^\beta \rangle + \langle \sigma^\beta_{\text{int}} v^\beta \rangle)$, where ρ^β is the dislocation density in this slip system. Here and in the following, angular brackets $\langle \cdots \rangle$ denote averages over the dislocation ensemble.

Let us for simplicity assume that no work is stored in the crystal, $\eta_{\text{st}} = 0$.[8] Then the energy dissipation on average equals the expended work, $\rho^\beta b \langle \sigma^\beta_{\text{eff}} v^\beta \rangle = \sigma^\beta_{\text{ext}} \langle \dot{\varepsilon}^\beta \rangle$. Using Orowan's relation $\dot{\varepsilon}^\beta = \rho^\beta b v^\beta$ yields

$$\langle \sigma^\beta_{\text{int}} v^\beta \rangle = 0, \tag{6}$$

hence the average work done by the internal stresses is zero. Using the definitions $\sigma^\beta_{\text{int}} := \langle \sigma^\beta_{\text{int}} \rangle + \delta\sigma^\beta_{\text{int}}$, $v^\beta := \langle v^\beta \rangle + \delta v^\beta$, and noting that $\delta\sigma^\beta_{\text{eff}} = \delta\sigma^\beta_{\text{int}}$, we obtain cross-correlations between the fluctuations of the effective stresses acting on the moving dislocations and of the dislocation velocities,

$$\langle \delta\sigma^\beta_{\text{eff}} \delta v^\beta \rangle = -\langle \sigma^\beta_{\text{int}} \rangle \langle v^\beta \rangle. \tag{7}$$

[6]The duration of such events can be assessed by surface observations. It corresponds to the time of activity of a slip line or slip band, which ranges from milliseconds to a few seconds [43].

[7]In a work-hardened crystal, this is mainly the stress field of other dislocations. However, the origin of the internal stresses is not crucial for the following argument.

[8]Taking into account energy storage modifies the results given in eqs (9) and (11) by a factor $(1 + \eta_{\text{st}}) \approx 1$.

At this point, it is important to avoid the pitfall of confusing spatial and ensemble averages and arguing that the average internal stress is zero. This holds for the *spatial average* but not for the *ensemble average* $\langle\sigma^{\beta}_{\text{int}}\rangle$ since dislocations spend most of the time in configurations where their motion is hindered by back stresses. When dislocations are highly mobile and when the strain rates are low, the dislocations spend practically all time in positions where the internal stress roughly balances the external stress, hence $\langle\sigma^{\beta}_{\text{int}}\rangle \approx -\sigma^{\beta}_{\text{ext}}$ holds.

In order to obtain the autocorrelations of the fluctuations of internal stress and dislocation velocity, one has to specify the stress dependence of the latter. We consider two cases:

(i) If the dislocation velocities v^{β} are mainly limited by phonon and electron drag effects, they are proportional to the effective shear stress acting on the dislocation,

$$v^{\beta} = b\sigma^{\beta}_{\text{eff}}/\mu, \tag{8}$$

where the dislocation drag coefficient μ is typically of the order of magnitude of 10^{-4} Pa s [59]. Accordingly, the average dislocation velocity is $\langle v^{\beta}\rangle = [b/\mu]\langle\sigma^{\beta}_{\text{eff}}\rangle$. To get an estimate of the order of magnitude of the average effective stress $\langle\sigma^{\beta}_{\text{eff}}\rangle$, we use Orowan's relation and consider typical material parameters for copper, $\mu = 5 \times 10^{-5}$ Pa s [59] and $b = 2.56 \times 10^{-10}$ m. At the beginning of hardening stage II of single-glide orientated Cu single crystals, the resolved shear stress and dislocation density in the primary slip system are $\sigma_{\text{ext}} = 5$ MPa and $\rho^{\beta} = 2 \times 10^{12}$ m^{-2} [60]. For these parameters and a typical strain rate $\langle\dot{\varepsilon}^{\beta}\rangle = 10^{-3}$ s^{-1} one finds that the effective stress in the primary system is $\langle\sigma^{\beta}_{\text{eff}}\rangle \approx 8 \times 10^{-7}$ MPa, which is negligibly small in comparison with the applied stress. Hence, the relation $\langle\sigma^{\beta}_{\text{int}}\rangle = -\sigma^{\beta}_{\text{ext}}$ is well fulfilled.

Inserting eq. (8) into eq. (7) and using $\langle\sigma^{\beta}_{\text{int}}\rangle \approx -\sigma^{\beta}_{\text{ext}}$ yields autocorrelations of the effective-stress and dislocation velocity fluctuations:

$$\left\langle\left(\delta\sigma^{\beta}_{\text{eff}}\right)^2\right\rangle = \sigma^{\beta}_{\text{ext}}\left\langle\sigma^{\beta}_{\text{eff}}\right\rangle, \qquad \frac{\langle(\delta v^{\beta})^2\rangle}{\langle v^{\beta}\rangle^2} = \frac{\sigma^{\beta}_{\text{ext}}}{\langle\sigma^{\beta}_{\text{eff}}\rangle} = \frac{b^2\rho^2\sigma^{\beta}_{\text{ext}}}{\mu\langle\dot{\varepsilon}^{\beta}\rangle}. \tag{9}$$

It is seen that the fluctuation amplitude of the effective stress is the geometrical mean of the external and effective stresses, while the relative amplitude of the velocity fluctuations is the ratio of both quantities. Since the average effective stresses may be very small, the dislocation velocity fluctuations are large. In the last step in eq. (9), we have used Orowan's relation to express the dislocation velocity fluctuations in terms of internal stress, imposed strain rate, dislocation density, and drag coefficient. For the parameters given above for Cu at the beginning of hardening stage II, we find that the relative amplitude of dislocation velocity fluctuations in the primary system may be as high as $\langle(\delta v^{\beta})^2\rangle/\langle v^{\beta}\rangle^2 \approx 6 \times 10^6$. Measurements by Neuhäuser [43] indicate that local shear strain rates deduced from slip-line kinematography in α-brass exceed the imposed strain rates by factors between 4.8×10^7 and 6.9×10^6 for imposed strain rates between $10^{-5}\,\text{s}^{-1}$ and $10^{-3}\,\text{s}^{-1}$. When one assumes that the dislocations in a slip line or slip band have all the same velocity while dislocations outside slip lines or bands do not move, the ratio between the local and global shear strain rates coincides with the relative magnitude of dislocation velocity fluctuations. Hence the estimates deduced from eq. (9) are in agreement with Neuhäuser's findings.

(ii) In [57,58], Hähner considered thermally activated dislocation motion, where the dislocation velocity is governed by an Arrhenius law, $v^\beta = v_0 \exp[-H^{\mathrm{eff},\beta}/k_\mathrm{B}T]$ where k_B is Boltzmann's constant, T temperature and H^{eff} a stress-dependent effective activation enthalpy for dislocation motion. Differentially this can be written as

$$\frac{\partial v^\beta}{\partial \sigma^\beta_{\mathrm{eff}}} = \frac{v^\beta}{S^\beta}, \qquad S^\beta = \frac{k_\mathrm{B}T}{V^\beta_\mathrm{a}}, \tag{10}$$

where $V^\beta_\mathrm{a} := \partial H^{\mathrm{eff},\beta}/\partial \sigma^\beta_{\mathrm{eff}}$ is the activation volume. Under the additional assumption that the stress and velocity fluctuations can be approximated by Gaussian stochastic processes, one can use the Furutsu–Novikov theorem [61] according to which (monotonic and differentiable) functions $h(\psi)$ of a Gaussian stochastic variable ψ fulfill the relationship $\langle \psi h(\psi)\rangle = \langle \psi^2\rangle\langle \partial h/\partial\psi\rangle$. With the relations $\partial v^\beta/\partial\sigma^\beta_{\mathrm{eff}} = -v^\beta/S$ and $\langle\sigma^\beta_{\mathrm{int}}\rangle \approx -\sigma^\beta_{\mathrm{ext}}$ one obtains from eq. (7) the fluctuation autocorrelations

$$\left\langle\left(\delta\sigma^\beta_{\mathrm{eff}}\right)^2\right\rangle = \sigma^\beta_{\mathrm{ext}} S^\beta, \qquad \frac{\langle(\delta v^\beta)^2\rangle}{\langle v^\beta\rangle^2} = \frac{\sigma^\beta_{\mathrm{ext}}}{S^\beta}. \tag{11}$$

These relations are similar to eq. (9) with the only difference that in the 'thermal' case the strain-rate sensitivities S^β replace the average effective stresses. Since strain-rate sensitivities are generally small as compared to the external stress ($\sigma^\beta_{\mathrm{ext}}/S^\beta \approx 10^2$ in f.c.c. metals, [62]) one finds again that the relative fluctuations of the dislocation velocities are large. In b.c.c. metals at temperatures below the transition temperature, S may be much larger and accordingly dislocation motion is smoother.

While the velocity distribution of dislocations or dislocation segments in a deforming crystal is difficult to measure, it can be easily determined in dislocation-dynamics simulations. Since in simulations the stress–velocity law is fixed a *priori,* it is a useful exercise to compare predictions obtained from eq. (9) to simulation results. Figure 3 shows the velocity spectrum of dislocation segments in a 3D simulation by Kubin et al. [63]. The simulation pertains to symmetrical multiple slip with the tensile axis in a $\langle 100\rangle$ orientation, an imposed strain rate of 50 s^{-1}, and a linear stress-velocity law with a drag coefficient $\mu = 5 \times 10^{-5}$ Pa s. The strain was about 0.6%, and the flow stress about 6 MPa. The dashed line indicates the average velocity of the dislocation ensemble. From fig. 3 one finds a relative fluctuation magnitude $\langle(\delta v^\beta)^2\rangle/\langle v^\beta\rangle^2 \approx 56$. Equation (9) yields a value of about 50, i.e., a correct estimate of the velocity fluctuation magnitude observed in the simulation. One also notes that the fluctuation magnitude in the simulation is substantially smaller than in a typical experiment mainly due to the high imposed strain rate. This is unavoidable since the relative velocity fluctuation amplitude directly translates into numerical stiffness of the simulations, and therefore a high imposed strain rate which reduces the fluctuations is indispensable to run simulations within reasonable computation time. This implies that simulations necessarily tend to underestimate the

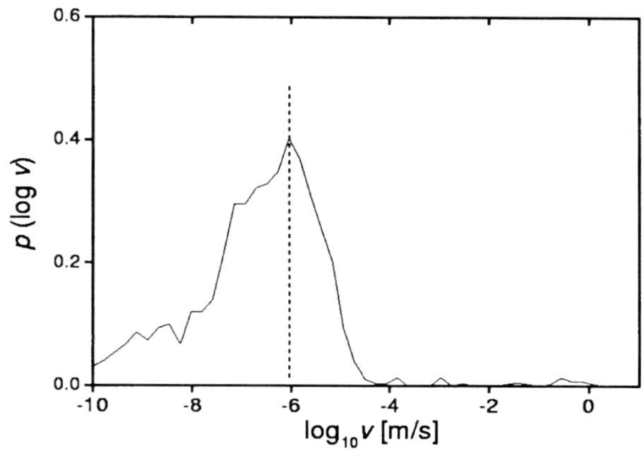

Fig. 3. Velocity spectrum of dislocation segments in a 3D simulation, strain 0.6%, flow stress 6 MPa; after Kubin and Devincre [63].

degree of intermittency of plastic flow and may not fully monitor the fluctuation effects that are observed by experiment.

2.5.2. Cooperative motion of dislocations

In f.c.c. metals, jerkiness and intermittency are not restricted to the microscopic level of single-dislocation motions but characterize the dynamics of plastic deformation also on mesoscopic scales. The observations of slip lines and slip bands indicate the simultaneous and correlated motion of many dislocations. Often a hierarchy of surface features is observed where, at the lowest level, one finds surface steps which are produced by the motion of small groups of dislocations and are denoted as slip lines. These lines may form clusters which are denoted as slip bands, and in turn those bands may aggregate into slip-band bundles [43]. In either case, the formation of the surface markings proceeds in a temporally correlated manner. The formation of a slip line is due to the correlated motion of several dislocations which typically takes a few milliseconds [43], the formation of a slip band is due to the correlated formation of several slip lines, and so on. It has been proposed by some investigators to characterize slip patterns in terms of fractal geometry [64–66], but no unambiguous evidence has been found that slip patterns are generically fractals. While slip-line patterns recorded in small strain intervals exhibit fractal characteristics, at large strains the slip-line distribution corresponds to a statistically homogeneous random pattern [65].

The cooperative character of dislocation motion has been noted already in the first systematic investigations of slip traces on the surface of deforming f.c.c. crystals [67] where the average slip-step height observed during Stage-II deformation indicates the collective motion of groups with an average number of about 20 dislocations. The corresponding slip events are confined in space, as may be seen from the fact that slip lines terminate on the surface. This observation has been interpreted in terms of confined ‘slip zones’ at the boundaries of which dislocations are blocked at least temporarily. The geometry of a slip zone can be qualitatively characterized in terms of three parameters, *viz*. (i) the extension ξ_g in the direction of dislocation motion (the slip-line length) which may depend on the dislocation character (edge or screw), (ii) the extension ξ_n normal to the glide plane (the width of a slip line or slip band), and (iii) the number n of dislocations involved in the 'slip event' leading to the formation of a slip zone or, equivalently, the characteristic strain $\hat{\varepsilon}$ produced by this event.

The observation of spatially confined slip zones has three important consequences:

- (i) The termination of a slip line implies storage of dislocations. This spatially inhomogeneous storage may play a rôle in the evolution of inhomogeneous dislocation microstructures.
- (ii) At the boundary of a slip zone there are large internal stresses which may be partly relaxed by slip on other systems or by the correlated formation of slip zones [68]. Again the resulting dislocation entanglements are distributed in an irregular and inhomogeneous manner.
- (iii) The heterogeneity of slip creates lattice rotations.

The preceding conclusions indicate that cooperative motions of dislocations are a crucial feature in work-hardening and microstructure evolution. However, there is no straightforward relationship between the parameters characterizing collective slip and those characterizing the inhomogeneous microstructure, as the slip-line lengths always exceed the characteristic 'wavelengths' of the dislocation patterns. In high-symmetry $\langle 100 \rangle$- and $\langle 111 \rangle$-orientated Cu single crystals, Ambrosi and Schwink report a ratio of slip-line length to cell size of about 3 [69]. For single-glide orientated copper, Mughrabi reports that the slip length of primary edge dislocations exceeds the distance between dislocation-dense bundles in the primary glide plane by a factor of about 6 [32] while Basinski and Basinski point out that this factor may be as large as 16 [62].

2.6. Microstructure evolution and work-hardening

Any theory of plastic deformation has to deal with three interrelated questions, namely (i) what determines the stress required to deform a solid with given microstructure (theory of the flow stress), (ii) how does plastic flow proceed on mesoscopic and microscopic scales (theory of the collective dynamics of dislocations), and (iii) what processes govern the evolution of the dislocation microstructure (theory of dislocation patterning and microstructure evolution).

2.6.1. Flow-stress models

While the flow stress of b.c.c. metals is well understood in terms of dislocation core properties (stress required to form kink pairs on screw dislocations), a long-standing controversy concerns the mechanism which controls the flow stress in f.c.c. metals. There is agreement that the flow stress is governed directly or indirectly by dislocation interactions, but the precise nature of the flow-stress-controlling interactions has been much disputed, and an abundance of mechanisms have been proposed in the literature of the fifties and sixties (for a compilation of such mechanisms, see [70]). Examples are (a) the stress required to activate a dislocation source, (b) the passing stress of two parallel straight dislocations, (c) the stress required to pass many-dislocation configurations creating long-range stresses (e.g., pile-ups), (d) the overcoming of attractive or repulsive interactions with forest dislocations, and (e) the stresses required to create and move jogs. All these mechanisms have in common that they predict the flow stress of a slip system to be proportional to the square root of some dislocation density ρ (Taylor relationship)

$$\sigma_{\mathrm{f}}^{\beta} = \alpha^{\beta} G b \sqrt{\rho}, \tag{12}$$

where G is an effective shear modulus, b the modulus of the Burgers vector, and α^{β} a non-dimensional parameter which varies for the different mechanisms typically between 0.2 and 0.5. The Taylor relationship is rather unspecific – irrespective of how the different mechanisms superimpose and which mechanism prevails in a given physical situation, the basic structure of eq. (12) is preserved. Not only do virtually all conceivable mechanisms lead to this relationship (with possible logarithmic corrections); it has also been found in both two- and three-dimensional dislocation dynamics simulations as well as in innumerable experimental investigations. In view of this situation one may think of adopting a pragmatic stance and accept the Taylor relationship as generic without going into the 'mechanistic' details.

While such a pragmatic approach may avoid fruitless controversies about the relevance of particular mechanisms, it is important to make a basic distinction which is crucial for theories of work-hardening and microstructure evolution. This concerns the *scale* of the flow-stress controlling mechanism. If we disregard mechanism (e), which involves dislocation core effects, all mechanisms mentioned above rely on elastic dislocation interactions. Mechanisms (a), (b), and (d) consider interactions between individual dislocations or dislocation segments. For these mechanisms, the range of the flow-stress controlling interactions is of the order of one dislocation spacing and therefore pertains to the microscopic scale of dislocation dynamics. Mechanism (c), on the other hand, relates the flow stress to long-range stresses caused by many-dislocation configurations with a large net Burgers vector. Such configurations are associated with mesoscopic incompatibilities of plastic flow such as terminations of slip lines. The characteristic range of these stresses is on the mesoscopic rather than on the microscopic scale. Here, mesoscopic stresses *are* significant, as shown by transmission electron microscopy or X-ray scattering (for a discussion of these observations, see section 4). The crucial question is how they influence the flow stress.

In the *composite models* formulated by Mughrabi and others [71–75], it is assumed that the flow stress is mainly governed by mechanisms on the microscopic scale such as dislocation–forest interactions. In this case it is possible to define a local flow stress for each mesoscopic volume element which depends only on the density and arrangement of the dislocations within this element. During loading, first the softest volume elements begin to deform plastically. Plastic flow in these elements then leads to mesoscopic strain gradients which, in turn, give rise to mesoscopic stresses that lead to a *stress re-distribution* until the sum of the mesoscopic stresses and the external stress matches the local flow stress everywhere. When this re-distribution is accomplished, the external stress (the macroscopic flow stress) is equal to the spatial average of the local flow stresses, while the mesoscopic stresses do not contribute. The crucial point of such composite models is

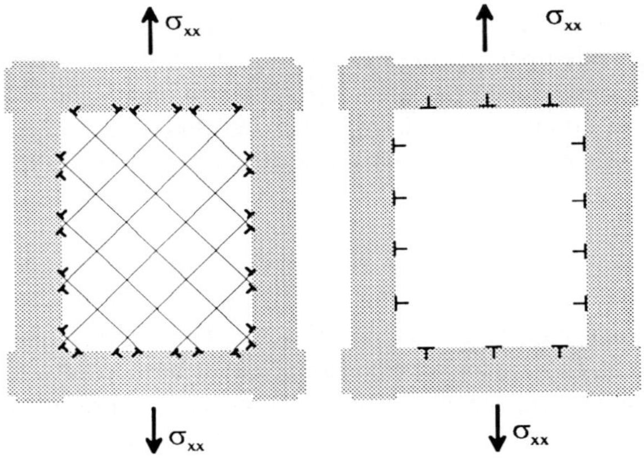

Fig. 4. Dislocation arrangement in a composite model, after Mughrabi [73]. The accumulated excess dislocations lead to a mesoscopic stress $\sigma_{xx}^{(me)} > 0$ in the cell walls and a mesoscopic stress $\sigma_{xx}^{(me)} < 0$ in the cell interiors. This internal-stress pattern directly monitors the pattern of local flow stresses, which are large in the cell walls and small in the interiors.

that, after an initial transient, plastic flow is supposed to proceed in a compatible manner on mesoscopic scales. Long-range stresses arise because they are required to establish compatible plastic flow in an inhomogeneous microstructure, i.e., they are a *consequence* of spatially varying local flow stresses and directly monitor the local flow-stress pattern. This is illustrated in fig. 4, which shows a cell structure in which the local flow stress is high within the dislocation-dense cell walls and low within the dislocation-depleted cell interiors. Dislocations moving in the cell interiors accumulate at the cell walls and thereby accommodate the strain gradient that develops as long as the walls do not yield. These dislocations create a mesoscopic internal stress field which enhances the tensile stress within the cell walls and reduces it within the cell interiors. Accumulation of dislocations continues until the mesoscopic stresses compensate the flow-stress difference between cell walls and interiors, so that that dislocations can move in both regions.[9]

According to long-range stress models [76–79], on the other hand, the mesoscopic stresses caused by incompatibilities of plastic flow on mesoscopic scale such as terminations of slip zones may substantially contribute to the flow stress. In this case, the flow stress of a mesoscopic volume element depends on the dislocation arrangement in the surrounding volume elements in a non-local manner. Such a situation is illustrated in fig. 5, which shows a group of dislocations immobilized by a strong obstacle. As in fig. 4, the accumulated dislocations create a 'forward' stress on the obstacle and a back stress in the obstacle-free region behind it. These stresses are mesoscopic in the sense that they vary on a scale that substantially exceeds the individual dislocation spacing. However, in contrast

[9]We note that, in addition to the mesoscopic stress field, the accumulated dislocations create local internal-stress fluctuations in the immediate vicinity of the dislocation lines. The range of these fluctuations is on the microscopic scale of one or two dislocation spacings only; they do not affect plastic flow over larger distances. However, as well as other kinds of short-range interaction, they may contribute to the local flow stress at the wall surface. In section 3.3 it will be shown that this contribution is negligibly small.

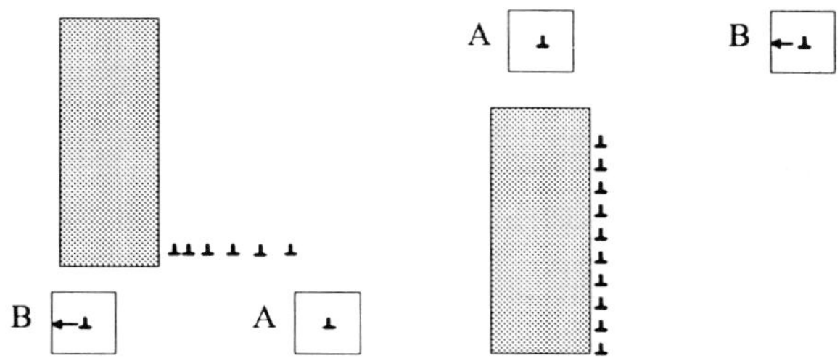

Fig. 5. Dislocation configurations with long-range internal stresses which cannot be envisaged within the composite framework: the long-range stresses do not monitor the local flow-stress pattern. While the local flow-stress contributions are the same in the points denoted by A and by B, the long-range stresses of the dislocation configurations (a pile-up and a terminated wall) hinder dislocation motion in the points A but assist dislocation motion in the points B.

to fig. 4 these mesoscopic stresses have a complicated spactial dependence that does *not* simply monitor the local flow-stress pattern. For instance, in the mesoscopic volume element indicated by A, dislocations may get trapped in spite of the fact that there are no local obstacles, while in the element indicated by B, they may move. The fact that the piled-up dislocations influence the motion of other dislocations in a non-local manner has important consequences. Arrangement of such pile-ups may give rise to a flow stress that exceeds the volume average of the local obstacle strengths. This can be seen by considering obstacles of large-but-finite strength and very small extension. The contribution of such obstacles to a volume-averaged strength is insignificant. Nevertheless, the obstacles act in a 'catalytic' manner when dislocations pile up against them. In this case, the motion of other dislocation is not only hindered by local obstacles, but also by the long-range stresses of the piled-up groups. Hence, it is not possible to 'disentangle' the flow stress from the mesoscopic stresses by separation of length scales. In calculating flow stresses in such situations, composite models fail and many-dislocation interactions on mesoscopic scales must be taken explicitly into consideration. This increases the difficulty of flow-stress calculations substantially unless very idealized dislocation configurations (e.g., isolated pile-ups which may be considered as 'superdislocations') are envisaged. The fact that possible sources of long-range stresses in real dislocation arrangements are much more complex [78,79] and can hardly be treated analytically may have contributed to a limited acceptance of long-range stress models. In the future, simulations may help to overcome this problem.

The dislocation arrangements evolving in f.c.c. single crystals orientated for single slip are rather complex and the corresponding internal-stress 'landscape' may not be fully described by either long-range stress or composite models. Within a slip zone, the dislocation microstructure is inhomogeneous since slip lines cross several cell walls [69] or dislocation entanglements [32]. Since these inhomogeneities act as dislocation obstacles, compatibility of deformation within the slip zone requires mesoscopic stresses which compensate the concomitant local flow-stress inhomogeneities. These stresses can be calculated from composite models. Terminations of slip lines, on the other hand, introduce mesoscopic incompatibilities of plastic deformation which can *not* be described within the composite framework which is built upon the assumption that plastic flow (after a transient) proceeds in a mesoscopically compatible manner. The flow stress is then a superposition of two components: (i) the stresses arising from the termination of slip zones, and (ii) the spatial average of the local flow-stress contributions within a slip zone that are due to dislocation interactions on the microscopic scale. The relative importance of both components may differ from case to case. When many slip systems are active such as in $\langle 100 \rangle$ or $\langle 111 \rangle$ orientated f.c.c. single crystals, plastic relaxation of mesoscopic stress concentrations is easy and 'microscopic' forest interactions are predominant. In these situations composite models yield a quantitative description of the relationships between mesoscopic internal stresses and dislocation microstructure [73]. On the other hand, in the early stages of deformation of crystals orientated for single slip long-range stresses may contribute significantly to the flow stress.

Experimental information about the scale of the flow-stress controlling mechanism and the nature of the mesoscopic stresses can be obtained by comparing the internal-stress patterns in the loaded and unloaded states. When the local flow stresses are governed by processes on the microscopic scale, the effective obstacle spacing is on the order of one dislocation spacing. In this case, one expects that during unloading the 'geometrically necessary' dislocations which create the mesoscopic stress pattern get trapped close to the positions which they have occupied under load. Hence the mesoscopic stresses do not change appreciably during unloading. If, by contrast, the flow stress is controlled by long-range stresses, the effective obstacle spacing is much larger and relaxation processes during unloading will substantially change the mesoscopic stress pattern (cf. section 4).

2.6.2. *Work-hardening*

Using the Taylor relationship, it is easy to formulate phenomenological hardening models in terms of differential equations for the evolution of dislocation densities. This has been the strategy pursued by the dislocation dynamics approach of the seventies and eighties. For instance, Kocks [80] proposed to describe dislocation accumulation by a phenomenological equation of the form

$$\frac{\partial \varrho}{\partial \varepsilon} = k_1 \sqrt{\varrho} - k_2 \varrho, \tag{13}$$

where ϱ is the total dislocation density, ε the total strain, and the phenomenological constants k_1 and k_2 are supposed to describe dislocation multiplication and dynamic recovery, respectively. With a Taylor relationship between ϱ and the (tensile) flow stress, $\sigma_{\text{ext}} = \alpha G b \sqrt{\varrho}$, this leads to a hardening law of the Voce type [81],

$$\frac{\partial \sigma_{\text{ext}}}{\partial \varepsilon} = \Theta_0 \left(1 - \frac{\sigma_{\text{ext}}}{\sigma_{\text{ext}}^{\infty}} \right), \tag{14}$$

where the initial hardening coefficient is $\Theta_0 = k_1 \alpha G b / 2$ and the asymptotic stress $\sigma_{\text{ext}}^{\infty} = \alpha G b k_1 / k_2$. This phenomenological model has been applied to characterize Stage-III hardening of single- and polycrystals.

Later this approach has been elaborated by distinguishing the densities ρ_{m} and ρ_{i} of mobile and immobile dislocations. This has proved useful to describe softening phenomena associated with rapid dislocation multiplication at yield [82]. More generally, the dislocation system may be described by a state vector $\boldsymbol{\rho} = [\rho_k]$ consisting of the densities of dislocations of different types, the evolution of which is determined by a set of ordinary differential equations

$$\partial_t \rho_i = \boldsymbol{f}\left([\rho_k], \left[\dot{\varepsilon}^{\beta}\right], [q_k]\right). \tag{15}$$

Here the q_k are control parameters such as temperature, materials parameters, and externally imposed strain rate. All terms in these equations which involve long-range dislocation motions (dislocation multiplication, reactions, dynamic recovery) scale in proportion with the respective dislocation fluxes, i.e., they are proportional to the shear strain rates $\dot{\varepsilon}^{\beta}$. When other processes (e.g., diffusion-controlled static recovery) are absent, this allows one to express the state of the dislocation system as a function of the plastic strain. Establishing a relation between the dislocation densities and the flow stress then again leads to a description of hardening.

2.6.3. *Dislocation patterning*

In the course of time, several shortcomings of the phenomenological density-based dislocation dynamics approach have become evident. Since the evolution of the dislocation system is described in terms of ordinary differential equations which do not include any spatial features, the approach cannot account for the spontaneous emergence of inhomogeneities in the dislocation arrangement, i.e., for dislocation patterning. While features of inhomogeneity can be introduced *a priori* into the equations of evolution, e.g. by assuming a cell pattern and writing separate evolution equations for the dislocation densities in the dislocation-rich cell walls and the dislocation-depleted cell interiors [83], this method cannot account for qualitative changes in the morphology of the dislocation arrangement. Here a modified theoretical framework is required.

Reaction–diffusion–transport equations A straightforward generalization of eq. (15) is to consider the fluxes $\boldsymbol{J}_k$ of the dislocation species ρ_k. This leads to reaction–transport equations of the type [84,85]

$$\partial_t \rho_i + \nabla \boldsymbol{J}_i = f_i\left([\rho_k], \left[\dot{\varepsilon}^{\beta}\right], [q_k]\right). \tag{16}$$

The fluxes $\boldsymbol{J}_k = \rho_k \boldsymbol{v}_k$ are envisaged as deterministic functionals of the dislocation densities $[\rho_k]$. As shown above, on microscopic and mesoscopic scales the dislocation velocities are strongly space- and time-dependent due to many-dislocation interactions. Physically motivated expressions for the fluxes $\boldsymbol{J}_k$ should monitor these interactions in an appropriate manner.

Several phenomenological models have been proposed in the literature to deal with this problem. In the work of Walgraef and Aifantis [86,87], space dependencies were introduced through diffusionlike terms of the form $\boldsymbol{J}_k = D_k \nabla \rho_k$. Kratochvil proposed to describe spatial interactions in terms of nonlocal expressions either for the flow stress evolution in general ('nonlocal hardening') [88] or, more specifically, for the flux of edge dislocation dipoles induced by their 'sweeping' by moving screw dislocations [89]. In one of the earliest papers on the subject, Holt [90] used an irreversible thermodynamics framework and assumed the dislocation fluxes to be proportional to the gradient of an energy functional. To obtain a well-defined energy functional, a phenomenological correlation function restricting the range of dislocation interactions had to be introduced, and the pattern wavelength obtained from the model turned out to be proportional to the spatial range of the correlations [90]. This observation points to a general problem: The stress fields of dislocations decay in space as $1/r$ and, hence, single-dislocation interactions do not exhibit any intrinsic length scale. All the above-mentioned models of patterning, on the other hand, introduce such scales in a phenomenological manner through diffusion coefficients or nonlocal interaction kernels. Since the results are pre-determined by the phenomenological 'input', it is doubtful whether this can solve the problem of length scale selection in dislocation patterning.

Experimentally, one observes that the more or less regular patterns of dislocation-rich and dislocation-depleted regions which develop in the course of deformation (for examples, see section 4) often obey a so-called 'law of similitude' [91]. This means that the characteristic length λ of these patterns scales in inverse proportion with the flow stress. As shown in a compilation of data by Raj and Pharr [92], relations of the type $\lambda \sim 1/\sigma_{\text{ext}}$ are obeyed in many materials and under widely varying deformation conditions. In conjunction with the Taylor relationship, this implies that the characteristic scale of dislocation patterns is proportional to, but in general much larger than, the average dislocation spacing [32,70]. In section 3 of the present work we demonstrate that pair correlations in the dislocation arrangement lead to screening of dislocation interactions on a scale which is proportional to the dislocation spacing and inversely proportional to the external stress, in agreement with the Taylor relationship and the 'law of similitude'. This indicates that the problem of length-scale selection in dislocation patterning can be solved without making a-priori-assumptions on the presence of characteristic scales if we go beyond the dislocation-density description and study the properties of correlations in the dislocation arrangement.

Stochastic dislocation dynamics Besides the problem of length-scale selection, the intrinsically intermittent and erratic nature of dislocation motions on mesoscopic scales poses a serious problem when deterministic reaction–diffusion–transport equations of the type of eq. (16) are used to formulate mesoscopic models of dislocation patterning. In chemical kinetics (see, e.g., the work of Prigogine [42]), the high-dimensional stochastic dynamics of the constituent atoms of a mixture can, on mesoscopic scale, be mapped on a dynamics of concentration fields with deterministic diffusion terms. This is possible because the characteristic length of fluctuations (range of diffusion jumps of single atoms) is much smaller than the wavelength of concentration patterns. In plastic flow, on the other hand, the high-dimensional dynamics of the individual dislocations leads to spatio-temporal fluctuations of the dislocation fluxes on scales that are *larger* than the wavelengths of dislocation patterns. This indicates that a deterministic reaction–diffusion–transport framework may not be fully adequate for describing the dynamics of the dislocation ensemble on mesoscopic scales.

In the models of work-hardening proposed by Seeger [76–79] and Hirsch [68,70], the accumulation of dislocations in a deforming crystal is explicitly related to the inhomogeneity of plastic deformation on mesoscopic scales that manifests itself through the formation of confined slip zones. These models consider the accumulation of dislocations on the macroscopic scale. Recently, Hähner [57,58,93,94] has re-formulated the idea of taking the inhomogeneous, intermittent and erratic nature of plastic flow as a starting point for the description of plastic deformation and dislocation patterning. To this end, he proposed to use a stochastic formulation for describing the evolution of dislocation microstructure on mesoscopic scales. In this *stochastic dislocation dynamics* approach the irregular and random character of many dislocation patterns (for examples, see section 4) is seen as a consequence of spatially and temporally fluctuating dislocation fluxes driving the evolution of an inhomogeneous dislocation microstructure. The basic framework of the density-based dislocation dynamics approach as defined by eq. (15) is preserved by stochastic dislocation dynamics. However, the plastic strain rates $\dot{\varepsilon}^{\beta} = \langle\dot{\varepsilon}^{\beta}\rangle + \delta\dot{\varepsilon}^{\beta}$ are now interpreted as the sum of a deterministic average plus a stochastic process with prescribed statistical properties. In this way, eq. (15) becomes a set of stochastic differential equations,

$$\partial_t \rho_i = f_i\big([\rho_k], \big[\langle\dot{\varepsilon}^{\beta}\rangle, \delta\dot{\varepsilon}^{\beta}\big], [q_k]\big). \tag{17}$$

In different mesoscopic volume elements one finds in general different realizations of the stochastic processes $\delta\dot{\varepsilon}^{\beta}(x,t)$ (i.e., different *local strain histories)* and this is seen as the major reason for the emergence of heterogeneous dislocation patterns. To obtain a closed description, eq. (15) must be supplemented by equations specifying the statistical properties of the strain-rate fluctuations on the different slip systems as functions of the dislocation densities and the external control parameters. Stochastic integration of eq. (17) then yields probability distributions $p(\boldsymbol{\rho}, t)$ which characterize the evolving microstructure in a probabilistic sense, i.e., the approach yields a statistical 'signature' of patterning but no direct information on how the dislocations are distributed in space.

The stochastic dislocation dynamics approach accounts in a natural manner for the observed irregularity and intermittency of plastic flow on mesoscopic scales. In this sense, it represents an attempt to introduce a realistic description of the dynamics of plastic flow into density-based models of dislocation patterning. The approach cannot directly account for the spatial features of the patterns which, of course, implies a significant loss of information. In view of the often irregular and random character of real dislocation patterns (section 4), this may be a minor shortcoming. On the other hand, the approach offers a conceptual framework which makes it possible to account for the interplay between inhomogeneous plastic flow, dislocation storage, dislocation patterning and the evolution of lattice rotations and misorientations. This will be illustrated in sections 5 and 6 by examples.

3. *Statistical characterization of dislocation arrangements and internal stress fields*

The intricate complexity of dislocation arrangements in plastically deformed crystals raises the basic question: How can dislocation arrangements and internal-stress fields be characterized on mesoscopic scale by a limited number of variables? The phenomenological dislocation-dynamics models discussed in section 2.6 formulate, in a semi-phenomenological manner, evolution equations for spatially averaged dislocation population densities. These models necessarily involve some arbitrariness, since systematic procedures for defining mesoscopic densities and deriving their evolution equations from the dynamics and interactions of discrete dislocations are not available. In this respect, an idea of Kröner [95] deserves attention. Almost three decades ago Kröner proposed to develop a statistical theory of plasticity based upon configurational averages rather than simple space-averaged densities. This idea has recently been taken up in a number of papers [96–98]. Interestingly, this has been partly motivated by discrete dislocation dynamics simulations in two and three dimensions [99,100]. They allow the investigation of the coupled dynamics of large numbers of dislocations or dislocation segments but, like all molecular dynamics simulations, require a statistical framework for interpreting the results.

The basic argument why a statistical theory of dislocation dynamics should not be formulated in terms of spatially averaged densities alone is simple: The evolution of the dislocation arrangement is due to the motion of dislocation segments, which is governed by the external and internal stresses acting on them. These stresses depend on the *configuration* of other segments in the surroundings, which is obviously influenced by the presence of the first segment – a fact which cannot be grasped by spatially averaged densities which represent probabilities for the spatial occurrence of a given type of dislocations (dislocation segments) irrespective of their arrangement relative to each other. Hence one has to ask for *conditional probabilities* to find other dislocations in a certain position relative to the dislocation under consideration, which mathematically implies the consideration of pair densities or pair correlations. The evolution of pair configurations depends, in turn, on the arrangement of third dislocations or dislocation segments, and so forth. In the present section we formulate a mathematical framework for the statistical characterization of mesoscopically inhomogeneous dislocation arrangements and their internal-stress fields based upon dislocation densities and correlations. To keep the presentation transparent, we restrict the development of this section to two-dimensional dislocation arrangements (i.e., parallel dislocations) while the extension to 3D dislocation systems will be briefly discussed later in section 6.

3.1. Statistical characterization of two-dimensional dislocation arrangements

3.1.1. *Mathematical formulation*

We consider an arrangement of N dislocations with line direction parallel to the z axis of a Cartesian coordinate system. The vectors $\boldsymbol{r}[i]$ $(i = 1, \dots, N)$ give the positions of the dislocations in the xy-plane. The slip system of the i-th dislocation is labeled by $\beta[i]$, and the sign of this dislocation by $s[i] \in [+, -]$. We focus on f.c.c. metals, where we may assume that the dislocation strength b (= the modulus of the Burgers vector) is the same for all dislocations. As opposed to the spatial coordinates $\boldsymbol{r}[i]$, we call the $s[i]$ and $\beta[i]$ *configurational coordinates* of the dislocation lines.

In principle, the evolution of the dislocation arrangement might be described by tracing the positions $\boldsymbol{r}[i]$ of all dislocations and accounting for dislocation reactions in terms of the corresponding changes in configurational coordinates. Unless otherwise stated, we confine ourselves to slip motion. When inertial effects can be neglected, the velocity of the i-th dislocation is

$$\begin{aligned} \boldsymbol{v}[i] &= s[i]\boldsymbol{e}_{\mathrm{g}}^{\beta[i]} v\big(\sigma^{\beta[i]}(\boldsymbol{r}[i])\big), \\ \sigma^{\beta[i]}\big(\boldsymbol{r}[i]\big) &= \sigma_{\mathrm{ext}}^{\beta[i]} + \sum_{k\neq i}^{N} s[k]\sigma^{\beta[i]\beta[k]}\big(\boldsymbol{r}[i]-\boldsymbol{r}[k]\big), \end{aligned} \tag{18}$$

where $\boldsymbol{e}_{\mathrm{g}}^{\beta}$ is a unit vector in the slip direction of dislocations of slip system β (for our 2D model system this is the intersection of the slip plane and the xy-plane), $\sigma_{\mathrm{ext}}^{\beta}$ is the external stress resolved in slip system β, and $\sigma^{\beta\beta'}(\boldsymbol{r}-\boldsymbol{r}')$ is the shear stress produced in this slip system at the position $\boldsymbol{r}$ by a (positive) dislocation of slip system β' located at $\boldsymbol{r}'$. These stresses are related to the external stress field and the dislocation stress fields via $\sigma_{\mathrm{ext}}^{\beta} = \sum_{kl}(\sigma_{\mathrm{ext}})_{kl} M_{kl}^{\beta}$, and $\sigma^{\beta\beta'}(\boldsymbol{r}) = \sum_{kl}\sigma_{kl}^{\beta'}(\boldsymbol{r}) M_{kl}^{\beta}$, respectively. In linear elasticity, the stress field of a (positive) dislocation of slip system β can be written as

$$\sigma_{kl}^{\beta}(\boldsymbol{r}) = Gb\frac{K_{kl}^{\beta}(\psi)}{|\boldsymbol{r}|}, \tag{19}$$

where G is an elastic (shear) modulus of the material and K_{kl}^{β} a dimensionless function of the angle ψ between the dislocation slip direction and the vector $\boldsymbol{r}$.

We consider an ensemble of statistically equivalent dislocation arrangements and define many-dislocation densities by the ensemble averages

$$\begin{aligned} \rho_{(1)}^{\beta_1,s_1}(\boldsymbol{r}_1) &:= \left\langle \sum_{i=1}^{N} \delta_{\beta_1\beta[i]}\delta_{s_1 s[i]}\delta\big(\boldsymbol{r}-\boldsymbol{r}[i]\big)\right\rangle, \\ \rho_{(2)}^{\beta_1\beta_2,s_1 s_2}(\boldsymbol{r}_1,\boldsymbol{r}_2) &:= \left\langle \sum_{i=1}^{N}\sum_{j\neq i}^{N} \delta_{\beta_1\beta[i]}\delta_{\beta_2\beta[j]}\delta_{s_1 s[i]}\delta_{s_2 s[j]}\delta\big(\boldsymbol{r}_1-\boldsymbol{r}[i]\big)\delta\big(\boldsymbol{r}_2-\boldsymbol{r}[j]\big)\right\rangle, \\ &\dots \end{aligned} \tag{20}$$

In eq. (20), $\rho_{(1)}^{\beta_1,s_1}(\boldsymbol{r}_1)$ denotes the density of dislocations of type $[\beta_1, s_1]$ at $\boldsymbol{r}_1$, and $\rho_{(2)}^{\beta_1\beta_2,s_1s_2}(\boldsymbol{r}_1,\boldsymbol{r}_2)$ the density of dislocation pairs of types $[\beta_1, s_1]$ and $[\beta_2, s_2]$ at the positions $\boldsymbol{r}_1$ and $\boldsymbol{r}_2$. Higher-order densities are defined accordingly. In the following, we shall often use an abbreviated notation, writing $\rho_{(1)}(1) := \rho_{(1)}^{\beta_1,s_1}(\boldsymbol{r}_1)$, $\rho_{(2)}(1,2) := \rho_{(2)}^{\beta_1\beta_2,s_1s_2}(\boldsymbol{r}_1,\boldsymbol{r}_2)$, etc.

Owing to the ensemble averaging, the many-dislocation densities are in general smooth functions of their arguments. Evolution equations for these densities may be obtained by ensemble averaging the equations of motion of the discrete dislocations. In slowly deforming f.c.c. metals, at each moment practically all dislocations are at rest and therefore many results can be obtained by assuming that most of the dislocations experience zero stress. In this case the single-dislocation densities fulfill the approximate equations

$$\sigma_{\text{ext}}^{\beta_1}\rho_{(1)}(1) = -\sum_{\beta_2,s_2} s_2 \int \rho_{(2)}(1,2)\sigma^{\beta_1\beta_2}(\boldsymbol{r}_1 - \boldsymbol{r}_2)\,\mathrm{d}^2 r_2. \tag{21}$$

For the pair densities one finds

$$\left[\sigma_{\text{ext}}^{\beta_1} + s_2\sigma^{\beta_1\beta_2}(\boldsymbol{r}_1 - \boldsymbol{r}_2)\right]\rho_{(2)}(1,2) = -\sum_{\beta_3,s_3} s_3 \int \rho_{(3)}(1,2,3)\sigma^{\beta_1\beta_3}(\boldsymbol{r}_1 - \boldsymbol{r}_3)\,\mathrm{d}^2 r_3, \tag{22}$$

and for the n-dislocation densities

$$\left[\sigma_{\text{ext}}^{\beta_i} + \sum_{j\neq i} s_j\sigma^{\beta_i\beta_j}(\boldsymbol{r}_i - \boldsymbol{r}_j)\right]\rho_{(n)}(1\ldots n)$$
$$= -\sum_{\beta_{n+1},s_{n+1}} s_{n+1} \int \rho_{(n+1)}(1\ldots n, n+1)\sigma^{\beta_i\beta_{n+1}}(\boldsymbol{r}_i - \boldsymbol{r}_{n+1})\,\mathrm{d}^2 r_{n+1}. \tag{23}$$

The equations for the n-dislocation densities contain integrals over the densities of next-higher order. Thus, we are confronted with an infinite hierarchy of equations. In practice, it must be truncated at some finite order, which necessarily involves a loss of information. To identify the information that is indispensable for characterizing the dislocation arrangement it is useful to study, in a first step, some scaling relations.

Scaling relations We investigate the scaling behaviour of eqs (21)–(23) by assuming that all single-dislocation densities are changed by a factor C_ρ, $\rho_{(1)}^{\beta,s} \to C_\rho\rho_{(1)}^{\beta,s}$. Further we require that the n-dislocation densities behave like products of n single-dislocation densities. Now we look for an appropriate transformation of the other quantities in eqs (21)–(23) such that these equations remain invariant upon the rescaling. One finds that there is exactly one transformation which fulfills this requirement; it is given by

$$\boldsymbol{r}_j \to C_r\boldsymbol{r}_j, \qquad \sigma_{\text{ext}} \to C_\sigma\sigma_{\text{ext}}, \tag{24}$$

where $C_r = C_\rho^{-1/2}$ and $C_\sigma = C_\rho^{1/2}$. Invariance under this transformation implies that any particular solution of (21)–(23) belongs to a one-parameter manifold of *similitude solutions* pertaining to different external stresses. These solutions may be parametrized, for instance, by the macroscopically averaged dislocation density $\varrho := \sum_{\beta,s}\{\rho_{(1)}^{\beta,s}(\boldsymbol{r})\}$. From a given solution the corresponding similitude manifold is generated according to eq. (24) by scaling all lengths in proportion with the mean dislocation spacing $1/\sqrt{\varrho}$ and the external stress in proportion with $\sqrt{\varrho}$.

This scaling behaviour is consistent with two basic empirical properties of dislocation systems discussed in section 2, *viz.* the Taylor relationship and the 'law of similitude'.

3.1.2. Pair correlations in quasi-static dislocation arrangements

The scaling property (24) gives an important guideline how to handle the infinite hierarchy of many-dislocation densities. The idea is to truncate this hierarchy at some level n by expressing densities of order $n+1$ in terms of densities of lower order. To identify the lowest order at which this is feasible, we require that truncation preserve the fundamental scaling property (24). From this requirement, one readily sees that it is not appropriate to describe the dislocation system in terms of first-order densities only, e.g. by using the mean-field approximation $\rho_{(2)}(1,2) \approx \rho_{(1)}(1)\rho_{(1)}(2)$ in eq. (21) as proposed in [96]. Equation (22) corresponds to the lowest order at which the property (24) is manifest, while from eq. (21) alone, only one of the scaling constants C_r and C_σ of eq. (24) can be determined. Hence, it is necessary to consider at least pair densities (or, equivalently, pair correlations) to account correctly for the scaling relations between stress, total dislocation density, and length scales of the dislocation pattern.

Screening property of pair correlations In the following, we focus on weakly correlated dislocation arrangements where distant dislocations behave in a statistically independent manner. Then the pair densities fulfill the asymptotic relationships

$$\rho_{(2)}(1,2) \rightarrow \rho_{(1)}(1)\rho_{(2)}(2) \quad \text{for } r_{12} := |\boldsymbol{r}_1 - \boldsymbol{r}_2| \rightarrow \infty, \tag{25}$$

and the third-order densities

$$\rho_{(3)}(1,2,3) \rightarrow \rho_{(2)}(1,2)\rho_{(1)}(3) \quad \text{for } \boldsymbol{r}_{13} \rightarrow \infty \text{ and } \boldsymbol{r}_{23} \rightarrow \infty. \tag{26}$$

We define the pair correlation function $D_{(2)}(1,2) := \rho_{(2)}(1,2) - \rho_{(1)}(1)\rho_{(2)}(2)$ with the boundary condition $D_{(2)}(1,2) \rightarrow 0$ for $\boldsymbol{r}_{12} \rightarrow \infty$ and study the asymptotic behaviour of eq. (22) at large $\boldsymbol{r}_{12}$. Using eqs (25), (26) and (21), and neglecting small terms, we obtain

$$\rho_{(1)}(2)\sigma^{\beta_1\beta_2}(\boldsymbol{r}_1 - \boldsymbol{r}_2) \approx -\sum_{\beta_3 s_3} s_2 s_3 \int D_{(2)}(2,3)\sigma^{\beta_1\beta_3}(\boldsymbol{r}_1 - \boldsymbol{r}_2)\,\mathrm{d}^2 r_3 \tag{27}$$

with the solution $D_{(2)}(2,3) \approx \widehat{D}^{23}(\boldsymbol{r}_2)\delta(\boldsymbol{r}_2 - \boldsymbol{r}_3)$. This means that the $D_{(2)}(2,3)$ are localized functions which, on large scales, can be approximated by δ-functions with weighting factors $\widehat{D}^{23}(\boldsymbol{r}_2) = \int D_{(2)}(2,3)\,\mathrm{d}^2 r_3$. These weighting factors fulfill the relation

$$s_2\rho_{(1)}^{\beta_2 s_2}(\boldsymbol{r}_2)\sigma^{\beta_1\beta_2}(\boldsymbol{r}_1 - \boldsymbol{r}_2) = -\sum_{\beta_3 s_3} s_3 \widehat{D}^{23}(\boldsymbol{r}_2)\sigma^{\beta_1\beta_3}(\boldsymbol{r}_1 - \boldsymbol{r}_2). \tag{28}$$

To interpret eq. (28) in physical terms, we consider the conditional probability $p(2|1)\,\mathrm{d}V_2$ to find in a volume element $\mathrm{d}V_2$ at $\boldsymbol{r}_2$ a dislocation of type $[\beta_2, s_2]$ provided there is a dislocation of type $[\beta_1, s_1]$ at $\boldsymbol{r}_1$. For a random dislocation arrangement, the coordinates of the dislocations are independent random variables and hence $p(2|1)\,\mathrm{d}V_2 = \rho_{(1)}(2)\,\mathrm{d}V_2$. For a dislocation arrangement with correlations, on the other hand, $p(2|1) = [\rho_{(2)}(1,2)/\rho_{(1)}(1)]\,\mathrm{d}V_2$. The difference $[p(2|1) - \rho_{(1)}(2)] = D_{(2)}(1,2)/\rho_{(1)}(1)$ may be interpreted as the density of 'excess dislocations' surrounding a dislocation in a correlated arrangement, and $\widehat{D}^{12}(\boldsymbol{r}_1)/\rho_{(1)}(1)$ is the average number of excess dislocations of type $[\beta_2, s_2]$ around a dislocation of type $[\beta_1, s_1]$ at $\boldsymbol{r}_1$. Anticorrelations are formally represented by negative excess dislocations.

Equation (28) then means that the shear stresses created by a dislocation of type $[\beta_2, s_2]$ in the slip system β_1 are, at large distances, completely balanced by the stresses of the surrounding excess dislocations of types $[\beta_3, s_3]$. We note that this screening implies that the sum of the Burgers vectors of the excess dislocations minus the Burgers vector of the first dislocation is zero.[10] Therefore, the screening property (28) holds not only for the resolved shear stresses in the active slip systems but for all components of the stress tensor.

Integral equation for the pair correlation functions: Kirkwood approximation To obtain a closed set of equations, we use the so-called Kirkwood approximation to express the third-order dislocation density functions in terms of pair densities:

$$\rho_{(3)}(1,2,3) = \frac{\rho_{(2)}(1,2)\rho_{(2)}(2,3)\rho_{(2)}(3,1)}{\rho_{(1)}(1)\rho_{(1)}(2)\rho_{(1)}(3)}. \tag{29}$$

(For a discussion of this approximation see [101].) Inserting into eq. (22) yields a set of non-linear integral equations for the pair correlation functions. Using the notation $d_{(2)}(1,2) := D_{(2)}(1,2)/[\rho_{(1)}(1)\rho_{(1)}(2)]$ we get

$$s_2\sigma^{\beta_1\beta_2}(\boldsymbol{r}_1 - \boldsymbol{r}_2) = -\sum_{\beta_3,s_3} s_3 \int \rho_{(1)}(3)d_{(2)}(2,3)\big[1 + d_{(2)}(3,1)\big]\sigma^{\beta_1\beta_3}(\boldsymbol{r}_1 - \boldsymbol{r}_3)\,\mathrm{d}^2r_3. \tag{30}$$

Asymptotic properties of the pair correlation functions are obtained by studying the asymptotic behaviour of this equation in Fourier space at large and small wave vectors. It follows that, in the Kirkwood approximation, the pair correlation functions $d_{(2)}(1,2)$ decay for large r_{12} faster than algebraically, while for small r_{12} the pair densities are expected to diverge as $1/r_{12}$.

A convenient method for determining pair densities or correlations consists in performing simulations of the discrete dislocation dynamics, exploiting the fact that because of the limited range of the pair correlations it is sufficient to simulate comparatively small systems. Figures 6 and 7 show pair densities (densities of pairs of the same and of opposite signs, respectively) obtained from simulations of dislocation dynamics in single glide [102]. Equal numbers of dislocations of both signs were initially placed at random, and pair densities were determined after relaxation of the dislocation system to an equilibrium configuration at zero external stress. Figure 6 reveals a tendency of dislocations of the same sign to form walls where they arrange perpendicularly above each other, while dislocations

[10] Suppose we have a dislocation arrangement with this property. Adding further excess dislocations without violating eq. (28) is possible if these form an arrangement which does not create shear stresses in the slip system β_1. Such an arrangement must either have zero net Burgers vector or consist of infinitely extended dislocation walls, which, however, would violate the boundary condition (25).

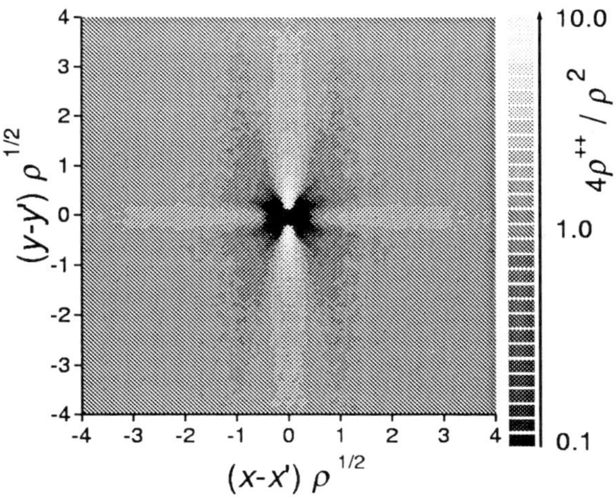

Fig. 6. Pair density of dislocations of the same sign, stationary dislocation arrangement at zero stress. The density function has been obtained by averaging over 50 2D simulations of systems consisting of 300 positive and 300 negative dislocations moving in a single slip systems. The pair density has been normalized by the square of the total dislocation density, the x and y coordinates are normalized by the average dislocation spacing. For a random distribution $4\rho_{(2)}/\rho^2 = 1$ holds; larger values imply correlations, smaller values anticorrelations.

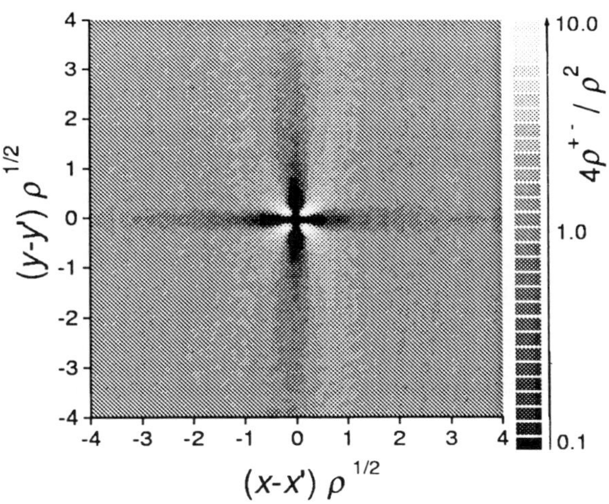

Fig. 7. Normalized pair density of dislocations of opposite signs; as fig. 6.

of opposite signs tend to form close dipoles with 45° orientations (fig. 7). The behaviour of the pair correlation functions at distances of about one average dislocation spacing indicates a tendency of the dipoles to arrange themselves vertically above each other. At larger distances, correlations decay rapidly and no indications of long-range ordering, e.g., in a Taylor lattice, are found. This is seen from fig. 8 showing the pair correlation function of

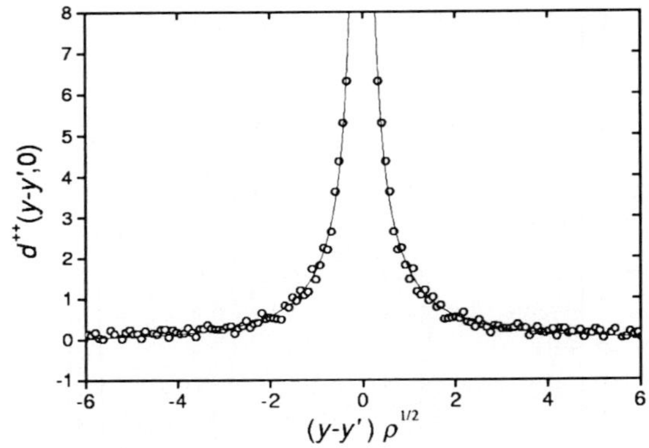

Fig. 8. Normalized pair correlation function $d^{++}_{(2)} = 4\rho^{++}_{(2)}/\rho^2 - 1$ of dislocations of the same sign arranged above each other in the direction normal to the slip plane; circles: simulation data; full line: fit function $f(y) = [2.48/y]\exp[-0.38y]$.

dislocations of the same sign arranged vertically above each other in the direction normal to the slip plane. The data are well approximated by a function $f(y) \propto (1/y)\exp[-0.38y]$, i.e., at large distances y, pair correlations decay exponentially such that their effective range is restricted to a few dislocation spacings.

While we have formulated our considerations for static dislocation arrangements, the qualitative conclusions carry over to dislocation arrangements in plastically deforming materials in which dislocation mobilities are high and deformation is slow. As discussed in section 2, in such materials deformation is brought about by rapid, highly correlated motions of groups of dislocations taking place in small parts of the crystal volume, while the overwhelming majority of dislocations are immobile at a given moment. Since the momentarily immobile majority determine the ensemble averages, the results of the present section apply to such quasi-static dislocation arrangements. It must be noted, however, that the dynamic *evolution* of the dislocation arrangement may be governed by those rapidly moving dislocations for which the present considerations fail.

3.2. The internal-stress pattern generated by two-dimensional dislocation arrangements

We consider an ensemble of statistically equivalent dislocation systems characterized by space-dependent densities and pair correlation functions and ask for the statistical signatures of the corresponding internal stress fields. For a particular N-dislocation system, the internal stress at the point $\boldsymbol{r}$ is given by

$$\sigma_{kl}(\boldsymbol{r}) = \sum_{i=1}^{N} s[i]\sigma_{kl}^{\beta[i]}\big(\boldsymbol{r} - \boldsymbol{r}[i]\big) = \sum_{\beta,s}\int\left[\sum_{i=1}^{N}\delta_{\beta\beta[i]}\delta_{ss[i]}\delta\big(\boldsymbol{r}' - \boldsymbol{r}[i]\big)\right]s\sigma_{kl}^{\beta}(\boldsymbol{r} - \boldsymbol{r}')\,\mathrm{d}^2r'. \tag{31}$$

Ensemble averaging and using eq. (18) yields the mean internal stress field

$$\sigma_{kl}^{(\mathrm{me})}(\boldsymbol{r}) := \langle \sigma_{kl}(\boldsymbol{r}) \rangle = \sum_{\beta} \int \left[\rho_{(1)}^{\beta,+}(\boldsymbol{r}_1) - \rho_{(1)}^{\beta,-}(\boldsymbol{r}_1)\right] \sigma_{kl}^{\beta}(\boldsymbol{r} - \boldsymbol{r}_1)\, \mathrm{d}^2 r_1. \tag{32}$$

This has the following properties:

(i) As $\sigma_{kl}^{(\mathrm{me})}(\boldsymbol{r})$ depends only on the difference of densities of dislocations of opposite sign, the mean internal stress field can be related to a non-vanishing value of the ensemble averaged Nye–Kröner dislocation density tensor $\boldsymbol{\alpha}$. For a single dislocation of type $[\beta, s]$ at $\boldsymbol{r}$ with Burgers vector $\boldsymbol{b} = s\boldsymbol{b}^{\beta}$ and unit tangent vector $\boldsymbol{t} = \boldsymbol{e}_z$, the Nye–Kröner tensor is $\boldsymbol{t} \otimes \boldsymbol{b}\, \delta(\boldsymbol{r}) = s\boldsymbol{e}_z \otimes \boldsymbol{b}^{\beta}\, \delta(\boldsymbol{r})$ [103]. Summation over all $[\beta, s]$ and ensemble averaging yields

$$\boldsymbol{\alpha}(\boldsymbol{r}) = \sum_{\beta} \boldsymbol{e}_z \otimes \boldsymbol{b}^{\beta} \left[\rho_{(1)}^{\beta,+}(\boldsymbol{r}) - \rho_{(1)}^{\beta,-}(\boldsymbol{r})\right]. \tag{33}$$

This allows us to establish relations between the formulation in terms of densities of dislocations of different signs and slip systems and the classical continuum theory of dislocations. For the present planar problem, only the components α_{zx}, α_{zy}, and α_{zz} of the Kröner tensor may differ from zero, and the mean stress field may be calculated from $\boldsymbol{\alpha}$ by the stress function method [104], which for isotropic materials reads

$$\sigma_{xx}^{(\mathrm{me})} = -\frac{\partial^2 \Psi}{\partial x^2}, \qquad \sigma_{yy}^{(\mathrm{me})} = \frac{\partial^2 \Psi}{\partial y^2}, \qquad \sigma_{xy}^{(\mathrm{me})} = \frac{\partial^2 \Psi}{\partial x \partial y},$$

$$\sigma_{yz}^{(\mathrm{me})} = -\frac{\partial \Phi}{\partial x}, \qquad \sigma_{xz}^{(\mathrm{me})} = \frac{\partial \Phi}{\partial y}, \qquad \sigma_{zz}^{(\mathrm{me})} = \nu\left[\sigma_{xx}^{(\mathrm{me})} + \sigma_{yy}^{(\mathrm{me})}\right], \tag{34}$$

$$\Delta^2 \Psi = \frac{G}{1-\nu}(\partial_y \alpha_{zx} - \partial_x \alpha_{zy}), \qquad \Delta \Phi = -G\alpha_{zz}. \tag{35}$$

Here, ν is Poisson's number and Ψ and Φ are the (ensemble averaged) Airy and Prandtl stress functions, respectively.

(ii) The 'wavelength' of the mean internal stress field is governed by the characteristic range λ of variations in the first-order dislocation densities. In the terminology of section 2, it therefore pertains to the mesoscopic scale. We take this into account by calling $\sigma_{kl}^{(\mathrm{me})}(\boldsymbol{r})$ the *mesoscopic internal stress* at $\boldsymbol{r}$.

The local internal stress $\sigma_{kl}(\boldsymbol{r})$ can be envisaged as the sum of the smoothly varying mesoscopic stress $\sigma_{kl}^{(\mathrm{me})}(\boldsymbol{r})$ and local internal-stress fluctuations with zero mean value,

$$\sigma_{kl}(\boldsymbol{r}) = \sigma_{kl}^{(\mathrm{me})}(\boldsymbol{r}) + \delta\sigma_{kl}(\boldsymbol{r}), \quad \text{where } \langle \delta\sigma_{kl}(\boldsymbol{r}) \rangle = 0. \tag{36}$$

While the mesoscopic stress is related to spatial variations in the *average* (first-order) dislocation densities, comparison of eqs (31) and (32) shows that the local stress fluctuations are a consequence of the *discreteness* of single dislocation lines. Therefore we speak, in the terminology of section 2, of *microscopic internal-stress fluctuations.* As shown in the following, the characteristic range of these fluctuations is governed by the pair correlations in the dislocation arrangement.

In a statistical sense, the spatial behaviour of the internal stress field is characterized by its spatial correlation $\langle\sigma_{kl}(\boldsymbol{r})\sigma_{kl}(\boldsymbol{r}')\rangle$. From eqs (20) and (31) this follows as

$$\begin{aligned}\big\langle\sigma_{kl}(\boldsymbol{r})\sigma_{kl}(\boldsymbol{r}')\big\rangle = &\sum_{\beta_1,s_1}\int \rho_{(1)}(1)\sigma_{kl}^{\beta_1}(\boldsymbol{r}-\boldsymbol{r}_1)\sigma_{kl}^{\beta_1}(\boldsymbol{r}'-\boldsymbol{r}_1)\,\mathrm{d}^2r_1\\ &+\sum_{\beta_1\beta_2,s_1s_2}\int\!\!\int s_1s_2\rho_{(2)}(1,2)\sigma_{kl}^{\beta_1}(\boldsymbol{r}-\boldsymbol{r}_1)\sigma_{kl}^{\beta_2}(\boldsymbol{r}'-\boldsymbol{r}_2)\,\mathrm{d}^2r_1\,\mathrm{d}^2r_2.\end{aligned} \tag{37}$$

Using the dislocation pair-correlation functions $D_{(2)}(1,2)$, eq. (37) is readily re-written as

$$\begin{aligned}\big\langle\sigma_{kl}(\boldsymbol{r})\sigma_{kl}(\boldsymbol{r}')\big\rangle = &\sum_{\beta_1\beta_2,s_1s_2}\int\!\!\int s_1s_2\rho_{(1)}(1)\rho_{(1)}(2)\sigma_{kl}^{\beta_1}(\boldsymbol{r}-\boldsymbol{r}_1)\sigma_{kl}^{\beta_2}(\boldsymbol{r}'-\boldsymbol{r}_2)\,\mathrm{d}^2r_1\,\mathrm{d}^2r_2\\ &+\sum_{\beta_1,s_1}\int \rho_{(1)}(1)\sigma_{kl}^{\beta_1}(\boldsymbol{r}-\boldsymbol{r}_1)\sigma_{kl}^{\beta_1}(\boldsymbol{r}'-\boldsymbol{r}_1)\,\mathrm{d}^2r_1\\ &+\sum_{\beta_1\beta_2,s_1s_2}\int\!\!\int s_1s_2D_{(2)}(1,2)\sigma_{kl}^{\beta_1}(\boldsymbol{r}-\boldsymbol{r}_1)\sigma_{kl}^{\beta_2}(\boldsymbol{r}'-\boldsymbol{r}_2)\,\mathrm{d}^2r_1\,\mathrm{d}^2r_2.\end{aligned} \tag{38}$$

The first term on the right-hand side is the product $\sigma_{kl}^{(\mathrm{me})}(r)\sigma_{kl}^{(\mathrm{me})}(r')$ of the mesoscopic stresses at $\boldsymbol{r}$ and $\boldsymbol{r}'$, while the second and third terms on the right-hand side yield the spatial correlation $\langle\delta\sigma_{kl}(\boldsymbol{r})\delta\sigma_{kl}(\boldsymbol{r}')\rangle$ of the microscopic stress fluctuations.

3.2.1. *Mean square stresses and elastic energy density*

The mean square stresses at the position $\boldsymbol{r}$ in the dislocated crystal, which are a measure of the 'amplitude' of the internal-stress field and its elastic energy density, are obtained by putting $\boldsymbol{r}'=\boldsymbol{r}$ in eq. (38). This yields $\langle\sigma_{kl}^2(\boldsymbol{r})\rangle=[\sigma_{kl}^{(\mathrm{me})}(\boldsymbol{r})]^2+\langle[\delta\sigma_{kl}(\boldsymbol{r})]^2\rangle$. For a given dislocation density distribution, the square of the mesoscopic stress is readily calculated within the continuum framework of eqs (33)–(35). In the following we focus on the microscopic stress fluctuations, the square of which follows from the second and third terms on the right-hand side of (38),

$$\big\langle[\delta\sigma_{kl}(\boldsymbol{r})]^2\big\rangle=\big\langle[\delta\sigma_{kl}(\boldsymbol{r})]^2\big\rangle_{(\mathrm{S})}+\big\langle[\delta\sigma_{kl}(\boldsymbol{r})]^2\big\rangle_{(\mathrm{I})}, \tag{39}$$

$$\big\langle[\delta\sigma_{kl}(\boldsymbol{r})]^2\big\rangle_{(\mathrm{S})}=\sum_{\beta_1,s_1}\int \rho_{(1)}(1)\big[\sigma_{kl}^{\beta_1}(\boldsymbol{r}-\boldsymbol{r}_1)\big]^2\,\mathrm{d}^2r_1, \tag{40}$$

$$\langle[\delta\sigma_{kl}(\boldsymbol{r})]^2\rangle_{(\mathrm{I})} = \sum_{\beta_1\beta_2,s_1s_2} \int\int s_1 s_2 D_{(2)}(1,2)\sigma_{kl}^{\beta_1}(\boldsymbol{r}-\boldsymbol{r}_1)\sigma_{kl}^{\beta_2}(\boldsymbol{r}'-\boldsymbol{r}_2)\,\mathrm{d}^2r_1\,\mathrm{d}^2r_2. \tag{41}$$

The terms $\langle[\delta\sigma_{kl}(\boldsymbol{r})]^2\rangle_{(\mathrm{S})}$ yield the contribution of the dislocation self-energies to the elastic energy density of the microscopic internal-stress field. Since $\sigma_{kl}^{\beta}(\boldsymbol{r}) \propto 1/|\boldsymbol{r}|$, the corresponding integral diverges both at small and at large values of $|\boldsymbol{r}-\boldsymbol{r}_1|$. While the divergence at $|\boldsymbol{r}-\boldsymbol{r}_1| \to 0$ can be removed by introducing an appropriate core radius $r_\mathrm{c} \approx b$ which accounts for the fact that on atomistic scale the expressions (19) for the dislocation stress fields cease to be valid, the divergence at large $|\boldsymbol{r}-\boldsymbol{r}_1|$ is limited only by the crystal size. With $C_{kl}^{\beta} := \int[K_{kl}^{\beta}(\psi)]^2\,\mathrm{d}\psi$ one obtains

$$\begin{aligned}\langle[\delta\sigma_{kl}(\boldsymbol{r})]^2\rangle_{(\mathrm{S})} &= \sum_{\beta_1,s_1}\int \rho_{(1)}(1)\left[\sigma_{kl}^{\beta_1}(\boldsymbol{r}-\boldsymbol{r}_1)\right]^2\mathrm{d}^2r_1\\ &= G^2b^2\sum_{\beta_1,s_1} C_{kl}^{\beta_1}\left(\{\rho_{(1)}^{\beta_1,s_1}\}\ln R_0 - \rho_{(1)}^{\beta_1,s_1}(\boldsymbol{r})\ln r_\mathrm{c}\right),\end{aligned} \tag{42}$$

where R_0 is proportional to the crystal diameter. Note that R_0 pertains to a macroscopic scale large as compared to the scale λ of dislocation density variations, such that in (42) the corresponding terms depend on the spatially averaged dislocation densities only.

The terms $\langle[\delta\sigma_{kl}(\boldsymbol{r})]^2\rangle_{(\mathrm{I})}$ in eq. (41) yield the interaction energies between dislocations. For a random arrangement of dislocations all correlation functions $D_{(2)}(1,2)$ are zero, i.e., these terms vanish and the mean square stresses and elastic energy density diverge in proportion with the logarithm of crystal size. This has been first recognized by Wilkens [105] who proposed to consider 'restrictedly random' distributions of dislocations where appropriate correlations remove this unphysical divergence (cf. below).

For weakly correlated dislocation arrangements, in which the range of correlations does not exceed a few dislocation spacings, the pair-correlation functions may be represented as $D_{(2)}(1,2) = \widehat{D}^{12}(\boldsymbol{r}_1)f_{\rho}^{12}(\boldsymbol{r}_1-\boldsymbol{r}_2) \approx \widehat{D}^{12}(\boldsymbol{r}_2)f_{\rho}^{12}(\boldsymbol{r}_1-\boldsymbol{r}_2)$. Here the $f_{\rho}^{12}(\boldsymbol{r})$ are normalized correlation functions ($\int f_{\rho}^{12}(\boldsymbol{r})\,\mathrm{d}^2r = 1$) with a 'microscopic' range of the order of magnitude of a few dislocation spacings. The weighting factors $\widehat{D}^{12}$ (cf. eq. (28)) are space dependent on the mesoscopic scale $\lambda \gg 1/\sqrt{\varrho}$ of dislocation density variations. For large values of r_{12} one may approximate $D_{(2)}(1,2)$ in eq. (37) by $\widehat{D}^{12}(\boldsymbol{r}_1)\delta(\boldsymbol{r}_1-\boldsymbol{r}_2)$. As a result, one finds that the diverging terms in $\langle[\delta\sigma_{kl}^2(\boldsymbol{r})]\rangle_{(\mathrm{S})}$ and $\langle[\delta\sigma_{kl}^2(\boldsymbol{r})]\rangle_{(\mathrm{I})}$ cancel each other when the condition

$$\sum_{\beta_1,s_1} C_{kl}^{\beta_1}\rho_{(1)}^{\beta_1}(\boldsymbol{r}) = -\sum_{\beta_1\beta_2,s_1s_2} s_1s_2C_{kl}^{\beta_1\beta_2}\widehat{D}^{12}(\boldsymbol{r}) \tag{43}$$

is fulfilled with $C_{kl}^{\beta_1\beta_2} := \int K_{kl}^{\beta_1}(\psi)K_{kl}^{\beta_2}(\psi)\,\mathrm{d}\psi$. It may be readily shown that all pair correlation functions satisfying the previously formulated screening condition (28) also fulfill (43).

When eq. (43) is fulfilled, the mean square of the microscopic internal-stress fluctuations may be written in the form

$$\left\langle\left[\delta\sigma_{kl}(\boldsymbol{r})\right]^2\right\rangle = G^2b^2\sum_{\beta,s} C^{\beta}_{kl}\rho^{\beta,s}_{(1)}(\boldsymbol{r})\ln\left[\xi^{\beta,s}_{\sigma}(\boldsymbol{r})/r_c\right]$$
$$= C_{kl}(\boldsymbol{r})G^2b^2\rho(\boldsymbol{r})\ln\left[\xi_{\sigma}(\boldsymbol{r})/r_c\right], \tag{44}$$

where $\rho := \sum_{\beta,s}\rho^{\beta,s}_{(1)}$, $C_{kl} := \sum_{\beta,s} C^{\beta,s}_{kl}\,\rho^{\beta,s}_{(1)}(\boldsymbol{r})/\rho(\boldsymbol{r})$, and $\ln\xi_{\sigma} := \sum_{\beta,s}\ln\xi^{\beta,s}_{\sigma}[C^{\beta}_{kl}\rho^{\beta,s}_{(1)}]/[C_{kl}\rho]$. The characteristic lengths $\xi^{\beta,s}_{\sigma}$ characterize the range of screening correlations around dislocations of type $[\beta, s]$; their calculation from the corresponding pair correlation functions is discussed below.

A simple model of a screened dislocation arrangement is obtained by Wilkens' construction of a 'restrictedly random distribution' of dislocations [105]. This construction proceeds as follows: The cross-section of the crystal is divided into areas $\mathcal{U}_i$ of size A and the dislocations are distributed in such a manner that each box contains the same number $N^{\beta,s}_{A}$ of dislocations of type $[\beta, s]$ while the arrangement of these dislocations within the boxes is random. The ensemble averaged dislocation densities are $\rho^{\beta,s}_{(1)} = N^{\beta,s}_{A}/A$, and the pair correlation functions for dislocations of the same kind are

$$D^{\beta\beta,ss}_{(2)}(\boldsymbol{r},\boldsymbol{r}') = \begin{cases} -\rho^{\beta,s}_{(1)}/A & \text{for } \boldsymbol{r},\boldsymbol{r}' \text{ in } \mathcal{U}_i, \\ 0 & \text{else} \end{cases} \qquad \widehat{D}^{\beta\beta,ss} = -\rho^{\beta,s}_{(1)}. \tag{45}$$

All other pair-correlation functions are zero. In physical terms, screening is brought about by the fact that the construction delimits the magnitude of possible dislocation density fluctuations. Formally, according to eq. (45) there is for each dislocation exactly one negative excess dislocation of the same type and sign within the same box. Hence (45) fulfills both screening conditions (28) and (43). It is interesting to note that the construction works even if only dislocations of only one type and sign are present, i.e., screening does not require that the dislocation arrangement is 'neutral' when averaged over large distances.

The restrictedly random distribution exhibits spatially constant densities $\rho^{\beta,s}_{(1)}$ and a common screening length which is proportional to the linear dimension of the sub-areas, $\xi^{\beta,s}_{\sigma} = \xi_{\sigma} = \zeta\sqrt{A}$. The proportionality factor ζ depends on the shape of the sub-areas and is typically on the order of 0.25 [105]. We note that the construction can be easily generalized to a spectrum of dislocation densities. To this end one covers the crystal cross-section with a distribution of sub-areas $\mathcal{U}_i$ of varying sizes A_i and distributes at random a fixed number N^{β}_{0} of dislocations over each sub-area. In this case eq. (43) still holds, i.e., the screening condition is fulfilled, while one has now a spectrum of local dislocation densities $N^{\beta,s}_{0}/A_i$ and screening lengths $\xi_{\sigma} = \zeta\sqrt{A_i}$.

3.2.2. *Spatial correlations in the internal-stress pattern*

Spatial correlations in the internal-stress pattern may be calculated from eq. (38). We consider pair correlation functions $D_{(2)}(1,2) = \widehat{D}^{12}(\boldsymbol{r}_1)f^{12}_{\rho}(\boldsymbol{r}_1 - \boldsymbol{r}_2)$ which fulfill the screening condition (28). We denote by $q^{\beta}_{kl}(\boldsymbol{k})$ the Fourier transform of $\sigma^{\beta}_{kl}(\boldsymbol{r})/[Gb]$ and by

$f_\rho^{12}(\boldsymbol{k})$ the Fourier transform of $f_\rho^{12}(\boldsymbol{r})$. The screening functions $f_\sigma^{\beta_1,s_1}(\boldsymbol{r})$, which characterize the screening of the stress fields of dislocations of type $[\beta_1, s_1]$ by surrounding dislocations, are the Fourier transforms of $f_\sigma^{\beta_1,s_1}(\boldsymbol{k}) := \sum_{\beta_2 s_2} s_1 s_2[\widehat{D}^{12}/\rho_{(1)}^{\beta_1 s_1}][q_{kl}^{\beta_1}(\boldsymbol{k})/q_{kl}^{\beta_2}(\boldsymbol{k})] \times f_\rho^{12}(\boldsymbol{k})$. Because of eq. (28), $\int f_\sigma^{\beta_1,s_1}(\boldsymbol{r})\,\mathrm{d}^2 r = 1$. We express the spatial correlation of the internal-stress fluctuations in terms of these screening functions. From the second and third terms on the right-hand side of eq. (38) we obtain

$$\begin{aligned}&\langle \delta\sigma_{kl}(\boldsymbol{r})\delta\sigma_{kl}(\boldsymbol{r}-\boldsymbol{r}')\rangle \\ &\quad = \sum_{\beta,s} \rho_{(1)}^{\beta,s}(\boldsymbol{r}_1) \iint \left[\delta(\boldsymbol{r}''-\boldsymbol{r}''') - f_\sigma^{\beta,s}(\boldsymbol{r}''-\boldsymbol{r}''')\right] \\ &\quad\quad \times \sigma_{kl}^{\beta}(\boldsymbol{r}''')\sigma_{kl}^{\beta}(\boldsymbol{r}''-\boldsymbol{r}')\,\mathrm{d}^2 r''\,\mathrm{d}^2 r'''. \end{aligned} \tag{46}$$

This may be written as

$$\langle \delta\sigma_{kl}(\boldsymbol{r})\delta\sigma_{ij}(\boldsymbol{r}-\boldsymbol{r}')\rangle = G^2 b^2 \sum_{\beta} \rho_{(1)}^{\beta,s}(\boldsymbol{r}) g_{kl}^{\beta,s}(\boldsymbol{r}'), \tag{47}$$

where the Fourier transforms of the stress correlation functions $g_{kl}^{\beta,s}(\boldsymbol{r})$ are

$$g_{kl}^{\beta,s}(\boldsymbol{k}) = \left[1 - f_\sigma^{\beta,s}(\boldsymbol{k})\right]\left|g_{kl}^{\beta}(\boldsymbol{k})\right|^2. \tag{48}$$

From eqs (47) and (48) one finds that the mean square of the microscopic stress fluctuations may be represented as $\langle[\delta\sigma_{kl}(\boldsymbol{r})]^2\rangle = G^2 b^2 \sum_{\beta,s} \rho_{(1)}^{\beta,s}(\boldsymbol{r}) g_{kl}^{\beta,s}(\boldsymbol{r}'=0)$. When computing $g_{kl}^{\beta,s}(\boldsymbol{r}'=0)$, the integration over $\boldsymbol{r}'''$ in eq. (46) must be restricted to values $|\boldsymbol{r}'''| > r_\mathrm{c}$. The screening lengths $\xi_\sigma^{\beta,s}$ are then obtained by comparing the result with eq. (44).

To assess the qualitative behaviour of the stress fluctuation correlations, we study some simple cases. We assume spherically symmetrical screening functions $f_\sigma^{\beta,s}(r)$ and calculate the angular average of $\langle\delta\sigma_{kl}(\boldsymbol{r})\delta\sigma_{kl}(\boldsymbol{r}-\boldsymbol{r}')\rangle$ over the directions of $\boldsymbol{r}'$,

$$\langle \delta\sigma_{kl}(\boldsymbol{r})\delta\sigma_{kl}(\boldsymbol{r}-\boldsymbol{r}')\rangle_\psi = G^2 b^2 \sum_{\beta,s} \rho_{(1)}^{\beta,s}(\boldsymbol{r}) \bar{g}_{kl}^{\beta,s}(|\boldsymbol{r}'|), \tag{49}$$

where $\bar{g}_{kl}^{\beta,s}(|\boldsymbol{r}|) := 1/(2\pi)\int g_{k,l}^{\beta,s}(\boldsymbol{r})\,\mathrm{d}\psi$. Using eq. (48), we obtain

$$\bar{g}_{kl}^{\beta,s}(|\boldsymbol{r}|) = C_{kl}^{\beta} \int \left[1 - f_\sigma^{\beta,s}(k)\right]\frac{1}{k} J_0\left(k|r'|\right)\mathrm{d}k, \tag{50}$$

where J_0 is the Bessel function of order zero. In deriving eq. (50), we have used that the angular average of $|q_{kl}^{\beta}(\boldsymbol{k})|^2$ is $C_{kl}^{\beta}/(2\pi k)^2$. The result may also be written as

$$\bar{g}_{kl}^{\beta,s}(r) = C_{kl}^{\beta} \int_r^\infty f_\sigma^{\beta,s}(r')\ln[r'/r]2\pi r'\,\mathrm{d}r'. \tag{51}$$

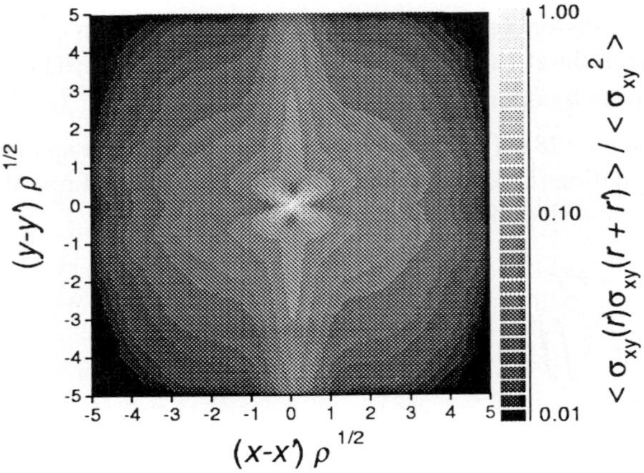

Fig. 9. Normalized correlation function of the internal shear stress created by a relaxed arrangement of edge dislocations of both signs; the pair density functions characterizing this arrangement are shown in figs 6 and 7; the Burgers vector points in the x direction.

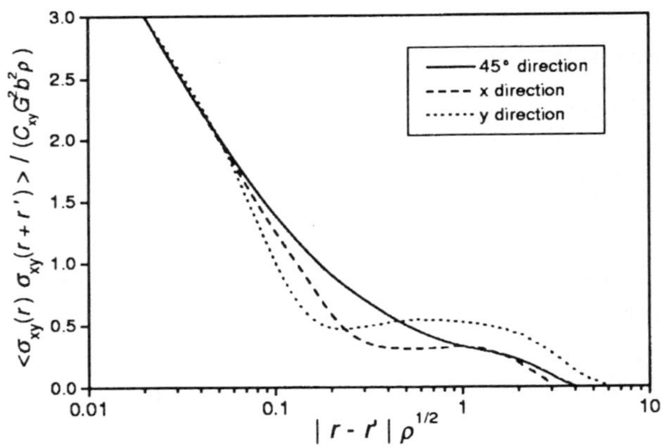

Fig. 10. Radial decay of the stress correlation function of fig. 9 for three different directions.

As an example, we consider the exponentially decaying pair correlation function $f_\sigma^{\beta,s}(r) = 1/(2\pi[r^{\beta,s}]^2)\exp[-r/r^{\beta,s}]$ with the characteristic range $r^{\beta,s}$. Equation (51) yields the corresponding radially averaged stress-correlation function, $\bar{g}_{kl}^{\beta,s}(\boldsymbol{r}) = C_{kl}^{\beta,s}[\exp(-r/r^{\beta,s}) - \mathrm{Ei}(-r/r^{\beta,s})]$, where $\mathrm{Ei}(x)$ is the exponential integral. At large r the stress correlation function monitors the exponential decay of the dislocation pair correlations, while at small r it exhibits a logarithmic divergence, $\bar{g}_{kl}^{\beta,s}(r) \approx C_{kl}^{\beta}(1 - \mathcal{C} + \ln[r^{\beta,s}/r])$, where $\mathcal{C}$ is Euler's constant. For $r \to 0$, this divergence must be truncated at the dislocation core radius. Inserting $r = r_\mathrm{c}$ and comparing the result with eq. (44), we find that the parameter $r^{\beta,s}$ of the dislocation pair correlation function and the corresponding screening length of the internal stresses are related by $\xi_\sigma^{\beta,s}/r^{\beta,s} = \exp[1 - \mathcal{C}]$. Doing the same exercise for a Gaussian correlation function $f^{\beta,s}(\boldsymbol{r}) = 1/[\pi(r^{\beta,s})^2]\exp[-r^2/(r^{\beta,s})^2]$ yields $\xi_\sigma^{\beta,s}/r^{\beta,s} = \exp[-C/2]$. The same relations have been derived by Krivoglaz [106] for an equivalent problem, *viz.* the effective cut-off radii determining the shape and width of X-ray diffraction lines.

In general, stress-correlation functions may be easily calculated numerically from the dislocation pair correlation functions using eqs (47) and (48). As an example, fig. 9 shows a 2D map of the two-point correlation function of the stress tensor component σ_{xy} calculated for a relaxed arrangement of edge dislocations of both signs at zero external stress (the pair density functions characterizing this arrangement are depicted in figs 6 and 7). Note that, for dislocation arrangements that are related by the scaling transformation (24), the same holds for the corresponding stress correlation functions. Therefore, a density-independent representation is obtained by expressing all lengths in units of the average dislocation spacing and scaling the stress correlation function in proportion to the dislocation density ρ. Figure 10 shows, in scaled coordinates and for three different directions, the radial decay of the stress correlation function depicted in fig. 9. Again one observes a logarithmic behaviour at small r ($r\sqrt{\varrho} \ll 1$). At large distances the stress correlation function becomes anisotropic and decays to zero on a scale of the order of magnitude of a few dislocation spacings. The numerical value of the screening length is less than one average dislocation spacing, $\xi_\sigma\sqrt{\varrho} \approx 0.55$, reflecting the dipolar character of the dislocation arrangement.

3.2.3. *Probability density function of internal stresses*

A method for calculating the probability density function of the internal stresses produced by the ensemble of dislocations, i.e., the probability $p(\boldsymbol{r}, \sigma_{kl})\,\mathrm{d}\sigma_{kl}$ to find at the position $\boldsymbol{r}$ an internal stress between σ_{kl} and $\sigma_{kl} + \mathrm{d}\sigma_{kl}$, has been proposed by Groma and Bako [97]. These authors use methods developed in the theory of X-ray diffraction by Krivoglaz et al. [107], Wilkens [108,109], and Groma et al. [110]. We briefly summarize this work in order to illustrate the close relation between the statistical distribution of internal stresses and the profile of X-ray diffraction lines which makes X-ray diffraction an ideal tool for monitoring statistical properties of internal-stress patterns and dislocation arrangements (section 4.2).

For a two-dimensional system of N dislocations, the (ensemble-averaged) probability density of internal stresses is given by

$$p(\sigma_{kl}, \boldsymbol{r}) = \frac{1}{N!}\sum_{\substack{\beta_1\ldots\beta_N\\ s_1\ldots s_N}}\int \rho_{(N)}(1\ldots N)\,\delta\left[\sigma_{kl} - \sum_{j=1}^{N} s_j\sigma_{kl}^{\beta_j}(\boldsymbol{r}-\boldsymbol{r}_j)\right]\mathrm{d}^2r_1\cdots\mathrm{d}^2r_N, \tag{52}$$

where $\rho_{(N)}$ is the N-dislocation density. The problem is now to express $p(\sigma_{kl}, \boldsymbol{r})$ in terms of dislocation densities and pair correlations. To this end, one considers the Fourier transform of eq. (52). Using the properties of the Dirac delta function, this is readily written as

$$A_{\sigma_{kl},\boldsymbol{r}}(n) = \frac{1}{N!}\sum_{\substack{\beta_1\ldots\beta_N\\ s_1\ldots s_N}}\int \rho_{(N)}(1\ldots N)\exp\left[\sum_{j=1}^{N} \mathrm{i}ns_j\sigma_{kl}^{\beta_j}(\boldsymbol{r}-\boldsymbol{r}_j)\right]\mathrm{d}^2r_1\cdots\mathrm{d}^2r_N, \tag{53}$$

where n is a scalar Fourier variable with dimension of reciprocal stress. One now considers $A_{\sigma_{kl},\boldsymbol{r}}(n)$ in the regime of small n. This yields the correct asymptotic behaviour of the probability distribution at large values of σ_{kl} as well as reasonable estimates for the flanks of the distribution. Defining $B_{kl}^{\beta,s}(\boldsymbol{r},n) := 1 - \exp[\mathrm{i}ns\sigma_{kl}^{\beta}(\boldsymbol{r})]$ and expanding eq. (53) into a power series of the B_{kl}^{β} yields

$$
\begin{aligned}
A_{\sigma_{kl},\boldsymbol{r}}(n) &= \frac{1}{N!}\sum_{\substack{\beta_1\ldots\beta_N\\ s_1\ldots s_N}}\int \rho_{(N)}(1\ldots N)\prod_{j=1}^{N}\bigl[1 - B_{kl}^{\beta_j,s_j}(\boldsymbol{r}-\boldsymbol{r}_j,n)\bigr]\,\mathrm{d}^2r_1\cdots\mathrm{d}^2r_N\\
&= 1 - \sum_{\beta_1,s_1}\int \rho_{(1)}(1)B_{kl}^{\beta_1,s_1}(n,\boldsymbol{r}-\boldsymbol{r}_1)\,\mathrm{d}^2r_1\\
&\quad + \frac{1}{2}\sum_{\substack{\beta_1\beta_2\\ s_1s_2}}\int \rho_{(2)}(1,2)B_{kl}^{\beta_1,s_1}(n,\boldsymbol{r}-\boldsymbol{r}_1)B_{kl}^{\beta_2,s_2}(n,\boldsymbol{r}-\boldsymbol{r}_2)\,\mathrm{d}^2r_1\,\mathrm{d}^2r_2\\
&\quad + \cdots .
\end{aligned} \tag{54}
$$

Equation (54) may be re-written in terms of dislocation pair correlations:

$$
\begin{aligned}
\ln A_{\sigma_{kl},\boldsymbol{r}}(n) \approx &- \sum_{\beta_1,s_1}\int \rho_{(1)}(1)B_{kl}^{\beta_1,s_1}(n,\boldsymbol{r}-\boldsymbol{r}_1)\,\mathrm{d}^2r_1\\
&+\frac{1}{2}\sum_{\substack{\beta_1\beta_2\\ s_1s_2}}\int D_{(2)}(1,2)B_{kl}^{\beta_1,s_1}(n,\boldsymbol{r}-\boldsymbol{r}_1)\\
&\qquad\times B_{kl}^{\beta_2,s_2}(n,\boldsymbol{r}-\boldsymbol{r}_2)\,\mathrm{d}^2r_1\,\mathrm{d}^2r_2.
\end{aligned} \tag{55}
$$

For a random dislocation arrangement, the second term in (55) as well as all higher-order terms vanish. The first term may be calculated using the stress field of a single dislocation, eq. (19). The result is

$$
\ln A_{\sigma_{kl},\boldsymbol{r}}(n) = \mathrm{i}\sigma_{kl}^{(\mathrm{me})}(\boldsymbol{r})n + \frac{C_{kl}(\boldsymbol{r})}{2}G^2b^2\rho(\boldsymbol{r})n^2\ln\left[\frac{\zeta_{kl}}{R_0}\sqrt{\frac{C_{kl}}{2}}Gbn\right] + \cdots . \tag{56}
$$

For an arrangement of dislocations of only one slip system β, the numerical factor ζ_{kl} is given by $\zeta_{kl}^{\beta} \approx \sqrt{2}\exp[-3/2 + C - [1/C_{kl}^{\beta}]\int_0^{2\pi}[K_{kl}^{\beta}(\psi)]^2\ln(K_{kl}^{\beta}(\psi)/\sqrt{C_{kl}^{\beta}})\,\mathrm{d}\psi]$. If σ_{kl} is the shear stress created by an arrangement of edge dislocations in their slip system, one finds $\zeta_{kl}^{\beta} \approx 0.331$, and for screw dislocations, $\zeta_{kl}^{\beta} \approx 0.261$. For a pattern of dislocations of several slip systems, one gets $\ln\zeta_{kl} = \sum_{\beta}(\rho_{(1)}^{\beta}C_{kl}^{\beta})/(\rho C_{kl})\ln[\zeta_{kl}^{\beta}\sqrt{C_{kl}^{\beta}/C_{kl}}]$.

The second term on the right-hand side of (56) diverges in proportion with the logarithm of the crystal size R_0. This monitors the divergence of the mean square stresses for a completely random dislocation arrangement, cf. eq. (42). When there are non-vanishing

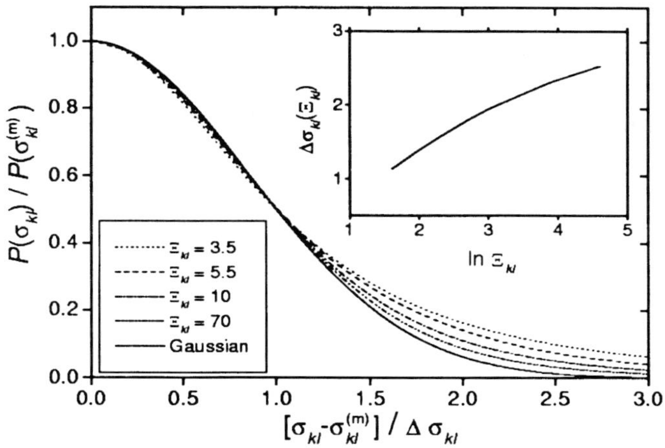

Fig. 11. Dependence of the probability distribution of internal stresses on the screening parameter Ξ_{kl}. All distributions have been normalized with respect to their peak value and half width.

correlations in the dislocation arrangement which fulfil the screening condition (43), the contribution of the second term on the right-hand side of (55) removes this divergence. In this case, in (56) the crystal diameter R_0 is replaced by the same screening length ξ_σ, which also appears in the expression for the mean square stresses, eq. (44).

The stress probability density function $p(\sigma_{kl}, \boldsymbol{r})$ is obtained by Fourier transformation of (56). It has the following properties:

(i) Its first moment is the mesoscopic internal stress $\sigma_{kl}^{(\mathrm{me})}(\boldsymbol{r})$.
(ii) The asymptotic behaviour of $p(\sigma_{kl}, \boldsymbol{r})$ at large $|\sigma_{kl}|$ depends only on the local dislocation density and is given by [97]

$$p(\sigma_{kl}, \boldsymbol{r}) = \frac{G^2 b^2}{2} C_{kl}(\boldsymbol{r})\rho(\boldsymbol{r})\frac{1}{|\sigma_{kl}|^3}. \tag{57}$$

This can be seen from a simple physical argument: The asymptotic tails of the stress probability distribution correspond to the high stresses in the immediate vicinity of single dislocations, where many-dislocation effects can be neglected. Using eq. (19) we find that the area $\mathcal{A}_{kl}^{\beta}(\sigma_{kl})\,\mathrm{d}\sigma_{kl}$ where a dislocation of slip system β produces stresses between σ_{kl} and $\sigma_{kl} + \mathrm{d}\sigma_{kl}$ is given by

$$\mathcal{A}_{kl}^{\beta}(\sigma_{kl})\,\mathrm{d}\sigma_{kl} = G^2 b^2 \left[\int_0^{2\pi} \left[K_{kl}^{\beta}(\psi)\right]^2 \mathrm{d}\psi\right] \frac{\mathrm{d}\sigma_{kl}}{|\sigma_{kl}^3|} = \frac{G^2 b^2}{2} C_{kl}^{\beta} \frac{\mathrm{d}\sigma_{kl}}{|\sigma_{kl}^3|}. \tag{58}$$

Now $p(\sigma_{kl}, \boldsymbol{r})\,\mathrm{d}\sigma_{kl}$ can be interpreted as the area fraction on which the stress is between σ_{kl} and $\sigma_{kl} + \mathrm{d}\sigma_{kl}$. Equation (57) follows by multiplication of eq. (58) with $\rho_{(1)}^{\beta}(\boldsymbol{r})$ and summation over all slip systems: $p(\sigma_{kl}, \boldsymbol{r}) \approx \sum_\beta \mathcal{A}_{kl}^{\beta} \rho_{(1)}^{\beta}(\boldsymbol{r})$ for $|\sigma_{kl}| \to \infty$.

(iii) The shape and width of the central part of the distribution $p(\sigma_{kl}, \boldsymbol{r})$ depend on the range of the screening correlations in the dislocation arrangement. A convenient representation is obtained by scaling $\sigma_{kl}\tilde{\sigma}_{kl}[\sqrt{(C_{kl}/2)}Gb\sqrt{\rho(\boldsymbol{r})}]$, denoting by $\tilde{n}$ the corresponding Fourier variable, and defining the nondimensional screening parameter $\Xi_{kl} := [\xi_\sigma\sqrt{\rho}]/\zeta_{kl}$. Replacing in eq. (56) R by ξ_σ and scaling all variables yields

$$A_{\sigma_{kl},\boldsymbol{r}}(\tilde{n}) = \exp\left[i\langle\tilde{\sigma}_{kl}(\boldsymbol{r})\rangle\tilde{n} + \frac{\tilde{n}^2}{2}\ln\frac{\tilde{n}}{\Xi_{kl}} + \cdots\right]. \tag{59}$$

The corresponding distribution $p(\tilde{\sigma}_{kl}, \boldsymbol{r})$ is symmetrical around $\tilde{\sigma}_{kl}^{(\mathrm{me})}$. Its shape and width depend on the screening parameter Ξ_{kl} (cf. the discussion of X-ray line profiles by Wilkens [109]). The half width $\Delta\tilde{\sigma}_{kl}(\Xi_{kl})$ of the scaled distribution $p(\tilde{\sigma}_{kl}, \boldsymbol{r})$ increases roughly in proportion with the logarithm of Ξ_{kl} (inset in fig. 11). The shape of the distributions can be seen from fig. 11, where distributions normalized with respect to peak value and half width have been compiled for different Ξ_{kl}. For $\ln\Xi_{kl} \gg 1$, the shape of the stress probability distribution approaches a Gaussian and the transition towards the asymptotic behaviour (57) occurs at large stresses only.

In general, no analytical expression for $p(\tilde{\sigma}_{kl}, \boldsymbol{r})$ is available. For small Ξ_{kl}, the analytical approximation [111] $p(\tilde{\sigma}_{kl}, \boldsymbol{r}) \approx 2K_{kl}/[(\tilde{\sigma}_{kl} - \tilde{\sigma}_{kl}^{(\mathrm{me})}(\boldsymbol{r}))^2 + K_{kl}]^{3/2}$ may be used where $K_{kl} = [\Delta\tilde{\sigma}_{kl}(\Xi_{kl})]^2/[2^{2/3} - 1]$. This yields an optimum fit for $\Xi_{kl} = 3.5$, where the analytical approximation practically coincides with the numerically calculated curve in fig. 11. Since $\zeta_{kl} \approx 0.3$, this corresponds to the realistic case when the screening length is approximately equal to one average dislocation spacing.

3.3. Generalized composite models of multiscale dislocation arrangements

The preceding discussion has demonstrated that an appropriate statistical characterization of dislocation arrangements must envisage both spatial variations in the (first-order) dislocation densities and dislocation pair correlations with a range which is typically of the order of magnitude of one average dislocation spacing. This is reflected by the internal-stress pattern: On the microscopic scale, the discreteness of single dislocation lines gives rise to internal-stress fluctuations $\delta\sigma_{kl}$ with a characteristic 'wavelength' that is governed by the range of dislocation pair correlations. These fluctuations are superimposed on a mesoscopic internal stress field $\sigma_{kl}^{(\mathrm{me})}(\boldsymbol{r})$ which stems from spatial inhomogeneities in the first-order dislocation densities. The previous paragraph has mainly dealt with the statistical properties of the 'microscopic' internal-stress fluctuations. In the present paragraph, we address the calculation of mesoscopic stresses in deforming crystals with inhomogeneous dislocation microstructure using composite models.

3.3.1. Mesoscopic stresses in inhomogeneous dislocation patterns

The spatially varying dislocation densities in an inhomogeneous dislocation arrangement lead to spatial variations of the flow stress. We consider situations where the local flow stresses $\sigma_{\mathrm{f}}^{\beta}(\boldsymbol{r})$ in the active slip system(s) can be related to microscopic features of the dislocation arrangement on the scale of the spacing between single dislocations, such as interactions between individual dislocations, junctions, etc. The local dislocation arrangement is, in a simplified manner, characterized by the local (first-order) dislocation densities $\rho_{(1)}^{\beta,s}$ and the screening lengths $\xi_{\sigma}^{\beta,s}$, which characterize the range of dislocation pair correlations. Compiling these variables into a state vector $\boldsymbol{\rho} := [\rho_{(1)}^{\beta_1,s_1}, \xi_{\sigma}^{\beta_1,s_1}; \rho_{(1)}^{\beta_2,s_2}, \xi_{\sigma}^{\beta_2,s_2}; \ldots]$, we can write the local flow stresses in the form $\sigma_{\mathrm{f}}^{\beta}(\boldsymbol{r}) = \sigma_{\mathrm{f}}^{\beta}(\boldsymbol{\rho}(\boldsymbol{r}))$. In this approximation, all mesoscopic volume elements $\mathrm{d}V_{\rho}$ with statistically equivalent local dislocation arrangements (same value of $\boldsymbol{\rho}$) show a similar deformation behaviour.

During loading, first the softest volume elements begin to deform plastically. Flow in these volume elements leads to mesoscopic plastic strain gradients which, in turn, give rise to a re-distribution of the internal stresses. We consider a representative volume ΔV_{ρ} over which the dislocation arrangement is constant and approximate the shape of this volume by an ellipsoid. According to Kröner [112], the mesoscopic internal stress in this volume is

$$\boldsymbol{\sigma}_{\rho}^{(\mathrm{me})} = -\Gamma G\big(\boldsymbol{\varepsilon}_{\rho}^{\mathrm{pl}} - \bar{\boldsymbol{\varepsilon}}^{\mathrm{pl}}\big), \tag{60}$$

where $\boldsymbol{\varepsilon}_{\rho}^{\mathrm{pl}}$ is the average plastic strain in the volume elements ΔV_{ρ} and $\bar{\varepsilon}^{\mathrm{pl}}$ the macroscopically averaged plastic strain. Γ is Eshelby's accommodation factor [113] which depends on the shape of the volume element ΔV_{ρ}.

Stress redistribution proceeds until the sum of the mesoscopic internal stresses and the external stress resolved in the active slip system(s) matches the respective flow stresses everywhere, such that further plastic deformation proceeds in a compatible manner. The mesoscopic stresses in the active slip systems are then given by

$$\sigma_{\rho}^{(\mathrm{me})}(\boldsymbol{r}) = \sigma_{\mathrm{f}}^{\beta}(\boldsymbol{\rho}) - \sigma_{\mathrm{ext}}^{\beta}. \tag{61}$$

Averageing over the macroscopic crystal volume V_0 shows that the external stress is the volume average of the local flow stresses,

$$\sigma_{\mathrm{ext}}^{\beta} = \{\sigma_{\mathrm{f}}^{\beta}\} = \frac{1}{V_0}\int_{V_0} \sigma_{\mathrm{f}}^{\beta}(\boldsymbol{r})\,\mathrm{d}^2 r, \tag{62}$$

since the macroscopic volume average of the mesoscopic stresses vanishes. Hence within the framework of a composite model these stresses do not contribute to the flow stress. Because of $\sigma_{\mathrm{f}}^{\beta}(\boldsymbol{r}) = \sigma_{\mathrm{f}}^{\beta}(\boldsymbol{\rho}(\boldsymbol{r}))$, eq. (62) may be written as [114]

$$\sigma_{\mathrm{ext}}^{\beta} = \int \sigma_{\mathrm{f}}^{\beta}(\boldsymbol{\rho})\, p(\boldsymbol{\rho})\,\mathrm{d}\boldsymbol{\rho}, \tag{63}$$

where $p(\boldsymbol{\rho})$ is a probability density function which characterizes the probability to find, in a mesoscopic volume element chosen at random, the local dislocation configuration $\boldsymbol{\rho}$.

The characteristic strains required for the initial stress re-distribution which establishes compatible plastic flow can be estimated as follows: The characteristic variation in the local flow stresses is of the order of the macroscopic flow stress. Putting $\sigma_\rho^{(\mathrm{me})} \approx \sigma_{\mathrm{ext}}^\beta$ we get from eq. (60) $\Delta\varepsilon^\beta := \bar{\varepsilon}^\beta - \varepsilon_\rho^\beta \approx \sigma_{\mathrm{ext}}^\beta / \Gamma G$. Since Γ is typically on the order of magnitude of unity ($\Gamma = 0.5$ for rectangular cells aligned along the stress axis in uniaxial straining [73]), the characteristic strains required for stress redistribution are typically on the order of 10^{-2}–10^{-3}.

Two remarks are appropriate at this point: (i) Even if it is possible to establish a local relationship between flow stress and dislocation configuration within a mesoscopic volume element, hardening involves exchange of dislocations between adjacent volume elements. On mesoscopic scales hardening is an intrinsically non-local process [115]. (ii) Redistribution of internal stresses to accommodate mesoscopic strain gradients necessarily implies redistribution of dislocations. The densities of these 'geometrically necessary' dislocations relate to the flow stress pattern in a non-local manner. Since such dislocations may, in turn, influence the 'local' flow stresses, the assumption of a local $\sigma_{\mathrm{f}}^\beta(\boldsymbol{\rho})$-relation is only consistent when the flow-stress contribution of the geometrically necessary dislocations is negligibly small. This entails the requirement that the geometrically necessary dislocations should make up only a small fraction of the total dislocation density. For mesoscopic strain gradients arising from spatial variations of the local flow stresses, it can be demonstrated by a simple estimate that this requirement is always fulfilled: Characteristic shear strain gradients are on the order of $\Delta\varepsilon^\beta/\lambda$ where λ is the 'wavelength' of the dislocation density pattern. The corresponding characteristic density ϕ of geometrically necessary dislocations is $\phi \sim \Delta\varepsilon^\beta/(\lambda b)$. When the tensile flow stress obeys a Taylor relationship, $\sigma_{\mathrm{ext}} = \alpha G b\sqrt{\varrho}$, where ϱ is the macroscopically averaged dislocation density, one finds that the ratio between the characteristic density of geometrically necessary dislocations and the total dislocation density is

$$\frac{\phi}{\varrho} = \frac{\alpha}{\Gamma\lambda\sqrt{\varrho}}. \tag{64}$$

Because of $(\alpha/\Gamma) \approx 1$ this is approximately the ratio between the average dislocation spacing and the wavelength of dislocation density patterns. If this ratio is small, the density of 'geometrically necessary' dislocations required to create the long-range stress field in a composite model can be neglected in comparison with the total dislocation density.[11] The condition $\lambda\sqrt{\varrho} \gg 1$ is fulfilled when the mesoscopic scale of dislocation density variations can be clearly distinguished from the microscopic spacing of individual dislocations.

3.3.2. Spatially averaged probability distribution of internal stresses

In section 3.2.3 we have considered the local probability distribution of internal stresses, which is defined with respect to a given point $\boldsymbol{r}$ and an ensemble of statistically equivalent

[11] At large strains, it is found that 'geometrically necessary' excess dislocations make up a substantial fraction of the total dislocation density. These dislocations are, however, mostly arranged in configurations which do not create long-range stresses but only stress-free lattice rotations. Hence they do not contribute to the mesoscopic stress field.

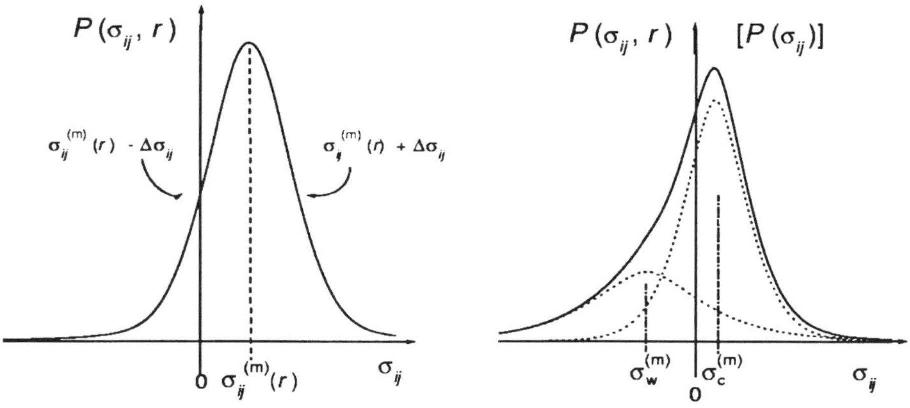

Fig. 12. Probability distributions of internal stresses; left: local probability distribution with mesoscopic stress $\langle\sigma_{ij}(r)\rangle$ and half-width $\Delta\sigma_{ij}(\boldsymbol{r})$; right: probability distribution for a two-phase composite model.

dislocation arrangements. Averaging over the macroscopic volume V_0, gives us the spatially averaged probability distribution (SAPD)

$$p(\sigma_{kl}) := \{p(\sigma_{kl}, \boldsymbol{r})\} = \frac{1}{V_0}\int_{V_0} p(\sigma_{kl}, \boldsymbol{r})\,\mathrm{d}^3 r \tag{65}$$

which is the density of the probability to find the internal stress σ_{kl} at a location chosen at random in the crystal volume. This distribution is of particular importance since it directly relates to the X-ray line profile and can therefore be determined experimentally (section 4.2). It may be considered as a superposition of local probability distributions which are according to eq. (59) centered around the local values of the mesoscopic internal stress $\sigma_{kl}^{(\mathrm{me})}(\boldsymbol{r})$ (fig. 12). The width and shape of the local probability distributions are governed by the local dislocation densities and screening lengths.

When a composite model can be used to establish a local relation between the mesoscopic stress field and the dislocation configuration $\boldsymbol{\rho}(\boldsymbol{r})$, the local internal-stress distributions can be written as $p(\sigma_{kl}, \boldsymbol{r}) = p(\sigma_{kl}, \boldsymbol{\rho}(\boldsymbol{r}))$ and the SAPD (65) can be expressed as

$$p(\sigma_{kl}) = \int p(\sigma_{kl}, \boldsymbol{\rho})\,p(\boldsymbol{\rho})\,\mathrm{d}\boldsymbol{\rho}. \tag{66}$$

Under these circumstances, virtually all information both about the statistics of internal stresses and about the macroscopic deformation properties is contained in the probability distribution $p(\boldsymbol{\rho})$. The theoretical calculation of such distributions using methods of stochastic dislocation dynamics will be discussed later in section 6.

3.3.3. An example: internal stresses in ⟨100⟩-orientated f.c.c. single crystals

As a concrete example of the application of composite models we consider the internal stresses developing in inhomogeneous dislocation arrangements (cell structures) of f.c.c.

single crystals deformed in tension or compression along a $\langle 100\rangle$ axis which we identify with the x-axis. Under these circumstances, eight active slip systems contribute symmetrically to the deformation while four systems remain inactive.[12] The external shear stresses in the eight active slip systems are $\sigma_{\text{ext}}^{\beta} = M\sigma_{\text{ext}}$, where σ_{ext} is the axial stress and the Schmid factor $M = 1/\sqrt{6}$. The dislocation densities are the same in all active slip systems. Inactive systems may be populated by reactions among dislocations of the active systems. This leads to a dislocation density in these systems which is a spatially constant fraction of the density in the active systems. Furthermore, the dislocation arrangements are 'neutral', i.e., dislocations of both signs have approximately equal densities. (Locally, there may be surplus dislocations which give rise to mesoscopic stresses but, for the reason discussed above, they make up only a small fraction of the total dislocation density.)

Under these circumstances, the dislocation arrangement in a mesoscopic volume element can be simply characterized by the total dislocation density $\rho(\boldsymbol{r})$. The local flow stresses in the active slip systems are $\sigma_{\text{f}}^{\beta}(\boldsymbol{r}) = \alpha^{\beta} Gb\sqrt{\rho(\boldsymbol{r})}$ where α^{β} is the same for all active slip systems. Since the composition of the dislocation arrangement is spatially constant, α^{β} is space-independent. Compatible plastic flow requires that the mesoscopic stress matches the flow stress everywhere, i.e., the mesoscopic tensile stress is given by

$$\sigma_{xx}^{(\text{me})}(\boldsymbol{r}) = \sigma_{\text{ext}} - \alpha^{\beta} Gb\sqrt{\rho(\boldsymbol{r})}/M. \tag{67}$$

This situation has been considered in a series of papers by Mughrabi et al. [73–75]. Here the simplifying assumption was made that the dislocations form a cell pattern consisting of two 'phases', namely dislocation-rich cell walls with dislocation density ρ_{w} and dislocation-depleted cell interiors with dislocation density ρ_{c}. (For experimental observations of such patterns, see section 4.) Similar two-phase models have also been applied to single-slip situations, where deformation is governed by slip in only one system, for instance by Pedersen et al. [71] for analysing the Bauschinger effect in copper deformed into hardening stage II and by Mughrabi [72] for analyzing mesoscopic internal stresses in the wall structure of persistent slip bands of cyclically deformed f.c.c. metals. In either case the internal-stress distribution corresponds to the situation depicted in fig. 12 (right): The SAPD of internal stresses is made up of two symmetrical sub-distributions centred around the values $\sigma_{\text{w}}^{(\text{me})} = \sigma_{\text{ext}} - \alpha^{\beta} Gb\sqrt{\rho_{\text{w}}}/M$ and $\sigma_{\text{c}}^{(\text{me})} = \sigma_{\text{ext}} - \alpha^{\beta} Gb\sqrt{\rho_{\text{c}}}/M$, corresponding to the cell walls or cell interiors. The relative 'weight' of these sub-distributions is determined by the volume fractions f_{w} and $f_{\text{c}} = 1 - f_{\text{w}}$ of cell walls and cell interiors. Macroscopic stress equilibrium is ensured by the flow stress law $\sigma_{\text{ext}}^{\beta} = \sigma_{\text{ext}}/M = \alpha^{\beta} Gb[f_{\text{w}}\sqrt{\rho_{\text{w}}} + f_{\text{c}}\sqrt{\rho_{\text{c}}}]$.

Within the framework of generalized composite models established by eqs (63) and (66), the two-phase composite models of Mughrabi, Pedersen and others correspond to a special case in which the local dislocation arrangement can be completely characterized by a single variable (the total dislocation density ρ). Furthermore, a specific assumption is made about the functional shape of the probability density $p(\rho)$, which is assumed in the form $p(\rho) = f_{\text{w}}\delta(\rho - \rho_{\text{w}}) + f_{\text{c}}\delta(\rho - \rho_{\text{c}})$. The generalization from two 'phases' to continuous

[12] TEM investigations of crystals deformed in a $\langle 100\rangle$ orientation indicate that under these deformation conditions the structure does not break up into blocks where only a few slip systems are active (cf. section 4.2.2).

probability distributions is straightforward. The theoretical calculation of such distributions for f.c.c. single crystals deformed along a ⟨100⟩ direction will be discussed in section 6.

4. Characterization of dislocation patterns and internal stress fields

4.1. Analysis of transmission electron microscopy images

Apart from birefringence and magnetic [116,117] techniques, both of which are restricted to limited classes of materials, for many years X-ray diffraction was the main tool for investigating internal stresses in crystalline materials. As dislocations are the dominant sources of internal stresses in deformed crystals, since the late 1950's the possibility to study dislocation arrangements in thinned samples by transmission electron microscopy (TEM) [118,119] has provided us with another, though admittedly indirect, access to internal stresses. It was soon realized, however, that this approach was beset by several pitfalls. These were related to the following three questions: (i) How and to what extent does the dislocation pattern change when the deformed crystals are unloaded before the thin sections required for TEM are prepared? (ii) To what extent are the observed dislocation arrangements affected by the unavoidable relaxation of the internal stresses accompanying the thinning? (iii) How can we extract from the limited fields of view of TEM micrographs relevant information on internal stresses and dislocation patterns on a mesoscopic scale?

Of these questions, (ii) was tackled first. Owing to the capability of screw dislocations to change their slip plane by cross-slip, it is the screw parts of the dislocations that are most likely to leave the sample during the thinning and thus make the dislocation patterns of thin sections unrepresentative of the bulk. In f.c.c. crystals the barrier against cross-slip may be increased by lowering the stacking-fault energy [120,121], hence materials with low stacking-fault energies should be less susceptible to 'thinning artefacts'. TEM studies on NiCo alloys, whose stacking-fault energy decreases strongly with increasing Co content as the f.c.c./h.c.p. phase boundary is approached, showed indeed dislocation pattern that were rather different from those of Cu or Ni, not to speak of Al with its particularly high stacking-fault energy [122]. At about the same time, Essmann [123] developed an approach based on the observation that metals are hardened by neutron irradiation. Irradiating a deformed sample in a nuclear reactor before thinning suppresses or, at least, strongly reduces the dislocation rearrangement during thinning. The dislocation patterns observed in, say, copper crystals treated in this way were indeed rather different from those seen in unirradiated crystals [124].[13] Question (i) was answered by an extension of this technique, viz. performing the neutron irradiation under load. With this technique, Mughrabi [126] was able to deduce from the curvature of dislocation lines that had been

[13] The realization that the thicker the TEM samples the better the chances to preserve the bulk microstructure aroused the interest of metallurgists and solid-state physicists in high-voltage electron microscopes. Before about 1960 the main incentives for developing high-voltage electron microscopy had come from cell biology [125]. However, when high-voltage microscopes did become available for physical-metallurgy work, the Essmann technique had solved the problem for all those materials that did not become intolerably radioactive under neutron irradiation.

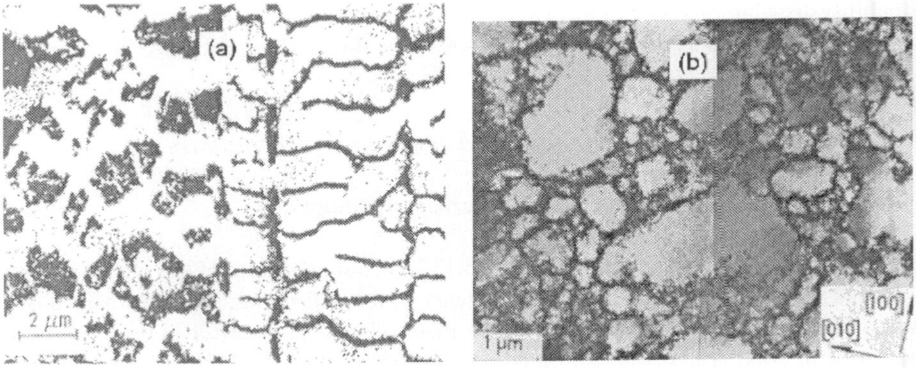

Fig. 13. (a) TEM image of Cu single crystal fatigued into saturation at room temperature, plastic shear-strain amplitude 4.1×10^{-3}, stress amplitude 28 MPa. Left-hand side: 'matrix' structure, right-hand side: periodic wall structure of persistent slip bands (PSBs) (courtesy Dr. U. Holzwarth). (b) Cell structure in a Cu single crystal deformed at room temperature in tension along [100] to a stress of 75.6 MPa, foil orientation [100], after Mughrabi et al. [73]

pinned by the irradiation (cf. below) the magnitude and distribution of shear stresses acting on dislocations in the primary slip system of Cu single crystals.

A first step in answering question (iii) was the construction of 'TEM maps' by putting together micrographs of adjacent areas (see, e.g., [123]). Since relevant features of mesoscopic dislocation patterns exhibit a range of scales from one to several hundreds dislocation spacings, this proved indispensable to decide which of these features are generic.[14] Such TEM maps have been used to characterize dislocation patterns by determining the characteristic shapes and extensions of dislocation-rich and dislocation-depleted regions or the properties of dislocation walls such as their net Burgers vector content and misorientation. While the earlier investigations focused mainly on determining average quantities such as mean cell sizes or average misorientations in cell structures, systematic investigations of the probability distributions of microstructural variables are rather recent [127–130]. There are several reasons for this development. A regular dislocation pattern such as a persistent slip-band structure (the ladderlike structure in fig. 13(a)) is adequately characterized by the average values of thickness and spacing of the dislocation-rich walls: the distribution of wall spacings shown in fig. 14 (top) exhibits a pronounced maximum close to the average spacing, and the same holds for the spacings of dislocation-rich 'veins' in the slightly more irregular 'matrix' pattern seen on the same micrograph (fig. 14, bottom). Hence one may discard the additional information contained in the statistical distribution of these quantities as incidental. If we consider, however, the cell pattern depicted in fig. 13(b), we find a very broad spectrum of cell sizes which makes it difficult to decide what is the 'correct' length scale to characterize the pattern. The corresponding cell-size distribution depicted in fig. 15 shows that the average cell size

[14] While it has been possible to observe, with sufficient experimental effort, most of the elementary dislocation reactions that have been proposed in the literature, to demonstrate that a particular feature is generic proved usually more difficult.

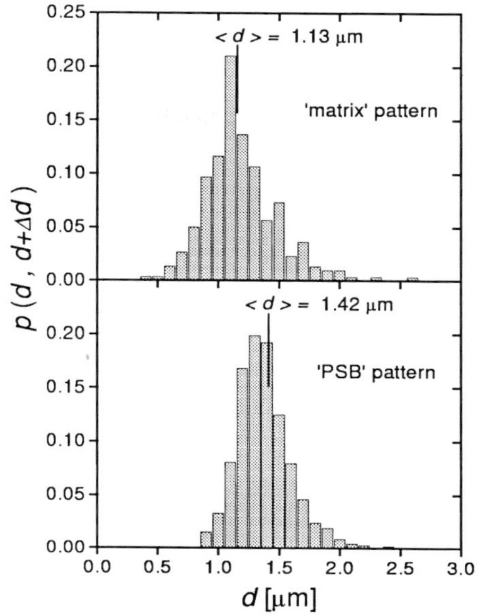

Fig. 14. Distributions of widths d of dislocation-depleted 'channels' in the microstructure of Cu fatigued to saturation at room temperature; shear stress amplitude 28 MPa; top: channel width distribution in a matrix structure [131]; bottom: distribution in a PSB structure [132].

$\langle\lambda\rangle$ is hardly representative of the overall pattern. Indeed, the question of a characteristic scale is ill-posed in this particular case: As will be discussed in detail in section 4.1.2, this pattern belongs to a class of self-similar cell structures which exhibit fractal scale-invariance. For such patterns virtually all relevant information on the microstructure resides in the parameters characterizing the functional shape of the distribution. The average cell size determined from a TEM micrograph may be incidental, being strongly influenced by parameters such as image size, image magnification, and foil thickness rather than representing an intrinsic property of the dislocation arrangement. A second reason for the interest in the statistical characterization of dislocation arragements stems from the fact that certain highly relevant processes are not governed by the average, 'normal' behaviour of the microstructural variables but by strongly deviant 'outliers'. An example is the rôle of cell or subgrain boundaries with very large misorientations which may serve as nucleation sites for recrystallization. This situation is, hence, governed by the statistics of extremes, which requires knowledge of the shape and, in particular, of the asymptotic tails of the misorientation distribution.

Recently, methods of quantitative image analysis have been systematically applied to bitmaps obtained from dislocation patterns [127,128,133,134]. This has been shown to constitute another powerful tool for characterizing dislocation arrangements on the mesoscopic scale. The method, which will be discussed in some detail in the following, yields both a qualitative characterization (long-range vs. short-range order, characteristic

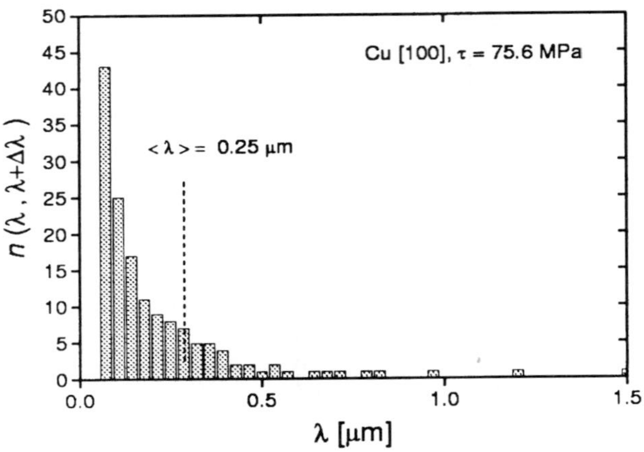

Fig. 15. Distribution of cell sizes in the cell structure of fig. 13(b).

scales vs. fractal scale invariance) and quantitative features ('wavelengths', fractal dimensions) of the patterns.

4.1.1. *Dislocation patterns with a single characteristic scale*

Determination of characteristic lengths Mesoscopic dislocation patterns often consist of dense dislocation entanglements separating almost dislocation-free regions. In such cases, dislocation-rich and dislocation-depleted regions constitute well-defined geometrical entities which can be characterized by their characteristic shapes and sizes. The TEM micrographs yield distributions of characteristic lengths but only little can be said about the dislocation densities in the entanglements, which on the micrographs appear simply as black areas (fig. 13). Depending on the morphology of the patterns, these characteristic lengths may be determined in different ways: For cellular patterns, in which dislocation-dense cell walls separate mutually non-connected, dislocation-depleted cell interiors as depicted in fig. 13(b), the cell size may be defined as the maximum diameter of a cell in a given direction or as the square root of the area of the cell interior. In either case, stereological information on the average cell shape can be used to obtain the distribution of cell sizes in three dimensions.

If the shapes of the dislocation-depleted areas are rather irregular or if these areas are mutually connected, characteristic lengths may be defined using a line-intercept method in which the spacing of dislocation-rich regions is determined along straight lines drawn through the micrograph. For instance, fig. 14 shows width distributions of dislocation-depleted 'channels' in matrix and PSB patterns of fatigued Cu single crystals. In both cases, the channel widths d determined in the slip direction exhibit unimodular distributions, indicating a preferred pattern wavelength. The maximum of the distributions (the most probable channel width) roughly coincides with the average channel width $\langle d \rangle$ [135]. While the average channel width is about the same for the matrix and the PSBs, the distribution is narrower for the PSBs, indicating a higher degree of order.

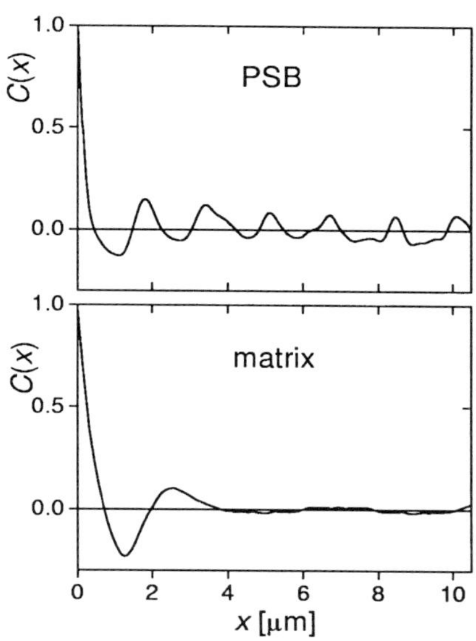

Fig. 16. Autocorrelation functions $C(x)$ of bitmaps of dislocation patterns observed in cyclic deformation [135], deformation conditions as in fig. 13(a). Top: persistent slip bands, bottom: matrix structure; the x-direction corresponds to the direction of slip.

Image analysis of bitmaps of TEM micrographs The information obtained from distributions of characteristic lengths may be complemented by quantitative image analysis of bitmaps obtained from TEM micrographs. To this end, the micrographs must be image-processed in order to eliminate all contrast features which are not due to the core contrast of the dislocations, such as brightness variations due to large-scale misorientations, foil-thickness variations, etc. This necessity limits the applicability of the method and, in particular, makes it difficult to analyze micrographs of cell structures with large misorientations. Technical details of the image processing procedure are given, e.g., in [127]. In the following we assume that the images have been processed in such a manner that dislocations or dislocation entanglements are represented by black pixels.

When a bitmap of the dislocation pattern can be created, important qualitative information is obtained from its autocorrelation function $C(\boldsymbol{r})$ [128]. We illustrate this for matrix and PSB patterns of fatigued Cu. For these patterns, the autocorrelation functions depicted in fig. 16 exhibit an oscillatory behaviour in the slip direction (this corresponds to the vertical direction in fig. 13(a)). The width of the *initial maximum* is proportional to the average thickness of the dislocation-rich regions, which amounts to about 1 μm for the matrix 'veins' and about 0.2 μm for the PSB 'walls', while the *first minimum* of the autocorrelation function relates to the preferred width of the dislocation-depleted channels in between. The wavelength of the oscillations, which is about 1.7 μm in the PSB and 2.4 μm in the matrix, corresponds to the sum of the average extensions of dislocation-rich and dislocation-depleted regions.

In the irregular matrix pattern, the correlations decay rapidly in space, indicating short-range order. In the PSBs, on the other hand, the autocorrelation function exhibits a well-

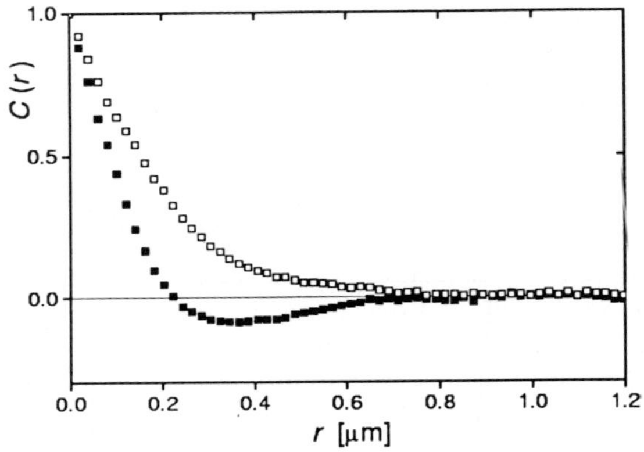

Fig. 17. Autocorrelation function $C(x)$ of bitmap obtained from a SEM micrograph of a Ni single crystal deformed at room temperature in tension to a flow stress of 117 MPa [136]. The crystal orientation was near 100. Full symbols: autocorrelation function in a direction of the temsile axis, open symbols: autocorrelation function in the direction normal to the tensile axis.

developed residual oscillation which extends over many wall spacings. This indicates that the wall arrangement in the PSBs has periodic long-range order in spite of a substantial scatter in the spacings between adjacent walls. The long-range order appears to be a specific feature of dislocation patterns forming in cyclic deformation. It is observed both in single and multiple slip conditions [128]. Cell patterns emerging in unidirectional deformation usually exhibit a much lower degree of order, and often the existence of a preferred cell size is indicated by a single minimum in the autocorrelation function only. For instance, the autocorrelation function shown in fig. 17 was obtained from the cell structure of a Ni single crystal deformed in tension along a near-⟨100⟩ direction [136]. In the direction of the tensile axis a single minimum indicates short-range order with a characteristic length of about 0.4 μm, while the structure is disordered in the direction perpendicular to this axis.

4.1.2. Fractal dislocation patterns

Patterns like that depicted in fig. 13(b) cannot be characterized by a dominant 'wavelength'. Rather, a certain self-similarity is suggested in this figure, and one may ask whether the pattern can be characterized in terms of fractal geometry. The possibility of fractal dislocation patterning was first recognized by Gil Sevillano and co-workers [133,137], who applied image analysis to cell structures of cold-worked Cu including the example shown in fig. 13(b). Later, analyses by Hähner, Bay and Zaiser [94,127] demonstrated the fractal character of dislocation cell structures in f.c.c. single crystals deformed in tension in ⟨100⟩ and ⟨111⟩ directions, and in NaCl polycrystals [134]. Independently, fractal dimensions of deformation-induced dislocation patterns were determined by Olemskoi et al. (see [138], and references therein).

Measures of the fractal dimension In the following brief description of methods that may be used to determine fractal dimensions of a pattern given in the form of a two-dimensional bitmap, we use the terminology proposed by Mandelbrot in [139–141]. (i) The 'box-counting dimension' is obtained by covering the bitmap with square grids of meshlength Δx and denoting by $N(\Delta x)$ the number of squares which contain at least one black pixel. A power-law behaviour $N(\Delta x) \propto \Delta x^{-D_B}$ defines the box-counting dimension D_B. (ii) The 'mass dimension' is defined as follows: Consider the average number $M(r)$ of black pixels contained within a circle of radius r around each black pixel. When $M(r)$ is proportional to some power of the radius r, $M(r) \propto r^{D_M}$, the exponent D_M gives the mass dimension of the bitmap. Note that $M(r)$ is the integral of the autocorrelation function $C(\boldsymbol{r})$ over a circle of radius r. (iii) For cellular patterns, a fractal dimension can be deduced from the distribution of cell sizes. When the number of cells above size Λ decreases according to $N(\lambda > \Lambda) \propto \Lambda^{-D_G}$, then D_G defines the so-called gap dimension [140]. We note that for self-similar fractal patterns, the different measures of fractal dimension listed above coincide and yield the Hausdorff dimension [140]. For self-affine fractals, on the other hand, different methods yield, in general, different values of the fractal dimensions. Thus simultaneous application of different methods allows us to decide whether or not a given pattern is self-similar.

Fractal analysis: examples and results The fractal analysis of self-similar dislocation patterns is exemplified for the cell structure depicted in fig. 13(b). Results of a box count performed on a bitmap obtained from this micrograph are shown in fig. 18. To enhance variations in slope in the double-logarithmic representation, $\log[N(\Delta x) \times \Delta x^2]$ has been plotted vs. $\log[\Delta x]$. The box-counting dimension is related to the slope $m = \partial \log[N(\Delta x) \times \Delta x^2]/\partial \log[\Delta x]$ by $D_G = 2 - m$. Three regimes can be distinguished: (i) at small Δx, the slope approaches zero. This is a consequence of the area-like character of the cell walls on the micrograph which shows up at small scales. (ii) At intermediate Δx, a scaling regime with slope $m = 0.21 \pm 0.01$ extends over almost two orders of magnitude. This corresponds to a box-counting dimension $D_B = 1.79 \pm 0.01$. (iii) At large Δx, the slope becomes zero when the largest cell within the analyzed area is covered completely. Determination of the mass dimension of the same pattern is demonstrated in fig. 19. Again, at small r, $M(r) \propto r^2$ due to the areal character of the cell walls, which is reflected by an asymptotically zero slope in the double-logarithmic $\log[M/r^2]$ vs. $\log r$ plot. The slope $m = -0.21 \pm 0.02$ of the intermediate scaling regime yields $D_M = 1.79 \pm 0.02$. This is consistent with the value of the box-counting dimension. For r above the maximum cell size on the micrograph, the scaling regime is again delimited by a regime of zero slope ($M \propto r^2$).

An ideal fractal is self-similar on all length scales. In reality, however, the range of scales where self-similarity is observed is always limited. The limits of the fractal scaling regime may be extrinsic or intrinsic. Intrinsic limits reflect a property of the structure; for instance, on the length scale of single dislocation segments (which are one-dimensional geometrical objects), dislocation arrangements necessarily cease to be fractal. Extrinsic limits, on the other hand, are related to the geometry of the analyzed sample. In figs 18 and 19, both the upper and lower bounds of the scaling regimes may be of extrinsic character. The lower bound can be related to the fact that a TEM micrograph represents the projection of a foil of finite thickness. Because of this, cells with size smaller than the foil thickness cannot be resolved [142]. The upper boundary, on the other hand, is governed by the size of the

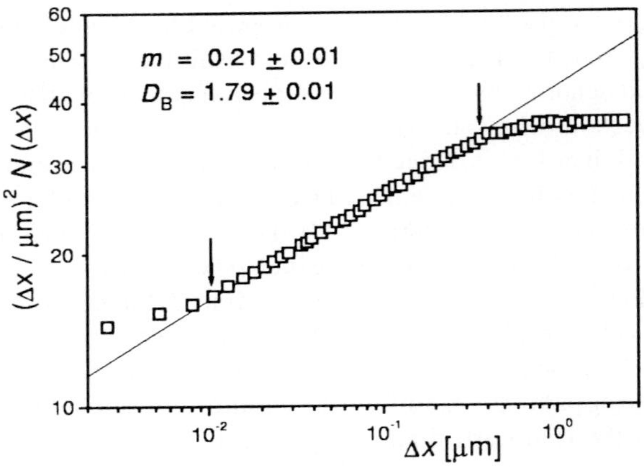

Fig. 18. Determination of the box dimension of the pattern in fig. 13(b). For details see text.

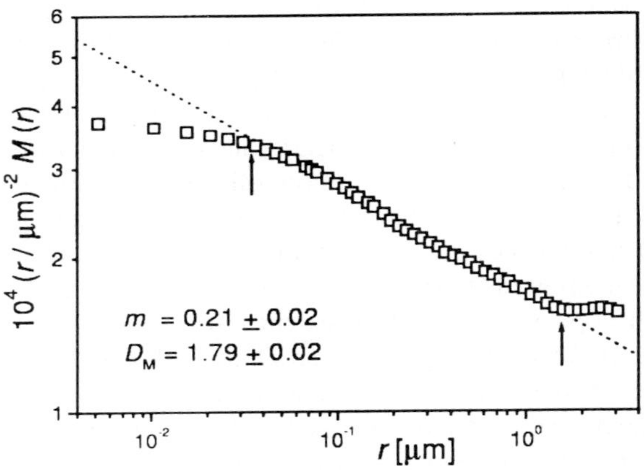

Fig. 19. Determination of the mass dimension of the pattern in fig. 13(b). For details see text.

largest cell visible on the micrographs. Within the intrinsic limits of the scaling regime, this size increases in proportion with the size of the micrographs, i.e., the scaling regime is extended when larger micrographs are analyzed.

The importance of finite-size effects is also seen when cell-size distributions are analyzed. In fig. 20, the full circles represent the cumulative cell size distribution $N(\lambda > \Lambda)$ obtained from fig. 13(b), which corresponds to an analyzed area of 40 μm^2. The distribution is hyperbolic, and the exponent yields a gap dimension $D_G = 1.78$. This consistent with the values obtained from the other methods. To illustrate the influence of micrograph size, we have included another distribution (open squares) obtained from a ⟨100⟩-orientated Cu3at%Mn single crystal deformed to approximately the same stress. In this case, a much larger area of about 410 μm^2 could be analyzed. The extension of the scaling regime and the size of the largest cell increase roughly in proportion with the

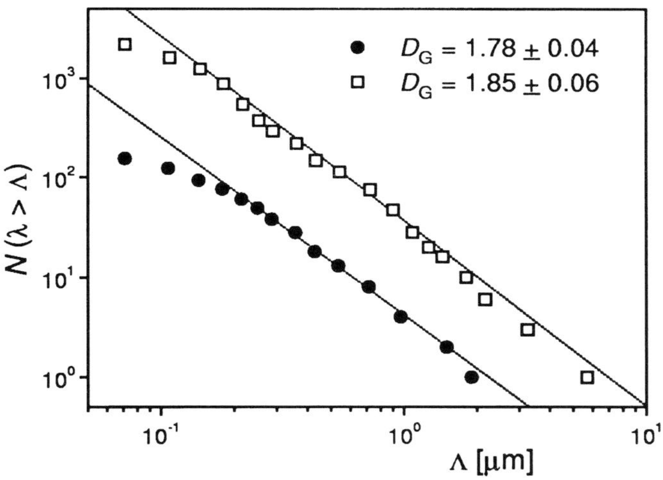

Fig. 20. Determination of gap dimensions. Full circles: number of cells above size Λ for the cell size distribution shown in fig. 15; open squares: the same for the cell size distribution in a Cu3at%Mn single crystal deformed in tension along [100] to a flow stress of 68.2 MPa (courtesy of U. Essmann).

linear extension of the analyzed area, as expected when one is analyzing samples of finite size which are below the intrinsic limits of the fractal scaling regimes. In [127] it was demonstrated that this dependence can be interpreted quantitatively in terms of a finite-size scaling relationship. This allows the prefactor C of the hyperbolic cell size distribution $N(\lambda \geqslant \Lambda) = C\Lambda^{-D_G}$ to be expressed in terms of the fractal dimension D_G and the area analyzed, A.

Micrographs taken from the same specimen as fig. 13(b) but with different foil orientations reveal an anisotropic (ellipsoidal) shape of the cells, which are slightly elongated in the direction of the tensile axis [73]. In spite of this anisotropy, fractal dimensions turn out to be independent of foil orientation. Together with the fact that dimensions deduced from different methods coincide within the limits of confidence, this supports the view that, within a scaling range extending over one to two orders of magnitude, this pattern is consistently characterized as a three-dimensionally self-similar fractal ('pore fractal'). Due to 3D self-similarity, the fractal dimension of the three-dimensional dislocation arrangement can be easily deduced from the sections represented by the electron micrographs: The dimension of a self-similar fractal in three-dimensional space is simply given by the dimension of its two-dimensional section plus one [139]. This relation has been used in the compilation of data in fig. 21.

Systematic fractal analyses have been performed for cell structures in high-symmetry oriented f.c.c. single crystals deformed in $\langle 100 \rangle$ and $\langle 111 \rangle$ directions [127]. At low to intermediate strains (resolved shear strain < 0.8), the fractal dimensions were found to increase with increasing flow stress (fig. 21). A similar finding was previously reported by Olemskoi [138] for dislocation patterns in ordered Ni_3Al. The observation of an increasing fractal dimension in $\langle 100 \rangle$ Cu was confirmed by Szekely et al. [143]. Gil Sevillano et al. [137] reported evidence for fractal cell patterning in cold-rolled Cu polycrystals up to

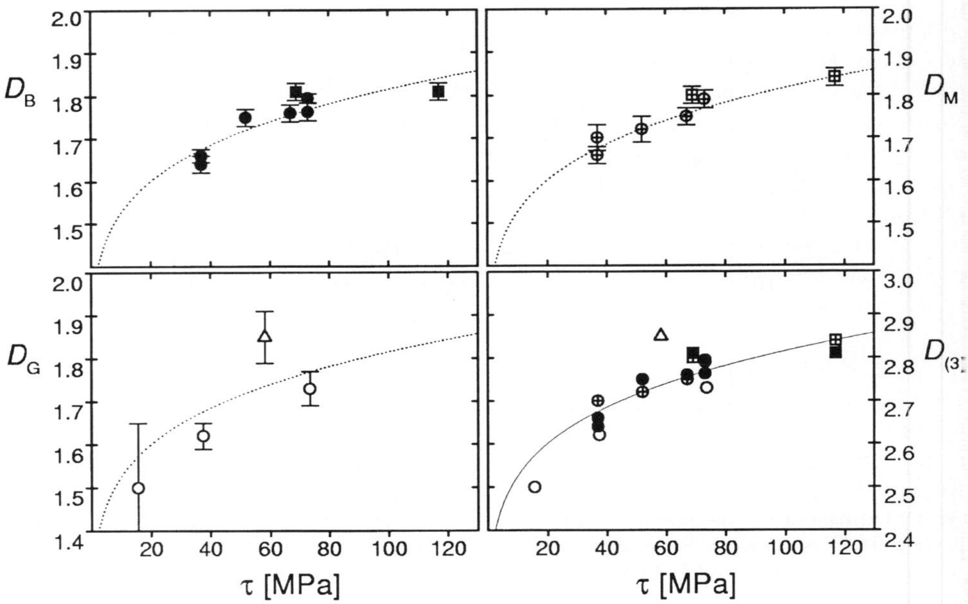

Fig. 21. Fractal dimensions of dislocation cell structures of ⟨100⟩ and ⟨111⟩-orientated f.c.c. single crystals deformed to different stresses. Circles: Cu ⟨100⟩, squares: Cu ⟨111⟩, triangles: Cu3at%Mn ⟨100⟩. Full symbols: box-counting dimensions, cross-centre symbols: mass dimensions; open symbols: gap dimensions; dotted line: guide to the eye.

rolling strains of the order of unity, while at larger strains their results indicate non-fractal patterns with a well-defined characteristic length scale. Cell structures of grains with ⟨100⟩ orientation in cold-rolled polycrystalline Cu have similar characteristics as those in single crystals; also in this case, a decrease of fractal dimension towards unity and a transition to non-fractal behavior is observed at large strains. However, more systematic analyses are needed to investigate whether such cross-overs are a genuine feature of large-strain deformation, as suggested in [63].

4.1.3. 'Self-similarity' versus 'similitude'

Statistical signatures of similitude Fractal dislocation patterns exhibit statistical self-similarity over a certain range of length scales. This property must be distinguished from the older notion of 'similitude' [91], which has been discussed in section 2 and refers to the fact that dislocation patterns observed after deformation under comparable conditions but up to different stresses, tend to show similar geometries with characteristic lengths that scale in inverse proportion with stress.

From a statistical viewpoint, microstructural variables are characterized by distribution functions, and similitude implies that these functions exhibit the scaling property that the distribution of a given variable normalized by its (stress- or strain-dependent) average is invariant in the course of deformation. When one is dealing with a characteristic length (cell size, wall spacing, etc.) this average is, in turn, expected to decrease in inverse proportion with the flow stress. The practical importance of such *scaling distributions* has been emphasized by Hughes et al. [129,130], who pointed out that the existence of scaling distributions considerably simplifies the modelling of statistically distributed microstructures, since they allow us to restrict ourselves basically to the calculation of averages. For a rather simplified statistical model of microstructure evolution during Stage II hardening in unidirectional deformation, the existence of a scaling distribution of dislocation densities with an average density that increases as the square of flow stress was demonstrated [114]. An improved version of this model which allows for fractal cell patterning has been proposed later by Hähner and Zaiser [144,145]. If will be discussed in more detail in section 6.

Implications of self-similarity in dislocation patterning In dislocation cell patters which can be characterized as self-similar fractals, the cell size distribution is well described by a power law over a certain range of scales. This has important consequences with regard to the interpretation of average quantities such as mean cell size and cell wall volume fraction. For a cell fractal with gap dimension D, the cell size probability distribution $p(\lambda)\,\mathrm{d}\lambda$ (probability for the size of a given cell to lie between λ and $\lambda+\mathrm{d}\lambda$) in 2D is given by

$$p(\lambda) = \mathcal{N}\lambda^{-D-1} = \left[D\lambda_{\min}^{D}\right]\lambda^{-D-1}, \tag{68}$$

where a lower cut-off size $\lambda_{\min}$ had to be introduced to normalize the distribution. From the distribution (68) it follows that the *average cell size* is proportional to the mimimum cell size,

$$\langle\lambda\rangle = \int_{\lambda_{\min}}^{\infty} \lambda p(\lambda)\,\mathrm{d}\lambda = \frac{D}{D+1}\lambda_{\min}. \tag{69}$$

When the distribution of characteristic lengths is determined by a line-intercept method, the fractal dimension is reduced by unity but eq. (69) remains valid. The proportionality of average and minimum cell size has important implications for the experimental determination of $\langle\lambda\rangle$, since in practice $\lambda_{\min}$ is strongly influenced by experimental parameters such as TEM magnification, foil thickness, and also by subjective factors. This may explain why there is often a large scatter in average cell sizes reported by different authors even for the same material and comparable deformation conditions [92].

When TEM micrographs are taken from the same microstructure but at different magnifications, because of eq. (69) fractal patterns exhibit an inverse proportionality between magnification and average cell size as long as the minimum resolvable cell size falls within the fractal scaling regime. By contrast, for Euclidean patterns characterized by a single characteristic length scale, averages are almost magnification-independent, provided the patterns can be resolved at all. This observation has been used [137] to distinguish fractal from non-fractal behaviour of cell structures. A negative unit slope in the double-logarithmic plot of average cell size vs. magnification, as reported [137], provides strong evidence for fractal behaviour though the method does not allow the reliable determination of fractal dimensions [127].

When micrographs of limited area are analyzed, in the case of a hyperbolic cell-size distribution the maximum cell sizes and cell wall volume fractions determined from such samples are affected by finite-size effects. As discussed in [127], for an ideal cell fractal the maximum size of a cell found, on average, on a micrograph of finite area A scales in proportion with the size of the micrograph, $\lambda_{\max} \propto \sqrt{A}$. The area fraction occupied by 'cell walls' is readily calculated from the hyperbolic distribution (68) by considering all features on the micrographs which fall below the lower cut-off size $\lambda_{\min}$ as parts of the cell walls. The result is

$$f_{\mathrm{w}} = \left[\frac{\lambda_{\min}}{\lambda_{\max}}\right]^{2-D}. \tag{70}$$

For a three-dimensional cell structure, the fractal dimension $D_{(3)}$ is equal to $D + 1$, where D is the dimension of the two-dimensional section. Calculating the wall volume fraction in three dimensions yields $f_{\mathrm{w}} = [\lambda_{\min}/\lambda_{\max}]^{(3-D_{(3)})} = [\lambda_{\min}/\lambda_{\max}]^{(2-D)}$, hence the wall volume fraction of the three-dimensional cell structure is the same as for its two-dimensional section. For $D \to 2$ ($D_{(3)} \to 3$), the area-filling character of the dislocation arrangement becomes manifest as $f_{\mathrm{w}} \to 1$. For smaller D, the wall area fraction depends through $\lambda_{\max}$ on the size of the analyzed area. For an ideal fractal and $D < 2$, f_{w} goes to zero as the analyzed area goes to infinity.

Equations (69) and (70) demonstrate that self-similar dislocation cell fractals exhibit characteristic length scales only through the upper and lower cut-off lengths which limit the fractal scaling regime. These cut-off lengths determine the values of important microstructural characteristics such as mean cell size or volume fractions of cell walls and cell interiors. In practice, these cut-off lengths may often be related to finite-size effects resulting from finite micrograph sizes and foil thicknesses, in which case they do not represent intrinsic characteristics but more or less incidental parameters of the experimental setting. For understanding the properties of self-similar microstructures it is, however, important to note that in nature there is no ideal fractal object, and to ask what are the *intrinsic* bounds of the fractal regime and what is their physical origin.

In [144] it was pointed out that a natural upper bound to the extension of a dislocation-depleted region in a deformed crystal is given by the slip-line length L. Experiment shows that in single crystals deformed in tension in $\langle 100\rangle$ and $\langle 111\rangle$ directions, L is inversely proportional to the flow stress [69]. A lower bound to the cell size can be defined by requiring that the interior of a 'cell' must be sufficiently large such that dislocations may cross it by a bowing mechanism. One can then estimate the minimum cell size by equating the corresponding Orowan stress to the flow stress of the cell walls: $Gb/\lambda \approx \alpha^{\beta} Gb\sqrt{\rho_{\mathrm{w}}}$, where ρ_{w} is the local dislocation density in the walls [144]. This yields $\lambda_{\min} = 1(\alpha^{\beta}\sqrt{\rho_{\mathrm{w}}})$, which is again inversely proportional to the flow stress. When both the upper and lower bounds of the scaling regime of cell fractals scale inversely to the flow stress, such patterns are consistent with the 'principle of similitude' as long as their fractal dimension is strain-independent. However, the experimental results indicate that this may not be the case (see fig. 21). When the changing fractal dimension of the cell structure is taken into account, cell wall volume fractions f_{w} calculated from eq. (70) for resolved shear strains $\gamma < 0.6$ show an increase of f_{w} with increasing flow stress [144]. This is consistent with experimental

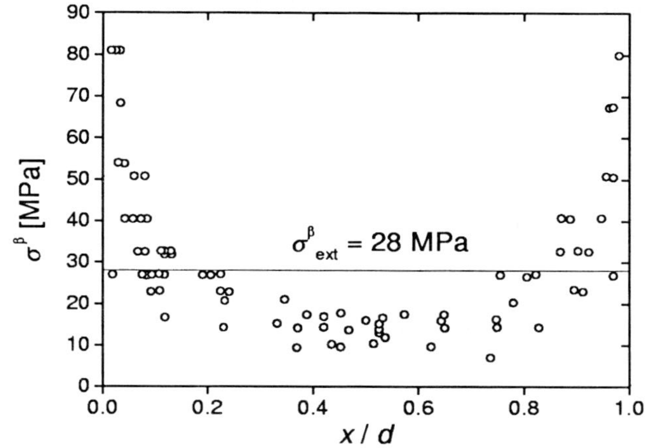

Fig. 22. Local stresses deduced from the curvature of dislocation lines in the persistent slip band structure of a fatigued Cu single crystal; stress applied state (resolved shear stress at the peak amplitude $\sigma_{ext} = 28$ MPa); x is the coordinate in the slip direction, d the width of a dislocation-depleted 'channel'. Dislocation-rich 'walls' are at $x/d = 0$ and 1. After Mughrabi [72].

determinations [142] of f_w by electron micrographs. Recently, the increase of f_w with increasing fractal dimension was confirmed in a work of Szekely et al. [143] where f_w was determined from X-ray line profile analysis.

4.1.4. Internal stresses as determined from TEM micrographs

Information on internal stresses may be obtained by transmission electron microscopy by determining the curvature of dislocation segments on TEM micrographs. Internal stresses determined in this manner allow us to directly link the internal stress and dislocation density patterns. To obtain a reliable picture, the dislocation structure in the sample must be stabilized by neutron irradiation before thinning [123,124,126]. This method has been applied to evaluate the internal-stress distribution not only in the unloaded state, but also under load (i.e., after irradiation in the stressed state).

An example pertaining to the internal-stress distribution in persistent slip bands of cyclically deformed Cu single crystals [72,146] is shown in fig. 22. In this case, the internal-stress distribution can be related to the dislocation arrangement in a straightforward manner: The observed internal stresses may be envisaged as mesoscopic stresses which increase the total stress acting on the hard, dislocation-rich 'walls' of the persistent slip band but decrease it in the soft, dislocation-depleted channels. As expected for mesoscopic internal stresses, their average over one 'wavelength' of the PSB pattern is about zero. During unloading, the mesoscopic stresses remain practically unchanged, which indicates that composite models may be used for their theoretical calculation [72] (section 3.3).

Internal-stress distributions have also been determined by Mughrabi for Cu deformed in tension into hardening Stage II [32,126,147] (fig. 23). An important finding which distinguishes uniaxial from cyclic deformation is that, in uniaxial deformation, the internal-stress pattern may change substantially during unloading. It is observed that the local

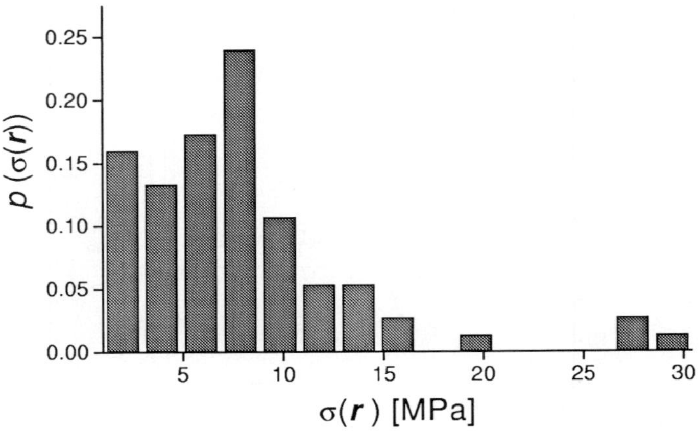

Fig. 23. Probability distribution of local stresses deduced from the curvature of dislocation segments in the primary glide plane of a Cu single crystal after deformation to 12 MPa; stress-applied state. After Mughrabi [147].

stresses in the stress-applied state range between zero and roughly twice the applied stress but partly relax upon unloading such that the mean internal stresses deduced from the curvature of dislocation segments in the unloaded state are only about the half of those in the loaded state. The effect is particularly significant in single-glide crystals where upon unloading a significant decrease in the internal stresses as well as a decrease in the number of free primary dislocations has been reported [126] and interpreted in terms of the relaxation of pile-up configurations. This observation indicates that it may be problematic to use composite models for relating internal stresses and dislocation densities in these situations. In this respect it is interesting to note that comparable observations have been reported recently by Borbely et al. [148], who performed X-ray diffraction measurements on both loaded and unloaded Cu single- and polycrystals. The general tendency reported in this work shows that during unloading there is a decrease both in the dislocation density and in the mesoscopic ('long-range') internal stresses.

4.2. Determination of internal stress and dislocation-density distributions from X-ray line profiles

TEM yields the *local* internal stresses acting on certain dislocation segments. This may be considered an advantage, since it is the internal stresses acting at the positions of the dislocations which determine the flow stress. Establishing statistically reliable experimental relations between the features of the internal-stress pattern and those of the overall dislocation arrangement requires, however, a substantial effort. While the internal-stress pattern seen in fig. 22 can be related to the characteristic features of a dislocation pattern consisting of regularly spaced dislocation-rich walls, it is much more difficult to establish similar relationships for dislocation patterns in unidirectionally deformed crystals. Due to the large variety of local features displayed by those patterns, the statistical significance of conclusions based on TEM observations is always disputable.

X-ray measurements, on the other hand, give access to the overall statistical properties of the internal-stress pattern. X-ray measurements average over sample regions containing many on dislocation configurations. Hence, relevant information is acquired different statistical properties of the dislocation arrangement. In particular, X-ray scattering gives direct access to the probability distribution of internal stresses and allows dislocation densities and correlation lengths to be determined. From the asymptotic behaviour of X-ray line profiles, the mean and mean square dislocation densities may be determined, while from the shape of the profiles the entire probability distribution of dislocation densities may be reconstructed in principle using generalized composite models.

4.2.1. X-ray scattering by inhomogeneous dislocation arrangements

We consider X-ray scattering by a set of planes characterized by the reciprocal lattice vector $\boldsymbol{g}$ and denote by $\boldsymbol{s} = \boldsymbol{\kappa} - 2\pi\boldsymbol{g}$ the difference between the actual scattering vector $\boldsymbol{\kappa}$ and the scattering vector from the undistorted lattice. Using kinematic scattering theory and retaining only first-order terms in $|\boldsymbol{\kappa}|/|\boldsymbol{g}|$, one finds that the scattered intensity $I_g(\boldsymbol{s})$ is given by [96]

$$I_g(\boldsymbol{s}) = \frac{C}{(2\pi)^3 V_0} \int\int_{V_0} \exp[\mathrm{i}\boldsymbol{s}\boldsymbol{r}'] \exp\big(2\pi\mathrm{i}\boldsymbol{g}\big[\boldsymbol{u}(\boldsymbol{r}+\boldsymbol{r}'/2) - \boldsymbol{u}(\boldsymbol{r}-\boldsymbol{r}'/2)\big]\big)\,\mathrm{d}^3r\,\mathrm{d}^3r', \tag{71}$$

where V_0 is the crystal volume, $\boldsymbol{u}(\boldsymbol{r})$ the displacement field, and C the atomic scattering factor. Henceforth, we consider intensity distributions which are normalized such that $\int I_g(\boldsymbol{s})\,\mathrm{d}^3s = 1$. In experiments, often only the dependence of I, on the diffraction angle (the X-ray line profile) is investigated. Denoting by s the component of $\boldsymbol{s}$ and by n the component of $\boldsymbol{r}'$ in the direction of the diffraction vector $\boldsymbol{g}$ and integrating over the normal directions, one obtains

$$\begin{aligned} I_g(s) &= \frac{1}{2\pi}\int A_g(n)\exp[\mathrm{i}ns]\,\mathrm{d}n, \\ A_g(n) &= \frac{1}{V_0}\int_{V_0} \exp\big(2\pi\mathrm{i}\boldsymbol{g}\big[\boldsymbol{u}(\boldsymbol{r}+n\boldsymbol{e}_g/2) - \boldsymbol{u}(\boldsymbol{r}-n\boldsymbol{e}_g/2)\big]\big)\,\mathrm{d}^3r, \end{aligned} \tag{72}$$

where $\boldsymbol{e}_g = \boldsymbol{g}/|\boldsymbol{g}|$.

We now specialize our considerations to planar dislocation systems and assume an arrangement of N dislocations characterized by the N-dislocation density function $\rho_{(N)}(1\ldots N)$. The variation in the displacement field $\boldsymbol{u}(\boldsymbol{r})$ over the distance n along $\boldsymbol{e}_g$ is characterized by the difference strain

$$\varepsilon_{n,gg}(\boldsymbol{r}) = \frac{1}{n}\int_{-n/2}^{n/2} \boldsymbol{e}_g\boldsymbol{\varepsilon}(\boldsymbol{r}+\boldsymbol{e}_g r')\boldsymbol{e}_g\,\mathrm{d}r', \tag{73}$$

which for $n \to 0$ reduces to the diagonal component $\varepsilon_{gg} := \boldsymbol{e}_g \boldsymbol{\varepsilon} \boldsymbol{e}_g$ of the (elastic) strain tensor $\boldsymbol{\varepsilon}$. Equation (72) is then readily re-written as

$$A_g(n) = \frac{1}{V_0} \int_{V_0} A_g(n, \boldsymbol{r})\, \mathrm{d}^3\boldsymbol{r} = \{A_g(n, \boldsymbol{r})\}, \tag{74}$$

$$A_g(n, \boldsymbol{r}) = \sum_{\substack{\beta_1 \dots \beta_N \\ s_1 \dots s_N}} \int \rho_{(N)}(1 \dots N) \exp\left[2\pi \mathrm{i} g n \sum_{j=1}^{N} s_j \varepsilon_{n,gg}^{\beta_j}(\boldsymbol{r} - \boldsymbol{r}_j)\right] \mathrm{d}^2 r_1 \cdots \mathrm{d}^2 r_N, \tag{75}$$

where $\varepsilon_{n,gg}^{\beta_j}(\boldsymbol{r} - \boldsymbol{r}_j)$ is the difference strain produced at $\boldsymbol{r}$ by a positive dislocation of type β_j located at $\boldsymbol{r}_j$. $A_g(n, \boldsymbol{r})$ is the Fourier transform of the local line profile $I_g(s, \boldsymbol{r})$ due to scattering from a mesoscopic volume element at $\boldsymbol{r}$, and $A_g(\boldsymbol{r})$ is the superposition of these local profiles.

As the stress tensor is a linear function of the strain tensor, eq. (75) would be completely equivalent to eq. (53) if we were allowed to replace the difference strain $\varepsilon_{n,gg}$ by the differential strain ε_{gg}. Indeed, the procedure for calculating the local scattering intensity distributions $A_g(n, \boldsymbol{r})$ from eq. (75) is completely analogous to the calculation of the local stress probability distribution $p(\sigma_{kl}, \boldsymbol{r})$, section 3.2.3. Since the procedure is based on the consideration of the Fourier transform $A(n, \boldsymbol{r})$ at small n where $\varepsilon_{n,gg} \to \varepsilon_{gg}$, results obtained for stress distributions directly carry over to X-ray line profiles and vice versa. In this respect, it is instructive to compare the papers published by Groma on both subjects [96,97].

The X-ray line profile is the superposition of the local line profiles and corresponds directly to the spatially averaged probability distribution of internal stresses, eq. (66). In particular, the following holds:

(1) For a completely random dislocation arrangement, the line profile is Gaussian with a width diverging in proportion with the logarithm of the crystal radius. This was first noted by Krivoglaz and Ryaboshapka [149].

(2) The asymptotic tails of the distribution are governed by scattering from the large strain fields in the immediate vicinity of single dislocations. They are, hence, independent of the total dislocation arrangement. As a consequence of the $1/r$ decay of the strains around dislocations, one finds a third-order power-law decay (cf. eq. (57))

$$I_g(s) = \lambda_g \langle \rho \rangle \frac{1}{s^3}, \tag{76}$$

where $\langle \rho \rangle$ is the average dislocation density and the pre-factor Λ can be calculated from the scattering of single dislocations [150]. An experimental example is shown in fig. 24.[15]

[15] In a recent paper by Levine and Thomson [151], the third-order power law decay has been doubted and a close-to fifth-order behaviour has been reported. Note, however, that the work of Levine and Thomson refers

(3) The Fourier transforms of the local line profiles $I_g(s, \boldsymbol{r})$ are given by

$$\ln A_g(n, \boldsymbol{r}) \approx 2\pi i g n \langle \varepsilon_{gg}(\boldsymbol{r}) \rangle + \lambda_g \rho(\boldsymbol{r}) n^2 \ln[n/\xi_g] + \cdots, \tag{77}$$

where $\xi_g \approx \xi_\sigma$ is a screening length (cf. eq. (56) for the stress probability distribution). The local line profiles $I_g(s, \boldsymbol{r})$ are centered at the values $2\pi g \langle \varepsilon_{gg}(\boldsymbol{r}) \rangle$ which are proportional to the mesoscopic strain at $\boldsymbol{r}$,

$$\langle \varepsilon_{gg}(\boldsymbol{r}) \rangle = \sum_{\beta, s} \int \rho_{(1)}^{\beta, s}(\boldsymbol{r}') s \varepsilon_{gg}^{\beta}(\boldsymbol{r} - \boldsymbol{r}') \, \mathrm{d}\boldsymbol{r}'. \tag{78}$$

Here $\varepsilon_{gg}^{\beta}(\boldsymbol{r} - \boldsymbol{r}')$ is the strain produced at $\boldsymbol{r}$ by a positive dislocation of type β located at $\boldsymbol{r}'$.[16] The profiles are symmetrical functions of s with respect to this centre. Their half width is proportional to the local dislocation density and depends on a shape factor Ξ_g in a similar manner as the half width of the local stress-probability distributions (cf. fig. 11).

4.2.2. Analysis of X-ray line profiles

The X-ray line profiles monitor in a quite straightforward manner the spatially averaged probability distribution of internal stresses. While until now we have considered the problem of calculating the X-ray line profile (the stress probability distribution) for a given dislocation arrangement, in the analysis of experiments we are confronted with the inverse problem: How to obtain from a given X-ray line profile parameters of the underlying dislocation pattern?

To this end, several procedures have been applied. Groma [96] proposed to determine statistical characteristics of the dislocation arrangement from an analysis of the asymptotic 'tails' of the line profile. While results obtained from a direct fit of eq. (76) to the asymptotic 'tails' of $I_g(s)$ are inaccurate because of the bad signal-to-noise ratio in the tails, where $I_g(s)$ is small (fig. 24), an improved procedure has been introduced in [96]. This is based upon considering the truncated moments of the line profile,

$$v_g^{(k)}(s) = \int_{-s}^{s} (s')^k I_g(s') \, \mathrm{d}s', \tag{79}$$

at large s. Due to normalization, for $s \to \infty$ we have $v_g^{(0)}(s) \to 1$. Since the mean strain produced by an arrangement of dislocations vanishes, $v_g^{(1)}(s) \to 0$. The higher-order $v_g^{(k)}$ in general diverge. One finds (the method of calculation is discussed in detail in [96]) that $v_g^{(2)}(s) \to 2\lambda_g \{\rho\} \ln(s/s_0)$ and $v_g^{(3)}(s) \to -6\{\rho \langle \varepsilon_{gg} \rangle\} \ln(s/s_1)$. $v_g^{(4)}(s)$ depends on $\{\rho\}$ and the spatial variance $\{\rho^2\} - \{\rho\}^2$. Accordingly, $\{\rho\}$, $\{\rho \langle \varepsilon_{gg} \rangle\}$ and $\{\rho^2\} - \{\rho\}^2$ govern the

to the intensity distribution in terms of the scattering vector $\boldsymbol{s}$, whereas the present consideration is based upon an integration over two directions in $\boldsymbol{s}$-space. Hence, there is no discrepancy between eq. (76) and the results of Levine and Thomson.

[16] This behaviour is analogous to the local internal-stress distributions which are centered at the local mesoscopic stress.

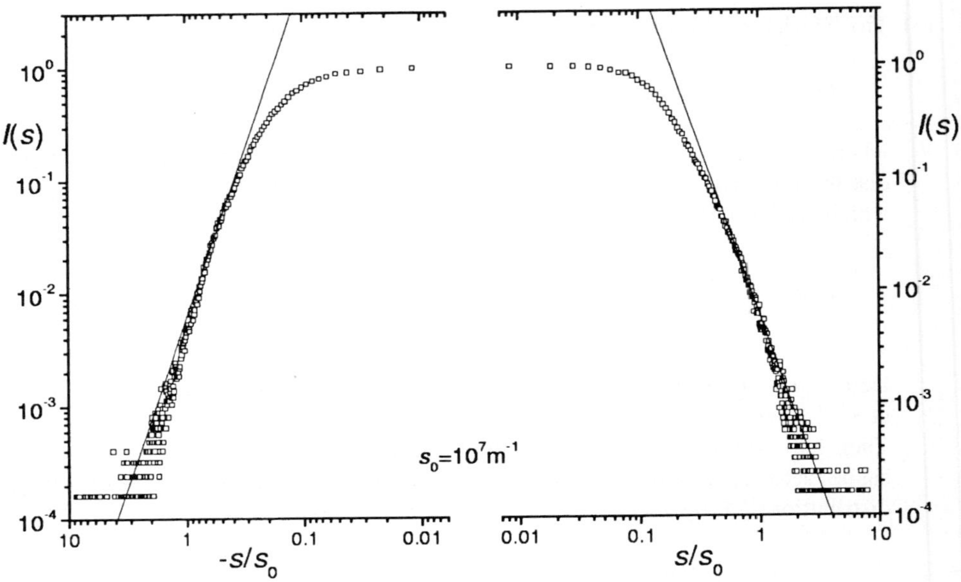

Fig. 24. X-ray scattering intensity distribution from a Cu single crystal deformed in tension to a resolved shear stress $\sigma_{\text{ext}} = 40$ MPa, $\varepsilon = (002)$. Data by Szekely et al. [143], lines: fit of the tails by eq. (4.8).

lowest-order terms in the asymptotics of $I_g(s)$ at large s. Fitting these expressions to the data gives access to the mean value and mean square fluctuation of the dislocation density, as well as to the correlation between the dislocation density and the mesoscopic strain. Recently, Szekely et al. [143,153] have conducted systematic investigations in which the dislocation density fluctuations were investigated as a function of flow stress for single and for $\langle 100\rangle$-symmetric multiple slip. Figure 25 shows results obtained for $\langle 100\rangle$ Cu single crystals; it is seen that the dislocation density fluctuation tends to decrease with increasing flow stress. Combined with TEM observations, such observations permit a quantitative characterization of the microstructure. For a microstructure consisting of dislocation-rich cell walls and dislocation-depleted cell interiors, the dislocation density fluctuation may be related to the cell wall volume fraction $f_{\text{w}} \approx \{\rho\}^2/\{\rho^2\}$ if the dislocation density in the cell interiors is much smaller than in the walls. A decreasing density fluctuation implies an increasing wall volume fraction. For fractal dislocation arrangements this may be related to the increase of the fractal dimension with increasing flow stress [143,144] (cf. eq. (70) and fig. 21).

The analysis of the line-profile asymptotics allows us to obtain statistical signatures of the dislocation arrangement without making *a priori* assumptions about the underlying microstructure. A complementary method is the analysis of the central parts of the line profiles. In analogy to the SAPD $p(\sigma_{ij})$, which is the superposition of the local stress probability distributions of regions with different *mesoscopic stress* (fig. 12), the X-ray line profile $I_g(s]$ may be considered as a superposition of line profiles from regions with different *mesoscopic strain* $\langle\varepsilon_{gg}(\boldsymbol{r})\rangle$. With assumptions on the distribution of local

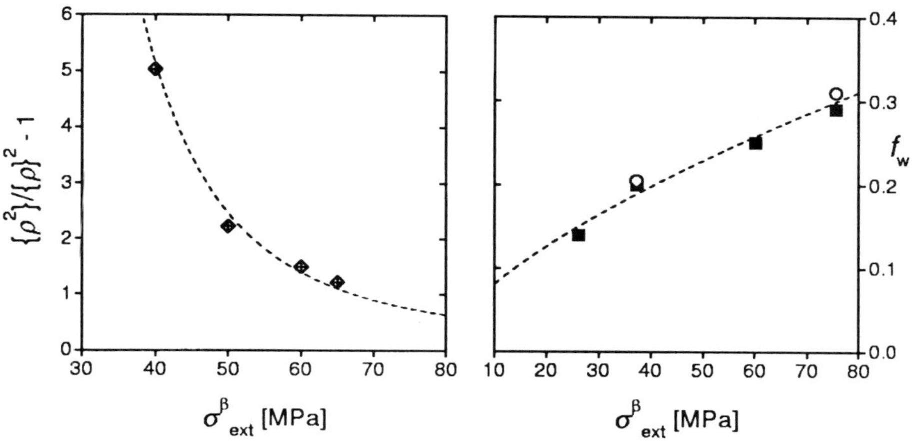

Fig. 25. Statistical characteristics of dislocation patterns in Cu deformed in tension along [100] as a function of flow stress; left: dislocation density fluctuation determined from X-ray line profiles, after Szekely et al. [143], right: cell wall volume fraction determined from X-ray line profiles (open points) and TEM micrographs (full points), after Mughrabi et al. [73].

dislocation densities and the correlations between dislocation densities and mesoscopic strain, parameters characterizing the dislocation arrangement may be extracted from the line profile. Generalized composite models as discussed in section 3.3 relate the mesoscopic stress in a volume element – and therefore also the mesoscopic strain – to the local dislocation densities. The X-ray line profiles may then be written as (cf. eq. (66))

$$I_g(s) = \int I_g(s, \boldsymbol{\rho}) p(\boldsymbol{\rho}) \, \mathrm{d}\boldsymbol{\rho}, \tag{80}$$

where the Fourier transforms of the local subprofiles are given by eq. (77) and depend on the local dislocation densities and screening lengths which are formally compiled into the state vector $\boldsymbol{\rho}$. Unfortunately, no general method exists for inverting the integral equation (80) in order to deduce from the intensity profiles the probability density $p(\boldsymbol{\rho})$. In practice, one has to assume a functional shape of $p(\boldsymbol{\rho})$ and may then derive the parameters characterizing the dislocation arrangement by fitting intensity profiles measured at different ε.

We illustrate the procedure just outlined for the uniaxial deformation of f.c.c. single crystals in ⟨100⟩, studied in section 3.3.3. In this highly symmetric deformation geometry, one may expect that the dislocation densities are the same in all active slip systems and the distribution of dislocations over the different Burgers vectors is space independent. Indeed, for Cu and CuMn single crystals deformed in ⟨100⟩ orientations, TEM investigations of Göttler [154] and of Neuhaus and Schwink [142] indicate that all Burgers vectors occur with approximately equal probability and that no large-scale lattice rotations (breaking up of the structure into blocks) are observed. Hence, the dislocation arrangement may be characterized by two variables, viz. the dislocation density ρ and the screening length ξ_σ.

The mesoscopic tensile strain $\langle \varepsilon_{gg}(\boldsymbol{r})\rangle$ corresponding to a [200] axial reflection is given by (cf. eq. (67))

$$\langle \varepsilon_{200}(\boldsymbol{r})\rangle = \left[\sigma_{\text{ext}} - \alpha^{\beta} G b \sqrt{\rho}/M\right]/E, \tag{81}$$

where E is Young's modulus. With regard to the functional shape of the density function $p(\boldsymbol{\rho})$, Mughrabi et al. [73] consider a two-phase structure consisting of cell walls and cell interiors and assume a fixed distribution of dislocations over the different Burgers vectors and line directions. The dislocation density and screening length in the cell walls are denoted by $\rho_{\text{W}}, \xi_{\sigma}^{\text{W}}$ and those in the cell interiors by $\rho_{\text{c}}, \xi_{\sigma}^{\text{c}}$. This leads to the density function

$$p(\rho,\xi) = f_{\text{W}}\delta(\rho-\rho_{\text{W}})\delta(\xi-\xi_{\sigma}^{\text{W}}) + (1-f_{\text{W}})\delta(\rho-\rho_{\text{c}})\delta(\xi-\xi_{\sigma}^{\text{c}}), \tag{82}$$

where f_{W} is the cell wall volume fraction. The density function (82) decomposes the asymmetric line profile into two symmetric subprofiles, and ρ and ξ_{σ} can be determined separately for each subprofile. The centres of the two subprofiles yield the mesoscopic strain in the cell walls and in the interiors; the relative areas under the subprofiles give the respective volume fractions. In this manner, the five parameters $[f_{\text{W}}, \rho_{\text{W}}, \xi_{\sigma}^{\text{W}}, \rho_{\text{c}}, \xi_{\sigma}^{\text{c}}]$ characterizing the density function as well as the Taylor parameter α^{β} may be determined. The consistency of the physical picture may be tested by determining these parameters for different reflections. Figure 25 (right) shows that the wall volume fractions determined by this method are consistent with TEM observations. It should be kept in mind, however, that TEM data may not always be reliable (see the discussion in section 4.1.3).

The analysis of the central parts of the line profile yields more specific information on the dislocation arrangement than an analysis based entirely on the asymptotics of the line profiles. However, in practice it has the drawback that one has to assume (i) the validity of a composite model and (ii) a prescribed functional shape of the probability density function $p(\boldsymbol{\rho})$. While the validity of assumption (i) can be tested by investigating the consistency of the results obtained from different reflections, 'input' from other methods such as TEM is required for choosing an appropriate probability density function.

It is expected that future high-intensity X-ray sources will enable us to combine the advantages of TEM (good spatial resolution) and of X-ray methods (direct monitoring of internal strain fields on mesoscopic scale). By using X-ray microbeams, local line profiles may be taken from regions with a size of less than 1 micron. In this way it should be possible to correlate dislocation patterns, long-range stresses, and local stress fluctuations directly. It is hoped that this will lead to an experimental clarification of the relations between the flow-stress, the mesoscopic and the microscopic internal stresses on the one hand, and the underlying dislocation patterns on the other hand.

5. *Stochastic dynamics of plastic flow: lattice rotations and mesoscopic internal stresses*

In this section we formulate a simple model of plastic flow which describes spatio-temporal heterogeneities of plastic deformation in terms of random fluctuations of slip. It is based on the observation that slip proceeds in discrete, spatially and temporally localized events. A consequence of the spatial localization is that excess dislocations (dislocation ensembles with nonzero net Burgers vector) are stored at the boundaries of the slip zones. These dislocations give rise to mesoscopic stresses and lattice curvature. For simplicity, we use a quasi-twodimensional formulation as in section 3. Generalizations to three dimensions will be discussed in section 6.

5.1. Plastic flow viewed as a stochastic process

The intrinsic spatio-temporal inhomogeneity of plastic deformation on mesoscopic scales is taken into account by writing the local strain rate in a mesoscopic volume element at $\boldsymbol{r}$ as a sum over discrete 'deformation events':

$$\dot{\varepsilon}^{\beta}(\boldsymbol{r},t)=\sum_{i}\hat{\varepsilon}_{i}^{\beta}(\boldsymbol{r})f_{i}(t-t_{i}). \tag{83}$$

In (83), $\hat{\varepsilon}_i^{\beta}(\boldsymbol{r})$ denotes the local strain produced by the i-th event, t_i the time at which this event is centred, and $f_i(t)$ a shape function which is normalized in such a way that $\int_{-\infty}^{\infty} f_i(t)\,\mathrm{d}t=0$ and $\int_{-\infty}^{\infty} t f_i(t)\,\mathrm{d}t=0$. During each deformation event, dislocation multiplication may take place and dislocations may be stored in sessile configurations. The evolution of the microstructure is brought about by the cumulative effects of many such deformation events.

In the simplest case, a 'deformation event' may be the motion of a single dislocation. In general, however, dislocation motion proceeds in 'avalanches' involving the collective motion of many interacting dislocations. Hence, an important characterization of the statistics of plastic flow is the avalanche size distribution $p(\hat{\varepsilon}^{\beta})$. Experimentally, such distributions may be assessed by acoustic emission measurements: In the high-frequency regime, the acoustic emission amplitude during a given 'event' is proportional to the number of dislocations moving collectively [155]. Thus, the distribution of acoustic emission amplitudes gives us information on the size distribution of dislocation avalanches. Experiments as well as recent dislocation dynamics simulations [156] have demonstrated that the ensuing statistics may exhibit features that are typical of avalanche phenomena in slowly driven non-equilibrium systems [157]. In such systems, the distribution functions of the avalanche sizes often exhibit a power-law decay which is truncated at large avalanche sizes. This type of behavior is illustrated in fig. 26, which shows the amplitude distribution of acoustic emission events recorded during plastic deformation of ice single crystals [156].

To obtain the distribution of strains following from eq. (83), we consider first the case where all events have the same size $\hat{\varepsilon}^{\beta}$. Then the number of events which have taken place in a mesoscopic volume element is $n=\varepsilon^{\beta}/\hat{\varepsilon}^{\beta}$, where ε^{β} is the local strain. The average number of events in such a volume element is $\langle n\rangle=\langle\varepsilon^{\beta}\rangle/\hat{\varepsilon}^{\beta}$. The distribution of n is Poissonian. According to the Moivre–Laplace limit theorem [158], for large $\langle n\rangle$ it approaches a Gaussian with average and variance both equal to $\langle n\rangle$. If we denote Gaussian distributions by $\mathcal{G}(a_1;a_2,a_3)$, where a_1 is the random variable, a_2 its average and a_3 its variance, the distribution of n is $\mathcal{G}(n;\langle n\rangle,\langle n\rangle)$. By change of variables, it follows that in the

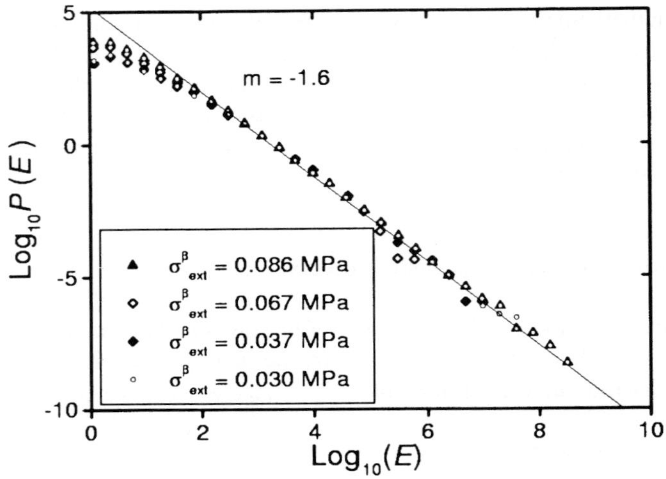

Fig. 26. Distribution of energy releases in acoustic emission 'events' recorded during creep deformation of ice single crystals. Frequency range of the acoustic transducer 0.1–1 MHz, temperature $T = 263$ K, resolved shear stresses on the basal plane as indicated in the inset. Over several orders the data follow a power law $P(E) \sim E^{-1.6}$. After Miguel et al. [156].

present case the distribution of strains is asymptotically given by the Gaussian distribution $\mathcal{G}(\varepsilon^\beta; \langle\varepsilon^\beta\rangle, \langle\varepsilon^\beta\rangle\hat{\varepsilon}^\beta)$ with average $\langle\varepsilon^\beta\rangle$ and variance $\langle\varepsilon^\beta\rangle\hat{\varepsilon}^\beta$.

Now assume that different event sizes $\hat{\varepsilon}^\beta_i$ occur with probability p_i. The average number of events that have taken place in a mesoscopic volume element is $\langle n\rangle = \langle\varepsilon^\beta\rangle/\langle\hat{\varepsilon}^\beta_i\rangle$, and the average number of events with size $\hat{\varepsilon}^\beta_i$ is $p_i\langle n\rangle$. The strain ε^β_i produced by these events is a random variable with asymptotic distribution $\mathcal{G}(\varepsilon^\beta_i; [p_i\hat{\varepsilon}^\beta_i\langle n\rangle], [p_i(\hat{\varepsilon}^\beta_i)^2\langle n\rangle])$. The distribution of a sum of random variables which are Gaussian-distributed is itself a Gaussian, its average being the sum of the averages of the individual variables and its variance the sum of the variances. It follows that the distribution of $\varepsilon^\beta = \sum_i \varepsilon^\beta_i$ is asymptotically given by $\mathcal{G}(\varepsilon^\beta; \langle\varepsilon^\beta\rangle, [\langle\varepsilon^\beta\rangle\varepsilon^\beta_{\mathrm{corr}}])$ with

$$\varepsilon^\beta_{\mathrm{corr}} = \frac{\langle[\hat{\varepsilon}^\beta_i]^2\rangle}{\langle\hat{\varepsilon}^\beta_i\rangle}. \tag{84}$$

The argument may be generalized to continuously distributed event sizes $\hat{\varepsilon}^\beta$, provided that the first and second moments $\langle\hat{\varepsilon}^\beta\rangle := \int \hat{\varepsilon}^\beta p(\hat{\varepsilon}^\beta)\,\mathrm{d}\hat{\varepsilon}^\beta$ and $\langle[\hat{\varepsilon}^\beta]^2\rangle := \int (\hat{\varepsilon}^\beta)^2 p(\hat{\varepsilon}^\beta)\,\mathrm{d}\hat{\varepsilon}^\beta$ of the event-size distribution exist.

When the event size distribution varies slowly with time (i.e., on a scale that is very large compared with the separation between individual events), a further generalization leads to

$$p(\varepsilon^\beta, t) = \mathcal{G}\left(\varepsilon^\beta; \langle\varepsilon^\beta\rangle, \left[\int_0^t \left[Q^\beta_\varepsilon\right]^2 \mathrm{d}t'\right]\right)$$

$$= \frac{1}{(\pi \int_0^t [Q_\varepsilon^\beta]^2 \, dt')^{1/2}} \exp\left[-\frac{(\varepsilon^\beta - \langle \varepsilon^\beta(t) \rangle)^2}{\int_0^t [Q_\varepsilon^\beta]^2 \, dt'} \right], \tag{85}$$

where the effective 'noise amplitude' at time t is given by

$$\left[Q_\varepsilon^\beta(t)\right]^2 = \frac{\langle [\hat{\varepsilon}^\beta]^2 \rangle_t}{\langle \hat{\varepsilon}^\beta \rangle_t} \langle \dot{\varepsilon}^\beta \rangle =: \varepsilon_{\rm corr}^\beta(t) \langle \dot{\varepsilon}^\beta \rangle. \tag{86}$$

An estimate of $\varepsilon_{\rm corr}^\beta$ based on physical reasoning is given below in section 5.3.1.

The fact that the microstructure evolution is dominated by the cumulative effects of large numbers of deformation events permits us to develop a simplified description in which the statistical accumulation of local strains is characterized by the asymptotic distribution (85). Any additional information contained in the event size distribution $p(\hat{\varepsilon}^\beta)$ characterizing plastic flow on short time scales (small strain increments) is lost at strains $\varepsilon^\beta = \langle \dot{\varepsilon}^\beta \rangle t \gg \varepsilon_{\rm corr}^\beta$. Therefore, on this time scale, eq. (83) may be replaced by a simpler stochastic differential equation which leads to the same asymptotic distribution (85). In this simplified equation,

$$\partial_t \varepsilon^\beta(\boldsymbol{r}, t) = \langle \dot{\varepsilon}^\beta \rangle + \delta \dot{\varepsilon}^\beta, \quad \text{where } \delta \dot{\varepsilon}^\beta := Q_\varepsilon^\beta \dot{w}^\beta, \tag{87}$$

$\dot{w}^\beta$ is a normalized Gaussian white-noise process. This means that its temporal correlation function is $\langle \dot{w}^\beta(t) \dot{w}^\beta(t') \rangle = \delta(t - t')$. The first term in eq. (87) characterizes the deterministic increase of the average strain $\langle \varepsilon^\beta \rangle = \langle \dot{\varepsilon}^\beta \rangle t$, while the fluctuating second term leads to a scatter of the local strains around this value. The properties of the event size distribution $p(\hat{\varepsilon}^\beta)$ enter only through its first and second moments, which according to eq. (86) determine the 'noise amplitude' Q_ε^β.

Stochastic integration of eq. (87) (for technicalities see, e.g., [159]) leads to a Fokker–Planck equation for $p(\varepsilon^\beta)$

$$\frac{\partial p(\varepsilon^\beta)}{\partial t} = -\langle \dot{\varepsilon}^\beta \rangle \frac{\partial p}{\partial \varepsilon^\beta} + \frac{[Q_\varepsilon^\beta]^2}{2} \frac{\partial^2 p}{\partial [\varepsilon^\beta]^2}. \tag{88}$$

The first term on the right-hand side ('drift term') stems from the deterministic term in eq. (87); it shifts the average of the distribution $p(\varepsilon^\beta)$ at the rate $\langle \dot{\varepsilon}^\beta \rangle$ towards larger strains. The second term ('diffusion term') stems from the white-noise process in (87); it causes a diffusionlike spreading of the distribution. The distribution (85) is the solution of this Fokker–Planck equation for initially zero strain (initial condition $p(\varepsilon^\beta, t = 0) = \delta(\varepsilon^\beta)$) and so-called natural boundary conditions. Hence eq. (84) is indeed compatible with the asymptotic distribution (85).

The spatial extension of a slip event may be characterized by two correlation lengths, viz., a length $\xi_{\rm g}^\beta$ in the direction of dislocation motion (on the surface, this corresponds to the slip line length) and a length $\xi_{\rm n}^\beta$ normal to the slip plane (this corresponds to the typical thickness of a slip line or slip band).[17] The product $\xi_{\rm n}^\beta \xi_{\rm g}^\beta := V_{\rm corr}^\beta$ will be called the (two-

[17] For individually moving dislocations, $\xi_{\rm g}^\beta$ is the average slip distance, $\xi_{\rm n}^\beta = b$, and $\varepsilon_{\rm corr}^\beta = 1$.

dimensional) 'correlation volume'. The spatial organization of collective slip is statistically characterized by the spatial correlation function $\langle\hat{\varepsilon}^\beta(\boldsymbol{r})\hat{\varepsilon}^\beta(\boldsymbol{r}')\rangle$ of the strains produced by the 'slip events'. In the following we assume a Gaussian decay of the correlations in the slip direction. Since, normal to the slip direction, correlations decay much faster than in the slip direction ($\xi_\mathrm{n}^\beta \ll \xi_\mathrm{g}^\beta$), the dependence of the correlation function on the normal coordinate may be approximated by a delta function. Under these assumptions the spatial correlation function of the slip events is given by $\langle\hat{\varepsilon}^\beta(\boldsymbol{r})\hat{\varepsilon}^\beta(\boldsymbol{r}')\rangle = \langle[\hat{\varepsilon}^\beta]^2\rangle f_\varepsilon^\beta(\boldsymbol{r}-\boldsymbol{r}')$ where

$$f_\varepsilon^\beta(\boldsymbol{r}) := \exp\left[-\pi\left(\frac{r_\mathrm{g}}{\xi_\mathrm{g}^\beta}\right)^2\right]\xi_\mathrm{n}^\beta\,\delta(r_\mathrm{n}). \tag{89}$$

Here $r_{\mathrm{g,n}} = \boldsymbol{r}\boldsymbol{e}_{\mathrm{g,n}}^\beta$ are the components of the vector $\boldsymbol{r}$ in the slip direction and the direction of the slip plane normal, respectively. The function (89) determines also the spatial correlations of the strain-rate fluctuations in eq. (87),

$$\left\langle\delta\dot{\varepsilon}^\beta(\boldsymbol{r})\delta\dot{\varepsilon}^\beta(\boldsymbol{r}')\right\rangle = \left[Q_\varepsilon^\beta\right]^2 f_\varepsilon^\beta(\boldsymbol{r}-\boldsymbol{r}'). \tag{90}$$

The spatio-temporal correlations of the Gaussian white noise processes $\dot{w}^\beta$ in eq. (87) are $\langle\dot{w}^\beta(\boldsymbol{r},t)\dot{w}^\beta(\boldsymbol{r}',t')\rangle = \delta(t-t')f_\varepsilon^\beta(\boldsymbol{r}-\boldsymbol{r}')$.

For the following considerations, the assumption that the pair correlation function $f_\varepsilon^\beta(\boldsymbol{r})$ has Gaussian shape is not crucial. Similar results are obtained for any correlation function with the following properties: (i) $f_\varepsilon^\beta(\boldsymbol{r})$ is everywhere at least once differentiable, (ii) it decreases monotonically in the radial direction, (iii) it is a 'localized' function with ranges ξ_g^β in the slip direction and $\xi_\mathrm{n}^\beta \ll \xi_\mathrm{g}^\beta$ in the normal direction,[18] and (iv) it is normalized such that $f_\varepsilon^\beta(0) = 1$ and $\int_V f_\varepsilon^\beta(\boldsymbol{r})\,\mathrm{d}\boldsymbol{r} = V_\mathrm{corr}^\beta$.

5.2. Lattice rotations and misorientations

5.2.1. Dislocation density tensor and lattice curvature

In order to establish relationships between spatial inhomogeneities of plastic flow, the concomitant storage of excess dislocations on the one hand and the resulting lattice rotations and misorientations on the other hand we consider the (mesoscopically averaged) Nye–Kröner dislocation density tensor $\boldsymbol{\alpha}$. This tensor may be expressed in terms of the densities ϕ^β of excess dislocations. These are, in turn, related to the shear-strain gradients on the different slip systems: $\phi^\beta = [\rho_{(1)}^{\beta,+} - \rho_{(1)}^{\beta,-}] = [-1/b]\nabla^\beta\delta\varepsilon^\beta$, where ∇^β is the directional derivative in the glide direction of the β dislocations. We obtain (cf. eq. (31))

$$\boldsymbol{\alpha}(\boldsymbol{r}) = -\sum_\beta[\boldsymbol{t}^\beta \otimes \boldsymbol{b}^\beta]\frac{1}{b}\nabla^\beta\varepsilon^\beta = \sum_\beta \boldsymbol{\alpha}^\beta\phi^\beta, \tag{91}$$

[18] This means that $f_\varepsilon^\beta(\boldsymbol{r})$ decreases for $r_\mathrm{n}/\xi_\mathrm{n}^\beta \gg 1$ or $r_\mathrm{g}/\xi_\mathrm{g}^\beta \gg 1$ faster than algebraically.

where $\boldsymbol{t}^\beta = \boldsymbol{e}_z$ is the dislocation line direction and $\boldsymbol{\alpha}^\beta := [\boldsymbol{t}^\beta \otimes \boldsymbol{b}^\beta]$.

The evolution of the excess densities ϕ^β is described by the stochastic differential equation

$$\partial_t \phi^\beta = -\frac{1}{b}\nabla^\beta\big[\langle\dot{\varepsilon}^\beta\rangle + \delta\dot{\varepsilon}^\beta\big]. \tag{92}$$

In the absence of macroscopic strain gradients $[\langle\dot{\varepsilon}^\beta\rangle = \mathrm{const.}(\boldsymbol{r})]$, the evolution of ϕ^β depends only on the gradient of the strain-rate fluctuations. In eq. (87), we describe these fluctuations by a white-noise process. Since this process has continuous spatial correlations, taking the spatial derivative $\nabla^\beta\delta\hat{\varepsilon}^\beta$ leads to a process which can again be approximated by a Gaussian white-noise process, but with different 'amplitude' and spatial correlations. Accordingly, the excess densities ϕ^β and therefore also the components of the dislocation density tensor are Gaussian random variables with zero mean and strain dependent variance which evolve in a diffusionlike manner.

The relation between excess densities and lattice rotations is established by considering Nye's lattice curvature tensor $\boldsymbol{\kappa}$, which for small rotations is connected to the dislocation density tensor by the linear relationship

$$\boldsymbol{\kappa} = (1/2)\mathbf{I}\,\mathrm{Tr}\,\boldsymbol{\alpha} - \boldsymbol{\alpha}, \tag{93}$$

where $\mathbf{I}$ is the unity tensor. Nye's tensor determines the differential rotation

$$\delta\boldsymbol{\omega} = \boldsymbol{\kappa}\,\delta\boldsymbol{r} \tag{94}$$

of the distorted lattice along $\delta\boldsymbol{r}$.[19] According to eq. (93), it may be written as

$$\boldsymbol{\kappa} = \sum_\beta \boldsymbol{\kappa}^\beta \phi^\beta, \tag{95}$$

where the tensors $\boldsymbol{\kappa}^\beta := (1/2)\mathbf{I}\,\mathrm{Tr}\,\boldsymbol{\alpha}^\beta - \boldsymbol{\alpha}^\beta$ characterize the lattice curvature created by excess dislocations of type β.

5.2.2. *Misorientations associated with cell walls*

A characteristic feature of dislocation cell structures is that inside the cells the dislocation densities and hence, according to eq. (95), the lattice curvature is small. Excess dislocations are stored in the cell walls. Orientation changes take place in a quasi-discontinuous manner across these walls. We consider a cell of volume V and calculate the net rate of the storage of excess dislocations within the cell volume.

The rate of excess dislocation storage in V is given by the difference of the fluxes $\boldsymbol{J}^{\beta,\pm} = \pm\rho_{(1)}^{\beta,\pm}\boldsymbol{v}^\beta$ ($\boldsymbol{v}^\beta$ is the velocity of a positive dislocation of slip system β) of positive and negative dislocations into this volume, integrated over the volume surface. Formally,

[19] Note that $\delta\boldsymbol{\omega}$ describes only the rotations which arise from the plastic distortion. It is a total differential only when the tensor of the plastic incompatibility is zero, i.e., when elastic distortions on mesoscopic scale are absent [103].

we may obtain this rate by integrating eq. (92) over V. When $\langle\dot\varepsilon^\beta\rangle$ is position-independent, we obtain

$$\partial_t \phi_V^\beta = -\frac{1}{bV}\int_V \nabla^\beta \delta\dot\varepsilon^\beta(\boldsymbol r)\,\mathrm d\boldsymbol r = -\frac{\boldsymbol e_{\mathrm g}^\beta}{bV}\int_{\partial V}\delta\dot\varepsilon^\beta(\boldsymbol r)\boldsymbol n(\boldsymbol r)\,\mathrm d\boldsymbol r. \tag{96}$$

Here ϕ_V^β is the excess dislocation density averaged over V, ∂V the surface of the cell volume, and $\boldsymbol n(\boldsymbol r)$ the unit vector normal to this surface. The relation to the dislocation fluxes is established by writing the shear strain rate in terms of the dislocation fluxes, $\dot\varepsilon^\beta = \boldsymbol e_{\mathrm g}^\beta b[\boldsymbol J^{\beta,+} - \boldsymbol J^{\beta,-}] = \langle\dot\varepsilon^\beta\rangle - \delta\dot\varepsilon^\beta$.

Each 'deformation event' leaves dislocations within the volume V. Their accumulation is characterized by the stochastic process on the right-hand side of eq. (96). By the argument of section 5.1, we can approximate this process by a white-noise process $Q_\phi^\beta(V)\dot w^\beta$ whose 'amplitude' $Q_\phi^\beta(V)$ follows from (90) and (96) as

$$\left[Q_\phi^\beta(V)\right]^2 = \frac{[Q_\varepsilon^\beta]^2}{V^2 b^2}\int_{\partial V}\int_{\partial V'}\left[\boldsymbol n(\boldsymbol r)\boldsymbol e_{\mathrm g}^\beta\right]\left[\boldsymbol n(\boldsymbol r')\boldsymbol e_{\mathrm g}^\beta\right] f_\varepsilon^\beta(\boldsymbol r - \boldsymbol r')\,\mathrm d\boldsymbol r\,\mathrm d\boldsymbol r'. \tag{97}$$

Since the cell interior is almost dislocation-free, virtually all excess dislocations are actually located in the cell walls, with a linear density that is proportional to the projection $|\boldsymbol n\boldsymbol e_{\mathrm g}^\beta|$ of the cell-wall normal $\boldsymbol n$ on the glide direction (fig. 27). Accordingly, the linear density of excess dislocations in a cell wall with orientation $\boldsymbol n$ is given by $\vartheta_{V,\boldsymbol n}^\beta = \phi_V^\beta|\boldsymbol n\boldsymbol e_{\mathrm g}^\beta|N_V$. The normalization factor $N_V = V[\int_{\partial V}|\boldsymbol n(\boldsymbol r)\boldsymbol e_{\mathrm g}^\beta|\,\mathrm d\boldsymbol r]^{-1}$ ensures that $\int_V \phi_V^\beta\,\mathrm d\boldsymbol r = \int_{\partial V}\vartheta_{V,\boldsymbol n}^\beta\,\mathrm d\boldsymbol r$. It follows from eqs (96) and (97) that the temporal evolution of $\vartheta_{V,\boldsymbol n}^\beta$ given by

$$\begin{aligned}&\partial_t \vartheta_{V,\boldsymbol n}^\beta = Q_\vartheta^\beta(V,\boldsymbol n)\dot w^\beta,\\ &\left[Q_\vartheta^\beta(V,\boldsymbol n)\right]^2 = \left[Q_\phi^\beta(V)\right]^2|\boldsymbol n\boldsymbol e_{\mathrm g}^\beta|^2 V^2\left[\int_{\partial V}|\boldsymbol n(\boldsymbol r)\boldsymbol e_{\mathrm g}^\beta|\,\mathrm d\boldsymbol r\right]^{-2}.\end{aligned} \tag{98}$$

There is no deterministic drift term in eq. (98), hence the corresponding Fokker–Planck equation has the form of a diffusion equation. For an initially excess-dislocation-free crystal the solution of this Fokker–Planck equation is therefore a Gaussian, $p(\vartheta_{V,\boldsymbol n}^\beta) = \mathcal G(\vartheta_{V,\boldsymbol n}^\beta; 0, [\int[Q_\vartheta^\beta(V,\boldsymbol n)]^2\,\mathrm dt])$.

The excess dislocations cause a lattice rotation across the cell wall. The rotation vector follows from eqs (94) and (95):

$$\boldsymbol\omega_{\boldsymbol n}^\beta = \boldsymbol\omega_0^\beta(\boldsymbol n)\vartheta_{\boldsymbol n}^\beta,\quad \boldsymbol\omega_0^\beta(\boldsymbol n) = \boldsymbol\kappa^\beta\boldsymbol n = 1/2\big(\boldsymbol t^\beta\boldsymbol b^\beta\big)\boldsymbol n - \big(\boldsymbol b^\beta\boldsymbol n\big)\boldsymbol t^\beta. \tag{99}$$

The rotation around the wall normal is due to the screw components of the excess dislocations, and the rotation around the dislocation line direction is due to the edge components.

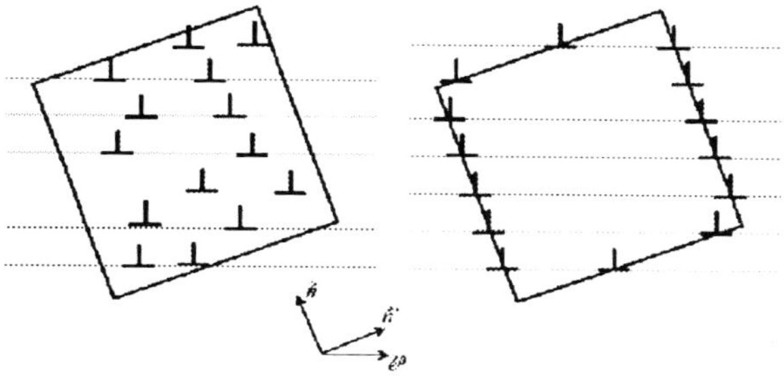

Fig. 27. Distribution of excess dislocations over the surfaces of a cell with normal vectors $\boldsymbol{n}$ and $\boldsymbol{n}'$ (schematically). Left: excess dislocations of slip system β statistically accumulated within the volume, right: the excess dislocations are actually 'condensed' in the cell wall. The density in a wall of orientation $\boldsymbol{n}$ is proportional to the projection $\boldsymbol{n}\boldsymbol{e}^{\beta}$.

The total misorientation of a cell wall with orientation $\boldsymbol{n}$ is $\boldsymbol{\omega}_{V,\boldsymbol{n}} = \sum_{\beta} \boldsymbol{\omega}_0^{\beta}(\boldsymbol{n})\vartheta_{V,\boldsymbol{n}}^{\beta} b$. Since they derive from a statistical accumulation of excess dislocations, the components of the rotation vector $\boldsymbol{\omega}_{V,\boldsymbol{n}}$ are Gaussian random variables. Their correlation matrix is given by

$$\langle \boldsymbol{\omega}_{V,\boldsymbol{n}} \otimes \boldsymbol{\omega}_{V,\boldsymbol{n}} \rangle = \sum_{\beta} \big[\boldsymbol{\omega}_0^{\beta}(\boldsymbol{n}) \otimes \boldsymbol{\omega}_0^{\beta}(\boldsymbol{n})\big] \big\langle \big[\vartheta_{V,\boldsymbol{n}}^{\beta}\big]^2 \big\rangle. \tag{100}$$

The mean square misorientation angle $\langle \boldsymbol{\omega}_{V,\boldsymbol{n}}^2 \rangle$ is equal to the trace of this matrix.

5.3. Statistical accumulation of long-range internal stresses

Statistical storage of excess dislocations leads to an accumulation of mesoscopic stresses. The fluctuating mesoscopic stress field associated with the excess dislocation densities $\phi^{\beta}(\boldsymbol{r})$ is given by (cf. eq. (32))

$$\sigma_{kl}^{(\mathrm{me})}(\boldsymbol{r}) = \sum_{\beta} \int \phi^{\beta}(\boldsymbol{r}')\sigma_{kl}^{\beta}(\boldsymbol{r} - \boldsymbol{r}')\,\mathrm{d}\boldsymbol{r}' = \frac{1}{b} \int \delta\varepsilon^{\beta}(\boldsymbol{r}')\tilde{\sigma}_{kl}^{\beta}(\boldsymbol{r} - \boldsymbol{r}')\,\mathrm{d}\boldsymbol{r}'. \tag{101}$$

In the last step we have performed an integration by parts, using $\phi^{\beta} = [1/b]\nabla^{\beta}\delta\varepsilon^{\beta}$ and defining $\tilde{\sigma}_{kl}^{\beta}(\boldsymbol{r}) := \nabla^{\beta}\sigma_{kl}^{\beta}(\boldsymbol{r})$. Temporal differentiation yields the stochastic differential equation

$$\partial_t \sigma_{kl}^{(\mathrm{me})}(\boldsymbol{r}) = -\sum_{\beta} \frac{1}{b} \int \delta\dot{\varepsilon}^{\beta}(\boldsymbol{r}')\tilde{\sigma}_{kl}^{\beta}(\boldsymbol{r} - \boldsymbol{r}')\,\mathrm{d}\boldsymbol{r}'. \tag{102}$$

We may approximate the stochastic process on the right-hand side of this equation by

$$\partial_t \sigma_{kl}^{(\mathrm{me})}(\boldsymbol{r}) = Q_{kl}^{\beta}\dot{w}_{kl}^{\beta}(\boldsymbol{r}, t), \tag{103}$$

where w_{kl}^{β} is again a normalized Gaussian white-noise process. The fluctuation amplitude Q_{kl}^{β} is given by

$$[Q_{kl}^{\beta}]^2 = \frac{1}{b^2}\int \langle \delta\dot{\varepsilon}^{\beta}(\boldsymbol{r})\delta\dot{\varepsilon}^{\beta}(\boldsymbol{r}')\rangle \tilde{\sigma}_{kl}^{\beta}(\boldsymbol{r})\tilde{\sigma}_{kl}^{\beta}(\boldsymbol{r}')\,\mathrm{d}\boldsymbol{r}\,\mathrm{d}\boldsymbol{r}' = \langle(\delta\dot{\varepsilon}^{\beta})^2\rangle G^2 \widetilde{C}_{kl}^{\beta}\frac{\xi_{\mathrm{n}}^{\beta}}{\xi_{\mathrm{g}}^{\beta}}. \tag{104}$$

The non-dimensional factors $\widetilde{C}_{kl}^{\beta} = [\xi_{\mathrm{g}}^{\beta}/(G^2b^2\xi_{\mathrm{n}}^{\beta})]\int f_{\varepsilon}^{\beta}(\boldsymbol{r}-\boldsymbol{r}')\tilde{\sigma}_{kl}^{\beta}(\boldsymbol{r})\tilde{\sigma}_{kl}^{\beta}(\boldsymbol{r}')\,\mathrm{d}\boldsymbol{r}\,\mathrm{d}\boldsymbol{r}'$ are defined in such a manner that, for correlation functions f_{ε}^{β} with the properties discussed in section 5.1, they are independent of the characteristic lengths $\xi_{\mathrm{g,n}}^{\beta}$. Evaluation in Fourier space gives us

$$\widetilde{C}_{kl}^{\beta} = \frac{\xi_{\mathrm{g}}^{\beta}}{(2\pi)^2\xi_{\mathrm{n}}^{\beta}}\int f_{\varepsilon}^{\beta}(\boldsymbol{k})k_{\mathrm{g}}^2|q_{kl}^{\beta}(\boldsymbol{k})|^2\,\mathrm{d}\boldsymbol{k}, \tag{105}$$

where $k_{\mathrm{g}} := \boldsymbol{k}\boldsymbol{e}_{\mathrm{g}}^{\beta}$ is the component of $\boldsymbol{k}$ in the glide direction, $q_{kl}^{\beta}(\boldsymbol{k})$ the Fourier transform of the non-dimensional function $\sigma_{kl}^{\beta}(\boldsymbol{r})/[Gb]$, and $f_{\varepsilon}^{\beta}(\boldsymbol{k})$ the Fourier transform of the strain correlation function $f_{\varepsilon}^{\beta}(\boldsymbol{r})$, eq. (89).

We evaluate the spatio-temporal correlation of the stochastic processes on the right-hand side of eq. (103) under the assumption that strain fluctuations on different slip systems are uncorrelated. Using eq. (90), we obtain

$$\langle \dot{w}_{kl}^{\beta}(\boldsymbol{r},t)\dot{w}_{kl}^{\beta}(\boldsymbol{r}',t')\rangle = \delta_{\beta\beta'}\delta(t-t')\big[g_{kl}^{\beta}(\boldsymbol{r}-\boldsymbol{r}')/g_{kl}^{\beta}(0)\big], \tag{106}$$

where the Fourier transform of the correlation function $g_{kl}^{\beta}(\boldsymbol{r})$ is $g_{kl}^{\beta}(\boldsymbol{k}) = f_{\varepsilon}^{\beta}(\boldsymbol{k})k_{\mathrm{g}}^2|q_{kl}^{\beta}(\boldsymbol{k})|^2$ (cf. eq. (48)). As an example, we consider the shear stress σ_{xy} resolved in a slip system in which edge dislocations glide in the x direction. In this case, $k_{\mathrm{g}} = k_x$ and $|q_{xy}^{\beta}|^2 = k_x^2k_y^4/[|\boldsymbol{k}|^8(1-\nu)^2]$. The Fourier transform of the correlation function f_{ε}^{β} (eq. (89)) is $f_{\varepsilon}^{\beta}(\boldsymbol{k}) = \xi_{\mathrm{n}}^{\beta}\xi_{\mathrm{g}}^{\beta}\exp[-(k_x\xi_{\mathrm{g}}^{\beta})^2/4\pi]$. With these functions, eq. (105) gives $\widetilde{C}_{xy}^{\beta} \approx [0.0623/(1-\nu)^2]$. The corresponding correlation function $g_{xy}^{\beta}(\boldsymbol{r})$ is shown in fig. 28. This function characterizes the decay of correlations in the mesoscopic stress field, providing a measure of the 'range' of the mesoscopic stresses. Figure 28 shows that this range is of the order of magnitude of the 'longitudinal' correlation length $\xi_{\mathrm{g}}^{\mathrm{n}}$ not only in the glide direction, but also perpendicular to it.

5.3.1. The contribution of fluctuating stresses to the flow stress

The fluctuating mesoscopic stress field $\sigma_{kl}^{(\mathrm{me})}$ arises from the incoherent superposition of many independent stress increments, each of them exhibiting long-range spatial correlations as shown in fig. 28. Stochastic integration of eq. (103) yields a Gaussian

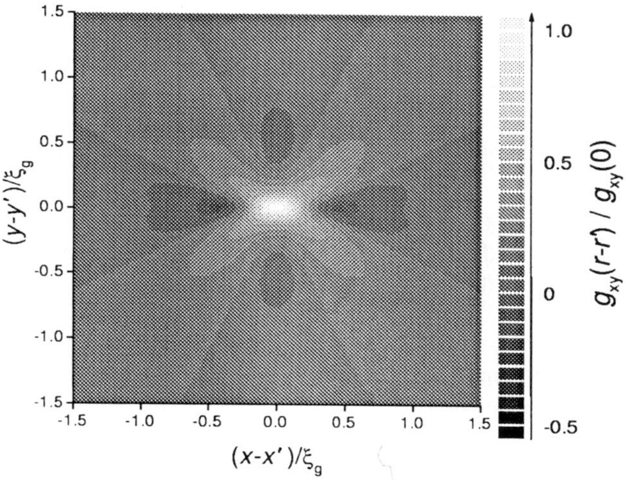

Fig. 28. Normalized correlation function of stresses created by statistical superposition of 'deformation events' with extension ξ_g parallel and $\xi_n \gg \xi_g$ perpendicular to the slip plane. The correlation function was calculated for the shear stress component σ_{xy} and a system of edge dislocations. For details see text.

distribution of the mesoscopic stress, the mean square of which follows from eq. (104) as

$$\langle(\sigma_{kl}^{(\mathrm{me})})^2\rangle = G^2 \sum_\beta \widetilde{C}_{kl}^\beta \int \frac{\xi_\mathrm{n}^\beta}{\xi_\mathrm{g}^\beta} \varepsilon_\mathrm{corr}^\beta \, \mathrm{d}\langle \varepsilon^\beta \rangle. \tag{107}$$

The motion of dislocations is controlled by regions on the slip plane where back stresses (internal stresses of sign opposite to the external stress) oppose the dislocation motion. Because of their mutual interaction, the moving dislocations average over the back stresses within the slip zone. Therefore, the contribution $\sigma_\beta^{(\mathrm{me})}$ of the mesoscopic stresses to the flow stress of slip system β is *not* governed by the peak value, but rather by the average of the back stresses within the slip volume V_corr^β. Calculating this average yields

$$[\sigma_\beta^{(\mathrm{me})}]^2 = \frac{1}{4} \sum_{\beta'} \frac{1}{(V_\mathrm{corr}^\beta)^2} \int \int_{V_\mathrm{corr}^\beta} \langle \sigma_{\beta\beta'}^{(\mathrm{me})}(\boldsymbol{r}) \sigma_{\beta\beta'}^{(\mathrm{me})}(\boldsymbol{r}') \rangle \, \mathrm{d}\boldsymbol{r} \, \mathrm{d}\boldsymbol{r}' =: \sum_{\beta'} \frac{A_{\beta\beta'}}{4} \langle (\sigma_{\beta\beta'}^{(\mathrm{me})})^2 \rangle. \tag{108}$$

Here the subscript $\beta\beta'$ denotes the resolved shear stress created by β' dislocations in the slip system β. The averaging reduces the mean square stresses by a factor $A_{\beta\beta'} < 1$. The factor $1/4$ in eq. (108) arises from the fact that only back stresses are considered.

Differentiation of eq. (108) with respect to $\langle \varepsilon^{\beta'} \rangle$ and using (107) yields the hardening matrix

$$\Theta^{\rm me}_{\beta\beta'} := \frac{\partial \sigma_\beta^{(\rm me)}}{\partial \langle \varepsilon^{\beta'} \rangle} = \frac{A_{\beta\beta'} \widetilde{C}^{\beta'}_{\beta\beta'}}{8} \frac{G^2}{\sigma_\beta^{(\rm me)}} \frac{\xi_{\rm n}^{\beta'}}{\xi_{\rm g}^{\beta'}} \varepsilon^{\beta'}_{\rm corr}. \tag{109}$$

In the following we calculate the hardening matrix for different special cases, *viz.* (i) individually moving dislocations on a single slip system, (ii) collectively moving dislocations on a single slip system, and (iii) collectively moving dislocations in symmetrical multiple slip.

Dislocations moving individually on a single slip system We consider dislocation sources of a given slip system emitting pairs of edge dislocations of opposite sign in an uncorrelated manner. In this case, $\varepsilon_{\rm corr} = 1$, $\xi_{\rm b}^\beta = b$, and $\xi_{\rm g}^\beta = 2L$, where L is the average slip distance of a dislocation. We distinguish two cases: (i) In the extreme case of slip distance that are comparable with the crystal diameter ($L \to \infty$) as in Stage-I hardening of f.c.c. single crystals [160], mesoscopic strain inhomogeneities are not built up and the corresponding hardening contribution is zero. In this limiting case, hardening is exclusively due to the 'microscopic' interactions between individual dislocations, (ii) In hardening stage II, the slip distance is limited by interactions with dislocations of secondary glide systems. When the dislocation arrangement complies with the 'law of similitude', one expects L to be inversely proportional to the flow stress. The hardening coefficients following from (109) are constant in this case. With $L = KGb/\sigma^\beta_{\rm ext}$, where $K \approx 10$–50 [70], and assuming $\sigma^\beta_{\rm ext} = \sigma_\beta^{(\rm me)}$, we obtain $\Theta^{(\rm me)}_{\beta\beta}/G = A_{\beta\beta}\widetilde{C}^\beta_{\beta\beta}/(16K)$. Numerically averaging the stress correlation function g^β_{xy} over the slip length yields $A_{\beta\beta} = 0.69$. With $\widetilde{C}^\beta_{\beta\beta} = \widetilde{C}^\beta_{xy} = 0.062/(1-\nu)^2$ and the typical values $\nu = 0.3$ and $10 < K < 50$ we obtain $5 \times 10^{-5} < \Theta^{(\rm me)}_{\beta\beta}/G < 2.5 \times 10^{-4}$. Comparing this to typical hardening coefficients in hardening stage II, *viz.* $\Theta_{\rm II} \approx 3 \times 10^{-3} G$, demonstrates that mesoscopic stresses arising from the random accumulation of individually moving dislocations cannot account for the hardening in this stage.

Dislocations moving collectively on a single slip system Significantly larger hardening than in the case of individually moving dislocations is may result if a sufficiently large number of dislocations move in a correlated manner. A classical example is the 'dislocation pile-up' created by n dislocations which are emitted by a single source during a slip event. For a group of n dislocations, we have $\varepsilon_{\rm corr} = n$, and again $\xi_{\rm b}^\beta = b$, $\xi_{\rm g}^\beta = 2L$. Obviously this increases the hardening by a factor of n. The ratio n/L may be estimated independently by assuming that a source is rendered inactive when the back stress of the previously emitted dislocations exceeds a certain fraction of the external stress [76,77,160].

The pile-up model may be generalized by allowing for activation of multiple sources and storage of the 'redundant' dislocations (which have zero net Burgers vector) within the slip zone. Termination of a slip event necessarily implies storage of excess dislocations. We emphasize that these dislocations do not have to be arranged in the specific configuration of a pile-up against a single obstacle, or on the same plane. At large distances, a group of n dislocations of the same sign spread over a distance ξ_{n}^{β} normal to the slip plane acts in a similar manner as a pile-up: For such a 'distributed pile-up', $\varepsilon_{\mathrm{corr}}^{\beta} = n$ must be replaced by $\varepsilon_{\mathrm{corr}}^{\beta} = n[b/\xi_{\mathrm{n}}^{\beta}]$ while the product $\varepsilon_{\mathrm{corr}}^{\beta}\xi_{\mathrm{n}}^{\beta} = nb$ and therefore the hardening matrix (109) remains unchanged. The ratio $\varepsilon_{\mathrm{corr}}^{\beta}\xi_{\mathrm{n}}^{\beta}/\xi_{\mathrm{g}}^{\beta}$, which according to eq. (107) governs the accumulation of mesoscopic stresses, is evaluated as follows: We consider a 'slip event' initiated in a source region around the origin and compute the back stress created by this slip event at the origin. Slip is assumed to terminate as soon as this back stress exceeds a fraction f^{b} of the external stress. For edge dislocations in an elastically isotropic material we obtain with the correlation function f_{ε}^{β} (eq. (89)) the result

$$\varepsilon_{\mathrm{corr}}^{\beta}\frac{\xi_{\mathrm{n}}^{\beta}}{\xi_{\mathrm{g}}^{\beta}} = \frac{f^{\mathrm{b}}\sigma_{\mathrm{ext}}^{\beta}(1-\nu)}{\sqrt{2}G}. \tag{110}$$

Inserting eq. (110) into eq. (109) and putting $\sigma_{\mathrm{ext}}^{\beta} \approx \sigma_{\beta}^{(\mathrm{me})}$ yields a constant hardening coefficient

$$\Theta^{(\mathrm{me})} = G\frac{A_{\beta\beta}\widetilde{C}_{xy}^{\beta} f^{\mathrm{b}}(1-\nu)}{8\sqrt{2}} \approx \Theta_{\mathrm{II}}. \tag{111}$$

Inserting $A_{\beta\beta} = 0.69$, $\widetilde{C}_{xy}^{\beta} = 0.062/(1-\nu)^2$, $\nu = 0.3$ yields $G/185 > \Theta^{(\mathrm{me})} > G/370$ for $1 > f^{\mathrm{b}} > 0.5$ [160]. These values are in the typical range of Stage-II hardening coefficients. Hence we find that long-range stresses which originate from mesoscopic strain incompatibilities associated with collective dislocation motions have the right order of magnitude to account for the Stage-II hardening of f.c.c. crystals deforming in single slip.

The results of section 5.2 allow us to relate internal stress fields and lattice rotations. During single-slip deformation, edge dislocation walls ('kink bands', [161]) emerge normal to the glide plane. These walls have a spacing which is of the order of the slip-line length [161] and a typical height normal to the slip plane which exceeds their spacing by a factor of 3–4 [162]. They lead to lattice rotations with an axis of rotation which is parallel to the line direction of the edge dislocations. The magnitude of these rotations can be calculated from eqs (97) and (98). We consider a rectangular volume of width ξ_{n} and height $4\xi_{\mathrm{n}}$. Evaluation of eq. (97) yields

$$\left[Q_{\phi}^{\beta}(V)\right]^2 = \left[Q_{\varepsilon}^{\beta}\right]^2\frac{\pi\xi_{\mathrm{n}}^{\beta}}{2b^2(\xi_{\mathrm{g}}^{\beta})^3}. \tag{112}$$

Excess dislocations are stored within the boundaries of this volume that are normal to the slip direction. We obtain from eq. (98)

$$\left[Q_{\vartheta}^{\beta}(\boldsymbol{e}_{\mathrm{g}}^{\beta}, V)\right]^2 = \frac{\pi}{8b^2}\frac{\xi_{\mathrm{n}}}{\xi_{\mathrm{g}}}\varepsilon_{\mathrm{corr}}^{\beta}\langle\dot{\varepsilon}^{\beta}\rangle. \tag{113}$$

The mean square rotation angle is then

$$\langle\omega^2\rangle = \frac{\pi}{8}\int \frac{\xi_{\mathrm{n}}^{\beta}}{\xi_{\mathrm{g}}^{\beta}}\varepsilon_{\mathrm{corr}}^{\beta}\,\mathrm{d}\langle\varepsilon^{\beta}\rangle. \tag{114}$$

Equation (114) may be evaluated using eq. (110). With $\sigma_{\mathrm{ext}}^{\beta} = \Theta_{\mathrm{II}}\langle\varepsilon^{\beta}\rangle$, $\Theta_{\mathrm{II}} = G/300$, $f^{\mathrm{b}} = 0.5$ and $\nu = 0.3$ we obtain $\sqrt{\langle\omega^2\rangle} = 4.6^{\circ}\langle\varepsilon^{\beta}\rangle$. Measurements by Ottenhaus [152] indicate that the characteristic rotation angle ω of kink bands increases linearly with strain in the regime $0.2 < \langle\varepsilon^{\beta}\rangle < 0.5$ according to $\omega \approx 3^{\circ}\langle\varepsilon^{\beta}\rangle$. This result is in reasonable agreement with the present estimate.

Collectively moving dislocations, symmetrical multiple slip We now consider tensile or compressive deformation of f.c.c. single crystals in $\langle 100\rangle$ orientation. Here 8 equivalent slip systems contribute equally to the total strain. The shear stress in each system is $\sigma_{\mathrm{ext}}^{\beta}$. From eq. (109) follows the total hardening coefficient

$$\Theta^{\mathrm{me}} = \frac{\partial\sigma_{\mathrm{ext}}^{\beta}}{\partial\langle\varepsilon\rangle} = \frac{1}{8}\sum_{\beta'}\Theta_{\beta\beta'}^{(\mathrm{me})} = \frac{\Theta_{\mathrm{II}}}{8}\sum_{\beta'}\frac{\Theta_{\beta\beta'}^{(\mathrm{me})}}{\Theta_{\beta\beta}^{(\mathrm{me})}}. \tag{115}$$

In the second step we have made use of the fact that the self-hardening of each slip system may be obtained using the same arguments as in the previous paragraph and therefore $\Theta_{\beta\beta}^{(\mathrm{me})} \approx \Theta_{\mathrm{II}}$ is given by eq. (113). For hardening caused by long-range stresses, the latent hardening ratios $\Theta_{\beta\beta'}^{(\mathrm{me})}/\Theta_{\beta\beta}^{(\mathrm{me})}$ have been calculated by Stroh [163], who obtained a ratio of 0.5 for coplanar systems and values around 0.8 for non-coplanar systems. Summing over all slip systems according to eq. (115), we find $\Theta^{(\mathrm{me})} \approx 0.75\Theta_{\mathrm{II}}$. It follows that the typical hardening contribution of the long-range stresses is on the order of $2 \times 10^{-3}G$. This must be compared with an initial hardening rate of $\langle 100\rangle$ Cu single crystals of $\Theta = 1.1 \times 10^{-2}G$ [142]. The comparison demonstrates that long-range stresses can only account for a small fraction of the hardening (and accordingly of the flow stress) of crystals deforming in symmetrical multiple slip. Moreover, the preceding consideration overestimates the hardening coefficient, since screening by the correlated activation of deformation events on different slip systems further reduces the long-range stresses (see also below). In symmetrical multiple slip the forest dislocation density for each slip system is large while the hardening contribution of long-range stresses is small. This leads to the conclusion that under these deformation conditions a forest hardening mechanism prevails.

5.3.2. Discussion

In section 5.3.1 it has been shown that long-range stresses can account for most of the Stage-II hardening of f.c.c. single crystals orientated for single slip. The model which we use can be considered a generalization of the 'pile-up model' of Stage-II hardening [76,77,160]. The main difference is that hardening is not related to large stress concentrations caused by specific dislocation configurations (passing stress of a pile-up) but to a mesoscopic internal-stress 'landscape' created by the incoherent superposition of long-range stresses arising from the terminations of many slip zones.

The central assumption in the model is that the individual slip events may be considered as statistically independent. One has therefore to ask to which extent this assumption is justified, and whether its relaxation will substantially change the numerical estimates. Let us first consider correlations between slip events in different slip systems which may reduce the mesoscopic stresses. Slip in secondary systems can efficiently screen stress concentrations close to an individual dislocation pile-up. However, in the present model we consider the superposition of the long-range 'tails' of the stress fields which are caused by terminations of slip zones. Our calculations demonstrate that these stress fields have a correlation length which is on the order of the slip-line length (fig. 28). Therefore they can be compensated by slip on secondary systems only when the total strains on secondary and primary systems are comparable. In Stage II of f.c.c. single crystals orientated for single slip, secondary systems contribute less than 10% of the total strain and therefore an effective screening is not possible. In crystals deforming in symmetrical multiple slip, on the other hand, screening may be effective but in this deformation mode long-range stresses are not relevant for the hardening anyway.

Screening of long-range stresses can also be due to the correlated activation of 'slip events' on the same slip system: In a kind of relay-race mechanism, multiple slip events may take place in such a manner that the stresses arising from the slip-line terminations are mutually screened. Such a mechanism has been envisaged by Hirsch [70]. In the terminology of the present model, such a mechanism would increase the 'longitudinal' correlation length ξ_g^β. However, this length does not show up explicitly in the hardening coefficient. If it is increased by correlated activation of multiple events, the characteristic strain $\varepsilon_{\rm corr}^\beta$ increases as well, since the back stress acting on the 'source region' of the first event is reduced by mutual screening of the long-range stresses. Therefore, the ratio $\varepsilon_{\rm corr}^\beta/\xi_g^\beta$ and the hardening matrix given by eqs (109) and (110) remain unchanged.

In the present model, the correlation length ξ_g^β may be eliminated in the calculation of the work-hardening slope. Nevertheless, it is somewhat unsatisfactory that the physical mechanism which governs the termination of slip events remains undetermined. The present model shares this weakness with other models of work-hardening. In particular, as emphasized by Hirsch [70], it is unclear how the observed termination of slip events might be understood within a forest-hardening model. This problem applies also to crystals deforming in high-symmetry orientations, where slip-line lengths have been reported to substantially exceed the dislocation cell sizes [69]. The classical explanation of slip-line blocking by Lomer–Cottrell locks [76,77] has been repeatedly doubted in view of the possibility to 'unzip' such locks in three dimensions at relatively low stresses. Recent large-scale discrete dislocation dynamics simulations have not shown evidence for strong locks which may lead to slip-line blocking and have, in general, failed to explain the experimentally observed slip-line lengths. Hence, the theoretical explanation of this basic observation remains an open issue.

We conclude the present section by some general remarks. The idea of using a 'Gaussian' description of the stochastic dynamics of plastic flow has been formulated already by Mott [164], who considered the accumulation of internal stresses in terms of

Gaussian statistics. More recently, Nabarro used a similar approach to estimate the random accumulation of excess dislocations in dislocation cell walls and the corresponding growth of misorientations [165]. This idea has been further elaborated by Pantleon [166] for two-dimensional and by Zaiser [50] for three-dimensional cell structures (cf. section 6). While a Gaussian approach towards the statistics of plastic deformation is attractive because of its simplicity, it is important to point out some limitations. We use Gaussian statistics for characterizing the cumulative effect of many elementary 'events' which themselves are not Gaussian distributed. This approach will fail when extreme events, so-called 'outliers', are relevant. An analogy of this situation is found in economics, where the price fluctuations of an asset result from a complex dynamics which gives rise to substantially non-Gaussian statistics on short and intermediate time scales. Predictions of the statistical evolution of prices based on Gaussian models usually work well on large time scales but may have disastrous consequences in the case of a financial crash [167]. The same may be true in plasticity where two examples of possible 'outliers' are (i) subgrains with large size and rotation where anomalous grain growth during recrystallization may be initiated, and (ii) large local stress concentrations leading to cracking.

6. *Work-hardening and dislocation microstructure evolution in symmetrical multiple slip*

6.1. Characterization of dislocation systems and plastic flow in three dimensions

Aiming at the generalization of the concepts developed in the previous sections to three-dimensional situations and their application to work-hardening and dislocation patterning in f.c.c. single crystals oriented for $\langle 100 \rangle$ symmetrical multiple slip, we consider dislocation dynamics in multiple slip, where several slip systems contribute to plastic deformation. In three dimensions, dislocation segments can have arbitrary directions within their slip planes (we do not consider climb). We distinguish dislocation segments of different orientation by the angle θ between their tangent and Burgers vectors. The tangent vector of a dislocation segment of slip system β and orientation θ is labeled $\boldsymbol{t}^\beta(\theta)$. The slip direction of this segment is $\boldsymbol{e}_{\mathrm{g}}^\beta(\theta) = n^\beta \times \boldsymbol{t}^\beta(\theta)$. For a system of parallel dislocations, this description reduces to the quasi-twodimensional formulation used in previous sections, since in this case θ can assume only two values, θ_0 or $\theta_0 + \pi$. These may be identified with the two possible 'signs' $s \in [+, -]$ of a straight dislocation.

On the mesoscopic scale, the dislocation arrangement is now characterized by scalar segment densities [98] $\rho^\beta(\theta, \boldsymbol{r})$, which can be interpreted as the length per unit volume of dislocation segments with orientation θ. The equations of evolution for these densities are given by

$$\partial_t \rho^\beta(\theta, \boldsymbol{r}) = -\nabla \boldsymbol{J}^\beta(\theta, \boldsymbol{r}) + K^\beta\big(\theta, \big[\rho^\beta(\theta, \boldsymbol{r})\big]\big). \tag{116}$$

Here $\boldsymbol{J}^\beta(\theta) = \rho^\beta(\theta)\boldsymbol{e}^\beta v^\beta(\theta)$ is the flux of dislocation segments of type $[\beta, \theta]$, $v^\beta(\theta)$ is the velocity of these segments, and $K^\beta(\theta)$ is the rate of change of the population $[\beta, \theta]$ due to dislocation multiplication and dislocation reactions which, in general, involve dislocations of different slip systems and orientations. The present formulation of dislocation dynamics in terms of segment densities is strongly inspired by the equations for the evolution of dislocation ensembles which have been proposed by El-Azab [98]. There is, however, one important difference between El-Azab's work and the present approach: El-Azab distinguishes dislocation segments not only by their spatial and configurational coordinates but also according to their velocity. The present approach uses a simplified stochastic formulation in which dislocation velocity variations are treated a summarily in terms of fluctuations of the dislocation fluxes or, equivalently, the shear strain rates.

The contribution of segments of orientation θ to the shear strain rate in the slip system β is $\dot{\varepsilon}^{\beta}(\theta) = \rho^{\beta}(\theta) b v^{\beta}(\theta)$, and the total shear strain rate in this slip system may be written as

$$\dot{\varepsilon}^{\beta} = b \int_0^{2\pi} \boldsymbol{J}^{\beta}(\theta, \boldsymbol{r}) \boldsymbol{e}_{\mathrm{g}}^{\beta}(\theta)\, \mathrm{d}\theta. \tag{117}$$

As in section 5.1, we envisage the shear strain rate as a sum over discrete deformation events which have the following properties: (i) On a given slip system, deformation events take place randomly in space and in time. (ii) During each event, multiple sources are activated and segments of different orientations move collectively. Only a small fraction of these segments are stored as 'excess dislocations' at the boundaries of the region where slip takes place during the event (the 'slip zone'), while the larger part remain as 'redundant dislocations' of zero net Burgers vector within the slip zone. (iii) On the time scale of microstructure evolution, the fluctuations of plastic flow may be approximated by Gaussian white-noise processes. Hence in generalization of eq. (87) we write

$$\dot{\varepsilon}^{\beta}(\theta) = \langle \dot{\varepsilon}^{\beta} \rangle + \delta\dot{\varepsilon}^{\beta}(\theta, \boldsymbol{r}), \tag{118}$$

where $\delta\dot{\varepsilon}^{\beta}(\theta) \approx Q_{\varepsilon}^{\beta}(\theta)\dot{w}^{\beta}$, $Q_{\varepsilon}^{\beta}(\theta) = Q_{\varepsilon}^{\beta} q^{\beta}(\theta)$. Here $q^{\beta}(\theta)$ is the relative strain-rate contribution of the segments with orientation θ, and Q_{ε}^{β} is given by eq. (86).

The spatio-temporal correlations of the stochastic processes $\dot{w}^{\beta}$ are assumed to have a similar form as in section 5.1: $\langle \dot{w}^{\beta}(\boldsymbol{r}, t)\dot{w}^{\beta}(\boldsymbol{r}', t')\rangle = \delta(t - t') f_{\varepsilon}^{\beta}(\boldsymbol{r} - \boldsymbol{r}')$. In three dimensions, three correlation lengths are required to characterize the spatial extension of the slip zones. In addition to the correlation length ξ_{n}^{β} normal to the slip plane, there are now two correlation lengths $\xi_{\mathrm{e,s}}^{\beta}$ in the slip directions of edge and screw segments which characterize the respective slip-line lengths. Instead of eq. (89) we therefore write

$$f_{\varepsilon}^{\beta}(\boldsymbol{r}) := \exp\left[-\pi\left[\left(\frac{r_{\mathrm{s}}}{\xi_{\mathrm{s}}^{\beta}}\right)^2 - \left(\frac{r_{\mathrm{e}}}{\xi_{\mathrm{e}}^{\beta}}\right)^2\right]\right] \xi_{\mathrm{n}}^{\beta} \delta(r_{\mathrm{n}}), \tag{119}$$

where r_{n}, r_{e}, and r_{s} are the components of the vector $\boldsymbol{r}$ in a coordinate system spanned by the glide plane normal and the glide directions of edge and screw segments, respectively. The average volume in which slip takes place during a 'deformation event' is $V_{\mathrm{corr}} = \xi_{\mathrm{e}}^{\beta} \xi_{\mathrm{s}}^{\beta} \xi_{\mathrm{n}}^{\beta}$.

6.2. Lattice curvature and misorientations in 3D dislocation systems

6.2.1. General relationships

We first consider the evolution of lattice curvature and misorientations. The treatment of this problem in section 5.2 can be generalized to three-dimensional dislocation systems in a straightforward way. For 3D dislocation systems, the dislocation density tensor is given by [85,98]

$$\boldsymbol{\alpha}(\boldsymbol{r}) = \sum_{\beta} \int_0^{2\pi} \rho^{\beta}(\boldsymbol{r},\theta)\big[\boldsymbol{t}^{\beta}(\theta)\otimes\boldsymbol{b}^{\beta}\big]\,\mathrm{d}\theta. \tag{120}$$

The evolution of this tensor is governed by the evolution of the segment densities, eq. (116). A considerable simplification comes from the fact that in dislocation reactions and multiplication, the 'Burgers vector content' of a mesoscopic volume element is preserved. Therefore, reaction and multiplication terms drop out when eqs (116) are summed up according to (120) in calculating the evolution of $\boldsymbol{\alpha}$ [98]. Since only the drift terms contribute, we may write

$$\partial_t\boldsymbol{\alpha}(\boldsymbol{r},t) = \sum_{\beta}\int_0^{\pi} \partial_t\phi^{\beta}(\theta)\boldsymbol{\alpha}^{\beta}(\theta)\,\mathrm{d}\theta, \qquad \partial_t\phi^{\beta}(\theta) = \nabla\big[\boldsymbol{J}^{\beta}(\theta) - \boldsymbol{J}^{\beta}(\theta+\pi)\big], \tag{121}$$

where $\boldsymbol{\alpha}^{\beta}(\theta) = [\boldsymbol{t}^{\beta}(\theta)\otimes\boldsymbol{b}^{\beta}]$ and $\boldsymbol{t}^{\beta}(\theta+\pi) = -\boldsymbol{t}^{\beta}(\theta)$ have been used. By analogy with section 5, we can interpret $\phi^{\beta}(\theta)$ formally as a density of excess dislocation segments of orientation θ.[20] In terms of $\phi^{\beta}(\theta)$, the dislocation density and lattice curvature tensors can be written as (cf. eqs (91) and (95))

$$\boldsymbol{\alpha}(\boldsymbol{r}) = \sum_{\beta}\int_0^{\pi} \phi^{\beta}(\theta)\boldsymbol{\alpha}^{\beta}(\theta)\phi^{\beta}(\theta)\,\mathrm{d}\theta, \qquad \boldsymbol{\kappa}(\boldsymbol{r}) = \sum_{\beta}\int_0^{\pi} \phi^{\beta}(\theta)\boldsymbol{\kappa}^{\beta}(\theta)\,\mathrm{d}\theta, \tag{122}$$

where $\boldsymbol{\kappa}^{\beta}(\theta) = [1/2]\mathbf{I}\,\mathrm{Tr}\,\boldsymbol{\alpha}^{\beta}(\theta) - \boldsymbol{\alpha}^{\beta}(\theta)$. The contribution of segments $[\beta,\theta]$ to the strain rate is given by $\dot{\varepsilon}^{\beta}(\theta) = b\boldsymbol{e}_{\mathrm{g}}^{\beta}\boldsymbol{J}^{\beta}(\theta)$. Therefore the evolution of $\phi^{\beta}(\theta)$ may be written as

$$\partial_t\phi^{\beta}(\theta) = -[1/b]\nabla_{\theta}\big[2\langle\dot{\varepsilon}^{\beta}(\theta)\rangle + \delta\dot{\varepsilon}^{\beta}(\theta) + \delta\dot{\varepsilon}^{\beta}(\theta+\pi)\big], \tag{123}$$

where we have split the strain-rate contributions into their mean values and fluctuations and used that $\langle\dot{\varepsilon}^{\beta}(\theta)\rangle = \langle\dot{\varepsilon}^{\beta}(\theta+\pi)\rangle$. The operator $\nabla_{\theta} = \boldsymbol{e}_{\mathrm{g}}^{\beta}(\theta)\nabla$ takes the directional derivative in the slip direction of the segments of orientation θ. As does eq. (92), eq. (123)

[20] In the absence of dislocation multiplication and reactions, $\phi^{\beta}(\theta)$ is identical with the actual population difference $\rho^{\beta}(\theta) - \rho^{\beta}(\theta+\pi)$. When reactions take place, $\phi^{\beta}(\theta)$ may differ from the actual population difference, but one may replace $\rho^{\beta}(\theta) - \rho^{\beta}(\theta+\pi)$ by $\phi^{\beta}(\theta)$ as far as the calculation of the dislocation density and lattice curvature tensors is concerned. Therefore we call it a 'formal' excess dislocation density.

contains a deterministic contribution which arises when the average dislocation fluxes are space dependent (e.g., when different slip systems are active in different regions or when there are macroscopic deformation gradients), plus a stochastic process which is obtained by applying the operator ∇_θ to the fluctuating strain-rate contributions $\delta\dot{\varepsilon}^\beta(\theta)$ and $\delta\dot{\varepsilon}^\beta(\theta+\pi)$. In the white-noise approximation (118), the excess densities ϕ^β and therefore also the components of the dislocation density and lattice-curvature tensors evolve in a diffusionlike manner; hence their distribution is Gaussian.

We now generalize the treatment of misorientations to three dimensions. Since the derivation follows very much the same lines as in section 5.2, we only pinpoint the crucial differences and give the results:

- If $\phi^\beta_V(\theta)$ denotes the excess segment density $\phi^\beta(\theta)$ averaged over a 3D cell volume V, the evolution of $\phi^\beta_V(\theta)$ is given by (cf. eq. (96))

$$\begin{aligned}\partial_t\phi^\beta_V(\theta) &= -\frac{1}{bV}\int_V \nabla_\theta\big[\delta\dot{\varepsilon}^\beta(\theta)+\delta\dot{\varepsilon}^\beta(\theta+\pi)\big]\,\mathrm{d}^3r\\ &= -\frac{\boldsymbol{e}_{\mathrm{g}}\beta(\theta)}{bV}\int_{\partial V}\big[\delta\dot{\varepsilon}^\beta(\theta,\boldsymbol{r})+\delta\dot{\varepsilon}^\beta(\theta+\pi,\boldsymbol{r})\big]\boldsymbol{n}(\boldsymbol{r})\,\mathrm{d}^2r.\end{aligned} \tag{124}$$

- All excess dislocation segments are stored in the cell walls. A cell wall contains only dislocation segments whose line direction is parallel to the wall $[\boldsymbol{t}^\beta(\theta)\boldsymbol{n}=0]$. The linear excess dislocation density is $\vartheta^\beta_{V,n}=\phi^\beta_V(\theta)|\boldsymbol{n}\boldsymbol{e}^\beta_{\mathrm{g}}(\theta)|\delta(\boldsymbol{n}\boldsymbol{t}^\beta(\theta))N_V$, with normalization factor $N_V = V[\int_{\partial V}|\boldsymbol{n}(\boldsymbol{r})\boldsymbol{e}^\beta_{\mathrm{g}}(\theta)|\delta(\boldsymbol{n}\boldsymbol{t}^\beta(\theta))\,\mathrm{d}\boldsymbol{r}]^{-1}$ (cf. the corresponding relations in section 5.2).
- The fluctuating strain rates which govern the evolution of ϕ^β_V and $\vartheta^\beta_{V,n}$ may be approximated by Gaussian white-noise processes. In this approximation, the temporal evolution of ϕ^β_V and $\vartheta^\beta_{V,n}$ is given by

$$\begin{aligned}&\partial_t\phi^\beta_V(\theta) = Q^\beta_\phi(V,\theta)\dot{w}^\beta,\\ &\big[Q^\beta_\phi(V,\theta)\big]^2 = \frac{[Q^\beta_\varepsilon(\theta)]^2}{V^2b^2}\int_{\partial V}\int_{\partial V'}\big[\boldsymbol{n}(\boldsymbol{r})\boldsymbol{e}^\beta_{\mathrm{g}}(\theta)\big]\big[\boldsymbol{n}(\boldsymbol{r}')\boldsymbol{e}^\beta_{\mathrm{g}}(\theta)\big]f^\beta_\varepsilon(\boldsymbol{r}-\boldsymbol{r}')\,\mathrm{d}\boldsymbol{r}\,\mathrm{d}\boldsymbol{r}',\end{aligned} \tag{125}$$

$$\begin{aligned}&\partial_t\vartheta^\beta_{V,n}(\theta) = Q^\beta_\vartheta(V,\boldsymbol{n})\dot{w}^\beta,\\ &\big[Q^\beta_\vartheta(V,\boldsymbol{n})\big]^2 = \big[Q^\beta_\phi(V,\theta_n)\big]^2\big|\boldsymbol{n}\boldsymbol{e}^\beta_{\mathrm{g}}(\theta_n)\big|^2V^2\bigg[\int_{\partial V}\big|\boldsymbol{n}(\boldsymbol{r})\boldsymbol{e}^\beta_{\mathrm{g}}(\theta)\big|\delta\big(\boldsymbol{n}\boldsymbol{t}^\beta(\theta)\big)\,\mathrm{d}\boldsymbol{r}\bigg]^{-2},\end{aligned} \tag{126}$$

  where θ_n is the orientation for which $\boldsymbol{t}^\beta(\theta)\boldsymbol{n}=0$. The distributions are Gaussian, $p(\phi^\beta_V(\theta)) = \mathcal{G}(\phi^\beta_V(\theta);0,\int[Q^\beta_\phi(V,\theta)]\,\mathrm{d}t)$ and $p(\vartheta^\beta_{V,n}) = \mathcal{G}(\vartheta^\beta_{V,n};0,\int[Q^\beta_\vartheta(V,\boldsymbol{n})]^2\,\mathrm{d}t)$.
- The lattice rotations across the cell wall are given by

$$\boldsymbol{\omega}^\beta_{V,\boldsymbol{n}} = \vartheta^\beta_{V,n}\boldsymbol{\omega}^\beta_0(\boldsymbol{n}),\qquad \boldsymbol{\omega}^\beta_0(\boldsymbol{n}) = \frac{1}{2}\big(\boldsymbol{t}^\beta(\theta_n)\boldsymbol{b}^\beta\big)\boldsymbol{n}-\big(\boldsymbol{b}^\beta\boldsymbol{n}\big)\boldsymbol{t}^\beta(\theta_n). \tag{127}$$

The components of $\boldsymbol{\omega}_{V,n}$ are Gaussian random variables with the correlation matrix

$$\langle \boldsymbol{\omega}_{V,n} \otimes \boldsymbol{\omega}_{V,\boldsymbol{n}} \rangle = \sum_{\beta} \big[\boldsymbol{\omega}_0^{\beta}(\boldsymbol{n}) \otimes \boldsymbol{\omega}_0^{\beta}(\boldsymbol{n})\big] \big\langle \big[\vartheta_{V,n}^{\beta}\big]^2 \big\rangle. \tag{128}$$

The mean square misorientation angle $\langle \boldsymbol{\omega}_{V,\boldsymbol{n}}^2 \rangle =: \langle \zeta_{V,n} \rangle$ is equal to the trace of this matrix.

- In experiments, often only the modulus $\omega_{V,n} = \sqrt{\zeta_{V,n}}$ of the misorientation angle is measured [129,130]. The distribution function $p(\omega_{V,n})$ can be calculated from the distribution of $\zeta_{V,n}$. To determine this distribution, we note that the correlation matrix (128) is symmetric and can therefore be diagonalized by choosing an appropriate coordinate system. A diagonal correlation matrix implies that the components $\omega_{V,n}^{(l)}$ of the misorientation vector $\boldsymbol{\omega}_{V,n} = [\omega_{V,n}^{(l)}]$ are statistically independent Gaussian variables with mean square $\langle (\omega_{V,n}^{(l)})^2 \rangle = \langle \zeta_{V,\boldsymbol{n}}^{(l)} \rangle$. The probability distribution of $\zeta_{V,n}^{(l)}$ follows from the distribution $p(\omega_{V,n}^{(l)}) = \mathcal{G}(\omega_{V,n}^{(l)}; 0, \langle (\omega_{V,n}^{(l)})^2 \rangle)$ by change of variables, and the distribution of the sum $\zeta_{V,n} = \sum_l \zeta_{V,n}^{(l)}$ is obtained from the distributions p_l of the $\zeta_{V\boldsymbol{n}}^{(l)}$ by a double convolution,

$$p_l(\zeta) = \frac{1}{\sqrt{2\pi \zeta \langle (\omega_{V,n}^l)^2 \rangle}} \exp\left[-\frac{\zeta}{2\langle (\omega_{V,n}^l)^2 \rangle} \right],$$

$$p(\zeta_{V,n}) = \int_0^{\zeta_{V,n}} p_1(\zeta_{V,n} - \zeta') \int_0^{\zeta'} p_2(\zeta'') p_3(\zeta' - \zeta'') \, \mathrm{d}\zeta'' \, \mathrm{d}\zeta'. \tag{129}$$

The probability distribution of the modulus of the misorientation angle, $\omega_{V,n} = \sqrt{\zeta_{V,n}}$, is obtained from this expression by a second change of variables, $p(\omega_{V,n}) \, \mathrm{d}\omega_{V,n} = p(\zeta_{V,n}) \, \mathrm{d}\zeta_{V,n}$.

6.2.2. *Misorientations in f.c.c. single crystals deformed in a $\langle 100 \rangle$ direction*

In the following we consider the particular case of f.c.c. single crystals deformed in tension or compression along a $\langle 100 \rangle$ direction. Eight equivalent $\langle 110 \rangle \{111\}$ slip systems β are active with the same Schmid factor $M = 1/\sqrt{6}$ and strain rate $\langle \dot{\varepsilon}^{\beta} \rangle = \dot{\varepsilon}/8M$ ($\dot{\varepsilon}$ is the externally imposed strain rate). We assume that segments of all orientations contribute equally to the average strain rate, i.e., $\langle \dot{\varepsilon}^{\beta}(\theta) \rangle = \langle \dot{\varepsilon}^{\beta} \rangle / (2\pi)$, $\delta \dot{\varepsilon}^{\beta}(\theta) = \delta \dot{\varepsilon}^{\beta} / (2\pi)$, and that the correlation lengths $\xi_{\mathrm{s}}^{\beta} = \xi_{\mathrm{e}}^{\beta} = \xi_{\mathrm{g}}^{\beta}$ are equal for all active slip systems. Furthermore, we assume that deformation events on different slip systems are statistically independent, so that the stochastic processes $\dot{w}^{\beta}$ in eqs (125) and (126) are uncorrelated for different β. Since correlations tend to reduce the misorientations, our results yield an upper estimate for rotations in $\langle 100 \rangle$ crystals. If slip on all equivalent slip systems took place in a fully correlated manner, there would be no misorientations at all. Uncorrelated fluctuations of the slip activity in the different slip systems lead to a statistical accumulation of excess dislocations and the emergence of misorientations. This must be distinguished from the *deterministic* growth of misorientations across boundaries between blocks in which different slip systems are active *on average* ('geometrically necessary boundaries' in the terminology of Hughes and Hansen [129]). Such boundaries are not typical for ⟨100⟩-orientated crystals [142].

We consider spherical cells of diameter λ and calculate the rate of excess dislocation accumulation. Evaluating eqs (125) and (126) yields

$$\left[Q^{\beta}_{\phi}(V,\theta)\right]^2 = \frac{12\pi}{5b^2}\left[Q^{\beta}_{\varepsilon}(\theta)\right]^2 \frac{\xi_{\mathrm{n}}}{\lambda\xi_{\mathrm{g}}^2}, \tag{130}$$

$$\left[Q^{\beta}_{\vartheta}(V,\boldsymbol{n})\right]^2 = \frac{3\pi^3}{20b^2}\left[Q^{\beta}_{\varepsilon}(\theta)\right]^2 \frac{\xi_{\mathrm{n}}\lambda}{\xi_{\mathrm{g}}^2}\left[1-(\boldsymbol{n}\boldsymbol{n}^{\beta})^2\right]. \tag{131}$$

Using eq. (86) and noting that $Q^{\beta}_{\varepsilon}(\theta) = Q^{\beta}_{\varepsilon}/(2\pi)$, we find that the mean square density of excess segments in a wall of a cell of size λ is given by

$$\left\langle\left[\vartheta^{\beta}_{\lambda,\boldsymbol{n}}\right]^2\right\rangle = \frac{3\pi}{80b^2}\int \varepsilon^{\beta}_{\mathrm{corr}}\frac{\xi_{\mathrm{n}}}{\xi_{\mathrm{g}}}\frac{\lambda}{\xi_{\mathrm{g}}}\left[1-(\boldsymbol{n}\boldsymbol{n}^{\beta})^2\right]\mathrm{d}\langle\varepsilon^{\beta}\rangle. \tag{132}$$

Using eq. (128), we may now calculate correlation matrices $\langle\boldsymbol{\omega}_{V,n}\otimes\boldsymbol{\omega}_{V,\boldsymbol{n}}\rangle$ for different wall orientations. With the notation

$$\Omega^2_{\lambda} = \frac{\pi}{40}\int \varepsilon^{\beta}_{\mathrm{corr}}\frac{\xi_{\mathrm{n}}}{\xi_{\mathrm{g}}}\frac{\lambda}{\xi_{\mathrm{g}}}\,\mathrm{d}\langle\varepsilon^{\beta}\rangle, \tag{133}$$

we obtain for $n = [100]$

$$\langle\boldsymbol{\omega}_{V,n}\otimes\boldsymbol{\omega}_{V,\boldsymbol{n}}\rangle = \Omega^2_{\lambda}\begin{bmatrix} 1/2 & 0 & 0 \\ 0 & 2 & 0 \\ 0 & 0 & 2 \end{bmatrix} \tag{134}$$

and for $\boldsymbol{n} = [010]$

$$\Omega^2_{\lambda}\begin{bmatrix} 1 & 0 & 0 \\ 0 & 5/4 & 0 \\ 0 & 0 & 1 \end{bmatrix}. \tag{135}$$

The mean square misorientation angles are $\langle\omega^2_n\rangle = (5/2)\Omega^2_{\lambda}$ for $\boldsymbol{n} = [100]$ and $(13/4)\Omega^2_{\lambda}$ for $\boldsymbol{n} = [010]$ and $[001]$. The misorientation-angle distributions exhibit a scaling property in the sense that the average misorientation $\langle\omega_n\rangle$ increases with increasing strain, while the distribution $p(\omega_{\lambda,\boldsymbol{n}}/\langle\omega_{\lambda,\boldsymbol{n}}\rangle)$ of the normalized angles is strain independent. This follows from the fact that the correlation matrices $\langle\boldsymbol{\omega}_{\lambda,\boldsymbol{n}}\otimes\boldsymbol{\omega}_{\lambda,\boldsymbol{n}}\rangle$ depend on strain only through the common scalar factor $\Omega^2_{\lambda}(\varepsilon)$. Scaled distributions for the cell-wall orientations [100] and [010] are plotted in fig. 29 together with the average of $p(\omega_{\lambda,\boldsymbol{n}}/\langle\omega_{\lambda,\boldsymbol{n}}\rangle)$ over all $\boldsymbol{n}$. The experimental data points in fig. 29 have been determined by Hughes et al. [130] for a multitude of polycrystalline Cu, Al and Ni specimens deformed to different strains and under different loading conditions and temperatures. These distributions represent averages

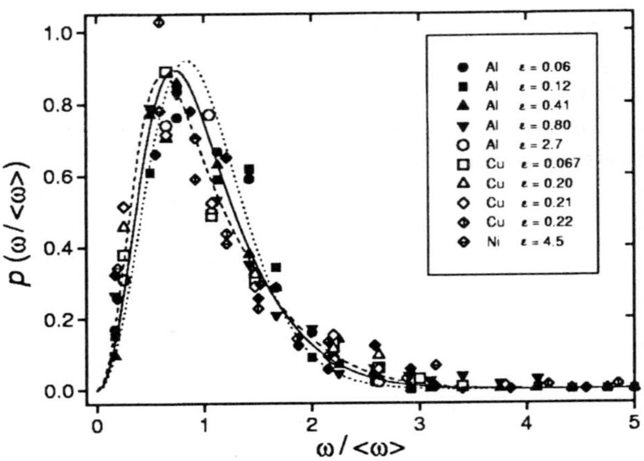

Fig. 29. Scaled distributions of misorientation angles; data points after Hughes et al. [130]. Theoretical curves for [100] symmetrical multiple slip. Dotted line $\boldsymbol{n} = [100]$, dashed line $\boldsymbol{n} = [010]$, full line: average over all $\boldsymbol{n}$; the angles have been normalized by their average value.

over different wall and grain orientations. It is seen that the scaled misorientation angle distributions fall approximately on a common curve. This implies that scaling is observed not only for different strains but also for different materials and loading conditions. The present calculations yield curves that are qualitatively consistent with the data of Hughes et al. However, we find that the shape of the misorientation distribution should in general depend on the activity of slip systems and the orientation of cell walls with respect to the Burgers vectors and glide plane normals. Further experimental data are required to clarify this point.

The magnitude of the misorientations and its evolution in the course of straining are determined by the strain dependent factor Ω_λ^2. To evaluate this factor, we note that the ratio $\varepsilon^\beta_{\mathrm{corr}}\xi_{\mathrm{n}}/\xi_{\mathrm{g}}$ can be estimated in a similar manner as in section 5.3. Using eq. (110) and averaging over all orientations of $\boldsymbol{n}$ and all cell sizes λ, we obtain the mean square misorientation angle

$$\langle \omega^2 \rangle = \frac{\sqrt{27}\pi}{320} \frac{\langle \lambda \rangle}{\xi_{\mathrm{g}}} f_{\mathrm{b}}(1-\nu) \int \frac{\sigma^\beta_{\mathrm{ext}}}{G} \, \mathrm{d}\langle \varepsilon \rangle. \tag{136}$$

At large strains, the flow stress $\sigma^\beta_{\mathrm{ext}}$ is almost strain-independent. Using the parameters $f_{\mathrm{b}} = 0.5$, $\nu = 0.3$, a ratio between the slip-line length and the average cell size $\xi_{\mathrm{g}}/\langle \lambda \rangle \approx 5$ [69], and a saturation stress $\sigma^\beta_{\mathrm{ext}} \approx 200$ MPa, we find that at large strains $\sqrt{\langle \omega^2 \rangle} \approx 1.51^\circ \sqrt{\langle \varepsilon \rangle}$. This is in reasonable agreement with experimental data quoted in [166] which indicate that the average misorientation angle increases like $1.8^\circ \sqrt{\varepsilon}$.

6.3. Dislocation-cell patterning and work-hardening

During the initial stages of deformation, excess dislocations which lead to lattice rotations and misorientations make up only a small fraction of the total dislocation density, while the larger part of the dislocation population has zero net Burgers vector. The spatial distribution of these 'redundant' dislocations determines the dislocation patterns and, when a forest hardening mechanism prevails, they are responsible for most of the flow stress.

In the following we are interested in the evolution of the total dislocation density. This may be deduced from eq. (116) by integrating over all angles and summing over all slip system. Dislocation segments of all orientations are assumed to contribute approximately equally to the local strain rates. In particular, the fluxes of dislocations of opposite 'signs' are approximately equal but have opposite directions, $\boldsymbol{J}(\theta,\boldsymbol{r}) \approx -\boldsymbol{J}(\theta,-\boldsymbol{r})$. Hence, the sum $\nabla \boldsymbol{J}(\theta,\boldsymbol{r}) + \nabla \boldsymbol{J}(\theta+\pi,\boldsymbol{r})$ is small and the flux terms cancel approximately when integrated over θ.[21] We are left with the terms describing dislocation multiplication and reactions. The rates of dislocation multiplication and reactions which govern the accumulation of these dislocations, may be obtained by two strategies: (i) Reaction cross-sections between dislocation segments of different slip systems and orientations may be deduced from systematic computer simulations of dislocation intersection processes [30]. In this manner one may determine the reaction terms $K^{\beta}(\theta,[\rho^{\beta}(\theta,\boldsymbol{r}])$ in eq. (116). This leads to a set of coupled integrodifferential equations for the dislocation populations $\rho^{\beta}(\theta,\boldsymbol{r})$ [50,98]. (ii) Instead of following the 'mechanistic' approach, one may pursue a 'holistic' top-to-down approach where energy storage and dissipation arguments are used to determine the basic structure of the equation of evolution of the total dislocation density. Since our main interest is in the basic processes which govern work hardening and dislocation patterning rather than the mechanistic details, we follow the second approach.

6.3.1. *Dislocation multiplication and dislocation reactions*

Dislocation multiplication By 'dislocation multiplication' we denote all processes in which mechanical work is expended to increase the dislocation length in the deforming crystal. Besides the large-scale expansion of dislocation segments and the operation of dislocation sources, this includes also the breaking of dislocation junctions. The rate of dislocation multiplication in a mesoscopic volume element is governed by the work done by the local stress. The multiplication rate in slip system β (integrated over all segment orientations) is

$$K^{\beta,+}(\boldsymbol{r}) = \frac{\eta_0^{\beta}}{E_{\mathrm{L}}}[\sigma_{\mathrm{ext}}^{\beta} + \sigma_{\mathrm{int}}^{\beta}(\boldsymbol{r})]\dot{\varepsilon}^{\beta} = \frac{\eta_0^{\beta}}{E_{\mathrm{L}}}\sigma_{\mathrm{eff}}^{\beta}(\boldsymbol{r})\dot{\varepsilon}^{\beta}, \tag{137}$$

where E_{L} denotes the dislocation line energy and η_0^{β} the fraction of the mechanical work that is not instantaneously dissipated in moving the dislocations. $\sigma_{\mathrm{ext}}^{\beta}$ and $\sigma_{\mathrm{int}}^{\beta}$ are the external and internal stresses acting on the dislocations of slip system β in a mesoscopic volume element at $\boldsymbol{r}$.

[21] This is does not hold so for the flux *difference* $\boldsymbol{J}(\theta,\boldsymbol{r}) - \boldsymbol{J}(\theta+\pi,\boldsymbol{r})$, the gradient of which governs excess dislocation accumulation, eq. (123) ff.

Both the internal stress $\sigma^{\beta}_{\rm int}$ (which must be understood as the internal stress acting *on the dislocations,* cf. section 2.5.1) and the shear strain rate $\dot{\varepsilon}^{\beta}$ fluctuate in space and time, $\sigma^{\beta}_{\rm int} = \langle\sigma^{\beta}_{\rm int}\rangle + \delta\sigma^{\beta}_{\rm int}$ and $\dot{\varepsilon}^{\beta} = \langle\dot{\varepsilon}^{\beta}\rangle + \delta\dot{\varepsilon}^{\beta}$. To assess the magnitude of these fluctuations, we use an argument similar to section 2.5.1: The average work done by the internal stresses on the moving dislocations is zero, $\langle\sigma^{\beta}_{\rm int}\dot{\varepsilon}^{\beta}\rangle = 0$. It follows that the fluctuation cross-correlation is $\langle\delta\sigma^{\beta}_{\rm int}\delta\dot{\varepsilon}^{\beta}\rangle = -\langle\sigma^{\beta}_{\rm int}\rangle\langle\dot{\varepsilon}^{\beta}\rangle$ (cf. eq. (7)). For a linear relation between local stress and strain rate, the identity $\delta\dot{\varepsilon}^{\beta}\langle\sigma^{\beta}_{\rm eff}\rangle = \delta\sigma^{\beta}_{\rm int}\langle\dot{\varepsilon}^{\beta}\rangle$ holds, and the stress fluctuation amplitude follows as $\langle(\delta\sigma^{\beta}_{\rm int})^2\rangle = \langle\sigma^{\beta}_{\rm eff}\rangle\langle\sigma^{\beta}_{\rm int}\rangle$ (cf. eq. (9)). When the dislocation mobility is high, the average effective stress $\langle\sigma^{\beta}_{\rm eff}\rangle$ is much smaller than the average internal stress $\langle\sigma^{\beta}_{\rm int}\rangle \approx -\sigma^{\beta}_{\rm ext}$. By inserting these relations into eq. (137) and approximating the second-order fluctuation term $\delta\sigma^{\beta}_{\rm int}\delta\dot{\varepsilon}^{\beta}$ by its average $\langle\delta\sigma^{\beta}_{\rm int}\delta\dot{\varepsilon}^{\beta}\rangle = -\langle\sigma^{\beta}_{\rm int}\rangle\langle\dot{\varepsilon}^{\beta}\rangle$, we obtain

$$K^{\beta,+}(\boldsymbol{r}) = \frac{\eta_0^{\beta}}{E_{\rm L}}\big[\sigma^{\beta}_{\rm ext}\langle\dot{\varepsilon}^{\beta}\rangle + 2\langle\dot{\varepsilon}^{\beta}\rangle\delta\sigma^{\beta}_{\rm int}\big] \approx \frac{\eta_0^{\beta}}{E_{\rm L}}\sigma^{\beta}_{\rm ext}\langle\dot{\varepsilon}^{\beta}\rangle. \tag{138}$$

In the last step, we have neglected the fluctuations since their relative magnitude with respect to the deterministic contribution is of the order of magnitude $\sqrt{\langle\sigma^{\beta}_{\rm eff}\rangle/\langle\sigma^{\beta}_{\rm ext}\rangle} \ll 1$. For materials with high dislocation mobilities this contribution may be safely neglected, and hence dislocation multiplication depends only on the average work $\sigma^{\beta}_{\rm ext}\langle\dot{\varepsilon}^{\beta}\rangle$ expended per unit volume. A similar argument has been given in [93] for the case when the strain rate depends exponentially on stress. Summing eq. (138) over all slip systems yields the total rate of dislocation multiplication,

$$K^{+} = \frac{\eta_0^{\beta}}{E_{\rm L}}\sigma^{\beta}_{\rm ext}\langle\dot{\varepsilon}\rangle, \tag{139}$$

where $\langle\dot{\varepsilon}\rangle = \sum_{\beta}\langle\dot{\varepsilon}^{\beta}\rangle$ and $\eta_0 = \sum_{\beta}\eta_0^{\beta}\langle\dot{\varepsilon}^{\beta}\rangle/\langle\dot{\varepsilon}\rangle$. For crystals deforming in tension or compression along [100], $\langle\dot{\varepsilon}\rangle = 8\langle\dot{\varepsilon}^{\beta}\rangle$ and $\eta_0 = \eta_0^{\beta}$.

Dislocation reactions and annihilation In multiple slip, the evolution of dislocation microstructure is accompanied by a continuous re-structuring of the dislocation arrangement: The motion of dislocations on several intersecting slip systems goes along with reactions between dislocation segments that lead to the formation of junction segments with new Burgers vectors and orientations such that the total dislocation length is reduced. As stored energy is released as heat, junction formation acts as a dissipation mechanism. The mean free path travelled by a dislocation before it undergoes a reaction is proportional to the meshlength of the dislocation network, and the total rate of dislocation density reduction by junction formation is thus

$$K_{\rm R}^{-} = \frac{\eta_{\rm R}}{b}\sqrt{\rho}\big[\langle\dot{\varepsilon}\rangle + \delta\dot{\varepsilon}\big]. \tag{140}$$

Here $\delta\dot{\varepsilon} = \sum_\beta \delta\dot{\varepsilon}^\beta$ is the fluctuating contribution to the total strain rate. Because of the spatio-temporal fluctuations of the strain rate, the dissipation by junction reactions proceeds in a heterogeneous manner in space and in time. Furthermore, it depends on the local dislocation density, being large in dislocation-dense and small in dislocation-depleted regions.

Another process of stored-energy dissipation is the annihilation of dislocations of the same slip system by cross-slip of near-screw and by climb of edge dislocation segments. In the following, we consider annihilation by cross-slip only. Annihilation by cross slip may take place when a near-screw segment (a segment that is within a critical angle $\Delta\theta$ around a screw orientation) is formed sufficiently close to a pre-existing near-screw segment of opposite sign such that their interaction is strong enough to induce cross slip before one of the segments becomes again activated, or disappears due to reactions with other segments. The rate at which screw segments are getting entangled is $2(\Delta\theta/(\pi b))\sqrt{\rho}\langle\dot{\varepsilon}\rangle$, the density of annihilation partners is $\Delta\theta\rho/(8\pi)$, and the probability to find an annihilation partner within a distance l_{cs} is approximately $2l_{\mathrm{c}}[\Delta\theta\rho/(8\pi)]^{1/2}$. The critical length l_{c} is estimated as follows: We equate the interaction stress of both segments to the stress σ_{cs} which is required to induce cross slip within the characteristic lifetime $t_{\mathrm{cs}} \sim [b\sqrt{\rho}\langle\dot{\varepsilon}\rangle]^{-1}$ of the configuration. This yields an annihilation distance $l_{\mathrm{cs}} = Gb/(4\pi\sigma_{\mathrm{cs}})$. Annihilation takes place when (a) two segments are closer than this distance and (b) the length of the annihilated segments is larger than the length of the edge segments which are left on the cross-slip plane as a result of the annihilation. The former length is on average $1/\sqrt{\rho}$, the latter l_{cs}. Since both conditions (a) and (b) must be fulfilled for annihilation by cross slip to take place, $l_{\mathrm{c}} = \min[l_{\mathrm{cs}}, 1/\sqrt{\rho}]$. Combining these relations, we obtain

$$K_{\mathrm{A}}^{-} = \eta_{\mathrm{A}}\rho\left[\mathrm{H}(\rho_{\mathrm{c}} - \rho) + \left(\frac{\rho_{\mathrm{c}}}{\rho}\right)^{1/2}\mathrm{H}(\rho - \rho_{\mathrm{c}})\right]\langle\varepsilon^\beta\rangle, \quad \text{where } \eta_{\mathrm{A}} = \left[\frac{\Delta\theta}{2\pi}\right]^{3/2}\frac{G}{\pi\sigma_{\mathrm{cs}}}, \tag{141}$$

H is the unit step function, $\mathrm{H}(x) = 1$ for $x > 0$ and $\mathrm{H}(x) = 0$ elsewhere, and $\rho_{\mathrm{c}} = 1/l_{\mathrm{cs}}^2$. We note that annihilation of screw segments on one slip system not only produces a 'drain' of the screw segment population but, since the populations of segments of different slip systems are coupled through reactions, also a net reduction of the populations of non-screw segments on other slip systems.[22] Therefore, we think that it is not necessary to consider a separate recovery mechanism for non-screw segments.

To formulate the dynamics of microstructure evolution and work hardening, we proceed in two steps: First, we use a stochastic formulation to calculate the probability distribution of dislocation densities which characterizes the inhomogeneous microstructure. Once the functional shape of this distribution is known, we may use it to calculate averages, i.e., *moments* of the distribution, the evolution of which gives us the work-hardening behaviour.

[22]For Cu single crystals deformed in ⟨100⟩ orientations, Göttler [154] reports an isotropic distribution of dislocations over the different line directions up to shear strains of about 0.4.

6.3.2. *Probability distributions of dislocation densities and cell sizes*

Combining the dislocation multiplication and reaction terms, we arrive at a stochastic differential equation for the total dislocation density,

$$\partial_t \rho = \frac{\eta_0}{E_\mathrm{L}} \sigma_\mathrm{ext}^\beta \langle \dot{\varepsilon} \rangle - \frac{\eta_\mathrm{R}}{b} \sqrt{\rho} \big[\langle \dot{\varepsilon} \rangle + \delta \dot{\varepsilon} \big] - \eta_\mathrm{A} \rho \langle \dot{\varepsilon} \rangle. \tag{142}$$

In the following it is convenient to introduce dimensionless variables defined by

$$\rho = \left(\frac{\eta_0 \sigma_\mathrm{ext}^\beta b}{E_\mathrm{L} \eta_\mathrm{R}} \right)^2 \tilde{\rho}, \qquad t = \frac{\eta_0 \sigma_\mathrm{ext}^\beta b^2}{E_\mathrm{L} \eta_\mathrm{R}^2 \langle \dot{\varepsilon} \rangle} \tilde{t}. \tag{143}$$

This leads to the non-dimensional stochastic differential equation

$$\partial_{\tilde{t}} \tilde{\rho} = 1 - \sqrt{\tilde{\rho}} - \tilde{\rho} T(\tilde{\rho}) + Q_\rho \sqrt{\tilde{\rho}} \dot{w}. \tag{144}$$

The function $T(\tilde{\rho})$ in eq. (144) is given by

$$T = T_1 + T_2 \left[\mathrm{H}(\tilde{\rho}_\mathrm{c} - \tilde{\rho}) - \mathrm{H}(\tilde{\rho} - \tilde{\rho}_\mathrm{c}) \sqrt{\frac{\tilde{\rho}_\mathrm{c}}{\tilde{\rho}}} \right],$$

$$T_1 = \frac{2 \eta_0 b^2}{E_\mathrm{L} \eta_\mathrm{R}^2} \frac{\partial \sigma_\mathrm{ext}^\beta}{\partial \langle \varepsilon \rangle}, \qquad T_2 = \frac{\eta_0 G b^2}{\pi E_\mathrm{L} \eta_\mathrm{R}^2} \left(\frac{\Delta \theta}{2\pi} \right)^{3/2} \frac{\sigma_\mathrm{ext}^\beta}{\sigma_\mathrm{cs}}. \tag{145}$$

The first term stems from the fact that the external stress is time dependent and therefore the scaling constant of the dislocation density increases, which formally introduces an additional loss term into the evolution of $\tilde{\rho}$. The second term characterizes dislocation annihilation by cross slip, cf. eq. (141).

The fluctuation term in eq. (144) comes from the addition of the strain-rate fluctuations on the eight active slip systems. Under the assumption that all ‘slip events’ are statistically independent, the non-dimensional fluctuation strength is

$$Q_\rho^2 = \frac{\eta_\mathrm{R}^2 E_\mathrm{L}}{\eta_0 \sigma_\mathrm{ext}^\beta b^2} \varepsilon_\mathrm{corr}^\beta. \tag{146}$$

Statistical information on the dislocation structure is obtained by calculating the distribution function $p(\tilde{\rho})$ of the total dislocation density. This is a solution of the Fokker–Planck equation

$$\partial_{\tilde{t}} p(\tilde{\rho}, \tilde{t}) = -\partial_{\tilde{\rho}} \left[1 - \sqrt{\tilde{\rho}} - \tilde{\rho} T(\tilde{\rho}) + \frac{Q_\rho^2}{4} \right] p(\tilde{\rho}, \tilde{t}) + \partial_{\tilde{\rho}}^2 \left[\frac{Q_\rho^2}{2} \tilde{\rho} p(\tilde{\rho}, \tilde{t}) \right]. \tag{147}$$

The steady-state solution $p(\tilde{\rho})$ of this eq. (137)[23] is given by

$$p(\tilde{\rho}) = \mathcal{N}\tilde{\rho}^{-[1-4/Q_\rho^2]/2}\exp\left[-\frac{2}{Q_\rho^2}h(\tilde{\rho})\right], \quad h(\tilde{\rho}) = 2\sqrt{\tilde{\rho}} + \int T(\tilde{\rho})\,\mathrm{d}\tilde{\rho}. \tag{148}$$

In eq. (148), $\mathcal{N}$ is a normalization constant which must be chosen such that $\int p(\tilde{\rho})\,\mathrm{d}\tilde{\rho} = 1$. The distribution $p(\tilde{\rho})$, which statistically characterizes the inhomogeneous dislocation arrangement, depends on three parameters Q_ρ^2, T_1 and T_2. The qualitative shape of the distribution, however, is governed exclusively by the fluctuation strength Q_ρ^2 while the parameters $T_{1/2}$ are only relevant for the exponentially decaying 'tail' of the distribution at large dislocation densities. From fig. 30, which shows the probability distributions for various noise intensities,[24] two limiting cases are distinguished: (i) In the deterministic limit, $Q_\rho^2 \to 0$, $p(\tilde{\rho}) \approx \delta(\tilde{\rho} - \langle\tilde{\rho}\rangle)$ gets sharply localized around its mean value. This situation corresponds to a more or less homogeneous dislocation network in which the only characteristic length is the mean dislocation spacing, (ii) In the strong-noise limit, $Q_\rho^2 \to \infty$, the distribution of dislocation densities exhibits a power-law decay with a stretched exponential cutoff at high densities. This strongly non-Gaussian behaviour is a consequence of the nonlinearity introduced by the dislocation reactions and the multiplicative character of the stochastic term in eq. (144). Between these extremes, there is a critical fluctuation strength $Q_c^2 = 4$ where peaked dislocation density distributions give way to monotonically decreasing spectra which, for $Q_\rho^2 > Q_c^2$, diverge as $\tilde{\rho}^{-[1-4/Q_\rho^2]/2}$. In the language of nonlinear phenomena, we may speak of a noise-induced phase transition [168], since the qualitative change in shape of the probability distribution of dislocation densities indicates a transition between qualitatively different dislocation patterns and the control parameter governing this transition is the effective fluctuation intensity.

To get a clearer understanding of the nature of this transition, we use the phenomenological relation $\lambda = C_\lambda/\sqrt{\rho}$ where λ is the size of a dislocation-depleted region on a glide plane, and C_λ is a constant of the order of 5–15. By this change of variables we obtain from the dislocation density distribution a distribution of cell sizes,

$$p_s(\lambda) \sim \lambda^{-(D+1)}h(\lambda). \tag{149}$$

The function $h(\lambda)$, which is obtained by substituting $\rho = [C_\lambda/\lambda]^2$ into $h(\rho)$, eq. (148), truncates the power-law distribution (149) at small cell sizes. The distribution (149) indicates that, on larger scales, the cells form a fractal pattern since the exponent $D = 1 + 4/Q_\rho^2$ can be interpreted as a fractal dimension (gap dimension) of the dislocation network (cf. section 4). For noise levels above the critical value $Q_\rho^2 = 4$, fractal dimensions range between the physically meaningful limits $2 > D > 1$. When $Q_\rho^2 > 4$, the dislocation network ceases to fill the specimen homogeneously and a spongy pattern of gaps with

[23]The scaling (143) reflects Taylor's relation $\sigma_{\text{ext}}^\beta \propto \sqrt{\rho}$. Since the stress $\sigma_{\text{ext}}^\beta$ increases in the course of deformation, a constant $\tilde{\rho}$ corresponds to a dislocation density that increases in proportion with the square of the stress.

[24]For the parameters given below in section 6.3.3, the influence of T_1 and T_2 on the distribution $p(\rho)$ is small; they have therefore been put to zero in fig. 30.

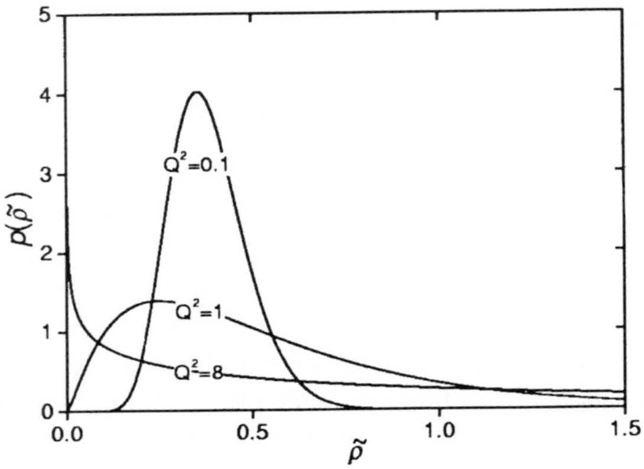

Fig. 30. Scaled distributions of dislocation densities for different noise amplitudes Q^2; the scaled hardening coefficient T has been taken equal to zero (see text).

a hyperbolic size distribution emerges. $Q_\rho^2 > 4$ can therefore be understood as a critical condition for the formation of fractal cell structures.

Fractal dislocation patterning constitutes a self-organization process which according to this condition is directly connected to energy dissipation by dislocation reactions. The control parameter which governs patterning is the 'noise level' Q_ρ^2, which is dependent on the ratio between the energy storage and dissipation parameters η_0 and η_R; large dissipation promotes fractal patterning.

To calculate the evolution of the fractal dimension D, we have to estimate the energy storage and dissipation parameters: The net stored energy fraction η_{st} is of the order of 15% at the onset of deformation [40], and accordingly about 85% of the expended work are ultimately dissipated. In symmetrical multiple slip, a fraction η_L of about 15% of the work is required to move dislocations through fluctuations of the long-range stress field;[25] this work is dissipated when dislocations move 'downhill' in the internal-stress landscape. The remaining work fraction is dissipated in dislocation reactions, i.e., we have $\eta_0 = 1 - \eta_L = 0.85$ and $\eta_R = \eta_0 - \eta_{st} = 0.7$. Using these parameters, and estimating the characteristic strain ε_{corr} in eq. (146) using eq. (110), we calculate the evolution of the fractal dimension D in the course of deformation. This yields

$$D = 1 + \frac{4\sqrt{2}\eta_0 b^2}{E_L f^b (1-\nu)\eta_R^2} \frac{\xi_n^\beta}{\xi_g^\beta}. \tag{150}$$

According to section 3, the dislocation line energy depends on the range of the dislocation–dislocation correlations, which is on the order of magnitude of one dislocation spacing. We use the average value of the dislocation line energy derived by Schoeck and Frydman [69], $E_L = 0.119Gb^2 \ln[1/(b\sqrt{\rho})]$, and approximate the expression in the logarithm by its average. This average is connected to the flow stress by the Taylor relation (cf. below), $1/(b\langle\sqrt{\rho}\rangle) = \alpha^\beta G/\sigma_{ext}^\beta$, where $\alpha^\beta \approx 0.38$ [169].

[25] This fraction corresponds to the contribution of the long-range stresses to the flow stress, cf. section 5.3.

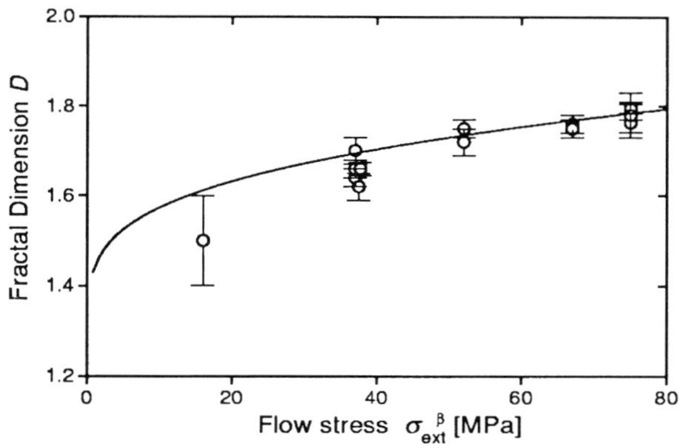

Fig. 31. Evolution of the fractal dimension in the course of deformation. Full line: curve calculated from eq. (150), for parameters see text. Data points: fractal dimensions determined for cell structures of Cu single crystals deformed in tension along ⟨100⟩ (see fig. 21).

The ratio ξ_n/ξ_g corresponds to the ratio of the dislocation spacing to the slip-line length, which for f.c.c. crystals deformed in tension in a ⟨100⟩ orientation was reported to be about 50 [69]. With these parameters and $f^b = 0.5$, $\nu = 0.3$, the evolution of fractal dimension as a function of flow stress has been calculated (fig. 31). One finds that the observed increase in fractal dimension with increasing flow stress is well reproduced both qualitatively and quantitatively.[26] This increase according to eq. (70) corresponds to an increase in the cell-wall volume fraction, which has been observed independently both by TEM [142] and by X-ray methods [73,143] and found to be in quantitative agreement with the evolution of D shown in fig. 31 [144].

6.3.3. *Work hardening*

In f.c.c. crystals uniaxially deformed in a ⟨100⟩ direction or, more generally, under deformation conditions where several slip systems are active, the flow stress is governed mainly by the stress required for cutting forest dislocations. In the terminology of section 2, this is a process which takes place on the *microscopic* scale. Hence, on the *mesoscopic* scale of dislocation density fluctuations the local flow stress is a function of the local dislocation density. Under the assumption that segments are distributed isotropically over all orientations, the local flow stress is given by $\sigma_f^{\beta}(\rho) = \alpha^{\beta} Gb/\sqrt{\rho}$, where according to Schoeck and Frydman $\alpha^{\beta} \approx 0.38$ [169]. Under these conditions, a generalized composite model (section 3.3) can be used. It follows that the total flow stress is simply the average of the local flow stresses,

$$\sigma_{ext}^{\beta} = \alpha^{\beta} Gb\langle\sqrt{\rho}\rangle. \tag{151}$$

[26] We note, however, that quantitative agreement is easy to achieve since the second term on the right-hand side of eq. (150) contains parameters which are known only approximately; this term may therefore readily be 'tuned' by a factor of about 2.

To determine the rate of work hardening, it is therefore necessary to calculate the evolution of the moment $\langle\sqrt{\rho}\rangle$ of the dislocation density distribution. To this end, we consider the evolution of $\sqrt{\rho}$ as a function of the average strain $\langle\varepsilon\rangle$. From eq. (142) we obtain by substituting $\mathrm{d}\langle\varepsilon\rangle = \langle\dot{\varepsilon}\rangle\,\mathrm{d}t$

$$\frac{\partial\sqrt{\rho}}{\partial\langle\varepsilon\rangle} = \frac{\eta_0\sigma^{\beta}_{\mathrm{ext}}}{2E_{\mathrm{L}}\sqrt{\rho}} - \frac{\eta_{\mathrm{R}}}{2b}\left[1 + \frac{\delta\dot{\varepsilon}}{\langle\dot{\varepsilon}\rangle}\right] - \frac{\eta_{\mathrm{A}}}{2}\left[\sqrt{\rho}\mathrm{H}(\rho-\rho_{\mathrm{c}}) + \sqrt{\rho_{\mathrm{c}}}\mathrm{H}(\rho_{\mathrm{c}}-\rho)\right]. \tag{152}$$

An equation for the flow stress is obtained by using eq. (151) and averaging over the distribution of ρ. Upon averaging, the additive fluctuation term vanishes, and we obtain the work-hardening slope $\Theta := \partial\sigma^{\beta}_{\mathrm{ext}}/\partial\langle\varepsilon\rangle$:

$$\frac{\Theta}{G} = \frac{\alpha^{\beta}}{2}\Bigg[\frac{\alpha^{\beta}Gb^2\eta_0}{E_{\mathrm{L}}}\langle\rho^{1/2}\rangle\langle\rho^{-1/2}\rangle - \eta_{\mathrm{R}} - \eta_{\mathrm{A}}\left[\frac{\sigma^{\beta}_{\mathrm{ext}}}{G}P(\rho<\rho_{\mathrm{c}}) + \frac{4\pi\sigma_{\mathrm{cs}}}{G}\left[1-P(\rho<\rho_{\mathrm{c}})\right]\right]\Bigg]. \tag{153}$$

The main difference with respect to previous work-hardening models, as for instance proposed by Kocks [80], is that the properties of the inhomogeneous microstructure enter eq. (153) in the form of averages over the dislocation density distribution: The moments $\langle\rho^{1/2}\rangle$ and $\langle\rho^{-1/2}\rangle$ are calculated from eq. (148) via $\langle\rho^{\pm1/2}\rangle = \int p(\rho)\rho^{\pm1/2}\,\mathrm{d}\rho$. The probability $P(\rho<\rho_{\mathrm{c}})$ is given by $P(\rho<\rho_{\mathrm{c}}) = \int_0^{\rho_{\mathrm{c}}} p(\rho)\,\mathrm{d}\rho$.

Stages III and IV: Voce-type hardening and its exhaustion The full lines in fig. 32 show the work-hardening slope Θ calculated from eq. (153) as a function of the resolved shear stress $\sigma^{\beta}_{\mathrm{ext}}$. The parameters used in the calculation were $\alpha^{\beta} = 0.38$ and $E_{\mathrm{L}} = 0.82Gb^2$ as given by Schoeck and Frydman [169], $\eta_0 = 0.85$ and $\eta_{\mathrm{R}} = 0.7$ (cf. the discussion preceding eq. (150)), an average value $D = 1.7$, and $\Delta\theta/(2\pi) = 0.15$. The logarithmic dislocation-density dependence of α^{β}, E_{L}, and D, which arises from the dislocation-density dependence of the dislocation line energy has been neglected. The critical stress σ_{cs} controlling cross-slip may be identified with the stress at the onset of Stage III in crystals deformed in single-slip orientations.

In Fig. 32, the work-hardening slope decreases first linearly with stress. In this Stage-III hardening regime (note that hardening stages I and II are absent in symmetrical multiple slip [142]), the hardening behaviour can be described by the phenomenological Voce equation, eq. (14). At higher stresses, however, a constant hardening slope is approached. This Stage-IV-type behaviour is, in the present simple model, not related to any new mechanism but simply to the 'exhaustion' of dislocation annihilation by cross-slip when the dislocation density is large. This exhaustion is due to the fact that, at high densities, only short segments can bow out and annihilate, and therefore the annihilation distance decreases. Since exclusively recovery by cross-slip is considered, all stresses scale like the critical stress σ_{cs} for cross slip, which we identify with the stress at which Stage III sets in in single-slip orientations. With $\sigma_{\mathrm{cs}} \approx 25$ MPa, which corresponds to the onset of Stage III in Cu single crystals deformed in single slip at room temperature [170], we find

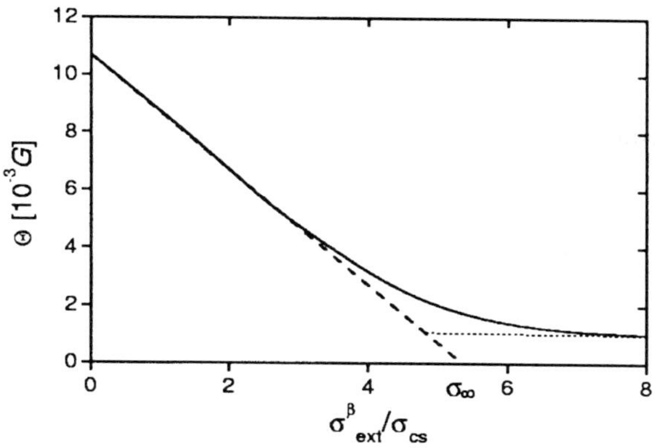

Fig. 32. Work-hardening slope vs. stress as calculated from eq. (153) using the $D(\sigma_{\rm ext}^{\beta})$ curve given in fig. 31. Dashed line: linear decrease of work-hardening slope when exhaustion of dislocation annihilation at dislocation densities $\rho > 1/l_{\rm cs}^2$ is neglected, dotted line: constant asymptotic hardening slope; for parameters see text.

good agreement with the work-hardening curves reported in [142] for Cu deformed in tension in a $\langle 100\rangle$ orientation at room temperature [142], where an initial hardening slope of $1.1 \times 10^{-2}G$ and an apparent saturation stress σ_∞ of about 100 MPa were observed. The transition between the initial Voce-type behaviour and the constant 'Stage IV' hardening depends on the properties of the dislocation density distribution. For a homogeneous microstructure where $p(\rho) = \delta(\rho - \langle\rho\rangle)$ and $P(\rho < \rho_{\rm c}) = {\rm H}(\rho_{\rm c} - \rho)$, an abrupt transition from Voce behaviour to a constant 'Stage IV hardening' is predicted (dotted line in fig. 32). For an inhomogeneous microstructure, the transition is smeared out (full line in fig. 32). While, according to the present model, hardening stages III and IV are governed by the same hardening mechanism, a new and qualitatively different mechanism must be invoked to account for the final decrease of the work-hardening slope to zero in hardening Stage V. This is in full agreement with the conclusions drawn from stored-energy measurement on polycrystalline Cu [11] and self-organization arguments [171].

6.4. Discussion and conclusions

In materials in which the dislocation motion is governed by dislocation–dislocation interactions, dislocation patterning and the accumulation of misorientations can be related to a common underlying mechanism, *viz.* the spontaneous emergence of large spatio-temporal fluctuations of the dislocation fluxes which can be characterized as a series of 'dislocation avalanches'. The formation of spatially inhomogeneous dislocation patterns results from localized and random dislocation reactions which continuously produce 'holes' in the dislocation network. Hence, the dislocation network is in a highly dynamic, dissipative state in which dislocation-depleted cells continuously form and disappear. Dissipation through dislocation reactions is the driving force for this process, whereas the storage of mechanical energy in form of dislocation lines tends to homogenize the dislocation arrangement. The interplay of these two mechanisms, together with the coupling of the fluctuating dislocation fluxes to the local dislocation densities by reaction terms, can lead to complex, multiscale dislocation arrangements which follow statistics that are essentially non-Gaussian. They are characterized by power-law distributions, which are typical of fractal patterns. In f.c.c. single crystals deformed in ⟨100⟩ orientations, fractal dislocation patterning persists throughout hardening stage III (and presumably into hardening stage IV) as shown by various investigations using both image analysis of TEM micrographs [127,128,133,137] and statistical analysis of X-ray line profiles [143]. The evolution of lattice rotations and misorientations (and of fluctuating mesoscopic stresses) is associated with the *spatial* heterogeneity of the randomly fluctuating dislocation fluxes which leads to a random accumulation of excess dislocations. The accumulation of excess dislocations is governed by inhomogeneities in the dislocation fluxes rather than by dislocation reactions, and therefore the statistics is Gaussian on large time scales.

In the present work, the statistics of plastic flow are described within a 'Gaussian' framework. This approximation to the real dynamics of collective dislocation motions is feasible on the time scale of dislocation microstructure evolution, which is the cumulative result of many elementary slip events. On shorter time scales, however, the statistics of slip is, as far as the limited experimental data indicate, essentially non-Gaussian as dislocation motion proceeds in discrete 'slip avalanches' with size distributions that exhibit a power-law decay. The theoretical understanding of the complex dynamics of interacting dislocations which creates these avalanches is still in its infancy [156]. We hope that the present work may stimulate future investigations into this direction, which may pave the way for a deeper understanding of the dynamics of plastic flow and the associated non-equilibrium phenomena.

References

[1] O. Mügge, Göttinger Nachrichten 1899, p. 56.
[2] W. Friedrich, P. Knipping and M. Laue, Münchner Sitzungsberichte 1912, p. 303.
[3] W. Friedrich, P. Knipping and M. Laue, Ann. Physik 41 (1913) 971.
[4] M. Laue, Münchner Sitzungsberichte (1912) p. 383.
[5] M. Laue, Ann. Physik 41 (1913) 989.
[6] H. Hort, Z. Verein Deutsch. Ing. 50 (1907) 1831.
[7] G.I. Taylor, Trans. Faraday Soc. 24 (1928) 121.
[8] W.S. Farren and G.I. Taylor, Proc. Roy. Soc. London A 107 (1925) 85.
[9] S. Satō, Science Reports Tôhoku Imperial University 20 (1931) 140.
[10] W. Rosenhain and V.H. Stott, Proc. Roy. Soc. London A 140 (1933) 9.
[11] G.I. Taylor and H. Quinney, Proc. Roy. Soc. London A 143 (1934) 307.
[12] H. Quinney and G.I. Taylor, Proc. Roy. Soc. London A 163 (1937) 157.
[13] L.M. Clarebrough, M.E. Hargreaves, D. Mitchell and G.W. West, Proc. Roy. Soc. London A 215 (1952) 507.
[14] A.L. Titchener and M.B. Bever, Progr. Metal Phys. 7 (1958) 247.
[15] R. Gross, Z. Metallkde. 16 (1924) 344.
[16] E. Schiebold, Z. Metallkde. 16 (1924) 462.
[17] K. Yamaguchi, Sci. Papers Inst. Phys. Chem. Research, Tokyo, 11 (1929) 151.
[18] E. Schmid, in: Proc. Internat. Congr. Appl. Mech. (Delft, 1924) p. 342.
[19] W. Meissner, M. Polanyi and E. Schmid, Z. Physik 66 (1930) 477.
[20] A. Seeger, in: Handbuch der Physik VII/2 (Springer, Berlin, 1958) p. 1.

[21] L. Prandtl, Z. Angew. Math. Mech. 8 (1928) 85.
[22] U. Dehlinger, Metallwirtschaft 28 (1928) 1172.
[23] E. Orowan, Z. Physik 89 (1934) 605.
[24] M. Polanyi, Z. Physik 89 (1934) 660.
[25] G.I. Taylor, Proc. Roy. Soc. London A 145 (1934) 363.
[26] V. Vitek, Cryst. Lattice Defects 5 (1974) 1.
[27] S. Ismail-Beigi and T.A. Arias, Phys. Rev. Lett. 84 (2000) 1499.
[28] M.S. Duesbery, Modelling Simul. Mater. Sci. Eng. 6 (1998) 35.
[29] U. Essmann and H. Mughrabi, Phil. Mag. A 40 (1979) 731.
[30] D. Rodney and R. Phillips, Phys. Rev. Lett. 82 (1999) 1704.
[31] A.T. Paxton and Y.Q. Sun, Phil. Mag. A 78 (1998) 85.
[32] H. Mughrabi, in: Constitutive Equations in Plasticity, ed. A.S. Argon (MIT Press, Cambridge, MA, 1975) p. 199.
[33] A. Seeger, Z. Metallkde. 72 (1981) 369.
[34] F. Ackermann, H. Mughrabi and A. Seeger, Acta Metall. 31 (1983) 1353.
[35] H.-R. Trebin, in: Introduction to Quasicrystals, eds J.B. Suck et al. (Springer, Heidelberg, 1999).
[36] E. Heyn and O. Bauer, Int. Zeitschr. Metallogr. 1 (1911) 16.
[37] G. Masing, Z. Techn. Physik 6 (1925) 569.
[38] A. Seeger, J. Mech. Behavior Mater. 10 (1999) 31.
[39] F.C. Frank and W.T. Read, Phys. Rev. 79 (1950) 722.
[40] L. Kaps, Dissertation TU Braunschweig 1996.
[41] I. Prigogine, From Being to Becoming (Freeman, San Francisco, 1980).
[42] G. Nicolis and I. Prigogine, Self-Organisation in Non-Equilibrium Systems (Wiley, New York, 1987).
[43] H. Neuhäuser, in: Dislocations in Solids, Vol. 6, ed. F.R.N. Nabarro (North-Holland, Amsterdam, 1984) p. 319.
[44] D. Kuhlmann-Wilsdorff, Mat. Res. Innovat. 1 (1998) 265.
[45] P.W. Bridgman, Rev. Mod. Phys. 22 (1950) 56.
[46] A. Seeger and W. Frank, Solid State Phenomena 3–4 (1988) 125.
[47] V. Frette, K. Christensen, A.M. Malte-Sørenssen, J. Feder, T. Jøssang and P. Meakin, Nature 397 (1996) 49.
[48] P. Bak, C. Tang and K. Wiesenfeld, Phys. Rev. Lett. 59 (1987) 381.
[49] R. Thomson and L.E. Levine, Phys. Rev. Lett. 81 (1998) 3884.
[50] M. Zaiser, Mater. Sci. Engng. A 309–310 (2001) 304.
[51] J. Weiss and J.R. Grasso, J. Phys. Chem. B 101 (1997) 6113.
[52] J. Weiss, F. Lahaie and J.R. Grasso, J. Geophys. Res. 105 (2000) 433.
[53] H. Leschhorn, T. Nattermann, S. Stepanov and L. Tang, Ann. Physik (Leipzig) 6 (1997) 1.
[54] M. Kardar, Phys. Rep. 301 (1998) 85.
[55] Y. Estrin and L.P. Kubin, Res. Mech. 23 (1988) 197.
[56] M. Zaiser and P. Hähner, Mater. Sci. Engng. A 238 (1997) 399.
[57] P. Hähner, in: Proc. 8th Int. Symp. on Continuum Models and Discrete Systems, ed. K.Z. Markov (World Scientific, Singapore, 1996) p. 514.
[58] P. Hähner, Appl. Phys. A 62 (1996) 473.
[59] V.I. Alshits and V.L. Indenbom, in: Dislocations in Solids, Vol. 7, ed. F.R.N. Nabarro (North-Holland, Amsterdam, 1986) p. 43.
[60] U. Essmann, Phys. Stat. Sol. 17 (1966) 725.
[61] U. Frisch, Turbulence: the Legacy of A.N. Kolmogorov (Cambridge Univ. Press, Cambridge, 1995) p. 43.
[62] S.J. Basinski and Z.S. Basinski, in: Dislocations in Solids, Vol. 4, ed. F.R.N. Nabarro (North-Holland, Amsterdam, 1979) p. 261.
[63] L. Kubin and B. Devincre, in: Proc. 20th Risø Int. Symp. on Materials Science: Deformation-Induced Microstructures – Analysis and Relation to Properties, eds J.B. Bilde-Sørensen et al. (Risø National Laboratory, Roskilde, 1999) p. 61.
[64] B. Sprusil and F. Hnilica, Czech. J. Phys. B 35 (1985) 897.
[65] T. Kleiser and M. Boček, Z. Metallkde. 77 (1986) 582.

[66] H. Neuhäuser, Res. Mech. 23 (1988) 113.
[67] S. Mader, Z. Physik 149 (1957) 73.
[68] P.B. Hirsch and T.E. Mitchell, Canad. J. Phys. 45 (1967) 663.
[69] P. Ambrosi and Ch. Schwink, Scripta Metall. 12 (1978) 303.
[70] P.B. Hirsch, in: The Physics of Metals, Vol. 2, ed. P.B. Hirsch (Cambridge University Press, Cambridge, 1975) p. 232.
[71] O.B. Pedersen, L.M. Brown and W.M. Stobbs, Acta Metall. 29 (1980) 1843.
[72] H. Mughrabi, Acta Metall. 31 (1983) 1367.
[73] H. Mughrabi, T. Ungár, W. Kienle and M. Wilkens, Phil. Mag. A 53 (1986) 793.
[74] H. Mughrabi, Mater. Sci. Engng. 85 (1987) 15.
[75] H. Mughrabi, Rev. Phys. Appl. 23 (1988) 367.
[76] A. Seeger, in: Dislocations and Mechanical Properties of Crystals (Wiley, New York, 1957) p. 232.
[77] A. Seeger, J. Diehl, S. Mader and H. Rebstock, Phil. Mag. 2 (1957) 323.
[78] A. Seeger and M. Wilkens, in: Realstruktur und Eigenschaften von Reinstoffen (Akademie-Verlag, Berlin, 1967) p. 29.
[79] A. Seeger, in: Work Hardening, eds J.P. Hirth and J. Weertman (Gordon and Breach, New York, 1968) p. 27.
[80] U.F. Kocks, J. Engng. Mater. Tech. 98 (1976) 76.
[81] E. Voce, J. Inst. Metals 74 (1948) 537.
[82] L.P. Kubin and Y. Estrin, Rev. Phys. Appl. 23 (1988) 573.
[83] Y. Estrin, L.S. Tóth, Y. Molinari and Y. Bréchet, Acta Mater. 46 (1998) 5509.
[84] E.C. Aifantis, Mater. Sci. Engng. 81 (1986) 563.
[85] E.C. Aifantis, Int. J. Plasticity 3 (1987) 211.
[86] D. Walgraef and E.C. Aifantis, Int. J. Engng. Sci. 23 (1985) 1351, 1359, 1365.
[87] D. Walgraef and E.C. Aifantis, J. Appl. Phys. 58 (1985) 688.
[88] J. Kratochvil, Scripta Metall. 20 (1986) 1625.
[89] J. Kratochvil, Scripta Metall. Mater. 26 (1993) 113.
[90] D.L. Holt, J. Appl. Phys. 41 (1970) 3197.
[91] N. Hansen and D. Kuhlmann-Wilsdorf, Mater. Sci. Engng. 81 (1986) 141.
[92] S.V. Raj and G.M. Pharr, Mater. Sci. Engng. 81 (1986) 217.
[93] P. Hähner, Acta Mater. 44 (1996) 2345.
[94] P. Hähner, K. Bay and M. Zaiser, Phys. Rev. Lett. 81 (1998) 2470.
[95] E. Kröner, in: Inelastic Behavior of Solids, eds M.F. Kanninen et al. (McGraw-Hill, New York, 1969) p. 137.
[96] I. Groma, Phys. Rev. B 57 (1998) 7535.
[97] I. Groma and B. Bako, Phys. Rev. B 58 (1998) 2969.
[98] A. El-Azab, Phys. Rev. B 61 (2000) 11956.
[99] L.P. Kubin, G. Canova, M. Condat, B. Devincre, V. Pontikis and Y. Bréchet, in: Non-Linear Phenomena in Materials Science II, eds G. Martin and L.P. Kubin (Trans Tech Publications, Aedermannsdorf, 1992) p. 137.
[100] I. Groma, M. Verdier, P. Balogh and B. Bako, Modelling Mater. Sci. Eng. 7 (1999) 795.
[101] E. Kotomin and V. Kuzovkov, Modern Aspects of Diffusion-Controlled Reactions (Elsevier, Amsterdam, 1996).
[102] C. Miguel, A. Vespignani, S. Zapperi, J. Weiss and J.-R. Grasso, Mater. Sci. Engng. A 309–310 (2001) 324; data by courtesy of C. Miguel.
[103] E. Kröner, Kontinuumstheorie der Versetzungen und Eigenspannungen (Springer, Berlin, 1958).
[104] E. Kröner, in: Physics of Defects, eds R. Balian, M. Kleman and J.-P. Poirier (North-Holland, Amsterdam, 1981) p. 219.
[105] M. Wilkens, Acta Metall. 15 (1967) 1412, 1415; 17 (1969) 1155.
[106] M.A. Krivoglaz, O.V. Martynenko and K.P. Ryaboshapka, Phys. Metals. Metallog. 55 (1983) 1.
[107] M.A. Krivoglaz, X-Ray and Neutron Diffraction in Nonideal Crystals (Springer, Berlin, 1996).
[108] M. Wilkens, in: Fundamental Aspects of Dislocation Theory II, eds J.A. Simmons, R. de Wit and R. Bullough (US Nat. Bur. Stand. Spec. Publ. 317, Washington, 1970) p. 1195.

[109] M. Wilkens, Phys. Stat. Sol. a 2 (1970) 359.
[110] I. Groma, T. Ungar and M. Wilkens, J. Appl. Cryst. 21 (1988) 47.
[111] B. Bako and I. Groma, Phys. Rev. B 60 (1999) 122.
[112] E. Kröner, Acta Metall. 9 (1961) 155.
[113] J.D. Eshelby, Proc. Roy. Soc. London A 241 (1957) 376.
[114] M. Zaiser, Mater. Sci. Engng. A 249 (1998) 145.
[115] M. Berveiller, D. Muller and J. Kratochvil, Int. J. Plasticity 9 (1993) 633.
[116] A. Seeger, J. Physique III 27 (1966) 68.
[117] H. Kronmüller, in: Proc. 5th Risø Int. Symp. on Metallurgy and Materials Science: Microstructural Characterization of Materials by Non-Microscopical Techniques (Risø National Laboratory, Roskilde, 1984) p. 79.
[118] W. Bollmann, Phys. Rev. 103 (1956) 1588.
[119] P.B. Hirsch, R.W. Horne and M.J. Whelan, Phil. Mag. 1 (1956) 677.
[120] G. Schoeck and A. Seeger, in: Rept. of a Conf. on Defects in Solids (Physical Society, London, 1955) p. 340.
[121] H. Wolf, Z. Naturforschg. 15a (1960) 180.
[122] S. Mader, in: Electron Microscopy and Strength of Crystals, eds G. Thomas and J. Washburn (Interscience, New York, 1963) p. 143.
[123] U. Essmann, Phys. Stat. Sol. 3 (1963) 932.
[124] H. Mughrabi, Phil. Mag. 8 (1968) 1211.
[125] A. Seeger, J. Electr. Microscopy 48 (1999) 301.
[126] H. Mughrabi, Phil. Mag. 23 (1971) 896, 897, 931.
[127] M. Zaiser, K. Bay and P. Hähner, Acta Mater. 47 (1999) 2463.
[128] M. Zaiser and P. Hähner, in: Proc. 20th Risø Int. Symp. on Materials Science: Deformation-Induced Microstructures – Analysis and Relation to Properties, eds J.B. Bilde-Sørensen et al. (Risø National Laboratory, Roskilde, 1999) p. 547.
[129] D.A. Hughes, Q. Liu, D.C. Chrzan and N. Hansen, Acta Mater. 45 (1996) 105.
[130] D.A. Hughes, D.C. Chrzan, Q. Liu and N. Hansen, Phys. Rev. Lett. 81 (1998) 4664.
[131] H. Mughrabi and F. Ackermann, unpublished data, courtesy of H. Mughrabi (University of Erlangen).
[132] U. Holzwarth and U. Essmann, Appl. Phys. A 57 (1993) 131.
[133] J. Gil Sevillano, Physica Scripta T49 (1993) 405.
[134] M. Zaiser, S. Marras, E. Nadgorny, H.P. Strunk and E.C. Aifantis, Phys. Stat. Sol. A 185 (2001) R4.
[135] U. Holzwarth and U. Essmann, Appl. Phys. A 58 (1994) 197.
[136] Courtesy of A. Schwab, Dresden.
[137] J. Gil Sevillano, E. Bouchaud and L.P. Kubin, Scripta Metall. Mater. 25 (1991) 355.
[138] A.I. Olemskoi, Phys. Rev. 18 (1995) 1.
[139] B. Mandelbrot, The Fractal Geometry of Nature (Freeman, New York, 1983).
[140] B. Mandelbrot, Physica Scripta 32 (1985) 257.
[141] B. Mandelbrot, in: Fractals in Physics, eds L. Pietronero and E. Tosatti (North-Holland, Amsterdam, 1986) p. 3.
[142] R. Neuhaus and Ch. Schwink, Phil. Mag. A 65 (1992) 1463.
[143] F. Szekely, I. Groma and J. Lendvai, Phys. Rev. B 62 (2000) 3093.
[144] M. Zaiser and P. Hähner, Mater. Sci. Engng. A 270 (1999) 299.
[145] P. Hähner and M. Zaiser, Mater. Sci. Engng. A 272 (1999) 443.
[146] H. Mughrabi, in: Proc. 4th Int. Conf. on the Strength of Metals and Alloys (Aachen, 1979) p. 1615.
[147] H. Mughrabi, J. Microsc. Spectrosc. Electron. 1 (1976) 571.
[148] A. Borbély, W. Blum and T. Ungár, Mater. Sci. Engng. A 276 (2000) 186.
[149] M.A Krivoglaz and K.P. Ryaboshapka, Phys. Metals Metallogr. 15 (1963) 14.
[150] M. Wilkens, Phys. Stat. Sol. 3 (1963) 1718.
[151] L.E. Levine and R. Thomson, Acta Cryst. A 53 (1997) 590.
[152] D. Ottenhaus, Dr.rer.nat. thesis Universität Stuttgart 1987.
[153] F. Szekely, I. Groma and J. Lendvai, Scripta Mater. 45 (2001) 55.
[154] E. Göttler, Phil. Mag. 28 (1973) 1057.

[155] D. Rouby, P. Fleischmann and C. Duvergnier, Phil. Mag. A 47 (1983) 671.
[156] M.-C. Miguel, A. Vespignani, S. Zapperi, J. Weiss and J.-R. Grasso, Nature 410 (2001) 667.
[157] J.P. Sethna, K.A. Dahmen and C.R. Myers, Nature 410 (2001) 242.
[158] B.V. Gnedenko and A.N. Kolmogorov, Limit Distributions for Sums of Independent Random Variables (Addison-Wesley, Cambridge, MA, 1954).
[159] H. Haken, Synergetics (Springer, Heidelberg, 1983).
[160] H. Kronmüller, in: Moderne Probleme der Metallphysik, ed. A. Seeger (Springer, Berlin, 1965) p. 126.
[161] M. Wilkens and K. Eckert, Z. Naturforschg. 19a (1964) 459.
[162] S. Mader and A. Seeger, Acta Metall. 8 (1960) 513.
[163] A.N. Stroh, Proc. Roy. Soc. London B 66 (1953) 2.
[164] N.F. Mott, Phil. Mag. 44 (1953) 741.
[165] F.R.N. Nabarro, Scripta Metall. Mater. 32 (1994) 1085.
[166] W. Pantleon, Acta Mater. 46 (1998) 451.
[167] J.-P. Bouchaud and M. Potters, Theory of Financial Risks (Cambridge University Press, Cambridge, 2000).
[168] W. Horsthemke and R. Lefever, Noise-Induced Transitions (Springer, Berlin, 1984).
[169] G. Schoeck and R. Frydman, Phys. Stat. Sol. b 53 (1972) 661.
[170] J. Diehl, Z. Metallkde. 47 (1956) 411.
[171] A. Seeger, Phil. Mag. Letters 81 (2001) 129.

CHAPTER 57

Collective Behaviour of Dislocations in Plasticity

L.P. KUBIN

LEM, CNRS-ONERA (OM)
29, Av. de la Division Leclerc, BP 72
92322 Châtillon Cedex
France

C. FRESSENGEAS

LPMM, CNRS – Université de Metz
Ile du Saulcy, 57045 Metz Cedex 01
France

and

G. ANANTHAKRISHNA

Materials Research Centre
Indian Institute of Science
Bangalore 560 012
India

Dislocations in Solids
Edited by F. R. N. Nabarro and M. S. Duesbery

Contents

1. Introduction

Plasticity is inherently a dynamical phenomenon, since it arises from the motion of dislocations under stress. Traditional investigations of plasticity in terms of the theory of dislocations involve computing the velocity of one dislocation segment under stress in the presence of various obstacles. In the presence of a large density of strong localised obstacles, dislocation–dislocation interactions can be comparatively weak and plastic flow may result from the sum of a large number of uncorrelated events. As the properties of the interaction with one isolated defect induce those of the bulk material, a connection between the microscopic and the macroscopic scales can be performed in a straightforward manner by simple averaging provided the density of mobile dislocations is known. This situation is met for instance in materials exhibiting a large lattice friction, like b.c.c. metals at low temperatures, or large Peierls–Nabarro forces like tetrahedrally bonded covalent crystals.

Such a connection is, however, not as straightforward when the mutual interactions of dislocations are significant. This type of situation occurs sooner or later during plastic flow and with increasing dislocation densities, as exemplified by the case of cyclically deformed b.c.c. metals at low temperatures. The transition between the micro and macro scales cannot be carried out by simple averaging because dislocation interactions and couplings induce a collective behaviour involving a large number of correlated events. The collective behaviour manifests itself in the form of dislocation patterns that emerge at an intermediate length scale called the mesoscopic scale. These patterns are structures exhibiting some degree of order due to the collective organisation of individual objects within a characteristic length scale. There are many examples of mesoscopic patterns in plasticity. Dislocation cells are observed by Transmission Electron Microscopy (TEM) in thin foils of many materials deformed in multislip. Strikingly regular structures are formed during cyclic deformation with low plastic strain amplitudes. Slip traces are observed at the surface of deformed single crystal specimens and result from the collective properties of the mobile dislocations. Patterns can also be inferred from serrations or from stress plateaus in the stress–strain curves of certain alloys. In the last case, the spatial structure of slip in the deforming sample is, to some extent, reflected in the fine structure of the stress vs. time curve.

To deal with such situations, one has to realise a micro–meso connection first, before attempting a meso–macro connection. In loose terms, this means going first from the properties of one individual dislocation to the collective properties of a dislocation population, or to the stress, strain and strain rate by averaging in a small representative volume of a sample. Intuitively, it is clear that the corresponding characteristic size should be much larger than the average distance between individual dislocations, but much smaller than the characteristic length scale of the emerging structures. Such averaging procedures inside a representative volume are inherent to continuum mechanics representations of materials behaviour. Next, these collective properties have to be integrated in constitutive forms which necessarily include length scales stemming from the mesoscopic level.

In this transition between scales, it was recognised only recently that the appropriate objects to be considered are dislocations populations, averaged over some elementary volume in order to be treated in a continuum manner, as well as their interactions and couplings. Thus, models have to be elaborated that involve sets of Partial Differential Equations (PDEs) governing the evolution in time and space of the dislocation densities. These models will be referred to as "kinematical" or "dynamical theories of dislocations" in what follows. Dynamical theories have, as discussed below, the property of showing bifurcations in space and/or time, a feature not shared by kinematical theories. The consideration of sets of PDEs raises the issue of the length scale of the domain over which they apply. Clearly, the latter has to be much larger than the elementary volume over which dislocation populations are defined, but shorter than or equal to the length scale of the sample. Full compatibility with the mechanical macroscopic boundary conditions should in principle be achieved, but since the edges of the sample are usually considered as remote, overall mechanical constraints on the sample may be applied in a relaxed form at some intermediate length scale, or may enter these sets of PDEs as control parameters only. Therefore, a full multiscale analysis from microscopic through macroscopic scale is not attempted in this chapter, but a discussion of the mechanical boundary conditions is provided.

From the point of view of modelling, a typical sequence of events might be described as follows: for low values of the control parameter, typically the stress or the plastic strain, uniform steady state dislocation densities are obtained. When the value of the control parameter is increased, the spatially uniform solution becomes unstable and non-uniform densities emerge. The critical value of the control parameter for which the slip regime switches from uniform to non-uniform is the hallmark of what is called a bifurcation point [1–3]. Beyond this point, the bifurcated solution may itself become unstable at larger values of the control parameter, giving way, after a second bifurcation, to an even more complex solution. When the material element is not driven too far from its uniform equilibrium state, its evolution is governed by linear laws (akin to all the linear laws relating fluxes and forces like the Fick's or Fourier's laws) deriving from thermodynamical extremum principles. In the situations of interest in this chapter, the "distance" from thermal equilibrium is large and the non-linearity of the governing laws is dominant. In such conditions, non-uniform solutions may exhibit features which are typical of dissipative systems like self-organised, periodic or multi-periodic patterns. Even more complex types of behaviour are presently drawing lot of attention, namely deterministic chaos and self-organised criticality. There is presently a growing number of examples associated with the dynamics of dislocation populations and involving either an infinite number of degrees of freedom (self-organised criticality) or strikingly small number (at last three) of them.

In the past decades, the study of these non-linear effects in various systems, physical, chemical, biological or other, has been accompanied with a considerable development of the tools suited for the study of dynamical systems. The application of these concepts and methods to materials science and more specifically to dislocation populations is relatively recent (cf. [3–5]). This will be illustrated in the present chapter by the examination of several sets of PDEs.[1] In the latter, the spatial term is often a gradient (actually a Laplacean) or sometimes a non-local form, i.e., an integral expression incorporating the fact that a

[1] Here we discuss exclusively deterministic models for patterning. A stochastic approach to patterning is treated in this volume by M. Zaiser and A. Seeger (Chapter 56, p. 1).

property at a given position depends on the properties of an extended neighbourhood. Numerical simulations provide an alternative approach to the question of patterning and multiscale modelling [6], as will be shown on a few examples.

In the first part of this chapter, the focus is on the simplest and most well known dislocation patterns, those emerging from the collective behaviour of mobile and immobile dislocations during cyclic deformation at low plastic strain amplitude. The next part is concerned with slip patterning, more particularly with yield point avalanches and band propagation phenomena, as in the (Piobert-)Lüders or Lüders-like phenomena and in the Portevin–Le Chatelier effect. Extended reviews of the experimental work describing these patterns are readily available and only brief accounts of the experimental situation will be given here. Therefore, the emphasis will be on the modelling efforts undertaken in the last decade. In the Portevin–Le Chatelier phenomenon, these models describe bands of plastic deformation that emerge periodically in time and propagate at constant velocity. The recordings of unstable stress–strain curves may, however, reveal considerably more complex evolution in time. The last part of the chapter is, therefore, devoted to a nascent topic, namely the examination of serrated flow in terms of collective behaviour of dislocations possibly leading to deterministic chaos, or to other types of complex behaviour such as self-organised criticality.

From this introductory material, it is readily seen that an understanding of the collective properties of dislocations involves several difficult steps, as well as some problems due to the need to connect descriptions at various scales of length. For instance, the connection between the theory of dislocations considered as discrete objects and a continuum theory of dislocations raises the issue of the size of the material element to be considered as representative of the local behaviour (coarse-graining length). Models that will be discussed in what follows may be considered as attempting a step in the still desirable connection between dislocation theory and continuum mechanics. Finally, some use will be made of the framework currently employed in the domain of non-linear dynamics. Some basic concepts that are necessary for an understanding of present models will be explained in some detail when necessary. It is the object of this review to show that significant advances have been recorded in the past decade in this somewhat composite domain. The spatial and temporal aspects of collective dislocation behaviour are now no longer a field beyond reach, although the question is very far from being settled.

2. *Collective dislocation behaviour in cyclic deformation*

2.1. Introduction

During plastic flow, the spatial arrangement of the dislocation structure essentially derives from a competition between two factors. When it exists, the interaction of dislocations with strong and dense localised obstacles other than dislocations (small clusters and precipitates, lattice friction and Peierls forces...) tends to induce rather uniform dislocation distributions. In such conditions, the plastic flow properties of the bulk material may simply reflect the behaviour of isolated mobile dislocations. In contrast, the mutual interactions of dislocations, both local and long ranged, become all the more important as dislocations multiply. Thus, the contribution of dislocation interactions to the flow stress increases during plastic flow. At a certain stage, characterised by a critical stress, strain, or dislocation density, collective dislocation behaviour sets in which is characterised by the emergence of dislocation-rich and dislocation-poor regions, with sometimes rather well defined spatial periodicities. This is usually referred to as dislocation patterning.

Dislocation patterning, its causes and its dependence on various material or defect properties has been the object of numerous experimental studies and is still the object of many debates (see, for instance, the reviews [7,8]). However, it is now increasingly recognised that this phenomenon is of dynamical origin. As a consequence, recent theoretical investigations have focused on the derivation of dynamical expressions describing the evolution of the dislocation densities in time and space during plastic flow. The dynamical models do not assume *a priori* that a certain structure is formed with characteristic geometrical dimensions. In contrast to former models, they attempt to describe the emergence of an organised structure from either an initial random distribution of dislocations or from a different, preexisting, organised structure.

This is best illustrated by studies devoted to the simplest possible situation, that of patterning during the cyclic deformation of single crystals oriented for easy glide. In such conditions, only one Burgers vector is active and only a few distinct dislocation populations need to be considered: mobile screw and edge dislocations, of either sign, and dipolar configurations of essentially edge character. The occurrence of very characteristic patterns appears to be favoured during cyclic deformation because very large dislocation densities are involved and immobile configurations must be stable with respect to both signs of the applied stress. In what follows, the experimental facts are first recalled without going too much into details. Three main theories developed in the past decade are then discussed in three steps: first, a presentation of the basic ingredients of the dynamical equations, then a brief summary of the solutions and, finally, a critical discussion. Indeed fatigue patterning has been the cornerstone of all dynamical models and it is instructive to compare their merits and shortcomings by considering how they deal with a few critical problems.

2.2. A brief description of the patterns formed in cyclic deformation

2.2.1. The matrix and Persistent Slip Band structures

Dislocation arrangements formed during the cyclic deformation of principally f.c.c. and b.c.c. metals and some of their alloys have been reviewed several times [9–14]. The most interesting configurations are found in the domain of low-cycle fatigue, at low and moderate temperatures (below $0.5T_{\mathrm{m}}$) and for small plastic strain amplitudes (lower than e.g., 10^{-2}). With increasing plastic strain amplitude, f.c.c. single crystals (Cu, Ni, Ag) oriented for single slip may exhibit two types of organised structures (cf. fig. 1) which successively emerge from an initial random microstructure: the matrix structure and the so-called Persistent Slip Bands (PSBs). Each structure corresponds to a well-defined stage in the cyclic stress–strain curve. The latter, which is shown in fig. 2, is a plot of the cyclic saturation stress vs. the plastic strain amplitude per half-cycle. Other structures formed in duplex or multiple slip, like labyrinth and cell structures, will not be considered here.

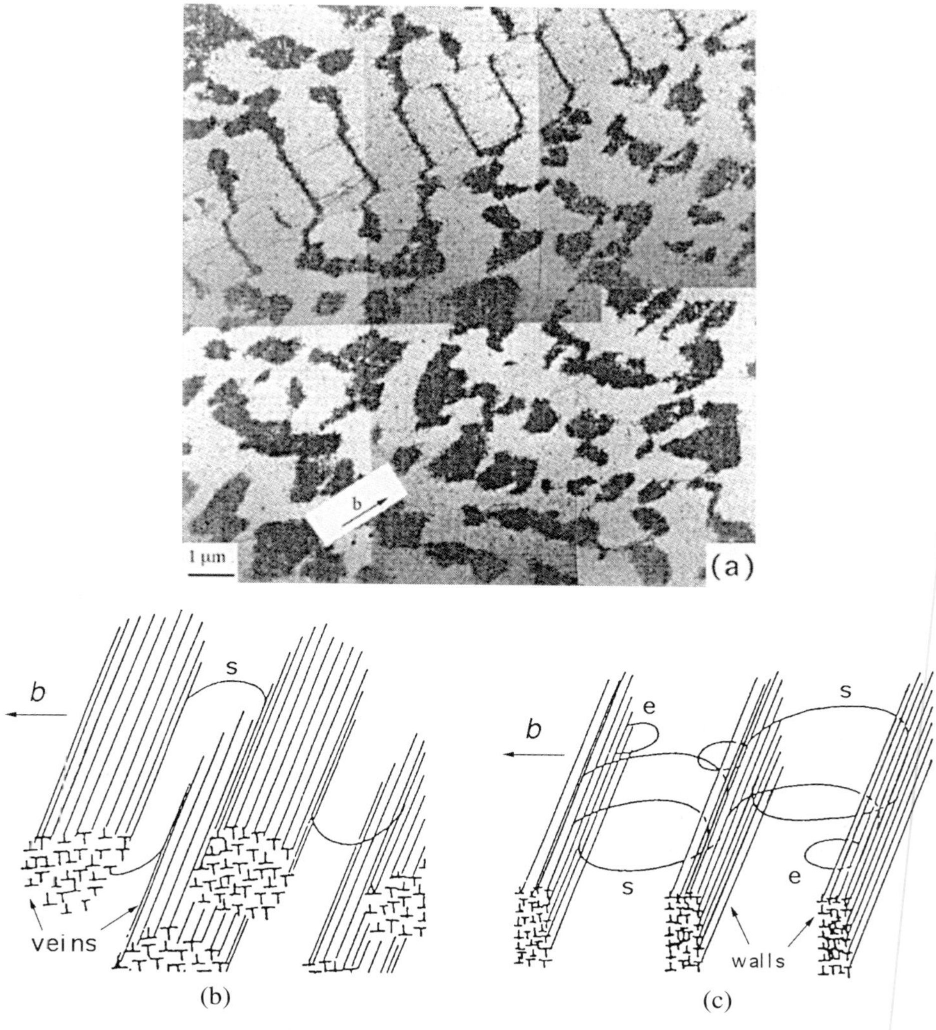

Fig. 1. (a) A Persistent Slip Band (top), with its characteristic ladder structure, embedded into a matrix structure in a copper crystal cycled at room temperature with a plastic strain amplitude of 2.2×10^{-3}. This $(1\bar{2}1)$ view contains the Burgers vector and is perpendicular to the edge direction (courtesy of U. Holzwarth and U. Essmann [15]). (b) and (c) Schematic dislocations arrangements in the matrix and the PSBs structures, respectively. Gliding screw segments (s) and edge segments bowing out from the walls (e) are shown in the channels (courtesy of H. Mughrabi [9]).

Quite generally, the patterns that we examine in this section are characterised by an unmixing of the screw and edge characters, the latter being organised in the form of dense dipolar configurations. Edge dislocations mutually annihilate with more difficulty than screw dislocations and are therefore always found with larger average density. The dense dipolar arrangements delimit empty channels in which screw dislocations can shuttle, thus

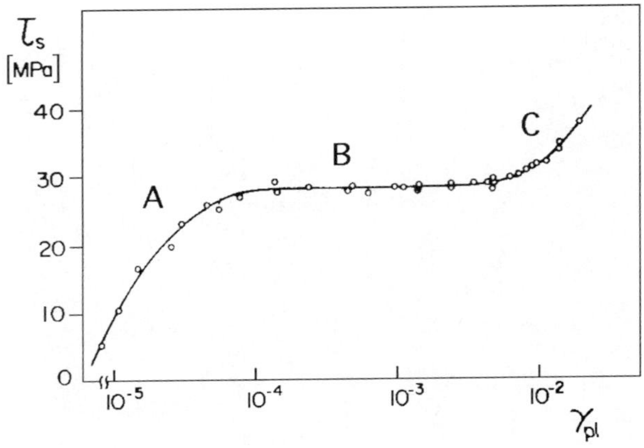

Fig. 2. The three-stage cyclic stress–strain curve of copper single crystals at 295 K (after Mughrabi [9]). The matrix structure is formed during stages A and B and the PSBs are formed with increasing volume fraction along the saturation plateau B.

carrying a large fraction of the plastic strain rate (cf. fig. 1(b), (c)). Schematically, the matrix structure (cf. figs 1(a) and 3) is formed as soon as the mean free path of edge dislocations during one-half cycle exceeds their mean separation. Then stable dipoles with heights such that their passing stress is smaller than the applied stress can be formed. This structure consists of an irregular arrangement of thick dipolar veins (with volume fraction up to 50%) and channels with a typical width in the micrometer range. Its occurrence is observed during the first stage of the cyclic stress–strain curve (A in fig. 2).

When the plastic strain amplitude is increased, a gradual transition occurs from the matrix structure to a quite fascinating structure, that of the PSBs. These bands are thin lamellae of thickness typically 1–2 μm parallel to the active slip planes, which may cross through the whole crystal. They owe their name to the fact that after having been removed by surface polishing they reappear at the same place during continued cycling, producing a local strain of about 10^{-2}. Their formation occurs via a local destabilisation of the matrix structure during the saturation plateau of the cyclic stress strain curve (B in fig. 2). The volume fraction of the PSBs is given by Winter's rule of mixtures [16] for the partition of the plastic strain amplitude between the coexisting matrix and the PSBs. As the plastic strain amplitude increases, in order to accommodate the imposed plastic strain rate, new PSBs are nucleated within the matrix structure under a constant saturation stress. In fig. 1(a) is shown the edge-on view of a PSB embedded into a matrix structure, with its very regular periodic structure of thin dipolar walls (the mean wall spacing is about 1.4 μm at room temperature). The saturation of the cyclic stress that occurs when the PSBs are activated suggests that a microstructural dynamical steady state has been reached, or at least that the PSBs are able to function during a large number of cycles without any significant change in their structure. The specimen becomes completely filled with PSBs when the plastic strain amplitude reaches a value of about 10^{-2}. Beyond this value, PSBs can no longer accommodate the imposed plastic strain and become unstable. In the third stage of the

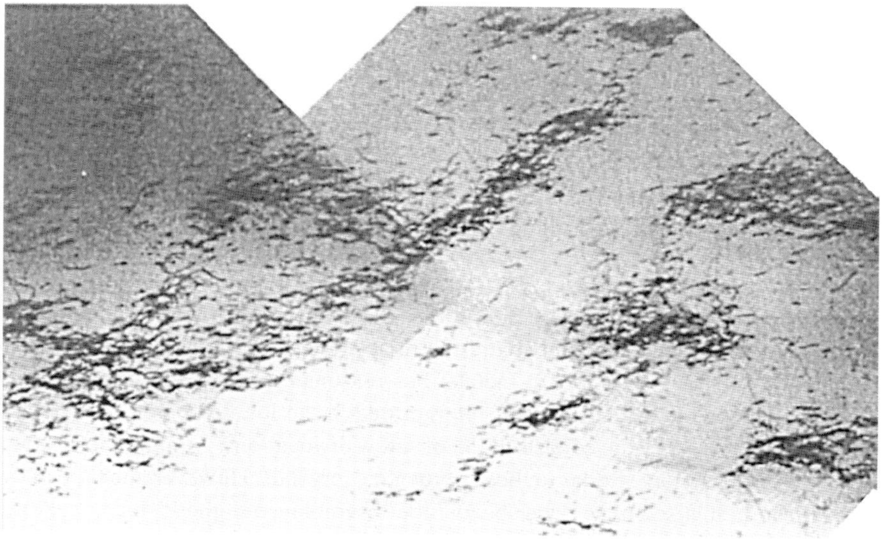

(a)

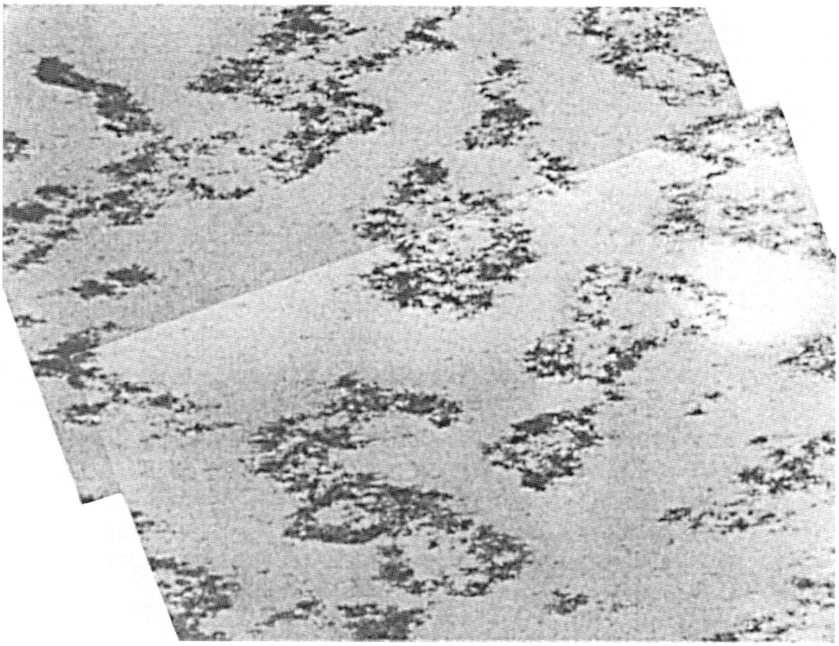

(b)

Fig. 3. Dislocation microstructures in Fe–0.9wt.%Si single crystals after cyclic deformation at room temperature with a constant plastic strain amplitude of 2×10^{-3} and a cumulative plastic strain of 1.6 (S. Libovicky, unpublished work, courtesy of A. Franek et al. [20]). (a) Section parallel to the $(\bar{1}01)$ glide plane; (b) section perpendicular to the slip plane and parallel to the slip direction. The veins are seen edge-on.

cyclic stress–strain curve (C in fig. 2), which is not discussed here, secondary slip systems are activated leading to the formation of a cell structure.

The occurrence of a dynamical steady state associated with the formation of a very regular structure with only one active Burgers vector and under a constant characteristic stress has made the PSBs very attractive in terms of modelling. The behaviour that has been described above is typical to the f.c.c. crystals. It is worth noting, however, that the case of aluminium appears to be a special case as the PSBs do not exhibit in general a distinct ladder structure in this material [17]. The reason for this peculiarity is not clearly understood.

The situation is less well documented for crystal structures others than f.c.c., except perhaps for the b.c.c. metals (cf. [18] and the review [11]). In the latter, a vein structure is also formed like that in the f.c.c. metals. For reasons that have not been understood yet, no observations of the PSBs have been reported in the low-temperature range where plastic flow is governed by the core structure of screw dislocations. At higher temperatures, plastic properties are more similar to those of the f.c.c. crystals. However, the type of cyclic strain localisation that seems to occur bears little resemblance to the PSBs [18]. A model b.c.c. material is Fe–Si which, like many alloys, exhibits an extended Stage I in monotonic deformation. This dominant feature allows one to study patterning in easy glide during cyclic deformation as well as during monotonic deformation [19]. In the latter case, the microstructure is not so different from the vein and wall structures observed in fatigue and shown in fig. 3. Kratochvil's model, that we discuss below, stresses the analogy between the two types of patterns.

2.2.2. *Dislocation mechanisms in PSBs*

Experimental studies on the PSBs have yielded certain results that have helped to understand the dislocation processes. During the saturation stage, the global pattern remains nearly unchanged and the densities of screws and of dislocations in dipole configuration remain constant as a result of a balance between generation and annihilation. Screw dislocations are produced by loops emitted from the walls (cf. fig. 1(c)) and mutually annihilate by cross slip. In the studies dealing with the dynamics of the PSBs, the annihilation of screw dislocations is represented by a critical annihilation distance (y_S) such that two screw segments whose slip planes distance is smaller than y_S mutually annihilate by cross-slip (in copper and at room temperature, $y_S \approx 45$ nm [21]). This length scale cannot be predicted by the available cross-slip models, which are based on the elastic theory of dislocations. However, simulations at the atomic scale (see [22] and references therein) can now address this question.

Edge segments are deposited on the walls by the motion of screw segments and since their density cannot grow indefinitely, an annihilation mechanism must exist in addition to their well known mechanism of destruction under stress. Even at temperatures where climb is not activated, the disintegration of dipoles of sufficiently small heights into a row of point defects is possible. It has been suggested that it can occur by a purely mechanical and temperature-independent process. When the elastic interaction stress becomes comparable to the theoretical limit of the crystal, a breakdown of the dipole is thought to occur. Actually, a "critical annihilation distance" for edge dipoles, y_e, has been measured in copper at room temperature [21] and in nickel between 77 K and 750 K [23]. Its value is of about 1.5–2.0 nm at room temperature, but the exact mechanism by which edge dipoles annihilate remains a bit mysterious. A recent calculation of this quantity at atomic scale [24] confirmed the value $y_e \approx 1.5$ nm in aluminium but not in copper, for reasons that are not well understood.

These dislocation mechanisms are incorporated into conventional models to establish relations between the dislocation densities, the saturation stress and various characteristic lengths (cf., e.g., the classical model by Essmann, Mughrabi and co-workers [18,21,25]). However a number of difficulties need to be mentioned. First, the length scale that is searched for in the present context, the channel width, seems to be inversely proportional to the saturation stress, i.e., the constant stress associated with the permanent regime [10]. Both quantities are strongly temperature-dependent, but there is no general agreement on the mechanism at the origin of this dependence. The latter may stem from the mobility of the screw dislocations interacting with various obstacles and debris in the channels or from the cross-slip probability via changes in the value of the critical annihilation distance y_s [14,15,26]. In addition, it must be noted that the macroscopic saturation stress is an average over spatially non-uniform stresses. As the walls do not carry much plastic strain, they almost behave as elastic inclusions, so that strain continuity induces the presence of compatibility stresses of opposite sign in the walls and in the channels and of strong stress gradients inside the channels [27]. Finally, the cyclic stress–strain curves of the f.c.c. metals and the occurrence of the PSBs can perhaps be rationalised in terms of a reduced stacking-fault energy, indicating again that cross slip may play a role in the formation of the PSBs. This can possibly explain the particular behaviour of aluminium, the f.c.c. metal where cross slip is the easiest. In summary, the characteristics of the PSBs, which has been the object of so many studies, depend on several material properties in a manner that is still under debate. As the formation of a matrix structure seems to be a more general phenomenon than that of the PSBs, it is probably less sensitive to the details of dislocation properties and to the crystallography of the material. The matrix structure is also interesting, as far as modelling is concerned, because it is directly formed from an initial random distribution of dislocations. It should, therefore, provide an easier check for dynamical models.

Dynamical models that are reviewed in the following sections adopt a simplified view by assuming that patterning emerges naturally from the conjunction of a few elementary processes like, among others, the mutual trapping of edge dislocations and the cross slip of screw dislocations. The way such features are, or are not, incorporated into the models provides an easy check of their physical content. We see from the above discussion, however, that even in the simplest case examined here, some of the underlying mechanisms have not been fully elucidated. The least that can be hoped for is that a model captures the typical wavelength of the ladder structure, its stress dependence and perhaps its temperature dependence. Somehow, the difference in behaviour between the f.c.c. and b.c.c. crystals should also be rationalised with the help of the model's output. Without anticipating the following sections, it can be mentioned at this point that one is still far from reaching such a stage.

2.3. General aspects of dynamical modelling

In what follows we give an account of the various steps involved in establishing dynamical equations for patterning in cyclic deformation. Major difficulties are illustrated with reference to the three different models that have been elaborated in the last decade. These three models are, in chronological order, those of Walgraef and Aifantis (WA), Kratochvil and co-workers (KC), and Differt and Essmann (DE) (cf. [28–30] for the basic reference of each model). After a short description of the evolutionary laws that have been developed and of the main results derived from their solutions, an assessment will be attempted of the basic assumptions involved in each model.

2.3.1. Evolutionary equations

The objective of dynamical modelling is to build up evolutionary laws for the relevant dislocation densities as a function of time and space. The three models that we discuss below can be considered as combining two different approaches that have been developed in the past into a single one. One is the continuum approach of dislocations (cf. [31,32] for an account) that was essentially developed to attempt a connection between continuum dislocation theory and mechanical approaches. The other, the kinetic approach, is more phenomenological and describes the time evolution of the average dislocation densities.

Continuum description of dislocation density

The spirit behind these theories is that they assume the existence of a coarse-graining length scale ℓ not appearing in the relevant equations, over which local dislocation densities of various kinds can be defined. In dislocation dynamics, reaction length scales for different transformations are different. For instance, the critical annihilation distance is not the same for dislocations of screw and edge character. As a consequence, ℓ should be larger than the length scales associated with all the assumed reactions occurring between individual dislocations. Other length scales can be identified which are much larger than coarse-graining length scale, and should therefore be relevant in studies of the spatial heterogeneity of slip. One such length scale is the mean free path of the mobile dislocations, which transport dislocation densities from one material element to the next, and is often taken as a "diffusive length scale". In most of the existing models, the reasons offered for the emergence of a diffusive spatial coupling between neighbouring elements are context-dependent. Double cross slip, which is schematically depicted in fig. 4, can be cited as an example. Here, a screw dislocation leaves its slip plane, for instance, through the effect of a repulsive stress from a dislocation of like sign, and cross slips back into the initial slip plane. The cross-slip height h_{cs} is a length scale that may be relevant in studying the spatial heterogeneity of slip, since double cross slip produces fresh dislocation sources in non-active slip planes and enhances the spreading out of slip activity. As further discussed in section 3.3.2, when this mechanism is associated with a pseudo-diffusion coefficient, one should have $h_{cs} > \ell$. Finally, the long-range interactions between dislocations provide a spatial coupling mechanism that can sometimes be treated without being reduced to a diffusive length scale (see sections 2.3.4 and 3.3.2).

Within the traditional dislocation approach, dislocation densities are averaged in space over the largest length scale corresponding to the dimension of the specimen. Hence, no

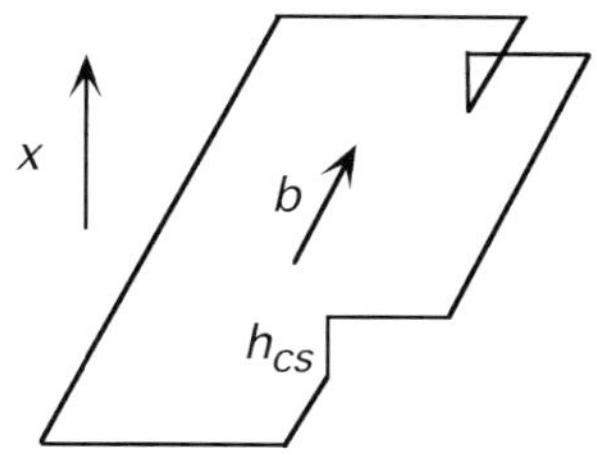

Fig. 4. Double cross slip of a screw portion of a dislocation loop.

patterning is considered. With the help of Orowan's law, time is replaced by a uniform plastic strain variable in the evolutionary laws. A typical example of a one-population model in terms of strain reads:

$$d\rho_f/d\varepsilon = s\rho_f^{1/2} - r\rho_f. \tag{1}$$

The total rate of increase of the dislocation density is the sum of a creation and a loss term. This famous equation [33] gives the evolution of the density of forest (i.e., immobile) dislocations (ρ_f) as the result of creation by storage (s) events and loss by recovery (r) events. Storage occurs by the immobilisation of mobile dislocations which form sessile locks with other dislocations, while recovery may occur by climb for edges or cross slip for screws. It is easy to see that there is a stable steady state corresponding to a saturation of the forest density. A more sophisticated set of equations involving two populations (forest and mobile dislocations, the latter with a density ρ_m) can also be written [34].

$$d\rho_m/d\varepsilon = m - a\rho_m - s\rho_f^{1/2}, \tag{2}$$

$$d\rho_f/d\varepsilon = s\rho_f^{1/2} + a'\rho_m - r\rho_f. \tag{3}$$

The terms at the right-hand side of such equations are called reaction terms, although they do not necessarily describe a thermally activated mechanism. In eqs (2), (3), the various mechanisms that have been incorporated are: multiplication (m), annihilation plus dipolar trapping that remove dislocations from the mobile density (a), the formation of dipoles and debris leading to a creation term, (a') in eq. (3), storage (s) and recovery (r). The physical justification of each term can be found in [34] and a full account of this framework can be found in [35]. Evolution equations written with respect to time have also been used in other contexts [36]. It is instructive to note that such equations have stable steady state solutions. These expressions can be considered as deriving from a more complex system that contains additional spatial terms. Then, what would be described here corresponds to the uniform solution of such extended systems. In the present context, spatial terms usually have a stabilising effect with respect to uniform flow. One may think, for instance, of a diffusion-like term that would smoothen out any local increase in the density. Hence, the reaction terms written above must be modified in such a way as to destabilise the uniform solution beyond certain critical conditions, thus leading to stable non-uniform solutions.

To achieve this, one has to take into account the fact that certain mechanisms are modified with increasing stress or dislocation density.

Within the classical continuum approach of dislocations, the dislocations accounted for are those that statically accommodate the strain gradients inside the specimen (cf. section 3.3.2). Other stored dislocations, mobile dislocations, as well as their mutual interactions, are ignored. It follows that within the classical continuum theory, reaction terms are neglected while spatial aspects are included through conservation-transport laws in the form

$$\partial\rho/\partial t + \operatorname{div} J = 0. \tag{4}$$

This equation expresses the conservation of the dislocation flux $J = \rho v$, where v is a dislocation velocity (here and in what follows, all equations are written in uni-dimensional and scalar form for the sake of simplicity). It is, of course, tempting to make use of some kind of Fick's law of the form $J = -D\nabla\rho$, where D is a pseudo-diffusion coefficient, in order to eventually obtain:

$$\partial\rho/\partial t - D\nabla^2\rho = 0. \tag{5}$$

Such diffusion-like forms have attracted lot of attention. However, the transition between eq. (4) and eq. (5) is usually not straightforward as it is not proven in general terms that dislocations may behave in a diffusive manner. The problem may be summed up as follows. To reach a form akin to the Fick's law, one has first to define a dislocation mobility M such that $v = M\tau^*$, where the effective stress τ^* is the algebraic sum of the internal stress τ_μ and of the applied stress. The latter, being uniform, has no divergence and what remains in eq. (4) is of the form $\operatorname{div}(\rho\tau_\mu)$. By definition, the internal stress is the sum on a given dislocation of all its elastic interactions stresses with other dislocations in the microstructure. Hence it depends on the unknown distribution of the densities of plus and minus dislocations. To estimate τ_μ one may, for instance, start from the following form giving the elastic interaction energy $E(r)$ at position r:

$$E(r) = \int \rho(r')f(r')I(|r - r'|)\,\mathrm{d}^3r'. \tag{6}$$

The function f describes the relative arrangement of dislocations of both signs and I is their pair interaction stress, the integration being carried out over the whole microstructure. As the calculation of this non-local term involves considerable difficulties, there have been attempts to simplify it by assuming a cut-off distance for the elastic interactions. Then, the first non-zero term (less often the first two terms) of a Taylor expansion of the energy near $r = r'$ is retained. Such a procedure is used in the early work of Holt [37] and further by Walgraef and Aifantis [28]. It can be criticised on the grounds that it artificially introduces a length scale, the cut-off distance, that induces the periodic nature of the solutions. Within this approximation, the elastic energy is found at first order to be proportional to the local density $\rho(r)$. The internal stress is then proportional to the negative gradient of the elastic energy, and hence to the local density. Using this, eq. (4) can then be rewritten

$$\operatorname{div} J = -\alpha \operatorname{div}[\rho\nabla\rho] \neq -D\nabla^2\rho, \tag{7}$$

where α is a positive constant. We see that further approximation is necessary to define a diffusion coefficient that, in any case, should depend on the local dislocation density. A more detailed discussion of this problem is given by Malygin [38] who reaches the same conclusion, namely that without the introduction of a cut-off radius one cannot reach a diffusion-like form. As will be further discussed in section 2.5.3, no formulation exists to date that leads rigorously or within acceptable approximations to a method for bypassing the non-local form of the internal stress.

Finally, the sets of equations currently used to describe dislocation patterning somehow mix up the kinetic and continuum approaches by combining both reaction terms (like those of eqs (1)–(3) and either a transport term or a diffusion-like term (eqs (4) and (5), respectively). As always, the starting point must be a conservation law for the relevant quantities, here the dislocation densities. At this step, it is worth mentioning that the approach presented here is a simplified one, as the derivation of conservation laws at different scales and their mutual consistency may pose rather deep problems (cf. section 2.5.1 and [39]).

2.3.2. *Reaction–transport: the Differt–Essmann (DE) model*

The starting point for the DE model is a transport equation which takes into account the flux through the surfaces of a small material element and the production rate (i.e., creation plus losses) in the same element. Quite naturally, we obtain:

$$\partial\rho_i/\partial t + \mathrm{div}[J_i] = g(\rho_i, \rho_j, c). \tag{8}$$

In this reaction–transport equation, the reaction term g takes into account both the creation and annihilation events of the population (i) and its coupling with other populations (j). c represents a set of control parameters. As indicated above, these quantities describe the external constraints imposed on the system, typically here the maximum stress during a fatigue cycle. Their magnitude governs the distance from equilibrium and, hence, the transition between the various types of spatio-temporal behaviour.

To take an example, the reaction–transport equations used in the DE model [27,40] to deal with the coupled behaviour of mobile dislocations and dipoles in the PSBs are reproduced in eqs (9)–(11). These equations refer only to the mechanisms taking place in the walls; they are complemented in a second part [41] by a consideration of the dynamical properties of screw dislocations in the channels. Three dislocation populations are considered in this first part of the model, mobile edge dislocations of both signs, ρ_1 and ρ_2, with velocities $v_1 = v = -v_2$ respectively, and an immobile population of dipoles ρ_{d}.

$$\partial\rho_1/\partial t + \mathrm{div}(\rho_1 v_1) = -4\rho_1\rho_2 v y_{\mathrm{d}}, \tag{9}$$

$$\partial\rho_2/\partial t + \mathrm{div}(\rho_2 v_2) = -4\rho_1\rho_2 v y_{\mathrm{d}}, \tag{10}$$

$$\partial\rho_{\mathrm{d}}/\partial t = 8\rho_1\rho_2 v(y_{\mathrm{d}} - y_{\mathrm{e}}) - 2y_{\mathrm{e}}\rho_{\mathrm{d}} v(\rho_1 + \rho_2). \tag{11}$$

Two length scales are introduced, an upper distance between slip planes y_{d}, below which a dipole is formed and beyond which a dipole is no longer stable under stress, and y_{e} the critical annihilation distance of edges (cf. section 2.2). The meaning of the creation terms is then clear: dipole formation is a loss term in eqs (9) and (10) and a creation term in eq. (11). Dipoles can also be destroyed by their encounter with mobile dislocations which results in a loss term, the last term at the right-hand side of eq. (11). Thus, all the reaction terms represent well-defined mechanisms and couplings, governed by length scales that are reasonably understood. One may note that no reaction term accounts for the multiplication mechanism of the mobile dislocations. This last mechanism occurs in the channels, whose properties are dealt with separately in the second part of the model [41]. As a consequence, the boundary conditions, in one dimension, account for two incoming edge dislocation fluxes on the wall sides by imposing $\rho_1(-\infty) = \rho_2(\infty) = \text{const.}$, $\rho_1(\infty) = \rho_2(-\infty) = 0$, with a uniform plastic strain rate $\dot{\gamma}/b = v(\rho_1 + \rho_2)$ and, hence a uniform dislocation velocity, v. Finally, the objective of this model is restricted to the description of the dynamical stationary state in the PSBs. Since the conditions of formation or destruction of these structures are not examined, no control parameter is introduced.

2.3.3. *Reaction–diffusion: the Walgraef–Aifantis (WA) model*

As shown in section 2.3.1, reaction–transport equations can be further developed into a reaction–diffusion frame, provided that simplifying assumptions are made. Reaction–diffusion equations have been widely used to study instabilities in non-linear systems. They were used in the domain of chemistry to model complex coupled chemical reactions leading to various types of spatio-temporal instabilities [42]. There, the physical meaning of "diffusion" and of "reaction" is perfectly clear and, for two species in concentrations $\rho_{\rm i}$ and $\rho_{\rm m}$, a typical set of equations is written:

$$\partial \rho_{\rm i}/\partial t - D_{\rm i}\partial^2 \rho_{\rm i}/\partial x^2 = g(\rho_{\rm i}) - b\rho_{\rm i} + \gamma \rho_{\rm m}\rho_{\rm i}^2, \tag{12}$$

$$\partial \rho_{\rm m}/\partial t - D_{\rm m}\partial^2 \rho_{\rm m}/\partial x^2 = b\rho_{\rm i} - \gamma \rho_{\rm m}\rho_{\rm i}^2. \tag{13}$$

These two coupled partial differential equations are also those used by Walgraef and Aifantis in their study of patterning in fatigue [28,43–50]. Thus, the peculiarity of the WA model is that it reproduces closely a well known set of reaction–diffusion equations, referred to as the "Brusselator", which models chemical patterning [51]. Although the analogy yields very interesting results, the physical relevance of the various terms with respect to dislocation properties has been the object of some criticism, which will be summed up in the next sections. The WA model is, however, the first full attempt to establish a dynamical theory of patterning recovering the succession of microstructures that have been described in section 2.2.2. The two populations are mobile (m) dislocations, whose character is not specified, and almost immobile (i) dipoles. The corresponding effective diffusion coefficients are written here in a one-dimensional frame, but the model has also 2- and 3-D extensions. These coefficients are such that $D_{\rm i} \ll D_{\rm m}$, and this is one of the conditions needed for obtaining patterning solutions.

The reaction mechanisms taken into account are:

- the destruction of dipoles ($-b\rho_{\rm i}$) when the stress exceeds a threshold value. Hence, the parameter b is related to the stress or the plastic strain rate and plays the role of a control parameter.

- A cubic term $\gamma \rho_{\mathrm{m}} \rho_{\mathrm{i}}^2$ representing the reaction between a mobile dislocation and a dipole, and
- a dislocation generation/loss function $g(\rho_{\mathrm{i}})$ for the immobile density, whose initial value is zero. The derivative $g(\rho_{\mathrm{i}})$ with respect to time is positive at low densities where the dipole density increases and is negative at high densities due to increased dipole destruction.

The set of eqs (12), (13) (and similar forms) has a large variety of solutions in unbounded media: the uniform solution may bifurcate to periodic temporal solutions (through the so-called Hopf bifurcation) or to several types of spatio-temporal patterns (via a Turing instability) according to the parameter values (for a derivation and study of these solutions, see [52,53]). For numerical solutions, the boundary condition adopted [47] is that of a domain of finite size, with zero dislocation flux at the boundaries (cf. the discussion in section 2.5.3).

2.3.4. *Non-local form: the model by Kratochvil and co-workers (KC)*

To conclude this preliminary survey of dynamical patterning equations and before attempting a comparison of their respective merits and weaknesses, a last equation is written that stems from the one-dimensional version of the KC model [54] (see also [19,20, 29,55–58]). In this model two populations are considered: – a uniform population of mobile screw dislocations, ρ, that produce the plastic strain rate, – a distribution of elongated dipolar loops, ρ_{d}. The evolutionary law of the latter is written:

$$\partial \rho_{\mathrm{d}}/\partial t + D\frac{\partial^2}{\partial x^2}\left[\int_{-\infty}^{+\infty} \Phi(x'-x)\rho(x')\,\mathrm{d}x'\right] = G(\rho_{\mathrm{d}})|\dot{\gamma}|. \tag{14}$$

This expression, which is now discussed in detail, is at the core of the KC model. Equation (14) contains a single reaction term at the right-hand side but is not a pure reaction–diffusion form like eqs (12), (13) since it contains no Laplacean of the dislocation density, nor a reaction–transport form like eqs (9), (10) of the DE model. Here, a non-local form involving an influence function Φ accounts for the long-range elastic interactions. This term has been successively attributed two meanings. In the earlier version of the model, the non-local form and the diffusion coefficient in front of it were interpreted as stemming from non-local hardening [54,55], more specifically from long-range elastic stresses. Then, this term should be of same nature as in eq. (6) except that stresses are computed instead of energies. The corresponding physical picture is that small, elongated, dipolar loops are produced at random places in the crystal, as reported by numerous TEM observations of the early stages of cyclic deformation (cf., e.g., [59]). It is assumed that patterning is principally produced by the ordering of such loops which is also confirmed by recent TEM studies on nickel [23,60]. In the drift model [19,56], the loops drift along their glide cylinder under the effect of the stress gradients produced by the dislocation structure. An alternative and possibly more effective interpretation of the non-local form has been given later on [29,58,59], which is based on the sweeping mechanism initially proposed by Sharp and Makin [61]. In this mechanism, bent screw dislocations that move in neighbouring slip planes interact elastically with the small loops. Under the applied

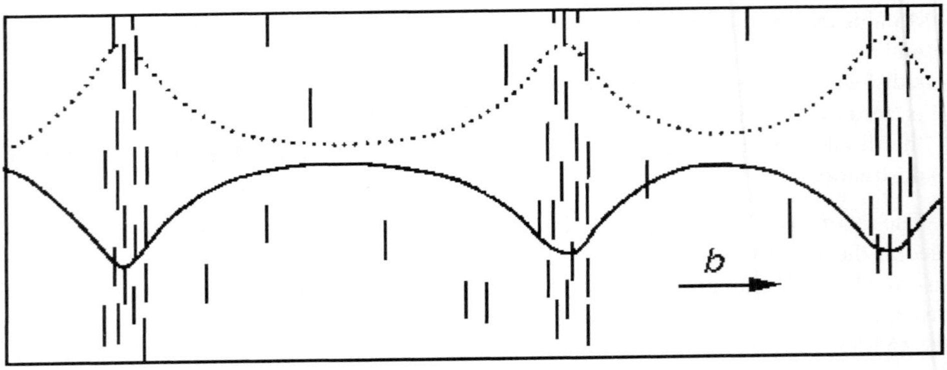

Fig. 5. The sweeping mechanism: mobile screw dislocations (full line and dotted line for a dislocation of opposite Burgers vector) interacting elastically with a distribution of small dipolar loops of same Burgers vector b sweep them along their glide cylinder. The dipolar loops eventually condense into clusters parallel to the edge direction. (Courtesy of J. Kratochvil and M. Saxlova [29].)

stress, the dipolar loops are swept [62] and tend to cluster, as illustrated by fig. 5 and fig. 9, below. From this viewpoint, the function Φ accounts for the change in shape of the gliding dislocations due to their interaction with individual dipoles or dipolar clusters (cf. section 2.4).

To compute the periodic solutions of this model, it is necessary to know the shape of the influence function Φ in eq. (14). The latter is not determined from theory and it is assumed that it has a bell-shaped form [19]. This introduces a length scale into the problem (the half-width of Φ), but no cut-off since Φ is defined over the entire real space. Therefore, this approximation is certainly less strong than a truncation of the long-ranged elastic stresses. On the other hand, since the form of Φ is assumed and not calculated, the geometry of the patterns can be discussed only within an unknown length scale.

The reaction term at the right-hand side of eq. (14) is justified as follows. The generation of the dipolar loops may involve mechanisms like jog dragging and/or cross slip and the generation rate should be proportional to the strain rate, $\dot{\gamma}$. The latter is related to the resolved shear stress through a power law relationship. The function $G(\rho_{\mathrm{d}})$ plays the same role as the function $g(\rho_{\mathrm{i}})$ in the WA model. It is positive at small densities, to represent loop generation, and negative at high densities to account for the increased rate of dipole destruction and annihilation. It is worth noting that dipoles cannot drift in uniform stress fields, as implicitly assumed in the WA model, but only under stress gradients. In the KC model, a diffusion coefficient D is associated with the drift or sweeping mechanism and it can be formally calculated without any assumption in the case of drift [54,55].

As emphasised by Kratochvil [54], mechanical equilibrium has to be achieved. While discrete dislocations have, by definition, stress fields that fulfil the condition of stress equilibrium and the requirement of compatibility between slipped and non-slipped areas in a crystal, it is not so for continuous dislocation densities. Hence, the KC model is complemented by the conditions of stress equilibrium and strain continuity inside the crystal. Within such a frame, elasticity has to be included in the crystal behaviour for a complete definition of the stress and strain fields. The unbounded crystal considered is then submitted to plane strain conditions, in agreement with the assumption made that the dipoles are of infinite length.

To summarise, three different sets of PDEs have been proposed for modelling dislocation patterning in cyclic deformation. All these models derive from dynamical balance equations and yield information about patterning despite significant differences in their basic assumptions. Before examining critically these differences, a brief account is given in the next section of the main results obtained from each model. For the sake of completeness, it must finally be mentioned that there have been several early attempts to model patterning from conservation laws, the earliest one being to the authors' knowledge the one by Vladimirov and Pegel [63].

2.4. Results

Figure 6 shows the bifurcation diagram obtained by Walgraef and Aifantis [28] from the three-dimensional version of their model. Here, the amplitude A of the spatial modulation of the dislocation density is plotted as a function of the control parameter, i.e. the cyclic saturation stress. With increasing saturation stress, the structures discussed in section 2.2.1 are successively obtained: first a uniform structure, then a rod structure (1 in fig. 6) similar to the vein structure and finally a structure of periodic walls (2 in fig. 6) akin to that of the PSBs. One may note that the last two structures can coexist within a certain range of values of the control parameter, as is the case for the vein and PSB structures. The behaviour of the latter have been examined in detail, both theoretically and numerically. For a full account, the reader is referred to the very detailed work by Schiller and Walgraef [47,64].

In fig. 7 is shown the one-dimensional stationary solution of eqs (9)–(11) of the DE model relative to the edge dipolar walls. The density of these walls is plotted along a direction perpendicular to them and their width constitutes a first length scale coming out

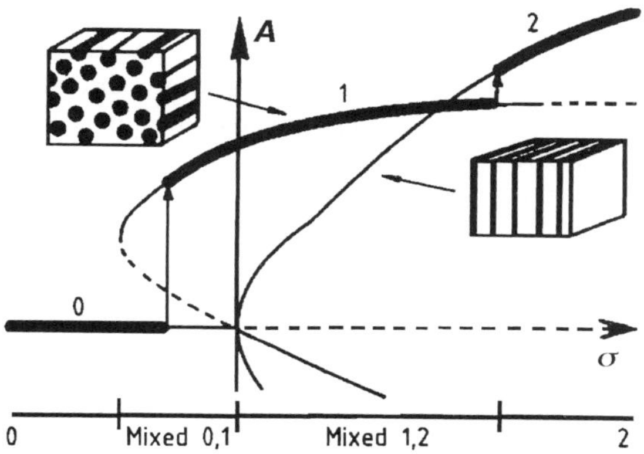

Fig. 6. Bifurcation diagram in the WA model (after [28]). A is the amplitude of the modulation of the immobile dislocation density and is the stress at saturation during cycling. The heavy lines indicate the range of stability of each of the three possible solutions, a uniform structure (0), a "vein" structure (1) and a "ladder" structure (2).

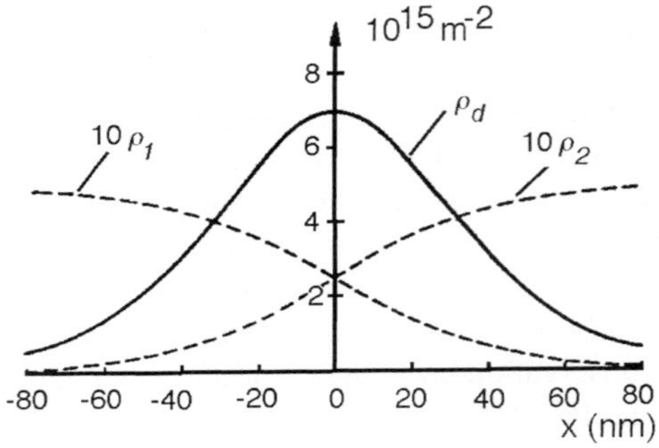

Fig. 7. The profile of a stationary edge wall in the DE model. The densities of the three dislocation populations ρ_1, ρ_2 and ρ_d are plotted along a direction perpendicular to the walls. The width of the wall is governed by the two length scales y_d (here, 24 nm for a resolved stress of 27 MPa) and $y_e = 1.6$ nm. (Courtesy of U. Essmann and K. Differt [40].)

of the model. The distance between walls, i.e., the channel width, cannot be estimated, however, without additional consideration of the dynamics of the screw dislocations in the channels.

In fig. 8 are shown two views of the structure obtained in the KC model through a bifurcation from the initial uniform solution. This structure consists in a periodic distribution of veins which is shown in fig. 8. Comparing with the two micrographs of fig. 3, which are presented in the same orientation, we see that the KC model captures rather well the geometry of this vein structure, within an unknown length scale, however. Thus, one may note that the WA and KC models reproduce rather well the structures obtained in the cyclic deformation of the f.c.c. and b.c.c. metals, respectively, although their input properties do not seem to be specific to a particular class of materials. For instance, such basic properties as the stacking-fault energy and cross-slip properties in f.c.c. crystals, or the thermally activated lattice friction on the screw dislocations in b.c.c. metals at intermediate and low temperatures, do not appear explicitely in these models.

The sweeping mechanism which, according to the KC model, is at the origin of dislocation patterning has been simulated with the help of a three-dimensional simulation of dislocation dynamics and interactions (an account of the numerical method can be found in [65]). This is illustrated by fig. 9. In fig. 9(a) is shown the initial microstructure which consists of a spiral dislocation source of initially screw character, pinned in the centre of a model f.c.c. crystal of cubic shape and of size 15 μm. The source segment glides in a (111) slip plane and its Burgers vector is $\frac{1}{2}[\bar{1}01]$. During its rotation, it interacts with elongated, prismatic, dipolar loops with the same Burgers vector. The latter are randomly distributed in neighbouring slip planes, but do not intersect the slip plane of the loop. After one rotation of the source, the dipolar loops have been swept up along their glide cylinder and released, forming clusters that delimit a channel parallel to the edge direction (cf. fig. 9(h)). This simulation shows that sweeping occurs via the local formation of tripolar configurations, this mechanism being all the more efficient as the character of the source has locally a strong edge component (cf. [62,66] for more detail). As emphasised in [65–67], such results suggest that the process by which edge dipolar loops are formed, presumably by

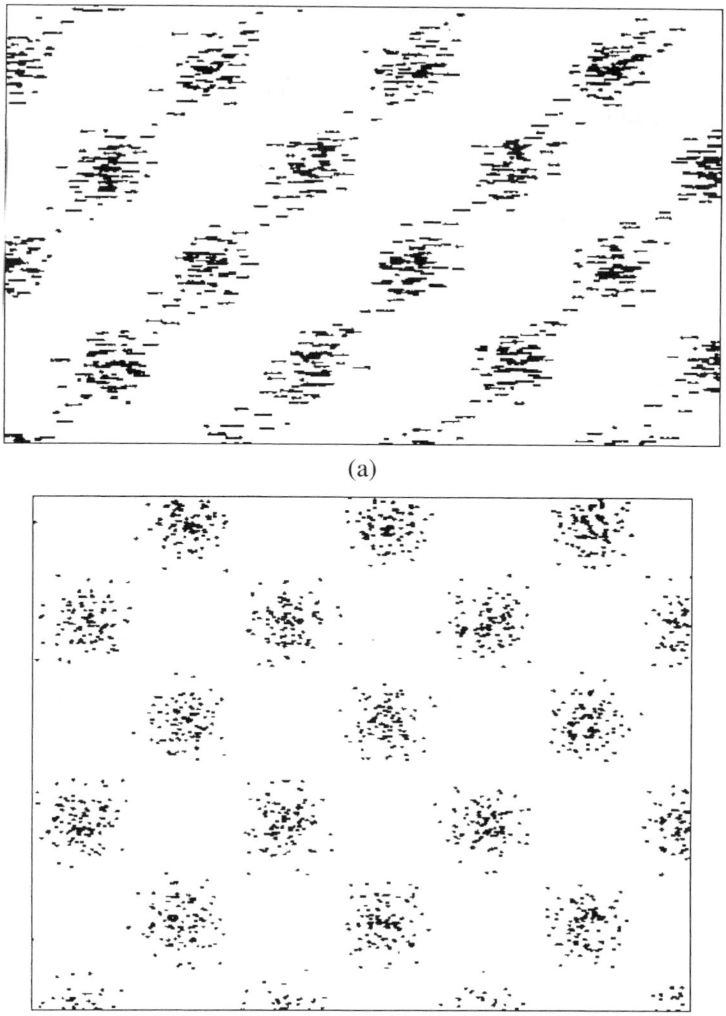

(a)

(b)

Fig. 8. The geometry of the vein structure in the KC model. The two figures are oriented like figs 3(a) and (b), respectively to allow for a comparison. (Courtesy of A. Franek et al. [20]). (a) Section parallel to the slip plane; (b) section perpendicular to the vein direction.

cross-slip, may play a significant role in fatigue patterning. Thus, the sweeping of dipolar loops may contribute to the early formation of clusters during the first cycles of fatigue. It may also play a role during the dynamical steady state of the PSBs. The loops formed in the middle of a channel are not swept by the gliding screw segments because the elastic interaction is too weak. Those formed at some distance from the middle interact with gliding segments which have a non-screw character, and they can be swept up towards the walls [41].

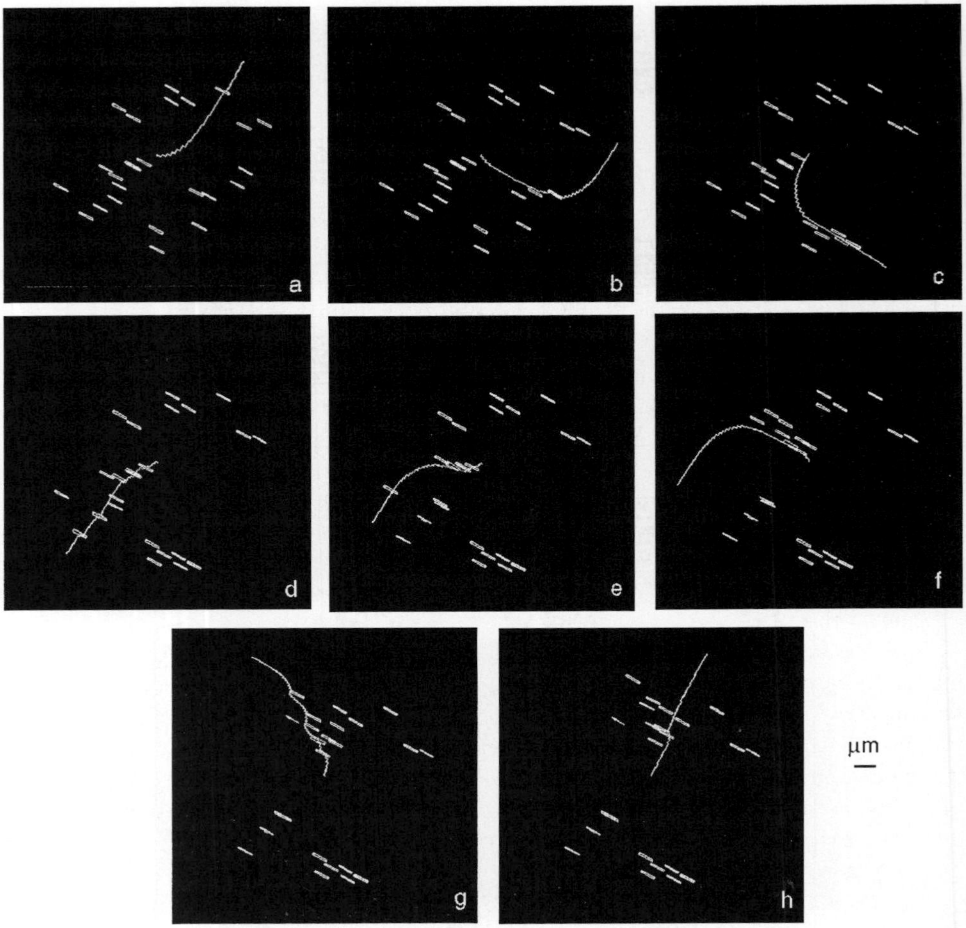

Fig. 9. Simulation of the sweeping mechanism, showing the dynamic interaction of a spiral source with a random distribution of edge dipolar loops, of dimension 1 µm × 0.1 µm, randomly distributed below and above its slip plane. The initial configuration is shown in (a) and the configuration after one rotation of the source in the clockwise direction is shown in (h). The dipoles have been swept, eventually forming clusters that delimit a channel parallel to the edge direction (after [66]).

2.5. Discussion

In this section, we come back to the three models and critically discuss several issues in their derivation. We focus, in particular, on the possible justifications for the various reaction terms, on the treatment of the long-range interaction stresses and on the additional equations or conditions that must be introduced in each model to put it in closed and solvable form.

2.5.1. *Dislocation populations and reaction terms*

As already mentioned, the main populations to be considered are the density of mobile dislocations, which can be of two different signs, and a density of weakly mobile or immobile dipoles or dipolar loops. According to the model considered and its objectives in terms of pattern description, there are large variations from one model to the other. We consider first the dipolar density. Considering the behaviour of the PSB walls in the DE model [30,40], the dipoles are assumed to be immobile and the purely reactional form of eq. (11) effectively takes into account the creation and destruction of dipoles, as governed by well-defined length scales. One may note that these terms are quadratic as they account for the probability of mutual encounter of two mobile dislocations or of one mobile dislocation and one dipole, the latter being considered as an entity.

In the WA model, the physical meaning of the reaction terms for the weakly mobile density, eq. (12), seems ambiguous. For example, the cubic term at the right-hand side of eq. (12), $\gamma\rho_{\mathrm{m}}\rho_{\mathrm{i}}^2$, has in principle the same meaning as the quadratic production term in eq. (11) of the DE model, $8\rho_1\rho_2 v(y_{\mathrm{d}} - y_{\mathrm{e}})$. This cubic non-linearity, which originates from the Brusselator (cf. section 2.3.3) plays a central role in the pattern-forming properties of the WA model. Despite some attempts to justify its occurrence [47], it seems that the latter implies an important assumption (see Salazar et al. [68] for a discussion).

As far as mobile dislocations are concerned they can be of either sign. In the WA model, these two densities are replaced by their sum, the total density $\rho_{\mathrm{m}} = \rho^+ + \rho^-$ that appears in eq. (13), and their difference, $\rho^* = \rho^+ - \rho^-$. The evolutionary equation for this last quantity is adiabatically eliminated [28,47]. This way, the equations of the model are brought into a form close to that of the Brusselator, at the expense of a few simplifying assumptions. In the DE model, the densities of plus and minus dislocations are treated explicitly and the loss term is consistent with a mechanism of dipole formation.

The DE model is interesting in several respects. First, the meaning of the reaction terms is physically founded and second, the wall properties are investigated only in terms of fluxes and densities. The velocity of the mobile dislocations, its relation to shear stress, the stress heterogeneities and long range stresses are not specified. These are strong simplifying assumptions and whether or not they can be justified and used to model emerging dislocation structures certainly deserves further consideration. Indeed, the DE model uses non-linearities but does not produce a bifurcation from uniform states and, hence, does not truly deal with pattern formation. In the KC model, the mobile density is assumed uniform and Orowan's law is used to describe its behaviour. The velocity of the mobile dislocations is described by a stress vs. velocity law, in the phenomenological form of a power law. No distinction is made between plus and minus mobile dislocations since the evolution of the dipole density is not related to the mobile density but to the strain rate.

This comparison raises questions about the level of sophistication needed for a sufficiently accurate description of the mobile density. The discussion of the reaction terms shows that the models under examination have no uniformity in their approach, although they deal with very similar properties. It is also clear that some pitfalls are now better understood than they were a few years ago.

2.5.2. Treatment of the spatial terms

In the three models, the spatial terms are treated in quite different ways. The DE model only considers fluxes (cf. eqs (9)–(11)) and, as seen just above, involves a few interesting but strong simplifying assumptions. In the KC model, the term expressing the divergence of the flux is expanded into a nonlocal form (eq. (14)), but the latter involves an influence function that has not yet been consistently calculated (cf. section 2.3.4). Finally, the WA model makes use of a Taylor expansion limited to the first non-zero term of the non-local form, thus assuming the existence of a cut-off distance.

We now turn to the discussion of the diffusive-like behaviour of the two populations of mobile and immobile dislocations. As far as mobile dislocations can be treated in a diffusive-like manner, one has to distinguish between their mobility in the slip plane and their spreading out from one slip plane to the other. The second mechanism, which is also central to the question of slip patterning (cf. section 3), has been modelled by several authors in terms of double cross slip (cf. section 2.3.1). In such a case, an effective diffusion coefficient can be defined provided one assumes that interactions between slip planes are limited to nearest-neighbours. The various (positive) pseudo-diffusion coefficients thus obtained differ slightly from one author to the other. They are stress or dislocation density-dependent and contain an adequate combination of cross-slip probability per unit time or glide area, cross-slip height and dislocation velocity (cf. [69] and the reviews [8,38]). A derivation of one of such diffusion coefficients is given in section 3.3.2. The spreading of slip activity in a slip plane and the propagation of an emerging structure can also, with some approximation, be expressed in diffusion-like form as discussed above in section 2.3.1 [44, 53]. The corresponding diffusion coefficient D_m appears in eq. (13). Alternatively, one can say that fluctuations of the mobile dislocation density do not necessarily relax according to Fick's law. Thus, it may seem much simpler, and perhaps no less correct, to assume either a uniform density of mobile dislocations (as in the KC model), or to take the flux or strain rate as a variable (as in the DE model).

The treatment of the immobile dipolar density also raises problems. It may seem paradoxical to look for the mobility properties of apparently immobile configurations. However, as noted by Differt and Essmann [30,40], experiments show that the periodicity of the wall structure is imperfect in freshly formed PSBs and becomes more regular upon further cycling. This question of wall mobility is elegantly solved in the DE model through a consideration of the balance of fluxes on both sides of a wall. An imbalance of the fluxes produces a translation of the walls, the latter being destroyed on the one side and reconstructed on the other (in fig. 7, above, the wall is stationary because the two fluxes are equilibrated). This may explain how a pattern adjusts to a change in stress or temperature. In the WA model which, unlike the two others, strictly follows a reaction–diffusion frame, it is thus necessary to define a diffusion coefficient D_i for the immobile or weakly mobile species. An additional constraint is that D_i must be sufficiently small compared to D_m in order to obtain pattern-forming instabilities. A derivation of D_i can be found in [28] but, as mentioned in section 2.3.1, it involves several assumptions (see also [8,68] for a discussion). Thus, the treatment of the immobile density within a reaction–diffusion frame has essentially a heuristic value. Finally, the non-local approach of the KC model also necessitates further elaboration in order to derive the exact form of the influence function (Φ in eq. (14)) in terms of the dislocation interactions.

We can conclude from this study of the spatial terms that the applicability of a pure reaction–diffusion framework to dislocation patterning in cyclic deformation is not straightforward. In addition, it is not clear to date whether or not the reaction–transport approach can be applied to such patterning problems. The non-local approach is the most rigorous of all, but no general solution has presently been elaborated in closed form.

2.5.3. Deformation and boundary conditions

The set of PDEs describing the evolution in time and space of the dislocation densities includes a specification of the straining conditions. The cycling conditions are in principle defined through a maximum stress or plastic strain amplitude per half-cycle, the cycling frequency not being a critical parameter. Both in the KC and WA models, the following argument is found. The dislocation density does not vary much during one cycle, so that its evolution is essentially governed by the maximum stress reached during a cycle. Then, this last quantity naturally plays the role of a control parameter. As a consequence, the cyclic conditions do not differ much from the conditions of monotonic deformation, at least in a single glide. As emphasised in the KC model, it follows that the dislocation structures formed in both situations (vein structure and extended Stage I of alloys, respectively) should exhibit large similarities. At the dislocation scale, this is equivalent to the assumption that in cyclic conditions the dislocations experience a unidirectional stress during their lifetime. This has been checked to be valid for screw dislocations [40], which carry most of the plastic strain rate. In contrast, the lifetime of the edge dislocations in the walls is about 2.5 cycles [30].

Orowan's law relates the plastic strain rate $\dot{\gamma}$ to the density and velocity of the mobile dislocations:

$$\dot{\gamma} = \rho_{\rm m} b v. \tag{15}$$

Its use is hardly avoidable in any consistent model of dislocation patterning since $\rho_{\rm m} v$ is a dislocation flux. Although Orowan's law is not included in the WA model, it appears in both KC and DE models. In the latter, it is used to express boundary condition specifying the dislocations fluxes on each side of the edge walls. In such conditions, it is necessary to distinguish between the two signs of the Burgers vectors because incoming dislocations of a given sign approach a wall from a given side.

Within a continuum frame, it is necessary to account for the conditions of stress equilibrium and strain continuity, which is only done in the KC model (cf. section 2.3.4). Thus, in principle, this model is potentially able to deal with the mechanical instabilities that are usually treated within the framework of continuum mechanics as well as with the material instabilities discussed in the present part. This led the authors of the KC model to put forward a somehow controversial hypothesis, according to which certain types of patterns, like the PSBs or cell structures, are only of mechanical origin [20,54,70]. Another important feature related to the boundary conditions, and which may condition the nature of the solutions of the PDEs, is the linear size of the domain considered. This point was considered by Glazov and Laird [71], who revisited the PSBs solutions of the WA model with a variable domain dimension, L. The latter was assumed to represent the grain size of a cyclically deformed polycrystal. As long as L is typically one order of magnitude larger than the width of the PSB channels, the solutions obtained are those of a medium of infinite size. This was for instance the case with the numerical solutions computed by Schiller and Walgraef [44] for the emergence of PSBs. For smaller values of L, the system gradually loses its ability to produce dislocation patterns.

Finally, the part of the DE model that completes it by treating the properties of screw dislocations [40] cannot be put in analytical form. This is essentially due to the present lack of understanding of which are the leading ones among several processes affecting the behaviour of screw dislocations in the channels. The possible processes are: the coupling with edges, the strong heterogeneity of the stress field due to the compatibility of deformation between the hard walls and the soft channels, the friction stress due to dislocation debris or point defects agglomerates, the sweeping of small prismatic loops and the annihilation by cross slip of screws of opposite sign.

2.6. Conclusion

The three models discussed in this part exemplify the advances recorded in the past decade in the field of patterning in cyclic deformation. The most ambitious of all is that of Walgraef and Aifantis which attempts to treat the succession of emerging structures met in the f.c.c. crystals with the same set of dynamical equations. In contrast, the model of Differt and Essmann makes much more reference to experimental reality but treats essentially the kinetic behaviour of non-screw dislocations in the PSBs. A mid path consists in modelling only the first bifurcation, from the uniform structure to the matrix structure, as attempted by Kratochvil and co-workers, but with the objective of predicting the relevant length scales.

In terms of non-linear dynamics, it is clear that many sets of equations can be constructed that may lead to the desired result. Again, the relevance to dislocation properties must be the cornerstone of any model. The applicability of the reaction–diffusion frame to dislocation patterning is still a matter of debate but gradient plasticity has, nevertheless, a wide potential for applications as exemplified in the next section. The recent stochastic model [72], as applied to patterning in cyclic deformation [73], provides an alternative manner to deal with the internal stresses due to the long-range elastic interactions. The problem is greatly simplified by treating the relevant quantities as stochastically fluctuating ones, whose average values can be calculated. The consequence is that the position variable does not appear explicitly in the model.

Emergent structures come out naturally as mathematical solutions of non-linear partial differential equations. In addition, it is certainly desirable to construct, whenever possible, simple physical pictures of the patterning processes, as done for example with the sweeping mechanism. It is hoped that, in the future, dynamical simulations will bring a useful complement to these theoretical studies and to the essentially static experimental observations.

3. Propagating band patterns

3.1. Introduction

The localisation of the plastic deformation of metals and alloys into slip bands and deformation bands has been investigated for more than a century. Various situations are encountered: the slip patterns can be either permanent and non-propagative, as in the Persistent Slip Bands (cf. section 2.2), or transient and propagative, as in the Lüders bands observed in uniaxial tension, or even repetitive and propagative, as in the Portevin–Le Chatelier (PLC) bands. Usually in the Lüders phenomenon, a single deformation front travels along the tensile specimen (cf. fig. 10(a)), while repetitive bands with well defined front and rear edges are associated with the PLC effect (cf. fig. 10(b)). Slip propagation is a particularly striking phenomenon, since it induces a spontaneous breaking of the symmetry of the specimen, and may result in slip markings often seen with the naked eye. In the present section, the focus is on the collective dislocation modes associated with propagative Lüders and PLC bands.

In both cases, the phenomenon has a damaging influence on forming properties and metallurgists have been looking for ways to avoid its occurrence. The mechanisms leading

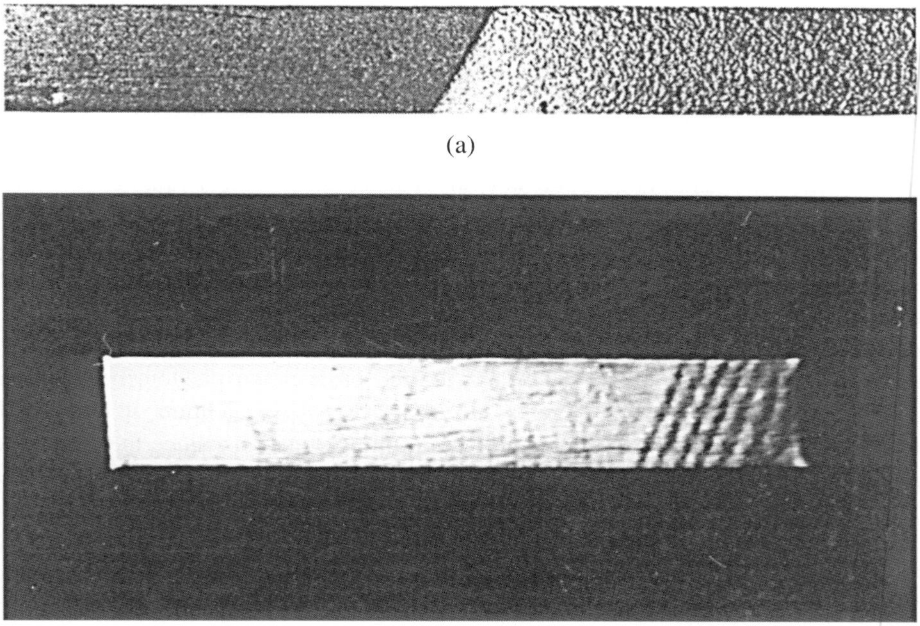

(a)

(b)

Fig. 10. (a) Lüders band in a mild steel tensile specimen, after C. Crussard [74]; (b) PLC band of type B in a tensile specimen of Al–5%Mg alloy, after Chihab et al. [75].

to bursts in slip activity are strain softening in the case of the Lüders phenomenon and strain-rate softening for the PLC bands. They are now reasonably well understood in term of the dynamics of the participating defect populations, but the physical origin of the propagation of slip patterns has not yet been fully elucidated. Early theories focused their attention on the local temporal evolution of mobile dislocations initially pinned by solute atoms (Lüders phenomenon) or interacting with the diffusing solute atoms (PLC effect). The conventional approaches to the PLC effect have undergone further refinements and extensions, and they are now able to successfully explain numerous effects of strain, strain rate and temperature on the occurrence of the propagating bands. In the Lüders phenomenon, early models [76] have remained dormant, until it was fully realised that conventional theories without any kind of spatial dependence could not quantitatively describe propagative patterns [77].

When plastic deformation is uniform, the associated length scale of the field is of the order of the sample dimensions. Therefore, it is usually large compared to any internal material length scale. Under these conditions, conventional constitutive formulations of plastic flow, which do not include internal length scales, are sufficient to describe the deformation process. However, gradients of plastic strain may appear when strain localisation occurs, due to some collective dislocation behaviour. When the plastic slip activity localises into deformation bands, the characteristic length scale of the deformation field becomes of the order of the width of the active glide bands. In a number of situations, the latter becomes small enough to be of the order of some internal length scale originating in the material structure, such as the grain size in polycrystals. In such conditions, internal length scales must be used when an account of the detailed structure of the strain field is undertaken. An appealing way to do so is to include them in the constitutive laws, by assuming that the flow stress depends on strain or strain gradients (see section 3.3).

Additional field-length scales are provided by the slip band spacing, whose range can span several orders of magnitude. When it occurs, slip propagation indicates the existence of a strong spatial coupling between neighbouring elements, which also provides a characteristic length scale. Therefore, gradients and length scales have to be incorporated into the formulations in order to account for the propagation phenomena. However, the physical origin of the coupling is still controversial. In addition, it may not be straightforward to obtain the solutions of the resulting equations, since the latter often admit a continuum of possible velocities. The problem, then, consists in selecting the velocity actually picked up by the deforming sample within the range.

It is the purpose of this part to present current work on the Portevin–Le Chatelier and Lüders (or Lüders-like) propagative plastic instabilities. In each case, the phenomenon will be described, and an introduction will be given to local and non-local constitutive formulations, the former describing the behaviour of a representative material element not interacting with its neighbourhood, the latter involving such interactions. Then, the mechanisms that have been proposed for the velocity selection and the predicted results will be discussed. A discussion of some of these aspects can also be found in recent reviews [78–80].

3.2. Propagating plastic instabilities

3.2.1. The Lüders phenomenon in polycrystals

In tensile samples loaded at a constant cross-head velocity, the Lüders phenomenon usually consists of a single band of plastic deformation travelling along the specimen, with a well defined angle with respect to the specimen axis, typically 50–55°, depending upon the sample geometry. The band nucleation, usually occurring at one grip, is associated with the elastic relaxation of the rest of the sample and a drop in stress, from the Upper Yield Point (UYP) to the Lower Yield Point (LYP). The plastic zone then spreads at constant LYP stress level. A clear demarcated front separates the plastically deformed area from the undeformed one into which it propagates, until the specimen is uniformly deformed at the so-called Lüders strain. Afterwards, the plastic deformation proceeds uniformly with positive strain hardening (cf. fig. 11).

The phenomenon, first reported in the early works of Piobert [81] and Lüders [82], occurs in mild steels, impure iron or b.c.c. polycrystals at room temperature. It is also currently observed in aluminium substitutional alloys [83]. In b.c.c. crystals, it is usually associated with the pinning of dislocations by interstitial impurities. Around room temperature, initially locked dislocations are collectively unpinned at the UYP, usually at one grip. No further repinning can occur, as it does in the PLC effect, as the time scale for diffusion of the impurity atoms is much larger than the total duration of the experiment. Since the dislocation propagation stress is much lower than the UYP, an abrupt multiplication of dislocations takes place. This surge of a large density of mobile dislocations locally produces strain softening. The characteristic stress plateau at the LYP level is then associated with the propagation with a constant velocity of a band of freed dislocations. From the end of the plateau onwards, plastic deformation proceeds uniformly in the specimen. When the temperature is increased, the interstitial atoms become weakly mobile by diffusion and a smooth transition with the PLC effect can sometimes be noticed.

This unpinning mechanism includes a strong grain-size effect, i.e., the yield-point phenomenon and the Lüders plateau decrease as the grain size is increased. Both vanish above a critical value amounting to a few hundreds microns [84]. The well-known Hall–Petch relation, which gives the dependence of the LYP on the grain size, originates from the

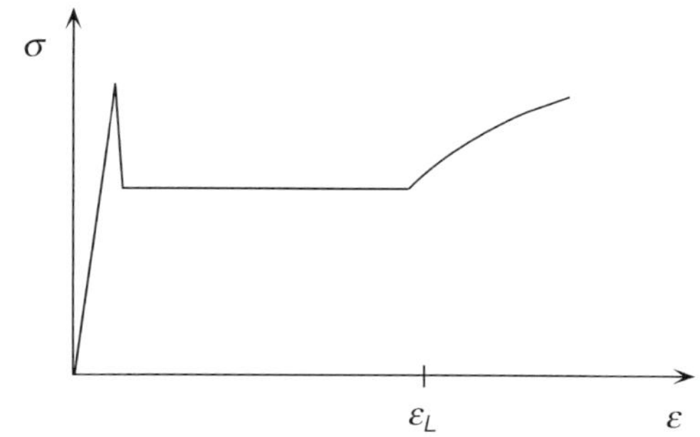

Fig. 11. Sketch of a typical stress–strain curve associated with the Lüders phenomenon in a mild steels. ε_L is the Lüders strain.

study of this effect [85]. According to the Cottrell assumption [86], the UYP is not reached until plastic slip is able to cross grain boundaries by activation of new dislocation sources in the neighbouring grains. This critical stress is attained once the stress concentration due to dislocations pile-ups reaches a critical value. In this modified mechanism, spatial interactions are thought to be responsible for slip propagation. They are intergranular in nature, since they originate from plastic slip gradients between neighbouring grains.

The Lüders band phenomenon may be further complicated by lattice rotations as, according to Jaoul [84], each grain deforms in single glide. Textural softening may therefore play an additional role in the instability, but it is unlikely that the phenomenon could be reduced to this single mechanism. The occurrence of single slip may also explain the band orientation. Actually, the average direction of the most favoured $\langle 111 \rangle$ slip direction in the grains of a randomly oriented b.c.c. polycrystal is 51°, which is precisely the observed orientation in flat specimens. As mentioned above, however, the sample geometry and the related stress state may also determine the value of the observed angle. A consistent model for the Lüders bands propagation should therefore combine the avalanche-like effect of the unlocked dislocations on the one hand, and on the other hand, the space gradients able to account for band propagation and size effects. Lattice rotations could be accounted for in a further refined model.

In Hähner's model [87], the features of dislocation dynamics in polycrystals are described by the densities of three different dislocation species: mobile dislocations, piled-up dislocations and immobile dislocations that are not piled up. Piled-up dislocations play a significant role in propagating the plastic strain across the grain boundaries and are therefore considered separately. Interactions between these populations include the generation of mobile dislocations by the pile-up mechanism, as well as the loss of mobile dislocations via their piling up at grain boundaries or by storage due to forest interactions. A relaxation mechanism of the pile-ups is also included. This local dynamics results in a set of nonlinear ordinary differential equations for the time evolution of the three unknown dislocation densities. Since this system does not exhibit any steady-state solution, it results in a hardening/softening behaviour. Assuming constant pile-up and mobile dislocation densities, the flow stress features a hardening behaviour. It is assumed that such conditions prevail in the trailing edge of the Lüders front. At the leading edge, the slip activity is initiated by the avalanche-like multiplication and motion of initially pinned dislocations, producing a strong strain softening. As a simplifying assumption, the immobile dislocation density is neglected, while it is again assumed that the pile-up dislocation density remains steady. As such, the model describes the local hardening/softening of a polycrystalline material and includes the Hall–Petch type of behaviour of the flow stress. The spatial aspects of the model will be described later on in section 3.3, and the front velocity determination will be discussed in section 3.4.4.

3.2.2. "Lüders-like" phenomena in single crystals

The feature common to the class of "Lüders-like" phenomena in single crystals is the occurrence of propagative slip bands associated with strain softening, sometimes with load serrations, or with an extended Stage I. The possible softening mechanisms are very diverse and, as the experimental observations have been extensively reviewed by Luft [88], Neuhäuser [89], and Neuhäuser and Hampel [90], we just recall here the main mechanisms.

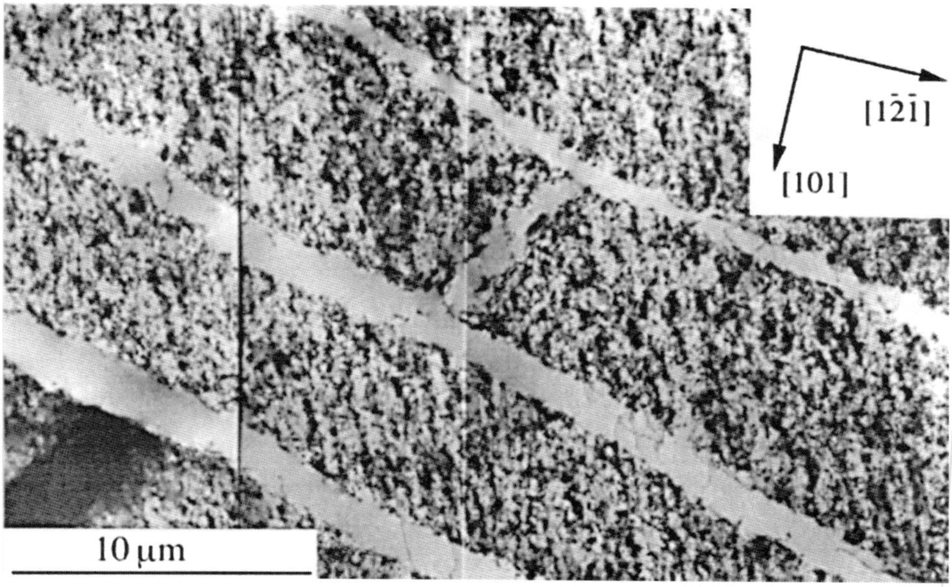

Fig. 12. Slip channels in a [941] molybdenum single crystal strained at 173 K and restrained at 378 K. Foil plane normal to the primary Burgers vector (1 MeV TEM). Courtesy A. Luft, after [88].

In semiconducting crystals of Si and Ge [91,92], extensive dislocation multiplication is responsible for the yield point behaviour at low temperatures. In other materials, the most important softening mechanism originates from the destruction of obstacles by the moving dislocations and the subsequent decrease of the glide resistance. The obstacles to be destroyed can be short-range ordered clusters, defects produced by quenching or by irradiation, or dislocation structures produced by a prestrain at a different temperature or along a different path of deformation. The shearing of small coherent precipitates also induces strain softening and slip localisation, but it is not clear whether propagation occurs in that case. It seems that slip propagation is only observed beyond a certain threshold in the softening of the glide resistance. TEM observations of the strain localisation regions [88] reveal that they consist of channels which have been cleared out of the pre-existing obstacles by the moving dislocations (cf. fig. 12).

Rather complex dislocation processes, almost invariably involving cross-slip, take place during strain softening. Strain hardening occurs next, as a result of dislocation interactions and accumulation of dislocation debris. When the glide resistance becomes again comparable to that of the virgin parts of the material, slip activity ceases in the strain-hardened region and is transferred to another place. At the specimen surfaces, the channels appear as slip bands of width typically in the μm range, containing fine slip lines with typical spacing in the 10 nm range. As emphasised by Neuhäuser in the case of neutron-irradiated bulk copper crystals [89], the observation of surface steps on strained specimens also reveals two other length scales. The slip bands are arranged in the form of bundles with typical thickness several tenth of microns and typical spacing between bundles in the

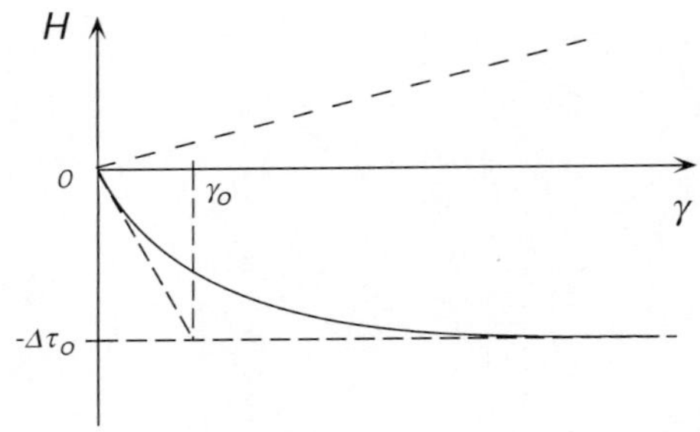

Fig. 13. Plot of the dependence of the hardening/softening function H in eq. (16) vs. the shear strain γ. H is the sum of a linear hardening term (dotted) plus a contribution which decays with a characteristic strain γ_o, inducing a total softening $\Delta\tau_o$.

range of several hundreds of microns. This hierarchical structure of length scales therefore involves different spatial couplings which will be discussed further in section 3.3.

Any mesoscopic modelling for the yield point effect and strain softening by avalanches can be lumped into a constitutive formulation by using two main parameters: the total amount of strain softening $\Delta\tau_o$ and the total plastic strain y_o required to remove a certain type of obstacles, thereby softening the crystal by $\Delta\tau_o$. In a simple way [93], the softening/hardening behaviour of the material can be reflected by taking the glide resistance in the form

$$\tau = \tau_o + H(\gamma) = \tau_o + h\gamma - \Delta\tau_o\big[1 - \exp(-\gamma/\gamma_o)\big]. \tag{16}$$

Here, τ_o is the initial glide resistance when $\gamma = 0$ and the function $H(\gamma)$ includes the softening behaviour in the form of a decaying exponential (cf. fig. 13).

In conditions of easy glide, strain hardening originates from the interactions between dislocations or pile-ups of opposite sign gliding on parallel slip planes. The associated hardening coefficient h typically amounts to the Stage I hardening value, $h \cong 3.10^{-4}G$ for the f.c.c. single crystals, or $h \cong 3.10^{-3}G$ when several active slip systems are present (Stage II). Softening occurs when the dimensionless quantity $\Delta\tau_o/h\gamma_o$ is larger than one. It is then possible to define the Lüders strain γ_L such as $H(\gamma_L) = H(0) = 0$. The effective stress τ^* in a glide plane is usually weakly strain-rate-dependent. In order to describe a weakly rate-dependent behaviour, associated with a very fast motion of dislocations at medium temperatures, use can be made of the simple relationship $\tau^* = \vartheta\dot{\gamma}$, where ϑ is a constant strain rate hardening coefficient. Therefore, the local part of the shear stress on a slip plane is obtained by including a strain rate-dependent contribution into eq. (16):

$$\tau = \vartheta\dot{\gamma} + \tau_o + H(\gamma) = \vartheta\dot{\gamma} + \tau_o + h\gamma - \Delta\tau_o\big[1 - \exp(-\gamma/\gamma_o)\big]. \tag{17}$$

As such, this model reproduces the softening behaviour associated with "Lüders-like" phenomena in single crystals, but not, obviously, the coupling between neighbouring active zones (see sections 3.3 and 3.4.5).

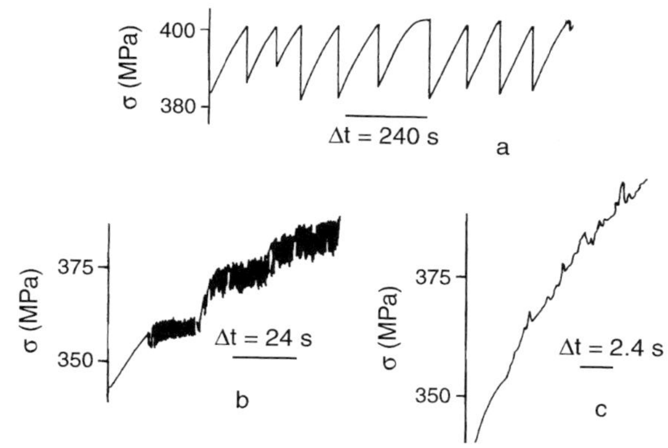

Fig. 14. Serrated stress-strain curves of an Al–5%Mg polycrystal deformed at room temperature with different imposed strain rates $\dot{\varepsilon}_a$. (a) $\dot{\varepsilon}_a = 5 \times 10^{-6}\,\mathrm{s}^{-1}$, type C serrations; (b) $\dot{\varepsilon}_a = 5 \times 10^{-4}\,\mathrm{s}^{-1}$, type B serrations; (c) $\dot{\varepsilon}_a = 5 \times 10^{-3}\,\mathrm{s}^{-1}$, type A serrations. After Chihab et al. [75].

3.2.3. The Portevin–Le Chatelier effect

The Portevin–Le Chatelier (PLC) effect consists in the repeated generation and often propagation of plastic deformation bands with sharp edges along a tensile specimen (see fig. 10(b)). In tests carried with a constant applied strain rate, each of these events is accompanied by an elastic relaxation of the sample and, therefore, by a drop in external load (cf. fig. 14). During experiments performed with a constant applied loading or stress rate, strain bursts are recorded.

The first detailed experimental studies were those of Le Chatelier [94] and Portevin and Le Chatelier [95] on mild steels and duralumin alloys respectively. With the advent of the dislocation theory, Cottrell proposed his basic model of pinning and unpinning of dislocations by impurity clouds [86]. In this model, dislocations were assumed to move continuously. Later on, McCormick [96] and van den Beukel [97] described their motion as a jerky one, the dislocations breaking through forest obstacles with the help of thermal activation and moving very fast between obstacles. It is now widely accepted that the PLC bands stem, on the dislocation scale, from Dynamic Strain Ageing (DSA) in solution hardened alloys. The mechanism of DSA is the additional pinning of mobile dislocations temporarily arrested at obstacles by diffusing solute atoms. Whereas the resistance of obstacles to dislocation motion increases with increasing dislocation velocity in a conventional thermally activated mechanism, the glide resistance increases with decreasing velocity when solute diffusion is effective during the waiting tune. When the strain rate and temperature are such that the waiting time of dislocations and the diffusion time of solute atoms are of the same order of magnitude, this inverse effect is at its maximum. When it becomes predominant, the competition may result in a negative Strain Rate Sensitivity (SRS), and in the occurrence of the PLC effect. Dislocations are successively locked and unlocked from the pinning solute atoms in a synchronised collective manner [98].

In samples deformed with a constant strain rate, three different types of load drops (or serrations) are observed, usually referred to as types A, B and C. Type C serrations

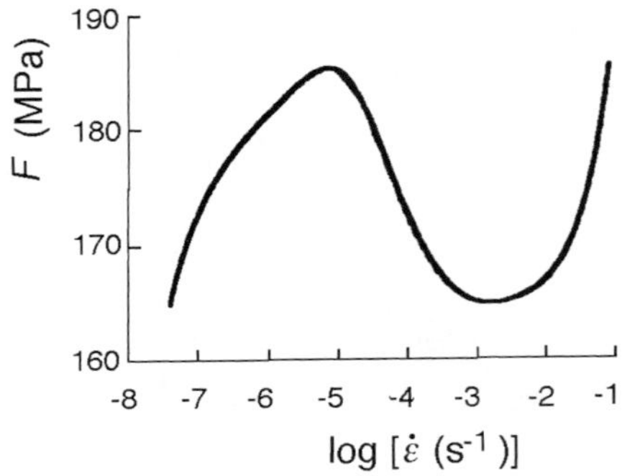

Fig. 15. The viscoplastic Penning's function $F(\dot{\varepsilon})$ for an Al–Mg alloy. In the intermediate range of strain rates, the portion of negative slope of F describes the negative strain-rate sensitivity associated with the PLC effect.

are observed at low strain rates (or high temperature). After the first band has initiated at one edge of the sample and propagated along the entire gauge length, other bands form at random locations, usually after some significant reloading time. Type B serrations are characteristically observed at intermediate strain rates (cf. fig. 10(b)). The associated load drop is typically smaller than in type C bands, and the band nucleation events are now correlated in space. Once the first band has formed, the next one forms at an adjacent location, and the plastically deformed area spreads in a relay-race manner. Type A serrations are obtained at larger strain rates (or low temperature). The load drops and reloading times display more dispersion than in type B or C serrations, the smaller ones being the more numerous. The correlation between bands is again stronger: they occur in a fast sequence, each band moving continuously as a steady-state profile from one end of the specimen to the other. The amplitude of these serrations and their strain rate or temperature dependencies are understood as reflecting the difference between the critical stresses of dynamically aged and free dislocations.

Macroscopically, when the dislocations properties are averaged over a cross-section, the local constitutive equation which incorporates this situation is of the Penning type [99]

$$\sigma = h\varepsilon + F(\dot{\varepsilon}). \tag{18}$$

Here σ denotes the uniaxial stress in the cross section of the sample, ε is the local the plastic strain and $\dot{\varepsilon}$ the local plastic strain rate. h is a constant hardening rate and $F(\dot{\varepsilon})$ a characteristic sigmoidal-shaped non-linear (viscoplastic) function with three strain-rate regimes (fig. 15). In the low strain-rate range, dislocations have to overcome the thermally activated obstacles and to drag along their solute atmosphere; the SRS is positive. In the high strain-rate regime, dislocations are free from solute atmospheres; the SRS is still positive but smaller. In the intermediate range, the dislocation velocity decreases with increasing stress due to increased pinning, which leads to a negative macroscopic SRS. In this last regime, dislocations must be either pinned or unpinned. Thus the interpretation is that they jump to (and from) the stable portions of the characteristic curve where dislocation velocity increases with stress. In parallel, at a macroscopic scale, linear stability analysis shows that uniform plastic deformation is unstable [100]. Uniform flow, then, gives way to spatio-temporal patterning. In a cross-section, oscillations in time are observed which reflect a spatial pattern organised in propagating bands of localised deformation. In the F vs. $\dot{\varepsilon}$ phase space, the system describes a periodic orbit (cf. fig. 19, below) including two discontinuous jumps between the slow/fast stable branches of the viscoplastic function F [100]. Numerical estimates for this function can be obtained within the framework of the current models for the dynamical interaction of dislocations and solute atoms (cf. [101] and section 4.3.3 for further discussion).

Despite a number of deficiencies in the above constitutive law (18), which remains admissible for the small strains involved in a single band, the Kubin–Estrin model [101] outlined above offers an adequate qualitative description of the temporal aspects of the PLC effect. A wealth of refinements have since been offered. In addition to the existence of two critical strain rates between which the PLC effect occurs, the model consistently explains numerous experimental features such as the occurrence of a critical incubation strain [101] and the decrease of load drop or strain burst amplitude as the applied rate is increased [102]. Unsolved problems in this model are those related to the spatio-temporal aspects of strain localisation. More specifically, the band velocity remains undetermined in the analysis. In the absence of any coupling mechanism between neighbouring material elements, no mechanism is defined by which the plastic deformation could be allowed to propagate along the sample. Models based on the current phenomenology of the PLC effect and involving different types of spatial couplings are discussed sections 3.3 and 3.4. A quite different type of model, which emphasizes the dynamical aspects of the phenomenon, will be discussed in section 4.2.2.

3.3. Modelling gradient effects

The space gradients involved in the plasticity formulations may be classified into two distinct groups. They originate either from mechanical effects related to the loading mode or triaxiality of the stress field, or from intrinsic material properties, at the scale of dislocation mechanisms or at the grain level. When they are of material origin, they are accounted for in the constitutive relations themselves.

3.3.1. Space coupling by mechanical effects: stress field triaxiality

Necking, i.e., the Considère instability [103], is a typical plastic strain localisation process. It occurs in tensile tests when the uniaxial nominal (engineering) stress vs. strain curve falls beyond a maximum. For those materials whose tensile curve falls uniformly from that point, the neck forms in some narrow region of the specimen remaining fixed with respect to the material; further plastic deformation of the sample is localised in the neck. Some materials however, particularly polymers, exhibit a nominal stress–strain curve with an upturn of the nominal stress as further straining occurs, while no softening of the material is observed in the true stress vs. strain curve. In such materials, the neck initiates much in the same way as in typical metallic materials, but the upturn of the nominal stress forces the propagation of the neck along the entire length of the specimen. In the transition zone between the necked and unnecked regions, substantial deviations of the stress field from the uniaxial approximation usually occur, and neck propagation can be explained from its full three-dimensional analysis [104]. It is the triaxiality of the stress field connecting the uniaxial states on either side of the transition area that provides the spatial coupling responsible for the propagation. Approximate analyses of stress triaxiality effects in the strain localisation regions often use the triaxiality factor F_t defined as $F_t = \sigma_e/\sigma$, where σ_e is the three dimensional effective stress, here the von Mises stress, and σ the uniaxial stress. F_t depends on the geometry of the sample and on the material properties. For rigid plastic materials, it has been given by Bridgman [105] in the form

$$F_t = 1/\left[(1 + 2R_c/R)\log(1 + R/2R_c)\right], \tag{19}$$

where R is the local radius of the specimen cross-section and R_c the local radius of curvature of the cross-section profile. In the linear approximation, valid for low curvatures,

$$\frac{1}{R} \cong \frac{\partial^2 R}{\partial x^2}, \tag{20}$$

and the Bridgman factor eventually reads (see [106,107])

$$F_t \cong 1 - \frac{A_0}{2\pi}\frac{\partial^2 R}{\partial x^2}\mathrm{e}^{-\varepsilon}. \tag{21}$$

A_0 is the initial cross-section of the specimen and ε the local plastic strain. Although this result is valid for an ideally plastic material only, it has been used unaltered for more complex model materials. Thus, models including space gradients have been proposed [106,108] that could yield quantitative predictions for the band velocity (see section 3.4).

In single crystals, stress field triaxiality stems from the effects of slip steps. In tensile specimens with fixed grips, any local shear invariably produces a bending moment along the specimen length. Therefore, shear stress gradients arise between the opposite edges of the specimen. However, two-dimensional finite element calculations performed for a flat model sample with a few slip steps [90] indicate that these gradients do not govern slip propagation. Further, it was assumed that the stress field induced by a slip step at the band front is shielded by the back stress of a piled-up dislocation group developed in the slip plane of the step, by analogy with the plastic zone developed around a moving crack. The long-range stresses of these dislocations induce a shear stress in the neighbouring slip planes of the virgin material which exhibit two maxima as a function of the distance to the slip step. In the model situation investigated, the slip step height is 0.1 μm and the maxima occur at distances 0.2 μm and 25 μm ahead of the band front. The smaller value corresponds well to the observed distances between slip lines and the larger to the distance between slip bands. Therefore, it has been suggested [90] that this slip step effect may induce the generation of new slip lines and bands, and govern the Lüders-like propagation in single crystals.

3.3.2. *Space coupling by intrinsic material effects*

Spatial coupling by material effects may originate either from the microstructure, i.e., from dislocation mechanisms in a single crystal, or from the “mesostructure”, more specifically from compatibility mechanisms between grains in a polycrystal. The microstructural mechanisms can be of two types: the spreading out of dislocation glide from an active region can be triggered either by double cross-slip or by long-range elastic stresses stemming from dislocation interactions.

Double cross-slip

Clearly, the double cross-slip of mobile screw dislocations is a coupling mechanism between adjacent active slip planes (cf. section 2.5.1 and fig. 9). Let us consider a simple one-dimensional model of slip activity, where slip planes are normal to the x direction. Each slip plane loses dislocations by cross-slip and receives dislocations that have cross-slipped from the two neighbouring slices at distance $\pm h_{\mathrm{cs}}$, where h_{cs} is the cross-slip height. If v is the average dislocation velocity and P the cross-slip probability, the rate of change of the mobile dislocation density ρ_{m} by cross-slip at position x and time t is:

$$\frac{\partial \rho_{\mathrm{m,cs}}(x,t)}{\partial t} = Pv\left\{-\rho_{\mathrm{m}}(x,t) + \frac{1}{2}\left[\rho(x+h_{\mathrm{cs}},t) + \rho(x-h_{\mathrm{cs}},t)\right]\right\}, \tag{22}$$

where the factor of $1/2$ stems from the fact that a cross-slip event can occur in either the positive or negative direction. This relationship implies that the number of cross-slip events is proportional to the cross-slip probability and to the glide area covered by the dislocations per unit time. Using a power expansion of the mobile dislocation density ρ_{m} limited to the second-order term, one obtains to the second order of approximation:

$$\frac{\partial \rho_{\mathrm{m,cs}}(x,t)}{\partial t} = \frac{1}{2}Pvh_{\mathrm{cs}}^2\frac{\partial^2 \rho_{\mathrm{m}}(x,t)}{\partial x^2} = D\frac{\partial^2 \rho_{\mathrm{m}}(x,t)}{\partial x^2}. \tag{23}$$

The *local* rate of change of the mobile dislocation density in a slip plane, $\partial\rho_{\mathrm{m}}/\partial t$, as due to dislocation reactions, is the *total* rate of change in the same slip plane $(\partial\rho_{\mathrm{m}}/\partial t)_{\mathrm{total}}$, minus the rate of change due to cross-slip, that is the non-local contribution from eq. (23). Thus,

$$\frac{\partial \rho_{\mathrm{m}}(x,t)}{\partial t} = \left(\frac{\partial \rho_{\mathrm{m}}(x,t)}{\partial t}\right)_{\mathrm{total}} - D\frac{\partial^2 \rho_{\mathrm{m}}(x,t)}{\partial x^2}. \tag{24}$$

Similar expressions involving a "diffusion-like" constant $D \propto Pvh_{\mathrm{cs}}^n$ (with n between 2 and 4) have been obtained by several investigators ([38,69,109,110], see also section 2.5.1). Such expressions can be used to transform conventional constitutive forms (not including space) into spatio-temporal relationships by replacing the total rate of change by its local counterpart. For instance, this procedure can be applied to the Orowan equation by averaging eq. (24) over a specimen cross-section. The new Orowan equation including the gradient term reads

$$\frac{\partial \varepsilon}{\partial t} - D\frac{\partial^2 \varepsilon}{\partial x^2} = \rho_{\mathrm{m}}bv = \dot{Y}, \tag{25}$$

where $\dot{Y}$ is the local strain rate, and $\partial\varepsilon/\partial t$ the total strain rate. The cross-slip model, originally suggested by Estrin and Kubin [101], includes gradients in the analysis by

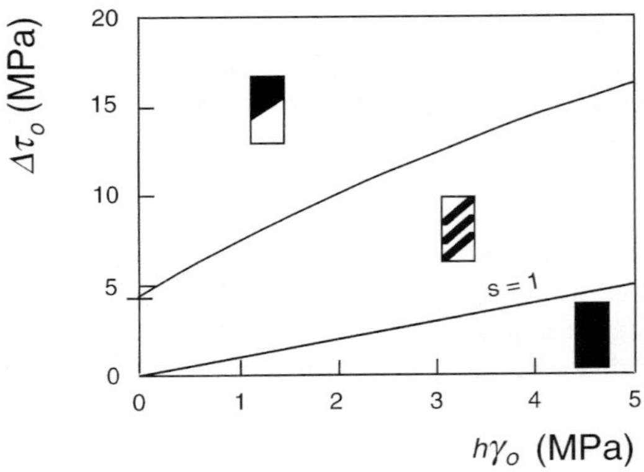

Fig. 16. Instability map showing the domains of uniform, localised and propagative flow as a function of the two softening parameters $\Delta\tau_o$ and $h\gamma_o$. Flow occurs in the black regions (after Bréchet et al. [93]).

substituting the local strain rate $\dot{Y}$ in eq. (25) for the total strain rate in the Penning's eq. (18). This leads to a gradient plasticity type of constitutive relation

$$\sigma = h\varepsilon + F\left(\frac{\partial \varepsilon}{\partial t} - D\frac{\partial^2 \varepsilon}{\partial x^2}\right). \tag{26}$$

The cross-slip distance, h_{cs}, is the natural length scale of this mechanism. As a double cross-slip event (cf. fig. 4) can be triggered by a repulsive interaction, the cross-slip height is governed by the distance travelled by the screw dislocation in the cross-slip plane, before it cross-slips back to the initial slip plane. Although no theoretical prediction is available for this quantity, the latter bears some analogy with the critical annihilation distance y_s defined in section 3.2.2. The cross-slip distance is much too short to account for Lüders band propagation, although it might play a role in the refinement of the slip pattern in the wake of the front. As far as the PLC effect is concerned, double cross slip is a possible candidate for band propagation mechanism (cf. section 3.4.4).

Elastic interactions of dislocations

The coupling of active slip planes via internal stresses has been observed in a three-dimensional simulation of dislocation dynamics and interactions (see section 2.4). In the case of a copper alloy, the simulated specimen is a cube of size 15 μm deforming in single or duplex glide. To mimic the organisation of slip activity during strain softening phenomena, a rule specifies the average strain softening effect on the slip plane, in the form of eq. (16) [93]. Three different situations can be met according to the relative values of $\Delta\tau_o$ and γ_o (cf. fig. 16).

By increasing the value of the dimensionless ratio $s = \Delta\tau_o/h\gamma_o$, one observes first uniform deformation, followed by a static strain localisation and finally a propagative localisation. In the latter, it is seen that the long-range elastic stresses of the mobile dislocations at the band front activate new sources ahead of the front, thus inducing a propagation of the active zone. The coupling stress is such that interacting dislocations (or

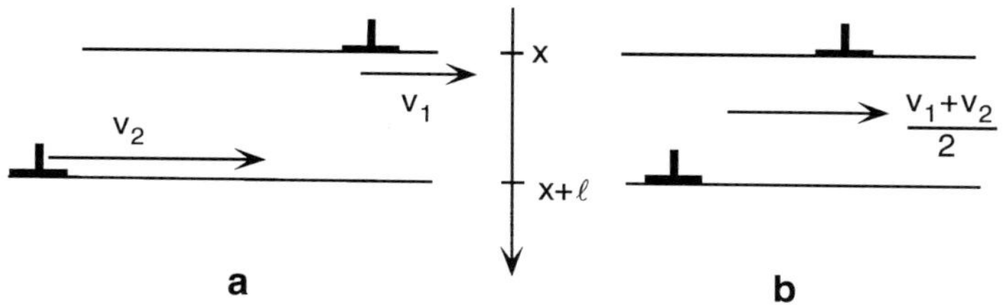

Fig. 17. (a) Two dislocations move with different velocities in neighbouring parallel slip planes, the faster being located at the band front; (b) under the effect of their elastic interaction, and if the passing stress is not exceeded, the two dislocations tend to adopt a common average velocity: the faster dislocation is slowed down and the slower one is accelerated.

pile-ups) of the same sign moving at different velocities tend to move with closer velocities during the activation event, the most active being slowed down and the slowest one being accelerated (cf. fig. 17).

These interactions can be considered as originating from gradients of either plastic strain rate or of plastic strain. This leads to two different non-local coupling stresses.

$$\tau_{\mathrm{nl}}(x) = \int G(x'-x)\big(\dot{\gamma}(x') - \dot{\gamma}(x)\big)\,\mathrm{d}x', \tag{27}$$

$$\tau_{\mathrm{nl}}(x) = \int M(x'-x)\big(\gamma(x') - \gamma(x)\big)\,\mathrm{d}x'. \tag{28}$$

In both cases, it is assumed that the stress at position x is influenced by the distant state at position x' through influence functions G and M (with different dimensions). It is natural to assume that these functions are even (and tend to zero at infinity). This can be seen from fig. 17, where the fast dislocation is slowed down whatever the sign of the distance between the two dislocations along the x axis. The slow dislocation is above or below it. Note the use of a similar influence function, Φ, in eq. (14).

By neglecting long-range interactions beyond a cut-off distance ℓ, for example an average spacing between active slip planes, and expanding in power series the influence functions and the slip (or slip rate), it is readily shown that the non-local parts of the stresses reduce to gradient forms

$$\tau_{\mathrm{n}\ell}(x) = \frac{1}{3}G(0)\ell^3\frac{\partial^2\dot{\gamma}}{\partial x^2} = \eta\ell^2\frac{\partial^2\dot{\gamma}}{\partial x^2}, \tag{29}$$

$$\tau_{\mathrm{n}\ell}(x) = \frac{1}{s}M(0)\ell^3\frac{\partial^2\gamma}{\partial x^2} = \zeta\ell^2\frac{\partial^2\gamma}{\partial x^2}. \tag{30}$$

In these relations, ℓ defines a characteristic length scale for the interaction. The sign of the constants η and ζ depends on the assumed coupling mechanism. Consider, for instance, in a tensile test, a neighbourhood where the strain or strain-rate profile undergoes positive fluctuation and, hence, exhibits a negative curvature. The resulting non-local stresses are negative if the constants η and ζ are positive. Thus, they describe a coupling mechanism that induces a softening of the glide resistance in this neighbourhood. Conversely, with negative values of the coupling constants, a positive glide resistance, i.e., a back stress, is generated. Further, the coupling term may have a stabilising or destabilising influence on uniform plastic flow, depending on the mode of loading. In a constant strain-rate test, softening (η or $\zeta > 0$) is destabilising, as strain rate tends to become localised in the region considered, whereas hardening has a stabilising effect. In tests performed with a constant loading rate, there is no global condition on the strain rates and no possible relaxation of stress associated with strain localisation: stress must increase in the localisation region. Softening non-local coupling (η or $\zeta > 0$) has therefore a stabilising effect on fluctuations, as it helps preserve mechanical equilibrium and material continuity. Thus non-local softening appears as a stabilising factor in constant loading rate and a destabilising one in constant strain rate. Several models will be considered in the next sections and in section 4.3.3, that involve different signs of the coupling constants, based on different physical arguments.

Compatibility stresses in polycrystals

During the plastic deformation of polycrystals, strain or rotation discontinuities can develop at the grain boundaries. Within a continuum approach, these discontinuities may be approximately replaced by gradients of the order of the jumps divided by the grain size d. Introducing gradients in this way proves necessary when the strain or rotation fields vary significantly over a distance of the order of the grain size. In such a case, the requirement of continuity of the deformation fields at grain boundaries induces additional elastic and plastic accommodation stresses in the constitutive formulations.

Compatibility stresses can be derived in a simple way if lattice rotations are not allowed. Consider a one-dimensional discrete model of slip activity in which the material is cut into parallel slices. Once again, the coupling stress can be written in the non-local form given by eq. (28). With cut-off conditions on the influence function M, we obtain the Laplacean form (30). If the interaction is limited to the nearest neighbours, then ℓ becomes the average grain size d. For dimensional reasons, the coupling constant ζ is scaled by an elastic modulus of the material. Although it is not based on the elastic interactions mechanism, Hähner's model [87] for the Lüders bands in polycrystals also uses a gradient dependence of the flow stress in the same Laplacean form. The use of a diffusion-like mechanism of the mobile dislocations is tentatively justified from the consideration of the granular structure and the random orientations of the activated slip systems. Full three-dimensional plasticity constitutive laws which extend the classical relations by including interaction stresses of the Laplacean type have been proposed by Mülhaus and Aifantis [111].

During the plastic deformation of polycrystals, discontinuities in the rotation rates may also develop at the grain boundaries. Within a continuum framework, these discontinuities can be described by rotation rate gradients of significant magnitude at the scale of the grain size. In the conventional theory of continuous media, the transmission of loads through a surface element dS is assumed to be completely described by the elementary stress vector $\mathbf{T}$ acting on that surface, while the conjugate state of deformation is characterised by the displacement gradients only. When rotation gradients must be included in the description of the state of deformation, a surface element transmits the conjugate couple stress vector $\mathbf{Q}$ in addition to the stress vector $\mathbf{T}$. This is the rationale of the so-called couple stress theory proposed early in the last century by Cosserat [112] and further developed by Toupin [113] and Mindlin and Tiersten [114] in an elastic context, and of a gradient plasticity theory recently proposed by Fleck et al. [115]. In a gradient plasticity model, one needs constitutive formulations relating the couple-stress field and the rotation-rate field. This step requires putting forward a length scale for which sound physically based justifications may be difficult to obtain. Considerable effort is presently devoted to that field, (cf., e.g., [116–118] among many others) but this work will not be further discussed here, as it is not relevant to the types of instabilities and localisations discussed in this part.

Concluding remarks
In this section, we have discussed different approaches that have been proposed to account for the nonlocal effects exhibited by various phenomena. Since all of these proposals involve the introduction of strain or rotation gradients in the formulations, there is clearly a need for additional boundary conditions. The equilibrium and constitutive relations should be supplemented by boundary conditions that preserve the well-posedness of the problem, that is, the existence and uniqueness of solutions. This point will be discussed further in the following sections.

3.4. Velocity selection mechanisms

In this part, all the phenomena under investigation are characterised by the occurrence of an unstable uniform state of deformation followed by the emergence of well-developed bands or fronts of plastic activity propagating throughout the sample and separating different states of plastic deformation. Usually, the plastic strain undergoes a step of finite size across the front. Therefore, propagation is usually governed by non-linear partial differential equations, from which stationary waves travelling at a constant velocity are to be extracted. However, the procedure may not be straightforward since these equations often admit a continuum of possible velocities. The problem then consists in selecting the actual velocity within this range. In a number of situations, the propagation is governed by the same mechanism of growth as the one controlling the small perturbations. In such cases, investigating the linearised behaviour turns out to be of physical relevance in the understanding of front propagation. In this section, we first concentrate on the linearised theory of front propagation that leads to the so-called Marginal Stability Hypothesis (MSH), first introduced by Dee and Langer [119]. Since the validity of the MSH cannot be proven in general, due to non-linear effects possibly becoming significant, we eventually turn to the specific non-linear selection mechanisms which have been devised for the phenomena under consideration.

3.4.1. Marginal stability hypothesis
The marginal stability theory is based on the assumption that most properties of the propagating front can be obtained from those of the leading edge of the front profile, i.e., the region where the deviations from the unstable state are small enough to ensure the validity of a linearised analysis. The dynamics of the non-linear part of the front just follows, as if dictated by the linear dynamics. The analysis focuses on the properties of

propagative solutions written in the form $\gamma(\xi)$, $\xi = x - ct$, where c is the constant front velocity. Boundary conditions at the edges of the sample are replaced by conditions in the limits $t \to \infty$, $x \to \infty$. The dynamics of the leading edge is that of a Fourier superposition of linearised deformation modes, of the symbolic complex form:

$$\delta\hat{\gamma} = \delta\gamma_{\mathrm{e}} \exp(ikx + \omega t), \tag{31}$$

where k is the mode real wave number and $\omega(k)$ the growth rate associated with a spatio-temporal fluctuation $\delta\hat{\gamma}$ of (small) amplitude $\delta\gamma_{\mathrm{e}}$. The relation $\omega(k)$ is called the linear dispersion relation. Seen from a frame moving at the velocity c, the long-term behaviour of the profile will be dominated by the mode k^* with the maximum growth rate, i.e., such that

$$\frac{\partial}{\partial k}(ikc + \omega) = 0. \tag{32}$$

Therefore, only this fastest growing mode is considered in the long-term dynamical selection mechanism which follows. The MSH consists now in selecting the velocity c^* such that, after a long time t and at a large distance x, the fastest growing k^* mode remains marginally stable in the frame moving at the velocity c^*. To an observer in that frame, the disturbance just keeps up with the front, neither growing nor decaying. Thus the velocity c^* is such that

$$\mathrm{Re}\big[ik^*c^* + \omega(k^*)\big] = 0. \tag{33}$$

The relations (32) and (33) can also be written in the form

$$\mathrm{Re}\left[\omega(k^*) - k^* \frac{\partial\omega}{\partial k}\right] = 0; \qquad c^* = i\frac{\partial\omega}{\partial k}. \tag{34}$$

It can be shown that this selection mechanism is efficient if the initial front profile is sharply localised [119–121]. As shown by van Saarlos [121], a breakdown of the MSH theory may occur if the non-linear front profile itself becomes unstable against some particular mode with a velocity larger than the MSH velocity c^*. In any case, there is no way to prove the validity of the MSH framework but to test it against a fully non-linear analysis. Following Zbib and Aifantis [122], we shall now illustrate the use of the MSH by applying it to the PLC effect.

3.4.2. The Portevin–Le Chatelier effect: the interaction stresses model

In the model of Zbib and Aifantis [122], a second gradient in strain is introduced into the Penning's equation to account in a heuristic manner for the effect of the long-range elastic interactions between dislocations:

$$\sigma = h\varepsilon + \zeta\ell^2 \frac{\partial^2\varepsilon}{\partial x^2} + F(\dot{\varepsilon}), \tag{35}$$

where $\zeta\ell^2$ is assumed to be a positive, generic constant term. The nature of the underlying coupling mechanism is not specified. Formally, this type of equation (with $\zeta > 0$) has been used by several authors but within different contexts. Its solutions in constant applied strain rate and constant loading rate were discussed by Hähner [123] (see also the review [80] by Zaiser and Hähner), who attributed the origin of the coupling to intergranular incompatibility stresses. Here, attention is restricted to the constant stress rate loading delivered by a "soft" machine, in view of the much simpler resulting behaviour. The case of the constant strain rate loading with $\zeta < 0$, as resulting from a different coupling mechanism, will be considered in section 4 in relation with the model of Lebyodkin et al. [124]. Neglecting inertial terms, the model results in the partial differential equation

$$\dot{\sigma}_o t + \sigma_o = h\varepsilon + \zeta\ell^2 \frac{\partial^2 \varepsilon}{\partial x^2} + F(\dot{\varepsilon}). \tag{36}$$

Here, $\dot{\sigma}_o$ denotes the applied stress rate and σ_o is the initial stress. A uniform steady-state solution exists for the strain and strain rate:

$$\dot{\varepsilon}_e = \dot{\sigma}_o / h; \qquad \varepsilon_e = \dot{\varepsilon}_e t. \tag{37}$$

When $\dot{\varepsilon}_e$ falls within the range of negative SRS $(\mathrm{d}F/\mathrm{d}\dot{\varepsilon})_{\dot{\varepsilon}_e} < 0)$, this solution is unstable and a propagating pattern of deformation in the form of PLC bands occurs. In the model [122], the band velocity is selected in a simple manner by making use of the Marginal Stability Hypothesis. However, it must be kept in mind that linear stability analyses can hardly be justified in a context set out by finite fluctuations about a uniform state of deformation. Indeed, in the PLC effect there is no continuous path connecting the uniform state and the non-uniform branches of solutions.

Marginal stability hypothesis

Let us define again small disturbances of the uniform state of deformation in the form (cf. eq. (31))

$$\varepsilon = \varepsilon_e + \delta\hat{\varepsilon}, \quad \delta\hat{\varepsilon} = \delta\varepsilon_e \exp(ikx + \omega t), \ \delta\varepsilon_e \ll \varepsilon_e. \tag{38}$$

Inserting this expression into in the partial differential eq. (36), and selecting the first-order terms in $\delta\varepsilon$, one obtains the dispersion relation

$$\omega = \frac{h(1 - \ell^2 k^2)}{|F'(\dot{\varepsilon}_e)|}. \tag{39}$$

Disturbances with wave numbers $k < 1/\ell$ display a positive rate of growth ω. Therefore, the uniform state of deformation, as given by eq. (37), is linearly locally unstable against such perturbations. The relation (34) then provides the velocity of propagative disturbances according to MSH

$$c^* = \frac{2h\ell}{|F'(\dot{\varepsilon}_e)|}. \tag{40}$$

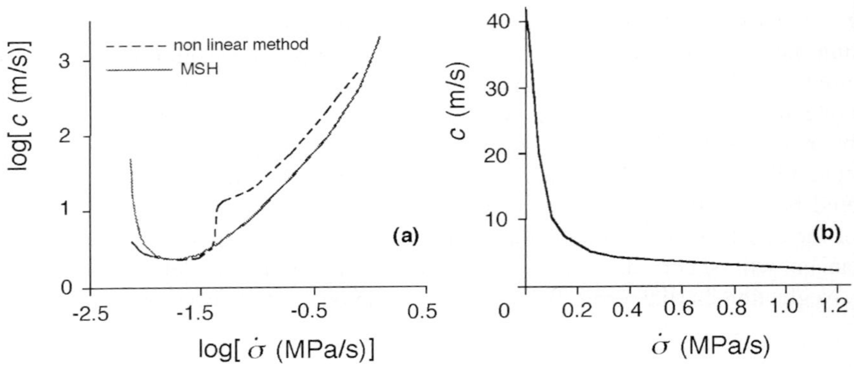

Fig. 18. PLC Band velocity c vs. stress rate as predicted by different models: (a) the internal stress model using the Marginal Stability Hypothesis (after [122]) and the internal stress model with a non-linear analysis (after [126]); (b) the cross-slip model (after [125]).

When $\dot{\varepsilon}_e$ tends towards the boundaries of the range of existence of the PLC effect, $F'(\dot{\varepsilon}_e)$ tends to zero. Therefore, c^* becomes infinite. In between, $F'(\dot{\varepsilon}_e)$ goes through a maximum. Thus the graph c^* vs. $\dot{\varepsilon}_e$ exhibits a U-shaped dependence (cf. fig. 18).

Whether or not this result is valid must be checked out by using non-linear techniques, as well as by comparison with experimental data. Useful non-linear methods, which are discussed now, can be designed by considering the system orbits in the appropriate phase space [124–127].

Analysis in the phase space
Let us introduce non-uniform finite disturbances of the plastic strain in the form

$$u = \varepsilon - \varepsilon_e, \tag{41}$$

where ε_e is the uniform state of deformation. One may rewrite eq. (36) as

$$\zeta \ell^2 \frac{\partial^2 u}{\partial x^2} + F(\dot{\varepsilon}_e + \dot{u}) - F(\dot{\varepsilon}_e) + hu = 0. \tag{42}$$

By appropriate scaling transformations, eq. (42) can be cast into the form

$$\kappa \frac{\partial^2 u}{\partial x^2} + f(\dot{u}) + hu = 0, \tag{43}$$

where κ characterises the strength of the gradient effects. The partial differential equation (43) is usually referred to as a Rayleigh equation (differentiating it with respect to time and setting $v = \dot{u}$ leads equivalently to a Liénard equation in v [128]). We look for solutions to eq. (43) in the form of stationary waves travelling with a constant velocity c:

$$u(x, t) = u(\xi), \quad \xi = x - ct, \quad \dot{u} = -cu', \quad \frac{\partial^2 u}{\partial x^2} = u'', \tag{44}$$

where the index ($'$) denotes the differentiation with respect to ξ eq. (43) then reduces to the ordinary differential equation

$$\kappa u'' + f(u') + u = 0, \tag{45}$$

where the function f depends implicitly on the velocity c. Note that eq. (45) can be interpreted as the equation of motion of a sliding particle of "mass" κ whose "position" u is a function of "time" ξ. The particle is submitted to the action of a linear spring, with a unit elastic stiffness and a non-linear friction force $-f(u')$. The determination of the velocity involves finding c (and f) such that a periodic motion exists for this particle. Using the phase variables (u_1, u_2) defined by

$$u_1 = u, \qquad u_2 = u' \tag{46}$$

the eq. (45) is equivalent to the autonomous set of equations

$$\begin{aligned} u_1' &= u_2, \\ \kappa u_2' &= -u_1 - f(u_2). \end{aligned} \tag{47}$$

Combining the two eqs (47), it follows that

$$\frac{1}{2}(u_1^2 + \kappa u_2^2)' = -u_2 f(u_2). \tag{48}$$

Integrating (48) along a path (γ) between two points (A, B) of the phase plane, one obtains

$$\left[\frac{1}{2}(u_1^2 + \kappa u_2^2)\right]_A^B = \int_\gamma u_2 f(u_2)\,\mathrm{d}\xi = -\int_\gamma f(u_2)\,\mathrm{d}u_1. \tag{49}$$

A necessary and sufficient condition to close the path (γ) between A and B is therefore [129]

$$\int_\gamma f(u_2)\,\mathrm{d}u_1 = 0. \tag{50}$$

Whenever the relation (50) is fulfilled, eq. (45) has a periodic solution and eq. (42) a propagative one. The velocity c is determined from eqs (47), (50) as follows. For a given value c, a path (γ) is obtained from the numerical integration of eq. (47). The consistency relation (50) is then checked and c is modified in eq. (47), if necessary, until the desired accuracy is obtained. Since a periodic orbit of eq. (45) is globally stable [129], the procedure converges rapidly. The results obtained from this model [126] are plotted in fig. 18, together with those from the MSH [122]. It can be seen that, despite strong non-linearities, the MSH applies quite successfully to this problem. Actually, its predictions are fairly close to those emerging from the non-linear analysis. A different type of stress vs. velocity law, with a propagation velocity decreasing with increasing loading rate, was obtained by Hähner (cf. [80] and references therein), based on a modified type of constitutive form.

3.4.3. The Portevin–Le Chatelier effect: the cross-slip model

Under constant stress rate loading conditions, the cross-slip model is summed up in the partial differential equation

$$Z = F(\dot{Y}), \quad \dot{Y} = \frac{\partial \varepsilon}{\partial t} - D\frac{\partial^2 \varepsilon}{\partial x^2}, \quad Z = \dot{\sigma}_o t + \sigma_o - h\varepsilon, \tag{51}$$

where F is the N-shaped Penning's function (see section 3.3.2, eq. (26)). Eq. (51) admits a uniform solution (37) in tension, which is rewritten

$$\dot{Y}_e = \dot{\varepsilon}_e = \frac{\dot{\sigma}_o}{h}, \qquad \varepsilon_e = \dot{\varepsilon}_e t, \qquad Z_e = \sigma_o = F(\dot{Y}_e). \tag{52}$$

Again, when $\dot{\varepsilon}_e$ falls within the range of negative SRS, this solution is unstable and a propagating pattern of deformation in the form of PLC bands occurs. Using the linear MSH would lead again to a prediction for the band velocity similar to that of eq. (40). Therefore, we jump at once to a non-linear analysis. Let us consider the non-uniform disturbances $(\dot{y}, z)$ of the uniform deformation $(\dot{Y}_e, Z_e)$ leading to the non-uniform state $(\dot{Y}, Z)$, as well as the perturbation in plastic strain u:

$$\dot{y} = \dot{Y} - \dot{Y}_e, \qquad z = Z_e, \qquad u = \varepsilon - \varepsilon_e. \tag{53}$$

In view of eq. (51), the evolution of the perturbation is governed by the partial differential equation

$$z = F(\dot{Y}_e + \dot{y}) - F(\dot{Y}_e). \tag{54}$$

In terms of the perturbation $(u, \dot{y})$, eq. (54) is written as $u = -z/h = f(\dot{y})$, or after inverting f separately on each stable branch of the function F:

$$\frac{\partial u}{\partial t} - D\frac{\partial^2 u}{\partial x^2} = g^i(u), \quad i = \mathrm{I}, \mathrm{II}. \tag{55}$$

The result is a couple of reaction-diffusion equations (cf. section 2.3.3) where the reaction term $g(u) = f^{-1}(u)$ alternatively pertains either to the slow regime (g^{I}) or to the fast regime (g^{II}) of dislocation glide (cf. fig. 19). Looking for solutions to eq. (55) in the form of the stationary waves travelling with the constant velocity c (cf. eq. (44)) leads to the ordinary differential equation:

$$Du'' + cu' + g(u) = 0. \tag{56}$$

With the same interpretation as in eq. (45), the particle is submitted to the action of a linear friction force – cu' and a non-linear spring – $g(u)$. The determination of the velocity involves finding c such that a periodic motion exists for this particle. Using the phase variables $u_1 = u,\ u_2 = u'$, eq. (56) becomes

$$Du_2\frac{du_2}{du_1} = -cu_2 - g(u_1). \tag{57}$$

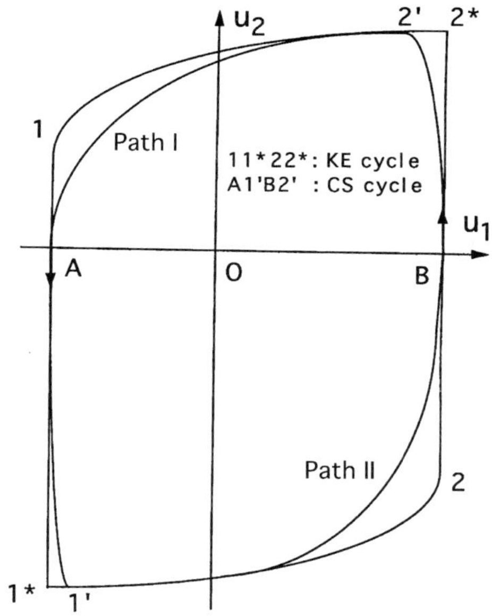

Fig. 19. The cross-slip (CS) model for PLC. The closed trajectory of the system in the phase space is a succession of slow and fast sequences. It is continuous, whereas discontinuous jumps in strain rate are allowed in the Kubin–Estrin (KE) model [102].

When the alternative determinations (g^{I}, g^{II}) are successively employed, the representative point follows a couple of paths, one slow and one fast, in the phase space (cf. fig. 19). The integration of eq. (57) along each path yields

$$\frac{1}{2}D\left[u_2^2(1') - u_2^2(2')\right] = -\int_{2'}^{1'} g(u_1)\,\mathrm{d}u_1 - c\int_{1'}^{2'} u_2\,\mathrm{d}u_1. \tag{58}$$

Since the left-hand side of eq. (58) does not depend on the path, the difference between the right-hand sides taken for path I and path II leads to the non-linear equation for the unknown velocity c

$$\int_{2'}^{1'} \left(g^{\mathrm{I}}(u_1) - g^{\mathrm{II}}(u_1)\right)\mathrm{d}u_1 = c\int_{2'}^{1'} \left(u_2^{\mathrm{II}} - u_2^{\mathrm{I}}\right)\mathrm{d}u_1. \tag{59}$$

When the closure condition (59) is satisfied, two consecutive paths are connected and a periodic orbit in the phase space exists (cf. fig. 19). This cycle results in temporal oscillations in a cross-section, and in the propagation of a stationary band profile along the sample with a constant velocity. Satisfaction of the closure condition means that the pseudo-diffusion effects are balanced over a cycle. This provides an exact equation for the band velocity, whose solution is obtained numerically [125]. The predicted cycle is continuous, but non-analytic at the connecting points. It can be thought of as a Kubin–Estrin cycle [100], regularised by diffusion effects. The predicted velocity vs. applied stress-rate behaviour is drawn in fig. 18 for the Penning's function F (cf. fig. 15) provided in Estrin and Kubin [102]. Note that the results are quite sensitive to the details of this viscoplastic function.

The comparison of the curves in fig. 18 shows that the interaction stresses on the one hand and Hähner's model [80] and the cross-slip model on the other hand yield conflicting predictions on the behaviour at large stress rates. According to the internal stresses model, the velocity should increase, whereas it should decrease according to the two other models. From the experimental point of view, the velocity of PLC bands is a well defined quantity during constant loading rate. However, the present experimental situation is quite confusing, as illustrated by fig. 20.

Karimi [10] measured band velocities at relatively low stress rates and at various temperatures in an austenitic stainless steel. The results show a monotonic decrease with increasing stress rate. More recently, a contradicting monotonic increase in the band velocity vs. loading rate dependence was reported by McCormick et al. [131] in an Al–Mg–Si alloy at room temperature, and at somewhat larger stress rates. Measurements by Bakir [132] of the band velocity in a Al–3%Mg alloy at 60°C yielded a more complex dependence of band velocity on loading rate. A decrease is observed at low stress rates, followed by an increase at large stress rates. Although it has been recognized recently that the measurements of band velocities require particular care regarding the surface condition of the samples [133], these results may explain the apparent discrepancy between earlier results. The latter may then have been dealing with too small ranges of stress rates. However, it is clear that more data on band velocities, for different materials and under various experimental conditions, would prove very useful. In particular, and as noted in [80], it should be desirable to measure the influence of grain size on band velocity. Understanding the reasons for an increase or a decrease of the band velocity with the loading rate is certainly a prerequisite to any progress in understanding the spatial couplings associated with the PLC effect.

3.4.4. The Lüders phenomenon in polycrystals

The generic reaction–diffusion equation (55), leading to the ordinary differential equation (56) for stationary waves propagating with a velocity c, as yet undetermined, is quite general. It applies to the Lüders bands model [87] as well as to various propagation phenomena in the field of chemistry or biology [120]. As in the previous section, it is instructive to interpret eq. (56) as the equation of motion of an analogue particle of "mass" D, submitted to the "linear viscous friction" $-cu'$, with a "damping constant" c, in the field of the non-linear "force" $g(u)$. The latter derives from a potential V such that

$$\frac{\mathrm{d}V}{\mathrm{d}u} = g(u). \tag{60}$$

Under a given applied stress, the Lüders potential $V(u)$ in [87] is assumed to exhibit two maxima associated with two stable uniform states of deformation on the rising branches of the stress–strain curve. The first maximum is set out on the elastic branch, prior to front propagation, the second one on the hardening part, once propagation is completed. A minimum of the potential occurs on the softening branch, and is associated to an

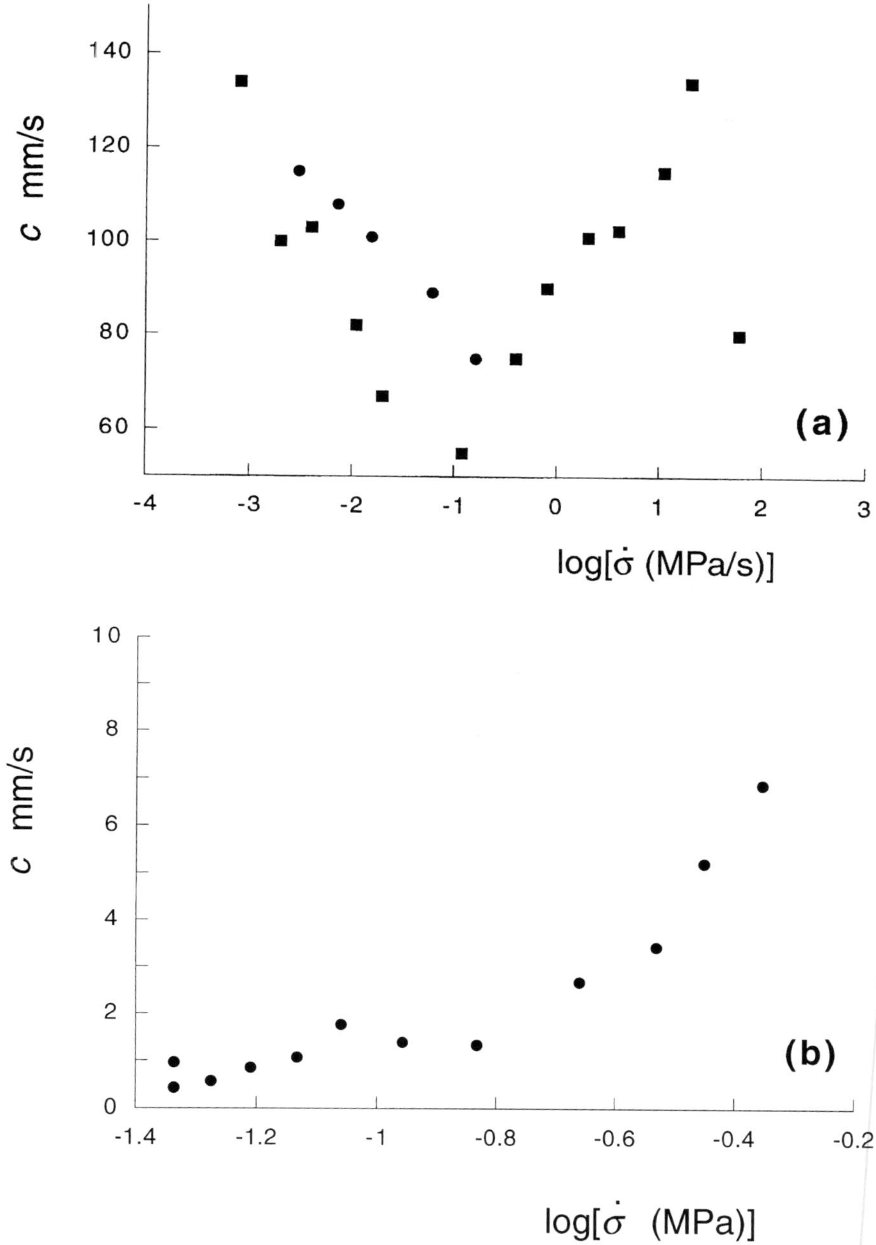

Fig. 20. Experimental measurements of the PLC band velocity vs. stress rate: (a) filled circles: stainless steel, $T = 850$ K (after Karimi [130]). Filled squares: Al–Mg alloy at 300 K (after Bakir [132]). Further studies [133] on specimens with polished surfaces seem to indicate that the velocity does not drop down at the highest stress rates. (b) Al–Mg–Si alloy, $T = 300$ K (after McCormick et al. [131]).

unstable (uniform) state. According to [87], eq. (53) thus belongs to the class of bistable reaction–diffusion equations, which are known to exhibit a front connecting the two stable steady states [120]. This framework is claimed to apply to the Lüders phenomenon, and therefore the yield point phenomenon is regarded as a finite excitation required to leave the metastable elastic ground state. In terms of the associated mechanical problem, the analogue particle is released from the largest potential maximum with zero initial velocity, and later on reaches the second stable steady state asymptotically with zero velocity. This is possible for only one value of the "damping constant" c. Thus, the front propagation velocity is uniquely determined. Multiplying eq. (56) by u' and integrating over the "time" span, one finds that c satisfies the equation

$$c = \frac{-[\Delta V]_{-\infty}^{+\infty}}{\int_{-\infty}^{+\infty} u'^2 \, \mathrm{d}\xi}. \tag{61}$$

Clearly, one does not need a velocity selection mechanism in this case since the velocity is uniquely determined, at least in principle. However, in order to establish a connection with the MSH selection method (see above section 4.1), consider the case where the system reaches its unstable state by some hypothetical elastoplastic process. Such an assumption would actually reflect more closely a dislocation avalanche than the bistability hypothesis. Then the analogue particle, released from the maximum of the potential $V(u)$, reaches the minimum with zero velocity for any value of the "damping constant" c, provided that the latter is large enough (say $c \geqslant c^0$). Clearly, the front velocity is undetermined and a selection mechanism is needed. In the underdamped case ($c < c_0$), the particle would oscillate when approaching zero velocity. In the MSH framework, the selected velocity c^* (cf. section 3.4.1 and eq. (33)) is the minimum possible value, c_0. Neglecting elastic relaxation, the Lüders bistable model [87] allows one to obtain the front velocity through an approximate evaluation of the integral term in eq. (61). Other characteristics, such as the Lüders strain, follow. The latter increases with the grain size and a minimum stress is required for propagation.

3.4.5. The Lüders-like phenomena in single crystals

Bréchet et al. [93] used the local constitutive relation (16), together with the spatial coupling

$$\tau_{\mathrm{nl}}(x) = \int M(x'-x)\big[H\big(\gamma(x')\big) - H\big(\gamma(x)\big)\big]\,\mathrm{d}x' \tag{62}$$

to write the gradient plasticity constitutive law in the form

$$\tau_{\mathrm{a}} = \eta\dot{\gamma} + \tau_{\mathrm{o}} + H(\gamma) - \ell^2 \frac{\partial^2 H(\gamma)}{\partial x^2}. \tag{63}$$

This expression was obtained via a limited Taylor expansion of $H(\gamma)$, thus introducing a cut-off. This spatial coupling is actually an extended form of the coupling via strain gradients (see eqs (28), (30)). Equilibrium conditions require that the applied resolved shear stress τ_a be not space-dependent. An additional constraint is that the applied shear stress be a constant along the plateau associated with band propagation. Thus eq. (63) is the starting equation for this problem. Let V be the velocity of the moving edge of the sample. If a stationary wave of the form $\gamma(\xi)$, $\xi = x - ct$, is assumed, this velocity is written

$$V = \int_{-\infty}^{+\infty} \dot{\gamma}\, \mathrm{d}\xi = -c \int_{-\infty}^{+\infty} \frac{\partial \gamma}{\partial \xi}\, \mathrm{d}\xi = c \Delta \gamma. \tag{64}$$

Eq. (64) is the so-called "machine equation", where $\Delta\gamma = \gamma_b - \gamma_\ell$ is the increment of shear strain achieved through the moving band, the value γ_b referring to the maximum strain at the rear edge of the band and γ_ℓ to a critical strain needed to initiate plastic flow at the leading edge via the activation of dislocation sources. The front moves forward via the activation of new sources by strain softening on its leading edge, while the same amount of sources stops operating at its rear edge under the effect of strain hardening. Thus, the strain rate and the strain gradient vanish at both the rear and the leading edges of the front. For $\gamma = \gamma_b$, the equilibrium equation (63) therefore reduces to

$$\tau_a = \tau_o + H(\gamma_b), \qquad H(\gamma_b) < 0. \tag{65}$$

Ahead of the front, it is shown again that

$$\tau_a = \tau_o + H(\gamma_\ell), \qquad H(\gamma_\ell) < 0. \tag{66}$$

Comparing eqs (65) and (66), it is seen that

$$H(\gamma_b) = H(\gamma_\ell). \tag{67}$$

Now, inserting eq. (66) into eq. (63), and using the relation $\dot{\gamma} = -c\gamma'$ between strain rate and strain gradient under steady state band propagation, one eventually obtains at the leading edge

$$\mu c \gamma_\ell' + \ell^2 \frac{\partial^2 H(\gamma_\ell)}{\partial x^2} = 0. \tag{68}$$

As in the MSH, an approximate solution method based on the behaviour of the leading edge of the deformation front is now set up. The strain at the leading edge increases by the approximate amount γ_ℓ over a distance of the order of the length scale ℓ. Therefore, the approximate values $\gamma_\ell' \cong -\gamma_\ell/\ell$, $\gamma_\ell'' \cong \gamma_\ell/\ell^2$ are used to write eq. (68) in the form

$$-\eta c \frac{\gamma_\ell}{\ell} + \frac{\partial^2 H(\gamma_\ell)}{\partial \gamma^2} \gamma_\ell^2 + \frac{\partial H(\gamma_\ell)}{\partial \gamma} \gamma_\ell = 0. \tag{69}$$

The approximate solution procedure is now as follows. Using eq. (64), the cross-head velocity V is shown to be constant during the time duration of the propagation plateau. In addition, it is uniquely related to the applied stress τ_a. The quantities of interest are therefore the applied stress τ_a, the plastic strains γ_b and γ_ℓ and the velocity c. They are linked by the independent relations ((64)–(66) or (67)) and (69). The results show that the normalised band velocity c/V is proportional to the applied strain rate and is a monotonically decreasing function of the hardening/softening parameter s. Its maximum value is obtained at the transition between the regimes of propagation and random localisation regimes. It decreases as softening increases, as should be expected. This result is in good quantitative agreement with data on Cu–Al and Cu–Mn single crystals [134]. Whether the nonlinear part of the front follows the propagation features obtained from assumptions on its leading edge remains to be checked.

As observed in the 3-D simulation of dislocation dynamics [65] referred to above in sections 2.4 and 3.2.2, front propagation is one of three possible modes of deformation. It occurs when the softening/hardening function $H(\gamma)$ exhibits a minimum, which happens when $s = \Delta\tau_o/h\gamma_o > 1$. When $s > 1$ and when a front propagates, new sources are activated at its leading edge, whose activation stress fluctuates around an average value. However, spatially uncorrelated sources may be activated if their critical stress is significantly smaller than the average. This competing mechanism leads to the possibility of slip being initiated at random locations instead of ahead of the front. According to eq. (66), the resolved shear stress needed for propagation is $\tau_a = \tau_o + H(\gamma_b)$. It is assumed that, away from the front, the most favoured source has a critical stress $\tau_o(1-\delta)$, $\delta < 1$, where the constant δ is related to the distribution of critical stresses for source activation in the specimen. Then, propagation occurs when $\tau_a < \tau_o(1-\delta)$. If not, random activation of sources prevails. A diagram of the domains of uniform, randomly localised and propagative regimes of deformation obtained from the above model is shown in fig. 16. It is in qualitative agreement with the results obtained from the 3-D simulation of dislocations dynamics [93].

3.5. Conclusion

The non-uniform patterns of deformation associated with the PLC effect, the Lüders and the Lüders-like phenomena have been modelled at macroscopic or mesoscopic scale, i.e., either in terms of dislocation populations or in terms of constitutive relations, by lumping the details of the dislocation dynamics into a limited number of parameters. The advantages are twofold. First, the models are generic in the sense that they encompass various physical situations. For example, various strain softening mechanisms are generically described via the two parameters $\Delta\tau_o$, γ_o, viz. the total amount of softening and the strain needed to achieve it. Second, mechanical boundary conditions and deformation field length scales are accounted for in the formulation. This is mandatory when non-local effects have to be considered.

Different scales of localisation have been encountered. According to the classification proposed by Neuhäuser [89], the slip morphology associated with strain softening in single crystals involves three different scales: nanometric or microscopic (slip lines), micrometric or mesoscopic (slip bands), millimetric or macroscopic (slip band bundles). The "true" Lüders fronts and the PLC bands pertain to the mesoscopic scale. Each scale may involve a different type of coupling mechanism. It is accepted that cross-slip is responsible for the smallest range of interactions, as shown by Luft [88] and observed in the 3-D simulations. The elastic interactions of dislocations may govern patterning at the mesoscopic scale, as suggested in [93] (cf. section 3.4.5). The coupling by compatibility stresses and mechanical effects should be held responsible, at least partially, for slip patterns at the largest scale. On the one hand, mechanical effects are likely to be involved in the PLC band patterns since the occurrence and morphology of bands are reported to depend on the sample's shape and size. The same holds for the Lüders phenomenon, since the angle of the Lüders bands with respect to the specimen axis is sensitive to the sample geometry. On the other hand, numerous experiments (cf. [75] for PLC bands and [135] for Lüders bands) indicate that the spatial correlation between bands is lost when the deformation is stopped by unloading the sample for a period of time larger than a critical relaxation time. This occurs whether or not the specimen surface has been repolished to remove the slip steps. This observation strongly suggests the involvement of an intrinsic material coupling, as well as the occurrence of some internal recovery or relaxation mechanism. As to the "true" Lüders bands, the situation is still unclear. Due to their polycrystalline nature, it is likely that compatibility stresses are at the origin of the spatial coupling needed for the propagation of the Lüders bands. However, a model accounting for these constitutive mechanisms is still to be worked out.

As to the PLC bands, the mesoscopic or macroscopic scale of the phenomenon seem to pinpoint the long-range elastic interactions of dislocations as being responsible for the spatial coupling. Still, no clear-cut check is available from experimental measurements of the velocity of the PLC bands under constant stress rate loading. Spatial coupling by the elastic interactions of dislocations acts globally as a stabilising factor for the propagating patterns in constant stress rate as well as constant strain rate loadings. As a matter of fact, it has been shown [129] (see also section 3.4.2) that, although small perturbations are locally linearly unstable (cf. eq. (39)), the periodic solutions associated with the propagation of PLC slip bands via dislocations interactions (cf. eqs (47), (50)) are globally stable under constant stress rate loading: the spatial coupling does relax any non-uniform disturbance of the periodic solution within a period of time. The situation is hardly as simple under constant strain rate loading. Actually, the observed behaviour rather exhibits a kind of limited instability, usually resulting in a complex spatio-temporal pattern of bands, and generating jerky flow with a seemingly random time dependence of the applied stress. Whether this complex regime can be characterised as being stochastic (i.e., random) or as involving some hidden order, is presently a problem of concern from both the experimental and theoretical points of view. Recent attempts to investigate this question are reported in the following part.

4. *Complex patterning*

In section 2, the spatio-temporal behaviour of weakly mobile dislocation ensembles was first studied at a scale corresponding to the characteristic length scale of the microstructure. Then, the collective properties of mobile dislocations during plastic instabilities were examined in section 3 leading to the construction of gradient constitutive laws. In this section, we again move up in length scales to that corresponding to the dimension of the deforming specimen and consider a different approach to patterning that has emerged in the past few years. It involves the measurement and analysis of purely temporal signals generated during deformation. Such signals are usually spatial averages over the entire specimen of a quantity sensitive to dislocation densities or mobilities, for example electrical resistivity or stress. They contain contributions from numerous spatially distributed events that have their own local dynamics. Thus, in jerky flow, stress is a complex time-dependent signal and a natural question that arises is whether it is purely random or has some internal structure. In the latter case, several possibilities exist and can indeed be observed, as will be shown below. The structure of time series has been extensively studied in different kinds of spatially extended systems and may in particular fall into two generic categories, viz., deterministic chaos and self-organised criticality. In this section, we focus our attention on the possible occurrence of such complex patterns in jerky flow. The temporal oscillations of the recorded stress will be analysed using methods that have become available in the recent past. The information thus obtained can be used to get new insight into the nature of the non-linearities present in the system.

4.1. Introduction

The experimental characterisation of complex time evolutions involving large numbers of individual events occurring at different spatial locations commonly uses spatially averaged signals. However, the capacity to monitor local events is also required. The underlying idea is to extract information on the correlations that are at the root of the patterning from the overall statistics, which arise at the scale of the sample. In practice, dislocation avalanches occurring during plastic instabilities can be investigated through their influence on various physical properties that are sensitive to the density and velocity of mobile dislocations.

For instance, the electrical response during discontinuous deformation at low temperatures has been studied in single crystals of aluminium and niobium [136–138]. In the latter case, mechanical twinning is the cause of the unstable flow, and in the former one, the plastic instability is due to a feedback between the heat generated by plastic flow and thermally activated glide of dislocations (thermomechanical instability). The electrical pulses recorded during drops in load are due to two main effects, which obey different time scales. One is the dragging of conduction electrons by the mobile dislocations; the other is an effect involving a coupling between local thermal and electrical properties. These investigations were performed along the same lines as the statistical analysis reported in section 4.3 and yielded very similar conclusions regarding the statistics of the stress drops and electrical pulses.

Acoustic emission (AE) is also a potentially powerful technique for the study of the cooperative movement of mobile dislocations. Although there have been other suggestions, it is known that AE activity is proportional to the rate of change of the mobile dislocation density as established by James and Carpenter [136]. Since then, there have been a number of attempts to establish a correlation between the acoustic activity and the PLC effect [140–148]. In the most recent studies, AE is followed with considerable precision and may involve simultaneous observation of slip line formation by fast cinematographic methods [147]. The acoustic sources can also be localised by measuring the difference in the arrival time of the AE signals recorded at two transducers located at the ends of the specimen [148]. In general terms, all the results obtained at various temperatures and strain rates and in various alloys indicate that there is a strong correlation between AE and the degree of heterogeneous deformation. Recent experimental results on ice crystals [149], which suggest that slip events have a self-organised character during globally uniform deformation, illustrate the enormous potential of this method.

A quantity which is directly related to the avalanche-like behaviour of mobile dislocations during serrated flow is obviously the stress recorded during a deformation test performed with a constant imposed strain rate. In what follows, several sets of stress time series are analysed. For the first set of experiments (section 4.2), a dynamical analysis is used to determine the type of order possibly underlying the oscillatory stress behaviour. The analysis shows that there is chaos in the regime of low and medium strain rates considered. In the second set of experiments (section 4.3), the statistics of stress drops are examined and characterised. The statistics display power-law distributions at high strain rates. Chaos and power-law statistics, of which the latter has come to be associated with Self-Organised Criticality (SOC), represent different dynamical states of the PLC effect. However, since these states have been characterised on two different materials, a third material is used for further analysis of the influence of strain rate. The idea is to look for a possible crossover from one dynamical state to the other as the strain rate is increased. This analysis will be presented in section 4.4.

4.2. Stochastic versus deterministic behaviour of dislocations

In section 3.4, it has been emphasised that spatio-temporal patterns emerge in the PLC effect due to strong non-linearities. One non-linear model for the PLC instability [150–152] predicts that the seemingly random stress vs. time records of constant strain rate loading could exhibit a chaotic deterministic behaviour. In order to lay a foundation for this analysis, some basic concepts from chaos are introduced in the present section, as well as a brief outline of the model by Ananthakrishna et al. [150,155].

4.2.1. Chaos versus randomness

Noise is often encountered both in science and engineering, and it will be hereafter referred to as stochastic noise, or random noise, or just noise. It is usually an undesirable element of experimental data and it is not considered a part of the phenomenon of interest. It can arise due to lack of control over external or internal parameters. For example, the measuring instruments themselves may produce noise due to their own vibrational frequencies interacting with the system, or there may be limitations on the accuracy of the measurement process. Another source of noise is the lack of knowledge of a large number of degrees of freedom contributing to the time evolution of the system. A standard example of such a noise is the Brownian motion of a particle wherein the randomness arises due to the innumerable collisions of the particle with the molecules of the reservoir. The basic feature of the Brownian motion is indeterminacy; a deterministic prediction of the future movement of the particle is impossible.

An apparently similar random behaviour can also arise in systems where a few degrees of freedom evolve deterministically, but interact in a non-linear way. Hereafter, this behaviour will be referred to as chaotic or deterministic noise as often in the literature. In the case of a deterministic evolution (non-linear or otherwise), there is no uncertainty in predicting the future once the initial conditions are given. Naively, one would think that determinism is equivalent to complete predictability, even in the case of non-linear systems, as was believed for a long time (for example in the study of celestial objects). However, the lack of predictability arises from an unexpected quarter, namely the sensitivity to initial conditions. No matter how small an indeterminacy in the initial conditions of such systems, it will explode at an exponential rate over some period of time, thereby rendering the prediction of the trajectories impossible. This sensitivity to initial conditions is one of the characteristic features of deterministic chaos. It also implies that, as for random noise, there is a loss of memory of the initial state. Thus, in both cases a statistical description becomes inevitable.

Loss of memory of the initial state

Consider the two sets of plots of the time series shown in figs 21(a) and (b). Both of them appear random and it is not easy to distinguish that these two time series are different at a fundamental level. Indeed, the trajectory in fig. 21(a) is an experimental stress vs. time curve which will be analysed below and will be shown to be of deterministic origin. In contrast, fig. 21(b) has been constructed out of a stochastic model. Such random looking functions would quickly decorrelate. The correlation function $C(\tau)$ of these time series is given by

$$C(\tau) = \frac{\langle \phi(t)\phi(t+\tau)\rangle - \langle \phi(t)\rangle^2}{\langle \phi(t)\rangle^2}, \tag{70}$$

where the symbol $\langle \cdots \rangle$ represents the time average and τ is the time interval. A plot of $C(\tau)$ corresponding to fig. 21(a) is shown in fig. 22(a). $C(\tau)$ is a rapidly decreasing function, meaning that there is a loss of memory of the initial state within a short time interval. The function $C(\tau)$ is highly oscillatory, implying that there is a large number of frequencies involved. Figure 22(b) shows the correlation function for the stochastic trajectory in fig. 21(b). This is also a rapidly decaying function, but it does not show any oscillatory behaviour. One can obtain the power spectrum of the signal from the correlation function, by taking the Fourier transform of $C(\tau)$. While the power spectrum for the deterministic signal shown in fig. 22(c) exhibits a broad band noise, that corresponding to the stochastic time series is indeed flat (cf. fig. 22(d)). This is one of the simplest, although tenuous, distinctions between a chaotic and a random signal.

Sensitivity to initial conditions – Lyapunov exponent

Consider two trajectories with slightly different initial conditions represented in the phase plane $z(x)$, as shown in fig. 23.

These trajectories are solutions of deterministic equations discussed in section 4.2.2 (cf. eqs (77)–(80)). Thus, they would be indistinguishable if they had the same initial conditions. However, since initial conditions are slightly different, the trajectories soon diverge. In contrast, for a random noise, two trajectories separate out even if they start with

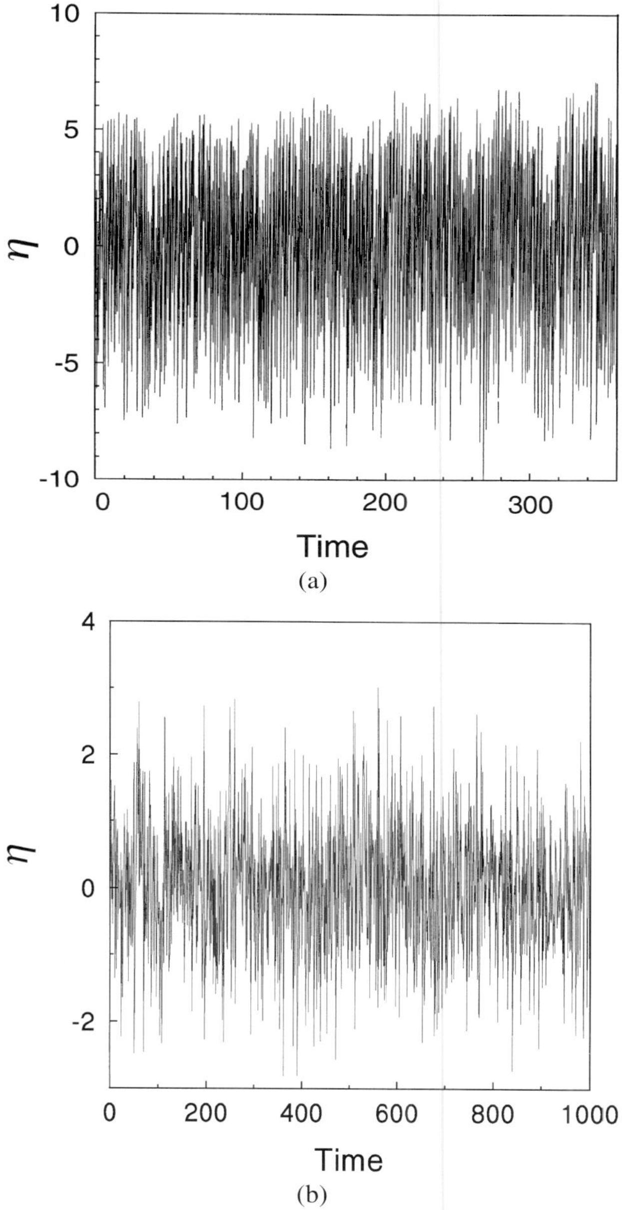

Fig. 21. (a) Plot of a time series of a deterministic signal, actually a stress vs. time series extracted from PLC data; (b) plot of a time series generated from a stochastic equation (Brownian motion).

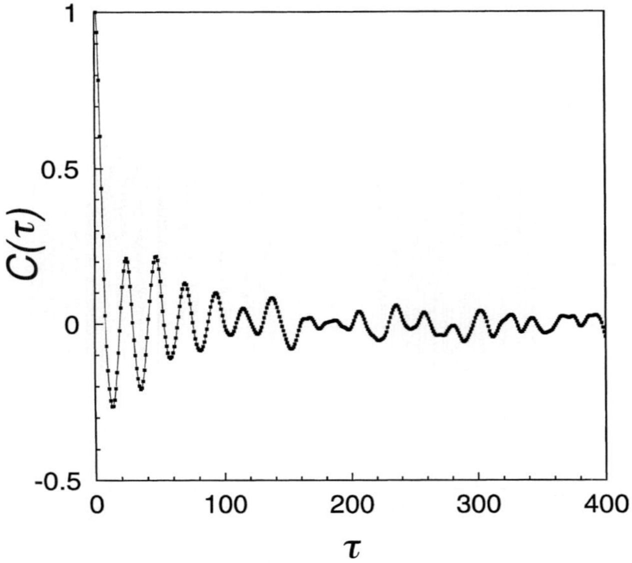

(a)

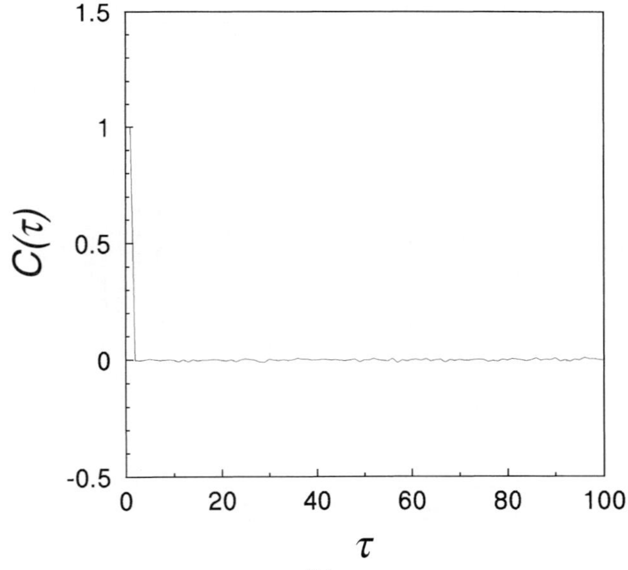

(b)

Fig. 22. (a) Plot of the correlation function $C(\tau)$ corresponding to the deterministic time series of fig. 21(a); (b) plot of the correlation function $C(\tau)$ corresponding to the stochastic signal of fig. 21(b); (c) power spectrum of the deterministic signal of fig. 21(a); (d) power spectrum of the stochastic signal of fig. 21(b).

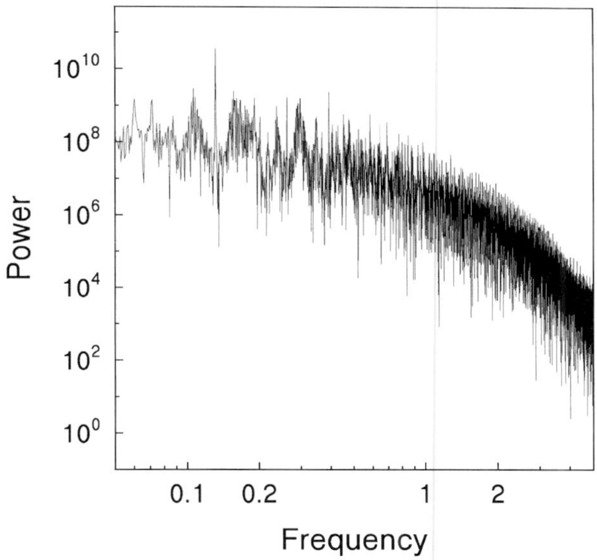

(c)

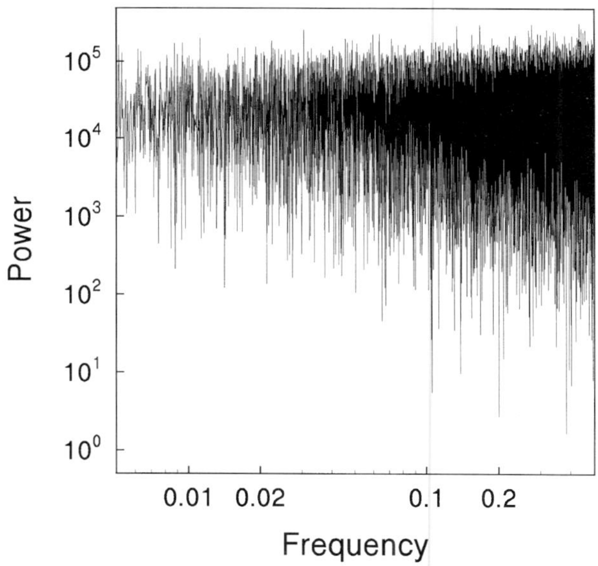

(d)

Fig. 22. Continued.

identical initial conditions. The rapid divergence of two nearby deterministic orbits further implies that there is a *stretching* in a particular direction of the phase space. Since for a dissipative system the total volume in the phase space is contracting towards some attractor

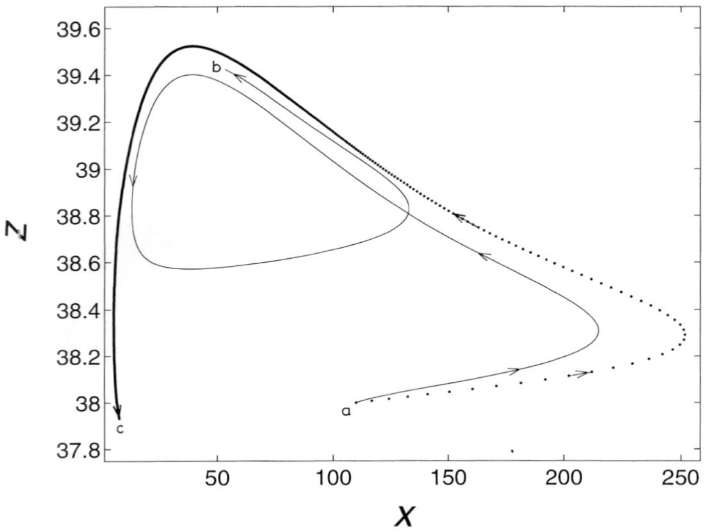

Fig. 23. A section (x, z) of the phase space diagram of the deterministic signal of the model [150], corresponding to the strange attractor shown below in fig. 25. Two orbits starting from nearly identical initial conditions (a) rapidly separate out. After some finite time the representative points (b), (c) may lie anywhere on the attractor.

(consider, for instance, a damped mechanical system, like a pendulum asymptotically tending to its stable equilibrium position), stretching in one direction implies that there should be contractions in other directions, *or folding*. Note that, in addition to dissipation, folding also stems from sheer non-linearity. As a simple example of stretching and folding, consider the conservative one dimensional map given by:

$$x_{n+1} = 2x_n \text{ (modulo 1)}, \tag{71}$$

where n is a discrete time index. Given the initial value x_0, successive values of the iterate are denoted by $x_1, x_2, \ldots, x_n$, etc. In non-linear systems, small changes in the input signals can produce very large outputs, unlike in linear systems. Consider two initial values $x_0 = 0.06249$ and $x'_0 = 0.06251$. The first few iterates of these two points are respectively: $x_1 = 0.12498, x_2 = 0.24996, x_3 = 0.49992, x_4 = 0.99984, x_5 = 0.99968$ and $x'_1 = 0.12502, x'_2 = 0.25004, x'_3 = 0.50008, x'_4 = 0.00016, x'_5 = 0.00032$. Clearly, though the points were initially as close as 2×10^{-5}, the fourth iterates of the two initial points are already far apart. This is entirely due to the fact that there is a stretching operation – the multiplicative factor of 2, and a folding one – the modulo unity operation, which is highly non-linear. In higher dimensions, an initial sphere of the phase space will soon evolve into an ellipsoid, with its long axis corresponding to the stretching direction and short axes to the contracting directions. This is schematically shown in fig. 24.

The magnitude of the rate of stretching when averaged over the entire phase space is a measure of the sensitivity to initial conditions of chaotic systems and goes by the name of *Lyapunov exponent*. Thus, an initial difference vector very quickly orients itself in the

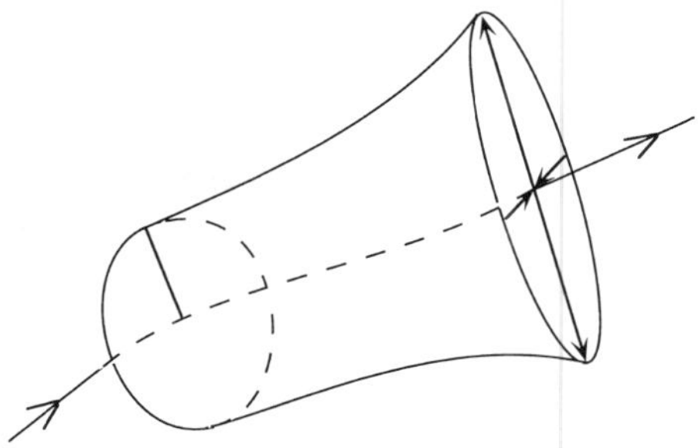

Fig. 24. Two-dimensional sketch showing the evolution of a neighbourhood along a phase space orbit. A circle becomes an ellipse by stretching along one direction and contracting along the other. The total area in the phase space is either conserved or decreases.

direction of maximum stretching and the corresponding exponent is positive. The values of other exponents corresponding to contracting directions should be negative. There should also be a zero exponent corresponding to the tangential direction, as there cannot be any change along the trajectory. From this discussion, it is seen that the existence of a positive Lyapunov exponent will guarantee the existence of chaos, while there would be no positive Lyapunov exponent for stochastic systems. In addition, the analysis implies that a system needs at least three degrees of freedom (one for stretching, one for contraction and one neutral) for the existence of a chaotic dynamics. One simple way to obtain a rough estimate of the positive Lyapunov exponent is to calculate the quantity

$$\lambda = \langle \log(d_{ij}(k)/d_{ij}(0)) \rangle / k\Delta t, \tag{72}$$

where $d_{ij}(k) = |\mathbf{X}_k(i) - \mathbf{X}_k(j)|$ is the distance between the vectors $\mathbf{X}_i$ and $\mathbf{X}_j$ (in a M-dimensional space) after k units of time increment Δt, starting with a distance $d_{ij}(0) = |\mathbf{X}_0(i) - \mathbf{X}_0(j)|$ at zero time. This time, the symbol $\langle \cdots \rangle$ represents the average over the entire phase space. Of course, if the evolution equations are known, the Lyapunov exponent can be calculated directly.

Strange attractors, self-similarity and fractal dimension

Consider a deterministic trajectory as a function of time similar to that shown in fig. 21(a), but now obtained from the model described below (cf. eqs (77)–(80). The corresponding phase plot $[\phi(t), x(t), z(t)]$, is shown in fig. 25. While the trajectory as a function of time is devoid of any pattern, this plot shows that the orbit organises itself into an ordered geometrical object called a "strange attractor". A strange attractor is a fractal object which exhibits a self-similar structure, namely, the appearance of the object is the same at various levels of resolution. Self-similarity is a direct consequence of the property of chaotic systems to *stretch* in some directions and *fold* in other directions. In contrast to this, a phase plot of a stochastic system shows no self-similarity and, instead, uniformly fills the space. Along with the sensitivity to initial conditions, self-similarity is an other basic property that distinguishes chaos from a random noise.

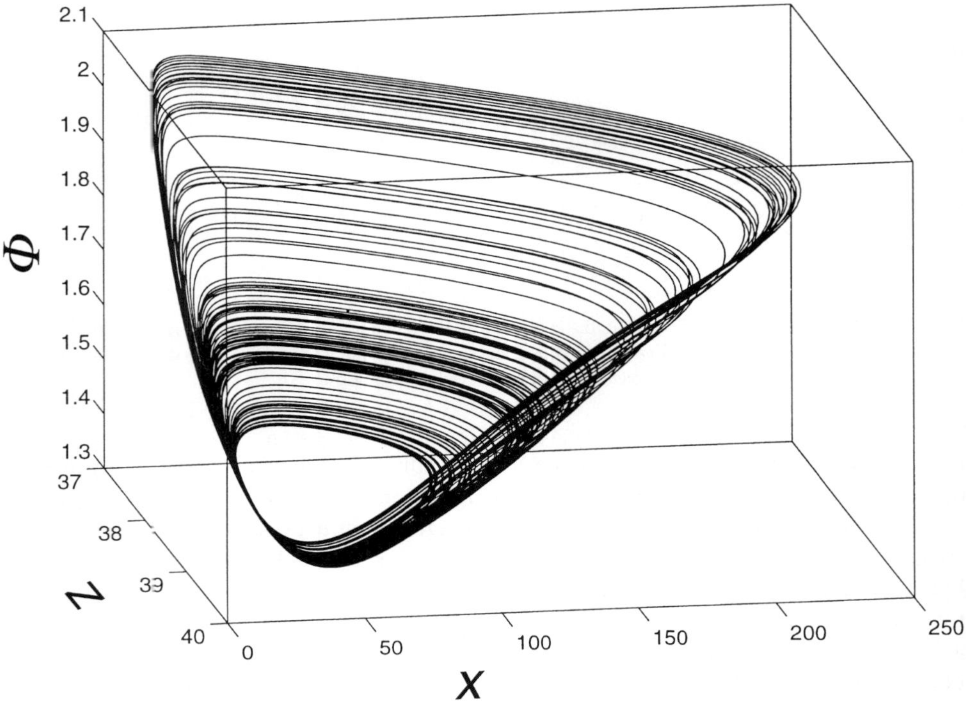

Fig. 25. Plot of an orbit in the phase space in the model [150] of the PLC effect, showing the structure of a strange attractor. A small section is shown in fig. 23.

Self-similarity is usually characterised by a quantity called the fractal dimension D_f of the attractor. An estimate of D_f is obtained by the box-counting method. In the latter, the attractor is covered by boxes of various sizes ℓ and one counts the number of occupied boxes $N(\ell)$. For a fractal object, $N(\ell) \approx \ell^{-D_\mathrm{f}}$. Thus, the slope of $\operatorname{Log} N(\ell)$ versus $\operatorname{Log} \ell$ is constant and equal to $-D_\mathrm{f}$, which is generally a non-integer. D_f is less than the number of degrees of freedom (or modes) of the system which corresponds to the dimension of the space in which the strange attractor is embedded. Thus, the lowest integer value of $(D_\mathrm{f}+1)$ is often taken as a lower bound of the number of degrees of freedom required for a dynamical description of the system. A more convenient method for estimating the fractal dimension of the attractor (or an equivalent quantity such as the correlation dimension) will be given later in section 4.2.3. Since all signals contain noise, the attractor is usually noisy. If the noise extent in the phase space is small as compared with the total extent of the attractor, then the attractor is itself robust in the sense that it is smeared only over the length scale corresponding to the amplitude of the noise. The attractor dimension remains unchanged, but the scaling region is restricted to larger length scales, as noise is usually found at small length scales. Experimental signals from chaotic systems would always have contributions from both deterministic noise and random noise. Given such time series, it is possible to quantify the extent of the contributions due to intrinsic non-linearity against those due to random noise (cf. section 4.2.3).

From the above discussion, it is clear that there are basic differences between chaotic and random types of behaviour. Several methods summarised below, both qualitative and quantitative, have been devised for distinguishing a chaotic signal from a stochastic one. Some of them, rather simplistic but easy to implement or to visualise in terms of plots, are quite useful in giving a clue about the possibility of chaos. Some others provide more quantitative estimates and one usually needs to carry out several of these to obtain a good level of confidence. The following is intended to be a check list:

(i) If a plot as a function of time of one or more variables of a non-linear system looks "noisy", and a phase plot (as sketched in fig. 23) has the appearance of an attractor, then there is a possibility of finding chaos.
(ii) The correlation function can be calculated to check if it decays rapidly and oscillates; the broad band nature of the calculated power spectrum is also considered an indicator of chaos.
(iii) If an attractor shows up in the phase plot, the correlation dimension can be calculated with the help of an appropriate method (cf. section 4.2.3).
(iv) Then, the Lyapunov exponents can be calculated to check the occurrence of a positive exponent (cf. section 4.2.3).

4.2.2. *A dynamical model for the PLC effect*

The model by Ananthakrishna et al. [150] for the PLC effect addresses the collective behaviour of dislocations. It basically attempts to recover the essence of the phenomenon at a more basic level than the models discussed in the previous part (cf. sections 3.4.2 and 3.4.3). The basic idea of the model is that most of the qualitative features of the PLC effect result from a non-linear interaction between the defect populations. Hence the model is dynamical in nature. In its simplest form, it consists in setting up rate equations for three types of dislocation densities [151] and solving them, taking into account the machine equation which defines the experimental conditions. The first versions of the model are carried out by ignoring the spatial aspect. The rationale behind this is that the variables (dislocation densities) correspond to the collective degrees of freedom of the spatially extended system. In this section, we briefly recapitulate the model and its principal results.

The model and its results

The model starts by setting up rate equations (73)–(75) for the evolution with time of three types of dislocation densities, namely mobile dislocations, ρ_{m}, immobile dislocations, ρ_{im} and another population ρ_{i} which may be regarded as dislocations interacting with "clouds" of diffusing solute atoms [86,151]. The machine equation (eq. (76)) describes the rate of change of the stress under a constant applied strain rate. The complete set of equations is written as:

$$\dot{\rho}_{\mathrm{m}} = \theta v \rho_{\mathrm{m}} - \mu \rho_{\mathrm{m}}^2 - \mu \rho_{\mathrm{m}} \rho_{\mathrm{im}} + \lambda \rho_{\mathrm{im}} - \alpha \rho_{\mathrm{m}}, \tag{73}$$

$$\dot{\rho}_{\mathrm{im}} = k \mu \rho_{\mathrm{m}}^2 - \mu \rho_{\mathrm{im}} \rho_{\mathrm{m}} - \lambda \rho_{\mathrm{im}} + \alpha' \rho_{\mathrm{i}}, \tag{74}$$

$$\dot{\rho}_{\mathrm{i}} = \alpha\rho_{\mathrm{m}} - \alpha'\rho_{\mathrm{i}}, \tag{75}$$

$$\dot{\sigma}_{\mathrm{a}} = \kappa(\dot{\varepsilon}_{\mathrm{a}} - b\rho_{\mathrm{m}}v). \tag{76}$$

The dots indicate derivatives with respect to time. The first term at the right-hand side of eq. (73) is the rate of production of dislocations due to double cross-slip with a rate constant θ. This term also includes the velocity of the mobile dislocations, $v = v_{\mathrm{o}}(\sigma/\sigma_{\mathrm{o}})^m$. The second term refers to two mobile dislocations either mutually annihilated and blocked with rate constants $(1-k)\mu$ and $k\mu$ respectively. The third term represents the annihilation of a mobile dislocation with an immobile one. The fourth term stands for the remobilization of immobile dislocations due to stress or thermal activation. The last term represents the immobilisation of mobile dislocations either due to solute atoms or to other pinning obstacles. If we choose to interpret this term to model the former, then α refers to the concentration of the solute atoms which participate in slowing down the dislocations. Once a certain concentration of solute atoms is gathered around the mobile dislocation, the latter is considered as belonging to another population, namely ρ_{i}. Thus, the loss term $\alpha\rho_{\mathrm{m}}$ in eq. (73) is a gain term for ρ_{i}. The second term in eq. (75) also represents a loss term. This term is introduced to mimic the fact that the dislocations ρ_{i} eventually become immobilised when the clouds of solute atoms reach large concentrations. The coefficient α' corresponds to the rate constant of this immobilisation. In eq. (76), κ is the effective elastic modulus and $\dot{\varepsilon}_{\mathrm{a}}$ is the applied strain rate. Clearly, every time the quantity $bv\rho_{\mathrm{m}}$ exceeds $\dot{\varepsilon}_{\mathrm{a}}$, a stress drop occurs. These equations can be cast into a dimensionless form by using the scaled densities x, y, z and the scaled stress ϕ:

$$\dot{x} = (\phi^m - a)x - bx^2 - xy + y, \tag{77}$$

$$\dot{y} = b_0(kb_0x^2 - xy - y + az), \tag{78}$$

$$\dot{z} = c(x - z), \tag{79}$$

$$\dot{\phi} = d(e - \phi^m x). \tag{80}$$

The parameter a is related to the concentration of the solute atoms, b_0 to the reactivation of immobile dislocations and c to the time scales over which the slowing down occurs, d refers to the scaled effective modulus and e is the scaled applied strain rate. The dependencies on stress and temperature are incorporated into the coefficients v, α and α'.

There is only one steady-state solution for these equations. There are ranges of the parameters a, b_0, c, d, m, k and e for which this steady state is linearly unstable, and a transition to a stable periodic state, referred to as a Hopf bifurcation, occurs as these parameters are varied. Limit cycle solutions are obtained for a range of physically relevant parameters a, b_0, c, d, m, k and e. In each of these ranges, the dislocations densities x, y and z are oscillatory. In physical terms, the most interesting parameter is the reduced strain rate e, which plays the role of a control parameter as further bifurcations occur when it is increased (see fig. 27, below). Many qualitative features of the PLC effect emerge naturally as a consequence of non-linear interaction between participating defects. For instance, the negative strain rate sensitivity of the flow stress and its dependence on plastic strain follow

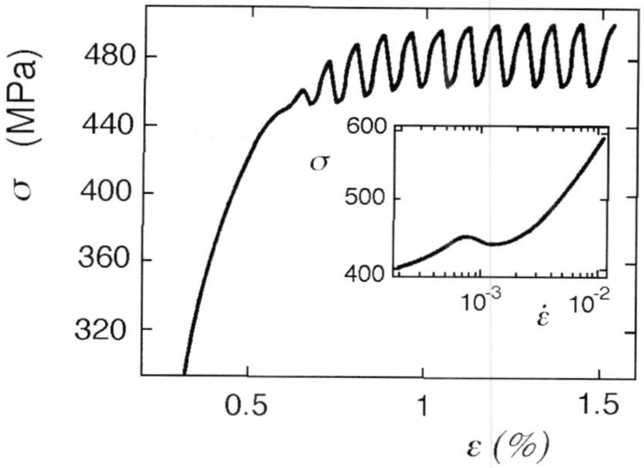

Fig. 26. Oscillatory stress vs. strain curve obtained as a solution of the set of eqs (77)–(80). The inset shows the strain rate dependence of the yield stress, obtained with the same value of the parameters, which exhibits a portion of negative strain rate sensitivity.

automatically [150]. Figure 26 shows a calculation of a serrated stress vs. time curve, with the inset showing the negative strain rate sensitivity of the flow stress.

The existence of two critical strain rates within which the phenomenon is observed [153] and the existence of a critical incubation strain are both predicted. Similarly, there are bounds on the concentration of solute atoms within which serrations are observed. Many other features consistent with experiments are predicted (cf. [150,152–154]). In the present form of eqs (73)–(76), the model ignores the spatial heterogeneity. A generalisation of the model has been proposed in which a spatial term is included by writing down the continuity equations for the three types of densities [155]. Then, band propagation is predicted in constant imposed strain rate, with a band velocity which increases with the applied strain rate [153]. All other results of the earlier model remain unaltered.

Chaos in the model and its consequences

One prediction of the model is that the stress drops exhibit a chaotic behaviour within a window of strain rates whose upper and lower values, typically from $10^{-5}\,\mathrm{s}^{-1}$ to $10^{-3}\,\mathrm{s}^{-1}$, depend on the set of parameters used [153,154]. The control parameter is the dimensionless applied strain rate e. For small values of e a periodic motion is found, say with a period T, through a Hopf bifurcation at $e = e_1$. As e is increased, this period-T orbit becomes unstable, giving way to a period-$2T$ orbit at e_2, and further to periodic orbits of $4T$ at e_3, $8T$ at e_4, etc. This process eventually culminates in chaos, which is aperiodic, at e_∞. This sequence of bifurcations is referred to as the period-doubling route to chaos [156]. One can calculate the rate of approach to e_∞ by calculating the so-called Feigenbaum exponent $\delta = (e_n - e_{n-1})/(e_{n+1} - e_n)$, where e_n is the value of the reduced strain rate at the n-th bifurcation. The value obtained from the above model is the same as that obtained for the textbook example referred to as the logistic map (cf. [1, p. 17]), namely, $\delta = 4.666$. One can construct in a simple way a bifurcation diagram by plotting, for example, the peak stress values during unstable flow as the control parameter e is increased (cf. fig. 27). In

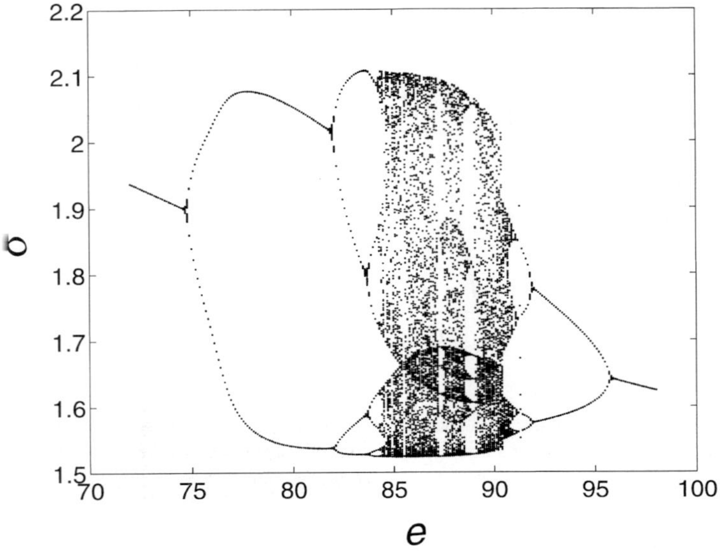

Fig. 27. Bifurcation diagram of the deterministic PLC model [150]. The peak stress during jerky flow is plotted vs. the control parameter e (the reduced applied strain rate). A sequence of bifurcations leads to chaos, followed by a reverse sequence leading to periodic motion. The two single lines at the extreme right and left represent the stable stress vs. strain rate regimes obtained outside the unstable domain.

addition to the sequence of bifurcations leading to chaos, the diagram exhibits a reverse bifurcation connected to the fact that the PLC effect is seen only in a window of strain rates. Just outside this range there are periodic solutions with a high period [153,154].

4.2.3. Experiments and time-series analysis

Experimental data

The first attempts to verify the prediction that the stress time series could be chaotic were undertaken by Ananthakrishna et al. [157], and further by several authors [158–161] (among which [158–160] used the same experimental data). The analysis was carried out on experimental stress signals obtained from $\langle 111 \rangle$ single crystals of Cu–14%Al loaded under an imposed strain rate of $3.34 \times 10^{-4}\,s^{-1}$ at a temperature of 294 K. Twelve files of 6144 points sampled at a rate of 242 points per second, corresponding to a kind of Stage I deformation with one active slip system and a rather low work-hardening coefficient (80 MPa) were analysed. The results, though preliminary, suggested that the time series could be chaotic. The methodology consisted of using the simplest algorithm for implementation, namely that due to Grassberger and Procaccia [162]. However, one limitation of this work was the shortness of the data sets, often containing in addition a high level of noise.

For this reason, specific experiments were performed to obtain reasonably long and accurate signals [163]. Al–2.8%Mg polycrystalline samples were obtained from a strip of thickness 0.5 mm in the form of parallelepipedal (length 50 mm, width 5 mm) tensile specimens. The samples were cold-rolled and annealed at 460°C for three hours, which

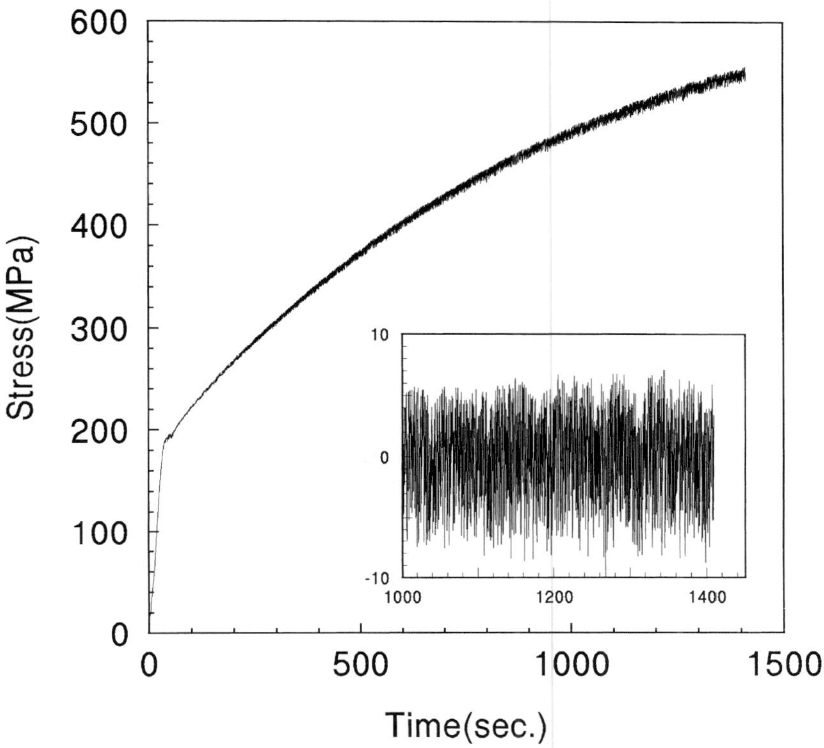

Fig. 28. Experimental stress vs. time plot in Al–2.8%Mg deformed at room temperature with a constant applied strain rate $\dot{\varepsilon}_a = 6.67 \times 10^{-5}\ s^{-1}$. The insert shows the last part of the curve after correction from the drift due to strain hardening.

yielded an average grain size of 250 μm. The deformation tests were carried out at room temperature with a constant imposed strain rate of $6.67 \times 10^{-5}\ s^{-1}$. The data were sampled at a frequency of 20 s^{-1}. The accuracy in the load measurement was up to 10^{-6} MPa, as compared with the 10^{-2} MPa accuracy of the files analysed in [157].

As illustrated by fig. 28, two features can be noticed at once from the stress vs. time plot of these data sets [163]. First, the stress–strain curve has a drift upward due to strain hardening. This background variation is considered as being without significance as far as the dynamical aspect of the PLC phenomenon is concerned. It is removed by either fitting the stress vs. time curve with a low-order polynomial or by constructing a moving average over a large number of points and subtracting this background value. The non-constancy of the amplitude of the stress drops also stems from the changes in the microstructure of the sample, which also implies non-stationarity of the time series. However, there is a saturation of this effect at large strains. The last part of the data exhibits a nearly constant serration amplitude and is, therefore, stationary. The analysis was mostly confined to this region and was carried out by subdividing the data into several sub-files to further minimise the effects due to changing amplitude. The subdivided data files on which the analysis was performed, referred to as files M13 to M15-16, contain 5000 points each, except M15-16 which has 8254 points. This last file is shown as an inset in fig. 28. Finally, a third data set obtained from Cu–10%Al single crystals will be examined from a global perspective of possible types of dynamical states in section 4.4.

Embedding and reconstruction of the attractor
The purpose of this section is to provide a working knowledge for the concepts introduced in section 4.2.1 in a form that will be directly used for the analysis of the experimental stress vs. time signals. Here, only one variable is monitored. However, the dynamics of the system is obviously governed by more degrees of freedom. Therefore, one attempts to reconstruct the original attractor by embedding the scalar time series in a higher-dimensional space and, in the process, obtain information about the number of degrees of freedom required for a proper dynamical description of the system. Consider the stress signal defined by

$$\left[\sigma(k),\ k=1,2,\ldots,N\right], \tag{81}$$

where the index k is in units of the time increment Δt. Let

$$\mathbf{X}(k)=\left[\sigma(k),\sigma(k+L),\ldots,\sigma\big(k+(M-1)\big)L\right],\quad k=1,\ldots,N, \tag{82}$$

be the M-dimensional vector defined by embedding this scalar signal, with a properly chosen time delay L . The "window length" t_{w} of the embedding is defined as $t_{\mathrm{w}} = (M-1)L$. In a M-dimensional space, referred to as the reconstructed phase space, the mapping $\mathbf{X}(k+1)=F[\mathbf{X}(k)]$ defines a trajectory (or an orbit) known as the reconstructed attractor. The idea behind the reconstruction is that one expects to obtain information about the actual attractor by studying the reconstructed attractor, as this mapping has been shown to mimic the dynamics of the true, but unknown, attractor [164,165]. If the dimension M of the reconstructed attractor is lower than the dimension of the actual attractor, two points on different orbits may appear as neighbours even though they are not actually neighbours on the attractor. Thus, increasing the embedding dimension M lowers the contribution arising from such "false" neighbours to various quantitative estimates, such as the correlation dimension, for example.

Correlation dimension
As mentioned earlier (cf. section 4.2.1), a measure of the self similarity of the strange attractor can be obtained by calculating the fractal dimension by the box-counting method. However, this method becomes prohibitive in terms of computation time when the dimension is more than three. A much simpler algorithm due to Grassberger and Procaccia [162] calculates another dimension called the correlation dimension ν, which has been shown to be an upper bound to D_{f}. This quantity is obtained from the correlation integral $C(r)$ which is defined as the fraction of pairs of points $[\mathbf{X}(i),\mathbf{X}(j)]$ in the M-dimensional reconstructed phase space whose distance is less than r. An expression for $C(r)$ is given by

$$C(r)=\frac{1}{N_{\mathrm{p}}}\sum_{i,j}H\left[r-\left|\mathbf{X}(i)-\mathbf{X}(j)\right|\right], \tag{83}$$

where $H(\cdot)$ is the Heaviside function and N_p is the number of pairs used in the sum. This quantity exhibits a scaling behaviour $C(r) \approx r^{\nu}$ if the orbit lies on a strange attractor. In low enough embedding dimension, even though one may observe a scaling regime, there are contributions from "false" neighbours and, therefore, the exponent will not be a true representative of the correct dimension of the attractor. As the embedding dimension M is increased, the contribution from false neighbours decreases and beyond a certain dimension it vanishes. Then, one expects that the dependence of the exponent ν on M saturates. In contrast, since noise is usually considered infinite dimensional, the fractal dimension of a purely stochastic set increases linearly with the dimension in which it is being embedded. Thus, if a scaling regime exists, the slope $\mathrm{dln}[C(r)]/\mathrm{dln}(r)$ tends, in the limit $r \to 0$ and $M \to \infty$, to a constant value ν taken as the correlation dimension. Actually, it has been shown that a proper estimate of ν is obtained when $M > \nu$, provided the series is long enough and noise-free [166]. In practical situations, there is always noise at small distances and the data is of limited length. Therefore, one looks for a scaling regime in the intermediate length scales for some values of the embedding dimension.

Curing of time series by the Singular Value Decomposition method
Most experimental signals are corrupted by random noise. It is therefore necessary to determine the contribution arising from intrinsic non-linearity as against that due to random noise. There are methods available in the literature which help to reduce the noise component. One such method is the so-called Singular Value Decomposition (SVD) [167, 168]. In addition to curing the time series of noise, the SVD method often provides an estimate of the dimension of the attractor. In this method, the trajectory matrix $\mathbf{A}$

$$ {}^{\mathrm{t}}\mathbf{A} = \frac{1}{\sqrt{N}}\big[\mathbf{X}(1), \mathbf{X}(2), \ldots, \mathbf{X}(N)\big] \tag{84} $$

is rotated onto the basis of its principal vectors. Here ${}^{\mathrm{t}}\mathbf{A}$ denotes the transpose of matrix $\mathbf{A}$. The M singular values of matrix $\mathbf{A}$ (in other words, the eigenvalues of the covariant matrix ${}^{\mathrm{t}}\mathbf{A} \cdot \mathbf{A}$) are positive; conventionally they are ordered in decreasing order of their magnitudes. If some principal values (say $q + 1, \ldots, M$) are zero, then the trajectory remains confined to the subspace spanned by the basis $(1, \ldots, q)$. This gives an estimate of the minimum number of variables required for a proper dynamical description of the phenomenon. In practice, noise prevents the eigenvalues from being strictly zero, but sometimes, there may be a set of eigenvalues which are "small" compared to the largest one. Then, the estimate of the embedding dimension may be taken to be the value of the dimension q beyond which there is an abrupt decrease of the eigenvalues to this "low" level. This distinction of some eigenvalues being much larger than others often permits the reduction of the noise level. After a rotation of the trajectory matrix $\mathbf{A}$ to $\mathbf{A}'$ in the basis of its principal vectors, the cured time series is obtained by keeping only the first p principal components of the rotated vectors $(1 < p \leqslant q)$ and back rotating the trajectory matrix $\mathbf{A}'$ to the initial basis. The SVD technique is particularly useful when using the algorithm of Grassberger and Procaccia with the matrix $\mathbf{A}'$ eventually limited to its first p components [168]. In the rotated M-dimensional embedding space, the correlation integral $C(r)$ can then be calculated using the cured time series.

Lyapunov exponents

In addition to the estimate of the (largest) positive Lyapunov exponent mentioned earlier, algorithms have been designed to calculate the whole spectrum of Lyapunov exponents [169–171]. However, this requires very long time series which are usually impractical to obtain. The spectrum of Lyapunov exponents can also be calculated by using a method suited for (not too) short time series [172], which is a modified algorithm of Eckmann et al. [169]. The algorithm relies on the construction of the matrices $\mathbf{T}_i$ such that

$$\mathbf{X}(j+k) - \mathbf{X}(i+k) = \mathbf{T}_i\big[\mathbf{X}(j) - \mathbf{X}(i)\big] \tag{85}$$

$\mathbf{T}_i$ maps the small difference vector $[\mathbf{X}(j) - \mathbf{X}(i)]$ to its counterpart $[\mathbf{X}(j+k) - \mathbf{X}(i+k)]$ after k units of time evolution. As mentioned earlier (cf. section 4.2.1), this difference vector evolves by stretching and folding. Thus, after a short time interval, it aligns itself in the direction of maximum stretching. The idea is to resolve the evolved difference vector, for a few neighbouring points lying within a certain shell in the various directions of elongation and contraction. This procedure determines the elements of the matrix $\mathbf{T}_i$. Since the amount of elongation and contraction depends on the position of the orbits on the attractor, an average is taken over the entire attractor. This procedure is adequate when the time series is long and the magnitude of noise superposed is small. However, most of the experimental time series are short and the superposed noise magnitude could be large in addition to being unknown. In order to smooth out the effects of noise, we designed an improved version of the Eckmann's algorithm wherein the size of the shell used for calculating the Lyapunov spectrum is varied. For a chaotic attractor, one should find a positive exponent whose values remain stable for a range of shell sizes. A stable zero exponent should also be found simultaneously. In a random time series, any positive Lyapunov exponent is expected to be a decreasing function of the shell size, in addition to the absence of a zero exponent.

4.2.4. Analysis of data on Al–Mg polycrystals

Experimental results

Following [163], characteristic results obtained with the help of the above algorithms will now be given for the file M15-16 relative to Al–Mg polycrystalline samples (the results for the other files on the same material are found to be similar). For the uncured files, the value of the autocorrelation time (the value at which the correlation drops to $1/e$ of its original value) is about five time units. Using this as the lag time L, one can obtain the correlation dimension ν. Figure 29 shows a plot of $\ln[C(r)]$ vs. $\ln(r)$. It is seen that the slope converges to 3.2 as the embedding dimension is increased to 5. Beyond this dimension, even though the slope remains the same, the extent of the scaling regime reduces. As shown in [166], this last feature is an artefact of short time series.

The SVD method has been used as a noise reduction technique to cure the data files. The spectrum of eigenvalues for the window lengths $t_w = 10$ and 12 is shown in fig. 30 where the relative amplitude of the eigenvalues, ordered in decreasing order of their magnitude, is plotted as a function of their index. No abrupt decrease to a base level can be noticed. Even so, the relative value falls by two orders of magnitude by the sixth component.

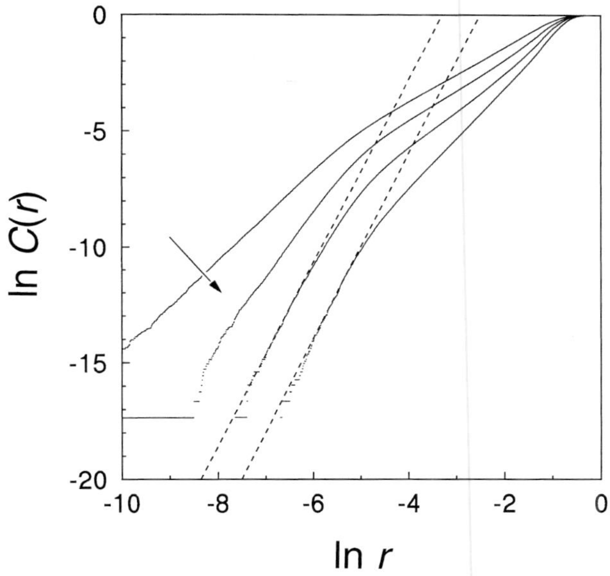

Fig. 29. Log–log plot of the correlation integral $C(r)$ as a function of distance r in the phase space for the uncured file M15-16 of [163]. Note the convergence of the slope to 3.2 (dotted line) as the embedding dimension is increased (arrow) from two to five.

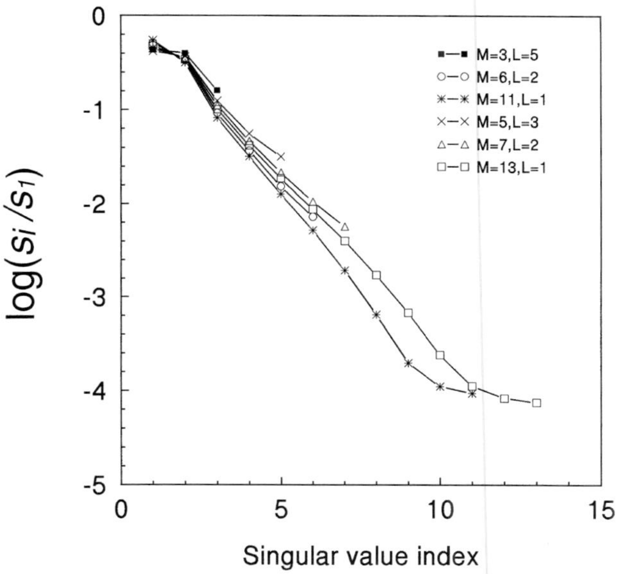

Fig. 30. Singular value spectrum for the cured file M15-16 of [163]. The window length is $t_W = (M - 1)L = 10$ or 12.

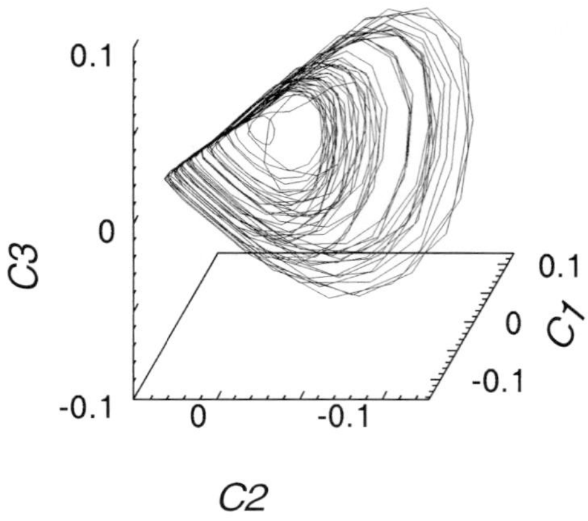

Fig. 31. Projected trajectories onto the first three principal components of the rotated phase space for the cured data file M15-16 of [163], containing only six principal components. For the sake of clarity, the three-dimensional plot contains only 10^3 points.

Therefore, the time series were cured by keeping only the first six principal components and back-rotating the trajectory matrix to the initial basis. The cured signal looks very much like the original time series. However, the cured data is expected to give a better visual picture of the strange attractor, due to reduced level of noise, as is evidenced by the plot of the rotated trajectory onto the first three principal components shown in fig. 31.

The calculation of the autocorrelation function of every cured file shows that the data quickly decorrelate; the characteristic autocorrelation time is less than four sampling units. The window length $t_w = (M-1)L$ is kept to be a few times the correlation time and both M and L are varied. As shown in fig. 32, it is clear that the slope $\mathrm{dln}[C(r)]/\mathrm{dln}(r)$ vs. $\ln(r)$ converges when M reaches the value 6 for $L = 2$, while for $L = 1$ the convergence is seen as M approaches 10. The slope saturates around $\nu \cong 3.1$ in the range $r \cong e^{-5} - e^{-4}$. The window length is roughly kept constant in these limits. This value of ν is slightly lower than that obtained with the uncured files, but it is still consistent with it. Thus, the correlation dimension ν can be taken to be 3.1–3.2, which suggests the existence of a strange attractor in the phase space.

Lyapunov exponents

Since Lyapunov exponents are unambiguous quantifiers of chaotic dynamics, the time series were also analysed for the existence of a positive Lyapunov exponent. The spectrum of Lyapunov exponents has been calculated for both the uncured and cured ($p = 6$) files M15-16, with the delay times of $L = 5$ and $L = 2$ respectively. Figures 33(a) and (b) show the plots of the Lyapunov spectrum as a function of time for $M = 5$. The convergence of the exponents is clear; the value of the largest exponent comes out to be 1.0 and 0.62 respectively. The existence of a positive exponent in both cases gives additional confidence in the chaotic nature of the time series.

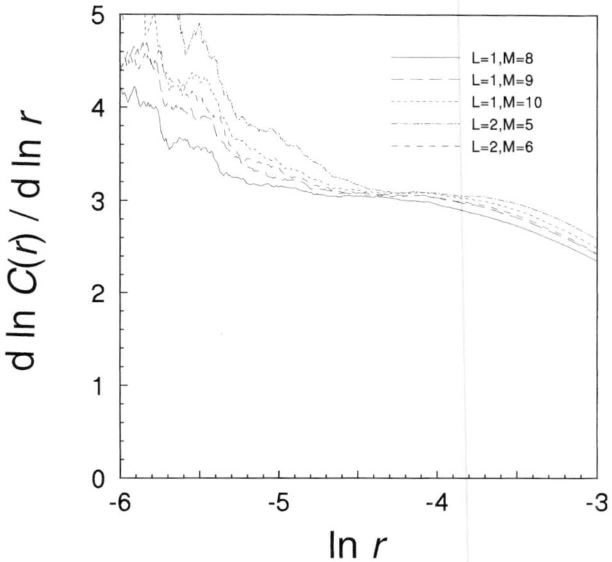

Fig. 32. Log–log plot of the slope of the correlation integral $C(r)$ as a function of distance r in the phase space for the cured file M15-16 of [162], with only six principal components. Note the convergence of the slope to 3.1 as the embedding dimension is increased to 10.

Surrogate data

Distinguishing low-dimensional dynamical behaviour from superimposed noise is not an easy task, as the interpretation of the data obtained from the correlation dimension and the Lyapunov exponents may involve some subjective judgement if the time series is short and noisy. If positive identification of chaotic dynamics is difficult, *a contrario* arguments are easier to find. For a low-dimensional dynamics, the Fourier phases play an essential role in the convergence of the correlation dimension when the embedding dimension increases [173]. Therefore, it is useful to consider stochastic surrogate data obtained by inverting a power spectrum identical to that of the data under study, but with random phases. If the correlation dimension obtained from the surrogate data is invariant in this randomising process, then the actual data is not indicative of chaotic dynamics. Note however that, conversely, a change in the correlation integral under phase randomisation does not necessarily imply in itself that the original time series is of non-linear origin. Thus, several surrogate data files generated by randomising the phases of the Fourier transform of the original data must be produced and analysed with the same methods. While for the original data the correlation dimension converges to 3.2 as the embedding dimension M is increased, the correlation dimension increases linearly with M for the surrogate data, as shown in fig. 34. The latter behaviour is characteristic of noise which is infinite dimensional. In the same way, the Lyapunov exponents have been calculated for the surrogate data. The results indicate that the largest one approaches zero as a the shell size increases. In contrast, the largest exponent for the real data is close to unity. Thus, the chaotic nature of the original signal is confirmed.

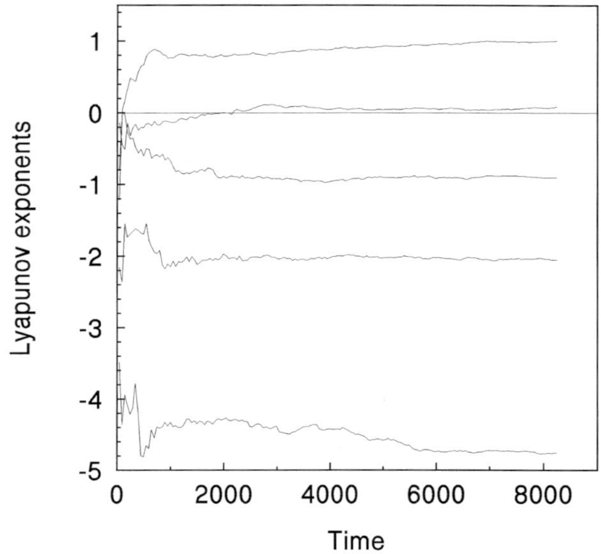

(a)

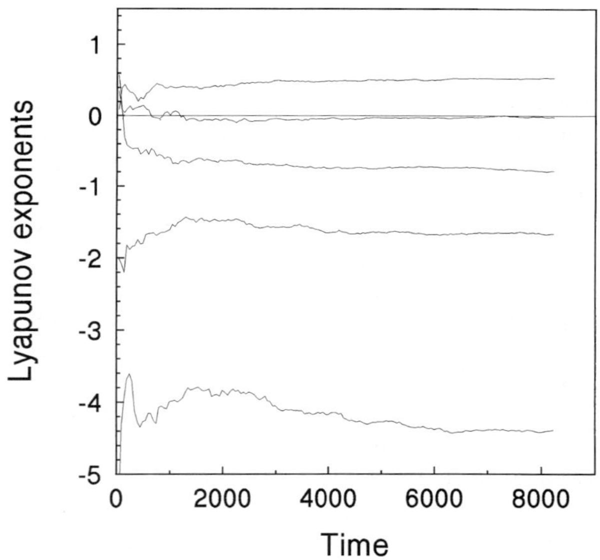

(b)

Fig. 33. (a) The Lyapunov spectrum vs. time for the uncured file M15-16. The evolving neighbourhood is a shell of size 12% of the attractor; (b) the Lyapunov spectrum vs. time for the cured file M15-16. The evolving neighbourhood is a shell of size 10% of the attractor.

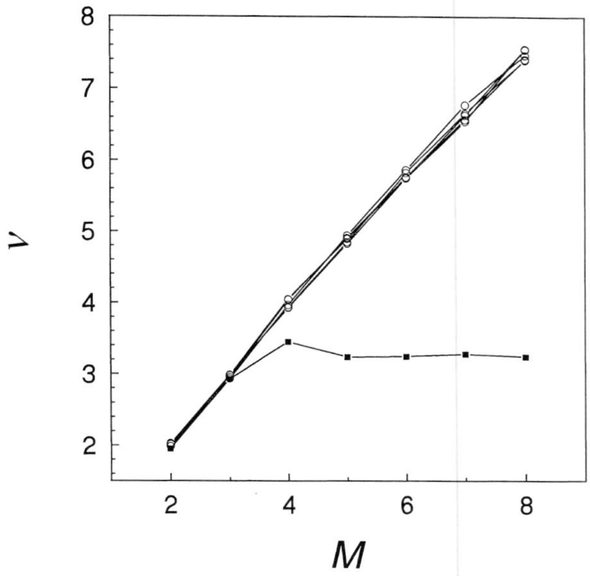

Fig. 34. The correlation dimension ν for the actual data and several sets of surrogate data as a function of the embedding dimension M. The points refer to the average slope of the correlation integral in the region $10^{-5} \leqslant r \leqslant 10^{-4}$. Note the convergence for the actual data as against the linear increase for the surrogate data.

4.2.5. Discussion

The positive results emerging from several complementary methods (such as noise filtering by the SVD method, the correlation dimension and Lyapunov spectrum estimations and a surrogate analysis), coupled with the physical insights into the phenomenon given by the existing model of the PLC effect, give confidence in the conclusion that the Al–Mg polycrystal data are indeed of low-dimensional non-linear origin. Further, the analysis suggests that only a few degrees of freedom, four or five, are needed to describe the dynamics of the PLC effect, consistent with the model [150–152].

A detailed account of the influence of the loading conditions (strain rate, temperature ...) and of the material parameters on the dynamical characteristics of stress vs. time plots is missing in the above analysis. It would be useful to view these results in a global perspective which might help to understand more clearly the intimate relations between the overall time evolution of the system and the spatial interactions involved in the band propagation. Put in other words, the question is about the relations between the spatial coupling, the spatial distribution of plastic slip events and the time series where one is actually measuring only homogeneous or spatially averaged quantities such as the stress level. Since the characteristic length scales of coupling are short as compared to the specimen dimensions, the formation of the PLC bands can be spatially correlated or uncorrelated. This must somehow be reflected in the ordered structure of the time series. Although, the relation is quite understandably a complex one, some insight can be obtained by analysing the stress-time series by a method which essentially addresses its statistical nature. This forms the objective of the next section.

4.3. Statistical analysis of the Portevin–Le Chatelier instability

4.3.1. The PLC effect, stick–slip and earthquakes

From a physical point of view, the PLC effect is a good example of a "stick–slip" situation [174]. The latter corresponds to a bistable behaviour, the system spending characteristically unequal times in the stick and slip states. During jerky flow the reloading period between two stress drops is much longer than the rather short duration of the jerks themselves. There are many other instabilities which also fall under the category of stick–slip situations. To name only a few, stick–slip due to solid friction [175], earthquakes [176], peeling of scotch tape [177], and flow of sand particles through an hour glass [177]. Such similarities have been suggested by several authors [178–182]. In all cases, the basic ingredient is that the systems exhibits a "negative flow rate characteristic", just as in the PLC effect.

Earthquakes are often considered as slip events on the interface of tectonic plates which are in relative motion caused by mantle-wide convection. In this context, the Burridge–Knopoff model has attracted considerable attention in the last few years [183,184]. This model describes the dynamical behaviour of a system consisting of a large number of coupled sliding blocks that interact with a fixed surface via solid friction. This is clearly an oversimplification of the actual tectonic plates interface, which can hardly reproduce sliding along a fault embedded in an elastic continuum [185]. However the model generates the observed Gutenberg–Richter's law [186] for the statistics of the earthquake events. According to this law, the histogram of the magnitudes of events is in the form of a decreasing power law with a characteristic exponent whose value is typically around unity.

The similarities between the PLC effect and stick–slip on faults goes much further, as has been discussed in some detail. In practice, this analogy suggests performing a statistical analysis of the stick–slip events in the PLC instability, in complement to the dynamical approach of the previous section. Following Lebyodkin et al. [124,181], such an analysis is presented in the next section. A numerical model which captures the essential features of the experimental results is then presented.

4.3.2. Statistics of stress drops in the PLC instabilities: an experimental study

From the above discussion, one expects to find situations where stress drops recorded during jerky flow exhibit a power law distribution instead of a random distribution. In order to verify this, specific experiments were performed on Al–4.5at.%Mg single crystals with stress axes $\langle 111 \rangle$ or $\langle 100 \rangle$ favouring multiple glide. No sizeable effect of orientation was observed, however. Tests were also performed on polycrystals, yielding tendencies similar to those recorded with single crystals. Here, we report the results obtained with an imposed strain rate $\dot{\varepsilon}_a$ varied in the range $3.2 \times 10^{-6}\,\mathrm{s}^{-1}$ to $1.3 \times 10^{-3}\,\mathrm{s}^{-1}$ at 300 K. The specimens exhibited serrations in this whole range of experimental conditions.

The influence of the changing material state as reflected in its strain hardening behaviour was eliminated by a normalisation accounting for the slow drift in the average stress level during straining. The magnitudes of the individual stress drops $\Delta\sigma$ were plotted as a function of the plastic strain ε on each deformation curve. A moving average stress drop value $\overline{\Delta\sigma} = f(\varepsilon)$ was then determined by a linear regression fit through the data points. The distribution of the non dimensional stress drops $\delta = \Delta\sigma/f(\varepsilon)$ was then investigated. Histograms with a statistical sample of 100–300 stress drops were extracted from the

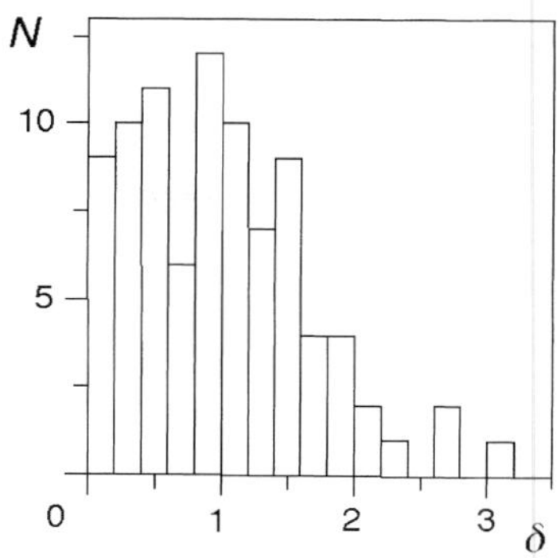

Fig. 35. Histogram of the stress drop statistics (number of drops N vs. the dimensionless stress drop magnitude δ) in the saturation region of the deformation curve for an Al–4.4%Mg single crystal deformed at $T = 300$ K with an applied strain rate $\dot{\varepsilon}_a = 1.3 \times 10^{-5}\,\mathrm{s}^{-1}$ (after [124]).

deformation curves. The latter exhibit a first region of strong strain hardening followed by a saturation plateau. Since this plateau corresponds to a stabilised microstructural state, it was used as a basis for comparison.

Figure 35 shows a peaked histogram obtained in the plateau region with a strain rate of $1.3 \times 10^{-5}\,\mathrm{s}^{-1}$. The effect of the imposed strain rate on the character of the stress drops statistics is illustrated in fig. 36. As the strain rate is increased to $1.3 \times 10^{-4}\,\mathrm{s}^{-1}$, the histograms become quite asymmetrical. This last type of distribution can be fitted by a power law of the form

$$P(\Delta\sigma) \propto \Delta\sigma^{-\alpha}, \tag{86}$$

with an exponent close to 1.1, bearing some resemblance with the Gutenberg–Richter law mentioned above in the context of earthquakes. Details of other experimental results can be found in [124] which confirm the observed trend.

4.3.3. A simple model for the PLC statistics

The model used by Lebyodkin et al. [124,181,182] involves a series of blocks coupled by linear springs which slide on a fixed surface with a strain rate-dependent solid friction F (cf. fig. 37). In what follows the set of blocks is pulled with a constant imposed strain rate. (Each element is characterized by its strain rate.)

This model is reminiscent of the coupled sliders with dry friction models which are used to mimic seismicity on crustal faults [184]. In one dimension, the position variable x represents the behaviour of one cross-section of the sample whose constitutive relation is of the Penning type (cf. section 3.2.3 and eq. (18)). As discussed in section 3, the study of propagative phenomena requires modelling of the spatial coupling between adjacent slices of the material. In the present case, it is assumed that the coupling stems from the elastic stresses resulting from plastic strain incompatibilities between slices. Each element is characterised by its strain and strain rate. Due to the one-dimensional approximation (local changes in the cross-sectional area are not taken into account) and in mechanical

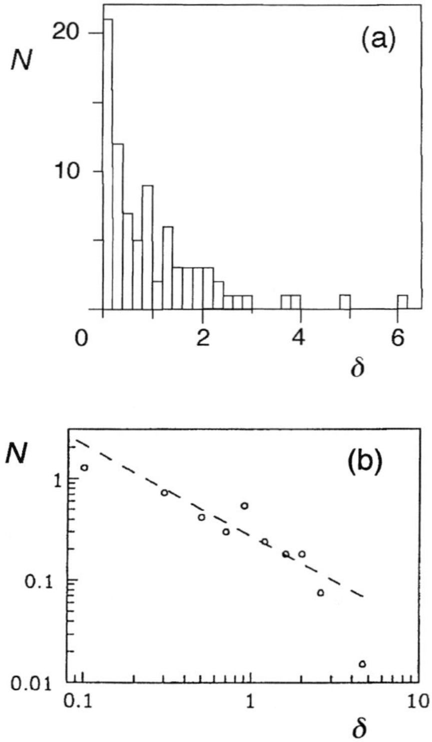

Fig. 36. (a) Histogram of the stress drop statistics for $\dot{\varepsilon}_a = 1.3 \times 10^{-4}\ s^{-1}$, showing a monotonically decreasing distribution (compare fig. 35). After [124]. (b) Log–log plot of this distribution.

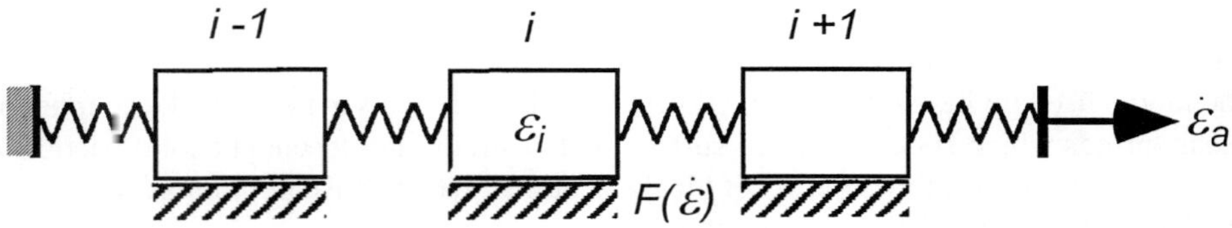

Fig. 37. Sketch of the model of coupled blocks connected by linear springs of stiffness k, sliding with a strain rate-dependent friction $F(\dot{\varepsilon})$ and deformed with a constant total imposed strain rate $\dot{\varepsilon}_a$.

equilibrium, the stress is constant throughout the system. When two slices deform at different strain rates, an internal stress develops which tends to homogenise the strain (see section 3 3.2). Using a cut-off distance, this internal stress is accounted for by coupling the slices in proportion to the strain difference between adjacent elements. Cutting the specimen into n slices perpendicular to the tensile axis (cf. fig. 37), each element i obeys, as a consequence, the discretised constitutive relation

$$\sigma = h\varepsilon + F(\dot{\varepsilon}_i) - k\big[(\varepsilon_{i-1} - \varepsilon_i) + (\varepsilon_{i+1} - \varepsilon_i)\big], \tag{87}$$

where $\dot{\varepsilon}_i$ is the local non-uniform strain rate, h is a constant strain hardening coefficient and $F(\dot{\varepsilon})$ is the N-shaped Penning-type characteristic function (cf. fig. 15). The constant k designates the strength of the coupling. Its adequate level, which depends on whether or not incompatibility stresses are plastically relaxed, can be obtained by varying its value between an elastic modulus and a plastic modulus (i.e., a strain hardening coefficient). Additional boundary conditions are used so that eq. (87) is no longer valid for the first and last slice, $i = 1$ and n, respectively. It is readily seen that the continuous version of this constitutive form involves a Laplacean with a minus sign, unlike some of the models discussed in the previous part (cf. section 3.4.2 and eq. (35)). This term induces non-local stabilising back stresses in constant strain rate loading (cf. section 3.3.2 and eq. (30)), in conformity with the present assumption that it stems from slip incompatibilities.

Under a constant imposed strain rate, the machine equation accounts for the possible heterogeneous deformation. In a discrete form, it is written

$$\dot{\varepsilon}_\mathrm{a} = \frac{\dot{\sigma}}{\mathcal{M}} + \frac{1}{n}\sum_{i=1} \dot{\varepsilon}_i . \tag{88}$$

Here, $\mathcal{M}$ is the combined elastic modulus of the specimen and the deforming apparatus. One peculiarity of the present model is the "dynamisation" of the $F(\dot{\varepsilon})$ curve, i.e. the fact that the shape of this characteristic curve depends on the applied strain rate. More specifically, by reconstructing the strain rate sensitivity from arguments based on the thermally activated motion of dislocations interacting with diffusing solute atoms, it has been shown that $F(\dot{\varepsilon})$ depends on the applied strain rate in a manner that can be modelled and numerically estimated [102]. Actually, it is found that an upward change in applied strain rate distorts the shape of this curve by reducing the range of the unstable domain and the amplitude of the stress drops.

The numerical procedure used for solving the set of eqs (87), (88) is described in [124]. The number of slices is $n = 300$. The $F(\dot{\varepsilon})$ curve was fitted to experiments representing the behaviour of the Al–Mg alloy as a function of strain rate. A value of $\mathcal{M} = 10^5$ MPa was selected for the combined elastic modulus and the strain hardening coefficient h was set to be either 10^2 MPa or zero in saturation conditions. The coupling constant was left to vary between $0.06\mathcal{M}$ and $\mathcal{M}$, the former value representing a plastic coupling between the slices and the latter an elastic coupling. Actually, this constant was tuned until a satisfactory agreement was obtained with the experimental statistics. The heterogeneities normally present in the specimen were accounted for by providing random initial fluctuations in the initial strain rate. There is no other randomness in the problem and the statistics is entirely of dynamical origin.

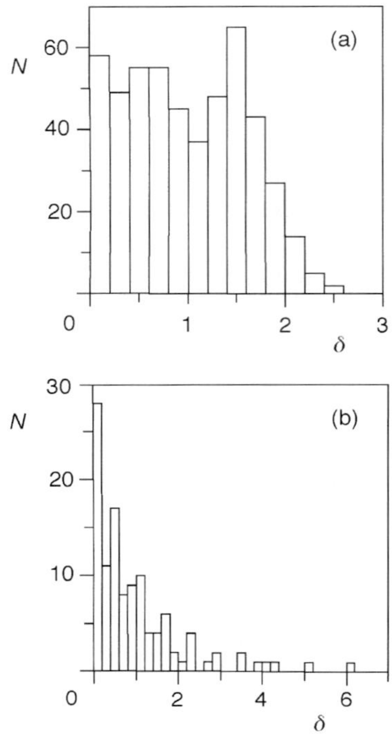

Fig. 38. Simulated stress drop statistics: (a) reference conditions: $k = 0.5\mathcal{M}$, $\dot{\varepsilon}_a = 10^{-4}\,\mathrm{s}^{-1}$; (b) same conditions but with an increased strain rate of $4 \times 10^{-4}\,\mathrm{s}^{-1}$. A monotonically decreasing histogram is obtained (compare fig. 36). After [124].

4.3.4. Results of numerical simulations

The strain hardening was set to zero in order to reproduce stabilised conditions; the introduction of a non-zero strain hardening has no marked effect. The results of the simulations were recorded in the form of stress vs. time curves and the statistical analysis of the stress drops was carried out. The results obtained in the simulation are in qualitative agreement with the observed features of the statistics of PLC instabilities. With a constant imposed strain rate of $\dot{\varepsilon}_a = 10^{-4}\,\mathrm{s}^{-1}$, the best agreement with experimental results was found for $k = 0.5\mathcal{M}$ (cf. fig. 38(a)), meaning that k represents, indeed, an essentially elastic coupling. These conditions were, therefore, taken as reference conditions. Then, with an increase in strain rate the distribution was found to change from a peaked or flat distribution to a monotonically decreasing one. This is illustrated by fig. 38(b) which shows the asymmetrical histogram obtained for a strain rate of $4 \times 10^{-4}\,\mathrm{s}^{-1}$. The monotonically decreasing distributions were found consistent with a power law form (cf. eq. (86)), with an exponent $\alpha \approx 1.1$, irrespective of the size n of the system. Within the model, these effects are due to the dependence of the Penning's function F on strain rate. A decrease of the value of the coupling constant k from its reference value also yields asymmetrical distributions with no preferred stress drop magnitude. This is understandable considering

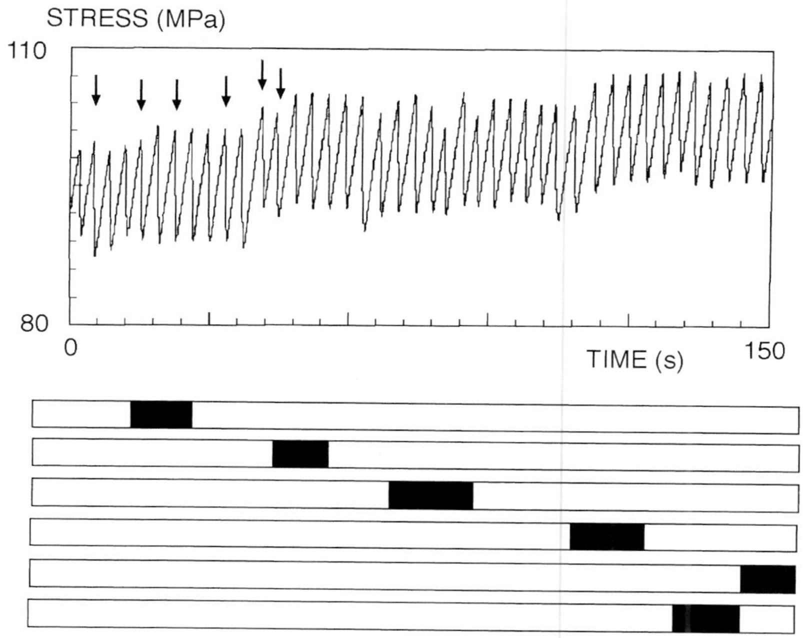

Fig. 39. Simulated stress vs. time curve of type B and the corresponding band pattern ($\dot{\varepsilon}_a = 5 \times 10^{-5}\,s^{-1}$). The arrows indicate the positions of the load drops during which snapshots of the deformation profiles have been taken. The latter are placed below each other in chronological order. The black regions represent the deformation bands which are clearly seen to occur in a propagating sequence. After [124].

that a strong coupling leads to a higher degree of correlation between the strain rate jumps of the blocks.

In addition, spatial patterns could be obtained that reproduce exactly the sequence of types of bands experimentally obtained with increasing applied strain rate (cf. section 3.2.3): spatially uncorrelated and static type C bands, and then spatially correlated hopping type B bands and propagative type A bands. As an example, fig. 39 shows a propagating pattern of type B bands and the corresponding stress vs. time curve.

4.3.5. Discussion

In this section, the statistics of stress drops in the PLC instabilities was shown to be consistent with a power-law distribution in a certain regime of high strain rates. The distributions however showed peaked distributions at small strain rates. This behaviour was qualitatively captured by a simple computer simulation. Although, the latter is phenomenological in the sense that it does not directly incorporate the behaviour of defect populations, it could also explain several features of the propagation of PLC bands and their velocities. This power-law behaviour is reminiscent of the Gutenberg–Richter law [186] which is another well-known power law describing the statistics of earthquakes events. Such statistical behaviour of dynamical systems has attracted considerable attention in the last decade and is associated with a new type of dynamical state referred to as Self-Organised Criticality (SOC) [187,188].

In the previous section we have seen how the time series obtained from a deforming specimen imply that only a few cooperative degrees of freedom are needed for a full description of its dynamics. Although the sample is a spatially extended system, there is an effective reduction in the number of degrees of freedom. However, there is another scenario of spatially extended non-linear systems exhibiting high degree of spatial and temporal correlation wherein the number of degrees of freedom cannot be reduced. One standard example is the sand-pile model originally formulated by Bak and co-workers [187]. This system evolves naturally to a critical state where avalanches of all sizes occur as the sand-pile is built up by the addition of sand particles. Since its introduction, the subject of SOC has been considerably developed (cf. [188] for a review). In particular, it has been shown that the steady state of such systems exhibit statistics which are described by power law distributions in both space and time. The results described above suggest that jerky flow might provide another example of SOC in a certain range of high strain rates. However, a detailed investigation of distributions of the magnitude and duration of the stress drops, power spectrum, and the interrelationship of the corresponding exponents would be required if a strong case is to be made out [189]. Such an attempt will be undertaken in the next section.

4.4. Cross-over from chaotic to self-organised dynamics in single crystals

The analysis presented in the last two sections shows that both chaos and SOC-type dynamics can be present in the PLC instability. Chaotic files analysed in section 4.2 at low and medium strain rates systematically showed peaked distributions for stress drops. At high strain rates (section 4.3), the statistics changed over to power law distributions compatible with SOC dynamics. The purpose of this section is to further investigate the possibility that, as the strain rate is increased from low to high values, a cross-over from chaos to SOC occurs in the dynamics of the PLC effect.

4.4.1. Experimental data

The intended analysis requires obtaining data over a wide range of strain rates. These are provided by files from single crystals of Cu–10%Al [189]. The specimens were initially homogenised for 36 h at 1230 K before mounting. The crystals were oriented for easy glide and the strain hardening coefficient was in the range 20 to 30 MPa. The deformation tests were carried out at 620 K under three different strain rates $3.3 \times 10^{-6}\,\mathrm{s}^{-1}$, $1.7 \times 10^{-5}\,\mathrm{s}^{-1}$ and $8.3 \times 10^{-5}\,\mathrm{s}^{-1}$. The sampling rate was 20 points per second and the corresponding files contained 4×10^4, 2×10^4 and 1.2×10^4 data points respectively. Clearly, these files are fairly long and we shall refer to them as PLC l, m and h (for "low", "medium" and "high"). Before we proceed further, a visual inspection of the time series is useful. A plot of a small portion of the three files is shown in fig. 40. While the PLC l and m files show numerous large yield drops, the file h shows a large number of small drops and much fewer large drops. Thus, the time series at low and medium strain rates are significantly different from those at high strain rates.

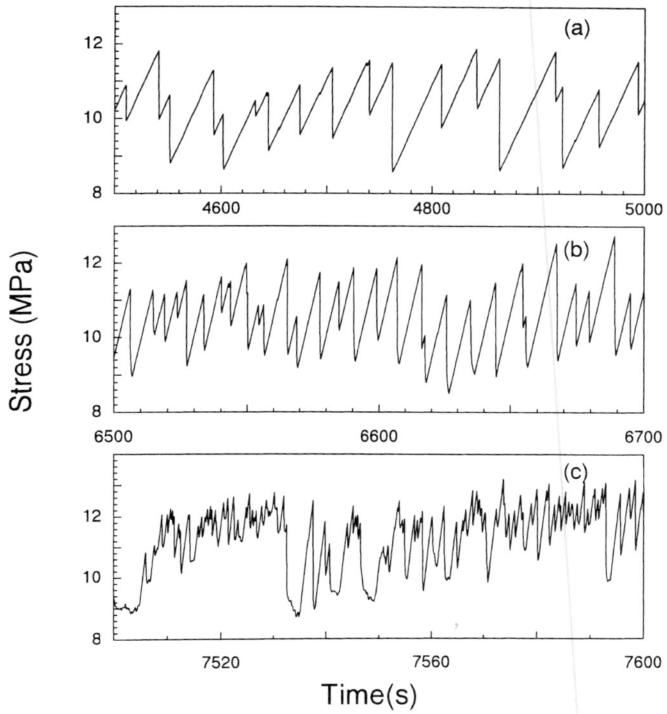

Fig. 40. Time series data of the PLC l, m and h files, (a), (b) and (c) respectively, showing the varying structure of the stress drops.

4.4.2. Results at low and medium strain rates

Since the methods used and the results obtained are very similar for the low and medium strain rate files PLC l and m, only results from file PLC m are shown. In both files, the distribution of stress drops is peaked. The autocorrelation function displays an oscillatory trend once it crosses the zero value with an autocorrelation time $t_c \approx 35$ units of sampling time $\Delta\tau$. Using a slightly smaller delay time $L = 20$, the correlation integral $C(r)$ is shown in fig. 41 for embedding dimensions from $M = 4$ to 9. A scaling region of two orders of magnitude can be seen in the interval $-4.0 < \ln(r) < -2.0$, with the slopes $\ln[C(r)]/\ln(r)$ converging as the embedding dimension approaches $M = 9$. The resulting correlation dimension is about 2.7. As a guide to the eye, dashed lines show the converged slopes for $M = 8$ and 9. A similar exercise was carried out on all surrogate data sets keeping the same time delay L. However, convergence of the slopes $\ln[C(r)]/\ln(r)$ was not found as the embedding dimension M was increased. Instead, the slopes kept increasing with M.

The Lyapunov spectrums of the PLC m file, calculated using an improved version of Eckmann's algorithm presented above in section 4.2.3 show one positive and one zero exponent, both stable in a range of values of the shell size. Similar analysis was carried out on 18 corresponding surrogate data sets. In this case, no stable positive and zero Lyapunov

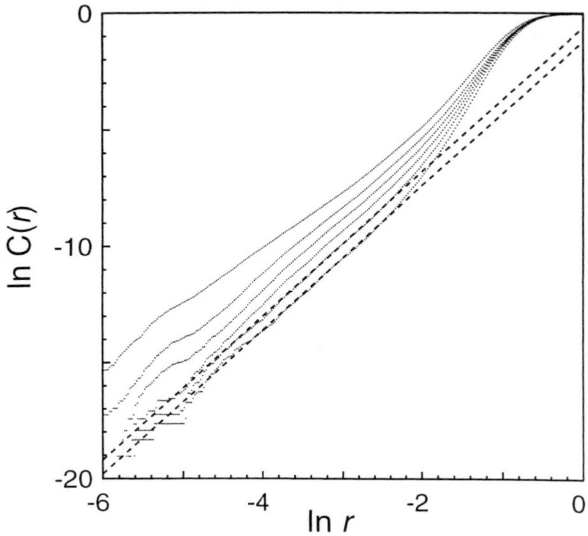

Fig. 41. Log–log plot of the correlation integral $C(r)$ vs. r for PLC m file. Dashed lines are shown as a guide to the eye with an exponent value $\upsilon = 2.7$.

exponent could be found. Thus, there is sufficiently strong evidence to conclude that the time series of the low (l) and medium (m) strain rate PLC files are of chaotic origin.

4.4.3. Results at high strain rate

The distribution of stress drops of the PLC h file is not peaked as was the case with the two previous files. Instead, it has the scaling form of eq. (86) with an exponent value $\alpha = 1.1$ (cf. fig. 42). The normalised distribution of the time duration Δt_{d} of stress drops (which is a multiple of the sampling time Δt) has also a scaling form

$$P(\Delta t_{\mathrm{d}}) \propto \Delta t_{\mathrm{d}}^{-\beta}. \tag{89}$$

In this case, there is a scatter for intermediate values of Δt_{d}. Actually, it is known that, even in numerical simulations, the distribution of the duration of the events is not as impressively scaled as that of the event sizes [187]. In the PLC h file, the scatter is largely due to the short time scale of the plastic relaxation time compared to the sampling time ($\Delta t \approx 0.05$ s). Even so, a rough estimate of the exponent β is still possible, yielding $\beta \approx 0.9$. Further, the magnitude of the events scales with their duration according to a power law given by

$$\Delta\sigma \approx (\Delta t_{\mathrm{d}})^{1/x}, \tag{90}$$

showing an exponent value $x = 1.25$ (cf. fig. 43). If power-law relations such as those of eqs (86) and (89)–(90) are satisfied, it can be shown that the three exponents are related through the scaling relation [190] $\alpha = 1 + x(\beta - 1)$. This is indeed the case here. However,

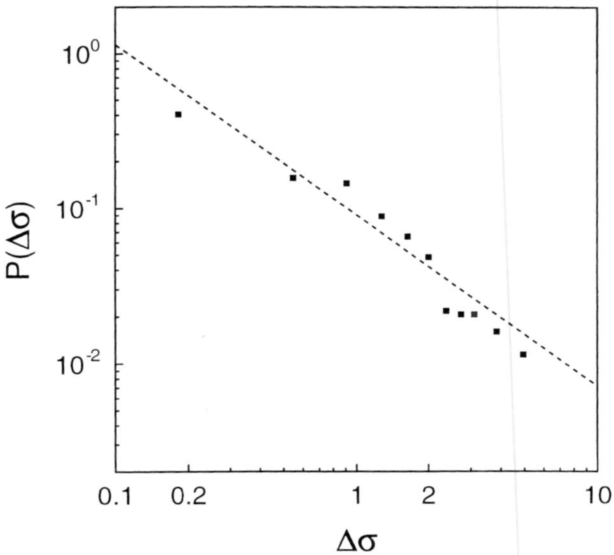

Fig. 42. Log–log plot of the distribution of the magnitudes of the stress drops for PLC h file, showing a power law distribution with an exponent value $\alpha = 1.1$.

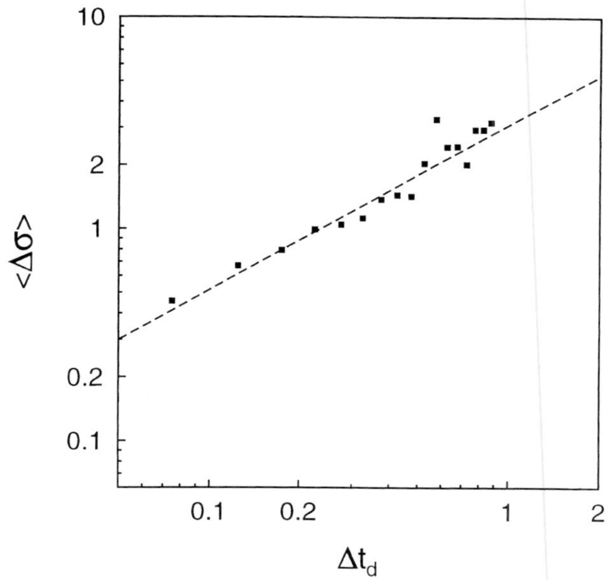

Fig. 43. Log–log plot of the mean of the stress drop amplitudes as a function of the stress drop durations for file PLC h, showing a scaling relation with an exponent value 1.25.

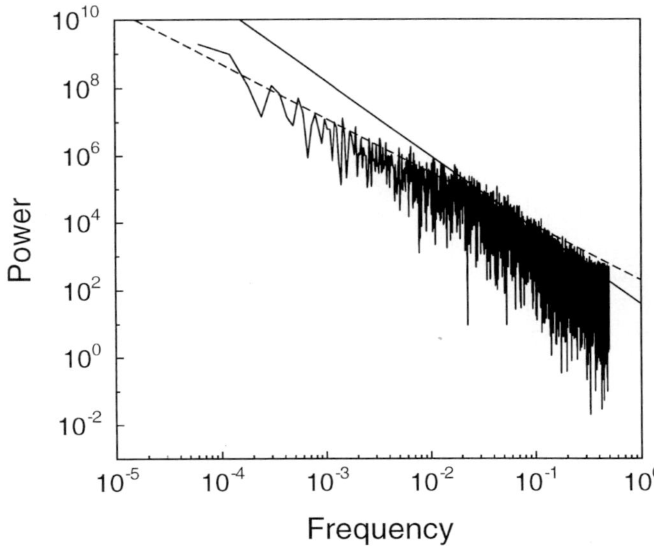

Fig. 44. Power spectrum of file PLC h. The dashed line fits the low frequency region with an exponent 1.55. The solid line fits the high frequency part and has an exponent 2.05.

since the confidence level in β is not high, an independent check of this scaling behaviour has to be carried out. For this purpose, an alternative scaling relation derived by Kertesz and Kiss [191] proves to be useful. This relation connects α and x to the low frequency exponent of the Fourier power spectrum of the time series. Assuming that the total energy dissipation stems from independent elementary events whose energy density spectrum is quasi-Lorentzian, these authors have shown that the low-frequency power spectrum $S(\omega)$ behaves as

$$S(\omega) \propto \omega^{-(3-\alpha)/x}, \tag{91}$$

provided the scaling exponents satisfy the inequality $2x + \alpha > 3$. Using the values obtained for α and x, it is seen that this inequality holds, and we get $S(\omega) \propto \omega^{-1.48}$. The low-frequency region of the power spectrum (shown in fig. 44) is found to have a scaling behaviour with an exponent value 1.55, reasonably consistent with the value 1.48 obtained from eq. (91). Therefore it is concluded that the scaling laws (86), (89)–(90) are consistent with a SOC-type dynamics.

The correlation dimension and Lyapunov spectrum are now calculated in order to understand these results from a dynamical point of view. For the file PLC h, the correlation time t_c is much larger than for the earlier two files. No convergent scaling region could be found in the correlation integral $C(r)$ as the embedding dimension is increased, which is in sharp contrast to the results obtained from the two other data sets [189]. Looked at from the point view of dynamics, a SOC signal is different from a chaotic one. There is a power law growth of any disturbance [191], as against an exponential growth in chaotic dynamics. Thus, we expect *no stable* positive Lyapunov exponent for a time series which exhibits SOC-type behaviour. For the PLC h file, the largest Lyapunov exponent shows a decreasing trend as a function of shell size. This is in sharp contrast to the stable values of the positive and zero exponents obtained for the PLC l and m data sets. These result should again be taken as supportive of the existence of a cross-over in the dynamics. Finally, both the correlation integral and Lyapunov spectrum were calculated for a number of surrogate data sets obtained from the PLC h. No converged value of the slope $\ln C(r)/\ln(r)$ and no stable positive or zero Lyapunov exponent could be found. Thus, the original signal and its surrogates are not distinguishable. This is again in marked contrast to the data sets PLC l and m, where the original files and the surrogates behave very differently, and should be taken as an additional support for the existence of a cross-over in the underlying dynamics as the strain rate is increased.

4.4.4. Discussion

We have shown in section 4.4.3 that the low and medium strain rate files PLC l and m, are chaotic. In contrast, the high strain rate file PLC h appears to be compatible with a SOC-type dynamics. Thus, a cross-over from chaos to a SOC-type dynamics has been identified as the applied strain rate is increased within the range of the PLC effect. This cross-over has been investigated recently in detail [193]. The physical factors that affect the observed dynamics are the instability mechanism itself, i.e., the existence of an anomalous force versus flux response of the dislocations, and the spatial coupling existing between the various defect populations. When the applied strain rate is in the range where the PLC effect occurs, the local plastic strain rate is bi-valued, with one low value corresponding to pinned dislocations, and the other, with a much higher value, to freed dislocations (see sections 3.2.3 and 4.2.2). Starting from the pinned state at any location in the sample, the access to the unpinned state requires a finite fluctuation which is provided by the gradual loading of the sample. Then, the local strain rate jumps to the higher level, eventually jumping back to the pinned state. The reloading time as well as the jump in the local strain rate depend on the applied overall strain rate. Both decrease due to the negative strain rate sensitivity as the applied strain rate increases. Spatial coupling is responsible for the correlation and propagation of the bands of localised plastic deformation. In single crystals, it may stem either from dislocation mechanisms or from strain incompatibilities between neighbouring slipped and unslipped regions of the crystal. At present, it is believed that these incompatibilities are accommodated elastically but can be also partly relaxed by plastic flow, with a certain relaxation time. The magnitude of the coupling constant is maximum when elasticity alone is involved; it decreases with increasing plastic relaxation, the latter being favoured by material or experimental conditions that lower the flow stress. Since the strain rate sensitivity of the flow stress is negative, the magnitude of the spatial coupling decreases as the applied strain rate increases. Thus, the plastic relaxation time increases with the applied strain rate.

The cross-over might be expected as the plastic relaxation time becomes comparable with the characteristic reloading time. At low strain rates, the reloading time is larger than the plastic relaxation time. In such conditions, incompatibilities are fully relaxed within the reloading time and, in space, within a characteristic relaxation length scale which allows for stress drop uniformisation. Within the spatial elements where the relaxation is complete, different types of dislocation populations themselves interact in a non-linear way leading to chaos. At sufficiently high applied strain rate, the plastic relaxation time becomes larger than the reloading time. Then, new small heterogeneities can form before plastic relaxation is complete. This picture results in a recurrence of partial relaxation processes, each time with a different magnitude, which in turn may provide stress drop distributions without a characteristic value and lead to SOC-type of dynamics. A full synthesis of all experimental results and of their interpretation can be found in a recent reference [193].

Finally, there is an enormous field opened for simple experimentation in the domain of jerky flow and possibly too in the analysis of small slip bursts during globally uniform deformation [149]. Experimental studies on collective dislocation behaviour usually yield rather complex data. This information can now be quantitatively analysed within a dynamical framework of non-linear dynamics. This allows for physical insights into stability phenomena, and also suggests new methods for quantification and modelling.

Acknowledgements

The authors are most grateful to many colleagues whose work is described therein, particularly S. Bakir, V. Jeanclaude, M. Lebyodkin, S.J. Noronha and L. Quaouire, as well as to Prof. J. Vergnol at University of Poitiers and Prof. Neuhäuser and his group at Braunschweig Technical University for providing data. The support of the Indo-French Centre for the Promotion of Advanced Research (IFCPAR/CEFIPRA) under the Research Project 1108-1, of JNCASR and CNRS under PICS Program n°657 is gratefully acknowledged. One of the authors (G.A.) would like to acknowledge the support of the University of Metz for his stays at Metz during 1997 and 1998.

References

[1] P. Bergé, Y. Pomeau and C. Vidal, Order in Chaos (John Wiley, New York, 1984).

[2] D. Walgraef (ed.), Patterns, Defects and Microstructures in Nonequilibrium Systems, NATO ASIE 121 (Martinus Nijhoff Publishers, Dordrecht, 1987).

[3] L. Kubin and G. Martin (eds), Non Linear Phenomena in Materials Science (Trans Tech Publications, Aedermannsdorf, 1988).

[4] D. Walgraef and N.M. Ghoniem (eds), Patterns, Defects and Materials Instabilities, NATO ASI E 183 (Kluwer Acad. Publ., Dordrecht, 1990).

[5] G. Martin and L. Kubin (eds), Non Linear Phenomena in Materials Science II (Trans Tech Publications, Aedermannsdorf, 1992).

[6] H. Kirchner, L. Kubin and V. Pontikis (eds), Computer Simulation in Materials Science, NATO ASI E 308 (Kluwer Acad. Publ., Dordrecht, 1996).

[7] L.P. Kubin, in: Stability of Materials, NATO ASI Series B 355, eds A. Gonis, P. Turchi and J. Kudrnovsky (Plenum Press, New York, 1996) p. 99.

[8] L.P. Kubin, in: Materials Science and Technology, eds R.W. Cahn, P. Haasen and E.J. Kramer, Vol. 6: Plastic Deformation and Fracture of Materials, ed. H. Mughrabi (VCH, Weinheim, 1993) Ch. 4, p. 138.

[9] H. Mughrabi, in: Strength of Metals and Alloys (ICSMA 5), Vol. 3, eds P. Haasen, V. Gerold and G. Kostorz (Pergamon Press, Oxford, 1979) p. 1615.

[10] L.M. Brown, in: Dislocation Modelling of Physical Systems, eds M.F. Ashby et al. (Pergamon Press, Oxford, 1981) p. 51.

[11] C. Laird, in: Dislocations in Solids, Vol. 6, ed. F.R.N. Nabarro (North-Holland, Amsterdam, 1983) p. 55.
[12] P. Neumann, in: Physical Metallurgy, Vol. 2, eds R.W. Cahn and P. Haasen (Elsevier, Amsterdam, 1983) p. 1553.
[13] H. Mughrabi, in: Strength of Metals and Alloys (ICSMA-CIRMA 7), Vol. 3, eds H.J. McQueen et al. (Pergamon Press, Oxford, 1985) p. 1917.
[14] Z.S. Basinski and S.J. Basinski, Prog. Mater. Sci. 36 (1992) 89.
[15] U. Holzwarth and U. Essmann, Appl. Phys. A 58 (1994) 197.
[16] A.T. Winter, Phil. Mag. 30 (1974) 719.
[17] T. Zhai, J.W. Martin and G.A.D. Briggs, Acta Mater. 44 (1996) 1729.
[18] H. Mughrabi, F. Ackermann and K. Herz, in: Fatigue Mechanisms, ASTM-STP 675, ed. J.T. Fong (Am. Soc. for Testing Materials, Philadelphia, 1979) p. 69.
[19] J. Kratochvil, Rev. Phys. Appl. 23 (1988) 419.
[20] A. Franek, R. Kalus and J. Kratochvil, Phil. Mag. 64 (1991) 497.
[21] U. Essmann and H. Mughrabi, Phil. Mag. A 40 (1979) 731.
[22] T. Vegge, T. Rasmussen, T. Leffers, O.B. Pedersen and K.W. Jacobsen, Phil. Mag. Lett. 81 (2001) 137.
[23] B. Tippelt, J. Brettschneider and P. Hähner, Phys. Stat. Sol. A 163 (1998) 11.
[24] A. Aslanides and V. Pontikis, Phil. Mag. A 80 (2000) 2337.
[25] U. Essmann, Phil. Mag. A 45 (1982) 171.
[26] O.B. Pedersen, Phil. Mag. 73 (1996) 829.
[27] K. Differt and U. Essmann, Mat. Sci. Eng. A 164 (1993) 295.
[28] D. Walgraef and E.C. Aifantis, J. Appl. Phys. 58 (1985) 688.
[29] J. Kratochvil and M. Saxlova, Scripta Metall. 26 (1992) 113.
[30] K. Differt and U. Essmann, Mat. Sci. Eng. A 164 (1993) 295.
[31] A.M. Kosevich, in: Dislocations in Solids, Vol. 1, ed. F.R.N. Nabarro (North-Holland, Amsterdam, 1979) p. 33.
[32] T. Mura, in: Micromechanics of Defects in Solids (Kluwer Acad. Publ., Dordrecht, 1993) p. 44.
[33] U.F. Kocks, J. Eng. Mat. Technology 98 (1976) 76.
[34] Y. Estrin and L.P. Kubin, Acta Metall. 34 (1986) 2455.
[35] Y. Estrin, in: Dislocation-Density-Related Constitutive Modelling (Academic Press, New York, 1996) p. 69.
[36] D. Sahoo and G. Ananthakrishna, J. Phys. D 15 (1982) 1439.
[37] D.L. Holt, J. Appl. Phys. 41 (1970) 3197.
[38] G.A. Malygin, Phys. Solid State 37 (1995) 1.
[39] S.B. Goryatchev, in: Computer Simulations in Materials Science, NATO ASI Series E 308, eds H.O.K. Kirchner, L.P. Kubin and V. Pontikis (Kluwer Acad. Publ., Dordrecht, 1996) p. 17.
[40] U. Essmann and K. Differt, Scripta Metall. 22 (1988) 1337.
[41] U. Essmann and K. Differt, Mat. Sci. Eng. A 208 (1996) 56.
[42] G. Nicolis and I. Prigogine, Self-Organisation in Nonequilibrium Systems (Wiley, New York, 1977).
[43] D. Walgraef and E.C. Aifantis, Int. J. Eng. Sci. 23 (1985) 1351, 1359, 1364.
[44] E.C. Aifantis, Mat. Sci. Eng. 81 (1986) 563.
[45] D. Walgraef, C. Schiller and E.C. Aifantis, in: Patterns, Defects and Microstructures in Nonequilibrium Systems, NATO ASI E121, ed. D. Walgraef (Martinus Nilhoff Publ., Dordrecht, 1987) p. 257.
[46] E.C. Aifantis, Solid State Phenomena 3–4 (1988) 397.
[47] C. Schiller and D. Walgraef, Acta Metall. 36 (1988) 563.
[48] D. Walgraef and E.C. Aifantis, Res. Mechanica 23 (1988) 161.
[49] D. Walgraef, in: Patterns, Defects and Materials Instabilities, NATO ASI E 183, eds D. Walgraef and N.M. Ghoniem (Kluwer Acad. Publ., Dordrecht, 1990) p. 73.
[50] E.C. Aifantis, Solid State Phenomena 23–24 (1992) 355.
[51] I. Prigogine and R. Lefever, J. Chem. Phys. 48 (1968) 1695.
[52] C. Misbah, Solid State Phenomena 3–4 (1988) 29.
[53] D. Walgraef, Solid State Phenomena 3–4 (1988) 77.
[54] J. Kratochvil, Solid State Phenomena 35–36 (1994) 71.
[55] J. Kratochvil, in: Basic Mechanisms in Fatigue of Metals, eds P. Lukas and J. Polak (Academia, Prague, 1988) p. 15.

[56] J. Kratochvil and S. Libovicky, Scripta Metall. 20 (1986) 1625.
[57] J. Kratochvil and M. Saxlova, Solid State Phenomena 23–24 (1992) 369.
[58] J. Kratochvil, M. Saxlova and R. Sedlacek, Key Engng. Mats. 97 (1994) 479.
[59] J.R. Hancock and J.C. Grosskreutz, Acta Metall. 17 (1969) 77.
[60] B. Tippelt, J. Bretschneider and C. Holste, Phil. Mag. Lett. 74 (1996) 161.
[61] J.V. Sharp and M.J. Makin, Phil. Mag. 10 (1964) 1011.
[62] L.P. Kubin and J. Kratochvil, Phil. Mag. A (1999) in press.
[63] V.I. Vladimirov and B. Pegel, Phys. Stat. Sol. B 56 (1973) K105.
[64] C. Schiller, Doctorate thesis (University of Brussels, 1988).
[65] B. Devincre and L.P. Kubin, in: Proc. ICSMA 11 (Prague, 1997), Mat. Sci. Engng. A 234–236 (1997) 8.
[66] L.P. Kubin and B. Devincre, in: Deformation-Induced Microstructures: Analysis and Relation to Properties, Proc. 20th Risø Int. Symposium, eds J.B. Bilde-Sørensen et al. (Risø National Laboratory, Roskilde, Denmark, 1999) p. 61.
[67] J. Kratochvil, M. Saxlova, B. Devincre and L.P. Kubin, in: Proc. ICSMA 11 (Prague, 1997), Mat. Sci. Engng. A 234–236 (1997) 318.
[68] J.M. Salazar, R. Fournet and N. Banai, Acta Metall. Mater. 43 (1995) 1127.
[69] Y. Bréchet, Doctorate thesis (University of Grenoble, 1987).
[70] J. Kratochvil and A. Orlova, Phil. Mag. 61 (1990) 281.
[71] M.V. Glazov and C. Laird, Acta Metall. Mater. 43 (1995) 2849.
[72] P. Hähner, Acta Mater. 44 (1996) 2345.
[73] P. Hähner, Appl. Phys. A 63 (1996) 45.
[74] C. Crussard, in: 6e Colloque de Métallurgie de l'INSTN (Presses Universitaires de France, Paris, 1962) p. 1.
[75] K. Chihab, Y. Estrin, L.P. Kubin and J. Vergnol, Scripta Metall. 21 (1987) 203.
[76] E.W. Hart, Acta Metall. 3 (1955) 146.
[77] U.F Kocks, Prog. Mater. Sci., Chalmers Anniversary Volume (Pergamon Press, Oxford, 1981) p. 185.
[78] Viewpoint Set No 21, Propagative Instabilities, Scripta Metall. Mater. 29 (1993) 1147.
[79] Y. Estrin and L.P. Kubin, in: Continuum Models for Materials with Microstructure, ed. H. Mülhaus (Wiley, Chichester, 1995) p. 395.
[80] M. Zaiser and P. Hähner, Phys. Stat. Sol. B 199 (1997) 267.
[81] A. Piobert, Mem. de l'artillerie 5 (1842) 525.
[82] W. Lüders, Dinglers Polytech. J. 155 (1860) 18.
[83] J. Balik, P. Lukac and L.P. Kubin, Scripta Mater. 42 (2000) 465.
[84] B. Jaoul, Etude de la Plasticité et Application aux Métaux (Dunod, Paris, 1965).
[85] E.G. Hall, Proc. Phys. Soc. London B 64 (1951) 747.
[86] A.H. Cottrell, Dislocations and Plastic Flow in Crystals (University Press, Oxford, 1953).
[87] P. Hähner and L.P. Kubin, Solid State Phenomena 23–24 (1992) 385.
[88] A. Luft, Progress in Materials Science 35 (1991) 97.
[89] H. Neuhäuser, in: Dislocations in Solids, Vol. 6, ed. F.R.N. Nabarro (North-Holland, Amsterdam, 1983) p. 319.
[90] H. Neuhäuser and A. Hampel, Scripta Metall. Mater. 29 (1993) 1151.
[91] H. Alexander and P. Haasen, Solid State Physics 22 (1968) 22.
[92] A. Moulin, M. Condat and L.P. Kubin, Acta Mater. 47 (1999) 2879.
[93] Y. Bréchet, G. Canova and L.P. Kubin, Acta Mater. 44 (1996) 4261.
[94] F. Le Chatelier, Rev. de Métallurgie 6 (1909) 914.
[95] A. Portevin and F. Le Chatelier, Trans. ASST 5 (1924) 457.
[96] P.G. McCormick, Acta Metall. 20 (1972) 35.
[97] A. van den Beukel, Phys. Stat. Sol. A 30 (1975) 197.
[98] P. Hähner, Acta Mater. 45 (1997) 3695.
[99] P. Penning, Acta Metall. 20 (1972) 1179.
[100] L.P. Kubin and Y. Estrin, Acta Metall. 33 (1985) 397.
[101] Y. Estrin and L.P. Kubin, J. Mech. Behav. of Mats. 2 (1989) 255.
[102] L.P. Kubin, K. Chihab and Y. Estrin, Acta Metall. 36 (1988) 2707.

[103] A. Considère, Annales des Ponts et Chaussées 9 (1885) 574.
[104] J.W. Hutchinson and K.W. Neale, J. Mech. Phys. Solids 31 (1983) 405.
[105] P.W. Bridgman, Studies in Large Plastic Flow and Fracture (Harvard Univ. Press, 1964).
[106] P. Hähner, Appl. Phys. A 58 (1994) 41.
[107] C. Fressengeas and A. Molinari, Acta Metall. 33 (1985) 387.
[108] Y. Estrin, C.P. Ling and P.G. McCormick, Acta Metall. 39 (1991) 2943.
[109] Y. Bréchet and F. Louchet, Solid State Phenomena 3–4 (1988) 347.
[110] Y. Estrin, Solid State Phenomena 3–4 (1988) 417.
[111] H.B. Mülhaus and E. Aifantis, Int. J. Solids Struct. 28 (1991) 845.
[112] E. Cosserat and F. Cosserat, Théorie des Corps Déformables (Hermann, Paris, 1909).
[113] R.A. Toupin, Arch. Rat. Mech. 11 (1962) 385.
[114] R.D. Mindlin and H.F. Tiersten, Arch. Rat. Mech. 11 (1962) 415.
[115] N.A. Fleck, G.M. Muller, M.F. Ashby and J.W. Hutchinson, Acta Metall. Mater. 42 (1994) 475.
[116] V.P. Smyshlyaev and N.A. Fleck, J. Mech. Phys. Solids 44 (1996) 465.
[117] H. Dai and D.M. Parks, in: Proceedings of Plasticity '97, ed. A.S. Khan (Neat Press, Fulton MD, 1997) p. 17.
[118] S. Forest, F. Barbe and G. Cailletaud, Int. J. Solids Struct., in press.
[119] G. Dee and J.S. Langer, Phys. Rev. Lett. 50 (1983) 383.
[120] D.G. Aronson and H.F. Weinberger, Partial Differential Equations and Related Topics, Lecture Notes in Math., Vol. 446, ed. J.A. Goldstein (Springer, Heidelberg, 1975).
[121] W. Van Sarloos, Phys. Rev. A 39 (1989) 6367.
[122] H.M. Zbib and E.C. Aifantis, Scripta Metall. 22 (1988) 1431.
[123] P. Hähner, Mat. Sci. Engng. A 164 (1993) 23.
[124] M. Lebyodkin, Y. Bréchet, Y. Estrin and L.P. Kubin, Acta Mater. 44 (1996) 453.
[125] V. Jeanclaude and C. Fressengeas, C. R. Acad. Sci. Paris II 316 (1993) 867.
[126] V. Jeanclaude and C. Fressengeas, C. R. Acad. Sci. Paris II 315 (1992) 7.
[127] V. Jeanclaude, C. Fressengeas and L.P. Kubin, Solid State Phenomena 23–24 (1992) 403.
[128] A. Liénard, Rev. Gen. de l'electricité 23 (1928).
[129] J. Lasalle, Quart. Appl. Math. VII 1 (1949).
[130] A. Karimi, Doctorate thesis (E.N.S. des Mines de Paris, 1981).
[131] P.G. McCormick, S. Venkadesan and C.P. Ling, Scripta Metall. Mater. 29 (1993) 1159.
[132] S. Bakir, Doctorate thesis (University of Metz, 1995).
[133] D. Thevenet, Doctorate thesis (University of Metz, 1999).
[134] A. Ziegenbein, C. Achmus, J. Plessing and H. Neuhäuser, in: Plastic and Fracture Instabilities in Materials, AMD-Vol. 200/MD-Vol. 57, ed. N. Ghoniem (ASME, 1995) p. 101.
[135] W. Sylvestrowicz and E.O. Hall, Proc. Phys. Soc. London B 64 (1951) 495.
[136] M.A. Lebyodkin, V.Ya. Kravchenko and V.S. Bobrov, Physica B 165–166 (1990) 267.
[137] M.A. Lebyodkin, V.Ya. Kravchenko and V.S. Bobrov, Mat. Sci. Eng. A 164 (1993) 252.
[138] M.A. Lebyodkin and V.S. Bobrov, Solid State Phenomena 35–36 (1994) 411.
[139] D.R. James and S.H. Carpenter, J. Appl. Phys. 42 (1971) 4685.
[140] R. Pascual, Scripta Metall. 8 (1974) 1461.
[141] H. Hatano, J. Appl. Phys. 47 (1976) 3873.
[142] P.F. Higgens and S.H. Carpenter, Acta Metall. 26 (1978) 133.
[143] C.R. Heiple, S.H. Carpenter and M.J. Carr, Metal. Sci. 15 (1981) 587.
[144] C.H. Caceres and H.R. Bertorello, Scripta Metall. 17 (1983) 1115.
[145] C.H. Caceres and A.M. Rodriguez, Acta Metall. 35 (1987) 2581.
[146] F. Zeides and I. Roman, Scripta Metall. 24 (1990) 1919.
[147] F. Chmelik, J. Dosoudil, J. PLessing, H. Neuhäuser, P. Lukac and Z. Trojanova, Key Eng. Mats. 97–98 (1994) 263.
[148] F. Chmelik, Z. Trojanova, Z. Prevorovsky and P. Lukac, J. Mat. Sci. Eng. A 164 (1993) 260.
[149] M.-C. Miguel, A. Vespignani, S. Zapperi, J. Weiss and J.R. Grasso, Nature 410 (2001) 667.
[150] G. Ananthakrishna and M.C. Valsakumar, J. Phys. D 15 (1982) L171.
[151] G. Ananthakrishna and D. Sahoo, J. Phys. D 14 (1981) 2081.

[152] G. Ananthakrishna and M.C. Valsakumar, Phys. Lett. A 95 (1983) 69.
[153] G. Ananthakrishna and T.M. John, in: Directions in Chaos, ed. Hao Bai-Lin (World Scientific, Singapore, 1990) p. 133.
[154] G. Ananthakrishna, Solid State Phenomena 23–24 (1992) 417.
[155] G. Ananthakrishna, Scripta Metall. Mater. 29 (1993) 1183.
[156] M.J. Feigenbaum, J. Stat. Phys. 19 (1978) 25.
[157] G. Ananthakrishna, C. Fressengeas, M. Grosbras, J. Vergnol, C. Engelke, J. Plessing, H. Neuhäuser, E. Bouchaud, J. Planès and L.P. Kubin, Scripta Metall. Mater. 32 (1995) 1731.
[158] L. Quaouire and C. Fressengeas, Solid State Phenomena 42–43 (1995) 293.
[159] G. Ananthakrishna and S.J. Noronha, Solid State Phenomena 42–43 (1995) 293.
[160] G. Ananthakrishna, S.J. Noronha, L. Quaouire and C. Fressengeas, Pramana J. Phys. 48 (1997) 705.
[161] S. Venkadesan, M.C. Valsakumar, K.P.N. Murthy and S. Rajasekar, Phys. Rev. E 54 (1996) 611.
[162] P. Grassberger and I. Procaccia, Physica D 9 (1983) 189.
[163] S.J. Noronha, G. Ananthakrishna, L. Quaouire, C. Fressengeas and L.P. Kubin, Int. J. Bifurcation and Chaos 7 (1997) 2577.
[164] N. Packard, J.A. Crutchfield, J.D. Farmer and R.S. Shaw, Phys. Rev. Lett. 45 (1980) 712.
[165] F. Takens, in: Dynamical Systems and Turbulence, Lecture Notes in Math., eds D.A. Rand and L.S. Young (Springer, Berlin, 1981).
[166] M. Ding, C. Grebogi, E. Ott, T. Sauer and J.A. Yorke, Phys. Rev. Lett. 70 (1993) 3872.
[167] D.S. Broomhead and G.P. King, Physica D 20 (1986) 217.
[168] A.M. Albano, J. Muench, C. Schwartz, A. Mees and P.E. Rapp, Phys. Rev. A 38 (1988) 3017.
[169] J.P. Eckmann, S.O. Kamphorst, D. Ruelle and S. Ciliberto, Phys. Rev. A 34 (1986) 4971.
[170] A. Wolf, J.B. Swift, H.L. Swinney and J.A. Vastano, Physica D 16 (1985) 285.
[171] M. Sano and Y. Sawada, Phys. Rev. Lett. 55 (1986) 1082.
[172] X. Zeng, R. Eykholt and R.A. Pielke, Phys. Rev. Lett. 66 (1991) 3229.
[173] J. Theiler, S. Eubank, A. Longtin, B. Galdrikian and J.D. Farmer, Physica D 58 (1992) 77.
[174] L.P. Kubin and J.P Poirier, Solid State Phenomena 3–4 (1988) 473.
[175] P.P. Bowden and D. Tabor, The Friction and Lubrications of Solids, Part I (Clarendon Press, Oxford, 1950).
[176] P. Bak and C. Tang, J. Geophys. Res. 94 (1989) 15635.
[177] D. Maugis, J. Mater. Sci. 20 (1985) 3041.
[178] X.-L. Wu, K.J. Maloy, A. Hansen, M. Ammi and D. Bideau, Phy. Rev. Lett. 71 (1993) 1363.
[179] L.P. Kubin and Y. Estrin, J. Physique 47 (1986) 497.
[180] G. Ananthakrishna and H. Ramachandran, in: Lecture Notes in Phys., Vol. 437, eds K.K. Bardhan, B.K. Chakrabarti and A. Hansen (Springer, Berlin, 1994) p. 78.
[181] M.A. Lebyodkin, Y. Bréchet, Y. Estrin and L.P. Kubin, Phys. Rev. Lett. 74 (1995) 4758.
[182] M. Lebyodkin, L. Dunin-Barkowskii, Y. Bréchet, Y. Estrin and L.P. Kubin, Acta Mater. 48 (2000) 2529.
[183] R. Burridge and L. Knopoff, Bull. Seismol. Soc. Am. 57 (1967) 341.
[184] J.M. Carlson and J.S. Langer, Phys. Rev. A 40 (1989) 6470.
[185] J.R. Rice, J. Geophys. Res.-Solid Earth (B6) 98 (1993) 9885.
[186] B. Gutenberg and C.F. Richter, Ann. Geophys. 9 (1956) 1.
[187] P. Bak, C. Tang and K. Wiesenfeld, Phys. Rev. Lett. 59 (1987) 381.
[188] P. Bak, C. Tang and K. Wiesenfeld, Phys. Rev. A 38 (1988) 364.
[189] G. Ananthakrishna, S.J. Noronha, C. Fressengeas and L.P. Kubin, Phys. Rev. E 60 (1999) 5455.
[190] J. Kertesz and L.B. Kiss, J. Phys. A: Math. Gen. 23 (1990) L433.
[191] K. Chen, P. Bak and S.P. Obukhov, Phys. Rev. 43 (1991) 625.
[192] M.S. Barathi, M. Lebyodkin, G. Ananthakrishna, C. Fressengeas and L.P. Kubin, Phys. Rev. Lett. 87 (2001) 165508.
[193] M.S. Bharathi, M. Lebyodkin, G. Ananthakrishna, C. Fressengeas and L.P. Kubin, Acta Mat. 50 (2002) 2813.

CHAPTER 58

Linear Work-Hardening and Secondary Slip in Crystals

L.M. BROWN

Cavendish Laboratory
Madingley Road
Cambridge CB3 OHE, UK

Dislocations in Solids
Edited by F. R. N. Nabarro and M. S. Duesbery

Contents

1. Introduction

This paper is an attempt to improve upon and enlarge somewhat an earlier paper of Brown [1]. It is based upon an unpublished article written nearly ten years ago to honour P.B. Hirsch's retirement, and I should like to think that this version might do the same for his 75th birthday.

The point of view is that the dislocation structures which cause the familiar work-hardening and fatigue behaviour of metals are *dissipative structures* in the sense of Prigogine [2], who emphasises the failure of principles based on equilibrium thermodynamics, such as 'minimum entropy production', to explain these structures. They are also Low Energy Dislocation Structures (LEDS) in the sense that when they are at rest they are in static equilibrium and so are metastable. But their evolution and the way they control the flow stress can only be understood by thinking of dissipative processes, mainly the bowing of dislocations between obstacles, followed by subsequent release. This point of view has recently been expounded by Nabarro [3], who shows that if cells are regarded as fluctuations in dislocation density driven by an external stress which can do work by making individual dislocations cross a cell from wall to wall, but resisted by internal forces derived from the elastic energy of the dislocation array, which normally favours a quasi-uniform distribution, then a cell size about thirteen times the average interdislocation spacing should be stable. He points out that such fluctuations cannot be treated by Maxwell–Boltzmann statistics, essentially because all dislocations are thermodynamically unstable and can be retained only by forces driving the structure far from equilibrium. This accords with Prigogine's point of view. What Nabarro calls an 'enthalpy' is simply a balance between dissipated energy and stored energy for an infinitesimal adjustment of cell size, not a quantity accessible to a heat sink in equilibrium with the crystal. Nabarro's result is very attractive, in that it refers back to earlier work by Staker and Holt [4] who take the view that cells are to be regarded as a 'spinodal instability' of the dislocation structures, and with whose experimental results Nabarro's theory agrees.

Our point of view is much more pragmatically based on observation. The hope is that if a consistent set of ideas and equations can be found, then one may see generalisations which might be incorporated into a sound theory of dislocation plasticity.

2. The model of a slip band

A slip band is conceived to be a blade-like ellipsoidal region, slightly inclined to the crystallographic slip direction. The aspect ratio is about 1 : 50, and the angles of inclination, randomly positive and negative with respect to the sense of rotation of the tensile axis, are about the same magnitude expressed in radians: that is to say, about one degree. The bands occupy a volume fraction of the crystal which is on average independent of strain. Between the bands are 'obstacles' – regions of high dislocation density, both primary and secondary.

The slip bands are the spatial and temporal product of a bifurcation process. An analogy which may help to understand the model is delta formation at the mouths of large rivers. As channels silt up, flow is confined to ever smaller rivulets between ever higher levees, until the flow is choked, the water level rises, and it breaks through a levee and a new channel is formed. Although the process of silting up is continuous, driven by water flowing from a distant elevated supply, the flow within the delta is subject to constant bifurcations and the mouth of the river, over a period of centuries, may suddenly migrate rather large distances to a less developed part of the delta. The elevated supply is so distant that it is unaware of the unsteady flows at the rivermouth, caused by the bifurcations: all that happens at the supply is a steady decrease in the flow of the river, caused by the progressive silting up: this is the work-hardening. Although this analogy is helpful, it is essentially two-dimensional, whereas the slip band structure is three-dimensional. Slip bands must be confined in three-dimensions.

The traditional model of a slip band has been that of a pile-up of dislocations on the crystallographic slip plane. It has, however, been widely agreed since Mott's early paper [5] that the pile-up produces intense stresses which must be relaxed by plastic flow. Hirsch and Mitchell [6] present detailed calculations of the stress field surrounding pile-ups in various configurations, and point out that the maximum back stress – the stress resisting further flow – is actually achieved on the plane of the pile-up itself. Under the applied stress, the dislocations in the pile-up experience zero stress – they are in equilibrium. This helps to clarify the point that what actually resists the dislocation motion is the obstacle at the head of the pile-up – not the 'back stress due to the pile-up', a phrase which can still be found in modern literature. However the pile-up is a treacherous starting point for a theory of work hardening because it occupies no volume, only an area. This makes it difficult to average the internal stresses correctly, and leads to inconsistencies in such theories [7].

Figure 1 shows various attempts to draw a slip band. Figure 1(a) shows an ellipsoid of aspect ratio about 1 : 2 and an angle of inclination about 30°. One must imagine this squashed normal to the slip planes, and extended parallel to them by about a factor of 25, to get close to an accurate depiction of the model. It is hard to draw. Figure 1(b) shows how the half-planes of the dislocations near the centre of the elongated band point all towards the inside – or, if the inclination were of the opposite sense, to the outside – of the band. This demonstrates that the inclination of the band produces a 'fibre stress' – a stress, either tensile or compressive, parallel to the long axis of the band. The fibre stress produces no force upon the primary dislocations in the walls making up the band, but it activates secondary slip, as recognised by Jackson [8]. Figures 1(c) and 1(d) show how the proposed picture relates to that drawn by Hirsch and Mitchell [6]. Figure 1(c) reproduces their drawing of the region of plastic relaxation near a pile-up, showing that the main stress caused by the crowded half-planes is in fact similar in origin to the fibre stress. Figure 1(d) is adapted from their fig. 10 which shows the pile-ups with associated plastic zones as ellipses – or close to it! – parallel to the crystallographic slip direction. They arrived at an ellipsoidal configuration by estimating theoretically the shape of the plastic zone within which secondary dislocations relieve what we are now calling the 'fibre stress'.

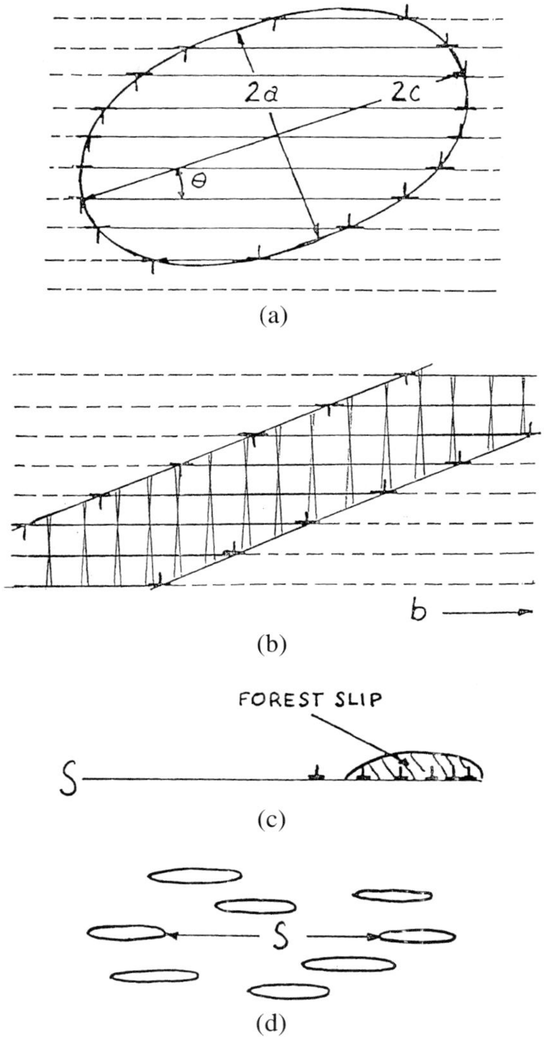

Fig. 1. (a) An inclined ellipsoidal slip band. The horizontal lines are the traces of the crystallographic slip planes. Slip occurs inside the inclined ellipsoid, and is blocked by obstacles around it. In reality, the angle of inclination is about 1 degree instead of 30, and the aspect ratio is about 25 rather than 2. (b) The disposition of primary dislocations in the walls of a long inclined band, near the middle of the band where the curvature due to its ellipsoidal shape is not noticeable. The fibre stress, compressive as drawn here, is relieved by secondary slip cross-crossing inside the band. (c) A diagram redrawn from Hirsch and Mitchell [6] showing the approximately ellipsoidal region at the tip of a planar pile-up where flow on secondary systems relaxes the stress. '*S*' refers to the position of the source of dislocations. (d) The disposition of ellipsoidal bunches of forest dislocations and sources, according to Hirsch and Mitchell [6].

It is important to recognise that the band is confined by obstacles at all its boundaries. If we look along the primary slip planes in the direction of the primary Burgers vector, the confinement is accomplished by edge dislocation dipoles front and back; by screw dislocation dipoles left and right; and by small angle twist and tilt boundaries top and bottom, as well as wherever reaction with secondary dislocations has occurred. The screw dipoles can be removed by cross slip and the edge dipoles by climb, but the boundaries are very resistent to recovery processes. For the moment, we confine ourselves to the situation where all types of obstacle are strong enough to resist penetration by the primary dislocations.

At this point, the reader will want some reassurance that the model of a slip band as an inclined ellipsoid is supported by some evidence. The evidence is indirect, but quite strong:

(i) Oppositely directed primary dislocations on closed spaced primary slip planes have been a feature of observations in the electron microscope since the earliest days.
(ii) Early studies of misorientations by Wilkens and co-workers [9] using X-ray topography proposed a model whereby secondary slip between the pairs of primary slip planes converts them to pairs of tilt boundaries enclosing crystal counter-rotated with respect to the sense of rotation of the tensile axis. This model is almost identical to the present one, differing from it only in the omission of twist boundaries with a rotation axis nearly parallel to the primary slip plane.
(iii) The fact that slip bands seem slightly inclined to the crystallographic slip plane is also a common observation. Basinski and Basinski [10] show that the model of a shallowly inclined band accounts for the geometry of surface features, because it relates the slip band length to the spacing.
(iv) By definition, when slip has just occurred within a band, the stress inside it must be uniformly zero, just as it must be in the planar pile-up model. A uniform stress is consistent with an ellipsoidal shape of the band, according to Eshelby's theory of elastic inclusions [11].
(v) Finally, it has been appreciated for many years that slip bands form suddenly throughout the stress–strain curve: the time of formation is measured in microseconds, decreasing strongly as the load on the crystal increases and the bands get smaller [12]. This feature, the temporal manifestation of bifurcation, accounts for much of the acoustic emission which accompanies plastic flow.

There has been to date no direct observation of an ellipsoidal slip band as postulated here. This is not surprising, in view of the experimental difficulties in imaging a highly-stressed object whose structure straddles the atomic scale and the macroscopic scale. However, the reader will find that the picture unifies much, if not all, the observational detail recorded in authoritative reviews [10,13].

3. *Work hardening in a slip band*

Inclined ellipsoidal bands produce linear work hardening. The main reason for this is that the fibre stress, which activates the secondary slip, is linearly proportional to the plastic strain in the band. Detailed calculations are presented in the earlier paper [1]. Here, we give a simplified account.

The angle of inclination of a slip band, θ, is statistically distributed over the bands, and is a constant characteristic of plastic deformation, unchanging with strain. As a result of the angle, however, the fibre stress in each band increases with plastic strain. The magnitude of the fibre stress caused by the extension of the band is, to first order in small quantities, the Young modulus multiplied by the fractional band extension: $E_Y\theta e_p$. The engineering shear strain in the band is e_p. However, a fraction of the fibre stress is relaxed by secondary slip across the band: that fraction is denoted β. The residual fibre stress drives secondary relaxation, which is impeded by the primary dislocations, seen by the secondary dislocations as a forest at the periphery of the band. The spacing between the primary dislocations L is $b/e_p\theta$, from geometry. These are 'geometrically necessary dislocations'. So we find

$$E_Y\theta e_p(1-\beta) = \text{const.}\,\alpha\mu b/L = \text{const.}\,\alpha\mu\theta e_p, \tag{1}$$

α here is the forest hardening constant, with a numerical value about equal to $1/3$. The value of the constant in eq. (1) can be determined exactly for any particular crystal system: for face-centered cubic crystals, one resolves the fibre stress and the source length (the primary dislocation spacing) onto the critical and conjugate systems. The constant is found to be $3/\sqrt{2}$. Thus one finds that the residual fibre stress is proportional to α:

$$1-\beta = 3\alpha\mu/E_Y\sqrt{2} = 3\alpha/2(1+v)\sqrt{2} \approx 0.8\alpha. \tag{2}$$

Taking Poisson's ratio v equal to $1/3$, we find that the fibre stress takes on a value of about $1/4$ its unrelaxed value, which essentially unloads the slip band and relieves the stresses. The principle of the calculation here is that of 'equilibrium relaxation' – the relaxation flow impeded by local forest hardening continues until the local flow stress is equal to the residual elastic stress. This principle gives satisfactory agreement with experiment in the case of dispersion hardening [14]. It is the point in the argument where an essentially heat-producing force is equated to a reversible elastic force.

All that is needed now is to calculate the flow stress on the primary system. The primary dislocations see the secondary dislocations as their forest, so one finds

$$\sigma_p = \alpha\beta\theta\mu e_p = \alpha(1-0.8\alpha)\theta\mu e_p. \tag{3}$$

Because of the smallness of the angle of inclination, θ, this gives the correct order of magnitude for the work-hardening. Equation (3) displays the interesting feature that the hardening rate, although determined entirely by forest hardening, does not depend markedly upon α, the forest hardening constant. The hardening would be zero if α were zero, because there would be no strength in the forest interactions, but it would also be zero if α were $1/0.8$, because then no relaxation could occur – the forest would be too strong. Since we expect α to lie between zero and unity, hardening will result, and for α around $1/3$, the hardening does not depend strongly on it. The hardening is maximised for $\alpha \approx 0.6$.

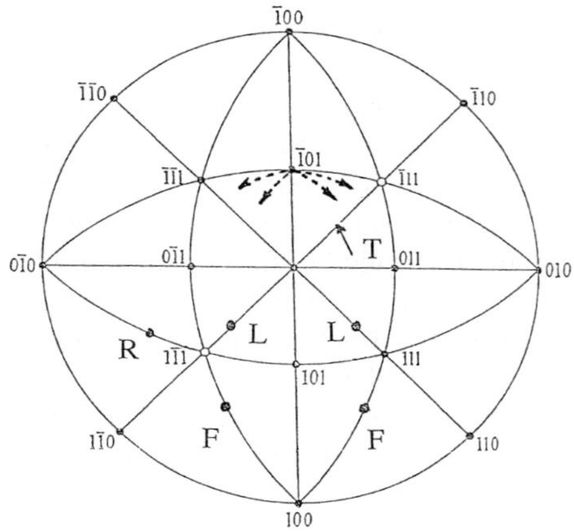

Fig. 2. A stereographic representation of the rotation of the tensile axis T caused by primary slip about R, the 'roller axis' (solid line) and secondary slip (dotted lines) which has four possible axes of rotation, F and L, depending upon the choice of secondary slip system. The primary system is $[\bar{1}01](111)$. Slip on the critical and conjugate planes, $(\bar{1}\bar{1}1)$ and $(\bar{1}11)$, is activated by the fibre stress, which is approximately purely tensile or compressive parallel to the primary slip direction. Depending upon the choice of slip direction, this produces the rotations around the roller axes L (Lomer–Cottrell-producing) or F. The average direction of the axis of secondary rotation is perpendicular to the primary slip plane, and to R. Although these angles of secondary rotation are small, they break the symmetry of the roller axis and are manifested in the misorientations of the dislocation cell structure, each orientation depending upon the local choice of slip system.

The concept of a dislocation arrangement in which the primary dislocations are impeded by the secondary ones, and *vice versa,* was first introduced by Kuhlmann-Wilsdorf and Comins [15], strongly influenced by Jackson and Basinski's [16] studies of latent hardening.

There are several details of the hardening mechanism worth listing:

(i) The rotations produced by the secondary flow are variable, but perpendicular to the roller axis – see fig. 2. There are four equivalent secondary systems, based on the critical and conjugate planes, two of which produce Lomer–Cottrell locks to harden the band. These produce equal and opposite rotations. Therefore we expect the misoriented cells produced by secondary flow to have alternating misorientations. The magnitude of the rotations is $\beta\theta e_{\mathrm{p}}$, usually less than one degree. The cell size should be about equal to the width of the primary slip band. There are many electron micrographs displaying this.

(ii) The slip distance of secondary dislocations is much shorter than the primary slip distance, and hence they produce little strain, although their density is comparable to the primary density.

(iii) The secondary systems activated are precisely those which interact strongly with the primary dislocations to produce Lomer–Cottrell locks.

Semi-quantitative comparison with experiment can be found in the earlier paper [1].

4. Slip bands and obstacles

What we call an 'obstacle' is really no more than a region of high dislocation density which separates two or more slip bands. The character of the obstacle is very different depending upon its location. If it is at the end of a band, it is a braid of densely packed primary dislocation dipoles, threaded by the secondary slip caused by the relaxation of the fibre stress. The very fine dipoles will lower their energy by exchanging edge dislocations to make themselves mostly vacancy type, so the dipolar braid will be a source of vacancies at temperatures where vacancies become mobile. If the obstacle is at the floor of a band, oriented nearly parallel to a crystallographic slip plane, it becomes a carpet of primary and secondary dislocations, and Lomer–Cottrell locks, knitted together in a network. Such structures have been commonly observed, and interpreted in this way, since the earliest observations in stainless steel by Whelan [17]. Throughout Stage II, on the sides of the band, there will also be screw dislocation dipoles, strongly reacted with secondary dislocations: these limit the slip distance of screws. As the stress level rises, the screw dislocations of opposite sign making up dipoles become ever more closely spaced, and by cross-slip can mutually annihilate. Above a certain stress level, all screw dislocation dipoles disappear, and the slip is now not confined in three dimensions – only in two. The slip can escape its confinement, surround the edge dipoles, and occupy a plane which crosses the specimen. The stress level at which this can occur is interpreted as the onset of Stage III and the end of Stage II.

Generally speaking, the obstacles are the regions which look dark and full of dislocations in electron micrographs. It seems likely that throughout the work-hardening process in Stage II, the volume fraction of obstacles remains constant, but the spacing between obstacles diminishes. It is not difficult to understand how this comes about, if obstacles are not permanent, but can form, disappear, and reform as the strain increases. Imagine an obstacle which forms at a certain level of stress and strain. As straining proceeds, it grows in size and the external wall of the obstacle must be strong enough to withstand the progressively increasing stress. Thus the obstacle is strong on the outside, but weak on the inside, which is shielded from the plastic strain. A concentrated forward stress builds up in the centre of the obstacle. At a certain point, as the level of stress continues to rise, the interior of the obstacle will give way, work softening will occur there, and a new slip band will form in the interior of the obstacle. This process allows the volume fraction of obstacles to remain constant, but the spacing of obstacles to diminish as straining proceeds.

The most compelling experimental evidence that this actually happens comes from the observation of structures resulting from cyclic plasticity: see, for example, Jin and Winter [18]. However, it is not difficult to imagine that the process is quite general. If so, it means that the nucleation of a new slip band does not occur just adjacent to an old one, where it would after all just be a continuation of the old one and effectively *increase* the length of an existing band. Instead, it suddenly is initiated some distance from the end of the band, where it forms a new, *shorter*, slip band appropriate to the current stress level, but hemmed in by the strong outer parts of existing obstacles.

If this view is accepted, it is not difficult to estimate the average volume fraction of obstacle. If we imagine a section through the structure, revealing the cross-sections of the slip bands, we see a random close-packed array of ellipses. If we imagine them derived by a shear transformation from a random close-packed array of circles, the volume fraction of space between them (the 'ullage') will be unaffected by the transformation, and equal to about 0.2. This is perhaps somewhat greater than what one might deduce from casual inspection of electron micrographs, but not inconsistent with such observations. Knowledge of the volume fraction enables one to estimate the average local strength of the dislocation forest in the obstacles, because all the external load is borne by them. Thus, if the volume fraction is f, we can write

$$\sigma = f\sigma_{\mathrm{loc}}. \tag{4}$$

The next point to note is that the fibre stresses must alternate between tension and compression, if a stress balance is to be achieved overall in the structure. So the slip bands are inclined by small angles which alternate from positive to negative as one travels perpendicularly to the bands. It follows from St. Venant's principle that the stress near the tip of a band will be appreciable only within a distance equal to the width of the band. If the length of the band is $2c$, and its width is $2a$, then the applied stress acting on the band – which is stress-free inside, and so acts elastically like a void – produces a stress intensity $\sigma\sqrt{c}$ and a concentrated stress distant a from the end of the band $\sigma\sqrt{c/a}$. We thus find that an obstacle is on the point of breaking up to produce a new band if

$$\sigma\sqrt{c/a} = \sigma_{\mathrm{loc}} = \sigma/f. \tag{5}$$

This condition effectively determines the aspect ratio of the slip bands:

$$a/c = f^2 = 1/25. \tag{6}$$

The angle of inclination is also determined, for if we imagine a band of length c being inserted into a space of width a, the angle through which it can turn is about a/c, so we write

$$\theta \approx a/c. \tag{7}$$

In the earlier paper, a much more elaborate argument based on random arrays of ellipses was given: it is not clear that it adds much to the one given here. What is needed is a computer model which tests the extent to which this semi-quantitative reasoning is valid. However, all the parameters of the problem are now determined, and one finds that to within a factor of about two they agree with experiment.

One final equation is required to determine the absolute length scale which fixes the size of the structure as a function of stress. Although the interior of the band is free from obstacles, so that the primary dislocations travel unimpeded within the band, nevertheless the band is very thin and the secondary dislocations must be able to penetrate a rather narrow channel. We assume therefore that the work softening at the centre of an obstacle progresses until the spacing between the walls of the slip band just permits the secondary dislocations to pass between them. At this point, the band is hardened, and the work softening stops. A further discussion of this condition is given in the next section. This determines the width of one band, $2a$, which is inversely proportional to the stress level, and thus linearly proportional to the strain. The resulting equation for the number of bands per unit length,

$$1/2a = 0.6 \times 10^5 e_p \ (\mathrm{cm}^{-1}) \tag{8}$$

is identical to equations dating back to Seeger's treatment of the problem, and agrees adequately with the experimental data, interpreting $2a$ as the slip plane spacing observed, for example, by replica techniques. The corresponding value of the slip-line length for edge dislocations is $2a/\theta$, which again agrees with observations. Basinski and Basinski [10] have given a careful comparison of these equations with data from several sources.

It is clear that the structure of the slip bands obeys the principle of 'similitude' as discussed by Kuhlmann-Wilsdorf [19]. Although the scale of the structure is reduced inversely proportional to the stress as straining increases, the characteristic appearance of the structure, suitably enlarged, is unchanged. The shape of the bands, their angles of inclination, and the volume fraction of obstacle remain constant. There are two points: the structural detail is subject to an almost biological variability, because of the random distribution of inclination angles, the random close-packed array of ellipsoids, and the random element in the sequence of bifurcation events leading to forest formation. Secondly, the principle of similitude does *not* apply to the slip-band offsets seen on the surface of the crystal: these remain on average constant throughout the deformation. It is always possible by observation alone to determine both the current stress level and current plastic strain.

The constancy of slip-step height is one of the earliest and most striking observations by Mader [20]. This was interpreted by Seeger as showing that the number of dislocations in a pile-up is constant throughout Stage II deformation. It is possible to re-interpret this if the bands are ellipsoidal. At the tip of a dislocation pile-up, the stress concentration which acts on the obstacle is just n, the number of dislocations in the pile-up. But for an ellipsoid, the stress concentration is $\sqrt{c/r}$, where r is the radius of curvature of the sharp end of the ellipsoid. This gives a stress concentration of c/a, which is in turn $1/\theta$, constant throughout the deformation. Of course, the band offsets are statistically distributed, reflecting the random close-packed nature of the structure.

Although the picture presented here seems sound, what it lacks is a model sufficiently detailed that one can make confident averages of the angles, stresses, etc., throughout the structure.

5. *The origin of work softening*

We now turn to the question of work-softening, which plays such a prominent role. As Nabarro [21] has pointed out, there is an elementary error in the exposition of the earlier paper [1], which undermines the argument put forward there that the basic cause of the work softening is the supposed outward sideways force which a bowing loop exerts on its ends. In fact, the force exerted by a bowing loop is always inwards. What follows is

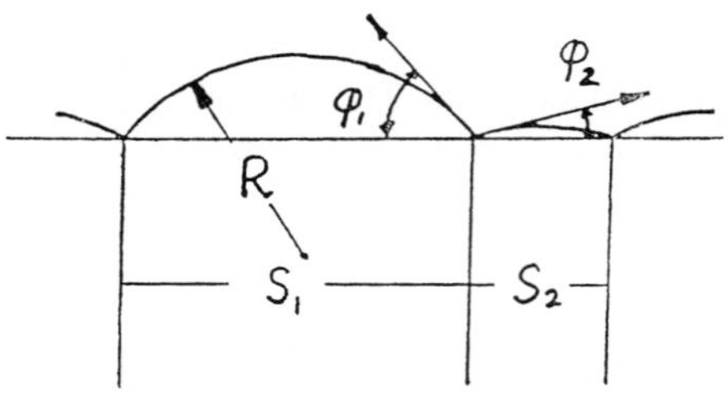

Fig. 3. A screw dislocation dragging dipoles spaced S_1 and S_2 apart, redrawn from Nabarro's book [22].

an attempt to correct the argument, in terms of the instability in the spacing of jogs on dislocation lines.

Figure 3, adapted from Nabarro's book [22], shows a jogged screw dislocation dragging dipoles. The force on each dipole is the resolved line tension, T, taken equal to $\mu b^2/2$. For a dipole separated from its neighbours by distances S_1 and S_2 as shown, the force is

$$F = T(\cos\phi_1 - \cos\phi_2) = T\left(\sqrt{1 - S_2^2/4R^2} - \sqrt{1 - S_1^2/AR^2}\right), \tag{9}$$

where

$$R = \mu b^2/2\sigma. \tag{10}$$

We can get some impression of the non-linear behaviour of the jogged dislocation with a small density of jogs on it if we denote the density by $n(x) = 1/S$ so that the change in spacing across a single jog is

$$\delta S = -\frac{1}{n^3}\frac{\mathrm{d}n}{\mathrm{d}x}\delta x \tag{11}$$

and the force on that jog is

$$F = T\frac{\partial}{\partial S}\left(\sqrt{1 - S^2/4R^2}\right)\delta S = T\frac{1}{\sqrt{4R^2n^2 - 1}}\frac{1}{2Rn^3}\frac{\mathrm{d}n}{\mathrm{d}x}. \tag{12}$$

Figure 4 shows how the force varies as a function of the jog density. The force is always directed towards a region of high density, so that under an applied stress the jogs will slide into clumps. The maximum force on a jog is T, rather than infinity as given by the approximate differential equation (12). When the jogs are spaced by the Frank–Read spacing, $S = \mu b/\sigma$, the maximum force is attained.

Now we imagine that in the interior of an obstacle there is a tangle of dislocations subject to the increasing applied stress. The effect is to redistribute such jogs as are mobile into clumps, leaving segments of line free and potentially mobile. Note that the jogs are not 'pushed' by the segments with large spacing, but are 'pulled' by the segments with small spacing! When the stress rises sufficiently, the jog-free segments attain the Frank–Read

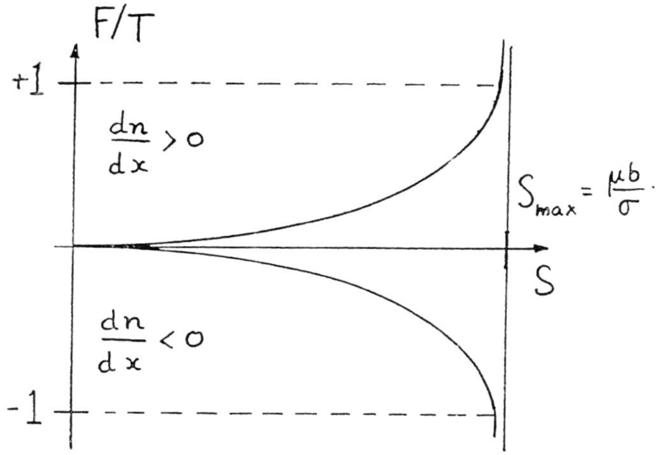

Fig. 4. The normalised force on a jog as a function of the gradient in jog density. The jogs are always pulled to a region of high density, with a maximum force equal to the line tension.

spacing, and the dislocation can move forward, such obstacles as it meets being pulled to one side. This is a possible process of work softening, which permits dormant obstacle interiors to participate again in the plastic flow, and to become the sites for the initiation of new slip bands. Of course, no sooner is the slip band initiated, than the fibre stress builds up, and secondary slip immobilises the band at a width and length appropriate to the current stress level.

6. The role of thermal activation

Although the reversible temperature dependence of the flow stress of a unidirectionally work-hardened metal is small, it is accurately measurable and it obeys the 'Cottrell–Stokes' law [23], which is an experimental finding that the ratio of the reversible temperature dependent part of the flow stress to the total work-hardened increment is constant, independent of plastic strain. This was interpreted by Mott as meaning that 'the important elastic forces on a glide dislocation come from its immediate neighbours, rather than from long-range stresses from piled-up groups; for then the elastic forces and the 'forest' forces acting on the dislocation can be both ascribed to the same thing, the presence of nearby dislocations' – quoted from Cottrell and Stokes [23]. Recently, this has been further interpreted by Nabarro [24] as meaning that the obstacles are of 'finite strength and range, and the dislocation advances rigidly across them'. In our terms, it is the temperature dependence of the forest hardening constant α itself which is responsible for the reversible temperature dependence. In effect, as the dislocations penetrate the forest, they are subject to fluctuating stresses which are caused by the thermal gas of phonons in the crystal. However, the overall structure is not comprehensible as a thermal fluctuation, only the

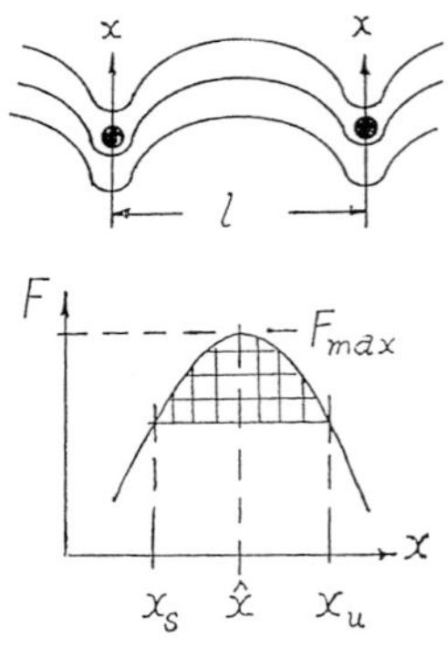

Fig. 5. A gliding dislocation bows between two secondary dislocations, whose points of crossing the glide plane form localised, short range, obstacles. The force on the gliding dislocation at each point goes through a maximum at the point $\hat{x}$. Thermal fluctuations must take the dislocation from its point of stability at x_s to the point where it is unstable and accelerates away, x_u.

overcoming of forest obstacles one by one. Figure 5 shows the situation. If we write the force at point x near its maximum in the close neighbourhood of a localised obstacle as

$$F = F_{\max}\left(1 - k^2(x - \hat{x})^2\right), \tag{13}$$

then the obstacle will be overcome in the absence of thermal fluctuations at a stress $\hat{\sigma}$, where

$$\hat{\sigma} bl = F_{\max}. \tag{14}$$

Here, we are using the 'hat' symbol, following the notation of Kocks et al. [25], for quantities appropriate to unactivated flow at absolute zero temperature. The algebra will be identical for attractive and repulsive obstacles, and is set up in recognition of the fact that thermal activation can be effective only over a very short range k^{-1} near the maximum force. For example, we anticipate that k will increase as the width of the dislocation core decreases, caused for example by increased stacking-fault energy.

The flow stress at temperature T is not greatly reduced from its value at absolute zero, so we write

$$\sigma = \hat{\sigma}\left(1 - \varepsilon(T)\right). \tag{15}$$

Then, following Nabarro [24], we can work out the temperature dependence by calculating the (free) energy[1] required to get the dislocation from its stable point of attack on the obstacle to its unstable point of release, and equating that to the thermal energy required to

[1] The only tricky point in this calculation is that the dislocation advances 'rigidly' over the obstacle: its curvature does not change. The thermodynamic justification for this is probably that the time constant for achieving equilibrium shape of the dislocation line is much greater than the period of atomic vibration. The line is frozen in shape during the activation. One can think of the atomic obstacle vibrating against the rigid dislocation.

overcome the obstacle in one second, nk_BT. Here n is the natural logarithm of the atomic vibration frequency, and k_BT is the usual exponent of the Boltzmann factor. The value of n can be taken to be about 30.

We find

$$\varepsilon = \left(\frac{9nk_BTk}{4\mu b^2}\right)^{2/3}, \tag{16}$$

where the line tension is, as usual, taken to be $\mu b^2/2$. This result exactly reproduces Nabarro's earlier result, if the constant k is taken to be equal to the reciprocal of Nabarro's informally defined obstacle width. As Nabarro first recognised, this is the basis of the Cottrell–Stokes law because the temperature dependence is the same whatever the spacing of the obstacles, and so will not depend upon plastic strain. In the context of this paper, it is the temperature dependence of the forest hardening which has been calculated, so we may write

$$\alpha = \hat{\alpha}\big(1 - \varepsilon(T)\big), \tag{17}$$

where the flow stress of the whole structure depends upon α through eq. (3). Thus the reversible temperature dependence is somewhat reduced, because the dependence of flow stress upon the forest hardening constant is reduced by the action of the internal stress. We find from eqs (3) and (17) that

$$-\frac{\Delta\sigma}{\sigma} = \frac{(1 - 1.6\hat{\alpha})\varepsilon(T) + 0.8\hat{\alpha}\varepsilon^2(T)}{(1 - 0.8\hat{\alpha})}. \tag{18}$$

If we insert reasonable numbers, taking $n = 30$, and the obstacle width k^{-1} equal to twice the Burgers vector, we find for aluminium $k_BT = \mu b^3/150$ at room temperature, and $\varepsilon(300\text{ K}) \approx 0.56$, giving for the Cottrell–Stokes ratio $\Delta\sigma/\sigma$ about -0.27, consistent with the experimental values. Figure 6 displays the original data to compare it with the theory. The agreement is good. Turning to other metals, only μb^3 depends upon the metal, and we find for the reversible reduction in flow stress between absolute zero and 300 K for Al, Ni, Cu, and Ag: 0.27, 0.13, 0.18, and 0.19, respectively. These are to be compared with the data given by Mitchell [13]: 0.27, 0.16, 0.10, 0.09, respectively. The agreement is satisfactory, and exact agreement can be achieved by changing the obstacle width slightly. These metals are in a sequence of decreasing stacking-fault energy, so one sees that the width of the obstacle, measured in Burgers vectors, increases slightly as the stacking-fault energy decreases and the partial dislocations get further apart. The thermally activated process is really an atomic process at the cores of the dislocations.

7. *Cross slip*

Finally, we turn to the problem of the screw dislocations, and the role of cross slip. It may have puzzled the reader that all the argument so far has concerned sections of the structure

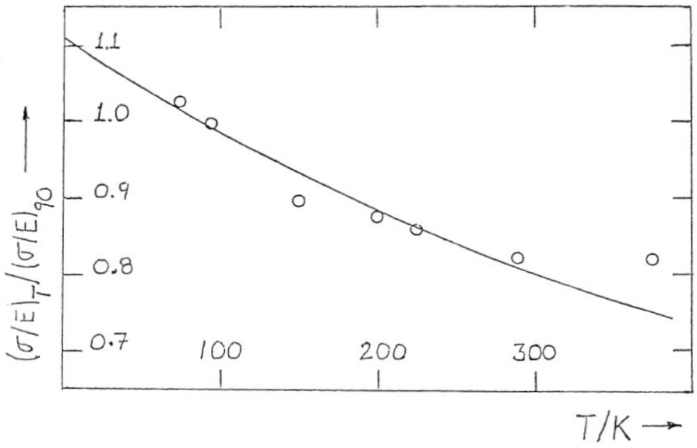

Fig. 6. The 'Cottrell–Stokes' ratio in aluminium as a function of temperature. The data (circles) from the original paper and the theory (line) have been normalised to 90 K, and the temperature dependence of the elastic constant has been factored out. There may be a discrepancy between theory and experiment for temperatures above room temperature.

drawn in a plane perpendicular to the slip plane and containing the Burgers vector. In reality, the screw dislocations travel perpendicular to this section, and trap one another. However, the resulting dipoles of screw dislocations, and the junctions that they form with secondary slip, have a maximum spacing inversely proportional to the stress if they are to be stable, so the pattern of these dipoles also obeys similitude and will produce linear hardening. Two closely spaced screw dislocations of opposite sign can cross slip and annihilate one another, a mechanism not available to edge dislocations. At low stresses, the screws will be widely separated, the dipoles will be stable and interact with the secondary dislocations, so any departure from similitude due to cross slip will be small. As the stress level rises, the spacing of screws in their dipoles decreases, the incidence of cross slip increases, until finally, above a certain stress, no further screw dislocation dipoles can be stored. This stress we identify with the onset of Stage III hardening, and noticeable departure from linear Stage II hardening because similitude has severely broken down. This stress is also equal to the saturation stress in cyclic plasticity, which is in turn approximately equal to the endurance limit for fatigue: the mean free path for screw dislocations at and above the saturation stress is constant, limited only by cross slip, and does not decrease further with increasing stress. This proposal is described in a recent paper [26], where the structure of persistent slip bands which control fatigue properties is discussed. It is notable that the fibre stress still plays an important role. It is not great enough to trigger secondary slip, or rather, if it does, new persistent slip bands form and the hardened ones become inactive. But where the persistent slip band meets the free surface, the fibre stress generates a logarithmic singularity in the surface stress, tension on one side of the band, and compression on the other. This mathematically infinite stress can initiate short cracks, the Stage I cracks of fatigue, which grow only slowly when they depart from the surface, but which can turn into Stage II cracks and ultimately cause the crystal to fail. Thus the fibre stress is an unexpected but centrally important feature of all slip bands.

The temperature dependence of the dislocation structures – the ladder spacing for example – in cyclically deformed copper is far greater than the temperature dependence of the forest hardening. These structures are controlled by cross slip, not by secondary slip. The length which determines the spacing is proposed to be the closest distance of approach which two screw dislocations of opposite sign can sustain before mutual annihilation by cross slip. Thus a strong *structural* temperature dependence is built in, which must be distinguished from the weak reversible temperature dependence manifested by the Cottrell–Stokes law.

8. Conclusions

The model outlined here is really to be thought of as a synthesis of the many models current in the literature. If the model is sufficiently self-consistent, one might at last begin to generate the differential equations of plasticity, and thereby achieve a useful, unified theory. However, it still seems very difficult to make the translation from the present semi-empirical – and no doubt, still controversial – picture to one of genuine engineering utility. The real question is: to what extent can dissipative structures, even simple ones such as are generated by dislocation plasticity, be predicted by current physical theory?

Acknowledgments

It is a pleasure to thank Frank Nabarro for suggesting that I could rescue this article from apparent oblivion and publish it in his *Dislocations in Solids* series; also for attending my lectures and commenting critically upon them as well as upon this paper.

References

[1] L.M. Brown, Metall. Trans. A 22 (1991) 1693–1708.
[2] I. Prigogine, From Being to Becoming (W.H. Freeman, San Francisco, 1980).
[3] F.R.N. Nabarro, Phil. Mag. A 80 (2000) 759–764.
[4] M.R. Staker and D.L. Holt, Acta Metall. 20 (1972) 569–579.
[5] N.F. Mott, Phil. Mag. 44 (1953) 742–765.
[6] P.B. Hirsch and T.E. Mitchell, Canad. J. Phys. 45 (1967) 663–706.
[7] L.M. Brown, in: Fundamentals of Deformation and Fracture, The Eshelby Memorial Symposium, eds B.A. Bilby, K.J. Miller and J.R. Willis (Cambridge Univ. Press, Cambridge, 1984) pp. 357–367.
[8] P.J. Jackson, Acta Metall. 32 (1985) 449–454.
[9] B. Obst, H. Auer and M. Wilkens, Mater. Sci. Eng. 3 (1968) 33–40. See also P. Gassenmeier and M. Willkens, Phys. Stat. Sol. 30 (1968) 833–843.
[10] S.J. Basinski and Z.S. Basinski, in: Dislocations in Solids, ed. F.R.N. Nabarro (North-Holland, 1979) pp. 260–362.
[11] J.D. Eshelby, Proc. Roy. Soc. A 241 (1957) 376–396.
[12] W. Schaarwachter and H. Ebener, Acta Metall. Mater. 38 (1990) 195–205.
[13] T.E. Michell, Progress in Applied Materials Research, Vol. 6, eds E.G. Stanford et al. (1964) pp. 116–237.
[14] L.M. Brown and W.M. Stobbs, Phil. Mag. 34 (1976) 351–372.
[15] D. Kuhlmann-Wilsdorf and N.R. Comins, Mater. Sci. Eng. 60 (1983) 7–24.
[16] P.J. Jackson and Z.S. Basinski, Canad. J. Phys. 45 (1967) 707–735.
[17] M.J. Whelan, Proc. Roy. Soc. A 249 (1958) 114–137.

[18] N.Y. Jin and A.T. Winter, Acta Metall. 32 (1984) 1173–1176.
[19] D. Kuhlmann-Wilsdorf, Metall. Trans. A 16 (1986) 2091–2108.
[20] S. Mader, Z. Phys. 149 (1957) 73–102.
[21] F.R.N. Nabarro, Scripta Met. Mat. 27 (1992) 277–278.
[22] F.R.N. Nabarro, Theory of Crystal Dislocations (Oxford Univ. Press, 1967) p. 389.
[23] A.H. Cottrell and R.J. Stokes, Proc. Roy. Soc. A 233 (1955) 17–34.
[24] F.R.N. Nabarro, Acta Metall. Mat. 38 (1990) 161–164.
[25] U.F. Kocks, A.S. Argon and M.F. Ashby, Thermodynamics and Kinetics of Slip, in: Progress in Materials Science, Vol. 19 (Pergamon Press, 1975).
[26] L.M. Brown, Materials Science and Engineering, Special issue in Honour of Prof. T. Mori A285 (2000) 35–42.

CHAPTER 59

The LES Theory of Solid Plasticity

Doris Kuhlmann-Wilsdorf

Department of Materials Science and Engineering
University of Virginia, Charlottesville, VA 22904

Dislocations in Solids
Edited by F. R. N. Nabarro and M. S. Duesbery

Contents

Preamble

This paper is divided into three parts comprising theoretical background information (part 1), the experimental evidence to be explained (part 2) and the LEDS theory (a special case of the LES theory) of work hardening (part 3). Each part can be read independently of the others. However, for proper appreciation of the theory (part 3) the first two parts are indispensable, since they lay out and make the case for the foundational assumptions on which the LEDS theory is built.

1. Theoretical background

1.1. Deformation occurs close to thermodynamical equilibrium – the LES hypothesis

Even while the plastic deformation of solids is one of nature's most effective means of converting mechanical work into heat, thereby increasing entropy, in regard to the interior structure of the deformed object it proceeds close to thermodynamical equilibrium [1–3]. This contention is encapsulated in the LES (*L*ow-*E*nergy *S*tructure) hypothesis as: "*In a material subject to mechanical stresses, that structure will be approached which has the lowest free energy among all structures which are in equilibrium with the tractions and are accessible to the system.*"

As a by-product of entropy generation, plastic deformation of solids produces ordered structures of various types. In the context of work hardening in dislocation-based deformation, these are seen as dislocation cell structures or deformation bands or slip bands, etc. However, once the tractions are reduced or removed those structures constitute excess free energy. As a result, on stress removal, deformation-induced patterns may be spontaneously released and converted into heat, or they may remain in metastable equilibrium and can be removed through annealing. Consequently, also, without annealing, deformation-induced internal structures are reversibly connected to the stresses that produce them, i.e. there exists a one-to-one correlation between the order produced and the highest previous stresses but not the strains.

The reason for the above is the requirement that (1) the second law of thermodynamics and (2) mechanical equilibrium, i.e. in regard to applied tractions Newton's third law, action equals reaction, must be obeyed at all times, and (3) that in the condition of minimum free energy specifically dislocations have mutually trapped into local energy minima. As dislocation structures generally do not reveal the direction from which the dislocations arrived, those relative energy minima are symmetrical against the inversion of the tractions that produced them. Hence, disregarding for the moment the Bauschinger effect (see section 1.10) and minor rearrangements accompanied by strains that are less than the elastic strains, a dislocation structure is typically stable on stress reduction and stress reversal up the magnitude of the previously highest.

Except for the accompanying irreversible shape changes associated with plastic deformation, the principle is not different for elastic deformation, as will become clear by contemplating the internal order in elastically stressed polycrystals. Here, again, the free energy of the system of tractions and deformed material is minimized. We are not used to think of plastic deformation in this way as we tend to be mesmerized by the irreversible external shape changes, e.g., a fender bender or a broken vase. Moreover, in materials science, theory has been unduly focused on the role of glide dislocations. As a result, the underlying principles in regard to the interior structure become much clearer when considering pattern formation on account of tractions in large structures. The best example here are stars and planets which, beginning with disordered gases or space debris, assume their generally spherical shapes and energy-minimum, onion-layer structures on account of gravity. It is deviations from thermodynamical equilibrium which, as second-order effects, trigger convections as in weather patterns or continental drift. Presumably these are treatable by Prigogine's thermodynamics of energy-flow-through systems, – but also these convections do *not* occur 'far from equilibrium' but rather on account of percentage-wise generally small deviations from equilibrium.

Thus it is concluded that in solid plasticity, the free energy of the 'system' of deformed material and deforming tractions is minimized independent of size, shape or chemical composition of the object considered. Besides the structure of astronomical objects just discussed, a most convincing example in support of this conclusion is the behavior of fatigue-cycled lamellar diblock co-polymers [1,4–7]. At modest amplitude above the glass transition temperature, constant-strain amplitude fatiguing establishes an ordered structure in initially disordered samples, wherein the lamellae are in the orientation with the lowest possible shear modulus relative to the applied fatigue stresses. This constitutes the structure predicted by the LES hypothesis, because at constant strain the stored elastic energy is proportional to the shear modulus. Next, above some critical fatigue amplitude applied to an initially ordered diblock polymer, kink bands develop with a morphology that minimizes the internal energy of the sample, as determined from the relationship between average kink-band length to width.

Similarly strong support for the LES hypothesis may be derived from the deformation-band structures of metals caused by unidirectional straining, to be discussed in greater detail in part 2. Their morphology similarly minimizes the stored energy, and the bands exhibit the same simple geometrical relationship of length to band width that is also found for the kinks in polymers. Most significantly, in that case the energy differentials involved can be shown to be so small that they would typically heat the samples through only about 0.01°C, – a close approach to equilibrium indeed.

The indicated reversibility of deformation structures is not obvious on account of their metastability already mentioned up to the magnitude of that stress level at which they were produced, and on stress reduction and reversal to that same stress magnitude. Reduction or reversal of stress level thus will not by itself reveal the inherent deformation structure reversibility predicted by the LES hypothesis. However, change of stress geometry, e.g., as in cross rolling, can cause work softening which is due to this effect, albeit not easily interpreted. Much more directly, through carefully controlled fatiguing at gradually reduced strain amplitude, Laird and coworkers [8,9] were indeed able to reverse the evolution of the structures that is normally observed on gradual amplitude increase.

Inherent structure reversibility lastly implies that creep at different temperatures generates those structures that otherwise are produced at the same stress and temperature in direct loading. That this is in fact so, is the conclusion drawn by this writer from scrutinizing, over many years, pertinent micrographs found in the literature. Albeit, straightforward evidence for this prediction of the LES hypothesis could not be located. Indeed, micrographs of dislocation structures produced by creep are comparatively scant in the literature. Among others they are not treated to any depth even in the recent book by Nabarro and de Villiers [10]. The best evidence for structure reversibility in creep that could be located is found in Kestenbach, Krause and da Silveira's study of creep in stainless steel [11] as reproduced in figs 1(a) and (b).

The above considerations will place in perspective the relative positions, in dislocation-based plasticity, taken in the LEDS (*L*ow-*E*nergy *D*islocation *S*tructure) and SODS (*S*elf-*O*rganizing *D*islocation *S*tructure) theories. Specifically, in opposition to the preceding facts, the SODS theory uses Prigogine's thermodynamics of energy-flow-through systems and assumes that plastic deformation occurs far removed from equilibrium. Pioneered by Aifantis [12–14] and Walgraef [15,16], the SODS theory was mechanistically derived from the work by Holt [17] and Staker and Holt [18] and is nowadays most closely associated with the name of Kubin [19–21]. The LEDS theory by contrast is strictly based on the second law of thermodynamics in conjunction with Newton's third law and was most recently summarized in [22] and [23]. It is but a special application of the LES hypothesis that according to the above introduction applies to all solid deformation independent of the specific mechanism.

The decisive observational support for the LES hypothesis outlined above and embodied in [1–3,22] and [23], rules out the SODS theory as well as all theories that are based on specific dislocation models, *except* if it could be shown that the proposed structures have the lowest energy among all those that would be accessible in principle. This condition is decidedly not fulfilled for dislocation pile-ups, the structure element previously most widely invoked in order to explain work hardening. Thereby the models by Kuhlmann et al. [24–26], Mott [27,28], Seeger et al. [29,30] and their many derivatives are ruled out. And this not only follows from abstract theoretical reasoning but results from the ready convertibility of dislocation pile-ups into LED structures by simple glide, without involvement of thermal activation.

Namely, as was first shown in 1952 [31], the illusion of the strong mutual blocking of dislocation sequences moving in opposite directions that was supposed to cause pile-ups and to represent a prime cause of workhardening, arises from focusing on edge dislocations only and neglecting the three-dimensional morphology. In actuality, in three dimensions, all mutually blocking dislocation sequences necessarily involve not only edge but also at least mixed dislocations, and none but pure edge dislocations can pile up because their position of maximum blocking stress arises not before but after the dislocations have passed each other. Specifically, at maximum blocking stress, pure edge dislocations on parallel slip planes meeting each other from opposite directions under a shear stress of, say, $^{\rightarrow}{}_{\leftarrow}$ type, have the relative positioning of "before passage" as $^{\perp}{}_{\top}$. By contrast, maximum blocking for all other dislocations occurs "after passage" as $_{\blacktriangle}{}^{\blacktriangledown}$, as clarified in fig. 2.

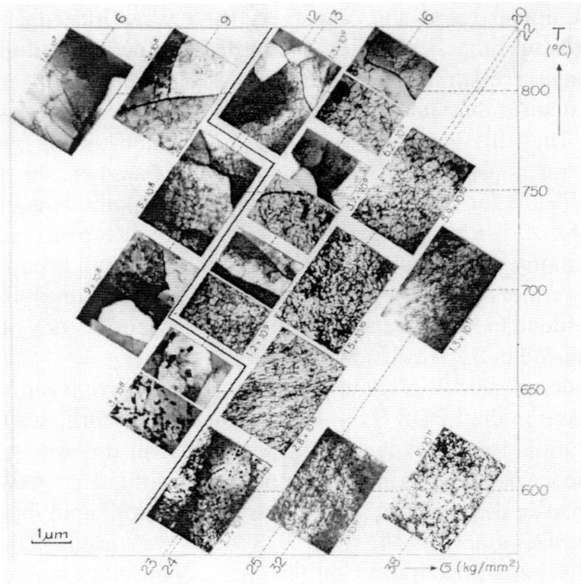

(a)

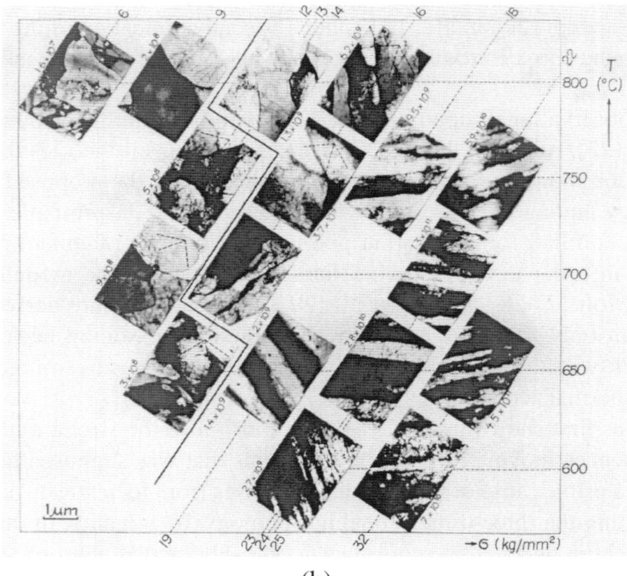

(b)

Fig. 1. Microstructures produced in 316 stainless steel during tensile creep at the indicated stresses and temperatures, (a) for random grain orientation and (b) for grains in which the stress axis was parallel to $\langle \bar{2}11 \rangle$, i.e. for which double glide was expected. The structures exhibit the features of planar glide in (a) and wavy glide in (b) and are believed to be substantially the same as for tensile tests at the same stresses and temperatures. (Figures 6 and 7 of Kestenbach, Krause and da Silveira [11].)

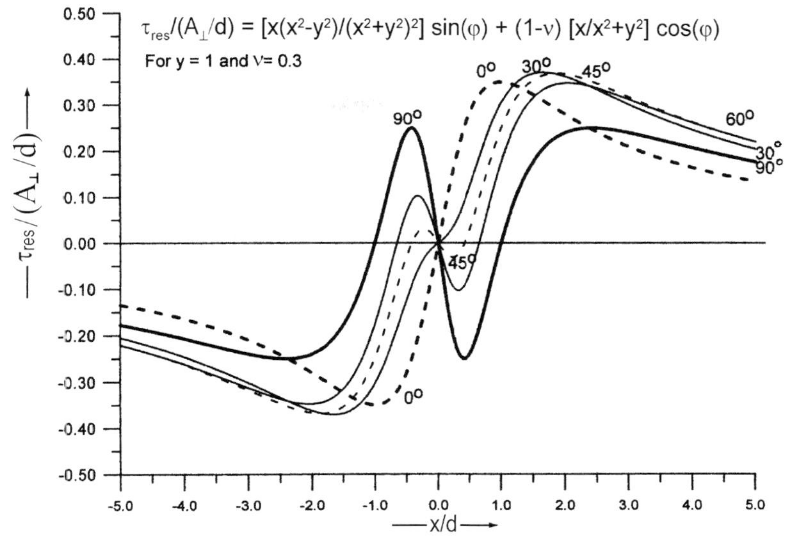

Fig. 2. The resolved shear stress, τ_{res} in units of $Gb/[2\pi d(1-\nu)]$, produced by a glide dislocation normal to the drawing with the indicated angles of the Burgers vector relative to its axis, on a similar dislocation on a parallel slip plane at distance d. This diagram is highly relevant to theories in which opposite pile-ups are supposed to block each other's progress. Namely, a stationary edge dislocation (90°) will maximally repel an opposite edge dislocation approaching, say, from the right, at the minimum in the boldly drawn curve, *before* the bypassing position at $x = 0$. By contrast, a fixed screw dislocation (0°) will attract an opposite screw at all positions and will block its progress through attraction only *after* bypassing, with the highest force at the curve minimum. In fact for all but edge dislocations the blocking stress *before* bypassing, if any, is smaller than that *after* bypassing. It follows [31] that pile-ups cannot possibly be mutually stabilized in three dimensions because there will inevitably be cross-over positions, and the pile-ups will convert into dipolar arrangements of greatly reduced strain energy, as in fig. 3, simply via moving such cross-over positions along the dislocation lines. (Figure 12 of [22].)

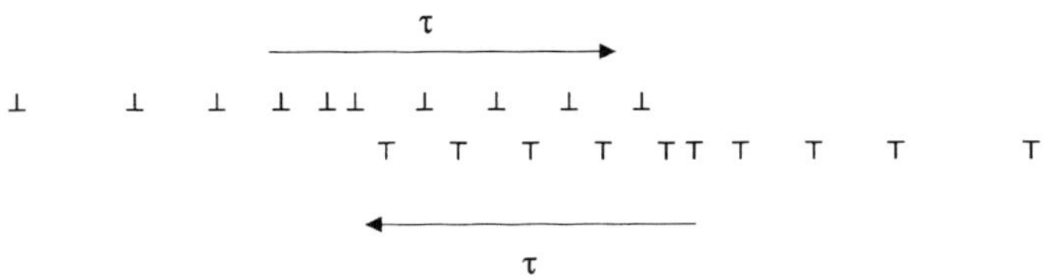

Fig. 3. The formation of a dipolar mat from edge dislocation trains meeting each other from opposite directions under shear stress τ as indicated. While the positive and negative edge dislocations are driven to the right and left, respectively, causing the pile-up distribution at either side of the growing dipolar mat (here exaggerated for clarification), the mat itself suffers no net force. The effect of the shear stress on it is only to mildly shift the $\perp$ and $\top$ dislocations from their zero-stress equilibrium positions. The indicated generation of dipolar mats is inevitable in any three-dimensional arrangement that in two-dimensional modeling would yield mutually blocking pile-ups.

In three dimensions, then, on account of the resulting inevitable 'cross-over' configurations that by their movement along the dislocation lines will shift the two dislocation sequences relative to each other, pile-ups will degenerate into sequences of dipoles as in fig. 3, simply by glide without the need for cross slip. Alas this decisive argument against pile-up models [31] was almost universally ignored, but it persuaded the present writer to rule out pile-ups as playing any significant role in workhardening. Other blocking-type dislocation configurations suffer the same defect of being three-dimensionally unstable against spontaneous rearrangement into configurations of lower energy, or certainly being unable to provide a significant source of work hardening.

1.2. The LEDS principle and general considerations on LES structures

In line with the preceding section, dislocation-mediated plastic deformation is governed by the LEDS hypothesis [2,22,23,32–34], i.e. the LES hypothesis for the particular case of glide dislocation involvement. According to it, among all microstructures which are in equilibrium with the applied tractions and are accessible to the dislocations, those are approached which minimize the stored energy.

The average specific line energy of dislocations is given by

$$U_D \cong Gb^2\{(1-\nu/2)/[4\pi(1-\nu)]\}\ell n(R/b) \approx 0.1Gb^2\ell n(R/b), \tag{1}$$

with G, b and ν (typically about 0.3), the shear modulus, Poisson's ratio and Burgers vector magnitude, respectively, and R the range of the dislocation stress field. It therefore follows that dislocations will arrange in positions in which they mutually screen each other stress fields to the level of τ_0, the friction force acting on dislocations, thereby limiting R. Any dislocation structure which fulfills this condition, to the effect that the upper cut-off radius, R, is comparable with the near-neighbor dislocation spacing, ℓ, is a LEDS (*L*ow-*E*nergy *D*islocation *S*tructure) by definition (see preface to [35]). The direct result of the LEDS hypothesis, and hence an integral part of the LEDS theory, is thus the "LEDS principle", namely that all dislocation structures formed in the course of plastic deformation be LEDS's. In fact, there are no known exceptions to the LEDS principle although, on account of the finite value of τ_0, the stress-screening is imperfect, and the various LED structures are not always well-formed and therefore not necessarily obvious. A considerable number of examples in support of the LEDS principle was presented in [36].

Next, in a LEDS comprising a more or less uniform mutually stress-screened dislocation distribution of density ρ, ℓ will be given by

$$\ell \cong 1/\sqrt{\rho}. \tag{2}$$

However, if the dislocations should cluster together, ℓ will be further reduced and with it the dislocation energy. If accessible, dislocations will for this reason assemble into planar LEDS arrangements. This is in fact the case, specifically very pervasively in the form of dislocation cell walls in mosaic block structures of mutually misoriented but otherwise more or less perfect volume elements, as first inferred from X-ray diffraction by Darwin [37].

The number of possible LEDS's is truly astronomical, in fact it is multiply infinite, rising with the number of available Burgers vector orientations. It is therefore very difficult indeed to verify the LEDS hypothesis by making predictions as to which LEDS to expect in any

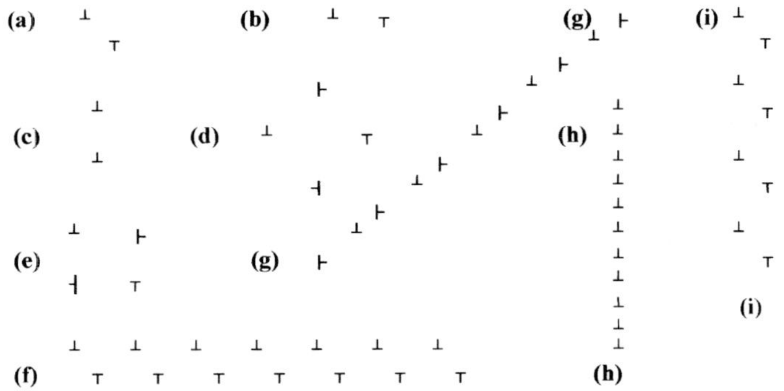

Fig. 4. The basic elements from which LEDS's may be constructed with only parallel edge dislocations of a single slip system ((a), (b, (c), (f) (h) and (i)) and of two systems with mutually perpendicular Burgers vectors ((d), (e) and (g)). In single glide *only* the possibilities (a), (b), (c), (f), (h) and (i) exist, and none other. Therefore, in Stage I as well as in single-glide low-amplitude fatigue, the LEDS theory predicts the formation of a selection of edge dislocation bundles of net zero Burgers vector content, of dipolar mats (f), of dipolar walls ((i), e.g., simple kink bands in uni-directional strain and the rungs in the ladder structure of persistent slip bands, PSB's), and of simple tilt walls (h), but no other dislocation structures. This is in agreement with experimental observation. The addition of parallel edge dislocations with a 90° rotated Burgers vector makes accessible all tilt walls in the zone defined by the edge dislocation lines, of which the 45° tilt wall (g) is an example. This, then, permits the formation of columnar but otherwise arbitrarily shaped dislocation cells delineated by tilt walls. All of these structures are unstable in climb. At vanishing friction stress, τ_0, the coplanar dipole (b), the quadrupole of (d) and the dipolar mat (f) are unstable in glide, also. Note that the edge dislocations "on top of each other" configuration (c) is the LEDS precursor of the tilt wall (h) but is not stress-screened according to the LEDS definition.

particular situation. However, retrospective analysis of dislocation structures is effective in supporting the hypothesis by revealing no difficulties. Moreover, for the case of single slip, with only a single Burgers vector direction available but in two opposite orientations, it so happens that *all* of the LEDS involve *only* edge dislocations, as indicated in fig. 4. It is therefore a simple matter to predict from the LEDS hypothesis which dislocation structures should form preferentially in single glide, meaning in Stage I and/or in low-strain-amplitude fatiguing. Namely in accordance with fig. 4 the expected dislocation structures in single glide are dipoles, edge dislocation bundles of net zero Burgers vector content, dipolar mats, dipolar walls and tilt walls. All of these are indeed characteristic for Stage I and for low-stress single-glide fatigue but are not observed otherwise.

The number of LEDS rises considerably if parallel edge dislocations with mutually perpendicular Burgers vectors are allowed additionally, also shown in fig. 4. Thereby tilt walls of arbitrary orientation become accessible of which the 45° wall of fig. 4(g) is a particular example, as well as dislocation quadrupoles and thence via assemblies of quadrupoles two-dimensional dislocation cells bounded by tilt walls but of arbitrary shapes [38–40]. Admitting additionally the presence of screw dislocations opens the way for the simple twist boundary of a screw dislocation cross grid fig. 5(a) and therewith the formation of three-dimensional cells. The range is still expanded when three coplanar Burgers vectors are allowed. In connection with close-packed planes including a low-energy stacking fault position, this introduces the hexagonal twist boundary (fig. 5(b)) as also dislocation extension into partials. For the particular case of a simple twist boundary

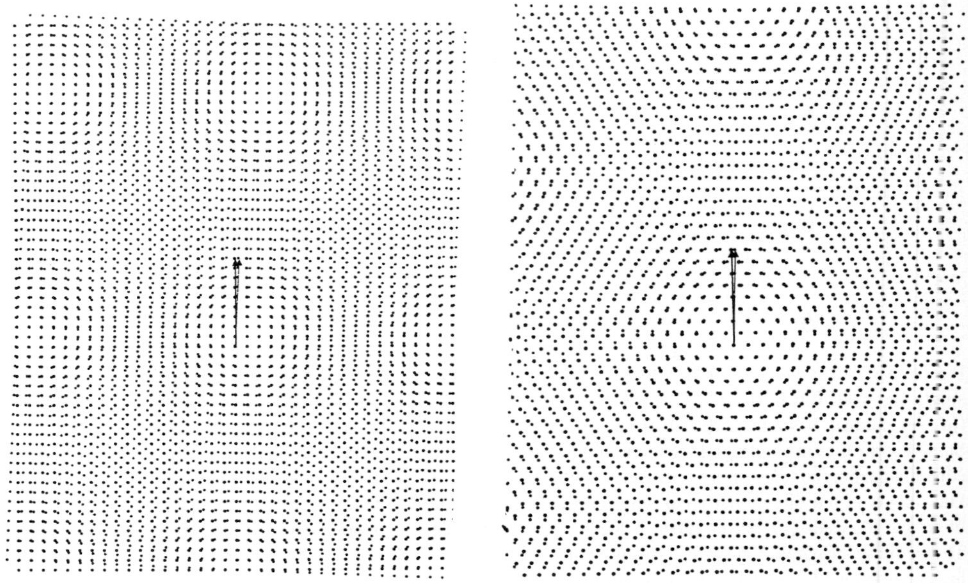

(a) (b)

Fig. 5. The simple twist boundary constructable from a cross grid of mutually perpendicular screw dislocations (a) and of three coplanar symmetrical Burgers vectors (b), modeled through the superimposition of two otherwise similar but slightly mutually rotated dot patterns that symbolize atoms on two adjacent crystal planes. Figures (a) and (b) model the (100) and (0001) planes of a simple cubic and simple hexagonal structure, respectively. Without the mutual rotation, indicated by the two arrows at the centre of each figure, the atomic positions in the two planes would coincide. The evident moire patterns generated by the rotation outline the expected screw dislocation geometry that will physically arise as by slight readjustments, the atoms concentrate the misfit into "dislocation cores" while everywhere else improving the fit.

the resulting hexagonal dislocation network is illustrated in fig. 6, using the notation of Thompson's tetrahedron when the plane of the boundary is (d), vector $\boldsymbol{b}_1$ is $\boldsymbol{AB}$, and $\boldsymbol{b}_2$ is either $\boldsymbol{BC}$ (fig. 6(a)) or $\boldsymbol{CB}$ (fig. 6(b)). We shall return to such formation of hexagonal networks in section 1.4. Importantly for the present discussion, note that this further reduction of stored energy through extended dislocations, especially in f.c.c. crystals, is not seen in freshly deformed so called "wavy glide materials" (see part 2) when straining took place below about half of the absolute melting temperature, but develops on mild annealing. This is so because it is obscured by "tangling" discussed in section 1.4, and takes place in response to only several percent of the dislocation line energy.

From here on matters become much more complex. Specifically, a considerable complication arises with the introduction of dislocations on non-orthogonal intersecting glide planes. For the particular case of edge and screw dislocations on the $\frac{1}{2}\langle 110\rangle\{111\}$ systems in f.c.c. the corresponding climb and glide stresses have been calculated in [41]. On average their magnitudes are only about one half to one third as for equal distances on the same system. However, the relevant distances, and thus the resultant forces acting between any two initially straight but non-parallel dislocations change, of course, along their axes. Thus quite generally dislocations on intersecting systems

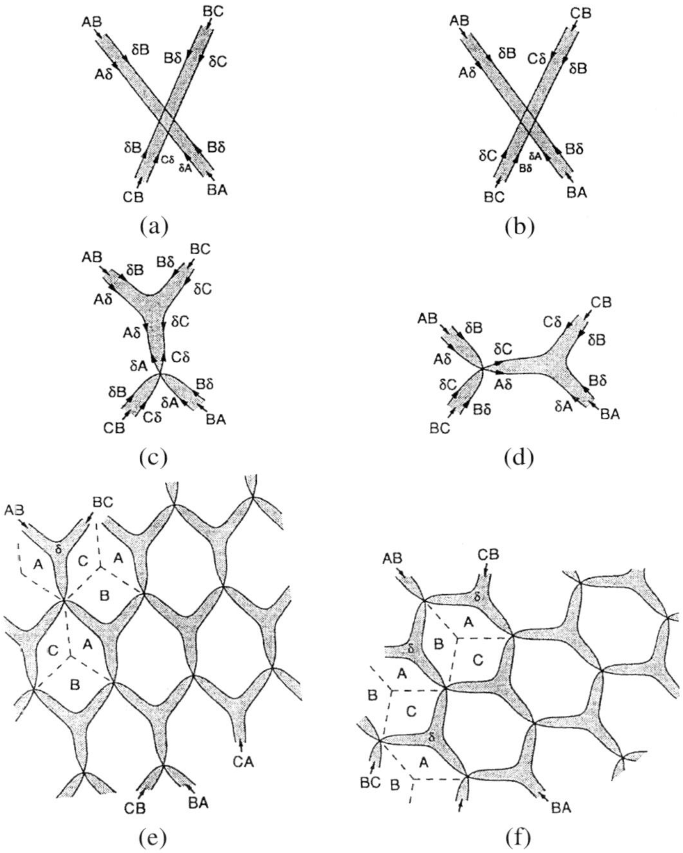

Fig. 6. Dissociation of four-fold nodes at the crossings of non-orthogonal, nominally "attractive" (***AB*** with ***BC***) and "repulsive" dislocations (***AB*** with ***CB***), into pairs of three-fold nodes. Given adequate dislocation flexibility, dissociation into three-fold nodes at dislocation crossing points takes place independent of the Burgers vectors involved, because this is lowers the free energy even without the additional extension into partials illustrated here that further decreases the energy. In the present example, $\frac{1}{2}\langle 110\rangle$ dislocations on the $\{111\}$ (d)-plane of an f.c.c. crystal with a lower intrinsic than extrinsic stacking fault energy are assumed. The same four-fold node intersections sketched in (a) and (b) that generate the three-fold nodes of (c) and (d), give rise to the networks (e) and (f) if not single dislocations but the corresponding dislocation sequences (or "grids") intersect.

interact so as to mutually bend. Finally, model calculations on LEDS dislocation cells including three coplanar Burgers vectors [42–47] have shown that energy minimization favors the formation of dislocation cells in a left/right/left/right checkerboard pattern of misorientation [46] as indeed seen in fatigued f.c.c. material [48].

With this we shall conclude the present general overview of the LEDS principle and characteristics of LEDSs and turn to a consideration of forces and energies among dislocations in greater detail. This is a step toward the ultimate challenge of predicting the lowest energy LEDS that can be formed under any particular straining conditions and that accordingly control the flow stress.

1.3. LEDSs versus MEDSs, HEDSs and NEDSs

1.3.1. Radial and tangential forces between parallel edge dislocations

The crucial difference between the LEDS and all other theories of plastic deformation is the insistence on energy minimization, whereas other models are content if they can demonstrate force equilibrium among the dislocations in the proposed models. That this is an unsound approach in principle may be demonstrated by considering the simplest and most important LEDS precursors, namely the two stable edge dislocation dipoles. We may readily understand them from the resolved shear stress of an edge dislocation extending along the z-axis with its Burgers vector, $\boldsymbol{b}$, parallel to the x-axis. Relevant text books, e.g., [49–53] give this in different forms as

$$\tau_{xy} = A_{\perp}\left[x\left(x^2 - y^2\right)/\left(x^2 + y^2\right)^2\right] = A_{\perp}\cos\phi/r = A_{\perp}\sin 4\phi/4d, \tag{3a}$$

where r is the distance from the dislocation axis, ϕ the angle between the radius vector $\boldsymbol{r}$ and the positive x-axis, d the distance between the slip planes of the two dislocations, and

$$A_{\perp} = Gb/2\pi(1-\nu), \tag{3b}$$

with G the shear modulus and ν Poisson's ratio. As seen from eq. (3a), glide force equilibria, i.e. positions of $\tau_{xy} = 0$, exist at $x = 0$, i.e. when the dislocations are "on top of each other", and at $x^2 = y^2$, i.e. on planes passing through the dislocation axis and inclined $\phi = \pm 45°$ to the glide plane either way. When the dislocations are "on top of each other" the equilibrium is stable for like dislocations (fig. 4(c)), which is the incipient form of a 90° tilt wall (fig. 4(h)), but unstable for opposite signs. The reverse is true on the 45° planes. On these, like dislocations are in unstable equilibrium but pairs of unlike dislocations form stable dipoles (fig. 4(a)). From $^{\perp}/_{\top}$ dipoles of type fig. 4(a) and/or dipoles moderately deviating from $\phi = 45°$ orientations, the "elementary quadrupole", dipolar mats (fig. 4(f)) and dipolar walls (fig. 4(i)) can be constructed, as well as Taylor lattices [54] as shown in fig. 7.

Even though these dipoles are stable in glide, they are unstable in climb. The basis of this apparent paradox is the enormously higher mobility in glide, at any temperature, as

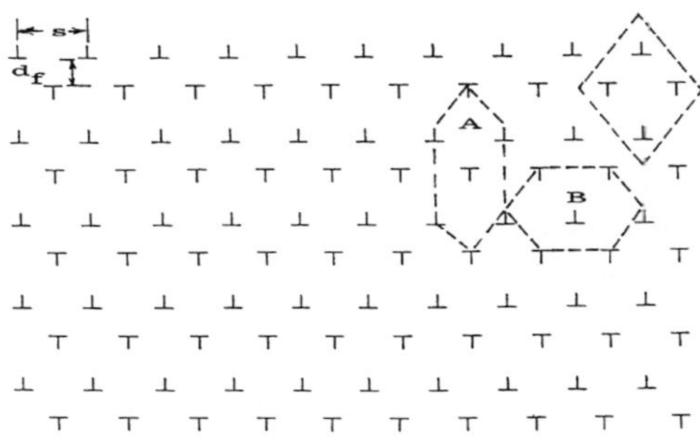

Fig. 7. The simple Taylor lattice as first proposed by G.I. Taylor [54].

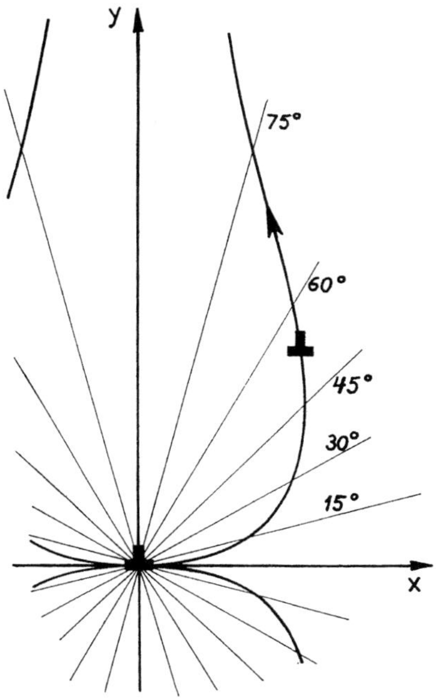

Fig. 8. Direction of the total force between similar parallel edge dislocations as indicated, due to glide force (in x-direction) and climb force (in y-direction) according to [55]. The glide force component is directed towards the y-axis for $45° < \phi < 135°$ (eq. (3a)). Correspondingly the dislocations attract in glide towards the "on top" position even though they are overall mutually repulsive. The force is inverted for dislocations of unlike signs. (Figure 1 of [55].)

compared to climb. The dislocations can therefore not respond to climb forces even while easily gliding in response to force components parallel to their respective glide planes. In this connection, the force vector distributions between parallel edge dislocations with parallel and normal Burgers vectors, respectively, given in figs 8 and 9 according to [55], are highly instructive. They are obtained from the radial (${}_{\mathrm{r}}F$) and tangential (${}_{\phi}F$) force components due to the dislocation at the origin. Both of these, in turn, are the vectorial sums of the respective glide and climb forces. Analytically the pertinent expressions are [55,56]

$$ {}_{\mathrm{r}}F_{\perp\perp} = bA_{\perp}/r, \tag{4a} $$

$$ {}_{\phi}F_{\perp\perp} = (bA_{\perp}/r)\sin 2\phi \tag{4b} $$

and

$$ {}_{\mathrm{r}}F_{\perp\dashv} = 0, \tag{5a} $$

$$ {}_{\phi}F_{\perp\dashv} = -(bA_{\perp}/r)\cos 2\phi. \tag{5b} $$

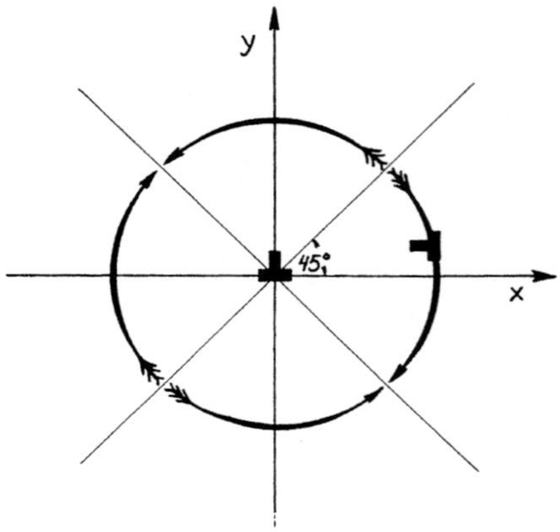

Fig. 9. As fig. 8 but for orthogonal Burgers vectors as indicated. Note the absence of any net attractive or repulsive forces between the dislocations even though at maximum they attract/repel with equal glide force as the dislocations in fig. 8 since $\tau_{xy} = \tau_{yx}$. (Figure 2 of [55].)

Evidently, the glide attraction between similar edge dislocations of like sign for $45° < \phi < 135°$ in accordance with eq. (3), which gives rise to the dipole of fig. 4(c) and thence the tilt wall, arises in spite of a strong overall repulsion among the two dislocations. By contrast, the similarly strong glide interaction between $\perp$ and $\dashv$ dislocations is free of radial force. The above is a specific illustration of the fact that net forces pull dislocations together or drive them apart in accordance with the square rule of Burgers vectors. Correspondingly, in the absence of the tractions that may have generated them, when dislocations are mobile in climb as well as in glide, complete mutual annihilation will result when the total sum of the Burgers vectors vanishes (as in all structures of fig. 4 except 4(c), 4(g) and 4(h)). Conversely, dislocations of one type and sign will be driven apart, eventually out of the sample, as is the case for all dislocation rotation boundaries. In either case net dislocation line length and dislocation energy will decrease, – theoretically to zero dislocation content or, as in the case of whiskers, one single dislocation passing through the specimen center. This, then, explains the effect of annealing.

1.3.2. The law of the square of Burgers vectors as a guide to dislocation structure energies With the enormous variety of possible LEDS's, and the difficulty of visually distinguishing LEDS's from non-LEDS's, the uninitiated is often tempted to believe that one structure is just as likely as another, provided the dislocations are in glide equilibrium. Over the years, the bulk of competing theoretical dislocation models have been constructed in this spirit. Yet, probabilities that LEDS's can be found by trial and error are astronomically small. Correspondingly, without exception it seems, all those models involve what have been dubbed MEDS's and HEDS's, for *M*edium- and *H*igh-*E*nergy *D*islocation *S*tructures, following an informal suggestion by F.R.N. Nabarro.

To provide a better distinction between those two ill-defined types versus LEDSs, we propose to add NEDS's for *N*eutral-*E*nergy *D*islocation *S*tructures, that incorporate no interaction energy, and proceed to examine the great conceptual difference between LEDS, on the one hand, and NEDS, MEDS and HEDS on the other hand. We shall go on to show that only LEDS's can form in nature, but none of the other types. To this end let us return once again to glide-stable edge dislocation dipoles of types $^{\perp}{}_{\top}$ (fig. 4(a)), $^{\perp}/_{\perp}$ (fig. 4(c)), and $\perp\dashv$ (top half of 4(d)). From eq. (3a), dissociating those dipoles in glide, by dragging or pushing one dislocation out of its stable position past the unstable position and thence guiding it beyond the reach of the other dislocation's influence, requires application of the same peak stress, namely

$$\tau_{\mathrm{dip}} = A_{\perp}/4d_{\mathrm{f}}, \tag{6}$$

if d_{f} is the distance of closest approach. The net work input or gain in the process of thus dissociating the dipoles is expended to increase the energy of the dislocation stress fields or is taken out of them, as they are converted from the condition in the dipole to the separated state. Conversely, the interaction energy, U_{int}, between the dislocation pairs in the respective dipoles is the net work gain or input, as the case may be, of work done on or by one of the dislocations as it is moved in glide from the widely separated state at distance R to its equilibrium position. One half of U_{int} may then be ascribed to each of the two dislocations in a pair.

On the basis of the above explanation, one may naively be tempted to believe that on account of the same τ_{dip} value involved, the dislocation energies in the three dipoles is the same or similar. This is very far from true, however. Namely, using eqs (4) and (5a) we can readily find the respective dislocation energies by using Cottrell's method. Thus, in thought, we bring the second dislocation from distance R on the $y = 0$ slip plane to its intended radial distance, r, from the dislocation at the origin against, or with the aid of, ${}_{\mathrm{r}}F$ and thence to its final position on an arc with or against ${}_{\phi}F$. Thus the interaction energies ${}_{ij}U_{\mathrm{int}}$ are for the three cases

$${}_{\perp\perp}U_{\mathrm{int}} = -\int_R^r {}_{\mathrm{r}}F_{\perp\perp}\,\mathrm{d}r - \int_0^{\phi} {}_{\phi}F_{\perp\perp} r\,\mathrm{d}\phi = bA_{\perp}\left[\ln(R/r) - \frac{1}{2}(1 - \cos 2\phi)\right], \tag{7a}$$

$${}_{\perp\top}U_{\mathrm{int}} = -{}_{\perp\perp}U_{\mathrm{int}}, \tag{8a}$$

$${}_{\perp\dashv}U_{\mathrm{int}} = -\int_0^{\phi} {}_{\phi}F_{\perp} r\,\mathrm{d}\phi = \frac{1}{2}bA_{\perp}\sin 2\phi. \tag{9a}$$

The resultant energy per unit length of dislocation line in the three cases is therefore

$${}_{\perp\perp}U(r,\phi) = U_{\perp} + \frac{1}{2}{}_{\perp\perp}U_{\mathrm{int}} = \frac{1}{2}bA_{\perp}\left[\ln(R/b) + \ln(R/r) - \frac{1}{2}(1 - \cos 2\phi)\right], \tag{7b}$$

$${}_{\perp\top}U(r,\phi) = U_{\perp} - \frac{1}{2}{}_{\perp\perp}U_{\mathrm{int}} = \frac{1}{2}bA_{\perp}\left[\ln(r/b) + \frac{1}{2}(1 - \cos 2\phi)\right], \tag{8b}$$

$$_{\perp\dashv}U(r,\phi) = U_\perp + \frac{1}{2}{}_\perp U_{\rm int} = \frac{1}{2}bA_\perp\left[\ln(R/b) + \frac{1}{2}\sin 2\phi\right], \tag{9b}$$

with $U_\perp = \frac{1}{2}bA_\perp \ln(R/b)$. At minimum, the dislocation energies per unit line length finally are

$$U_{\perp\perp} = \frac{1}{2}bA_\perp\left\{\ln[R(R/r)/b] - \frac{1}{2}\right\}, \tag{7c}$$

$$U_{\perp\top} = \frac{1}{2}bA_\perp\left[\ln(r/b) - \frac{1}{2}\right], \tag{8c}$$

$$U_{\perp\dashv} = \frac{1}{2}bA_\perp\left[\ln(R/b) - \frac{1}{2}\right]. \tag{9c}$$

Correspondingly, it is convenient to write all specific dislocation line energies in the form of

$$U_{jk} = \frac{1}{2}bA_\perp(\ln R^*_{jk}/b); \tag{10a}$$

if this is done we find

$$R^*_{\perp\perp} = R(R/r\sqrt{e}), \tag{7d}$$

$$R^*_{\perp\top} = r, \tag{8d}$$

$$R^*_{\perp\dashv} = R/\sqrt{e}. \tag{9d}$$

It should be added that through disregarding boundary conditions, eqs (7a) to (9c) are not reliable to the value of $\frac{1}{4}A_\perp$ although their relative differences are accurate. Similarly the factors $1/\sqrt{e}$ in eqs 7(d) and 9(d) are not significant.

Even though perhaps initially unexpected, the above results should not be surprising: To begin with, the maximum shear stress required to separate the dipoles (eq. (6)) reflects mainly the angular dependence of the interaction energy rather than net radial repulsion and attraction, if any. The cause of this is the large difference already indicated in responsiveness to glide and climb forces. In regard to the values of R^*, note the following: (i) At distances large compared to their separation r (where normally a great part of the dislocation energy resides), any assembly of parallel dislocation looks like one single dislocation with Burgers vector $\sum \boldsymbol{b}_j$; (ii) the dislocation stresses close to the axis are so large that they are not significantly affected at radial distances smaller than about one half the distance to their nearest neighbor. Thus, when in an arrangement of parallel dislocations $\sum \boldsymbol{b}_j$ vanishes, the far-away stress field also vanishes while the stresses within about half the nearest neighbor distance remain intact. In that case, therefore, the R^* value is bound to be comparable to the near-neighbor distance. That is the case for the $^\perp{}_\top$ dipole (eq. (8d)), and indeed for all LEDS's since in accordance with the LEDS definition the dislocations in them are stress-screened to the near-neighbor distance, more or less. Next, when $\sum \boldsymbol{b}_j$ does not vanish but the Burgers vectors are normal to each other, there is no change of energy according to the rule of the square of the Burgers vectors. Correspondingly, there is no mutual screening of stresses nor change in R^* from its prior value of R, as for the $\perp\dashv$ dipole (eq. (9d)). Finally, with any net increase of parallel Burgers vectors, as in the $\perp\perp$ dipole, the square rule of Burgers vectors indicates the corresponding increase in strain energy and thus an increased value of R so as to double the energy, as evidenced in eq. (7d).

1.3.3. LEDSs, NEDSs and HEDSs defined in terms of the upper cut-off radius

The above considerations naturally lead to a definition for LEDSs, NEDSs and HEDSs in terms of the respective R^* or R^*/r values, remembering that in actual dislocation structures r is typically related to the dislocation density ρ as

$$r = m/\sqrt{\rho} \tag{10}$$

with m not too far from unity. This is in line with eq. (2) since in LEDS's, by definition, near-neighbor dislocations mutually stress-screen so that R^* compares with the near-neighbor distance. Therefore *in LEDSs $R^*/r \approx 1$, i.e. R^* is small and shrinks with increasing dislocation density*. In neutral-energy dislocation structures, i.e. in statistical dislocation distributions expected for non-interacting dislocations [57–59], R^* is the non-dislocation stress-screening distance. Therefore *in NEDSs R^* is independent of dislocation density and equals R, the specimen diameter or grain size.* The label "HEDSs" (*H*igh-*E*nergy *D*islocation *S*tructures) is applied when dislocation energies are much higher than in the isolated state. Correspondingly, *in HEDSs $R^* > R$ and $R^*/r \gg 1$, rising with dislocation density.* The perhaps facetiously proposed term MEDS has yet to be defined but certainly implies R^* significantly larger than R. Pending any proposed definition, such as some mutually agreed-on value for R^*/R or R^*/r, it may best be dropped.

Non-LEDS theories of flow stress and/or work hardening in which dislocations are assumed to mutually block each other's motion, necessarily imply the formation of HEDSs, i.e. large R^*/r values. This follows because, in glide, dislocations solely respond to resolved shear stresses at their axes. If they are to impede each other's motions other than through mutual trapping, which is the basis of the LEDS theory, they therefore must be somehow forced together by the applied stress. The extra work imparted to the system thereby must be converted into extra stresses and these, in turn, are expressed in terms of $R^* \gg R$.

Dislocation pile-ups are the pre-eminent HEDS. Theory [60] shows that the interaction energy per unit length of dislocation line in an assembly of pile-ups consisting of $N \geqslant 10$ edge dislocations each is

$${}_{\text{pile-up}}U_{\text{int}} \approx 1.6(N-2)\left(\frac{1}{2}bA_{\perp}\right) \tag{11a}$$

for

$${}_{\text{pile-up}}U_{\text{int}}/U_{\perp} \approx 1.6(N-2)/\ln(R/b). \tag{11b}$$

Conversion of ${}_{\text{pile-up}}U_{\text{int}}$ into the form of eq. (7) is nonsensical since R^* would become millions and billions of light-years. In fact, depending on dislocation density, ${}_{\text{pile-up}}U_{\text{int}}/U_{\perp}$ hovers about unity for N between 5 and 10, and it is near 10 for $N \approx 40$.

An implication of the above considerations is that MEDS's and HEDS's cannot form in nature once the density of the dislocations is high enough that they predominantly interact among each other rather than with other crystal defects. The reasons are (i) the existence, and equal but opposite responses to stresses, of dislocations of opposite sign, (ii) the typical availability of multiple slip systems, and (iii) the flexibility of dislocations. To begin with (i), how can the approach among dislocations be impeded by "repulsive" stresses when those same stresses are attractive for dislocations of the opposite sign which ordinarily are freely available? As to (ii), even if there were situations in which glide on one system would be impeded through long-range dislocation stresses, surely those stresses would soon trigger glide on other systems. Finally (iii) as we already saw, non-parallel dislocations bend in each other's stress-fields. For two interacting dislocations it at most requires a local 90° bend of one, or 45° bends or less in each, to transform a repulsive interaction into an attractive one. What, then, could possibly compel dislocations to block each other's path through repulsion considering that they interact with forces that lead down energy gradients and are directed to induce bending to transform mutual repulsion into attraction? Conversely, in view of all of the above, what possibly could prevent dislocations *not* to react to their long- and short-range attractive forces and *not* mutually trap into low-energy positions, i.e. LEDSs? And as to NEDSs, they are unrealistic since the only case of mutual indifference are parallel edge and screw dislocations.

The problem with non-LEDS models is, and has been throughout, that alternative dislocation motions in accordance with (i), (ii) and (iii) above, which would destroy the supposed repulsive balances, are not considered. The preceding discussion of the pile-up (compare figs 2 and 3) is an instructive example to demonstrate the point. Mutually impeding pile-ups on intersecting slip systems are even less stable than those on the same glide system already discussed. Here crossover points are geometrically unavoidable except that one may think of invoking sessile barriers, especially the Lomer–Cottrell lock, ending up with sequences of long parallel dislocation locks. However, any such locks are easily unraveled from their ends [61,62].

In summary of this section, the rule of the square of Burgers vectors is a much better guide for assessing the stability of dislocation structures than considerations of local stress equilibria, – *except* for dislocation rotation boundaries to be discussed presently.

1.4. "Tangling" alias "mushrooming": dislocation–vacancy interactions

In "wavy-glide materials", comprising almost all pure metals and many alloys, especially at elevated temperatures, the underlying LEDS order is obscured through what has been, and still sometimes is, called "tangling". Figure 10 of [63] shows examples. The process occurs because in the course of gliding, all dislocations that are not firmly restricted to movements on their own crystallographic glide planes generate vacancies that rather rapidly build up to an estimated atomistic concentration of 10^{-4}. The morphological cause for the vacancy generation could be dragging of intersection jogs and/or "dislocation uncertainty" [64].

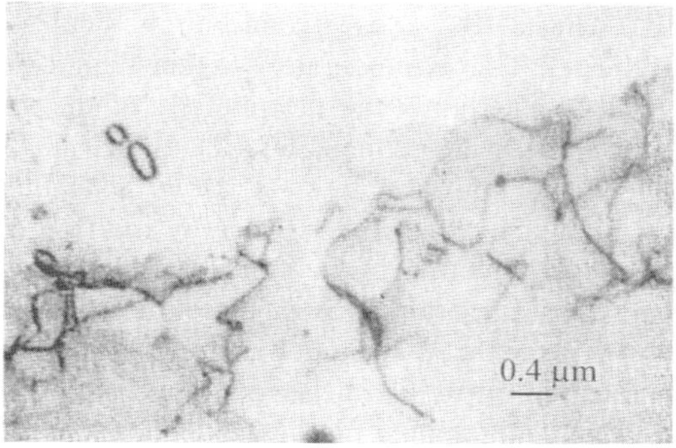

(a)

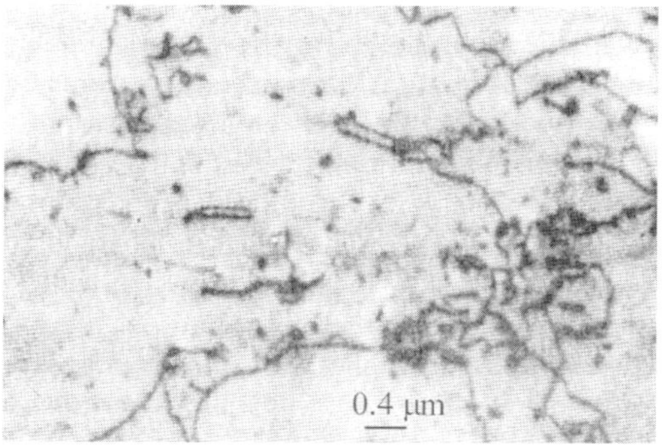

(b)

Fig. 10. Dislocation "tangles", ascribed to interactions of lattice vacancies with dislocations in the process of coming to rest, in pure aluminum mildly deformed (a) after annealing and (b) after annealing followed by water quenching. The higher density of prismatic loops and dislocation kinks in (b) is due to the extra lattice vacancies through quenching, supplementing those generated through dislocation glide ([63]).

Fast moving dislocations, e.g., at $v > 1$ μm/sec, are indifferent to those vacant lattice sites. However, at the end of their glide path, the dislocations slow down and come to rest in an environment of strongly supersaturated vacancies and in a complex mechanism deplete them [63–66]. That elimination of supersaturated point defects and of tiny vacancy clusters through dislocations takes the form of local climb, causing jogs and small dislocation loops appended to the dislocations. Driven by the local stress, free dislocation lengths between such pinning points bow out, cross-slip and double-cross-slip irregularly. In the process, the local stress distribution becomes increasingly complex on account of the interacting dislocation parts in varied orientations. On thus spreading, the dislocation tangle increases the volume out of which supersaturated defects are being drained, in line, once again, with the second law of thermodynamics. The whole process has therefore been descriptively dubbed "mushrooming". In this manner, then, dislocations which glided in fairly smooth configurations come to rest in the form of "tangles".

The driving energy for producing the extra dislocation line length is that of the supersaturated vacancies, e.g., about 1 eV as compared to a few eV for one Burgers vector length of dislocation line. Thus a 10^{-4} vacancy concentration could produce a dislocation concentration of $\approx 10^{10}/cm^2$. This is moderately higher than seen in even the densest 'tangles', e.g., as in [67], and certainly the morphology of the mushrooming structure conforms closely to the mechanism outlined [63–67]. Hence in their final movements, dislocations at the end of their gliding path so-to-speak "vacuum up" the vacancies in a roughly cylindrical volume about their axes. Thereby the free energy of the material decreases, as point defect energy is transformed into heat and partly into dislocation line energy, jogs and, perhaps most importantly, vacancy loops in all orientations. Additional support for this mechanism is the fact that about one half of the stored energy of deformed wavy-glide samples is released on mild annealing without significant effect on the flow stress but accompanied by a straightening out of the dislocations, as was first discovered by Bailey and Hirsch [68].

The importance of this process for work-hardening behavior can hardly be overestimated, since it makes available all possible Burgers vectors in potentially high concentrations and thus permits the formation of dislocation cell structures that cannot be generated with just one primary Burgers vector direction. This conclusion is additionally supported by a study by Wilsdorf and Schmitz [69] who have compared the tangling/mushrooming mechanisms in aluminum, copper and nickel. Finding no essential differences they concluded that ease of cross-slip cannot be a determining factor in the effect. They further found that the tangles emit new glide dislocations singly as well as provide the stopping points for them at the end of their path, thereby ruling out the models by Mott [27,28], Seeger et al. [70,71] and Hirsch and Mitchell [72] that in Stage I (and perhaps throughout) dislocations are emitted by multiply acting sources in groups of tens. Even so, the precise mechanisms involved in mushrooming urgently require elucidation since it takes place even at the lowest temperatures although with a greatly decreased radius of affected zone. Herein, the transient heating due to the energy release already discussed is probably responsible for the persistence of mushrooming to liquid helium temperatures. Whatever the answer to this problem may be, certainly profuse cross-slip, climb and nucleation of vacancy loops take place on a fine scale; and probably every possible Burgers vector is generated through vacancy loop nucleation and subsequent emission of glide dislocations from the loops.

In planar glide materials "mushrooming" does not take place, dislocations do not tangle, and cells by which Stage II is terminated do not form. Mostly, the difficulty appears to lie in insufficient three-dimensional dislocation mobility. Specifically in α-brass type alloys, the wide extension of the glide dislocations on account of low stacking-fault energies not only inhibits cross slip but also the statistical formation of point defects in glide and interactions with supersaturated point defects. However, in Al–Mg alloys the stacking-fault energy is high [73,74]. Even so they are of planar glide type [75–79] and do not form cells. In this case the problem appears to be lack of "unpredicted" Burgers vectors which as already indicated are otherwise provided through vacancy loop formation. It has been suggested [80] that vacancy loop nucleation is in this case prevented because Al–Mg alloys have a very high thermal equilibrium concentration of vacancies [81] so that the vacancies produced by gliding dislocations do not cause a significant supersaturation.

The alternative interpretation of "tangling" as due to the interaction of dislocations with different Burgers vectors lies close at hand since quite frequently primary glide dislocations slow down as they encounter non-primary dislocations. This alternative explanation was therefore tested by changing the concentration of point defects and point defect aggregates pre-existing before strain in various ways. Figures 10(a) and (b) are a case in point. Both pertain to lightly deformed pure aluminum, fig. 10(a) strained after annealing and fig. 10(b) similarly strained after water quenching from near the melting point. In fig. 10(b) the point defects generated through the gliding dislocations are thus supplemented by the quenched-in vacancies (already during quenching converted into invisible voids) whence the tangling in this micrograph is more severe than in fig. 10(a). Next, via extremely accurate density determinations, the destruction of the voids in the tangling process could be verified via the corresponding very small increase of mechanical density, both in aluminum [82] and in gold [83]. Also, tangling was studied through very careful observations of the dislocation geometry involved, including Burgers vector determinations and stereo micrography. Again the result was to confirm that point defects and not dislocation interactions are responsible for the effect [67]. Finally, in TEM foils, in which no vacancy supersaturations can build up, dislocations never tangle [84] although they may climb in response to image forces via vacancies entering the foils via pipe diffusion [85].

Tangling is very characteristic of pure metals because in them, dislocations have constricted cores which offer ready nucleation sites for point defect loops and permit easy cross slip. As a result, in pure metals the underlying regularities of dislocation structures are typically obscured by "tangling", as already mentioned at the start of this section. Metals with low stacking-fault energies contain dislocations extended into partials which do not exhibit tangling, being resistant to cross slip, jogging as well as nucleation of point defect platelets at them. These therefore reveal LEDS geometries much more clearly than pure metals. More importantly, they exhibit planar glide, the dislocations remaining largely confined to their glide planes. This is the mechanism by which stacking-fault energy affects overall mechanical behavior.

1.5. Dislocation walls

1.5.1. Why do dislocation cluster into walls?

One of the most obvious features in deformed materials is the clustering of dislocations into two-dimensional arrangements, commonly called "dislocation walls". In the framework of the LEDS theory, these must obviously represent LEDSs and can be of two kinds. Firstly, they may have zero net Burgers vector content like the dipolar mat (fig. 4(f)) or wall (fig. 4(i)), or, secondly, they may have a finite Burgers vector content and in that case are 'dislocation rotation boundaries' separating two mutually rotated crystal parts, like the tilt walls of fig. 4(g) and (h). Both of these types can involve more than the one or two 'dislocation grids' concerned in fig. 4, meaning sets of more or less uniformly spaced similar dislocations of same Burgers vector. In fact, conceptually, there is no limit to the number of ways in which one may construct dislocation walls with net zero Burgers vector content, i.e. walls in which a Burgers circuit would close about a representative small dislocation grouping. Depending on the number of Burgers vectors involved, these are called 'dipolar' or 'multipolar' walls. By contrast, 'dislocation rotation boundaries' are conceptually due to the joining together of two unstressed but relatively rotated crystals of same lattice structure. This joining operation would give rise to a defect boundary that is also analyzable in terms of dislocation grids. These, too, are free of long-range stresses and thus are LEDSs, but a Burgers circuit about a representative small dislocation grouping in them will not close but reflect the closure failure due to the relative lattice rotation across the wall. And, again, there is no limit to the variety of dislocation rotation boundaries that may be constructed in principle.

Dislocation walls of both types are very prevalent in deformed crystalline materials and one may ask why this should be so. There are three interrelated reasons:

(1) Glide dislocations have a tendency to assemble into planar arrays so as to reduce the screening distance of their long-range stress fields as discussed in connection with eqs (1) and (2). If the clustering dislocations have no net Burgers vector content, as would be the case for glide dislocations meeting each other from opposite directions as in Stage I and in plus-minus fatigue, dipolar or multipolar walls are expected. Indeed, the so-called 'maze structures' in fatigue, as studied in particular by Charsley [86,87] and by Dickson and coworkers [88–91] are of this kind. However, dislocation clustering for closer stress-screening among dislocations with significant net Burgers vector content will give rise to dislocation rotation boundaries, and thence to mosaic block, i.e. dislocation-cell structures.

(2) Plastic shear strains cause lattice rotations as a simple geometrical effect. For example in the very common case of tensile deformation, single glide will rotate the slip direction towards the tensile axis, and other strain configurations will cause their own peculiar rotations. In polycrystals, such reorientations cause textures. In principle the lattice rotations discussed should continue until a stable lattice orientation is reached that, in accordance with the Taylor criterion [92], is maintained by the simultaneous operation of at least five independent slip systems. However, as will be further discussed in part 2, the local flow stress rises with the number of simultaneously operating slip systems. Consequently, locally not the requisite five, but only one, two or perhaps three slip systems operate together, namely within small volume elements with different slip system selections. Among a grouping of two or more such volume elements, those slip systems will operate that overall produce the external shape change which is enforced by the imposed tractions. Hence at the boundaries between the individual volume elements operating with different slip system selections, the different shears and lattice rotations produce the corresponding rotation boundaries. The dislocation content in these boundaries is provided by those glide dislocations which were stopped at the boundary.

(3) On account of the 'law of the conservation of Burgers vectors', glide dislocations are stopped at dislocation rotation boundaries. If these dislocations arrive preferentially from one side, they will increase the misorientation.

1.5.2. Frank's formula

Frank's formula (eq. (12)) states the necessary and sufficient condition that dislocation 'walls' are free of long-range stresses, and thus are LEDSs, in compact but not entirely obvious form. It relates the Burgers vector content in the wall to $\boldsymbol{u}$, unit vector parallel to the axis about which the crystal parts on either side are rotated relative to each other, and Φ, the angle of misorientation, as follows [93]: Inscribe on the network arbitrary unit vector $\boldsymbol{v}$ (large compared to the dislocation spacing so as to intersect a representative sampling of the dislocations) and determine $\boldsymbol{B}$, the vectorial sum of the Burgers vectors of all dislocations intersected by $\boldsymbol{v}$ ($\boldsymbol{B}$ thus having the dimensions of length per unit length). The network will be an ideal LEDS if, and only if,

$$\boldsymbol{B} = \sum \boldsymbol{b}_j/d_j = 2\sin(\Phi/2)[\boldsymbol{u} \times \boldsymbol{v}] \approx \Phi[\boldsymbol{u} \times \boldsymbol{v}], \tag{12}$$

where d_j is the dislocation spacing in the j-th 'grid' (meaning set of parallel dislocations with same Burgers vector in the array). If $\sum \boldsymbol{b}_j/d_j$ is finite for any orientation of $\boldsymbol{v}$ and eq. (12) is not fulfilled, the wall has long-range stresses.

For $\boldsymbol{B} = 0$, i.e. no net Burgers vector content, Frank's formula describes a dipolar or multipolar wall or mat, not associated with any relative lattice rotation such as in figs 4(i) and (f). Rigid lattice rotations are associated with all walls obeying Frank's formula having finite values of $\boldsymbol{B}$. Let us consider in turn such boundaries consisting of one, two or more grids as follows.

(i) For one single grid, i.e. with $\boldsymbol{B}$ necessarily parallel to $\boldsymbol{b}$ and this normal to all orientations of $\boldsymbol{v}$, i.e. normal to the boundary, the 90° tilt wall (fig. 4(h)) is the only solution to eq. (12). (ii) The best-known two-grid rotation boundary is the cross-grid of all left-handed or all right-handed screw dislocations shown in fig. 5(a), defining a pure twist boundary. In this case $\boldsymbol{u}$ is normal to the boundary and $\boldsymbol{B}$ lies in the boundary for all directions of $\boldsymbol{v}$. (iii) The other simple type of two-grid rotation boundary is the family of tilt walls of parallel edge dislocations which was already discussed in connection with the 45° tilt wall (fig. 4(g)).

For the remainder, the vector analysis of eq. (12) is not trivial. It has been excellently reviewed by Amelinckx and Dekeyser [94]. The general solution for two-grid boundaries entails no restriction as to the Burgers vector components normal to the boundary. Therefore an arbitrary tilt can be superimposed on a twist rotation. If the Burgers vectors are crystallographically determined but both dislocation densities and axis directions are freely adjustable, two-grid boundaries can be constructed on any arbitrary plane. However, there is then no choice of the direction of the rotation axis.

With three independent dislocation grids and crystallographic Burgers vector directions, but freedom to choose dislocation spacings and axis directions, eq. (12) can be fulfilled for any arbitrary choice of both boundary plane and rotation axis orientation [94,95]. With $\boldsymbol{n}$ the normal to the boundary plane and $\boldsymbol{b}_j$ the Burgers vectors, the conditions for the axis directions, $\boldsymbol{r}_j$, are

$$\boldsymbol{r}_1 = [\boldsymbol{u}(\boldsymbol{b}_2 \times \boldsymbol{b}_3)] \times \boldsymbol{n} \tag{13}$$

and correspondingly for $\boldsymbol{r}_2$ and $\boldsymbol{r}_3$; and the conditions for the dislocation densities $(1/d_j)$ are

$$1/d_j = \left|\boldsymbol{n} \times (\boldsymbol{b}_j^* \times \boldsymbol{u})\right|\Phi \tag{14a}$$

with

$$\boldsymbol{b}_1^* = (\boldsymbol{b}_2 \times \boldsymbol{b}_3)/\boldsymbol{b}_1 \cdot (\boldsymbol{b}_2 \times \boldsymbol{b}_3). \tag{14b}$$

1.5.3. Graphical analysis of Frank's formula

An alternative to the above vector analysis of Frank's formula (which apparently cannot be extended to four-grid boundaries or higher) is a much simpler, physically more meaningful graphical treatment [96], applicable to any number of grids. It is based on the six equations represented by eq. (12) when rewritten in terms of vector components in an arbitrary x–y coordinate system inscribed on the boundary. Herein, when the dislocations of the j-th grid include angle ρ_j with the $+x$-axis, and their Burgers vector component parallel to the boundary ($_{\|}\boldsymbol{b}_j$) the angle α_j, then $_{\|}b_j \sin(\rho_j - \alpha_j)$ is the screw component of the dislocations in the j-th grid, and $_{\|}b_j \cos(\rho_j - \alpha_j)$ is the edge component parallel to the boundary. The component of the Burgers vector normal to the boundary may be designated $_N\boldsymbol{b}$ and, again, not the Burgers vectors but the "Burgers vector densities", $_{\|}\beta_j = {}_{\|}b_j/d_j$ and $_N\beta_j = {}_N b_j/d_j$, are the significant variables.

The graphical treatment under discussion represents Frank's formula in terms of three vectorial sums of the β's plotted as vectors in different directions in the said x–y coordinate system. Figure 11 illustrates the method by the example of a six-grid boundary. In figs 11(a) and 11(b) the $_{\|}\beta_j$'s are summed as vectors oriented at angles $(\rho_j - \alpha_j)$ and $(\rho_j + \alpha_j)$ against the x-axis, respectively, and in fig. 11(c), the $_N\beta_j$'s are similarly summed at angles ρ_j. According to fig. 11(a), Frank's formula demands that the resultant y-component, i.e. $\sum_{\|} \beta_j \sin(\rho_j - \alpha_j)$, vanish and shows that the resultant x-component, i.e. $\sum_{\|} \beta_j \cos(\rho_j - \alpha_j)$, equals $-2\Phi \boldsymbol{u}_N$ being the negative of twice the angle of twist.

As already indicated, those x- and y-components represent the screw- and the edge-component Burgers vector densities of the grids in the plane of the boundary. Physically, the import of fig. 11(a) is therefore, firstly, to forbid any net content of edge Burgers vector component in the boundary and, secondly, to state that the angle of twist is determined by the sum of the screw component densities. Evidently, the first is a very stringent but also eminently understandable condition: A net edge Burgers vector density in the plane of the boundary would generate the long-range dilatational and compressive stresses commensurate with the corresponding missing and extra atomic half-planes, and therefore is incompatible with Frank's formula. As to the screw-components, fig. 11(a) is simply descriptive, saying that the angle of twist, $\Phi \boldsymbol{u}_N$, is one half of the negative of the total screw component Burgers vector density.

Even simpler in physical interpretation is fig. 11(c), concerning the $_N\beta_j$'s. It is entirely descriptive, placing no restrictions on the normal components of the Burgers vector densities. In all dislocation rotation boundaries the $_N\beta_j$'s can thus be chosen without restriction but, once established, they determine the direction of the tilt axis and the angle of tilt, $\Phi \boldsymbol{u}_{\|}$, as indicated.

Figure 11(b), finally, is highly restrictive. Its function is to ensure freedom from long-range shear stresses via the permissible dislocation grid orientations. In regard to boundaries formed of two intersecting grids, this aspect of Frank's formula demands that each grid be normal to the other's Burgers vector component parallel to the boundary. That

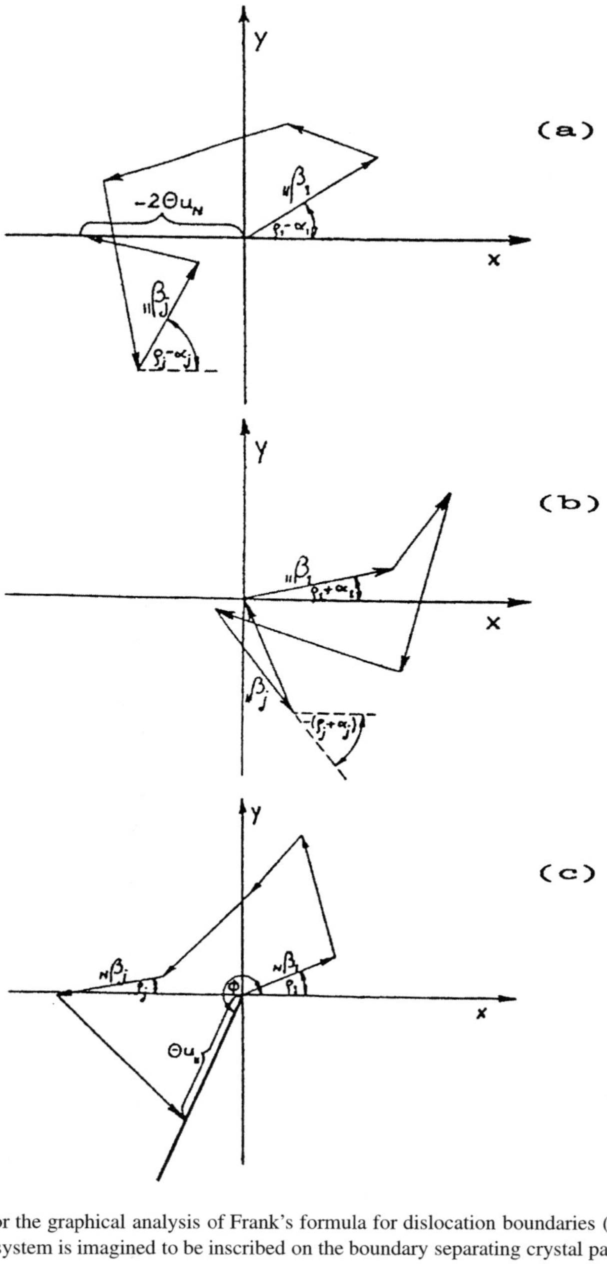

Fig. 11. Diagrams for the graphical analysis of Frank's formula for dislocation boundaries (eq. (12)), after [96]. The x–y coordinate system is imagined to be inscribed on the boundary separating crystal parts relatively rotated about $\boldsymbol{u}$ through angle Φ. In the diagrams, each vector represents one of the (in this example six) grids of "Burgers vector density" $\beta_j = b_j/d_j$ with d_j the dislocation spacing in the j-th grid. Subscripts $_{\parallel}$ and $_N$ designate components parallel and normal to the boundary. The positive x-axis makes angles ρ_j and α_j with the dislocation axis and Burgers vector components parallel to the boundary of the j-th grid, respectively. Hence $\rho_j - \alpha_j$ is the angle between dislocation axis and Burgers vector component in the boundary.

is evidently fulfilled for the cross-grid of screws already discussed but generally is a very restrictive condition.

In order to find specific solutions for three-grid boundaries in accordance with eqs (13) and (14), and for four-grid boundaries with crystallographic Burgers vectors, graphical methods are proposed in [96]. With more than four grids, simple adjustments of dislocation spacings permit satisfying Frank's formula under all conditions, and with six or more grids, solutions are no more unique but the same pre-selected arbitrary relative rotations on the same arbitrary planes can be generated in infinitely many different ways.

1.5.4. Energy of dislocation rotation boundaries

Accepting that the great majority of dislocation walls in uni-directionally strained materials are dislocation rotation boundaries which substantially obey Frank's formula and therefore may be classified as LEDS, the task of predicting which may form from case to case is far from achieved. This is due to the stupendous variety of possibilities all with similar energies for the same dislocation content. The dependence of the specific energy of low-angle boundaries (E_B) on lattice misorientation angle, Φ, is readily derived from the LEDS principle: Within the boundary the dislocations are stress-screened to about their near-neighbor distances, say, d, and therefore the dislocation energy per unit line length is given by eq. (1) with $R = d$, while the lattice misorientation angle is $\Phi \approx b/d$ and the dislocation line length per unit boundary area is $\approx 1/d$. Thus including the "redundancy factor" $M_r \cong 2$ that takes account of the excess dislocation content in the boundaries, mostly on account of mushrooming (see section 1.4) the boundary energy per unit area becomes

$$E_B \approx \Phi M_r Gb\{(1-\nu/2)/[4\pi(1-\nu)]\}\ell n(1/\Phi) \cong 0.1\Phi M_r Gb\ell n(1/\Phi). \tag{15}$$

This has a maximum at $\Phi = 21°$ when $\Phi\ell n(1/\Phi) = 1/e$ and in any event dislocation cores overlap. Thus for $\Phi \geqslant 21°$ we expect the asymptotic value of

$$E_{B,\Phi\geqslant 21°} \cong 0.035 M_r Gb. \tag{16}$$

The above derivation evidently leaves room for significant differences in the energies of boundaries constructable of the same dislocation densities and Burgers vectors. It is therefore impossible to predict, on the basis of the LEDS hypothesis in conjunction with eq. (15), which specific boundaries and hence dislocation cell structures will form under given deformation conditions. In general, a larger number of participating Burgers vectors permits more effective stress screening and should be favored for that reason. However, mutual interference among different slip systems argues in the opposite direction. Even so, two-grid boundaries are rather rare. They occur only when the two grids are either parallel, as in the family of two-grid tilt walls (fig. 4(g)), or normal to each other, as in the case of the cross grid of screw dislocations (fig. 5(a)). Both of these have been repeatedly observed (for a good example of the square cross grid see, for example, fig. 29 of [97]). The reason why two-grid boundaries are scarce is that non-orthogonal intersecting grids automatically react to generate a third grid. Namely, on account of their mechanical instability, individual four-fold dislocation nodes dissociate into pairs of three-fold nodes, drawing out between them a dislocation with a third Burgers vector as indicated in fig. 6(a) to (d). In the case of similar intersections among dislocation grids, a third grid is thereby generated as in figs 6(e) and (f), and this occurs even in "tangles", albeit the geometry is obscured.

Designating the Burgers vectors of the reacting dislocations by $\boldsymbol{b}_1$ and $\boldsymbol{b}_2$, the Burgers vectors of the new links are either $\boldsymbol{b}_{3+} = \boldsymbol{b}_1 + \boldsymbol{b}_2$ or $\boldsymbol{b}_{3-} = \boldsymbol{b}_1 - \boldsymbol{b}_2$, whichever is the smaller [98] according to the "law of the square of the Burgers vectors", i.e., $\boldsymbol{b}_3^2 < \boldsymbol{b}_1^2 + \boldsymbol{b}_2^2$. This choice is possible because the interacting dislocations can form links of either type as in fig. 6, i.e. the concept of 'attractive' and 'repulsive' dislocation intersections is a fallacy. The illustrated formation of alternating open and closed nodes in three-grid $\frac{1}{2}\langle 110\rangle\{111\}$ networks was first predicted by Frank [99] whose notation is used in figs 6(e) and (f). It has since been very widely observed, e.g., [94], – a testimony to the close approach to thermodynamical equilibrium in dislocation structures.

1.6. Cell structures

1.6.1. Scaling of rotation angles of cell walls and its relevance for similitude

For a long time it was doubted whether dislocation "walls" in cell structures were in fact dislocation rotation boundaries, as had for the first time been suggested in 1968 [100] and has ever since been a foundational tenet of the LEDS theory (initially still called the 'mesh length theory'). The first conclusive evidence for this contention was obtained by Langford and Cohen [101] who at the same time presented excellent measurements of the distribution of the associated misorientation angles and their evolution with straining. Since then there has been a very large body of evidence to affirm that cell walls are indeed rotation boundaries and to determine the rules governing their misorientation angles. A first survey of the corresponding experimental work, to a large extent accumulated by N. Hansen, D.A. Hughes and coworkers, was presented in [102]. It documented that cell walls are rotation boundaries obeying Frank's formula beyond any reasonable remaining doubt, thereby ruling out Mughrabi's model of work hardening [103,104]. Since then an additional wealth of experimental results has added proof to proof, e.g., [105–108].

Most decisively of all, Hughes, Chrzan, Liu and Hansen [109] have shown (i) that the frequency distribution of rotation angles across ordinary dislocation cell walls scales with the most frequent misorientation angle, Φ_{av}, (ii) that Φ_{av} in turn rises with strain (i.e. stress), and (iii) that the distribution function appears to be "universal", being independent of the deformation temperature, stacking fault energy, purity, ease of cross slip and ease of climb.

This is exactly the behavior expected from the LES hypothesis, since the energy of the cell structure is a function of morphology (i.e. dislocation cell size and shape distribution) and average rotation angle Φ_{av} (which for force equilibrium is bound to rise with stress), but for the remainder is proportional to $M_{\mathrm{r}}Gb$ in accordance with eq. (15). Thus there can be a dependence on all other variables only indirectly through M_{r}, G and b, and there is no reason for such dependence except perhaps through M_{r} which presumably can vary no more than between 1 and 2 or so. Besides, the gradual increase of Φ_{av} at same functional distribution of rotation angles, is a splendid example of "similitude", first enunciated in 1962 [110]. This is the concept that for a given type of LEDS, the structure will scale inversely with the effective stress, $(\tau - \tau_0)$, for the reason that the equilibrium is maintained by the interacting dislocation stress fields which decrease with the inverse of the distance from their axes.

Observationally, in TEM, increasing rotation angles give rise to a continuous evolution of the cell wall structure which those boundaries represent. At low dislocation densities, on account of mushrooming in conjunction with the initially small driving stresses, low-angle dislocation rotation boundaries, i.e. cell walls, tend to have significant width and to be composed of plentifully jogged and kinked dislocations without obvious regularity. At this stage their true nature may not be recognized and they are often mistaken for the kind of "dislocation tangles" characteristic for Stage I and shown in fig. 10. With increasing stress and resultant decreasing spacing, the dislocation interactions become stronger, and the boundaries sharpen but apparently without change of the admixture of 'redundant' dislocations, i.e. if M_r should change, it is not obvious in which direction [33]. Gradually, the individual dislocations in the networks become difficult to resolve and then they increasingly merge. At the stage in which individual dislocations can no longer be recognized, we tend to speak of the boundaries as sub-boundaries, and eventually the dislocation rotation boundaries are indistinguishable from ordinary grain boundaries whose appearance is independent of rotation angle.

Nowhere in the evolution described is there a sharp transition from dislocation rotation boundaries to grain boundaries, and the analysis in terms of dislocations remains valid at any value of Φ, albeit it becomes much more intricate [111,112]. Even so, LEDS considerations continue to apply, on occasion to amazing detail [113–116]. At the other end of the scale, i.e. small misorientations, the character of "tangles" as being parts of dislocation rotation boundaries obeying Frank's formula cannot be ascertained in TEM and is therefore often doubted, as already indicated. Here methods permitting observations on a larger scale may be invoked to prove the point. Examples found in [117] include [118] and [119]. It may be added here that besides the ordinary cell walls, increasing strains cause an increasing admixture of longer boundaries, dubbed GNBs for 'geometrically necessary boundaries' or CBBs (for cell block boundaries). These delineate 'cell blocks' each with their unique selection of slip systems short of the five which would be required for homologous deformation. More will be said about them in table 1 and in section 2.9.

1.6.2. Mosaic block structures form on account of wall-end stresses

Finite dislocation rotation boundaries, unlike their infinite counterparts discussed in connection with Frank's formula eq. (12), have long-range stresses, as has been known for a long time [49,52,94,120]. This is in accord with the previously discussed rule that at large distances the stress field of any assembly of parallel dislocations is that of a single dislocation with Burgers vector $\boldsymbol{b} = \sum \boldsymbol{b}_j$. Hence, at close range the dislocations in any boundary obeying Frank's formula are stress-screened to about the near neighbor distance, virtually independent of the extent of the wall, and at large distances individual cell walls have stresses as of one dislocation for each grid [49,52,94,120,121]. For tilt walls, the end-stresses as well as other properties are especially well known through Li's exhaustive study of infinite [122] and finite [121] tilt walls in isotropie media. Pertinent additional results in connection with simple models of dislocation cells were obtained in [42–47] and, more recently, by Jackson and Nabarro [123], but more needs to be done. One facet herein is the fact that the long-range end-stresses of terminating boundaries are somewhat relieved if the rotation is not strictly rigid but Φ is permitted to decrease toward the circumference, i.e. edge of the boundary. Several examples of this effect may already be found in [94].

From modeling results [45,46] it is evident that isolated cells, morphologically resembling pebbles in concrete, have a rather high energy as compared to cell aggregates. This is a powerful reason why they are not observed in nature but mosaic block structures instead. Besides, isolated cells are unstable since they can shrink into nothing by simple glide. In the process of shrinking the angular misorientation will increase, of course, as the dislocation distances in the boundaries decrease. The lowest energy for mosaic block structures was found for rectangular cells sharing the same rotation axis and with the same rotation angle, alternately rotated in a left-right-left-right checkerboard pattern [46], as in fact repeatedly (but certainly not universally) observed.

As will be argued in part 2, in cell structures the flow stress is controlled by the supercritical Frank–Read bowing of always the longest individual dislocation links in the walls. This mechanism does not depend on the cell size because within terminating boundaries the dislocations play the role of misfit dislocations even while their host walls may at long range act as stress centres. This dual aspect is readily understood by mentally making a boundary through a cutting, rotating and rejoining operation. For a bicrystal boundary, this entails slicing the specimen in two along the intended boundary plane, rotating the two sides relative to each other by angle Φ about the selected rotation axis $\boldsymbol{u}$, trimming the two parts so that they fit back together, and rejoining. Dislocations will arise spontaneously at the interface so as to accommodate the misfit, thereby lowering the free energy, – exactly in the same manner as dislocations in epitaxy and at phase boundaries. If so produced, the resulting boundary will be free of long-range stresses and the dislocation grids will obey Frank's formula (eq. (12)).

Long-range stresses other than due to wall-ends, and the corresponding deviations from Frank's formula, are caused by superimposed elastic strains that differ from one side to the other. For example, a simple rotation about the normal of a cut that passes completely through a crystal, would cause a twist boundary. If it is parallel to a $\{111\}$ plane in f.c.c., the result would be a hexagonal $\frac{1}{2}\langle 110\rangle\{111\}$ screw dislocation network as in fig. 5(b). On the other extreme, the same cut without relative rotation but with one side uniformly elastically compressed would also result in a hexagonal network but composed of edge dislocations. Geometrically the two networks would be indistinguishable in TEM. In fact the edge dislocation network would be similarly indistinguishable from an interfacial network on $\{111\}$ between, say, copper and silver. Either way, edge dislocations will accommodate the lattice constant difference. However, while an epitaxial edge dislocation network is stable, this is not so for the contemplated edge dislocation network since it has the stresses of three mutually rotated, uniformly spaced dislocation pile-ups.

Based on the above considerations, dislocation cell structures should be modeled by assembling pieces of dislocation rotation walls and searching for the patterns of minimum stored energy at given dislocation density. Mathematically, the stresses of area elements of rotation boundaries could be modeled as disclinations. Theorists are urged to explore this route toward modeling of dislocation cell structure in lieu of abortive SODS modeling.

1.7. Taylor lattices

1.7.1. Basic properties of extended Taylor lattices

The first to make a substantive contribution to our understanding of the Taylor lattice, after its introduction and use as the model for work hardening by Taylor [54], was Masing [124]. He pointed out that reversible displacements of the dislocations under low stresses would falsify the elastic constants, depending on the ratio s/d_f where s is the dislocation spacing within the same mat and d_f the distance between adjacent mats (see fig. 7). However Masing's numerical results, as well as Taylor's own on which they were based, were marred by a faulty solution for the stresses of the edge dislocation which Taylor had obtained from Love's treatise on elasticity (compare [54]). This error was corrected by Nabarro [125] who also derived the dependence of relative displacements between the $\perp$ and $\top$ sublattices as a function of applied stress and s/d_f. Nabarro's results were somewhat expanded, including to anisotropic elasticity, in [126] from which figs 12 to 14 are extracted.

The stress, τ_{TL}, with which an infinitely extended Taylor lattice resists relative displacements (x/s) of the $\perp$ and $\top$ sub-lattices is, physically, the equal and opposite of the applied stress minus the friction stress. Analytically it is given by [125]

$$\tau_{TL} = F\tau_{dip} = \tau_{dip} \sum_{n=-\infty}^{+\infty} \frac{\pi^2(x/d_f + ns/d_f)}{\{\cosh(\pi/2)(x/d_f + ns/d_f)\}^2}, \tag{17}$$

where $\tau_{dip} = A_{\perp}/4d_f$ is the dissociation stress of the dipole of eq. (6). In terms of the more suitable variable $\xi/s = (x - s/2)/s$, the function $F = \tau_{TL}/\tau_{dip}$ is shown in fig. 12, albeit in [126] and hence in figs 12 and 13 τ_{dip} is designated τ_{LF} for "loop flipping". The equilibrium position is at $\xi/s = 0$ and the "on-top" position of unstable equilibrium is at $\xi/s = 0.5$, i.e. at $x = 0$.

When the applied stress exceeds the friction stress by more than $(\tau_{TL})_{MAX}$, the extremum on each curve in fig. 12, the sublattices slide past each other irreversibly. As seen, and summarized in fig. 13, the corresponding value $F_{max} = (\tau_{TL})_{MAX}/\tau_{LF}$ is negligible for $s/d_f < 0.5$, then increases sharply to 2.5 at $s/d_f = 2$ and saturates at 2.83 for $s/d_f > 3$ Returning to fig. 12, note that with rising s/d_f the critical configuration at $(\tau_{TL})_{MAX}$ more and more closely approaches parallel but well separated dipolar walls, albeit with modified distances between the $\perp$ and $\top$ dislocations in each wall. Those distances may be extracted from fig. 12 or fig. 14, remembering that the "on top" position is $\xi/s = 0.5$.

The remaining unknown was the s/d_f value of minimum energy, i.e. the configuration which should be approached in accordance with the LEDS hypothesis. Intuitively, this had been assumed to be close packing of the dislocations at $s/d_f = 2/\sqrt{3} \approx 1.15$. In this form of close packing the hexagon B in fig. 7 would be even-sided with the dipolar angle 60° and $(\tau_{TL})_{MAX}/\tau_{LF} \approx 0.8$, as indicated in figs 12 and 13. However, according to Neumann's computer simulation [127], which will be more fully discussed in the next section, the minimum energy configuration approximates the other possible close packing arrangement for which hexagon A is even-sided, and indeed lies beyond it for Taylor lattices of restricted size. Thus, based on Neumann's simulation as well as the LEDS hypothesis, one will expect Taylor lattices to have $s/d_f \geqslant 2\sqrt{3} \approx 3.5$ with dipolar angles smaller than or equal to 30° and $F_{max} = 2.8$ or up to 2.831 in accordance with fig. 13.

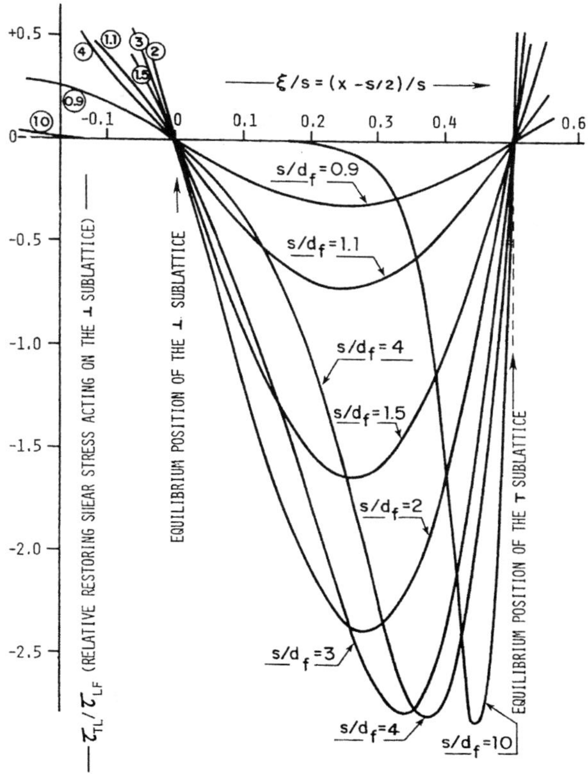

Fig. 12. The relative reactive shear stress τ_{TL}/τ_{LF} (where $\tau_{LF} = \tau_{dip} = A_{\perp}/4d_f$ of eq. (6) with which an infinitely extended Taylor lattice resists relative displacements (ξ/s) of the $\perp$ and $\top$ sub-lattices as a function of s/d_f (compare fig. 7). The stable equilibrium position is at $\xi/s = 0$ and the unstable "on-top" position at $\xi/s = 0.5$. When the relative magnitude of the applied stress minus friction stress exceeds $(\tau_{TL})_{MAX}/\tau_{LF}$, the extremum on each curve, the sublattices slide past each other irreversibly. (Figure 3 of [126].)

1.7.2. The Stage I/Stage II LEDS tansformation of Taylor lattices through "unpredicted" glide

The initial choice of $s/d_f = 2/\sqrt{3}$, i.e. the (B) form of close packing [126], was due to overlooking the second (A) form of close packing with $s/d_f = 2\sqrt{3}$ that was pointed out by Neumann [127]. While both of them, as indeed all Taylor lattices regardless of s/d_f, are LEDSs, being stress-screened to the near-neighbor distance, more or less, there can be no question that $s/d_f = 2\sqrt{3}$ is preferred by the LEDS hypothesis. This is so because for same dislocation density, $\rho = 1/sd_f$, form (A) is associated with the much smaller tensile/compressive stresses in the slabs between the $\perp$ and $\top$ mats, and thus lower energy per unit length of dislocation line, i.e. smaller R^* values. Moreover, it had already then been realized by others [128–131] that these extension/compression stresses trigger "unpredicted" glide in the same manner as primary glide would operate in a tensile or compression test, and that this is the cause for the Stage I to Stage II transition in

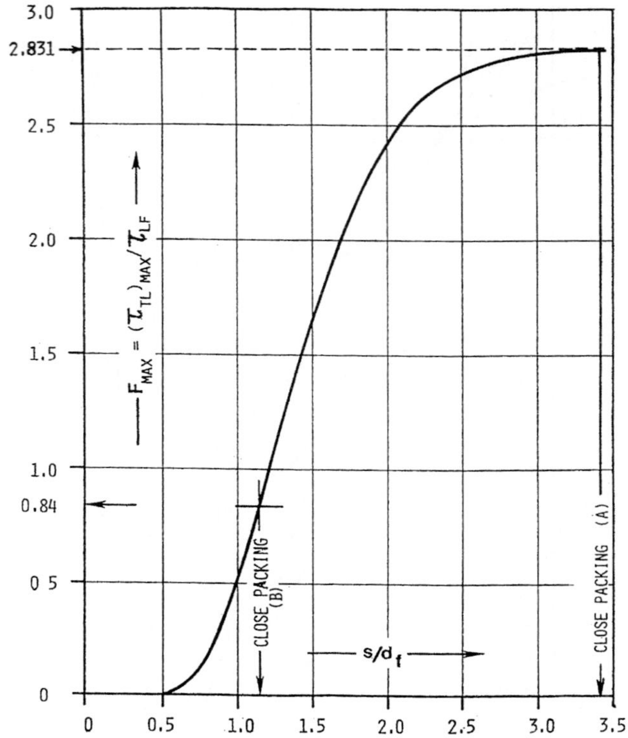

Fig. 13. The values of $F_{\max} = (\tau_{TL})_{MAX}/\tau_{LF}$ (i.e. the yield stress of an infinite Taylor lattice in units of $\tau_{LF} = \tau_{dip} = A_{\perp}/4d_f$ of eq. (6) as a function of s/d_f in accordance with fig. 12. The indicated close-packed arrangement at B (compare fig. 7), with a 60° dipolar angle at $s/d_f = 2\sqrt{3}$ and $F_{\max} \approx 0.84$, had erroneously been assumed to be stable. Neumann's computations [127] revealed the other close-packed arrangement, A of fig. 7, with 30° dipolar angle at $s/d_f = 2\sqrt{3} = 3.46$ and with $F_{\max} \approx 2.83$ to be stable. (Figure 4 of [126].)

f.c.c. metals. In terms of the LEDS theory, the corresponding LEDS transformation was examined in [132]. The underlying geometry is illustrated in fig. 15(a) and the resulting emergence of a rotated Taylor lattice through reactions between primary and unpredicted dislocations is shown in fig. 15(b).

Microscopically, the Stage II configuration of fig. 15(b) assumes different appearances, depending on the three-dimensional dislocation mobility. With a moderate or strong three-dimensional mobility (i.e. in marginal or full wavy glide), the dislocations somehow cluster together into 'carpets' as expected from the LES hypothesis in conjunction with eq. (1). In that case the aspect of parallel tilt walls implied in fig. 15(b) is accentuated, causing a structure of parallel slabs with alternating rotation about the axis normal to the primary Burgers vector as in fig. 15(b). This "Yamaguchi–Taylor roller misorientation" [133,134] was discovered, even before dislocations, through the resulting asterism (see pp. 196–199 of [135]). Examples of the carpet structure as characteristic of Stage II in wavy glide are easily found in the literature, e.g. among the earliest by Foxall, Duesbery and Hirsch [136],

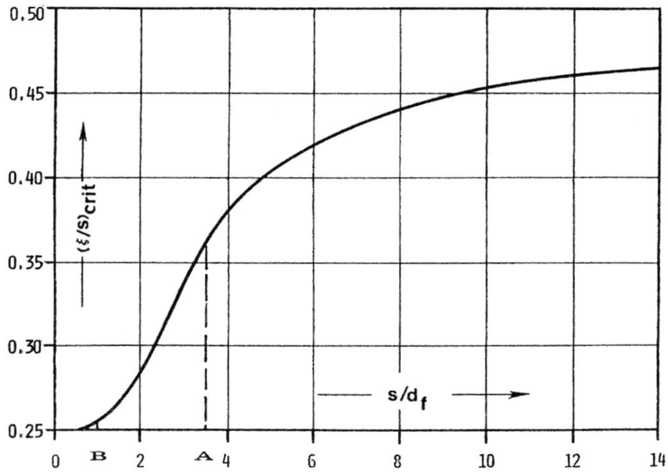

Fig. 14. The critical displacements of the sub-lattices, $(\xi/s)_{crit}$, i.e. ξ/s at $(\tau_{TL})_{MAX}$, as a function of s/d_f in accordance with figs 12 and 13. With increasing $(\xi/s)_{crit}$ the distortion of the Taylor lattice into a set of parallel dipolar walls becomes accentuated. (Figure 5 of [126].)

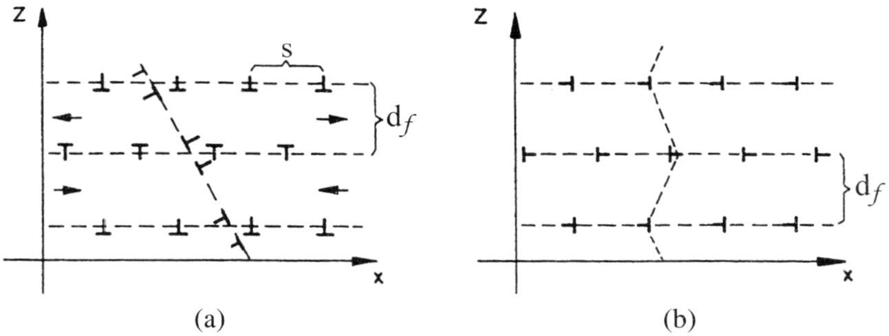

Fig. 15. Idealized geometry of the Stage I to Stage II LEDS transformation, namely from a simple Taylor lattice (a) into what may be seen either as a rotated Taylor lattice (b), or a set of tilt walls parallel to the active slip plane with alternating sense of rotation. The transformation is caused by "unpredicted" slip that is triggered by the alternating tensile and compressive stresses in the slabs of (a). Albeit, the geometry depicted is idealized since the intersection of the slip planes involved is not parallel to the edge dislocations so that networks rather than sets of parallel edge dislocations form. In a manner not yet well understood, with a modicum of three-dimensional mobility as in wavy-glide materials, the dislocations cluster into the 'carpet structure' of alternating tilt walls parallel to the slip plane and thereby decrease their screening distance eq. (1). In TEM this structure is seen as parallel strips of alternating contrast as at lower middle right of fig. 1(b). In planar glide, lacking 3-D mobility, the dislocations remain in the form of more or less well-formed rotated Taylor lattices (a). These, as all Taylor lattices, are commonly misinterpreted as 'tangled dislocations', of which the lower right of fig. 1(a) gives examples. (Figure 4 of [132].)

Moon and Robinson [137] and Steeds [138] in niobium, silver and copper single crystals, respectively, and in fig. 1(b) it is clearly seen at lower right centre. However, in pure planar glide, when three-dimensional mobility is too low, the structure retains its Taylor lattice aspect of more or less evenly distributed dislocations, as seen in fig. 1(a) especially below and to the right of the diagram diagonal.

The critical stress for the Stage I to Stage II transformation discussed is closely related to the friction stress, τ_0, at the start of the deformation. Following the derivation of [132] for the case of tension in mildly modified form, a first-order estimate of it is found as follows: The elastic strain in the slabs before the transformation, due to the extra and missing half-planes of atoms represented by the dislocations, is

$$\varepsilon_{\rm tc} = b/2s \tag{18}$$

for an associated equivalent resolved shear stress on the "unpredicted" glide plane of

$$\tau_{\rm bb} = m_{\rm c} E b/2s, \tag{19}$$

where E is Young's modulus and $m_{\rm c}$ is the Schmid factor of the unpredicted system relative to the tensile/compressive stress $\tau_{\rm bb}$. The unpredicted glide, and thereby the corresponding LEDS transformation, will be triggered at that applied stress, $\tau_{\rm trans}$, at which the resolved shear stress on the unpredicted system first exceeds the friction stress, τ_0, through the combined action of $\tau_{\rm bb}$ and the applied stress. Thus at the critical value of $\varepsilon_{\rm trans} = b/2s_{\rm trans}$ it must be (see [132])

$$\tau_{\rm trans} = \tau_0(1 - m_{\rm s}/m_{\rm p}), \tag{20}$$

with $m_{\rm s}/m_{\rm p}$ the ratio of the Schmid factors of the unpredicted and primary systems relative to the applied tensile stress. Therefore by combining eqs (19) and (20) we obtain

$$b/s_{\rm trans} = (2\tau_0/E)(1/m_{\rm c} - m_{\rm s}/m_{\rm p}m_{\rm c}). \tag{21}$$

The value of $s_{\rm trans}$ is related to $\tau_{\rm trans}$ through eq. (6) via its associated $d_{\rm f\,trans}$ and $F_{\rm max}$ values, i.e.

$$F_{\rm max} = (\tau_{\rm TL})_{\rm MAX}/\tau_{\rm LF} \equiv (\tau_{\rm TL})_{\rm MAX}/\tau_{\rm dip} = (\tau_{\rm TL})_{\rm MAX}(4d_{\rm f})/A_{\perp}. \tag{22}$$

Thus with eq. (21) and, in this case, $F_{\rm max} = 2.83$

$$\begin{aligned}\tau_{\rm trans} &= \tau_0 + (2.83/4)A_{\perp}/d_{\rm f\,trans}\\ &= \tau_0\{1 + [2.83G/\pi(1-\nu)E](1/m_{\rm c} - m_{\rm s}/m_{\rm p}m_{\rm c})\}.\end{aligned} \tag{23a}$$

Numerically, $m_{\rm c} \approx 0.4 \approx G/E$, $\nu \approx 0.3$ and $m_{\rm s}/m_{\rm p} \approx 2/3$, say, for

$$\tau_{\rm trans} \approx \tau_0(1 + 0.4). \tag{23b}$$

The physical import of the above is as follows: In simple Taylor lattices generated through unidirectional single glide, which are extensive and most likely restrained by confining domain boundaries (see section 1.7.4), unpredicted glide will be triggered at, say, $(1.4 \pm 0.3)\tau_0$. As a first, perhaps transient, step they are thereby transformed into Taylor lattices with two, or through reaction at the nodes three, Burgers vectors. In this transformation from one type of LEDS into another, the dislocation density remains substantially constant. Any associated increase in stored mechanical energy is provided by relief of τ_{bb} in conformity with the LEDS hypothesis, and the transformation reduces R^*, i.e. the energy per unit length of dislocation line. As already indicated in conjunction with [136–138] and fig. 1, and further discussed in part 2, the theoretically anticipated dislocation structures are in good accord with experimental observations, and so is eq. (23b) for the Stage I to start of Stage II transition [132].

1.7.3. Dissociation of unconfined Taylor lattices into dipolar walls

In his work already cited, P. Neumann [127] studied the stable configurations of various groupings of parallel edge dislocations with only one crystallographic Burgers vector and net zero Burgers vector content (i.e. $\sum \boldsymbol{B} = 0$ for any Burgers vector circuit about a representative group of neighboring dislocations). His ultimate motivation was to understand the "loop patch" or "vein" structures seen in fatigue. Neumann simulated the vein structure through diamond-shaped bundles. Making use of a previously available numerical algorithm he determined the dislocation equilibrium positions in bundles of up to 9×9 elementary quadrupoles (see fig. 16(a)). Neumann's computations revealed the

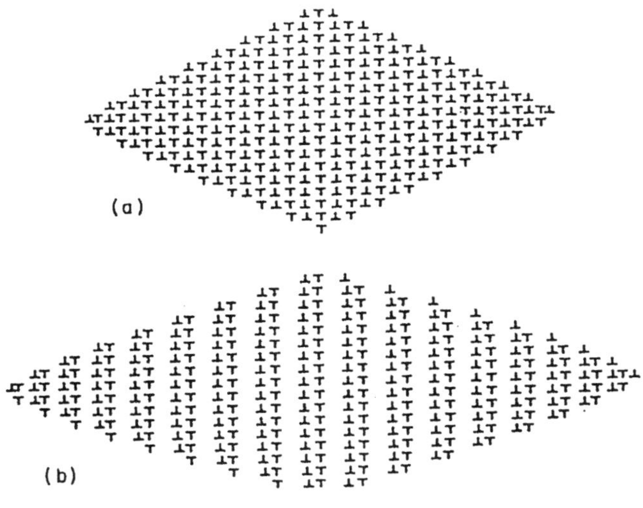

Fig. 16. Computer simulation of the equilibrium dislocation positions in a diamond-shaped Taylor lattice, modeling a fatigue loop patch vein (or bundle). It is composed of 81 elementary quadrupoles in a 9×9 configuration (i.e. both axes comprise 9 quadrupoles) according to Neumann [127]. (a) Configuration without applied stress; (b) the same assembly but under the critical applied stress of $\tau_{dis} = 0.425A_{\perp}/4d$. The increased energy (and thus R^* value) of the (b) configuration and its impending instability are graphically evident from the larger dislocation spacings compared to (a). The dipolar pairing of the dislocations directly before the dissociation of the Taylor lattice into the corresponding walls reflects the dislocation positions at the minimum of the curve for $s/d_f \approx 3.3$ in fig. 12, but is not numerically exact on account of the variations of tetragonality within the diamond. (Figure 6 of Neumann [127].)

ratio s/d_f, (in his nomenclature the "tetragonality" $t = 2s/d_f$) to be the most important parameter. Not only does the tetragonality vary moderately within any one bundle, but it is a rather sensitive function of applied stress and size of aggregate. With size of bundle as well as with increasing applied stress the tetragonality rises until it reaches a point of instability and the lattice disintegrates into dipolar walls (fig. 16(b)). Further, the yield strength of the bundles is always lower than the dipole dissociation stress, i.e. only about 40% thereof for the 9×9 bundle in fig. 16. These results are in stark contrast with the relatively high shear strength of the infinitely extended and/or confined Taylor lattice as reflected in the $F_{max}(s/d_f)$ curve of fig. 13, namely by the factor $F_{max}(s/d_f \geqslant 3) \approx 2.83$.

The following conclusions may be drawn from the evidence presented in figs 13 to 16.

- Unconfined Taylor lattices of small extent behave very differently from extended/confined Taylor lattices.
- The differences between these two types of Taylor lattices are due to the different boundary constraints, namely permitting or inhibiting lateral spread of the lattices. In summary: By minimizing free energy, in effect Neumann based his study on the LEDS hypothesis, i.e. that a fixed dislocation population will assume the configuration of minimum energy [127]. For practical reasons he restricted his study to diamond-shaped bundles and walls of similar parallel edge dislocations, both of which are LEDS's. On this basis he found that parallel edge dislocations of alternating sign will form bundles up to some critical size, depending on applied stress, and that beyond a combination of critical sizes and stresses these will dissociate into sets of parallel dipolar walls. The basic features of Neumann's analysis are in harmony with observations on the vein structure/ladder structure LEDS transformation in fatigue [127]. This, then, is a second LEDS transformation of Taylor lattices in addition to that in fig. 15.

1.7.4. Kinematics of Taylor lattice formation and densification

In a remarkable series of experiments, Mitchell and coworkers [139–145] studied the low-strain dislocation behavior in carefully prepared, tensile-tested single crystals of solid-solution Cu–Al. This alloy provides an extreme example of planar glide behavior, exhibiting no, or even negative [141] work hardening in Stage I. The initial motivation of the study had been to test, under the most favorable conditions possible, Frank's theory of dislocation reflection [146]. According to this theory, which predated the discovery of the Frank–Read source mechanism, dislocations multiplied when encountering glide dislocation obstacles at relativistic speeds. In that case, part of the excess kinetic energy was assumed to be converted into new dislocations of opposite sign. However, the results obtained by Mitchell and coworkers showed that dislocations reached at most about 10% of the speed of sound [141], with an insignificant relativistic energy increase of $(1 - v^2/c^2)^{-1/2} - 1 <\approx (1 - 0.01)^{-1/2} - 1 = 0.5\%$. Ergo, "dynamic" effects in which the kinetic energy of the dislocations is assumed to play a role are ruled out by experimental evidence. Mutatis mutandis, dislocations always move slowly enough to be constantly in momentary glide force equilibrium. Hence at constant stress, after virtually instantaneous glide equilibration, dislocation structures change only through (1) climb, (2) thermally activated changes in τ_0, or (3) non-dislocation structure changes, e.g., the formation or dissolution of precipitates.

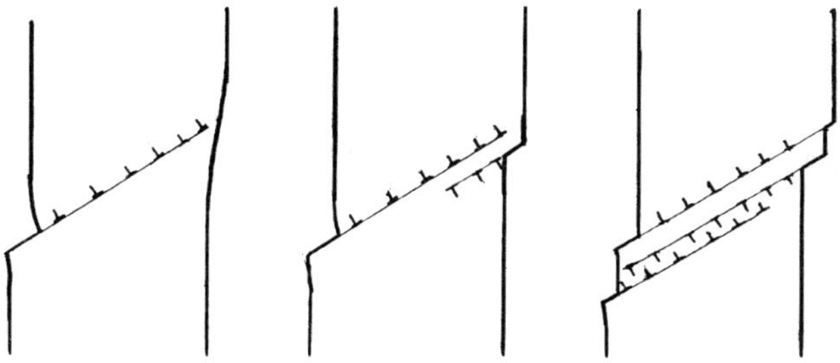

Fig. 17. Taylor lattice formation in a planar-glide single crystal according to Mitchell et al.: A train of edge dislocations entering the crystal from a 'terrace source' at lower left in (a), triggers the activation of a new terrace source as it approaches the free surface (b). The edge dislocations of opposite sign from the new source form a dipolar mat with those of the first train and in turn repeat the process on approaching the left free surface in (c). It is believed that the continuation of the process causes the spread of Lüders bands. However, the driving force in (c) is much less than in (b) and it is possible that preferentially pairs of dislocation trains form, rather than larger multiples, as also suggested by fig. 21.

Apart from this insight, the references cited contain the best ever record of the formation of dipolar mats and thence Taylor lattices, albeit lacking TEM dislocation images. Throughout Stage I, no evidence of interior sources of primary dislocations was found, but towards the end of Stage I a minor amount of "unpredicted" slip, as manifested by short slip lines (e.g., fig. 7 of [139]) and short sequences of etch pits (particularly studied by Hockey and Mitchell [140]), was observed. These are the tell-tale signs of the mechanism of fig. 15 but at the time remained unexplained.

As to primary glide, accepting the observations to have general applicability for planar-glide Stage-I single crystals, the dislocation behavior is entirely counter-intuitive, as illustrated in fig. 17: The process begins with the release of edge dislocations from a surface "terrace source" in large groups (e.g., ≈ 200). These move into and through the crystal, wherein several or many similar sequences, on planes too closely spaced for clear resolution, may follow each other as a long train of up to thousands of dislocations. Almost perversely, after traversing the crystal, a dislocation train approaching the opposite surface does not simply begin to spill out dislocations and thereby initiate a process of dislocations entering and emerging on pairs of matching, gradually deepening slip lines on opposite sides of the crystal. Instead, before actually reaching the surface, the stress concentration of the leading dislocations of a train (fig. 17(a)) triggers the generation of edge dislocations of opposite sign from a new "terrace source" on a parallel slip plane that is offset by the passing distance from the first train (fig. 17(b)) and thereby brings the first train to a halt. As the gradually lengthening train of new dislocations of opposite sign passes along the first train, the dislocations form low-energy dipoles via successively changing partners until the second train approaches the first side of the crystal. And, again, the dislocations may not spill out of the surface but by the same mechanism as before trigger a third terrace source (17(c)), although now with reduced driving force, and so on. The result is a gradually widening Lüders band within which slip by a (nearly) fixed amount occurs on each newly activated slip plane, and essentially all of the dislocations, once formed, remain trapped in the crystal.

The "terrace sources" appear to be nothing but surface steps. The generation of the dislocations against their theoretically very high surface image forces is assisted by stress concentrations at the steps and, even more so, by a "push" from the newly formed surface ledges because on account of oxygen adsorption the surface free energy has a negative value [145].

Hockey and Mitchell [140] concluded (using a now unfamiliar nomenclature instead of the term "Taylor lattice" that was coined only later [126]), that they obtained "the first clear example of a moving forest of edge dislocations of the type envisaged by Taylor". Regrettably, they could provide only indirect, even though very powerful, evidence for their mechanism. A more recent TEM micrograph by Müllner, Solenthaler, Uggowitzer and Speidel [147], reproduced as fig. 18, fills that gap. It shows a train of dislocations in a TEM foil of nitrogen-alloyed austenitic stainless steel coming from the lower left which has triggered a "terrace source" at upper right. Before the obvious three semicircular loops, the terrace source had already emitted a sequence of roughly thirty dislocations which have formed near-edge dipoles with the initial train. Interestingly, self-climb (facilitated by the nearby surface) has begun to locally pinch off some of the dipoles, most evidently near the left edge of the dipole sequence which in three dimensions would become part of a Taylor lattice, and to shorten the resultant presumed interstitial-type loops. This is much more readily possible in thin foils than in bulk on account of rapid pipe diffusion of vacancies from the surfaces along the dislocation cores, in a manner that is highly reminiscent of already mentioned earlier climb observations in TEM foils [66,85].

Figure 19 extracted from [85] shows that the transformation of HEDS dislocation sequences into LEDSs through the generation of the appropriate "forest" dislocations is not restricted to edge dipoles as in fig. 18 but also occurs for screws. In contrast to fig. 18, here the complementary screw dislocations belong to a different but coplanar slip system. They enter on the same atomistic glide plane and begin to form a twist boundary.

The above excursion into consideration of dislocation reactions in thin foils has yielded impressive examples of dislocation trapping under mechanical equilibrium conditions (i.e. while dislocation mobility is essentially limited to glide). In these, and similarly in all other cases, such trapping occurs via the formation of LEDS's of such low energy that the driving force for the elimination of the dislocations (in the case of fig. 18 even through the geometrically simple egress from close-by free surfaces) is insufficient for their removal. In fig. 18 that very low stored energy is manifested through the very close spacing of the dipoles. However, the examples also demonstrate the occasionally very powerful influence of free surfaces, whose role especially in the mechanism of Mitchell et al., as in figs 17 and 18, is central. Hence, slip-line evidence must be treated with great caution since it may not reveal conditions in the bulk of the material.

Yet work hardening is pre-eminently a property of bulk materials and we must understand Taylor lattices in these. Specifically, in bulk, the law of the conservation of Burgers vectors demands that sources must emit $\top$ and $\perp$ dislocations in equal numbers. How, then, do Taylor lattices arise in polycrystals? The answer proposed herewith, depicted in fig. 20, appears to be the only possible morphology by which interior Taylor lattices can be generated and perpetuated: Conceptually, the process begins with the emission of a net-zero-Burgers-vector sequence of dislocations from an interior source, S, within a volume element bounded by some imagined impenetrable surfaces normal to the slip plane. The

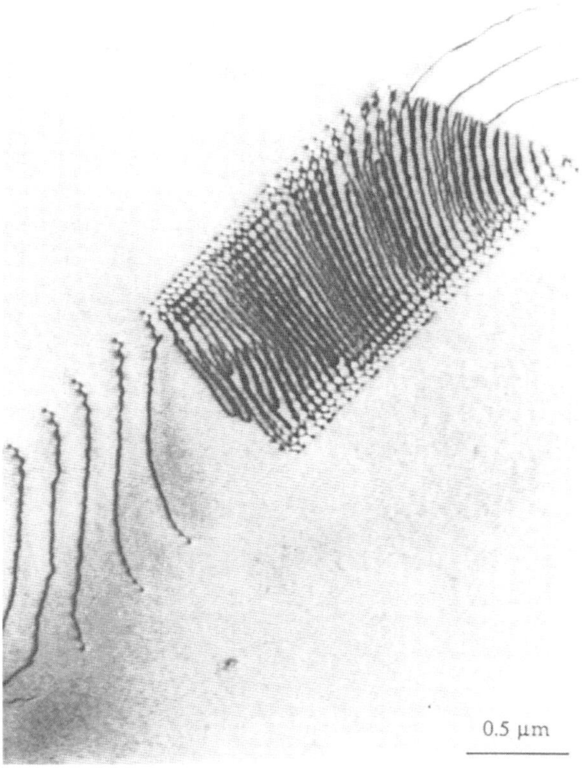

Fig. 18. Example of the process indicated in fig. 17(b) in nitrogen-alloyed 316-type stainless steel. The initial dislocation sequence in near-edge orientation progresses from the lower left and ends at the foil surface, indicated by the straight discontinuous cutoff near the upper right. The complementary dislocations of opposite sign are produced by the glide loop sequence which is emitted from the free surface on a parallel glide plane displaced approximately 0.1 μm from the first towards the observer. Five still unpaired primary dislocations are seen at lower left, followed by the dark area of closely spaced dipoles. The process indicated reduces the free energy of the sample but remarkably enough not through generation of opposite-sign dislocations on the same glide plane which would eliminate the dislocations, but through dipole formation. However, that final reduction of energy can take place in the stress-free state through climb. This has already pinched off the ends of some dipoles into prismatic loop configuration, causing them to draw away from the foil surface through climb, mainly at the lower edge of the track as seen, but a few also from the upper track edge. (Figure 2 of Müllner et al. [147].)

sequence would deform the volume element as exaggeratedly shown in fig. 20(a). If a similar sequence were to arrive from a neighboring volume element on a parallel slip plane (fig. 20(b)), the elastic strain at the imagined interface would be converted into two steps as the dislocations penetrate into each other's territory, while forming dipoles (fig. 20(c), comparable to fig. 17(b)).

The volume elements considered are "domains" and the dislocation sources (indicated by ⊗ symbols in fig. 20(d)) are believed to be situated in domain boundaries (indicated

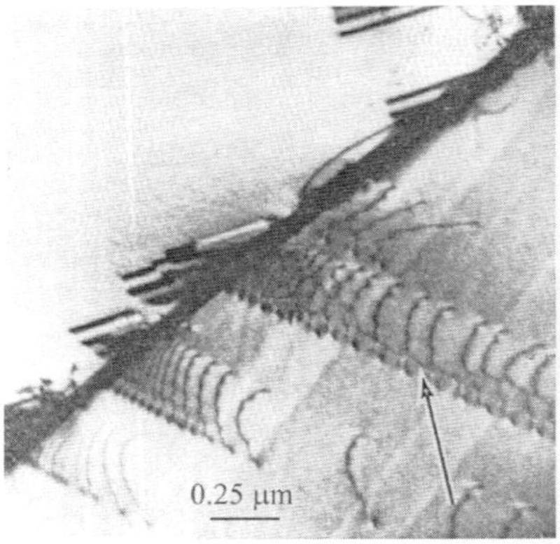

(a)

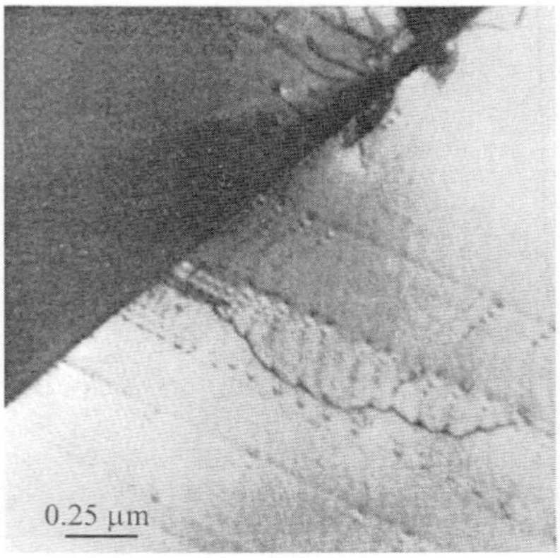

(b)

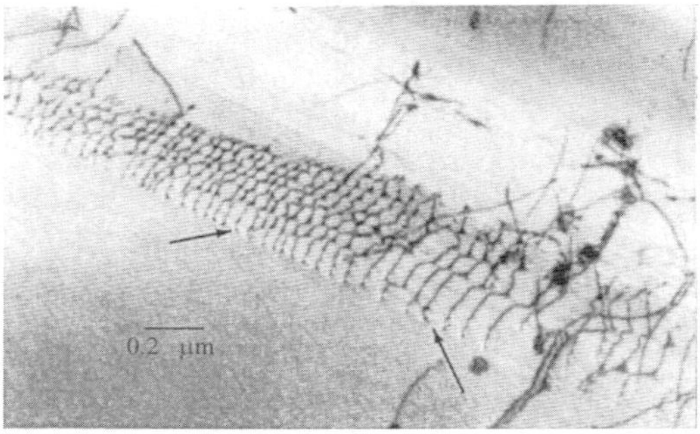

(c)

Fig. 19. The counterpart of fig. 18 but for the case of screw dislocations that form a twist boundary. The "forest dislocations" (as they would erroneously be called in non-LEDS theories) are drawn into the foil by the stresses of the HEDS pile-up of near-screw dislocations at a grain boundary in a TEM foil of 304 stainless steel. In (a) a single "forest" dislocation has entered from the edge of the foil (indicated by the arrow). Unlike the case of fig. 18, it lies on the same plane as the pile-up but has a different Burgers vector and thus belongs to a different but coplanar slip system. In (b) the "forest" dislocation has been brought into contrast through tilting which also reveals triangular stacking faults through the incipient dissociation of the four-fold into two three-fold nodes comparable to fig. 6. The same process at a later stage but in a different foil is shown in (c). The arrows here indicate the entering points at the surface of one of the "forest" dislocations. (Figures 15 and 16 of [85].)

by diagonal shading) that are normal to the active glide plane. If those domain boundaries should be either low-angle tilt walls or dipolar walls, as seems most likely, they could readily provide the required dislocation sources, and on account of their inherent mobility parallel to the glide plane would remain normal to the active glide plane independent of the shear caused by the Taylor lattice dislocations. With increasing stress, then, additional sources would be pair-wise activated in neighboring domain walls to fill in extra-wide gaps between rows with new dipole sequences. In fig. 20(d), the two next most likely sites for such extra dipolar dislocation sequences are indicated by dashed lines. However, since, in turn, any one source can emit dislocations only pair-wise, i.e. in opposite directions along

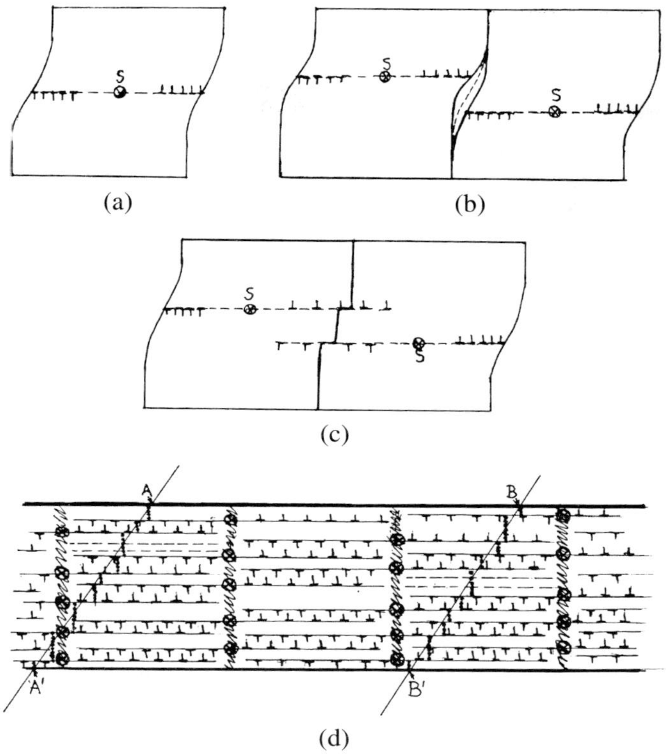

Fig. 20. Model for the formation of a Taylor lattice in the bulk, believed to describe actual Taylor lattices, e.g., as in fig. 22. (a) Schematic of the deformation of a cube through the activation of a dislocations source (S ⊗) at its center. (b) The coordinated shape change of two neighboring cubes with sources on parallel slip planes. This evidently unstable situation would be removed through the fusion of the two interior block surfaces and the passage of dislocations through them, as in (c). Now the centre of the two-block volume element is free of strain gradients, but the opposing positive and negative dislocations generate a tensile stress between them of the kind which causes the Stage I to Stage II transition clarified in fig. 7. (d) Expansion of the model to a sequence of neighboring domains, each with several rows of dislocations and the sources (⊗) situated in the domain walls. Before the generation of dislocations from any of the sources, lines AA′ and BB′ would have been vertical, parallel to the domain walls. They now consist of vertical sections (indicated by rows of dots •) that have been displaced by the passage of dislocations as shown.

the slip plane and therefore moving into neighboring domains, the gap-filling would ideally have to be coordinated among more than two adjoining domains in a process that would require a goodly degree of coordination.

One may rightfully expect such coordination to be rather imperfect in actual structures, and thus real Taylor lattices to fall significantly short of perfection. Figures 21(a) and (b), obtained from early to mid-Stage I of α-brass at room temperature, being figs 23(b) and (c) of [148], bear out this expectation. Noteworthy herein is, firstly, the considerable inhomogeneity of dislocation and strain distribution and, secondly, the evident tendency for the pair-wise formation of both slip lines and dislocation sequences. From this it

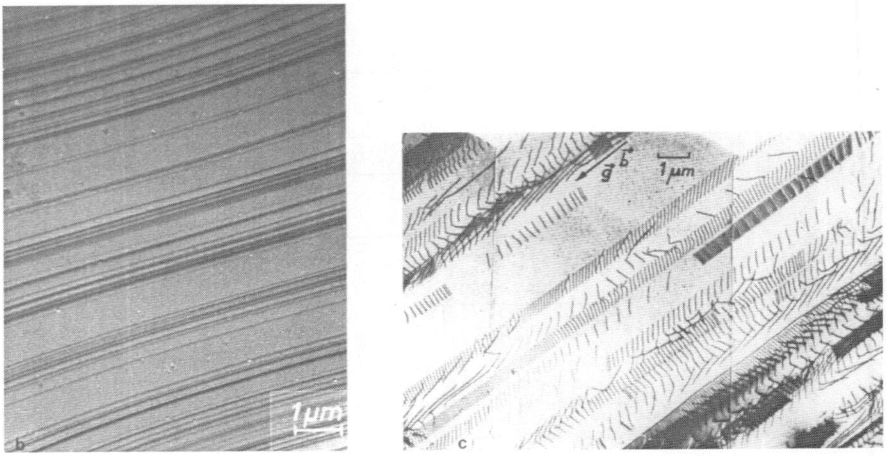

(a) (b)

Fig. 21. (a) Slip lines and (b) dislocation distribution of α-brass deformed into Stage I at room temperature. Note the frequent pairing of slip lines as well as dislocation sequences which support the model of fig. 17 but also suggest that commonly it may not proceed into the phase of fig. 17(c). (Figures 23(b) and (c) of [148].)

would seem that the process of fig. 17 may commonly stop with fig. 17(b). Yet, the relative positioning of the sequences is not as expected, and also in this area more research is certainly needed. But it should also be noted that fig. 21 pertains to Stage I before dislocations have filled the volume and begun to strongly interact with each other everywhere. Correspondingly we expect dislocation structures to become much more regular towards the end of Stage I and beyond.

Returning, then, to fig. 20, in spite of the strong strain inhomogeneity normal to the slip plane seen in fig. 21, the requirement of plastic strain compatibility in the sliding direction is rather stringent, namely within the range of less than the average long-range elastic strain. Correspondingly, the plastic shear strain can change only slowly along sequences of domains in the slip direction. This is indicated by the same inclination of the lines AA′ and BB′ which each would have been normal to the slip plane before the start of the Taylor lattice formation, namely as indicated by the columns of dots (•), but in which the passage of each dislocation has caused a horizontal kink of one Burgers vector magnitude.

What would make such a complex dislocation arrangement possible to begin with? The most obvious answer is the fact that this would be the configuration of lowest energy and hence to be expected based on the LEDS hypothesis. Besides, geometrically, Taylor structure formation is rather less demanding than appears at first sight: Specifically, similitude could be readily preserved while the dislocations densify with increasing stress, since every column or column pair of edge dislocations in the Taylor lattice is a potential domain boundary [149]. And by the same token, domain boundaries, being essentially but columns in the Taylor lattice, could be dissolved and rejoin the Taylor lattice structure. In fact we do not know the expected domain shape and size of minimum energy. The corresponding theoretical work to solve this problem would be highly welcome. On the positive side, the above model does not conflict with available micrographs (e.g., figs 8 to

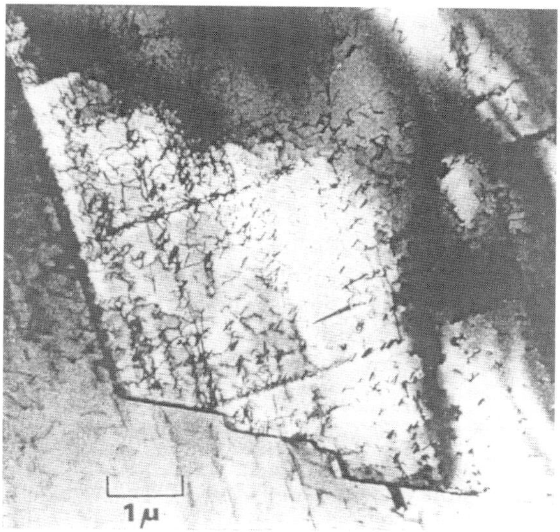

Fig. 22. Taylor lattice in strained planar glide material in α-brass. To the best possible present interpretation, it is still in Stage I, and the active glide plane slants steeply from top left to bottom right. The lines at right angles thereto, one just above the other below the centre of the figure, are interpreted as domain walls. The indistinct structuring parallel to the assumed domain walls, i.e. mildly slanting upwards from the horizontal, is interpreted as not-very-well-formed tilt walls, i.e. the columns in the Taylor lattice. By coalescing pair-wise, these could ripen into further domain walls in the process of refining the lattice in accordance with similitude. Interpreting this micrograph in terms of fig. 20, edge dislocations of opposite sign are emitted pair-wise, in opposite directions, from the dipolar domain walls. On arriving at the next domain wall, the dislocations will trigger the emission of new dislocation pairs, their respective opposites moving back into the initial domain so as to form dipolar mats with the arriving dislocations, and the opposite edge dislocations spreading the glide into the next domain. In the present micrograph, the number of dislocations between neighboring domain walls, which in this interpretation equals the glide path parameter g of eqs (35) and on, is about 20. More and better documented micrographs to reveal the architecture of Taylor lattices are urgently needed but do not seem to be available in the literature. (Figure 21 of Patterson and Wilsdorf [150].)

10 of [142]). Alas, fig. 21 due Patterson and Wilsdorf [150], reproduced here as fig. 22, is the one and only micrograph found in the literature that reveals the architecture of any Taylor lattice in the right orientation and on a large enough scale to recognize the pattern. In fig. 22 the rows and columns may be clearly seen, as is one very clear domain wall just above the centre of the figure.

Contemplating the present section, a certain uneasiness remains: Without a doubt there is a most urgent need for systematic micrographical studies of Taylor lattices. At a minimum we need to know much more about Taylor lattice evolution in Stages I and II of planar glide with particular concentration on the architecture of the domains.

1.8. The friction stress τ_0 and different dislocation locking mechanisms

The friction stress, τ_0, designates the lowest local resolved shear stress at which a dislocation will glide. Besides the intrinsic glide resistance of dislocations, minimally the

Peierls–Nabarro stress, τ_0 is due to jogs and various anchoring mechanisms. In turn, the local resolved shear stress is the sum of all stress contributions, besides the applied stress, of all other internal stresses, including that of all other dislocations and also that of the dislocation itself. As a result, since force equilibrium must be maintained at all levels, dislocations will not only move laterally if the local resolved shear stress exceeds τ_0 but, if pinned at blocking points, e.g., nodes, jogs or precipitates, will locally bow out between these until the local resolved shear stress has decreased to the level of τ_0. Consequently, within fairly narrow limits, during deformation the local value of τ_0 is also the local value of the resolved shear stress along every dislocation line everywhere, as first pointed out in 1962 [110]. From this also follows that an applied stress is always diminished by τ_0 to the effective stress $(\tau - \tau_0)$. It is for this reason that alloy hardening frequently simply adds a constant value to the flow stress, i.e. raises the stress strain curves by a constant stress increment. However, if alloying changes the dislocation structure morphology, e.g., from cells to Taylor lattices as is the case for α-brass type alloys, its effect is much more profound than simply adding to τ_0.

Technologically and scientifically important effects result if τ_0 changes in time, through aging or annealing with or without stress application. Typically such changes are due to the thermally activated movements of point defects, impurity or alloying atoms. These may give rise to changes of τ_0 on account of changing precipitate character, sizes and distribution, as in age hardening, and/or defects or atoms may diffuse to positions of lowered energy in or about dislocation cores, thereby locking them into position. The latter effects are particularly pronounced after, or in conjunction with, deformation, i.e. in “strain aging”. The required time intervals of aging depend on the particular mechanism involved and, of course, strongly decrease with annealing or aging temperature.

The best researched effect in this category is anchoring through Cottrell atmospheres, named in honour of Cottrell who first proposed the mechanism [151] and together with Bilby derived the applicable theory [152]. A sharp (‘upper’) yield point occurs when with increasing stress the dislocations suddenly break away from their atmospheres, followed by a strain interval (at the ‘lower yield point’) of as much as several percent without hardening while Lüders bands spread across the sample, after which ordinary work hardening begins. At sufficiently high temperatures the atmospheres move along smoothly with the dislocations through thermal diffusion and the yield point effect vanishes. In an intermediate temperature interval stop/go/stop/go movements result which cause the Portevin–LeChatelier effect of serrated stress–strain curves. Blue-brittleness in carbon steels between about 230°C and 370°C is the most important example here. However, the Portevin–LeChatelier effect can also occur through twinning and strain-induced phase transformations. An excellent exposition of the above subject may be found in Cottrell’s important early book [153].

A related mechanism is ‘Nabarro locking’ [154]. It is due to just a single atomistic jump of interstitial atoms from one to another orientation of tetragonal distortion, again most prominently of carbon in iron (see also Nabarro [49], pp. 418–420). The detailed theory is due to Cochardt, Schöck and Wiedersich [155]. Still another type of dislocation anchoring is ‘Suzuki locking’ [156]. It takes place in f.c.c. metals, preferably of planar-glide type, and typically also requires but one or two atomic jumps, namely of substitutional atoms into or out of the stacking faults of extended dislocations. Figure 23 due to Piercy, Cahn

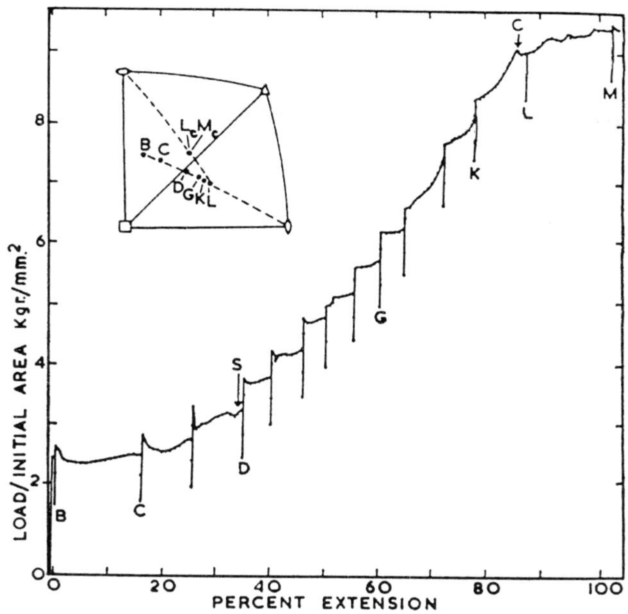

Fig. 23. Load-extension curve of 70%–30% α-brass (99.99% Cu, 99.999% Zn) with intermediate 2 h anneals at 200°C at positions A..G..M, and the associated lattice orientations shown in the insert. The secondary system begins acting at C near the top. The increases of flow stress due to the aging treatments, essentially independent of preceding strain and crystallographic orientation, are presumably due to Suzuki locking. (Figure 6 due to Piercy, Cahn and Cottrell [157].)

and Cottrell [157], in which a series of yield points was induced through aging at intervals of deformation in α-brass, is almost certainly an example of this type of age-hardening. In agreement with the above assertion that the effective stress is simply $(\tau - \tau_0)$, note here the remarkable constancy of the addition to the flow stress, i.e. of τ_0, caused by each anneal, independent of dislocation density.

Also vacancy locking in the course of mushrooming (section 1.4) adds to τ_0. It is very pervasive but so small that it often goes unnoticed. A good example may be seen in fig. 1 of Aaron and Birnbaum [158]. Hardening through vacancies becomes noticeable at high concentrations, specifically after quenching from an elevated temperature to give rise to 'quench hardening', or after energetic irradiation, i.e. radiation damage. A fine exposition of vacancy hardening may be found in [159] by Kimura and Maddin.

1.9. Internal stresses caused by strain inhomogeneities

The fact that, at all dislocation axes everywhere, the local resolved shear stress cannot significantly exceed τ_0, does not preclude the presence of large longer-range or long-range internal stresses. Dislocation pile-ups in Stage I would be an example here.

Internal stresses of any kind and on any scale that are directly caused by plastic deformation (as distinct from phase transformations), are due to strain inhomogeneities. On the smallest scale, the glide dislocation stress fields themselves are the best example. Namely, dislocations are the boundaries across which the shear displacement between adjacent atomic planes differs by one Burgers vector, $\boldsymbol{b}$, with the enormous local shear strain gradient of

$$\Delta\gamma/\Delta x = (b/d_g)/2r_0 = \frac{1}{2}r_0 \approx 10^7/\text{cm}, \tag{24}$$

where d_g and r_0 are the spacing of the adjacent atomistic glide planes and the radius of the dislocation core, respectively. The mathematical definition of dislocations (e.g., as in Love's book [160] whence it entered materials science through Burgers' fundamental paper [54], and which has already been used in section 1.3.2) bears this out: Mathematically, dislocations are the result of cutting, *rigid* displacement and rejoining operations, whereas disclinations are due to rotations before rejoining. However, the mathematical theory, e.g., [49–53], also shows that for isolated dislocations the stress field is independent of the positioning of the cut. It is for this reason that the stress fields of isolated straight dislocations do not reveal the direction from which they arrived, and also why in first approximation the strengthening due to LEDS's is independent of the sign of stress application, as already discussed in section 1.1. Yet, the *plastic strain* caused by moving dislocations is a physical property of the deformed materials morphology. This can be ascertained independent of the stress fields of the individual dislocations, e.g., most simply through the attendant external shape changes, or through the displacement of microscopic features such as grain boundaries or cut particles. The widespread neglect of this fundamental fact is one of the flaws of SODS modeling.

Larger-scale internal stresses may be traced to strain inhomogeneities on the corresponding scale, i.e. in dislocation-based plasticity to strain gradients caused by dislocation motions. Among others, this is the basic reason for end-stresses of dislocation rotation boundaries: The participating dislocations had to come from somewhere and they trailed with them their respective shear displacements of $\boldsymbol{b}$ over the areas swept out by their axes. This basic fact imparts to all dislocation rotation boundaries a dual nature, namely (1) as due to the mutual trapping of glide dislocations into LEDS's, thereby lowering the free energy of the deformed material in accordance with eq. (1) (the aspect that is responsible for the pseudo-epitaxial nature of rotation boundary dislocations outlined in section 1.6.2), and (2) of being 'geometrically necessary' as boundaries between mutually rotated crystal parts, namely the very rotation that is due to the strain gradient on account of the dislocations trapped in the boundary. This topic has been treated in somewhat greater detail in section 18 of [22].

As already discussed in section 1.6.2, the arrangement of the trapped dislocations into a mosaic block/cell structure minimizes, in fact virtually eliminates, the 'longer-range' internal stresses, i.e. on the scale of the average dislocation cell size. Further, in Taylor lattices 'longer-range' stresses (i.e. in this case on the scale of the domain size) are minimized by the cooperative operation of dislocation sources in neighboring domain walls (compare fig. 20). Yet, in that case, clearly strain inhomogeneities normal to the preferred glide plane(s) can be arbitrarily large without introducing significant internal stresses. This arises because an assembly of parallel slab-like volume elements, e.g., as modeled by a pack of playing cards, may undergo deformations that are strongly inhomogeneous normal to their slip planes without causing internal stresses, but not at right angles thereto. This fact is, of course, the underlying concept in 'relaxed constraints' texture modeling (compare [2, 3]).

The morphology of surface slip markings provides the most obvious evidence of strain gradients that are the cause of internal stresses. These very commonly form colonies of more or less uniformly spaced parallel slip bands or lines, at least at small strains. Herein, the inhomogeneity of strain distribution in the direction normal to the lines/bands, e.g., as in fig. 21(a), is clear from the very existence of the line/band structure. Yet, for elongated sets of parallel lines it is of little consequence as already indicated by the example of the pack of playing cards. Much more important are strain gradients in the direction of the lines/bands that are concentrated at the band ends. These are greatly aggravated because plastic deformation of single crystals at higher strains, and of polycrystals for all significant deformations, requires multiple glide on intersecting slip systems, thereby enforcing the increase of strain inhomogeneity (evidenced by the shortening of slip band sections) in glide direction. Even based on the observed slip distribution, then, significant internal stresses on the scale of the average slip-line length are unavoidable and are commonplace. That inhomogeneity is accentuated, or perhaps the very pattern of slip lines/bands is due to, the difficulty of the mutual intersection of slip systems.

The indicated resistance against the locally simultaneous operation of intersecting slip systems, i.e. 'latent hardening', is most obvious in planar glide where it causes 'overshooting'. This is the phenomenon that single glide on the primary system continues even after double glide should have been triggered based on relative resolved shear stress strength, followed by continuing single glide but now on the secondary system. Figure 23 demonstrates the effect (see the insert) and also that it is quite independent of dislocation locking. Overshooting was discovered in the 'twenties' by Elam [161] and Masima and Sachs [162], was thoroughly investigated by Piercy, Cahn and Cottrell [157] and extensively surveyed with a wealth of data by Mitchell [163]. In wavy-glide materials, the avoidance of mutual intersection of slip systems is not as obvious but it is present nonetheless and clearly recognizable in slip-band morphology as most recently discussed in [2].

Returning once again to slip-line morphology, note the fact that not strains but strain gradients give rise to internal stresses, while LEDSs do not ordinarily reveal from which direction the dislocations arrived. This gives rise to a curious disconnection between dislocation structures and slip lines/bands. Indeed, experimentally it is difficult and largely impossible to connect the two, slip line and dislocation structures, e.g., [164,165], except in Stage I. And this problem is aggravated by the already noted effect of surface energies on slip-line formation (section 1.7.4). Even so, it seems almost certain that in wavy glide, slip bands arise because slip is channeled between obstacles, specifically carpets in Stage II and DDWs ("dense dislocation walls") and CBBs ("cell block boundaries" as first defined in [166]) in Stages III and IV. At least partly, the latter are believed to be deformation band boundaries [2] (compare section 1.10).

1.10. Internal stresses, work hardening and the LES hypothesis

The strength of longer-range internal stresses is limited by the flow stress. On the largest scale, namely that of macroscopic strain inhomogeneities, observations on the Bauschinger effect prove this in any event obvious point, and on the level of structural inhomogeneities the same is true for composites. No doubt, also in all other circumstances the local flow stress limits the intensity of longer-range and long-range stresses. Moreover, internal stresses are of necessity balanced, with as much push as pull, left and right torsion, so that they arithmetically average to zero. Given, then, that the maximum value of internal stresses compares to the flow stress and their average value is zero, one thus expects (see [22,23,149]) that internal stresses average about one half of the flow stress. In fact the question of the magnitude of internal stresses has been researched very thoroughly by means of low-angle X-ray scattering principally by H. Mughrabi, T. Ungar and M. Wilkens, with the result that typically work hardening indeed causes internal stresses of about one half the momentary flow stress [167–172]. It follows that, contrary to the initial conviction of those researchers, work hardening cannot be due to the build-up of internal stresses but, instead, internal stresses are a byproduct of work hardening.

The free energy represented by internal stresses is part and parcel of the free energy that is minimized according to the LES hypothesis. This aspect of the LEDS theory had been overlooked until rather recently when it was discovered that it controls the formation and morphology of deformation bands in metals [2,3] as it does that of kink bands in fatigued polymers [1]. As already pointed out in section 1.1, the approach to energy minimum in terms of deformation bands is so close that the remaining stored energy would heat the material only through about 0.01°C, thereby providing very strong support for the LES hypothesis.

The impact of this insight on deformation texture modeling is potentially large since it substitutes the stored energy for the much larger plastic work done in accordance with the Taylor [92] or, equivalently [173], the Bishop and Hill [174] model. Even so, internal stresses do not affect work hardening except indirectly through slip-band and deformation-band morphology [2]. This is in line with long-held conviction beginning with [110] and [100], that long-range stresses can be neglected in theories of work hardening, as will be done also in part 2 of this paper.

2. *Evidence to be explained*

2.1. Fundamentals of work hardening curves

Based on the background information on dislocation properties presented in part 1, the task at hand is to explain the microstructures observed in different materials as a function of the work hardening stages, to derive therefrom the expected flow stresses and other mechanical properties, and to compare these with the experimental evidence. First and foremost that evidence includes the shape of the stress–strain curve as a function of lattice structure, alloying content and phases, and its strain-rate, temperature and grain-size dependence. The explanations will have to take account also of slip-line morphology and stored energy

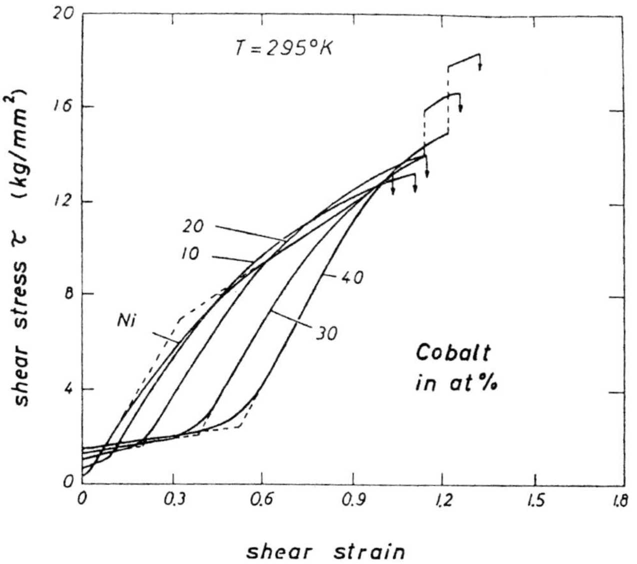

Fig. 24. Shear stress versus shear strain curves (resolved with respect to the momentarily active slip systems) of single crystals of nickel and solid-solution Ni/Co alloys with the indicated atomic weight percentages. While in wavy glide (Co concentrations to 20%), double glide begins on reaching the [100]/[111] symmetry line and proceeds with equal participation of the primary and secondary system, in planar glide (30% and 40% Co) overshooting is observed. Herein the primary system ceases to operate when glide starts abruptly on the secondary system only after its resolved shear stress has risen well beyond that of the primary system, as indicated by the vertical lines. Independent of this tell-tale feature, note the gradual transition from the wavy-glide characteristics of Ni, to the planar-glide of the α-brass-type, Ni–40wt%Co alloy, as evidenced by the lengthening of both Stages I and II. (Meissner, [175].)

as well as annealing behavior. It is therefore useful to give a brief account of the basic experimental evidence to be explained. Herein, and throughout this paper, the discussion is deliberately restricted to f.c.c. and b.c.c. metals, so as to permit focusing on dislocation behavior undisturbed by twinning which strongly modifies the behavior of materials of lesser crystallographic symmetry.

To begin with the stress–strain curves, these have been most carefully studied for pure f.c.c. metals and homogenous solid solutions (so-called 'α-brass type' alloys). Figure 24 due to Meissner [175] shows the transition of single crystal shear stress/shear strain curves from the typically 'wavy-glide' behavior of pure nickel, to the typical 'planar-glide' behavior of Ni/40at%Co alloy at room temperature. The evident planar-glide characteristics that develop with rising cobalt content are (i) the increasing yield stress, (ii) the lengthening and simultaneous flattening of Stage I (or 'easy glide'), (iii) the lengthening of Stage II, i.e. the gradually rising value of τ_{III}, the transition stress between Stages II and III, and (iv) the onset and increase in the magnitude of 'overshooting' (compare section 1.9 and insert in fig. 23) visible as the dashed vertical lines in the 30% and 40% curves that record the excess stress required for triggering the operation of the secondary (or 'conjugate') slip system.

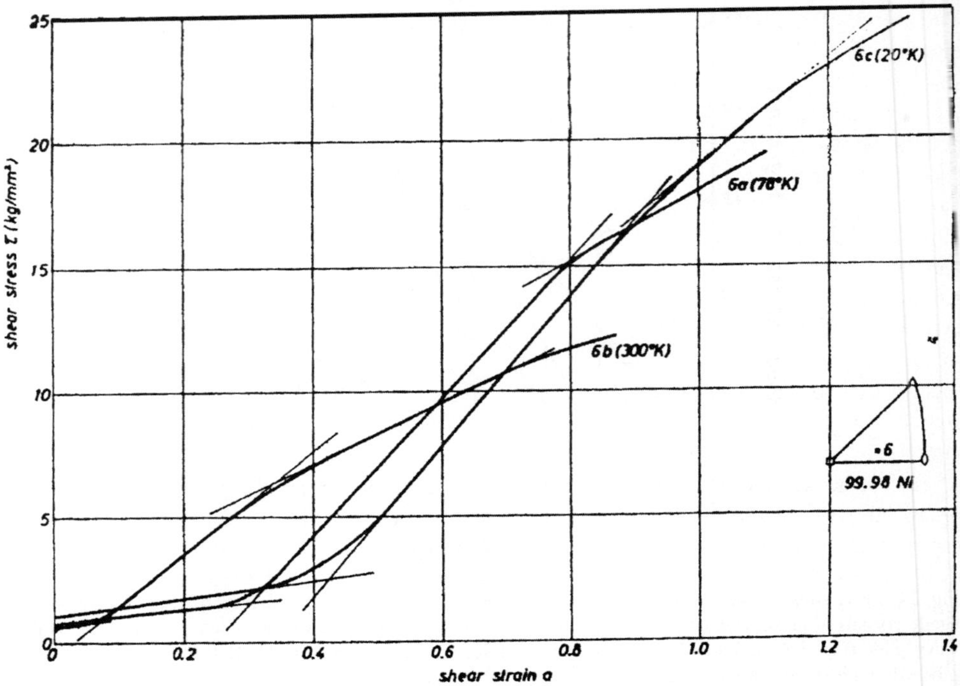

Fig. 25. Experimental three-stage work-hardening curves for single crystals of nickel, a wavy-glide material, at temperatures of 20 K, 78 K and 300 K, due to Haasen [177]. Note that the curves are somewhat idealized in that Stage II is not entirely linear and the transition to Stage III is in actual fact much less distinct than given in this figure.

Most remarkable is the almost constant value of Θ_{II}, the slope of Stage II, i.e. the slope of the almost linear, steepest part of the curve. Indeed, as has been known for a long time (e.g., Seeger [176]),

$$G/\Theta_{\mathrm{II}} \cong 250, \tag{25}$$

within a factor of two or so, is a hallmark of Stage II. The reason for this rather wide margin of variance is not only minor differences among materials but, more importantly, that in wavy glide Stage II is distinctly curved, while it is much more nearly linear in planar glide (compare the corresponding discussion in [100]).

A similar gradual evolution from wavy to planar-glide characteristics in the stress–strain curve takes place when the testing temperature is lowered, as shown fig. 25 by means of experimental, albeit somewhat idealized, curves for pure nickel due to Haasen [177]. Again, but now as a function of decreasing temperature instead of increasing solute alloying concentration, both the flow stress in, and extent of, Stage I increases. Again Stage II lengthens, i.e. τ_{III} rises, and its slope depends on temperature only through the shear modulus, G, i.e. in Stage II the relative flow stress, τ/G, is temperature independent.

The onset of macroscopic double glide occurs as the secondary slip system becomes equally stressed with the primary system and leaves no trace on the stress–strain curve. To some extent this is remarkable because, as we shall see later, slip systems avoid each other which results in different types of dislocation walls, dubbed incidental and geometrically necessary, which will be further discussed in different sections below [178].

As a counterpart to fig. 25, the temperature dependence of the single-crystal work hardening curve for planar glide is shown in fig. 26, namely for Ni/50at%Co between 78 K and 773 K, due to Pfaff [179]. Below and above ≈ 400°C (673 K), this is a planar- and wavy-glide material, respectively. The general features are similar to those in fig. 25, but the planar-glide regime is clearly distinct from wavy-glide through the considerable extension of Stages I and II and that on account of overshooting, double glide as well as Stage III is suppressed below 400°C. Again, in the planar-glide regime, the onset of secondary (conjugate) glide is indicated by the vertical sections in the curves, as glide switches to the more highly stressed secondary slip plane, but this effect vanishes in wavy glide.

The basic characteristics of the single-crystal three-stage stress–strain curves are remarkably persistent and are similarly found for b.c.c. metals, e.g., as shown in fig. 27 for niobium due to Foxall. Duesbery and Hirsch [136], although they can be masked by special effects. One such is, of course, the well-researched dislocation locking and unlocking discussed in section 1.8 (compare also fig. 23). Less well known but also well understood is the strain rate effect that arises when plastic deformation rapidly multiplies glide dislocations beginning from a very low concentration. Figure 28 due to Haasen [180] is an example.

In polycrystals, Stage I is eliminated and quite typically Stage II is masked by the distribution of friction stress and Schmid/Taylor factors among the grains. Moreover, on account of enforced polyslip, Stage II is drastically shortened if not also eliminated in wavy glide, especially at elevated temperatures. The temperature dependence of wavy-glide polycrystals is therefore almost completely controlled by that of Stages III and IV, the latter not seen in single-crystal tests because the requisite high strains are not attained. Besides on account of polyslip (as enforced by grain boundaries), Stage I is eliminated by disturbances of all kinds, e.g., impurities, precipitates and second phases, that strongly shorten the dislocation mean free path even at low dislocation densities, as first pointed out in [110].

The extra hardening due to grain boundaries has received a great deal of attention and matters are far from simple. In fact, within reasonable limits, grain refinement does not affect the work-hardening curves of wavy-glide materials. Put differently, seeing that grain boundaries eliminate Stage I and greatly shorten or eliminate Stage II, they evidently have no effect on Stages III and IV. The resulting fact that therefore the work-hardening curve of a polycrystal of impure aluminum is close to the average of work-hardening curves of single crystals of different orientations made of the same material, was noted as early as 1927 [181] (compare fig. 90 of [135]) and was well known to Taylor [54,92]. Similarly, a pure f.c.c. metal crystal oriented for polyslip, specifically in $\langle 111 \rangle$ orientation, has much the same stress–strain curve as a polycrystal of the same material. This aspect of polyslip was especially studied by Kocks [182], Thompson and Baskes [183] and Thompson [184]. Figure 29(a) due to Thompson [184] gives an example.

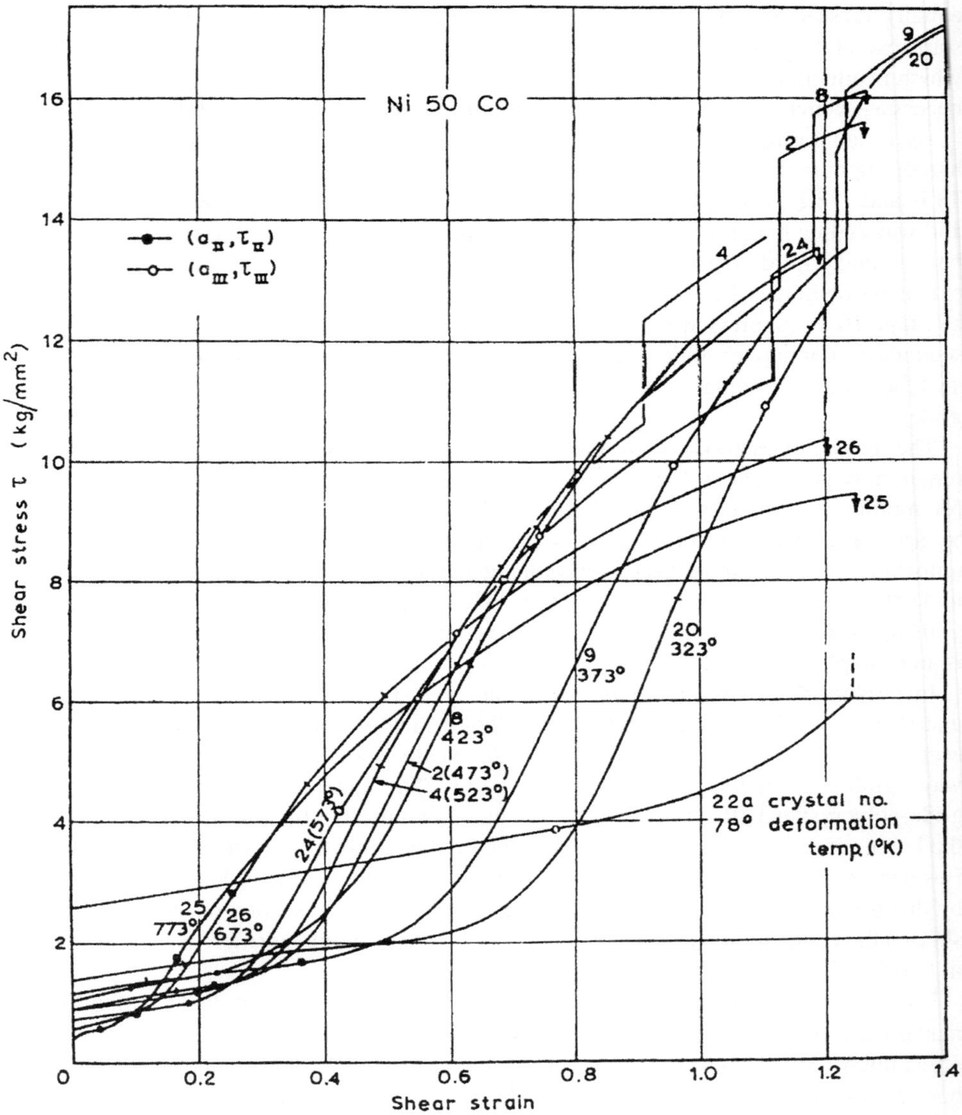

Fig. 26. Shear stress-shear strain curves for f.c.c. Ni-50at%Co crystals due to [179]. Below and above ≈ 400°C (673 K), this is a planar- and wavy-glide material, respectively. The general features are similar to those in fig. 25. In the wavy- as well as the planar-glide regime, both Stages I and II are greatly extended. The planar-glide regime is recognizable by over-shooting, indicated by the vertical sections in the curves, where glide switches from the primary to the more highly stressed secondary (or 'conjugate') slip plane. Thus, below 400°C, double glide as well as Stage III is suppressed. However, in the wavy-glide regime above 400°C, the onset of macroscopic double glide is smooth without any effect on the work-hardening curve. (Pfaff, [179].)

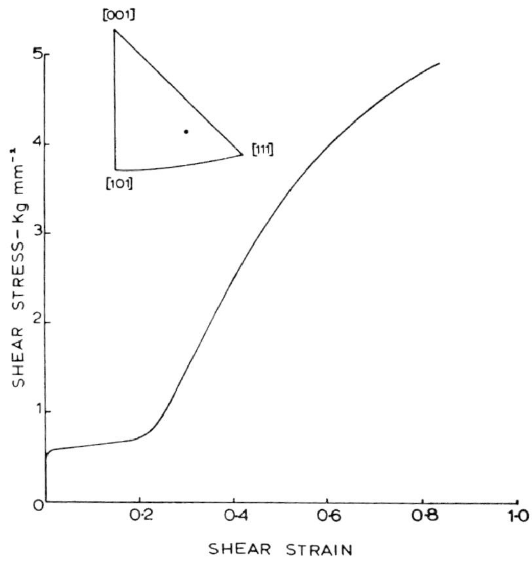

Fig. 27. A typical work-hardening curve of single-crystalline pure niobium, a b.c.c. wavy-glide metal, at room temperature. (Figure 1 of Foxall, Duesbery and Hirsch [136].)

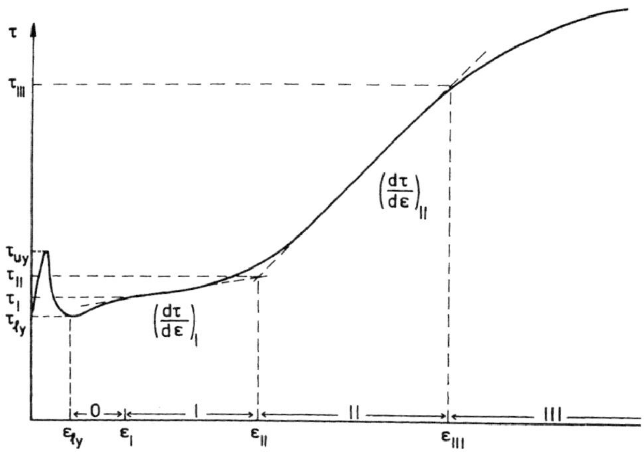

Fig. 28. Shear stress/shear strain curve of germanium showing an upper- and lower yield-point effect due to the initial scarcity of glide dislocations in germanium (Fig. 65 of Haasen [180]).

Matters are quite different in planar glide. Here grain-boundary hardening does indeed add an almost constant stress increment to the flow stress as first noted in [185] and shown in fig. 29(b) due to Thompson [184]. Albeit, the Hall–Petch relationship is not obeyed herein. It was specifically derived from, and has been verified for, the yield stress associated with dislocation locking by Cottrell atmospheres (compare section 3.12 and [22,23]).

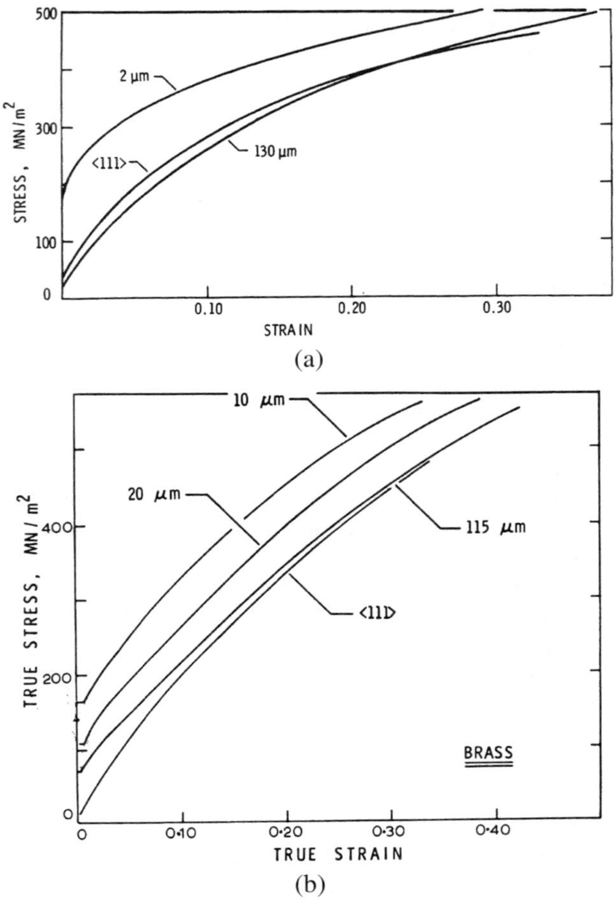

Fig. 29. The effects of grain size on the work-hardening curves of wavy- and planar-glide-materials. The examples here are polycrystals of the indicated grain sizes in comparison with ⟨111⟩ oriented single crystals, in (a) of pure nickel (wavy glide) and in (b) of planar-glide a-brass with 30 wt% Zn. As seen in these examples and still more clearly in many others, beginning with fig. 30 of [185] (see also fig. 14 of [149] and fig. 4(a) of [190]), in wavy glide, only the smallest grain sizes increase the level of the work-hardening curves while in planar glide, grain-size hardening is observed for virtually all grain sizes (figs 16 and 17 of [184]).

The most important remaining question, to be discussed further in the next section, is to what extent, if at all, the work-hardening stages can be recognized also in polycrystals.

2.2. The shape of Stage III

Visually, polycrystalline stress–strain curves are very varied and except for an occasional Stage I, mainly due to dislocation locking, give no evidence for division into stages. A different picture emerges from comparing plots of the work-hardening coefficients in terms of resolved shear stress versus shear strain ($\Theta = \mathrm{d}\tau/\mathrm{d}\gamma$) as a function of shear stress for single crystals, and of true tensile axial stress and strain ($\Theta_\sigma = \mathrm{d}\sigma/\mathrm{d}\varepsilon$) as a function of true axial stress for polycrystals. The two work-hardening coefficients are related through the Taylor factor, M, as

$$\Theta = \mathrm{d}\tau/\mathrm{d}\gamma = \left(1/M^2\right)\Theta_\sigma = \left(1/M^2\right)\mathrm{d}\sigma/\mathrm{d}\varepsilon. \tag{26}$$

This method was developed, it seems, by Mecking and Lücke [186,187] and the following examples are extracted from Mecking [188]. Specifically, figs 30 and 31 show such plots for single crystals of pure silver and polycrystals of pure copper, respectively, the latter credited to M. Biermann [189], both for a range of temperatures.

Evidently, these two figures are very dissimilar at low stresses. In fact the initial steep rise of the curves in fig. 30 is due to the transition from Stage I to Stage II, whereas the initial steep drop of the curves for the polycrystals in fig. 31 is due to the gradual onset of plastic strain at yielding. For the remainder, the similarity is unmistakable: Stage II is seen as a broad peak at low temperature that gradually sharpens with rising deformation temperature, namely on account of the shortening and ultimate elimination of Stage II. The top curve, drawn as a full dark line in fig. 30, indicates an extrapolation for continuing Stage II at the lowest temperatures. Importantly note that the highest slope of Stage II, indicated by the peaks of the curves in fig. 30 and of the plateaus in fig. 31, decreases mildly with deformation temperature, judging by fig. 5 of [188] so as to let the average of Θ_{II}/G remain constant for polycrystals.

Stage III, following Stage II after a short transition, is seen in figs 30 and 31 as a substantially linear decrease of Θ with stress, of the general form

$$\Theta = \Theta_\mathrm{s} - B\tau. \tag{27}$$

Characteristically, the negative slope, B, of the straight $\Theta(\tau)$ line in Stage III, is substantially constant. Next, except for the lowest temperatures, the $\Theta(\tau)$ lines are shifted to lower stresses by equal stress increments for equal temperature intervals. These two characteristic features, including also the much wider shift of the lines along the stress axis at the lowest temperatures, were observed also in an analysis performed on different stress/strain curves extracted from the literature before the author became aware of Mecking's prior work (see [190]). Among those examples were the stress–strain curves in [185] which had documented the absence of a grain-size effect in ordinary copper (see fig. 32). This, then, shows that not only changing temperatures but also changing grain sizes do not affect the slope of the $\Theta(\tau)$ lines in Stage III.

Eq. (27) represents the Voce curve interpretation of work-hardening curves [191,192]

$$\tau = \tau_\mathrm{s} + \Delta\tau e^{\frac{-\gamma-\gamma_\mathrm{S}}{\gamma_\mathrm{c}}} \tag{28}$$

that describes a curve segment starting with $\tau = \tau_\mathrm{s}$ at $\gamma = \gamma_\mathrm{S}$ and attaining the maximum stress of $\tau = \tau_\mathrm{s} + \Delta\tau$ asymptotically for $(\gamma - \gamma_\mathrm{S}) \gg \gamma_\mathrm{c}$. Namely, differentiation of eq. (28) yields

$$\Theta = \mathrm{d}\tau/\mathrm{d}\gamma = [(\Delta\tau + \tau_\mathrm{s})/\gamma_\mathrm{c}] - \tau/\gamma_\mathrm{c} = \Theta_\mathrm{s} - \tau/\gamma_\mathrm{c}. \tag{29}$$

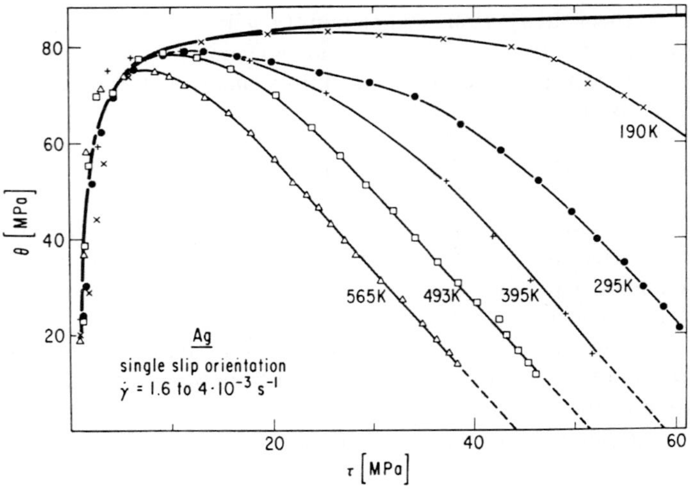

Fig. 30. The dependence of the tensile-test work-hardening coefficient, $\Theta = \mathrm{d}\tau/\mathrm{d}\gamma$, on the resolved shear stress, τ, for silver single crystals, at the indicated testing temperatures. The initial steep rise of $\Theta(\tau)$ is due to the Stage I to Stage II transition, followed by the plateaus or crests due to Stage II. As seen, the extent of Stage II decreases with rising deformation temperature while the level of Θ_{II} varies only slightly, apparently such as to keep Θ_{II}/G substantially constant. The decreasing straight lines are due to Stage III within which the work-hardening coefficient decreases linearly with flow stress. Except at the lowest temperature, Θ also decreases nearly linearly with testing temperature, as indicated by the shifting of the curves along the stress axis by about equal amounts for equal temperature intervals. The top curve given as a full dark line represents Stage II extrapolated to 0 K. A reason for the larger shift of the 190 K curve compared to the others is suggested in section 3.8.4. (Figure 2 of Mecking [188].)

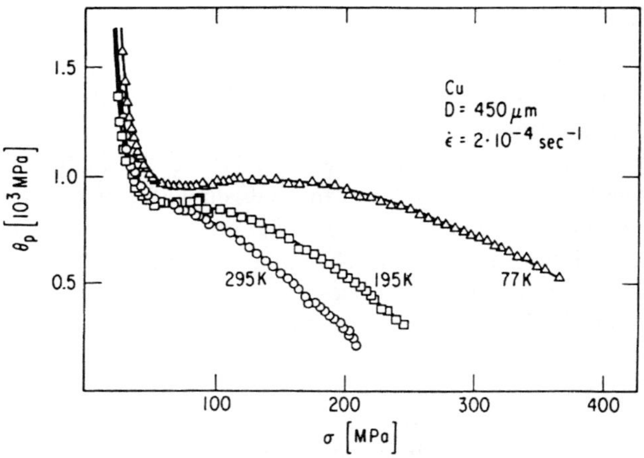

Fig. 31. As fig. 30 but for copper polycrystals, i.e. $\Theta_\sigma(\sigma) = M^2\Theta(\tau)$, see eq. (26), which is the reason for the higher stress levels along the axes. On account of the gradual onset of flow in polycrystals without a Stage I, the curves commence with a steep drop. Otherwise the features are quite similar to those in fig. 30, thereby indicating that Stage II can also exist in polycrystals, while Stage III does so routinely. (Figure 3 of Mecking [188].)

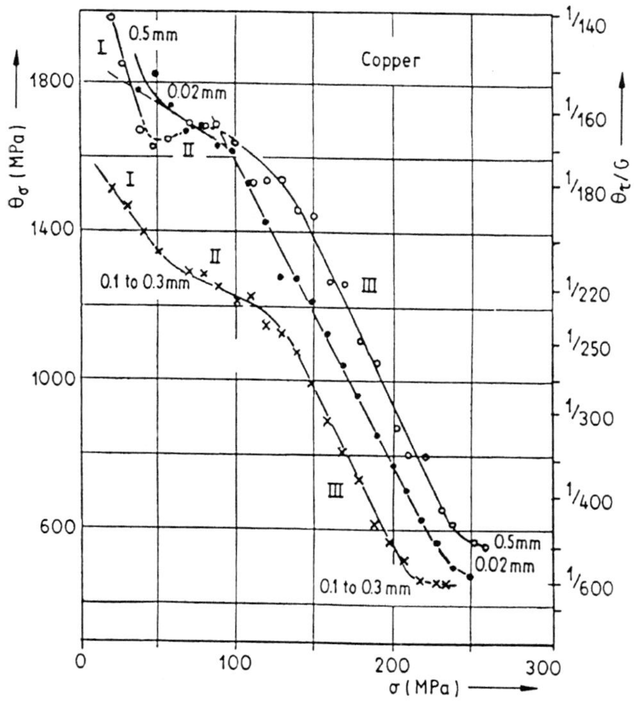

Fig. 32. As fig. 31 for copper polycrystals of different grain sizes as indicated. Note that the slope of the Stage III part is independent of grain size. (Figure 6 of [190].)

Hence the negative slope, B, of the straight part of the $\Theta(\tau)$ curve is the inverse of the parameter γ_c. This in turn is the strain increment at which the maximum stress of $(\Delta\tau + \tau_s)$ would be attained at the constant work-hardening coefficient Θ_s, i.e. the work-hardening coefficient at stress τ_s.

The evidence for Voce curves does not only rest on figs 30 to 32 and other examples in [188] and [190], nor on the fact that Kocks has championed Voce curves above other empirical fittings beginning with [193] plus many other examples in the literature (e.g., figs 33 and 34 of [194]). More persuasively yet, design engineers have routinely used Voce curves for decades, and stress–strain curves composed of piece-wise connected segments of Voce curves have been widely observed for aluminum and aluminum alloys [195].

2.3. Stage IV

Stage IV is not evident in normal tensile testing, for the reason that the requisite strains are not attained. However, it is prevalent in high-strain deformation as the next stage following Stage III [196–199]. Superficially, Stage IV appears to be linear and therefore it was initially mistaken for Stage II [200,201]. However, its slope is much smaller than of

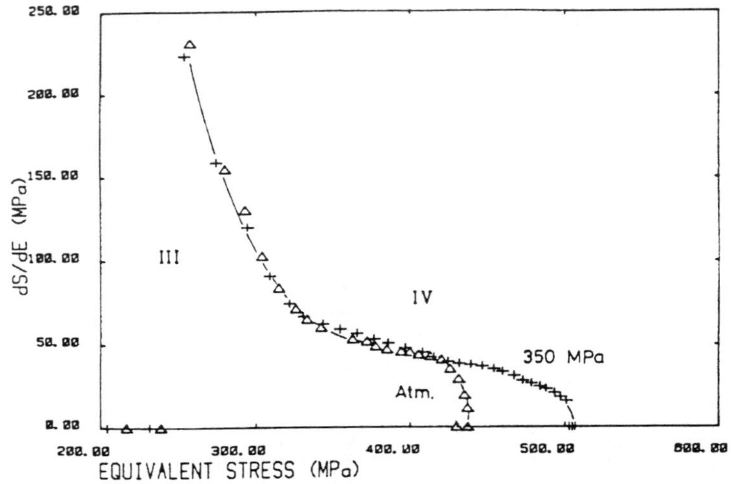

Fig. 33. Experimental plot of the work-hardening coefficient as a function of flow stress, i.e. $\Theta_\sigma(\sigma)$ in the manner of figs 31 and 32, for oxygen-free, high-conductivity copper at constant slow strain rate in torsion at ambient and 3500 atm. pressure, as indicated. Note again the linear decrease in Stage III, while the flatter part at higher stresses is Stage IV. At the highest stresses, the curves turn down in what may be Stage V (see table 1), at which stage the true strain amounts to hundreds. As seen, the imposition of a hydrostatic pressure has little effect except for perhaps adding modestly to the flow stress (i.e. shifting the curve a bit to the right) and for considerably lengthening Stage IV. (Figure 1 of [194], credited to Sturges, Parsons and B.N. Cole [202].)

Stage II and if is quite possibly not linear but is another Voce curve. This is indicated by the substantially linear decrease of the work-hardening coefficient with stress as seen in figs 33 and 34 reproduced from [194] and credited to Sturges, Parsons and Cole [202] and Gil Sevillano [203], respectively.

2.4. Correlation of dislocation structures with work-hardening stages

As already indicated in section 1.2, to best knowledge and belief all dislocation structures that are dominated by mutual dislocation interactions, in contrast to interactions with non-dislocation defects, are LEDSs. There, too, it was already mentioned that Stage I is dominated by LEDSs which can be constructed from dislocations belonging to only one slip system, i.e. the primary slip system, and it was shown by means of fig. 4 that all of such LEDSs comprise solely edge dislocations.

This simple basic condition of dislocation structures in Stage I is not immediately obvious for two reasons: (i) Equilibrium LEDSs are approached only to within the friction stress, τ_0, and (ii) in wavy glide, 'mushrooming' obscures the basic architecture, as illustrated in fig. 10 in comparison to fig. 21(b). Even so, there can be no question that dislocation bundles with net zero Burgers vector content (often called 'braids', e.g., fig. 12 of Moon and Robinson [137] and reproduced as fig. 2 in [22]), dipolar mats (e.g., fig. 21(b)), dipolar walls and tilt walls are the typical structures in Stage I.

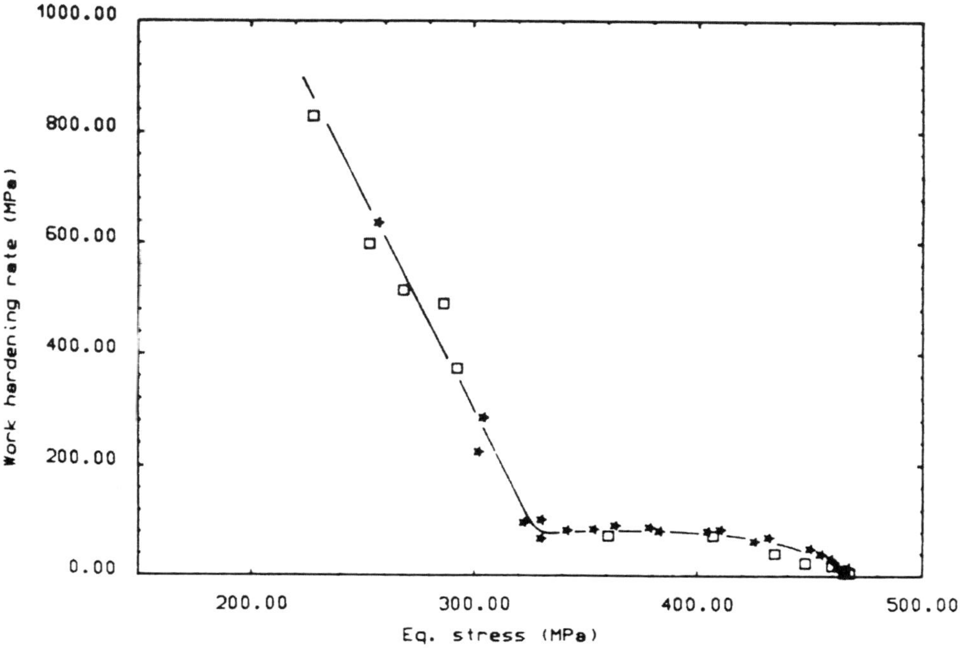

Fig. 34. As fig. 33 but for electrolytic tough pitch copper in torsion (⋆), and wire drawing with interspersed tensile tests (□). Note the great similarity with fig. 33. (Figure 2 of [194] credited to Gil Sevillano [203].)

For the remainder, in spite of an incredibly large volume of TEM micrographs, the correlation between dislocation structures and work-hardening stages is not as well documented as one might wish, in particular not in regard to Taylor lattices but also with deficiencies in regard to cell structures. To some extent this is due to protracted controversies in work-hardening theories, to the effect that even until quite recently and perhaps still now, dislocation cell structures were/are modeled as "*dislocation patterning*" that "*is now increasingly recognized... (as)... a dissipative system driven far from thermal equilibrium*" ([21], compare section 22 of [22]), failing to recognize that without a shadow of doubt cell walls are dislocation rotation boundaries (see section 1.6.1). And similarly, Taylor lattices have been or still are modeled as "*homogeneous microstructures*" that "*occur when dislocations strongly interact with a large density of localized extrinsic or intrinsic obstacles.... In such cases, plastic flow is thermally activated and the spatial distribution of dislocations is rather uniform since dislocation interactions are comparatively weak*" ([21], compare section 22 of [22]). Clearly, and as documented especially in section 1.7, this characterization of, and the role of thermal activation in, Taylor lattices is totally erroneous.

The actual role of thermal activation is very different, as further discussed in part 3.

Against this background, the relative scarcity of reliable correlations between work-hardening stages and dislocation structures arises because microscopists tend to overlook what they do not expect, whereas perhaps the majority of relevant investigations were

Table 1
Succession of microstructures and LEDS transformations in monotonic stress increase

Stage	Microstructure	Terminating LEDS transformation
I	$\perp\top$ LEDSs, including bundles, dipolar mats and walls, tilt walls, incipient primary Taylor lattices	Stress-relief in Taylor lattices, i.e. push/pull stresses, via "unpredicted" (forest) dislocations, causing new Taylor lattices or carpets
II	*In wavy glide*: "carpets" of tilt walls, mostly parallel to primary plane and sometimes noncrystallographic. *In planar glide*: multi-Burgers vector Taylor lattices	Cell formation from Taylor lattices or carpets. This requires some three-dimensional dislocation mobility, and therefore is inhibited in planar glide
III	Roughly equiaxed dislocation cells, composed of dislocation rotation boundaries. These are grouped into large "cell blocks" (CB's) each with its own selection of slip systems of fewer than the 5 for homologous deformation	Cessation of tangling? The size of the "cell blocks" shrinks. Their morphology is increasingly correlated with the external shape, e.g., filamentary in wire drawing and sheet-like in rolling. 'Tangling' in cell walls diminishes and perhaps ceases at the end of Stage III
IV	"Cell blocks" subdivided by equiaxed dislocation cells shrink in size and are more evident than in Stage III with more intricate morphology	Shear bands? "Cell blocks" continue to shrink, eventually perhaps to one "subgrain" each. Inhibition against simultaneous operation of five slip systems may cease and cause deformation by shear bands
V	"Steady-state flow" through equiaxed "clean" subgrains	On changing of straining conditions, any of the above structures can be re-established

devoted to finding supportive evidence for this, that or the other theoretical dislocation model. And lastly, trying to derive the dislocation architecture from etch pit patterns or TEM foils even with an unbiased mind, compares to deducing the architecture of buildings from typically randomly oriented sections that do not distinguish between ceilings, walls, stairs and furniture. Who would recognize his/her own home from such random sections? Still, one hastens to add that there are a number of very fine pertinent studies, noteworthy among them, in alphabetic order, Bowen, Christian and Taylor [204], Foxall, Duesbery and Hirsch [136], Keh and Weissmann [205], Mader [206], Moon and Robinson [137], Mughrabi [207], Steeds [138] and Swann [208]. From these and other literature, a table has been constructed in [190] which may now be tentatively updated to table 1.

The word "tentatively" above was used advisedly, since much uncertainty still persists especially in regard to the difference between Stages III and IV and the very existence of an independent Stage V. If Stage V does indeed exist, it is probably responsible for the final downturn of the curves in figs 33 and 34.

Examples of the different structures with a progression through Stages II and III from bottom to top, with cell structures predominating at the highest temperatures, may be found in fig. 1. The difference between fig. 1(a) and (b) regarding the Stage II structures, namely Taylor lattices with evenly distributed dislocations in fig. 1(a), and carpet structures of tilt walls with alternating sense of rotation revealed through alternating light/dark/light banding in fig. 1(b), is due to planar and wavy-glide behavior, respectively. From this and other figures in [11] it thus appears that the material, i.e. 316 stainless steel, transitions from planar to wavy glide with greater difficulty and therefore at moderately higher stresses and temperatures when it is in $\langle 100\rangle$ orientation, presumably on account of increased

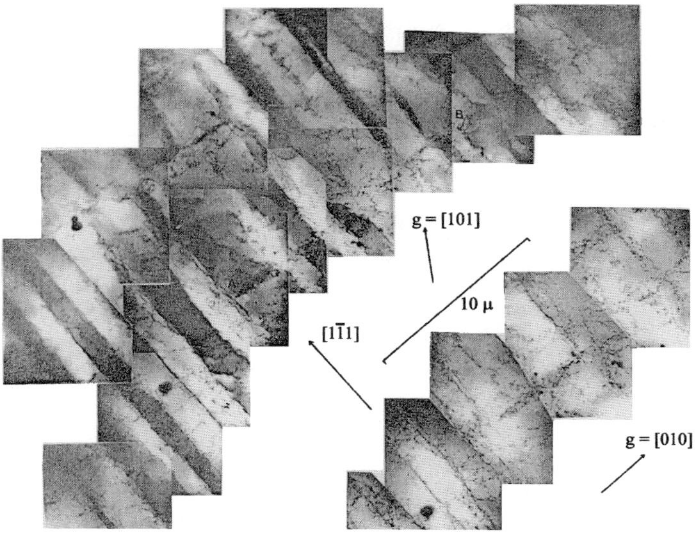

Fig. 35. An example of the characteristic Stage II structure of "wavy-glide materials", in this case an extended niobium single crystal, namely "carpets" parallel to the primary slip plane. These carpets constitute tilt boundaries, formed of primary dislocations that have reacted with "unpredicted dislocations" in accordance with fig. 15 to result in a rotation axis direction parallel to the primary slip plane, normal to the primary Burgers vector. The present micrograph represents a cut parallel to the primary slip direction. Since niobium is a b.c.c. 'pencil-glide' material, in cuts normal to the slip direction, the pattern is correspondingly disturbed into a wavy pattern, as shown by fig. 8 of [136]. The inset at lower right demonstrates that the tilt walls are largely composed of primary dislocations, in accordance with fig. 15. It is believed that the "hedges", seen in the inset to be only partly of dipolar primary edge dislocation walls, are the beginnings from which the Stage III cell structure develops. (Figure 9 due to Foxall, Duesbery and Hirsch [136].)

enforcement of multiple glide. The carpet structure, but in that case with the carpets in non-crystallographic orientations, for a stainless steel was also shown in fig. 7(b) of [190] and reproduced as fig. 5 in [23]. In good crystallographic alignment, carpet structures have been well documented, e.g., in [136–138] as already mentioned in section 1.7.2. An especially impressive example is given as fig. 35 due to Foxall, Duesbery and Hirsch [136]. This micrograph is of particular interest not only because the alternation of contrast due to the alternating sense of tilt wall rotation is seen over an unusually large sample width, as well as the imperfections in the structure, but the material, niobium, is of "wavy glide" type also in the traditional sense of exhibiting wavy slip lines on account of pencil glide. Correspondingly, in cuts at right angles to the slip direction, the pattern is disturbed into a wavy configuration, as documented by fig. 8 of [136]. The inset in fig. 35, documenting a relatively small contribution of primary dislocations to the carpets in spite of their formation in primary glide, further supports the mechanism of fig. 15.

Stage III is characterized by more or less equiaxed mutually mildly misoriented dislocation cells, as already variously discussed in part 1. Similarly cells are formed in fatigue, and for these, Laufer and Roberts [209] gave the first proof that dislocation cell walls are rotation boundaries composed of regular glide dislocations forming networks of

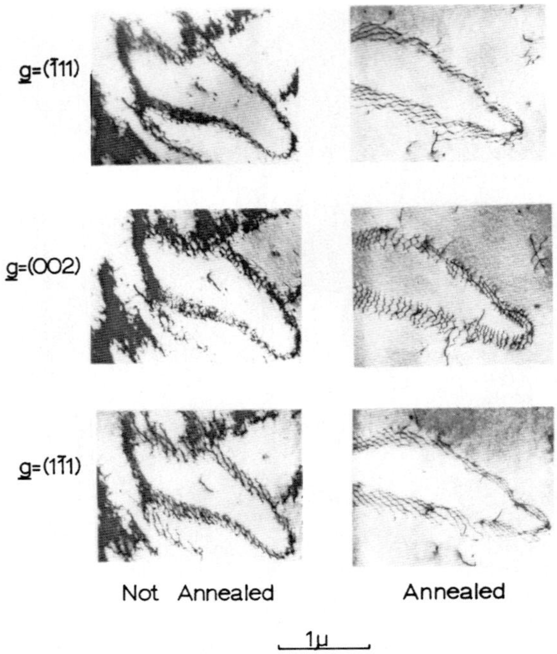

Fig. 36. Dislocation cells observed underneath a persistent slip band (PSB) in fatigued copper, illuminated with three different electron beams to reveal the Burgers vectors of which their boundaries are composed. Left, before annealing, the cell wall structure is obscured through dislocation 'mushrooming' (compare fig. 10), but at right, after mild annealing, the underlying structure of the type in fig. 6 is revealed. To be compared with figs 53 and 54. (Figure 1 of [210].)

the kind in fig. 6. They also obtained the same result as Bailey and Hirsch [68] that mild annealing does not change the cell morphology but removes redundant dislocations, i.e. reduces the redundancy-factor M_r (see 1.5.4) to near unity, as shown in fig. 36 from [210].

As expected from similitude (see section 1.6.1) the average cell diameter, D, shrinks inversely with the effective stress, i.e. empirically,

$$(\tau - \tau_0) = KGb/D, \tag{30}$$

where D is the average cell diameter and $K \approx 10$. This relationship has been known for a long time and a very extensive survey of literature data due to Raj and Pharr [211] has substantiated it for a great wealth of data, albeit with the exponent of D mildly variable and K decreasing with increasing stress.

Numerous TEM micrographs exist to show dislocation cell structures in Stages III and IV. Figure 37 extracted from [36] and based on micrographs by Langford in connection with Langford and Cohen's classical study [212] shows these in drawn iron wire. The cell sizes herein are in accord with eq. (30) but with a K value of near 6 [33]. The conclusion embodied in table 1 that Stage IV is not due to a phase transformation but that the 'cell block' morphology continues to evolve, will be further discussed in 2.6 and 3.8.4. At any

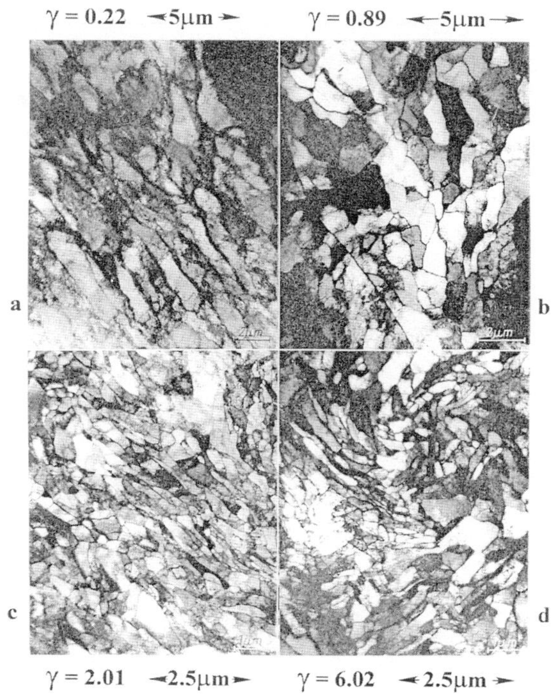

Fig. 37. Late Stage III (a) and Stage IV (b)–(d) dislocation structures in the cross-sections of severely drawn almost pure iron wire from the classic investigation by Langford and Cohen [212]. The true strains (γ) and magnifications are as indicated. At increasing drawing strain and thus flow stress, the structure shrinks in scale but remains otherwise similar to itself, – exhibiting "similitude". The abrupt changes of contrast across cell walls are due to the corresponding abrupt changes of lattice orientation, i.e. the walls are dislocation rotation boundaries. (Figure 9 of [36] based on micrographs kindly supplied by G. Langford and M. Cohen.)

rate, in view of the extensive studies most recently by Hansen, Hughes and coworkers already cited [105–109], there is no doubt left of the existence of 'geometrically necessary boundaries' [213], commonly called CBBs for 'cell block boundaries'. Note that cell blocks gradually shrink in size (see section 1.9) and deform each with their own selection of slip systems, always falling short of the five required in polycrystals according to the equivalent [173] criteria of Taylor [92] and Bishop and Hill [174].

2.5. Stored energy

Typically, only a few percent of the expended work is stored in the material. This percentage decreases with increasing strain down to only fractions of a percent [214,215]. Of particular interest in this connection is fig. 38 extracted from [194]. For one, it shows that the stored energy rises in proportion with the square of the flow stress. Considering that the greater part of the energy is stored in the form of dislocations whose line energy

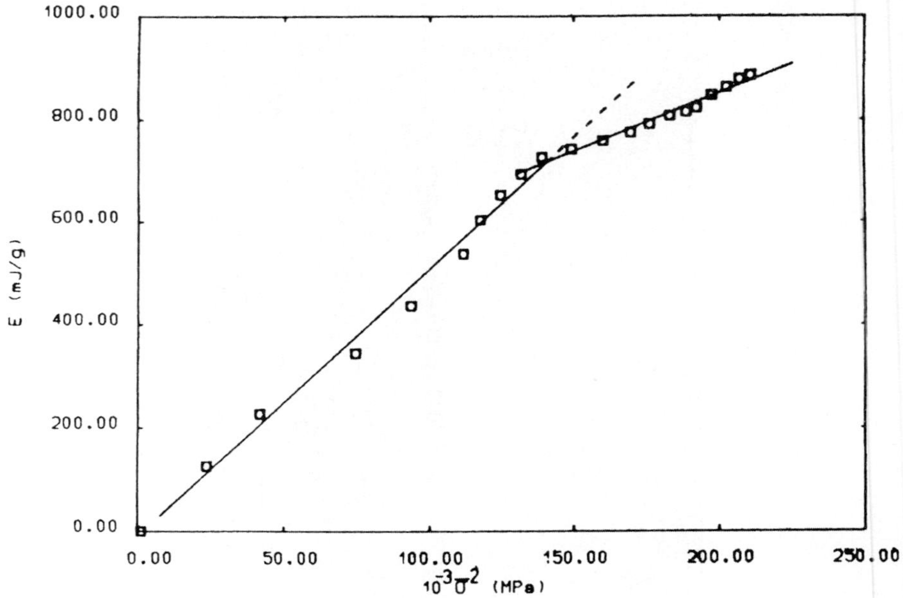

Fig. 38. Stored energy in rolled pure copper versus the square of the inferred flow stress. Note here the kink that almost certainly signifies the onset of Stage IV. Hence Stage IV is characterized by a lowered rate of energy storage as compared to Stage III. (Figure 3 of [194] credited to Gil Sevillano [203].)

does not change by more than a factor of two or so for all reasonable changes of the upper cut-off radius, R (eq. (1)), this means that the empirical correlation between flow stress and dislocation density, ρ, namely

$$(\tau - \tau_0) = \alpha G b \sqrt{\rho} \tag{31}$$

with $\alpha = 1/2$ within a factor of two, holds throughout Stages III and IV. Also the energy stored in the form of longer-range internal stresses is proportional to the square of the flow stress if their level is proportional to the flow stress itself, as argued in section 1.10. However, as will be seen in section 3.8.3 this part of the stored energy is typically negligible.

Since the slope of the stored-energy curve changes abruptly but there is no discontinuity in its magnitude, we may further conclude from fig. 38 that the Stage III/Stage IV transition is not due to a reorganization of dislocation structure (which is known in any event from the TEM evidence, e.g., fig. 37). This observation is a major reason for the conclusion already discussed in table 4 that the Stage III/Stage IV transition is not due to a LEDS transformation. A tentative explanation for the lowered slope for Stage IV in fig. 38 will be offered in section 3.8.4.

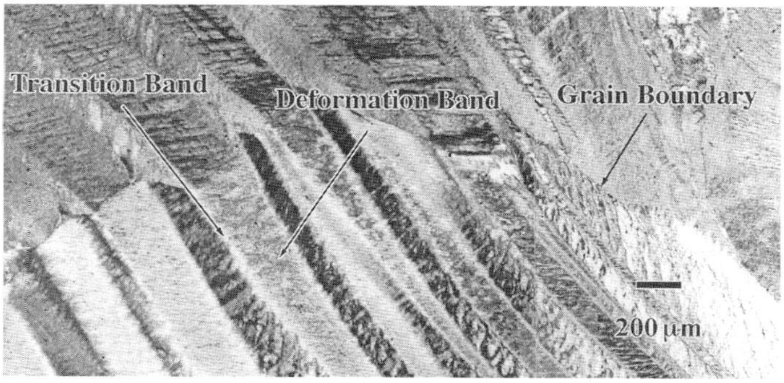

Fig. 39. 'Regular' deformation bands on the etched mid-plane of Al–0.5 wt%Cu after uniaxial room temperature compression at 1%/sec strain rate to 50% compression. Note the zigzag deformation of the grain boundary at left center showing (i) the overall uniformity of the strain within each primary deformation band and (ii) the strain-complementarity among the two sets of primary bands in the colony. This is a very unusually orderly structure exhibiting nested primary, secondary and tertiary DB's. Presumably the primary and secondary DB's formed after about 5% and 40% strain at $\tau_{DB} \cong 2 \times 10^7$ Pa and $\cong 4 \times 10^7$ Pa, respectively. (Figure 9 of [217] see also [216].)

2.6. Deformation bands

The mutual avoidance of slip systems that leads to a number of locally simultaneously acting slip systems below the Taylor/Bishop and Hill required number of five, and which is responsible for the break-up into cell blocks (delineated by CBBs, compare fig. 59), unavoidably causes internal stresses. Based on the LES hypothesis, the free energy associated with those internal stresses no doubt controls the intricacies of cell block morphology. Quite frequently, the cell blocks may have somewhat irregular shapes and form a kind of mosaic block structure of their own. If so, they are not evident in optical microscopy. However, on account of the lowering of energy that is possible through 'relaxed constraints' (compare section 1.9), a band-like morphology is favoured. And indeed there is plentiful evidence for slab- or band-like 'regular' deformation bands that occur in colonies of parallel bands, preferably with alternating strain deviations from the average imposed strain. Yet, whether or not the morphology is evidently band-like, all cell-block morphologies are now believed to be due to the same phenomenon of slip system avoidance [2,3] and therefore are subsumed under the name of 'deformation bands'.

In the course of deformation, initially large deformation bands are bound to refine, but their boundaries are fixed in the material since these delineate sizeable plastic strain discontinuities which caused the corresponding differences in lattice orientation that cannot be undone. Correspondingly, deformation-band boundaries deform along with their surroundings, but the number of bands per colony increases through the subdivision of pre-existing bands and, further, bands commonly subdivide into secondary bands and these again may divide into tertiary bands. The topic of deformation bands has been rather thoroughly investigated in recent papers [2,3,216–218] and will be taken up again in part 3. At this point, fig. 39, extracted from [216] and also included in [217], serves as the best example to date for illuminating the ideal deformation band morphology. Most notable

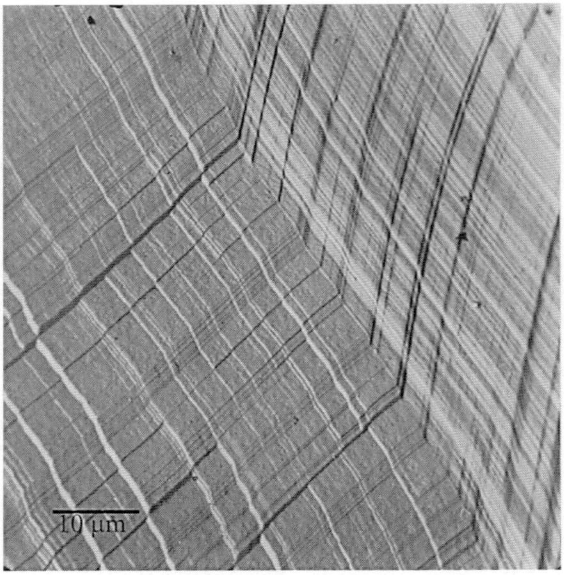

Fig. 40. Electron micrograph of a surface replica, obliquely shadowed with WO_3, of slip lines on either side of a twin boundary in moderately deformed polycrystalline α-brass, as characteristic of Stage II planar-glide. There are two sets of intersecting slip lines that evidently share the same slip direction which lies in the twinning plane. This follows because (i) one set of planes is parallel to the twin boundary and (ii) at numerous crossings some or all of the slip is transferred between intersecting lines, thereby forming "V" configurations. Such transference of glide from one slip plane to another with the same slip direction, two occurrences of which are here marked by arrows, is "cross slip". Note that the areas between the slip lines are free from surface markings and that the slip lines are long and sharply defined. In fact they are almost certainly due to glide on one single or a very few closely spaced atomic planes. (Courtesy of H.G.F. Wilsdorf.)

here is the already indicated complementary deviation of the local strain from the average among the two kinds of primary band that make up the colony, as seen from the shear displacement of the grain boundary at upper left.

2.7. Slip markings: slip lines and slip bands

The earliest high-resolution slip line/band studies were made on aluminum [219–222] and include a wealth of information up to moderate strains on single and polycrystals. This was followed by a comparison of the slip markings on different metals [185] wherein the difference between 'wavy' and 'planar' glide was discovered albeit not by those names. Specifically, throughout, planar glide materials exhibit rather sharp and long slip lines of which figs 21 and 40 are examples for a single crystal in Stage I and about 10% tensile strain in a polycrystal, i.e. Stage II, respectively. Note here the apparent absence of any glide at all between slip lines. The very existence of distinct slip lines and bands betokens the existence of strong strain gradients already briefly discussed in section 1.9. For slip on intersecting glide planes, the otherwise expected strong internal stresses are reduced when these share the same slip direction, as in fig. 40, i.e. are in cross-slip relationship and thus produce a kind of coarse pencil glide. In fact, cross slip was first discovered by means of slip lines on α-brass by Maddin, Matthewson and Hibbard [223,224].

The behavior is entirely different in wavy-glide materials. They exhibit an 'elementary structure' of long, very shallow more or less evenly spaced slip lines in stage I [219, 220] that was later renamed 'fine slip' by Mader [225]. According to Seeger et al. [29], Seeger [30, Chpt. VI] and Mader and Seeger [226], in Stage II the fine slip lines shorten and deepen, form loose groupings of higher and lower shear deformation, and are increasingly interspersed with "unpredicted" slip lines. The latter doubtlessly trace the slip which converts the Taylor lattice into carpets in accordance with fig. 15. In agreement with this conclusion note that the same kind of unpredicted slip lines are also seen in planar-glide α-brass (e.g., fig. 7 of Mitchell et al. [139]) as already mentioned. With similar certainty one will ascribe the variations of shear intensity that are indicated by the slip-line groupings, to channeling of glide dislocations between the carpets. In Stage III (and presumably also in Stage IV) rather well-defined, finely lamellated slip bands appear. In agreement with similitude, the slip-band length is inversely proportional to the stress but undoubtedly, throughout, it remains much longer than the average cell diameter. This may be seen in figs 41 and 42 which show the slip-band structure at the highest strains accessible in simple tensile tests, namely at $\cong 65\%$ shear in a pure copper single crystal due to Mader [225] and at $\cong 90\%$ shear in a dispersion-hardened copper crystal due to Ebeling [227]. The contrast and slip-line geometry indicates that in each case the slip systems did not operate simultaneously but alternated; and certainly there is no evidence for the five independent slip system that would have been expected based on the Taylor/Bishop and Hill criterion and on which modeling of textures is based (compare [2,3]). And even cross slip is not very evident, while it should profusely operate according to the Seeger pile-up model.

The lamellation in the slip bands in wavy glide has a similar spacing as that in the elementary structure [220]. Further, the depth of both the fine slip lines as well as of the lamellae within the slip bands does not change with stress or strain, but rather incongruously it depends on surface orientation [222]. Therefore, in agreement with Mitchell [145], it has been concluded [23] that the lamellation in wavy glide is a surface effect. Namely, the slip line structure is modified relative to the interior slip distribution due to the surface energy of the newly exposed surface ledges. Since this surface energy is typically negative through adsorption of gases on freshly exposed ledges, there is an energy gain on deepening a slip step, indeed accentuating with rising temperature. Yet that energy gain is reduced by edges, and therefore the LES hypothesis predicts some close-range coincidence of dislocation paths at surface intersections so as to widen already existing ledges instead of producing ever new ones. This conclusion is reinforced by the observation that in slow deformation at elevated temperatures the depth of individual slip steps on small single crystals can apparently grow without limit, e.g., figs 122 and 123 of Schmid and Boas [135], due to Boas and Schmid [228] and Masing and Polanyi [229], respectively, and fig. 43 extracted from [230], due to Tsien and Chow [231].

Evidently, then, the slip marking structure in wavy-glide materials is somewhat influenced by surface conditions. However, independent of such artifacts, it seems clear

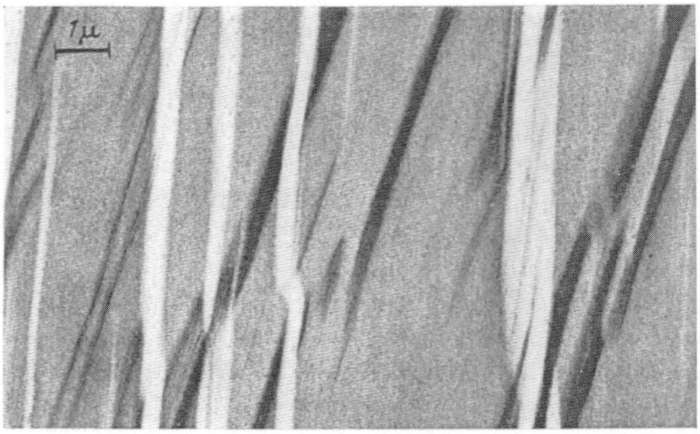

Fig. 41. Slip-band structure after ≅ 65% shear on a pure copper single crystal oriented so that double glide would have commenced at about 23% shear. Note the appearance of mutual avoidance of slip systems, even though probably the width and certainly the length of the slip bands is larger than the average, <∼ 1 μm cell diameter as judged from eq. (30). (Figure 15 of Mader [225].)

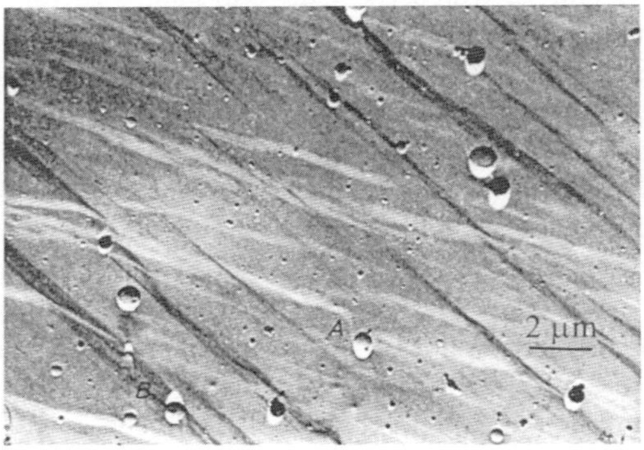

Fig. 42. Slip-band structure in a tensile-tested dispersion-hardened copper crystal strained to a shear of 90%. As in fig. 41, the band structure suggests the avoidance of momentary, local double glide on a scale rather larger than the average <∼ 1 μm cell diameter. (Figure 10 of Ebeling [227].)

that also in the bulk, shear is concentrated into zones, next that in Stage II slip-line lengths decrease inversely with the resolved shear strain [29,225,226,232–236] (meaning also inversely with the resolved shear stress!), that the number of slip bands per unit length of specimen increases approximately linearly with the resolved shear stress, and that in Stage III slip bands form that fragment and shorten considerably (e.g., figs 41, 42).

Over the years a number of very careful studies has been made also on planar glide slip lines. For the case of α-brass in single glide, these were shown by Wilsdorf and Fourie [237] to be due to sliding on either one single atomic plane or at most a few very closely spaced ones; and Fourie and Wilsdorf [238] determined that in course of straining (i) individual slip lines grow, (ii) that the gaps between them are gradually filled up with similar lines such that the number of lines rises linearly, but not proportionally, with stress,

ε = 28% @ 1000 degrees Celsius

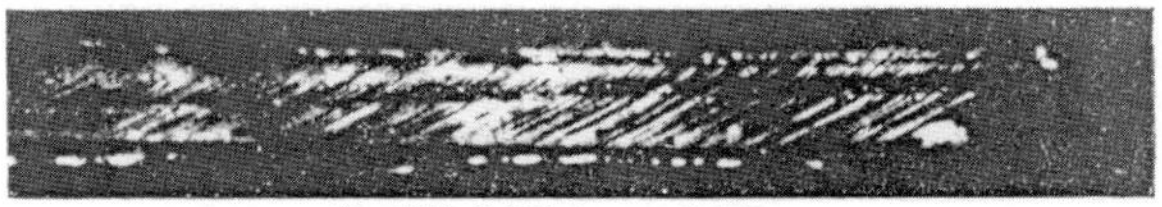

ε = 12% @ 300 degrees Celsius

ε = 12% @ 20 degrees Celsius

Fig. 43. The frequently observed increase of average slip band depth with increasing temperature and decreasing strain rate for the example of molybdenum single crystals, strained to (a) 28% elongation at 1000°C, (b) 12% elongation at 300°C and (c) similar elongation at 20°C. (Figures 6 to 8 extracted from [230] credited to Tsien and Chow [231].)

and (iii) that the average glide per line also rises linearly, but not proportionally, with the resolved shear stress, wherein the glide per line and the line spacing are almost independent of crystal orientation.

Perhaps most remarkably of all, unlike the case of planar glide, e.g., fig. 21, in wavy glide there is typically no recognizable connection between the surface markings and the underlying dislocation structures [239,240].

2.8. Alloy hardening

Besides perhaps changing the behavior from wavy to planar glide, as in α-brass, alloying most commonly simply adds to τ_0, or especially in wavy glide its effect may be much the same as prestraining, i.e. as if to eliminate the initial part of the stress–strain curve, in particular of course Stage I but also rather more than that. Examples have been given in [22] and [23], and this aspect of the theory will be discussed in part 3.

2.9. The Cottrell–Stokes law and three other types of work softening

'Work softening' is the reverse of work hardening, i.e. the decrease instead of the increase of the flow stress through plastic deformation. In addition to dislocation unlocking from some type of atmosphere already treated in section 1.8, work softening can have any of

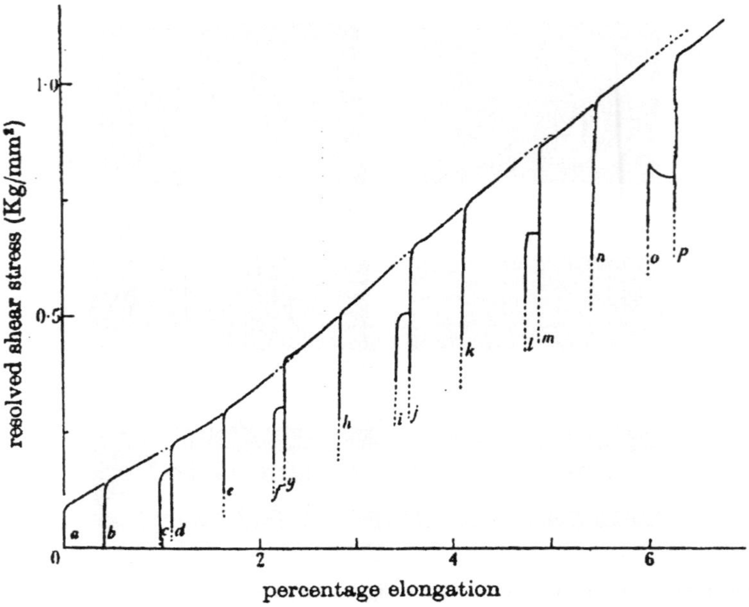

Fig. 44. Resolved shear stress/shear strain curve of a 99.992% Al crystal strained in tension at $2.5 \times 10^{-3}\%/\text{sec}$ strain rate, mostly in liquid nitrogen (90 K) and partly (curves c, f, i, ℓ and o) in iced water (273 K). Test increments in liquid nitrogen were preceded by 30 min rest periods at 273 K. These rest periods as well as the changes of temperature per se evidently did not affect the flow stress, as seen from sections a/b, d/e, g, h, j/k and m/n. The flow stress reduction compared to the 90 K curve, observed in the interspersed tests at 273 K, represent changes of the temperature-dependent part of the flow stress, τ_S, which may be seen to be proportional to the temperature independent part, τ_G. In curve ℓ the onset of work softening, i.e. reduction of flow stress through straining instead of the normal work hardening, is barely perceptible, while in o work softening is evident. From the data and description given by the authors it may be concluded that work softening occurs at strains which fall into Stage II at the low temperature but into Stage III at the higher straining temperature; i.e. the work softening bridges the difference between the flow stress appropriate to Stages II and III. (Figure 11 of Cottrell and Stokes [242].)

four causes. One is discontinuous changes of the straining temperature. The pioneering study of this effect is due to Stokes and Cottrell [241] and Cottrell and Stokes [242] as follows. In a tensile machine with a strain rate of $2.5 \times 10^{-3}\%/\text{sec}$, they elongated 99.992% pure aluminum crystals for several percent in a bath of some fixed temperature, then unloaded and resumed straining at a different temperature, after an additional strain interval making repeated changes of straining temperatures anywhere between 78 K and 423 K, i.e. -195°C and 150°C or $0.084T_M$ to $0.45T_M$. An example of their results is shown in fig. 44. They established the so-called "Cottrell–Stokes law", to wit that the flow stress consists of two parts, nowadays, following Seeger [30], mostly dubbed τ_G and τ_S. The first of these, τ_G, depends on temperature only by being proportional to the shear modulus while τ_S is the increment of flow stress effected through changes of straining temperature, and thus presumably is subject to thermal activation.

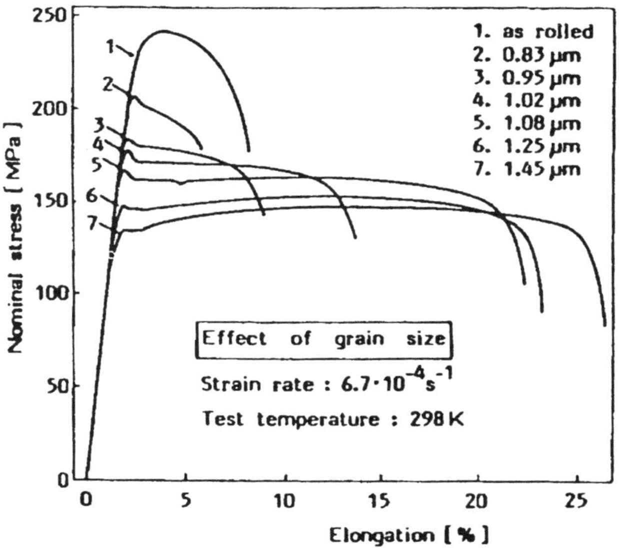

Fig. 45. Dependence of the room-temperature tensile stress-strain curves of an aluminum alloy on grain size. The associated averaged Hall–Petch constant drops from 191 MPa $\mu m^{1/2}$ for curve 2 to 175 MPa $\mu m^{1/2}$ for curves 3 and 4. While specifics change moderately from case to case, the general trend of the curves is typical for ultra-fine-grained material. (Figure 13 of [244] due to Westengen [245].)

Surprisingly, when corrected for the temperature dependence of the elastic modulus, τ_S/τ_G was found to be constant in the order of 10% (as, discounting thermal expansion of the specimens, it is in fig. 44) independent of strain and thus work hardening after the initial few percent, i.e. in Stages II and III. Hirsch and Warrington [243] found that the τ_S/τ_G ratio rises somewhat with further increases of straining temperature, i.e. in the range above $T_M/2$.

The second type of work softening, independent of impurity effects, was observed in the same study. It occurred when low-temperature straining in Stage II was followed by straining at an increased temperature such that, at the same strain, Stage III would prevail. In that case the first slight local deformation caused work softening that initiated the corresponding Lüders bands of about 0.1% shear strain. It is confidently concluded that this flow stress drop bridges the gap between Stages II and III (see fig. 44). Cottrell and Stokes contrasted the very rapid reduction of flow stress through mild straining at the higher temperature discussed, with the much more sluggish, only partial removal of work hardening through exposure to the same temperature without straining, saying: "Thus, although low-temperature work hardening can be partly removed by annealing, it can be removed more rapidly and more completely by plastically working the material during annealing" [242].

Next, materials of very small grain size are prone to a third type of work softening, illustrated in fig. 45 extracted from [244], due to Westengen [245]. Finally, changing of the straining geometry can sometimes also lead to work softening.

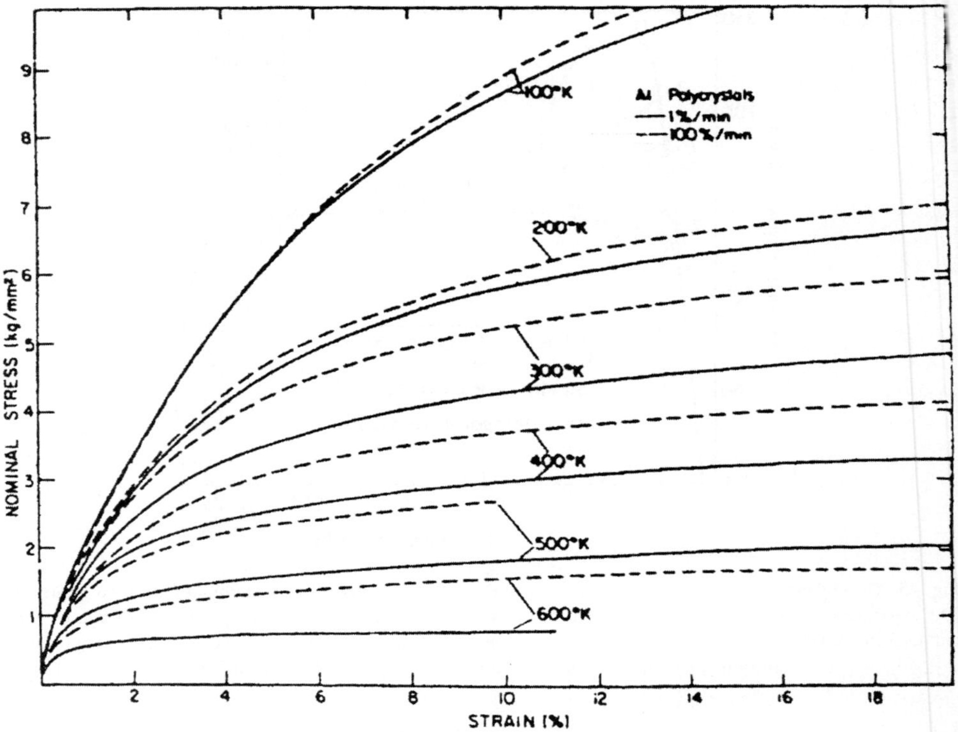

Fig. 46. The temperature dependence of the stress–strain curves of pure polycrystalline aluminum at different strain rates. (Figure 2 of Kocks, Chen, Rigney and Schaefer [246].)

2.10. The strain-rate dependence of work hardening and anelasticity

The observations in section 2.9 above are in line with the temperature and minor strain-rate dependence of work-hardening curves, in that shifts of the work-hardening curve through strain-rate changes must evidently be due to τ_s and be a rough measure of it. Work hardening, then, is principally caused by a rise of τ_G which is accompanied by a typically proportional rise of τ_s. In fact, strain-rate changes have only a momentary but not permanent effect on the flow stress. They amount to only several percent per decade increase of flow stress, e.g., as illustrated in fig. 46 due to Kocks, Schaefer and [246].

The combination of the facts that true LES equilibrium is not attained on account of τ_0, that τ_0 can never be exceeded along dislocation lines (section 1.8) and that τ_0 and τ_s are temperature dependent, i.e. subject to thermal activation, gives rise to numerous transitional effects, apart from the increasingly important addition of creep strains at temperatures above about one half of the melting point. Specifically, the slope of the initial part of a stress–strain curve, before any noticeable plastic deformation, is smaller than corresponds to elastic deformation. This is called the "ΔE effect" (see section 3.14). Further, on unloading and reloading a specimen, the downward and upward parts of the curve, which in first order should be determined by Young's modulus and therefore be coincident, form a narrow hysteresis loop, and there arises the corresponding crossing-over of the two parts when the previous highest applied stress is exceeded. Also, when a specimen is loaded at constant stress below the previous highest flow stress or unloaded after work hardening, there is some minor forward or back "creep", with total extensions not exceeding the elastic strain [247,248].

Most regrettably, judged with the wisdom of hindsight, those anelastic creep effects, although of scant technological import, have played an unduly great role by becoming the focus of attention. The mistake of expecting a close link between minor anelastic effects and work-hardening proper, originated with R. Becker's theory [249–251] and subsequent attempts to verify the theory by means of the anelastic creep effects discussed [247, 248,252,253]. Even though the result was negative [247] the error was never decisively corrected [248] and consequently was indirectly propagated, most lately in the form of SODS modeling. As a result, by diverting attention from the fundamental issues, the development of the theory of work hardening has been greatly impeded. Additional information on the historical development of the issues have been presented in [254] and in section 2.3 of [22].

2.11. Recovery

The previously mentioned elimination of redundant dislocations on mild annealing, first discovered by Bailey and Hirsch [68] and illustrated by means of fig. 36, reduces τ_0 but leaves the work-hardening curve on reloading unchanged [255]. This is the first step of 'recovery'. Fairly recently, Nes [256] has presented a literature review. Among the contributors to this area the following deserve special mention, listed in chronological order: Dix, Anderson and Shumaker [257], Michalak and Paxton [258], Sanders, Baumann and Stumpf [259], Bay and Hansen [260], Young, Headley and Lytton [97] and Furu, Örsund and Nes [261]. The major result from micrographs is that with increasing recovery times and temperatures, cell structures tend to coarsen, but apparently not uniformly [97].

Almost all flow-stress recovery data in the literature have been plotted versus the logarithm of recovery time, t, and a linear decrease as

$$R_{\mathrm{r}} = (\tau - \tau_0)/(\tau_{\max} - \tau_0) = 1 - A\ell n(1 + Bt), \tag{32}$$

with τ and $\tau_{\max}$ the momentary and initial yield stress, respectively, is often approximated. This relationship, with A proportional to the absolute temperature of annealing, is the characteristic log time law of aftereffects that results when the activation energy of a thermally activated process rises linearly with the associated property change [262]. This recovery time law was already proposed in 1948, namely for pure aluminum single crystals [263]. Since then, eq. (32) has been found to hold more or less accurately for the recovery of f.c.c. and b.c.c. metals, of pure metals and alloys, of single- and polycrystals, and of planar and wavy glide materials (see Nes [256]), albeit with deviations for small and large recovery times. Figure 47 gives an example for wavy-glide b.c.c. iron due to Michalak and Paxton [258]. Here the flattening of the recovery curves at long times is very

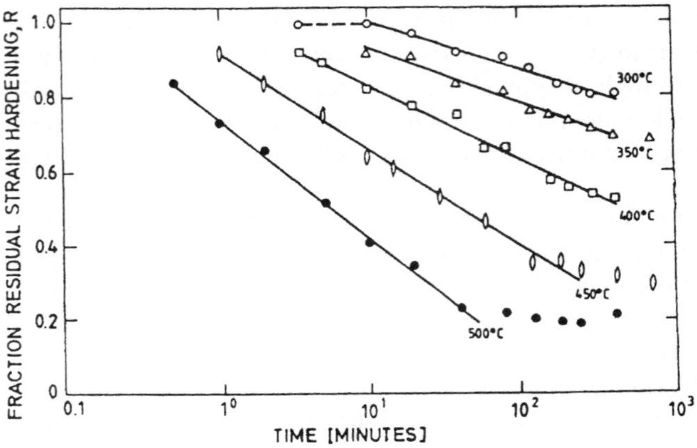

Fig. 47. Recovery curve, i.e. dependence of yield stress versus annealing time, of polycrystalline, wavy-glide 99.98% iron after 5% tensile strain at 0°C. Note that the negative slopes of the curves increase approximately linearly with the absolute temperature, as expected from the theory (see section 3.11). The 450°C and 500°C curves clearly indicate that recovery does not remove work hardening completely, as already observed also in [263]. In fact the remnant hardening after total recovery was found to rise with pre-strain. (Figure 3 of Michalak and Paxton [258].)

evident, and a similar flattening was also observed in [263]. Since, further, the percentage of flow stress that can be recovered decreases with increasing prestress, i.e. $\tau_{\max}$, one will suspect that there is in fact a mild overall curve rather than an abrupt kink. This surmise is strongly supported by the data of Young et al. [97] of which fig. 48 gives examples.

The simple behavior seems to be displayed in planar glide, at least judging by the most astounding proof of the fundamental nature and persistence of the recovery time law of eq. (32) illustrated in fig. 49, extracted from [256]. The corresponding data extend over a seventeen year period and pertain to planar-glide Al–Mg alloys [257,259]. The conclusion that certainly in this planar-glide example, recovery is due to a singly thermally activated process that derives from the most basic properties of dislocations, is inescapable.

2.12. LED structures in fatigue

Dislocation structures formed in long-term plus-minus constant-strain amplitude fatigue testing differ in three significant ways from those in unidirectional straining at same temperature. Firstly, dislocations with Burgers vectors of opposite sign tend to be more evenly mixed, secondly the structures are closer to thermodynamic equilibrium and, third, typically only one or two slip systems are activated so that effects of slip-system avoidance that cause cell blocks and deformation bands in unidirectional deformation (see section 2.6) do not enter. Therefore the LEDS nature of fatigue-induced structures is often clearly evident. In fact, the existence of Taylor lattices was first recognized in fatigued samples, namely as the so-called 'loop patch structure' [126]. Also Neumann's [127] analysis of the decomposition of unconfined Taylor lattices (alias the lancet-shaped strands of the

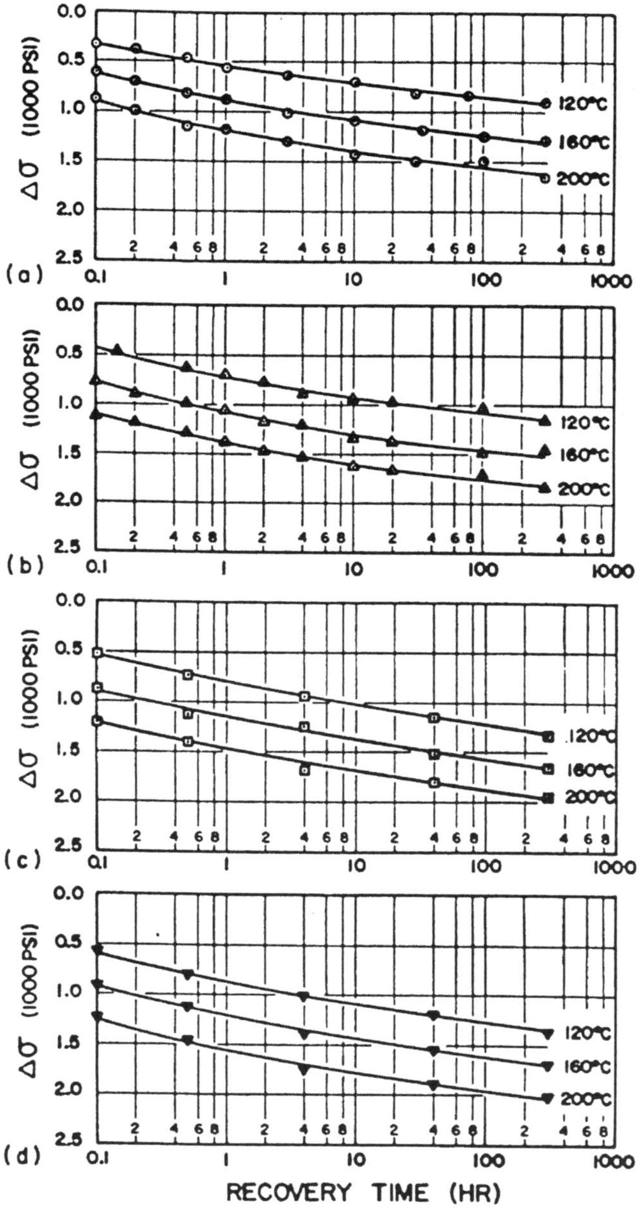

Fig. 48. Recovery curves for high-purity (99.99%) aluminum polycrystals. These curves are comparabe to the recovery curves in fig. 47. Note the small curvature that according to section 3.11.4 would be expected from an increasing grain size during recovery, as indeed observed. (Figure 2 of [97].)

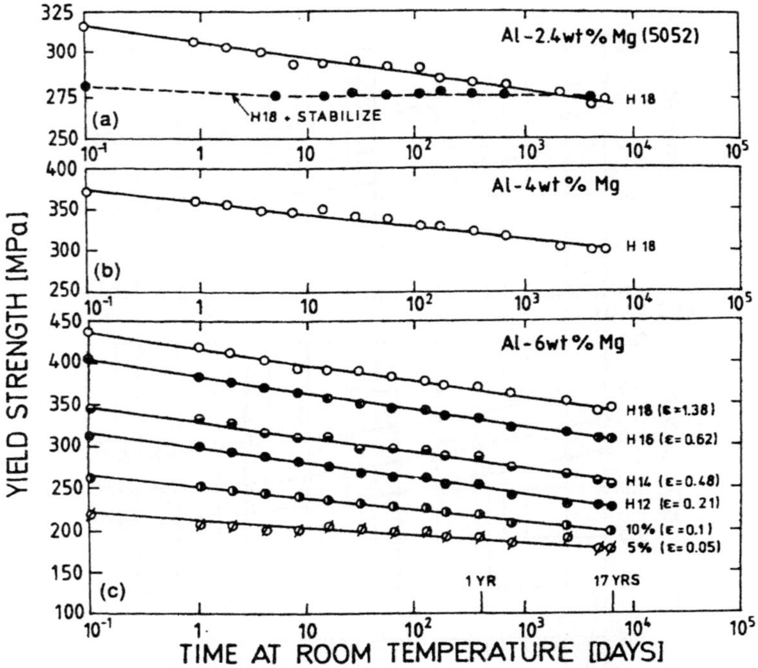

Fig. 49. Time-dependence of the yield stress of different Al–Mg alloys (being planar glide materials and hence comprising Taylor lattices) during recovery at room temperature. Note the extraordinary persistence of the logarithmic time law over a seventeen year (!) period. This time law is expected from the theory of after effects [262] and was first observed for wavy-glide aluminum single crystals in [263]. In wavy glide (e.g., also figs 47 and 48), i.e. in cell structures, the log time law is less accurately obeyed than in this diagram. (Figure 12 of Nes [256] based on [257,259].)

loop patches) that are separated by clear channels in which the glide dislocations move, into the ladder structure of dipolar walls of primary edge dislocations, has already been presented in section 1.7.3 and fig. 16. An example of the corresponding microstructure in a fatigued copper single crystal is shown in fig. 50 due to Winter [264]. Here the Taylor lattice in the form of loop patches is on the right, and the dipolar primary edge dislocation walls of the ladder structure into which it is transformed on the left. That ladder structure characterizes the material underneath PSBs (Persistent Slip Bands) in fatigued f.c.c. samples. Eventually, fatigue cracks nucleate in PSBs and spread along them. This, then, is another experimentally confirmed LEDS transformation by which a Taylor lattice can change into a lower-energy form.

Multiple glide causes ‘maze structures’, briefly discussed in section 1.5.1(1). Among these, maze structures with two mutually perpendicular sets of walls and due to the participation of two slip systems are rather frequent (compare Laird, Charsley and Mughrabi [265]). Figure 51 gives an example observed in fatigued f.c.c. Cu–Ni alloy, due to Charsley [266]. Surprisingly, in this case the walls are parallel to {100}, not {110}, planes. Yet they, too, are essentially edge dislocation-dipolar, i.e. are equivalent to fig. 4(i),

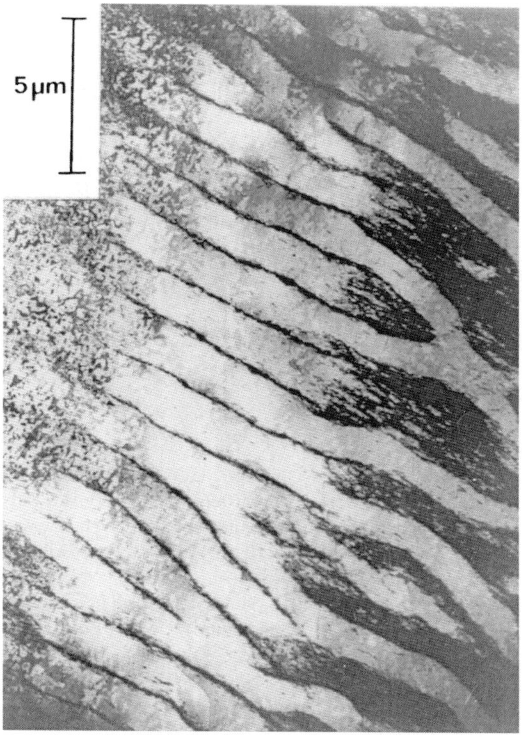

Fig. 50. LEDS transformation of an unconfined Taylor lattice as represented by a loop patch in a fatigued copper single crystal into the 'ladder structure' of parallel dipolar walls composed of primary edge dislocations. This figure is to be compared with fig. 16. The ladder structure underlies a persistent slip band (PDB) extending almost vertically from bottom left to top center. Since the open channels in the ladder as well as in the loop patch structure are normal to the primary slip direction, the plastic fatigue strain is accommodated by sequences of screw dislocations moving to and fro along the channels. (Courtesy of Winter [264].)

because the $\frac{a}{2}\langle 110\rangle$ dislocations from the two slip systems of which they are equally composed, are combined as $\boldsymbol{b}_1 + \boldsymbol{b}_2$ (e.g., $\frac{a}{2}[110] + \frac{a}{2}[\bar{1}10] = a[010]$) and as $\boldsymbol{b}_1 - \boldsymbol{b}_2$ (i.e. in this case $\frac{a}{2}[110] + \frac{a}{2}[1\bar{1}0] = a[100]$) to constitute compound Burgers vectors that are normal to each other and each perpendicular to their own wall.

However, this is not the only way in which a two-wall maze structure can form. Instead, both wall sets may be simple tilt walls comprising only one Burgers vector each but again with mutually perpendicular Burgers vectors, e.g., $\frac{a}{2}[110]$ and $\frac{a}{2}[\bar{1}10]$, as before. The difference determining which of the two possible forms arise, may well be the intimacy of local mixing of the two slip systems. Figure 52 due to Wang, Laird and Chai [267] appears to show an example of the second type. Here a simple ladder structure at left gives way to a simple two-wall maze via the introduction of a second Burgers vector through the activation of a second slip system at right, whose Burgers vector is normal to the first.

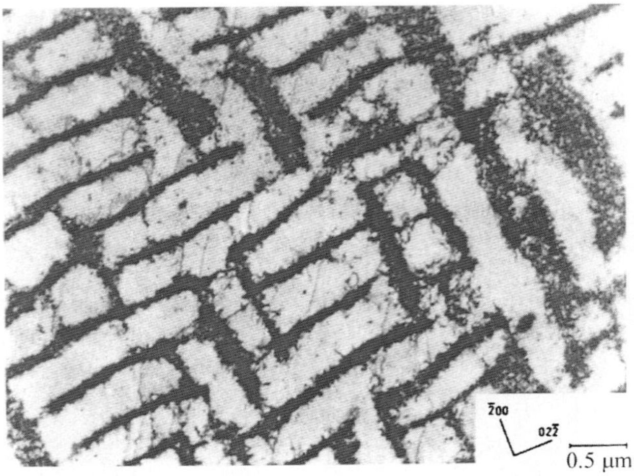

Fig. 51. A maze structure of mutually perpendicular, {100}-oriented dipolar walls in fatigued f.c.c. Cu–Ni alloy. Each wall contains two Burgers vectors in equal proportion but assembled as $\boldsymbol{b}_1 + \boldsymbol{b}_2$ (e.g., $\frac{a}{2}[110] + \frac{a}{2}[\bar{1}10] = a[010]$) and $\boldsymbol{b}_1 - \boldsymbol{b}_2$ (i.e. $\frac{a}{2}[110] + \frac{a}{2}[1\bar{1}0] = a[100]$. All of the walls are LEDSs with zero net Burgers vector content and hence free of mutual misorientation in accordance with Frank's formula (eq. (12)). The relative frequencies of free wall ends, L- and T-joints, are in reverse order of their estimated free energies, in harmony with the LEDS hypothesis. (Figure 1 of Charsley [266].)

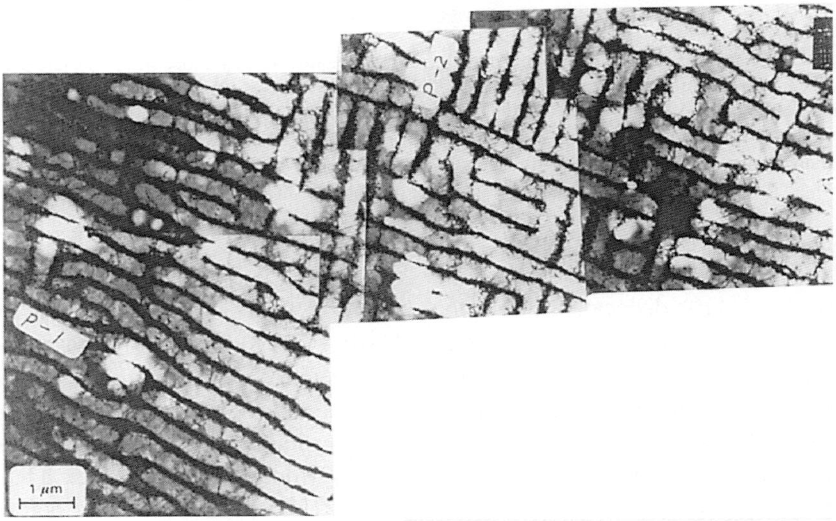

Fig. 52. A single-glide ladder structure at left adjoining a two-wall maze structure at right due to increasing secondary slip system operation from left to right. Unlike fig. 51, each wall is composed of edge dislocations from only one slip system. Note that the simultaneous action of the two participating slip systems is somewhat avoided since in the maze structure the walls do not simply intersect as in a rectangular network. (Wang, Laird and Chai [267].)

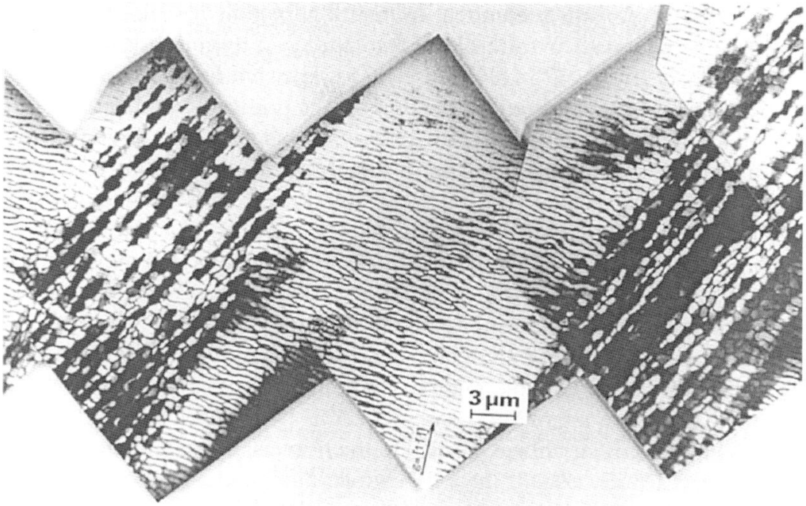

Fig. 53. Dislocation structure in a fatigued, multiple-slip-oriented copper single crystal. The geometry is clarified by means of fig. 54. (Figure 4 of Lepistö, Kuokkala and Kettunen [268].)

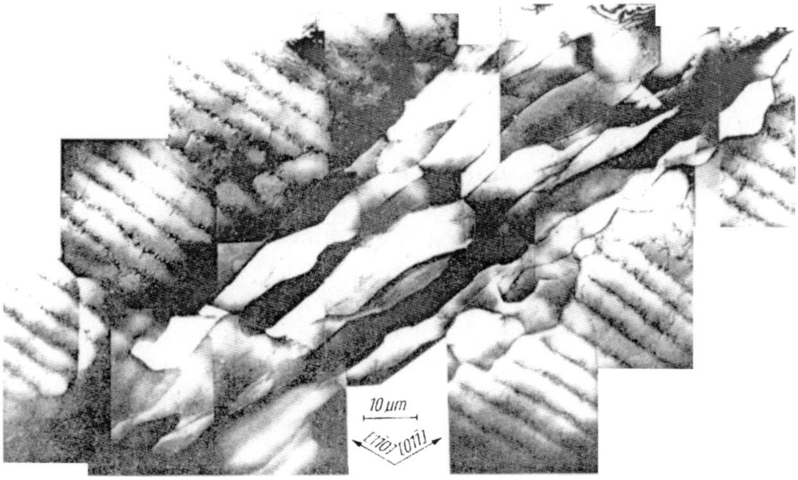

Fig. 54. A group of dislocation cells in a single-glide ladder structure, as in fig. 53, but in fatigued germanium. The band of cells extends parallel to the primary {111} plane, and with continued cycling widens so as to gradually consume the pre-existing ladder structure. The cells have alternating rotations of $\pm(1/4°)$ about an axis normal to the primary Burgers vector in the primary slip plane, constituting the so-called 'checker board pattern' [46]. Compare also fig. 36 which is a cell of the kind in this figure and shows the three participating Burgers vectors. (Figure 13 of [46] due to Scoble and Weissmann [48].)

From these arrangements it seems that mutually perpendicular dipolar walls have an especially low free energy. This inference is strengthened by microstructures such as in fig. 53 due to Lepistö, Kuokkala and Kettunen [268]. The cells are due to a secondary glide system whose $\frac{a}{2}\langle 110\rangle$ Burgers vector does not lie in the primary slip plane. In this case the maze walls, that would be inclined at 60° against the ladder walls, do not form. Instead the dislocations react in the manner of fig. 6 to generate the third Burgers vector on the primary plane. This, then, makes possible the formation of dislocation cells with relative misorientations among each other (of which also fig. 36 is an example), much as in the computer simulations of [44–47]. In these it had been shown that the minimum energy of rectangular cells occurs for alternating rotations about the same axis in a three-dimensional checker-board pattern [46]. Gratifyingly, this morphology was observed in the very first well-researched case, namely due to Scoble and Weissmann [48] shown in fig. 54. Here, as in fig. 53, the cells due to double glide extend along the PSB's, and in this case were determined to be rotated in alternating sense, presumably in a checkerboard pattern, about the axis parallel to the primary plane, normal to the primary Burgers vector.

In each of the cases illustrated, the dislocations have mutually trapped into LEDSs, and presumably embody (nearly) the lowest free energy accessible to them. Moreover, the micrographs chosen here are representative, thereby providing the corresponding further strong support for the LES hypothesis. In fact, Charsley and Kuhlmann-Wilsdorf [87] showed that the arrangement of the walls in a maze structure of the type in fig. 51, implies the minimization of wall-end stresses. Dickson et al. [88–91] successfully expanded and refined that semi-quantitative analysis to other maze structures, including rather complex patterns, and showed that these, too, minimize free energy to the extent that can be ascertained. The final, and perhaps most convincing support of the LES hypothesis that may be drawn from fatigue dislocation patterns, may be their reversibility by cycling at reduced amplitude, already mentioned in section 1.1.

3. *The LEDS theory of work hardening*

3.1. The theory of theories: Occam's razor quantified

Every theory whatsoever consists of conclusions drawn from a set of assumptions via mathematical manipulation and/or careful reasoning. Consequently, every theory has some finite probability of being false from two sources: (1) Flaws in reasoning and/or mathematical treatment and (2) flaws in the assumptions made. Accepting that flaws of mathematics and logic are relatively easily discovered and for competent investigators are rare, the major limitation in the reliability of theories therefore derives from the cumulative effect of the uncertainties in the underlying assumptions. This recognition is the basis for Occam's razor, to the effect that the theory involving the fewest and most reasonable assumptions is to be preferred.

One may quantify Occam's razor by assigning to each of the essential assumptions underlying a theory or part of a theory, labelled $j = 1, 2, 3, \ldots, n-1, n$, its specific probability of being correct, P_{jc}, or if one prefers of being false, $P_{jf} = 1 - P_{jc}$. It then follows that the probability of a (or of an affected part of a) theory being correct, P_c, is at best (i.e. free of all other errors)

$$P_c = \prod P_{jc} = \prod (1 - P_{jf}) \tag{33}$$

since a deduction cannot be correct unless *all* significant assumptions are correct.

Where to draw the line as to acceptable values of P_c in scientific research is matter of personal judgment, for theorists what they want to publish and for editors what they will accept for journals or conference proceedings. A reasonable rule would appear to limit publication to, say, $P_c \geqslant 80\%$, with theorists probably opting for a somewhat lower number and editors for higher. However, in the interest of the public, who after all foot the bill for the lion's share of all research, $P_c < 50\%$ would seem to be too low for anyone's standard unless the appropriate disclaimers are made and attention is drawn to the speculative nature of the theory. Yet, the rule discussed is quite commonly overlooked and thus violated by writers and editors alike. In this connection it is sobering to contemplate that already three plausible assumptions with 80% probability of being right place the end result at $P_c < 52\%$.

These considerations are of particular relevance to theories of work hardening, since the evidence to be explained is necessarily included among the underlying assumptions. Unquestionably, for example, a work hardening theory stands or falls with adequately understanding and taking into account the dislocation structures underlying the different work hardening stages, and similarly the fraction of the flow stress that is subject to thermal activation, plus a host of other experimental facts. It is for this reason that the essential observational facts regarding dislocation-based plastic deformation were so carefully laid out in part 2. By the same token, work hardening theories utterly depend on an acceptable and consistent application of dislocation theory. It is this precondition that is addressed in part 1, although a clean-cut division between the application of dislocation theory and observational fact is not always possible.

In the present part 3, therefore, the underlying assumptions laid out in parts 1 and 2 will be applied without further explanations. Doubtless the most important among these is the LES hypothesis. Indeed, for most of the evolution of the LEDS theory eq. (33) was kept in mind. Correspondingly, at least beginning with [110], dislocation models not based on firm experimental evidence were shunned in favor of generic considerations based on a minimum of specifics. At any rate, the number of superficially plausible models is so large that the probability of fortuitously hitting on the right model is $P_{jc} \ll 1\%$. As understanding grew and observational material increased, the theory was expanded step by step, always stopping when P_c appeared to drop below about 80%. And yet mistakes were made. Notably, for a long time it was believed that both Stages II and III were governed by dislocation cells and that cell refinement ceased at the onset of Stage III. Fortunately, those errors have long since been corrected, but others doubtless remain, to be hopefully discovered in the not-too-distant future. In this spirit the following presentation of the LEDS theory will follow the same path from the most general to increasing levels of complexity and increasing number of assumptions.

3.2. The general work hardening coefficient

3.2.1. The equation $(\tau - \tau_0) = \alpha G b \sqrt{\rho}$
The empirical proportionality between the effective flow stress, $\tau - \tau_0$, and the root of the dislocation density, $\sqrt{\rho}$, eq. (31), follows whenever dislocations mutually interact to form LEDSs. This is due to a combination of two basic facts. Firstly, the average distance of dislocation separation, ℓ, is bound to be related to the dislocation density as given in eq. (2), i.e. as

$$\rho = m/\ell^2 \approx 1/\ell^2 \tag{34}$$

with m not far from unity but depending on detailed morphology. Secondly, that the level of the dislocation stresses at distance ℓ, being proportional to Gb/ℓ, is bound to be a measure of the effective flow stress $(\tau - \tau_0)$. Much on the above line of reasoning eq. (31) was first derived in [110] and much earlier yet it was found by G.I. Taylor [54] for the Taylor lattice. In neither case was it stated in the form of eq. (31) but the associated α values, that depend somewhat on dislocation arrangement, always fall within the limits of a factor of two or so from $\alpha = 0.3$.

3.2.2. The general work hardening coefficient $(\mathrm{d}\tau/\mathrm{d}\gamma) \cong G\alpha\beta/2g$
Equation (31) implies that dislocation-based work hardening goes hand in hand with dislocation generation, while straining evidently is caused by moving dislocations. Provided similitude prevails within any one stage, their mean free paths, L, must be some fixed multiple of ℓ, i.e.

$$L = g\ell \cong g/\sqrt{\rho} = g\alpha G b/(\tau - \tau_0), \tag{35}$$

with g the so-called 'glide path parameter'. Further, of a newly generated glide dislocation density increment, $\mathrm{d}\rho_\mathrm{g}$, only a fraction $\mathrm{d}\rho$ will be trapped in the LEDS and contribute to its refinement, say,

$$\mathrm{d}\rho = \beta\, \mathrm{d}\rho_\mathrm{g}, \tag{36}$$

with β the 'dislocation retention parameter'. Hence a plastic shear strain increment, $\mathrm{d}\gamma$, is linked to the concurrent dislocation density increment, $\mathrm{d}\rho$, as

$$\mathrm{d}\gamma = g\ell b\, \mathrm{d}\rho/\beta = (gb/\beta)\, \mathrm{d}\rho/\sqrt{\rho} \tag{37}$$

However, eq. (31) yields

$$\mathrm{d}\rho/\sqrt{\rho} = (2/\alpha G b)\, \mathrm{d}\tau \tag{38}$$

leading to

$$\mathrm{d}\gamma \cong (2g/\alpha G\beta)\, \mathrm{d}\tau, \tag{39}$$

and the specific work-hardening coefficient

$$\Theta/G = [d(\tau - \tau_0)/\mathrm{d}\gamma]/G \cong \alpha\beta/2g. \tag{40}$$

As the above derivation does not depend on any particular model but makes appeal only to general LEDSs properties, eq. (40) is not specific to any stage. Constant work hardening coefficients in Stages I and II thus imply a constant value of $\alpha\beta/g$. Next, already knowing that τ_0 is typically only a small part of the flow stress and that α does not vary greatly from $\alpha = 0.3$, and that g is liable to be constant on the strength of similitude, one may conclude that within any one stage the stress dependence of Θ (figs 30–34) is that of the dislocation retention parameter β, provided similitude prevails.

3.3. Stage I

An early outline of the experimental facts regarding Stage I may be found in [110] which still seems as sound as then, including its qualitative explanation: In materials exhibiting a long Stage I, the dislocations are initially so few and far between that their mean free path is correspondingly long, i.e. g is large and the work-hardening coefficient is small. Initially, in fact, the free dislocation path is limited largely by obstacles other than dislocations. This explains occasional pile-ups, especially in early Stage I of planar glide (e.g. fig. 19(a)), because the scarcity of interfering dislocations permits the multiple operation of Frank–Read sources. That condition does not persist after dislocations have begun to interact principally among each other, i.e. at latest at the end of Stage I, and from then on multiply-acting dislocation sources are not expected, nor are they ever observed.

In thin homogeneous samples, L can exceed the specimen diameter, thereby causing a size-dependence of the length of Stage I. Conversely, the presence of non-dislocation obstacles, such as precipitates and inclusions, shortens L to the inter-obstacle spacing, thereby decreasing g and raising the work-hardening rate. This effect is particularly powerful for insoluble impurities in nominally "pure" f.c.c. metals, to the point of the suppression of Stage I by even small admixtures, as first recognized by Rosi [269]. It is for this reason that Stage I in wavy glide materials was discovered only when, after World War II, very pure metals became available. For the remainder, simply geometrically, mutual dislocation annihilation is virtually absent in Stage I and, apart from the size effect, the dislocation retention parameter, β, is essentially unity.

'Mushrooming' (see section 1.4) is held responsible for the increased slope of Stage I in wavy as compared to planar glide, since it increases the range within which dislocations mutually interfere. Yet, details in this respect as well as the control of g through non-dislocation obstacles are still unclear and much remains to be done. On the positive side, the stress at the transformation Stage I to Stage II derived in section 1.7.2, namely $\tau_{\mathrm{trans}} \approx \tau_0(1 + 0.4)$ following eq. (23(b)) is remarkably well supported. This may be verified in figs 24 and 26. In these (but not in the idealized fig. 25) the slope of Stage I is seen to decrease as its length increases and the total stress interval of Stage I approximates $0.4\tau_0$.

3.4. Stage II

3.4.1. Planar glide

As documented in parts 1 and 2, planar-glide Stage II is controlled by Taylor lattices, where also the critical flow stress of Taylor lattices and the pertinent dislocation kinematics were presented (fig. 20). In order to find the work-hardening coefficient according to eq. (40) the values of g, β and α are required. The first of these may be tentatively extracted from fig. 22 by counting the number of columns per domain. This yields $g \cong 20$. Namely, at release of each new dislocation from a domain wall, all of the g dislocations in the row extending to the next domain wall are moving a distance L/g if L designates the spacing of the domain walls, thereby yielding an effective mean free path of the new dislocation of L. Next β was already found as $\beta = 1$. This follows because, simply geometrically, mutual dislocation annihilations are bound to be very few compared to captures in the form of dipoles and higher aggregates.

Finally, α must be determined from the relation of flow stress to dislocation density via eq. (31). Albeit a difficulty now arises in that the Stage II Taylor lattice is rotated against that in Stage I. Even so it is assumed that the flow stresses are adequately represented for both orientations, – but again realizing that much more needs to be done.

Proceeding with the above assumption, one obtains an upper limit for α by using eqs (6) and (17) in conjunction with fig. 13 for the particular case of the stable configuration A, i.e. for $s/d_\mathrm{f} = 2\sqrt{3}$ and $F = 2.83$, to express the critical Taylor lattice stress, with $\nu = 0.3$, as

$$ {}_\mathrm{A}\tau_\mathrm{TL} - \tau_0 = F_\mathrm{A}\tau_\mathrm{dip} = F_\mathrm{A}Gb/[8\pi(1-\nu)d_\mathrm{f}] \cong 0.161Gb/d_\mathrm{f}. \tag{41} $$

Next introduce into eq. (41) the dislocation density in terms of d_f, namely

$$ \rho = (1/sd_\mathrm{f}) = (d_\mathrm{f}/s)/d_\mathrm{f}^2, \tag{42a} $$

or, specifically for the stable A-lattice with $(s/d_\mathrm{f})_\mathrm{A} = 2\sqrt{3}$,

$$ (1/d_\mathrm{f})_\mathrm{A} = \left[\rho(s/d_\mathrm{f})_\mathrm{A}\right]^{1/2} = 1.86\sqrt{\rho}. \tag{42b} $$

Thus one obtains

$$ \begin{aligned} {}_\mathrm{A}\tau_\mathrm{TL} - \tau_0 &= F_\mathrm{A}\left[Gb(s/d_\mathrm{f})_\mathrm{A}^{1/2}\right]/\left[8\pi(1-\nu)\right]\sqrt{\rho} \\ &= 0.161(s/d_\mathrm{f})_\mathrm{A}^{1/2}Gb\sqrt{\rho} = 0.30Gb\sqrt{\rho}, \end{aligned} \tag{43} $$

i.e. by comparison with eq. (31) find $\alpha_\mathrm{A} = 0.30$. This value is fortuitously close to the previously introduced empirical average value of α.

With these data and eq. (40) the work-hardening coefficient for the most stable Taylor lattice configuration A, becomes

$$ {}_\mathrm{A}\Theta_\mathrm{TL}/G \cong \alpha_\mathrm{A}\beta/2g \cong 0.3/40 \cong 1/130. \tag{44a} $$

By considering the most stable configuration, the upper limit of α was used above. One may obtain a reasonable lower limit, α_B, from configuration B of fig. 14, with $F = 0.84$ and $s/d_f = 2/\sqrt{3}$. Repeating, then, the previous calculation for the B-configuration yields $\alpha_B = 0.054$ for

$$_B\Theta_{TL}/G \cong \alpha_B\beta/2g \cong 0.054/40 \cong 1/740. \tag{44b}$$

These two extreme values for Θ_{TL}/G in eqs (44a) and (44b) comfortably bracket the most often cited average value for planar glide Stage II of $\Theta_{II}/G \cong 1/250$.

Experimentally there is a smooth transition of Stage II behavior between wavy and planar glide (e.g., figs 24 and 26), and similarly of the slip-line structures. No doubt there is also a continuous transition in the microstructures as indicated in fig. 1. However, there is a severe lack of systematic TEM micrographs that would document such a transition or even the detailed architecture of Taylor lattices. Correspondingly there is no evident way to improve on the above derivation of planar-glide Θ_{II}/G. In particular, the work-hardening curves of Al–Mg alloys exhibit planar glide behavior (see section 1.4) even though their microstructures do not show dislocation arrangements of the kind normally associated with planar glide.

3.4.2. Linear Stage II in wavy glide

The problem of accurately accounting for the work-hardening coefficient in Stage II is aggravated for wavy glide: The associated carpet structure evidently includes almost dislocation-free slabs between the alternating tilt boundaries that are schematically shown in fig. 15 and as TEM images in figs 1(b) and 35. Seeing that these slabs are also nearly free of internal stresses, judging from the micrographs, and none are expected on account of the tilt-wall carpets, the most likely obstacles that could account for the flow stress are the ‘hedges’ normal to the carpets, seen also in fig. 35. However, we have no data or micrographs from which to decide whether or not this is so, nor how the hedges together with the carpets transform into the Stage III cell structure at the end of Stage II. At this point even less well understood is the structure of non-crystallographic carpets found in polycrystalline wavy glide and/or in borderline planar-wavy glide (fig. 55).

On the positive side, the carpet structures are without a doubt LEDSs, and they are clearly related to Taylor lattices by basically the same Stage I/Stage II transformation, except that in this case cross slip appears to assume the role of the ‘unpredicted glide’ in Taylor lattice transformation (compare also [132]). This aspect was pointed out by P.J. Jackson and this writer [270].

All of the Stage II structures, including the variant of non-crystallographic carpets (fig. 55), are believed to share the characteristic that mutual dislocation annihilation is negligible so that $\beta \cong 1$. Further, at least in the Taylor lattices and in the planar carpet structures, similitude appears to obtain so that g is constant and of similar magnitude, thus accounting for the near-linearity of Stage II and the similar slopes of Stage II in these cases.

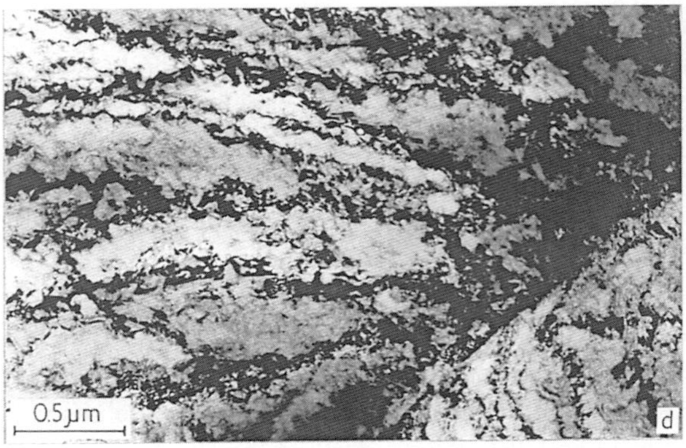

Fig. 55. TEM micrograph of a dislocation structure in polycrystalline AISI 310 stainless steel, tensile-tested at room temperature, to be compared with fig. 35. The somewhat irregularly alternating light/dark/light.. contrast reveals this to be a Stage II carpet structure, composed at each contrast change of similar dislocation rotation boundaries but with alternating sense of rotation. However, the carpets are clearly non-crystallographic and thus presumably contain dislocations from at least one additional slip system, besides the 'unpredicted dislocations' in accordance with the model of fig. 15. (Micrograph due to I. Shi, fig. 7(b) of [190].)

3.5. Stage III

3.5.1. α in Stage III

Stage III is rather well understood. Clearly, on account of the involvement of a larger number of Burgers vectors, and as verified by computer modeling (section 1.6.2), a mosaic block structure of more or less equiaxed cells has a lower energy than Taylor lattices or carpets. This fact drives the Stage II to III, Taylor lattice/carpet to mosaic block structure LEDS transformation as soon as three-dimensional dislocation mobility and availability of Burgers vectors are adequate. With the introduction of the cell structure, dislocation mean free paths are shortened and glide dislocation kinematics change dramatically to that indicated in fig. 56. No longer do the dislocations move in trains over relatively long mean free paths, but they move singly from cell to cell, their mean free paths ending at each cell wall whence one or more new loops bow out super-critically.

Rather detailed derivations of the parameter α for dislocation cell structures have been presented in [22,100,149] and [200]. In abbreviated form, it is this: The flow stress in a Stage III cell structure is controlled by super-critical bowing of always the longest links out of the cell walls as in fig. 56. For a link length l_s and with Poisson's ratio of $\nu \cong 1/3$, this requires the Frank–Read stress of [271]

$$\tau_{FR} \cong [[Gb(1-\nu/2)/2\pi(1-\nu)]/\ell_s]\ln(\ell_s/b) \approx Gb/\ell_s. \quad (45)$$

Namely, based on a simple geometrical argument that is supported by observation [110, 149,200], the longest representative links are by a factor of $n \cong 3$ longer than the average

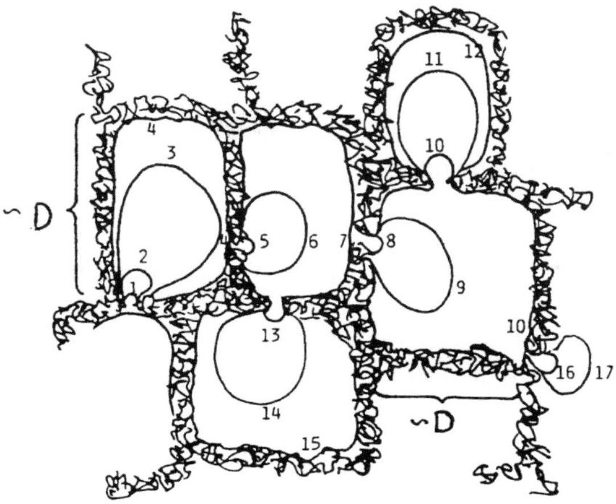

Fig. 56. Progression of a glide dislocation through a cell structure. Successive positions are indicated by consecutive numbering.

link length in the boundaries (compare eq. (83), section 3.9.3). Therefore in a Stage III cell structure

$$\ell_s = n\ell \cong 3\ell \cong 3/\sqrt{\rho}, \tag{46}$$

so that the flow stress becomes

$$\tau - \tau_0 = \alpha_C G b \sqrt{\rho}, \tag{47}$$

with $\alpha \cong 1/3$.

Importantly note that the penetration of dislocations through a cell structure illustrated in fig. 56, does not lead to any increase of rotation angle of the cell walls, since always the subsequent loop in the next cell deposits a dislocation of opposite sign at the cell wall from which its parent dislocation came. On account of various irregularities, e.g., those involved in mushrooming, the dislocations do not completely annihilate but tend to form close opposite dipoles. These contribute to the redundancy factor, M_r (section 1.5.4), but not to the flow stress except through a minor contribution to τ_0. Increases in the flow stress must therefore arise through another mechanism. That mechanism is believed to be the increase of average cell wall rotation angle in conjunction with simultaneous cell size refinement in accordance with eq. (30) that will be discussed below.

3.5.2. The Voce-curve shape of Stage III

Given eq. (40), given that α_C varies slowly at most by the factor of two, and given that g must remain substantially constant on account of similitude, the more or less linear decline of the Stage III work-hardening coefficient must be due to a similar linear decline of the dislocation retention parameter β. We may rationalize such a $\beta(\tau)$ relationship as follows:

Dislocation annihilation can take place, firstly, only among locally more or less parallel dislocation segments and, secondly, when these approach each other to within some critical distance, say Δl_j. Of interest in this connection are principally encounters among dislocations with different Burgers vectors, not among equal and opposite dislocations across ordinary cell walls from which no significant impact on flow stress can be expected as already indicated. Here j refers to the specific annihilation mechanism considered, e.g., cross slip, climb, self-climb or lattice breakdown. For any given dislocation path geometry, and with l the average link length, the chance of such an approach will be proportional to $\Delta\ell_j/l$.

But the average approach among dislocations is linked to the flow stress via eqs (31) and (34) so that one finds for the linear decrease of β,

$$\beta = (1 - p_j \Delta l_j / l) = 1 - (\tau - \tau_0)/\tau_j, \tag{48}$$

where p_j is the probability that two dislocations within distance Δl_j annihilate (e.g., cross slip affects only screws, segments may not be parallel, etc.) and τ_j stands for

$$\tau_j = \alpha G b / p_j \Delta l_j. \tag{49}$$

Integrating eq. (40) with eq. (48) yields the Voce curve

$$\tau = \tau_s + \tau_j \{1 - \exp[-G\alpha(\gamma - \gamma_s)/2g\tau_j]\}, \tag{50}$$

i.e. starting with $\tau = \tau_s$ at $\gamma = \gamma_s$ and attaining the maximum stress of $\tau = \tau_s + \tau_j$ asymptotically.[1]

3.5.3. *The temperature dependence of Stage III*

Since neither α nor g can be significantly temperature dependent, and in pure wavy-glide metals τ_0 makes only a minor contribution of the flow stress, the temperature dependence of Stage III demonstrated in figs 30 to 32 can only be due to the corresponding temperature dependence of β, i.e. of the dislocation annihilation mechanisms. Namely, the physically significant entity in mutual dislocation annihilation is not so much their separation distances, Δl_j, but the associated magnitude of the local stress which drives the dislocations towards each other. In first approximation, this driving stress is

$$\tau_{Dj} = Gb/2\pi \Delta l_j = p_j \tau_j / \pi. \tag{51}$$

Thus a temperature-dependent β indicates that the (perhaps partial) annihilation of the dislocations by any of the enumerated processes is facilitated through thermal activation. Moreover, the associated activation energy, U_β, is expected to decrease linearly with the local stress intensity acting through a characteristic activation volume, V. Hence, in the expected most simple case, for the j-th annihilation mechanism the pertinent activation energy will be

$$U_j = (U_{0j} - V_j \tau_{Dj}), \tag{52}$$

[1] Note that in [22,23] this equation contains a typographical error that was propagated from [190].

and the resulting dislocation annihilation parameter will be

$$\beta_j(\tau, T) = \beta_{0j} \exp\bigl[-(U_{0j} - V_j \tau_{Dj})/kT\bigr]. \tag{53}$$

Consequently, if a certain β-value, i.e. a certain dislocation annihilation rate due to mechanism j, is found under stress τ_{Dj} at temperature T, then the same β-value will be found for any constant ratio, C_j, of the stress-reduced activation energy $(U_{0j} - V_j \tau_{Dj})$ to kT. Thus, from

$$(U_{0j} - V_j \tau_{Dj}) = C_j kT \tag{54}$$

follows the temperature dependence of the critical stress

$$\tau_{Dj} = (U_j - C_j kT)/V_j = p_j \tau_j/\pi. \tag{55}$$

By the use of eqs (48) and (55) one then finds

$$\beta(\tau, T) = 1 - p_j(\tau - \tau_0)/\pi\tau_{Dj} = 1 - (\tau - \tau_0) p_j V_j/\bigl[\pi(U_j - C_j kT)\bigr]. \tag{56}$$

For the extrapolated flow stress at which β vanishes, i.e. for extrapolated vanishing work hardening at $\tau_{(\beta=0)} = \tau_s + \tau_j$ according to eq. (50) one finds from eq. (56)

$$(\tau - \tau_0)_{(\beta=0)} = (\pi U_j/p_j V_j) - (\pi C_j k/p_j V_j)T. \tag{57}$$

Hence for the intercepts with the τ-axis of the extrapolated straight declining β-lines of Stage III, eq. (57) predicts a linear decrease with temperature, i.e.

$$\mathrm{d}(\tau - \tau_0)_{(\beta=0)}/\mathrm{d}T = -(\pi C_j k/p_j V_j). \tag{58}$$

This is indeed observed in all relevant presently available $\beta(\tau, T)$ plots, i.e. not only in figs 30 to 32 but also in figs 5(a) and (b) of [190]. Further, from the second part of eq. (56) the slope of the Stage III decline of β with flow stress is

$$\mathrm{d}\beta/\mathrm{d}\tau = -p_j V_j/\bigl[\pi(U_j - C_j kT)\bigr], \tag{59}$$

i.e. slowly rising with increasing temperature, again as in figs 30 and 31 as well as in figs 5(a) and (b) of [190].

Certainly, this fine agreement between the above simple theory of $\beta(\tau, T)$ and all available pertinent evidence is very persuasive. The requisite detailed determination of the various parameters appearing in the theory has not yet been attempted but is urgently needed.

3.6. Stage IV

The origin of Stage IV is still in doubt. Certainly, in Stage IV there is a greatly increased incidence of the high-angle walls in the mosaic block structure, dubbed CBBs, that delineate volume elements across which the selection of active slip systems changes (compare 2.4 and table 1). Even so, and as further discussed in section 3.9.2, CBs are recognized as deformation bands or volume elements that share their essential nature and do not (significantly) affect work hardening. The low, almost stabilized value of β that may be inferred from figs 33 and 34 must have a different reason. Also, the decrease in the rate of stored energy versus stress squared shown in fig. 38 cannot be explained through CBBs.

It was previously believed probable that Stage IV, too, has Voce curve shape and that the theory outlined for Stage III also applies to Stage IV but with different values of the parameters [32]. However, the jury is still out on this question. Figure 34 suggests a linear Stage IV, and so does the quantitative analysis of the TEM data of the Langford and Cohen [212] micrographs in conjunction with their stress–strain curve in [200]. Therein, Stage IV was mistaken for Stage II on account of an almost linear shear-stress/shear strain relationship. However, the curve is not strictly linear since at constant values of α, g and β a minor curvature is imparted by the logarithmic term in the Frank–Read stress (eq. (45)). At any rate, on account of the success in [200], work hardening in Stage IV is unlikely to hold many surprises, especially since a powerful confirmation of the analysis of [200] was obtained in [33] as discussed in the next section 3.7.

Both a linear as well as a Voce curve Stage IV can be accounted for by a tentative explanation in terms of redundant dislocations in the cell walls. This will be presented in section 3.8.4.

3.7. Cell refinement through work hardening

In an equiaxed mosaic block structure of average cell diameter D, and misorientation angle Φ, the dislocation density, ρ, is

$$\rho \approx 3M_{\mathrm{r}}/Dl \approx 3M_{\mathrm{r}}\Phi/bD. \tag{60}$$

Consequently, for a fixed dislocation density, a growing cell size implies a shrinking near-neighbor distance and thus, following eq. (15) a decreasing specific dislocation line energy

$$U_{\mathrm{D}} \cong Gb^2\{(1-\nu/2)/[4\pi(1-\nu)]\}ln(1/\Phi). \tag{61}$$

If there were no other contributions to the stored energy, the LEDS hypothesis would therefore demand that the average dislocation cell diameter should grow without limit, yet empirically the cell size shrinks as $D \cong KGb/(\tau-\tau_0)$ (eq. (30)). The reason are strain gradients that are unavoidably associated with a mosaic block structure [33]. At flow stress τ these cause stress gradients of average magnitude $\tau/(D/2)$. However, the LES hypothesis demands that an elastic strain energy density be converted into the corresponding dislocations once these have a smaller energy. The principle is readily

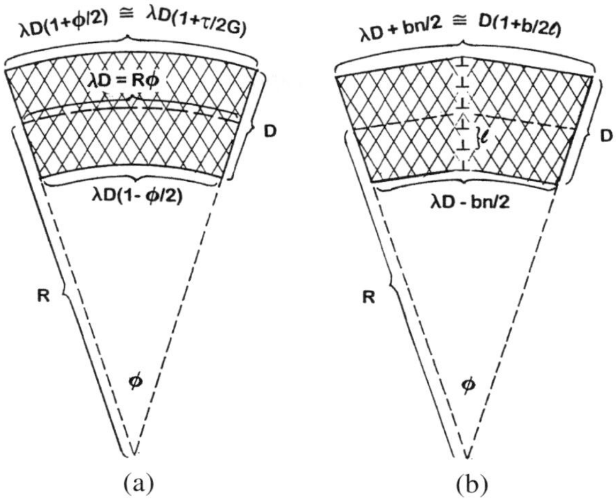

Fig. 57. The refinement of a cell structure with rising stress in accordance with similitude (e.g., fig. 37) occurs through the nucleation of new cell walls when unavoidable strain gradients, especially of bending type, are replaced by new dislocation boundaries of lower stored energy, as indicated. The empirically widely observed relationship $D = KGb/(\tau - \tau_0)$, eq. (30), is a consequence of this mechanism. (Figure 17 of [22].)

explained by the example of bending stresses which are converted into tilt walls as indicated in fig. 57. Specifically, the elastic bending of a cubic cell (i.e. with $\lambda = 1$ in fig. 57) through an angle

$$\Phi = \tau/G \tag{62}$$

can be relieved by the nucleation of a tilt wall of spacing

$$l_P = b\Phi = b\tau/G \tag{63}$$

containing

$$n_P = D/l_P \tag{64}$$

dislocations. While the bending is still purely elastic (fig. 57(a)), with the stress rising linearly from the neutral axis to a maximum of $\pm\tau$ at the most stressed cell surfaces, the strain energy density is, with eq. (62),

$$U_{el} \approx \tau^2/6G, \tag{65}$$

independent of cell size. After replacing the elastic bending by the tilt wall, including an excess of $(M_r - 1) \cong 1$ of "filler" dislocations, the energy density is, with eqs (61) and (62),

$$U_P \cong (D/\ell_P)M_r U_D/D^2 \cong 0.1 M_r (b/D)\tau \ell n(G/\tau). \tag{66}$$

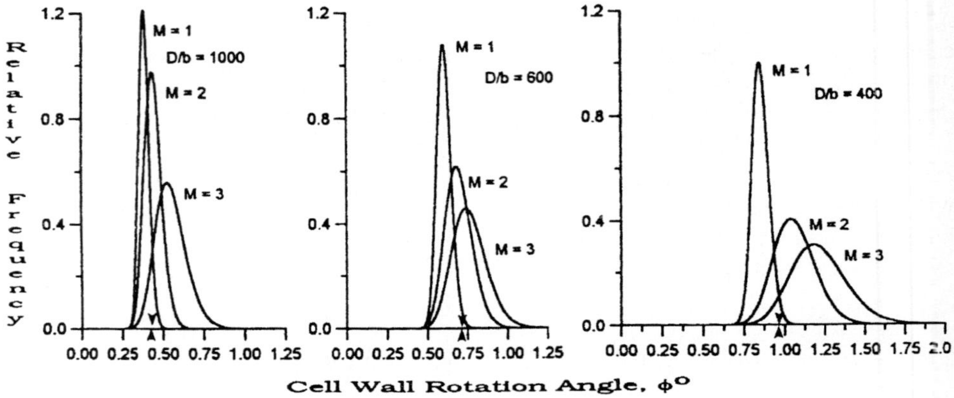

Fig. 58. Comparison of experimentally determined average cell rotation angles with those theoretically determined from the model of fig. 57, including bell-shaped curves modeled after fig. 61. Calculations have been made for 'redundancy factors', M_r (in the drawing M) of 1, 2 and 3. The vertical double arrows along the abscissa indicate the angles which had in [200] been deduced from measurements on micrographs such as fig. 37. Fortuitously, those measurements were made on the walls of 'ordinary' cells (not CB boundaries to which the theory does not apply), because only for these were the link lengths long enough to be microscopically resolved. The curves suggest $M = 2$. Gratifyingly, this is in the middle of the range of M_r-values that had in [200] been deduced from the early micrographs. (Figure 4 of [33].)

Based on the LEDS hypothesis, a new tilt wall as in fig. 57(b) is expected to replace the elastic bending when $U_P \leqslant U_{el}$ which following eqs (65) and (66) occurs at the cell size

$$D_{max} \cong [(0.6M_r Gb)\ell n(G/\tau)]/\tau \cong 4D/3. \tag{67}$$

The last part of eq. (67) arises because cell sizes will vary between D_{max} and $D_{max}/2$ before and after subdivision, respectively, for an average cell size of $D \cong (3/4)D_{max}$. Comparison of eq. (67) with eq. (30) then renders

$$K \cong 0.45M_r \ell n(G/\tau). \tag{68}$$

Numerically, with $10^{-4}G < \tau < 10^{-2}G$ and $M_r = 2$ one finds $K \cong 5$, in good agreement with empirical evidence considering the simplifications made in the theory.

The detailed analysis is in excellent accord with all known data [33,272]. Most convincingly, the expected rotation angles of newly formed walls were calculated for three wire-drawing stages based on Langford and Cohen's data [212], assuming $M_r = 1, 2$ and 3. These are indicated by the peaks of the bell-shaped curves in fig. 58, extracted from [33]. As seen, these are in close agreement with the corresponding three experimentally inferred cell rotation angles from the analysis of [200] that are shown by the vertical double arrow heads on the abscissae. This agreement is compelling because it links inferred rotation angles that were based on micrographs of another laboratory in 1970, long before any other researchers believed cell walls to be dislocation rotation boundaries, to a 1996 theory. It is especially gratifying because (i) already in [200] the most likely value of M_r (given as

M in fig. 58) was estimated as $M_r = 2$, and (ii) the analysis of [33] contains no adjustable constants. However, while at the time [200] it was thought that the observed angles were those of the *average* walls, we now know that they were those of ordinary cell walls, disregarding CBBs (compare section 3.8), that typically have much higher angles.

This error of 1970 is eminently understandable since it was then thought that all cells had more or less the same rotation angle, the difference between CBBs and ordinary cell walls was recognized only much later [178], and in any event only the ordinary cell walls had angles small enough to be determined by link length measurements. However, actual cell structures, at the least from late Stage III onward, contain CBBs (compare table 1).

3.8. Theory of slab-like deformation bands and stored energies

3.8.1. The morphology of deformation bands

Deformation bands (see section 2.6 and fig. 39) and of course also CBs are subject to the LEDS hypothesis. They lower the free energy by decreasing the flow stress through the reduction of the number of locally simultaneously operating slip systems, at the price of (i) overall deformation band boundary energy density, U_B, and (ii) energy density on account of band end stresses due to plastic strain mismatch, U_a. As already indicated (sections 1.9, 2.4 and 2.6) all cell blocks (CBs) are believed to be of the nature of deformation bands, whether primary, secondary or tertiary, although not necessarily of band-like geometry. For the case of band-like shape, the effect of energy minimization gives rise to a rather simple relationship between band length, L and width, W, derived here following [2] and [3]:

In line with the LEDS hypothesis, the deformation band morphology nearly minimizes $U_a + U_B$. The relevant equations are readily derived for bands of average length L, breadth T and thickness, W, when it is assumed that the flat sides of the bands (of dimension $L \times T$) fit together without mismatch, and that the elastic accommodation strain energy is concentrated within a zone of length $\lambda_a W$ at both ends of each band, where it causes the flow stress at the moment of band formation to rise from τ_{DB} to $\tau_{DB}(1+\delta_a)$. If so, the associated mismatch energy density is

$$U_a = (2\lambda_a W/L)\tau_{DB}^2\left[(1+\delta_a)^2 - 1\right]/2G \cong (2\lambda_a\delta_a W/L)\tau_{DB}^2/G. \tag{69}$$

Next, naming Γ_{DB} the average surface energy of deformation-band boundaries, we obtain

$$U_B = \Gamma_{DB}(1/W + 1/T + 1/L) \approx \Gamma_{DB}/W. \tag{70}$$

The energy minimum is found from $d(U_a + U_B)/dW = 0$ at band width

$$W = \left\{(\Gamma_{DB}LG)/\left[\lambda_a\left(2\delta_a + \delta_a^2\right)\tau_{DB}^2\right]\right\}^{1/2} \cong \left\{\Gamma_{DB}LG/\left[2\lambda_a\delta_a\tau_{DB}^2\right]\right\}^{1/2}. \tag{71}$$

Following eq. (16), one may write

$$\Gamma_{DB} \approx 0.035Gb, \tag{72}$$

on the argument that the average band boundary energy is only about one half of the saturation energy at maximum misorientation while at the same time $M_r \cong 2$ on account of redundant dislocations. Inserting eq. (72) into eq. (71) yields, with the most plausible values of $\delta_a \cong 1$ and $\lambda_a \cong 0.4$,

$$bL/W^2 \approx 29\left[\lambda_a\left(2\delta_a + \delta_a^2\right)\right](\tau_{DB}/G)^2 \approx 35(\tau_{DB}/G)^2. \tag{73}$$

The prediction from eq. (73) that the ratio of $(L/w^2)(G/\tau_{DB})^2$ is nearly constant for all sizes of deformation bands, whether primary, secondary or tertiary, and independent of other materials characteristics, was tested. Herein use was made of measurements obtained from fig. 39 (including primary, secondary and tertiary bands), almost all micrographs of deformation bands that could be found in the literature and for which the requisite data were available, plus a number of other micrographs by Kulkarni [216–218]. The results have supported eq. (73) remarkably well [2,3]. Further, the morphology of kink bands in diblock copolymers discussed in section 1.1 is subject to the same simple analysis except for a different, and still unknown, expression for Γ_{DB}. At any rate, at fixed stress, also for these L/W^2 was found constant for all sizes [1].

3.8.2. The energy density stored in deformation bands

The energy density differences that cause deformation banding, e.g., as in fig. 39 and the various deformation band images analyzed in [2] and [3], are tiny. This is not surprising because deformation bands are but a larger-scale structuring of Stage III and IV cell structures that only modestly changes their energy density. Namely, as explained, the deformation band patterning is driven by the decrease of flow stress, say, $\Delta\tau_{DB}$, when slip systems do not act locally together in accordance with the Taylor hypothesis but instead the material breaks up into volume elements (CBs) with a smaller number of active slip systems. The energy density, $U_{DB} = U_a + U_B$, stored in the bands is thus necessarily smaller than $\Delta\tau_{DB}^2/2G$, but it is not known how much smaller, nor is the value of $\Delta\tau_{DB}$ known. Even so, U_{DB} is bound to be a measure of the energy density changes to which the microstructure will still respond in its approach to thermodynamical equilibrium. It is therefore useful to evaluate U_{DB} in order to assess the degree to which deformation microstructures respond to the second law, i.e. the LES hypothesis, as already outlined in section 1.1.

The numerical evaluation of U_{DB} is simplified if it is realized that, in accordance with eqs (69) to (71), U_a and U_B must be quite similar in magnitude since U_B decreases inversely with W, while U_a is approximately proportional to W and $\mathrm{d}(U_a + U_B)/\mathrm{d}W \cong 0$. Hence

$$U_{DB} \cong 2\Gamma_{DB}/W \approx 0.07Gb/W. \tag{74}$$

In order to assess how much of a deviation from thermodynamical equilibrium U_{DB} represents, it is useful to translate it into the temperature rise that would result if it were converted into heat, i.e.

$$\Delta T_{DB} = U_{DB}/cd_m, \tag{75}$$

with c the specific heat and d_m the mechanical density. For most metals of technological interest the numerical value of G/cd_m at room temperature ranges around 15 000°C. Hence the equivalent temperature rise of deformation band formation is, with a safety factor of 2,

$$\Delta T_{DB} < 0.15Gb/Wcd_m \cong 2000b/W. \tag{76}$$

This means that even for the smallest deformation bands the equivalent temperature rise is less than one degree centigrade, and more typically it is 0.01°C [2,3].

3.8.3. Other forms of stored energy and energy expended in plastic straining

It is instructive to compare U_{DB} with U_{LEDS}, the free energy density stored in LEDSs at flow stress τ, which is fairly independent of detailed morphology. Namely, eq. (1) with $R = \ell$ and eq. (2), i.e. $\ell \approx 1/\sqrt{\rho}$, yields for the dislocation line energy in LEDSs

$$U_D \cong Gb^2\{(1-\nu/2)/[4\pi(1-\nu)]\}\ell n(1/b\sqrt{\rho}) \approx 0.1Gb^2\ell n\big(1/b\sqrt{\rho}\big). \tag{77}$$

With eq. (31) and with $\alpha \cong 1/3$, $\ell n(1/b\sqrt{\rho}) \cong 6$, Poisson's ratio $\nu = 0.3$, and $M_r = 2$, this yields

$$U_{LEDS} \cong M_r\rho Gb^2\{(1-\nu/2)/[4\pi(1-\nu)]\}\ell n\big(1/b\sqrt{\rho}\big) \cong 6\tau^2/G = 6G\gamma_{el}^2. \tag{78}$$

Here $\gamma_{el} = \tau/G$ is the overall elastic shear strain on account of the flow stress which produced the LEDS.

To this must be added the energy due to longer-range stresses which, as we saw, average $\tau/2$ and therefore can in this context be neglected. Also to be added is the energy of the deformation bands, i.e. U_{DB}. This is even smaller, and as a result the much slower increase of stored energy that is associated with large boundary rotation angles, as in CBBs, does not significantly impact the proportionality between stored energy and the square of the flow stress. This, then, is the reason for the rise of the stored energy in proportion with τ^2 illustrated in fig. 38 in both Stages III and IV.

As to numerical values: With $G/cd_m \approx 15\,000$°C as before, the equivalent temperature rise

$$\Delta T_{LEDS} = U_{LEDS}/cd_m \cong 6G\gamma_{el}^2/(cd_m) \approx 90\,000[°\mathrm{C}]\gamma_{el}^2. \tag{79}$$

For tensile testing, with $M \cong 2.8$ the Taylor factor, we find at the yield stress $\varepsilon_{el} = \gamma_{el}/M \cong 0.02\%$ for $\gamma_{el} = 0.00056$, while at fracture ε_{el} could be as high as $\varepsilon_{el} = 1\%$ for $\gamma_{el} \cong 0.028$. Correspondingly, ΔT_{LEDS} may range between

$$\approx 0.03°\mathrm{C} < \Delta T_{LEDS} <\approx 70°\mathrm{C},$$

but most commonly ΔT_{LEDS} will be smaller than 10°C.

By comparison U_{el}, the elastic strain energy density on account of the applied flow stress, is

$$U_{el} = (1/2)\varepsilon_{el}^2 E = (1+\nu)G\gamma_{el}^2/M^2 \cong 0.2G\gamma_{el}^2 = U_{LEDS}/30. \tag{80}$$

Thus U_{LEDS}, the energy stored in the dislocation structure, by far exceeds the simultaneous elastic strain energy due to the applied tractions, which will not exceed a very few degrees Centigrade even in extreme cases.

By contrast, U_{LEDS} is by far smaller than the plastic work expended by the tractions. Namely, U_{in}, the work-input density at plastic shear strain γ_{pl} is, very approximately,

$$U_{\mathrm{in}} \cong [\tau_0 + (2/3)\tau]\gamma_{\mathrm{pl}} = [\tau_0 + (2/3)G\gamma_{\mathrm{el}}]\gamma_{\mathrm{pl}} \approx \tau_0\gamma_{\mathrm{pl}} + (1/9)U_{\mathrm{LEDS}}(\gamma_{\mathrm{pl}}/\gamma_{\mathrm{el}}), \tag{81}$$

or for the fractional energy storage at $\tau \gg \tau_0$

$$U_{\mathrm{LEDS}}/U_{\mathrm{in}} \approx 9\varepsilon_{\mathrm{el}}/\varepsilon_{\mathrm{pl}}, \tag{82}$$

where ε_j designates the true tensile or compression strain. Correspondingly, and in agreement with observation, the fraction of plastic work input that is stored decreases steadily with straining and can attain very low levels, e.g., as in the case of drawn iron wire [200] at $\varepsilon_{\mathrm{el}} \approx 1/2\%$ and $\varepsilon_{\mathrm{pl}} \approx 7$ to below 1%. In such a case the work-input density would heat the material through a few hundred °C.

3.8.4. Energy stored in Stage IV, mushrooming and a tentative explanation for Stage IV
Finally, it may be useful to consider the reason for the decreased rate of energy storage in proportion with τ^2 observed in Stage IV. Most plausibly, this signifies a corresponding decrease in the coefficient preceding τ^2/G in eq. (78), and the only likely candidate here would seem to be a reduction in the magnitude of M_{r}, indeed its essential elimination, i.e. cessation of mushrooming. This explanation, i.e. that Stage IV is due to cessation of mushrooming, would be in general agreement with TEM micrographs which overall show a distinct lessening of wall tangling with increasing strain. But what would be the physical cause? The most plausible reason at this point is elimination of the requisite vacancy supersaturations through (i) reduced dislocation sliding distances that prevent build-up of point defects and (ii) nearness of cell walls that efficiently drain away those that have formed.

This interpretation is particularly attractive because the decrease of the slope in fig. 38 is not far from the factor of two, i.e. it has the right value to be explained by the elimination of M_{r}. And why should the work-hardening coefficient concurrently begin to decrease at a much slower rate, i.e. β become nearly constant in Stage IV (compare fig. 34)? Answer: Because the networks become more regular and the mutual bending together of intersecting dislocations as in fig. 6(c) and (d) will no longer be assisted by local climb in the course of mushrooming. Hence a more or less fixed fraction of the candidate dislocations for only partial mutual annihilation will be retained, yielding an almost constant value of β, and thereby of the work-hardening coefficient in accordance with eq. (40). However, those candidate dislocations cannot be of the same slip system, since their reactions cannot increase the 'geometrically necessary' dislocation content as illustrated in fig. 56. Thus the finite value of β must be ascribed to interference among dislocations of different slip systems.

As to the wall morphology, while the formation of new tangles would cease in tandem with mushrooming, there is no reason why M_r should suddenly plummet. Rather, throughout, mushrooming occurs only while dislocations move slowly. Climb of essentially stationary dislocations, including pre-existing redundant dislocations in cell walls, is expected to be very sluggish. Hence wall tangles would be only gradually eliminated.

One last very attractive feature of the present explanation is that by inversion it would account for the otherwise unexplained rise of the work-hardening coefficient at low temperatures, e.g., the 190 K curve in fig. 30, the 77 K curve in fig. 31 and the 100 K curve in fig. 46. In that case the increased point defect supersaturation would lead to a significant increase of M_r, thereby impeding mutual dislocation annihilation and hence raising β. However, a caveat is in order: Mindful of the theory of theories this is recognized as only a tentative proposal with perhaps a P_c in eq. (33) of 50%. Much more needs to be done to either confirm or rule out the hypothesis that cessation of mushrooming accounts for Stage IV. The requisite next steps include a detailed examination of the cell wall structure, and of the temperature and strain-rate dependence of the slope of Stage IV as well as of the stress at the start of Stage IV.

3.9. Theory of cell-structure work hardening and deformation banding

3.9.1. General considerations

Ultimately, the theory of cell structures and their behavior in plastic deformation will require determination of the cell structure of lowest free energy that would be stable under the applied stresses and accessible to the dislocations under the prevailing straining conditions. The most obvious approach to this goal will be expanded and refined computer modeling of the kind in [40,42–47]. Herein, individual pieces of cell walls could be modeled not as discrete dislocations but as disclinations in a continuum, as already suggested (section 1.6.2). From then on, in regard to the cell morphology, matters would be fairly straightforward as one will expect similitude to prevail. However, a number of other problems will still have to be solved, among them the detailed mechanism of mushrooming, the detailed mechanism of new cell wall generation and, conversely, cell wall merging which is prominently involved in wire drawing and rolling when the reduction of specimen size is more rapid than cell size shrinkage.

Most importantly, also, the theory of β (sections 3.5.2 and 3.5.3) needs to be greatly refined. In this task, the so far missing link is the manner in which individual dislocations from different slip systems move through cells and how they partially annihilate across cell walls. Namely, the picture that has now emerged from the various preceding considerations and culminating in section 3.8.4 above, is that dislocations of the same slip system encountering each other across cell walls (or indeed CB-walls) do not contribute to β (compare fig. 56). Yet, in spite of all of those needed refinements, it is not believed that any of the enumerated research will significantly change the expressions outlined above. In the spirit of the theory of theories (section 3.1) these were obtained without detailed models, and the very fact that the equations are so successful and that there are such wide-reaching similarities among the different metals, proves that only the most general

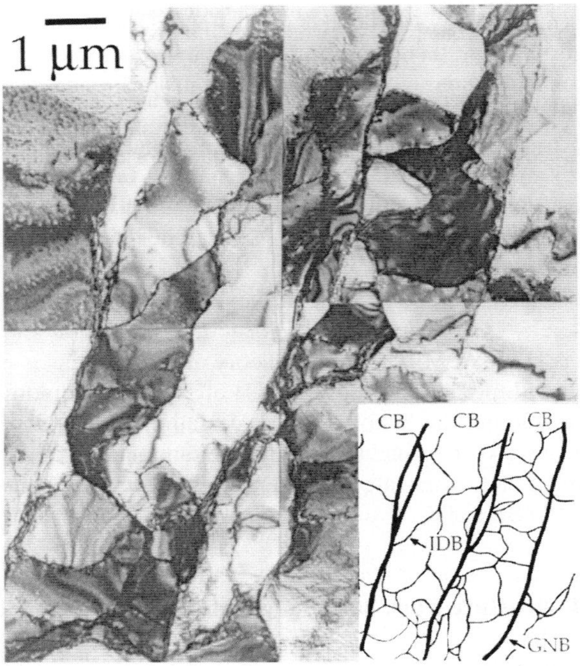

Fig. 59. Typical dislocation structure in rolled pure aluminum, being of the type from which the data of fig. 53 were obtained. It demonstrates the difference between CBBs (cell block boundaries, here labeled CB) that are believed to be deformation band boundaries (sections 2.4, 2.6 and 3.8.1), and 'ordinary' dislocation cell walls (labeled IDB for incidental dislocation boundaries [213]). Pure Al compressed to $\varepsilon = 0.6$. (Figure 1 of Hughes, Chrzan, Liu and Hansen [109].)

dislocation properties need to be considered for an understanding of the basic features of work hardening.

3.9.2. Deformation band boundaries or CBBs hardly affect the flow stress

The most important question at this point is that of the role of the boundaries between regions with different slip system selections, whether they are the boundaries of evident deformation bands or CBBs (cell block boundaries) delineating cell blocks of more irregular shape. Figure 59, extracted from [109] shows an example in which the distinction is rather obvious. The conclusion drawn from all of the evidence and theoretical considerations presented above is that deformation band boundaries/CBBs play much the same role as grain boundaries and therefore have almost no influence on the flow stress as long as they are significantly more widely spaced than regular cell walls (compare section 2.1, fig. 29(a)).

3.9.3. Link lengths in cell walls: distribution and reduction inversely with flow stress

Work hardening in Stage III, then, is determined by the supercritical bowing out of always the longest links from the cell walls within any one CB (fig. 56). Next, the

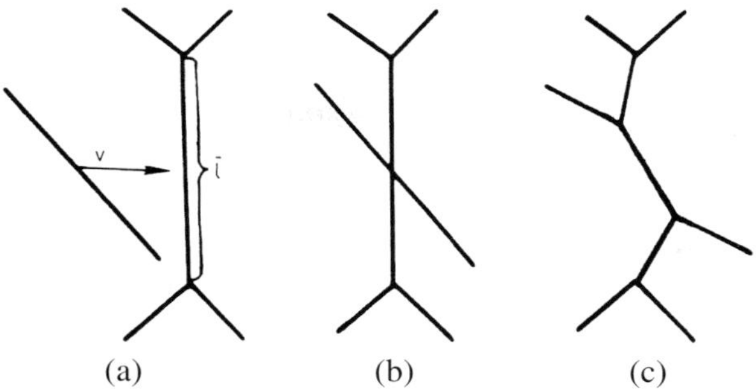

Fig. 60. Mechanism of the refinement of stationary link lengths through impinging glide dislocations with a different Burgers vector. In the drawing the glide dislocation arrives from the left. Manifold repetitions of this process cause the gradual shrinking of the dislocation link lengths in cell walls, and thereby cause work hardening, since flow requires the Frank–Read bowing of always the longest dislocation links out of, or through, cell walls. (Figure 4 of [274].)

considerations on the dislocation retention parameter, β, have shown that only encounters between dislocations with different Burgers vectors will increase the dislocation density. Furthermore, encounters between dislocations with different Burgers vectors cause a subdivision of the initial lengths of the obstacle dislocation in the cell walls into three parts to yield a new link length that is about one third of the initial one, as shown in fig. 60 extracted from [274]. This had already been dimly recognized in 1962 [110] where the ratio of the source link length (ℓ_s of eqs (45) and (46)) to the average link length, ℓ,

$$n = \ell_s/\ell \tag{83}$$

was estimated at $n \cong \pi$ (compare section 3.5.1).

The argument for $n \cong 3$ was formally explained in [200]. There also the first determination of n was made by direct observation (fig. 2 of [200]) which substantiated $n \cong 3$. Finally, in [149] and [274], the distribution of link lengths based on this mechanism was estimated to have the form of fig. 61. At the time of writing that paper, the only experimental determination of the link-length distribution function known to the author was that of Bilde–Sørensen [275], and this did not fit fig. 61 too well. However, Lin, Lee and Ardell [276] have made excellent determinations of the link-length distribution in single crystals of both rock salt and aluminum in different stages of high-temperature creep as reproduced in fig. 62. The close similarity among the three distributions is evident. In fact it is even closer than immediately obvious because the scale of the ordinate relative to that of the abscissa is moderately higher in fig. 61 than in fig. 62, making the former appear to be more steeply peaked. Importantly, too, these data strongly support the LES hypothesis, in connection with the conclusion in section 1.1 that "structure reversibility... implies that creep at different temperatures generates those structures that otherwise are produced at the same stress and temperature in direct loading."

In spite of this most gratifying agreement, a margin of uncertainty remains because Lin et al. made their measurements on etch pits without specific correlation with sub-boundaries and their rotation angles. Actually, if translated into rotation angles across

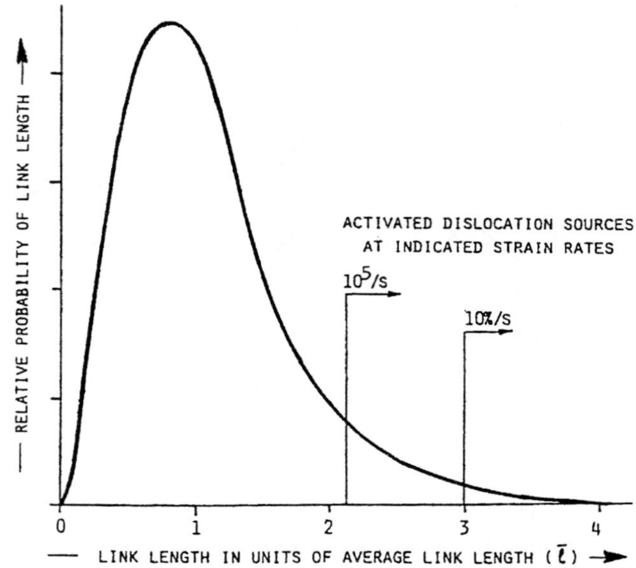

Fig. 61. The distribution function of link lengths in dislocation cell walls that, based on the LEDS theory, is semi-quantitatively expected from the mechanism in fig. 60. (Figure 1 of [274].)

cell walls, rather similar functions as in figs 61 and 62 are generated, as one may readily convince oneself. The missing connection to cell walls can therefore be made (albeit again not with 100% confidence) through the recent data on the distribution function of cell wall (*not* CBB!) rotation angles recently published by Hughes, Chrzan, Liu and Hansen [109] and reproduced in fig. 63.

As is the case for link lengths in fig. 62, so also the distribution of wall rotation angles is independent of applied stress, strain rate, temperature and a host of different materials characteristics that tend to be invoked in model-based theories of work hardening. As already indicated, the data in fig. 63 were all obtained from ordinary cell walls, i.e. IDBs (incidental boundaries [178]) that subdivide individual cell blocks, delineated by cell block boundaries, here labeled CB, as clarified in fig. 59. Thereby the earlier conclusion that work hardening in cell structures is solely a function of the ordinary dislocation cells is strongly supported.

Altogether, figs 61 to 63 (with incidentally fig. 61 having served as model for the bell-shaped curves in fig. 58) constitutes a powerful proof of the LES hypothesis, as they demonstrate the existence of an equilibrium dislocation structure that under comparable circumstances is presumed to be reversibly approached (compare section 1.1). And by being in full harmony with it, they similarly constitute powerful support for the LEDS theory.

3.9.4. Distribution of dislocation cell sizes

Still outstanding are data on the distribution of cell diameters. In this connection it is important to consider the origin of the dislocations that form the new cell walls. Either these

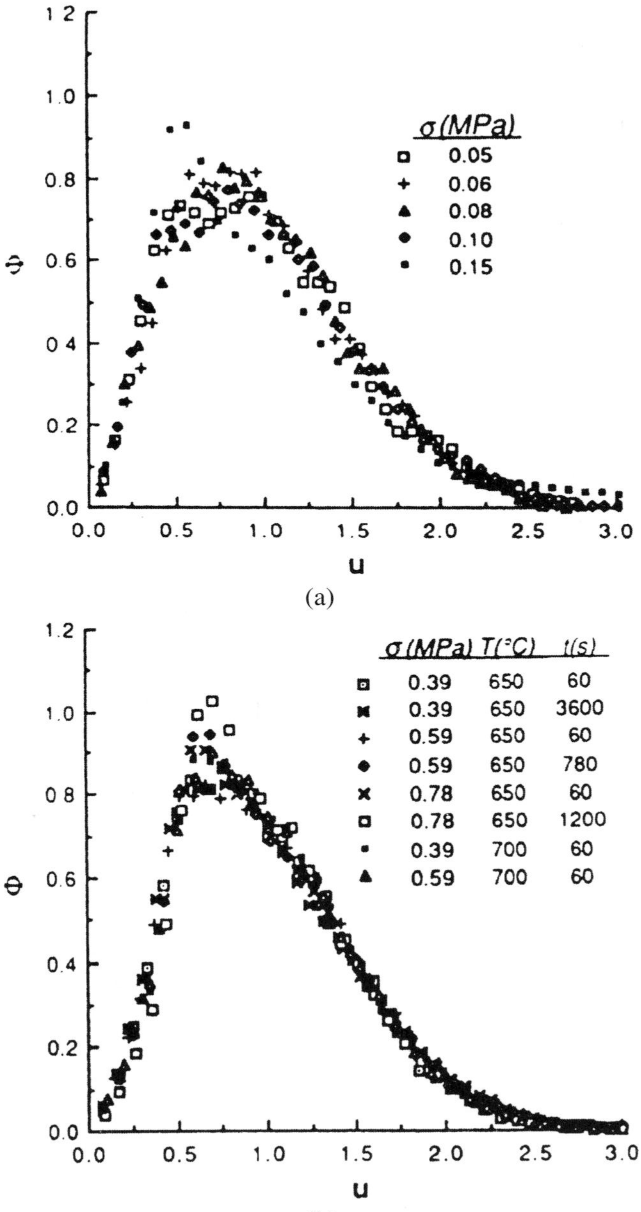

Fig. 62. Experimentally determined normalized distribution functions of link lengths in high-temperature compression creep, to be compared with fig. 61. Top: Al in steady-state creep at 647°C under the indicated stresses. Bottom: NaCl in primary creep at the indicated test conditions. In both cases data above ℓ/ℓ_{av} (here named $u = 3$) have been neglected, i.e. these curves in fact extend beyond $u = 3$, as does that in fig. 61. (Figures 6 and 7 by Lin et al. [276].)

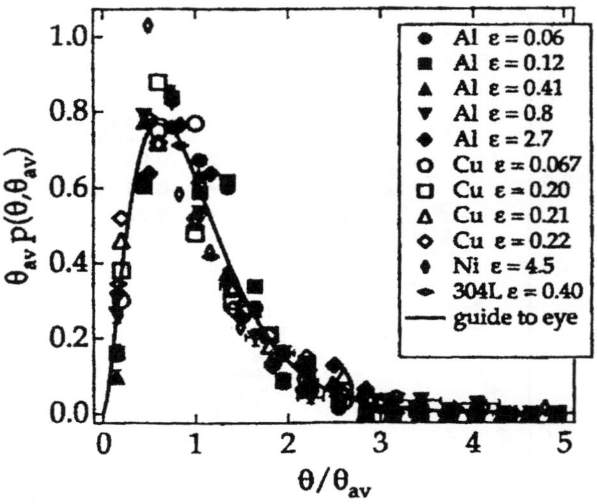

Fig. 63. Experimentally determined, normalized distribution function of the rotation angles of ordinary dislocation cells for a number of metals and strains as indicated. The existence of such a distribution is predicted by the LEDS theory (see fig. 61) and therefore this figure constitutes very powerful support for it. The Al and Ni data pertain to room temperature (being $T/T_M = 0.31$ and 0.15, respectively) while the Cu data (□) were obtained at 600°C, i.e. $T/T_M = 0.64$, and the 304L stainless steel data (—) at 1000°C, i.e. $T/T_M = 0.75$. (Figure 2 of Hughes, Chrzan, Liu and Hansen [109].)

must come from the deformation band (or cell block) boundaries to permit cell division as in fig. 57, or the walls must arise pair-wise so as not to violate the law of the conservation of Burgers vectors. This problem was previously considered in the section "Nucleation of New Cell Walls" in [277]. If pair-wise cell wall nucleation should be predominant, any one cell would be divided into three, reminiscent of the mechanism of link-length shortening in fig. 60. In that case the dislocation cell size distribution would be rather close to the shapes in figs 61 to 63. Otherwise the division would be into two cells so as to let cell sizes $> 2D_{AV}$ be scarce (rather than $3D_{AV}$ as resulting from division into three as in figs 61 to 63), and the distribution function would be correspondingly steeper. However, it should be noted that in a cell size distribution analysis, the 'small pancake-shaped cells' (PSBs see fig. 1 of [166]) which form transition bands (compare [2]) must be excluded from the count.

3.10. Strain-rate effects

A number of strain-rate effects exist that can be explained without recourse to work-hardening theory. These include (i) machine effects, (ii) initial strong increases of mobile dislocation density (whether due to break-away from various forms of anchoring, compare section 1.8, or paucity of grown-in dislocations such as in fig. 28), (iii) changes in the friction stress, τ_0 (among which Becker-type anelasticity has been prominent in past research, compare section 2.10), and (iv) oxidation or other chemical attack (e.g., as in fig. 43). However, the technologically most important form of strain-rate dependence, e.g., as seen in fig. 46, cannot be understood independent of work-hardening theory. It transiently raises the flow stress by several percent per decade of strain rate but the flow stress on further straining at the initial strain rate is not affected. Also, dislocation structures formed are not affected by rate of straining up to shear strain rates between 10^6/sec and 10^7/sec (see fig. 2 of [22] extracted from Murr and Kuhlmann-Wilsdorf [278]).

In spite of superficial appearances, the strain-rate effect documented in fig. 46 is not thermally activated. Rather, as shown first in outline [149] and then in greater detail [273], in wavy glide this basic strain rate effect is due to the distribution of dislocation links in cell walls documented in fig. 62. Just as a chain is as strong as its weakest link, so the flow stress in a cell structure is determined by the longest link lengths with the correspondingly lowest Frank–Read stress. At infinitely slow loading without thermal activation, the beginning of plastic deformation in a cell structure thus occurs with the super-critical bowing-out of the longest favorably oriented dislocation links. However, generally some finite strain rate is enforced which requires the simultaneous super-critical bowing of a number of links that rises with increasing strain rate. In planar glide, where stresses for mutual dislocation by-passing control plastic flow, variations in the spacing of rows in Taylor lattices play that same role. And no doubt the distribution function of bypassing distances in planar glide is also similar to fig. 62.

In either case, always the longest link lengths and the widest-spaced by-passing dipolar arrangements will be activated, and of these always as many as will accommodate the imposed plastic strain rate. Raising the strain rate thus requires the simultaneous activation of an increasing number of such events and therefore requires the corresponding increase of applied stress, as indicated in fig. 61. The structure becomes destabilized and brittle failure occurs when a too large fraction of the dislocations have to move simultaneously. Typically, and in line also with the already cited evidence in [278], this occurs at above 10^6/sec, moderately below the highest strain rate which can be attained on account of the limited velocity with which elastic waves are propagated.

For the quantitative treatment the reader is referred to [274]. It is made difficult because one does not know the stress-dependence of average dislocation velocities during the traverse of the cell interiors. Almost certainly this rises with stress, e.g., as found by Suzuki and Ishi [279]. This effect will be counteracted by some rise of τ_0. All in all reasonable numbers for the simultaneously acting sources and the flow stress rise as a function of strain rate were obtained. Albeit, the treatment is somewhat involved and tedious.

3.11. Theory of recovery

3.11.1. General considerations

LEDSs play an entirely different role under stress application than in the unstressed state (compare section 1.1): Under the stress that formed them they are in thermodynamic equilibrium, and in the course of their generation they have been the vehicle for entropy generation through the conversion of almost all of the plastic work into heat (compare section 3.8.3). After removal of the tractions they simply represent excess free energy. Correspondingly, at zero applied stress there is a driving force for their elimination.

However, as already explained in section 1.1, within LEDSs the individual dislocations occupy positions of relative energy minima in regard to glide motions, i.e. they suffer no glide forces, but the same is not true in regard to climb forces (compare section 1.3). Therefore, the driving forces for the removal of LEDSs in the stress-free state are bound to be climb forces, – in all stages, in cell structures as well as in Taylor lattices. Next, if well obeyed, as in fig. 49, the most often reported time law of recovery (section 2.11, and figs 47–49),

$$R_r = (\tau - \tau_0)/(\tau_{max} - \tau_0) = 1 - A\ell n(1 + Bt) \qquad \text{(see eq. (32))}$$

must be due to a singly activated process whose activation energy decreases linearly with the flow stress, i.e. is due to a relationship of the form

$$d\tau/dt = -\text{const} \times \exp[-(U_0 - V\tau)/kT], \tag{84}$$

where V is the activation volume ([262], compare section 2.11).

Qualitatively, a reason for eq. (84) is not hard to find since the climb forces that are acting to remove the LEDSs are proportional to the flow stress (compare figs 8 and 9), i.e.

$$\tau_{climb} \cong \frac{1}{2}G/[2\pi(1 - \nu)\ell] \cong (\tau - \tau_0) \cong \alpha Gb\sqrt{\rho}, \tag{85}$$

for all LEDSs. However, the direct effect is the reduction of dislocation density from which the flow stress decrease stems, and according to eq. (31),

$$d\rho = 2\sqrt{\rho}\, d\tau \tag{86}$$

so that eq. (32), implies

$$dp/dt = -C_1\sqrt{\rho}\exp[-(U_0 - V\tau)/kT]. \tag{87}$$

The remaining task is to identify the particular mechanism of recovery. Herein some models are easily ruled out: Certainly in neither cell structures nor Taylor lattices can the dislocations retrace the paths which brought them to their positions within the LEDSs. This is evident already from the simple observation that with minor exceptions (e.g., figs 18 and 19) LEDSs do not reveal the directions from which the dislocations in them arrived (e.g., figs 1, 21, 22, 35–37, 50–55 and 59). As a result, recovery is not accompanied by any significant back creep.

Also ruled out are models in which the “law of the conservation of Burgers vectors” is violated, i.e. internal dislocation removal can take place only in steps of zero net Burgers vector change, e.g., by pair-wise annihilation or in triplets of $\sum \boldsymbol{b}_j = 0$. Note that this poses no restriction on the removal of dislocation “kinking” and in general any extra dislocation line length due to mushrooming, since these extra line lengths have zero net Burgers vector content. It is for this reason that ‘tangling’ is so speedily removed (e.g., as demonstrated in fig. 36) and causes the corresponding rapid decrease of τ_0. However, the distances between the nodes that control the Frank–Read stress in cell structures, and the bypassing distances between parallel rows in Taylor lattices, are not changed thereby. Therefore, also, the work-hardening curve on reloading remains unchanged, except for the indicated transient decrease of τ_0 [255].

The remaining possibilities and resulting recovery-time laws are different for Taylor lattices and cell structures and will be discussed separately, as follows.

3.11.2. Recovery in Taylor lattices

For Taylor lattices, three possible recovery mechanisms exist: (1) Direct mutual annihilation among neighbor dislocations, e.g., as seen in the beginning elimination of dipoles in fig. 18. (2) Pairwise annihilation of dislocations at domain walls. (3) Elimination of individual dislocations at grain boundaries. Mechanism (2) moreover has two variants, namely with or without similitude. That opens the question whether similitude operates also during recovery. Specifically, does the domain size, or similarly in wavy glide the cell size, grow in tandem with the lengthening of average dislocation separation, l? For Taylor lattices, the answer is an unequivocal "NO". Namely, similitude is driven by the need for the dislocation structures to balance the applied stresses and preserve force equilibrium, whereas recovery occurs under zero applied force and removes that cause for similitude. At any rate, in terms of formalism, mechanism (2) without similitude is the same as (3) so that there are in fact only three basic recovery mechanisms.

The effect of similitude on recovery arises because it affects the "sink density", ${}_iN_{Sj}$, i.e. the number of locations per unit volume at which dislocations can be annihilated, with subscript i identifying the structure, whether cells (cell) or Taylor lattices (TL), and subscript j the case, whether without (O) or with (W) similitude. Namely, the rate of dislocation elimination is

$$\mathrm{d}\rho/\mathrm{d}t = -{}_iN_{Sj}P\exp[-(U_0 - V\tau)/kT], \tag{88}$$

where P is the probability factor (the number of 'attempts' per unit time) and $\exp[-(U_0 - V\tau)/kT]$ is the fraction of successful attempts that lead to the removal of (part of) a dislocation.

Specifically for *Taylor lattice dislocations being eliminated at grain or fixed domain boundaries,* the sink density is the length of "joint lines" between dislocation lattice rows and domain or grain boundaries per unit volume. Choosing for simplicity a square dislocation lattice with $s = d_f$, i.e. dislocation density $\rho = 1/l^2$, it is thus

$${}_{\mathrm{TL}}N_{\mathrm{SO}} = 1/Dl = \sqrt{\rho}/D, \tag{89}$$

yielding

$${}_{\mathrm{TL}}(\mathrm{d}\rho/\mathrm{d}t)_{\mathrm{O}} = -P\sqrt{\rho}/D\exp[-(U_0 - V\tau)/kT]. \tag{90}$$

With $\mathrm{d}\tau/\mathrm{d}t = (2\sqrt{\rho})\,\mathrm{d}\rho/\mathrm{d}t$ according to eq. (47), then,

$${}_{\mathrm{TL}}(\mathrm{d}\tau/\mathrm{d}t)_{\mathrm{O}} = -{}_{\mathrm{TL}}P/D\exp[-(U_0 - V\tau)/kT], \tag{91}$$

i.e. the form of eq. (84) that yields the logarithmic time law of eq. (32) observed in fig. 49.

If similitude were to operate (contrary to expectation, one may add) then one would have

$$ {}_{\mathrm{TL}}D_{\mathrm{SW}} = gl = g/\sqrt{\rho} \tag{92}$$

leading to

$$ {}_{\mathrm{TL}}(\mathrm{d}\tau/\mathrm{d}t)_{\mathrm{W}} = -{}_{\mathrm{TL}}P\big(\sqrt{\rho}/g\big)\exp[-(U_0 - V\tau)/kT]. \tag{93}$$

In that case, then, the pre-exponential factor, $\sqrt{\rho}/g$, would act to distort the ($\log t$) recovery law into a dish-shape, as with decreasing dislocation density the recovery is increasingly retarded.

Finally, for *direct annihilation among Taylor lattice dislocations*, one would have ${}_{\mathrm{direct}}N_{\mathrm{S}} = \rho$ leading to

$$ {}_{\mathrm{direct}}(\mathrm{d}\tau/\mathrm{d}t) = -{}_{\mathrm{direct}}P\sqrt{\rho}\exp[-(U_0 - V\tau)/kT]. \tag{94}$$

This, then, would yield the same dish-shape of the yield stress versus log of recovery time curves as eq. (93) due to dislocation annihilation at domain walls subject to similitude.

3.11.3. Recovery in dislocation cell structures

On account of the conservation of Burgers vectors requirement, in cell structures, all except "filler" dislocations can mutually annihilate only at nodal lines, where neighboring cell walls are joined. In fact, Young et al. [97] have documented this mechanism during the recovery of high-purity aluminum (see fig. 64). More problematic is answering the question whether similitude is to be assumed. While the previous argument that true similitude ceases with the removal of stresses leads to the answer "NO", yet at same dislocation content, larger dislocation cells have the smaller energy (see section 3.7) so that there is a driving force for cell coarsening that engenders an ambivalent "YES/NO" response. However, the elimination of dislocations from cell walls lowers the misorientation angles and this would give rise to longer-range stresses unless it occurred in a fairly coordinated fashion over several neighboring cells, thereby stimulating similitude.

More importantly yet, morphologically, cell growth cannot take place by the slow continuous mutual destruction of dislocations so far contemplated. Rather, cell coarsening necessarily requires the elimination of cell walls, just as cell refinement typically requires the formation of new cell walls (compare fig. 57), and wall elimination is bound to take the form of coalescence of cell walls. This mechanism has in fact been repeatedly reported, specifically by Langford and Cohen [212], as the cross sectional area of drawn wires decreased faster than the cell diameter, and by Young et al. in the course of cell size coarsening during the recovery of pure aluminum [97]. And finally on the "NO" side is the tendency for boundary anchoring through precipitates and impurities to further impede cell coarsening. On balance, due to these competing effects one will expect recovery curves that are bracketed between that of full similitude and fixed cell sizes.

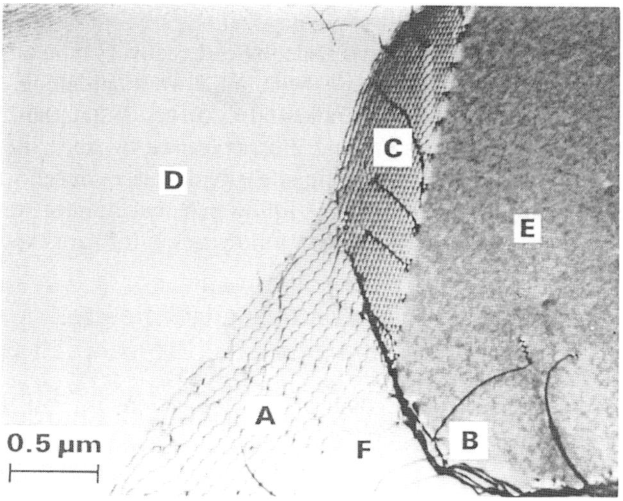

Fig. 64. Cell wall networks joined at a nodal line in pure Al polycrystal in the process of recovery. The regular interlocking of the adjoining networks at nodal lines permits the mutual dislocation annihilation during recovery that is responsible for the reduction of dislocation density and thus flow stress. (Figure 24 from Young, Headley and Lytton [97].)

Quantitatively, in a material with cubic cells of diameter D, the dislocation density is $\rho = 6/Dl$ and the node line density is $3/D^2$ for

$$_{\text{cell}}N_{\text{SO}} - 3/D^2 = \text{const.} \tag{95}$$

and

$$_{\text{cell}}N_{\text{SW}} = 3/D^2 = \rho/2g \tag{96}$$

without and with similitude, respectively. Accordingly,

$$_{\text{cell}}(\mathrm{d}\tau/\mathrm{d}t)_{\text{SO}} = -(1/\sqrt{\rho})\exp[-(U_0 - V\tau)/kT] \tag{97}$$

and

$$_{\text{cell}}(\mathrm{d}\tau/\mathrm{d}t)_{\text{SW}} \propto -\sqrt{\rho}\exp[-(U_0 - V\tau)/kT]. \tag{98}$$

Consequently, except for a 'fluke', the linear log time law of recovery, as in fig. 49, cannot be expected for dislocation cell structures since the pre-exponential $\sqrt{\rho}$ and $1/\sqrt{\rho}$ factors will distort it into a dish-shape or convex curvature.

3.11.4. Does recovery go on forever?

Contrary to practical experience, i.e. the indefinite persistence of work hardening in formed sheet such as in car bodies and of hard-drawn wires in steel cables, the above theory seems to permit recovery to continue until all of the dislocations are removed. Perhaps this could happen in planar, but certainly not in wavy glide. Indeed, recovery typically removes only a fraction of the work hardening that decreases with initial work hardening, as seen in fig. 47 by the abrupt final flattening of the 500°C and 450°C curves. What, then, causes recovery to cease? The most likely answer is internal stresses caused by steric hindrances through the lattice rotations that accompany dislocation density decreases in cell walls. For a rough estimate we use the results of section 3.7, as follows: A reduction of wall rotation angle by $\delta\phi$ is liable to cause strain gradients which, following eq. (65) are expected to store the elastic strain energy density of

$$U_{\delta\phi} = (G/6)\delta\phi^2. \tag{99}$$

Meanwhile the energy density on account of the dislocations is, with $\rho \cong 6\phi/Db$ and eq. (61), choosing Poisson's ratio $\nu = 0.3$ and $\phi = 1°$,

$$U_{\mathrm{D}} \cong \rho(0.4Gb^2) \cong (2.4Gb/D)(\phi_{\mathrm{max}} - \delta\phi). \tag{100}$$

The extra energy $U_{\delta\phi}$ will begin to overbalance the reduction in U_{D} at

$$\mathrm{d}(U_{\delta\phi} + U_{\mathrm{D}})/\mathrm{d}\delta\phi = (G/3)\delta\phi - 2.4Gb/D = 0, \tag{101}$$

i.e. at

$$\delta\phi_{\mathrm{max}} \cong 7.2b/D. \tag{102a}$$

However, if D remained constant at its initial value while $\tau = \tau_{\mathrm{max}}$, one will find with eq. (30), using $K = 10$ and remembering that $\phi \cong \tau/G$ (eq. (62))

$$\delta\phi_{\mathrm{max}} \cong 0.72\phi_{\mathrm{max}}. \tag{102b}$$

Following this very simple estimate, then, $\approx 28\%$ of the work hardening would be non-recoverable. A more stringent limit on recoverable work hardening, namely about 50%, follows from the still simpler argument that internal stresses due to steric mismatch cannot exceed the flow stress because, if so, they would cause local plastic deformation at greater energy expenditure than is gained by the recovery process. Additionally, these most basic limits on the asymptotic value of R_{r} of eq. (32), apply only to homogeneous materials. The presence of second phases and inclusions will further limit the possible reorientation of cells during recovery. And lastly straining causes non-recoverable contributions to the friction stress, other than through tangling, in accordance with the Cottrell–Stokes law (compare section 2.9), that further impede recovery.

This latter consideration regarding τ_0 applies equally also to planar glide, whereas none of the arguments based on lattice reorientation do apply. It is therefore not only possible but likely that recovery in planar glide can be more complete than in wavy glide.

3.11.5. *Conclusions regarding recovery in planar and in wavy glide*

On account of the term $\exp[-(U_0 - V\tau)/kT]$ in eqs (91), (93), (94), (97) and (98), recovery through all plausible mechanisms will approximate the linear (log time) law of eq. (32). However, in its pure form it is found *only* for planar glide through the mutual annihilation among neighboring opposite-sign dislocations at fixed sinks, either domain or grain boundaries (eq. (91)). Figure 49 thus suggests the operation of this mechanism. The pre-exponential factor $\sqrt{\rho}$ in eq. (93) (Taylor lattices with similitude), eq. (94) (direct annihilation among any neighbor dislocations in Taylor lattices) and eq. (98) (recovery in cell structures with similitude, i.e. increasing cell sizes) yields a concave or dish-shaped distortion. That is in fact seen in fig. 48 and is in agreement with the cell size growth concurrent with the recovery of the aluminum samples observed by Young et al. [97]. Lastly the $1/\sqrt{\rho}$ pre-exponential factor in eq. (97) for recovery in wavy glide at constant cell size would give a convex shape to the (log time) curve. Of this there is a hint in fig. 47, i.e. in the recovery of iron with its sizeable friction stress. However, in this case the two competing effects appear to have permitted some cell growth but well below similitude so that the curves are rather linear.

All in all, the simple models discussed are in excellent accord with flow-stress time curves and the non-recoverable work hardening fraction. In his admirably detailed and well-researched paper, Nes [256] also explains these but via complex models that raise concerns regarding eq. (33).

3.12. Theory of grain-size hardening (Hall–Petch hardening)

As already indicated (section 2.1, see fig. 29), grain-size, or colloquially speaking "Hall–Petch" hardening is weak or absent in *pure "wavy" glide*, unless the grain size begins to compete with the average cell size. Similarly also deformation band boundaries and CBBs do not significantly affect work hardening (see section 3.9.2), for the reason that the dislocation free paths are limited by cell walls no less than by CBB's or grain boundaries, coupled with the fact that they offer less glide resistance and therefore control work hardening.

The effect of grain boundaries to more or less add a constant stress increment to the whole stress–strain curve of *planar-glide materials*, which stress addition rises with decreasing grain size as in fig. 29(b), derives from the following very general features:

(1) Driven by an applied stress, glide dislocations line up roughly parallel to planar obstacles.
(2) The blocking range of an obstacle (L_i) compares with its lateral size, $\sqrt{A}$; i.e. for a boundary in a D_G grain size structure, it is $A = D_G^2$ and $L_i \cong D_G$.
(3) Resolved shear stresses, τ, cause glide forces $F_D = b\tau$ per unit dislocation line length, in the glide plane, always normal to the axes, independent of relative Burgers vector orientation.
(4) The obstacle must supply the reaction force, F_r, to the component of the glide forces normal to it, in excess of the friction force, acting on dislocations held up by it.
(5) The density of the dislocations whose progress is blocked by an obstacle, ρ_B, is proportional to $(\tau - \tau_0)^2$ in accordance with eq. (31).

From the above points it follows for planar glide, in which dislocations are distributed within grains rather than being confined to cell walls as in wavy glide, that a grain boundary must support the force exerted on it by the dislocations in front of it, i.e. is acted upon by the force

$$F_r \cong (1/2)AL_i\rho_B(\tau - \tau_0) = \text{const}\, D_G^3(\tau - \tau_0)^3, \tag{103}$$

and hence must withstand a distributed reactive shear stress of

$$\tau_r = F_r/A \cong (1/2)L_i\rho_B(\tau - \tau_0) = \text{const}\, D_G(\tau - \tau_0)^3. \tag{104}$$

Two reasonable cases present themselves, (i) If the boundary permits dislocation passage at a fixed value of τ_r, then the correlated flow stress is, from eq. (104),

$$(\tau - \tau_0) = K_{HP}^*/D_G^{1/3}. \tag{105}$$

Alternatively, if the boundary strength is proportional to the flow stress, i.e. gives way at the stress $\tau_r = \text{const.}(\tau - \tau_0)$, then the Hall–Petch relationship

$$(\tau - \tau_0) = K_{HP}/D_G^{1/2} \tag{106}$$

results.

For the combined data from fig. 29(b) and from fig. 6 of [185], eq. (105) rather than the Hall–Petch relationship of eq. (106) was found (compare fig. 14 of [23]). That result by no means disproves the Hall–Petch relationship [280,281] since that was proposed, and documented to exist, only for the value of the lower yield stress at which Lüders bands spread, as reviewed by Armstrong, Codd, Douthwaite and Petch [282]. In fact, the difference of the grain size effect in planar and wavy glide illustrated in fig. 29 has been repeatedly used to quantify the degree of glide planarity, among others by Thompson and Baskes [183], Marcinkowski and Fisher [283], and especially by Johnston and Feltner [284] who adduced data to show that the strain dependence of K_{HP} is correlated with the electron–atom ratio, with low e/a values giving rise to distributed glide and $e/a > 1.15$ to planar glide.

3.13. Theory of alloy hardening and brittleness

3.13.1. General theory

The superior strength of brass over copper, and similarly other planar-glide alloys over their respective pure metals, is due to (i) an increased value of τ_0 on account of the substitutional alloying and (ii) lack of three-dimensional dislocation mobility that almost totally inhibits mutual dislocation annihilation. As a result the annihilation parameter, β, is maintained close to unity in planar glide and steep work hardening in the form of Stage II continues. However, only a minority of alloys are of α-brass type, while the considerable majority, most prominently aluminum-copper alloys and steels, are of wavy-glide type.

Their superior strength also depends on two causes, including the same rise of τ_0 due to increased lattice friction, but secondly on a limitation of dislocation mean free paths through precipitates and/or second phases. This, too, can be most easily expressed as an increase in τ_0, and this is commonly, although as we shall see incorrectly, done. Indeed, phenomenologically the effect of limiting dislocation mean free paths to L, the average spacing of the obstacles to dislocation motion, resembles raising the stress strain curve by the corresponding Orowan stress,

$$\tau_{\mathrm{Or}} \cong Gb/2L. \tag{107}$$

An example of this effect is illustrated in fig. 65 extracted from [285]. However, simply to subsume τ_{Or} as part of τ_0, although serviceable, is incorrect. A deeper insight, including also the application to pearlitic steels, based on [149], follows.

Beginning with a dislocation density $\rho < 1/L^2$, a precipitation-hardened material with inter-particle spacing L will macroscopically yield when the dislocation density has risen to

$$\rho_{\mathrm{yield}} \cong 1/L^2. \tag{108}$$

At densities below ρ_{yield}, the dislocations are predominantly held up by the fixed obstacles in their path and the retention parameter will be $\beta \cong 1$. At substantially higher densities, however, the dislocations will predominantly interact with each other and ordinary work hardening will begin. In the interval between τ_0 and the macroscopic yield stress,

$$\tau_{\mathrm{yield}} = \tau_0 + \tau_{\mathrm{Or}} \cong \tau_0 + \alpha Gb(\rho_{\mathrm{yield}})^{1/2} \cong \tau_0 + \alpha Gb/L, \tag{109}$$

glide dislocations of rapidly increasing density $\rho < \rho_{\mathrm{yield}}$ move on average through mean free paths that are comparable to L. In the process they cause a plastic shear strain of

$$\gamma_{\mathrm{yield}} \cong \rho bL < \rho_{\mathrm{yield}} bL \cong b/L. \tag{110}$$

The average work-hardening coefficient up to the yield point is thus

$$\Theta_{\mathrm{yield}} = \tau_{\mathrm{yield}}/\gamma_{\mathrm{yield}} > \alpha G \cong G/3, \tag{111}$$

i.e. the pre-yield slope of the work-hardening curve of precipitation-hardened metals is so steep that it will be generally mistaken for the elastic slope. This is the major reason for the widely held belief that precipitation hardening simply adds the Orowan stress to the flow stress, i.e. raises the work-hardening curve along the stress axis by τ_{Or}. However, as seen, in fact the presence of precipitates or other obstacles essentially cuts off the first part of the work-hardening curve up to γ_{yield}, thereby shifting it along the strain axis by $-\gamma_{\mathrm{yield}}$. The effect is thus in good agreement with fig. 65 and is much the same as of pre-straining, illustrated in fig. 66 extracted from [286].

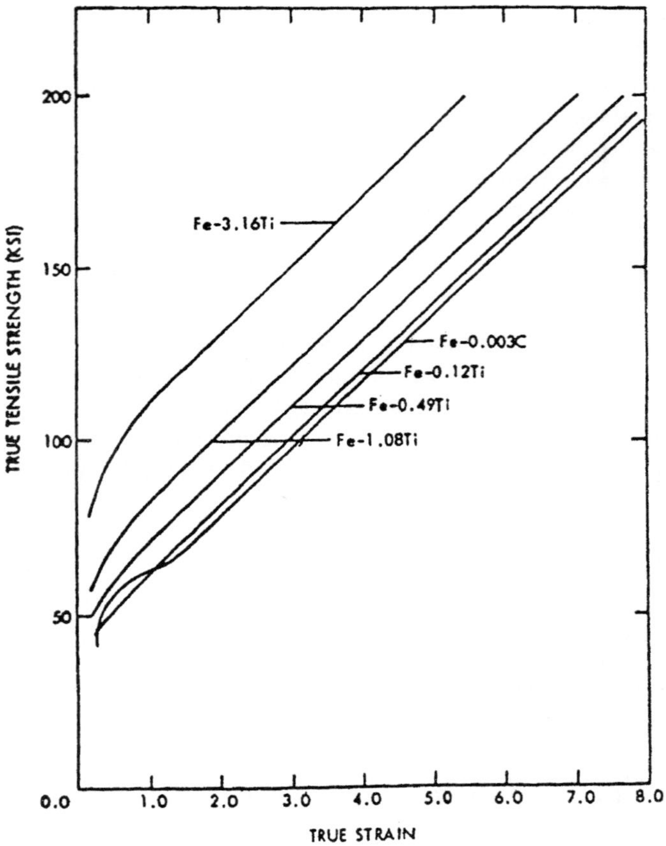

Fig. 65. True stress/true strain curves gained as the envelopes of tensile tests of iron wires drawn to various strains in the manner of Langford and Cohen [212], with alloying concentrations of Ti as indicated (in wt. percent). The lowest curve represents almost pure iron, and the next lowest curve, iron in which the interstitials are "gettered" with a small amount of Ti. In the next two higher curves the Ti is in solid solution, and in the top curve it has formed brittle precipitates. Evidently all five curves are much the same except for some disturbances at small strains and being displaced. Correspondingly the effect of alloying can be interpreted as raising τ_0 and/or displacing the curves along the strain axis as if beginning at higher strain. The indicated limitation to the same maximum stress favors the second interpretation. (Figure 2 of Rack and Cohen [285].)

3.13.2. Application to pearlitic steels

Continuing to follow [149] we note that available evidence [287–292] shows plastic deformation in steels to occur by the same dislocation mechanisms as in other metals and that it is pre-eminently the ferrite which deforms. Pearlitic steels may therefore be regarded as double composites of soft ferrite and hard pearlite, while the pearlite is composed of deformable ferrite lamellae separated by hard cementite platelets.

At very small deformations the plastic strain is accommodated by the ferrite [291], but extended deformation requires that ferrite and pearlite deform by very similar amounts.

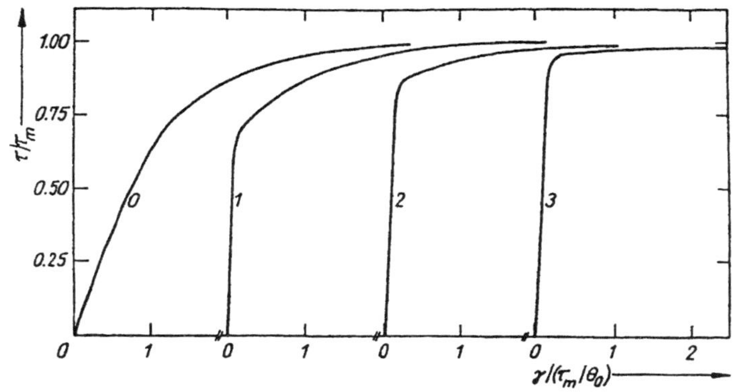

Fig. 66. Changes in a stress–strain curve through pre-strains of $\gamma/(\tau_m/\theta_0) = 1$, 2 and 3, respectively. The effect of hard particles is similar to pre-strain in that they typically eliminate Stage I, and may eliminate parts or all of Stage II and into Stage III, depending on particle density. Specifically, particles spaced L_p apart limit the mean free dislocation path to $\leqslant L_p$. Formally this causes an initial value of $g = L_p/\ell$ below unity and a commensurately high work-hardening coefficient (eq. (40)). Initially, therefore, the dislocations multiply rapidly without producing much strain and the stress-strain curve rises steeply, indeed quasi-elastically, until the dislocations are so densely spaced that they determine the value of g at its normal value, and ordinary work-hardening begins. (Figure 2 of [286].)

Within the pearlite, however, glide is (almost) completely restricted to the ferrite lamellae. In these, after initial generation of individual dislocations [290,291], ordinary dislocation cells are formed. The averaged dimensions of the dislocation cells in the ferrite lamellae are initially equal to the lamella thickness. The cells then refine with further strain much in accordance with the empirical law of eq. (30) [291]. However, since at, say, 0.6% C and above, the applied loads are principally carried by the pearlite, and in this practically only the ferrite lamellae deform, the strain is concentrated in these. In that case the work-hardening rate is that of ferrite multiplied with the inverse volume fraction of pearlite in the steel, and multiplied again with the inverse volume fraction of ferrite in pearlite.

It seems probable that the reported saturation of the cell size at higher strains [291,292] does not in fact strictly occur. Rather at the high stresses concerned, Stage IV will have set in with its very low work-hardening rate and thus very slow cell size reduction as a function of strain.

Equations (108)–(111) may be applied to the ferrite lamellae, wherein the ferrite lamella thickness S_p assumes the role of L. This agrees with the observations by Dollar et al. [290] that (i) the flow stress is inversely proportional to S_p rather than to S_p^2 that would have been expected from the Hall–Petch relationship (although the latter could be barely made to fit) and that (ii) the dislocation density rises in accordance with eq. (31). In summary, therefore, the evidence discussed is in good accord with the theory.

The details of the dislocation mechanisms involved are intricate and much work has been devoted to unraveling them, especially by Ashby [293–295], Hirsch and Humphreys [296], and Kocks [297–299], and several fine surveys are available [300–305]. Gratifyingly, also radiation damage contributes to τ_0 closely as expected from theory [306].

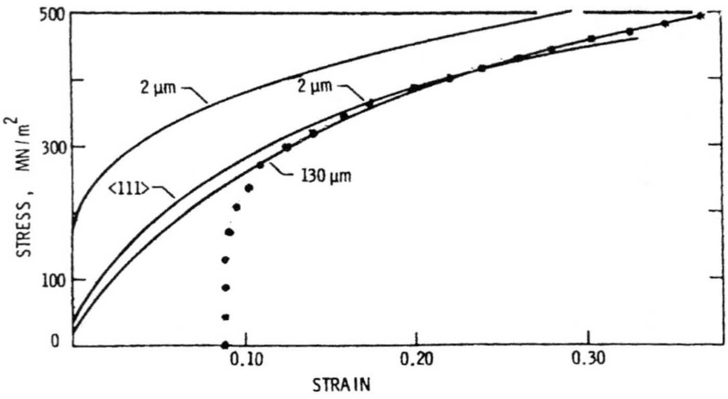

Fig. 67. As fig. 29(a) but with the 2 µm curve displaced along the strain axis (solid dots). The good fit demonstrates that the effect of grain sizes smaller than the average dislocation link length in the undeformed state, is much the same as of small particles and pre-straining (compare section 3.13.2 and fig. 66), i.e. to replace the early parts of the work-hardening curves by a pseudo-elastic strain.

3.13.3. Application to ultra-fine grain sizes and why very strong metals are brittle

For very small grain diameters, D, the condition of $1/\sqrt{\rho} \cong l > D$, i.e. average dislocation bowing length larger than the grain diameter, can occur. In that case there is a grain size effect also in wavy glide, e.g., as demonstrated in fig. 29. According to the above theory, one thus expects the same basic work-hardening curve also at ultra-small grain sizes but with an early part cut off. This prediction was tested by means of fig. 67 based on fig. 29(a). It will be seen to be well obeyed.

The general syndrome discussed, that obstacles in the path of glide dislocations do not lift the overall level of the work-hardening curve but eliminate early parts, to the greater extent the larger the extra yield stress, has the effect of reducing deformability in tandem with rising yield strength (see fig. 66). On the reasonable surmise that fracture occurs at some limiting cell refinement, this explains the effect that is well known in practice: The harder the material (meaning the higher its yield strength) the less ductile. On recovery, without change of grain size, the process is reversed, and as the yield stress drops, so the strain to fracture increases.

3.14. The ΔE-effect

3.14.1. Taylor lattices

Reversible dislocation movements are common in all LEDSs. Specifically, for Taylor lattices with the dislocation density $1/sd_{\mathrm{f}}$, a relative displacement by $+\xi/2$ and $-\xi/2$ of the two sub-lattices causes the shear

$$\delta\gamma = \xi b/2sd_{\mathrm{f}}, \tag{112}$$

while, from eq. (17) with eq. (112)

$$\tau_{TL} = F\tau_{dip} = FGb/d_f 4\pi(1-\nu) = \{FGs/\xi 4\pi(1-\nu)\}\delta\gamma. \tag{113}$$

Comparing the elastic strain $\gamma_{el} = \tau_{TL}/G$ with $\delta\gamma$ extracted from eq. (113) thus yields

$$\delta\gamma/\gamma_{el} = 4\pi(1-\nu)/[F/(\xi/s)] \tag{114a}$$

or for infinitesimally small strains

$$(\delta\gamma/\gamma_{el})_0 = 4\pi(1-\nu)/(\mathrm{d}F/\mathrm{d}\xi/s)_{\xi=0} \tag{114b}$$

and at the critical stress

$$(\delta\gamma/\gamma_{el})_{crit} = [4\pi(1-\nu)/F_{max}](\xi/s)_{crit}. \tag{114c}$$

Correspondingly, materials with Taylor lattices exhibit a quasi-elastic behavior with an apparent shear modulus, G^*, well below the actual modulus.

In accordance with eq. (114a) the relative apparent shear modulus at any stress is

$$G^*/G = \gamma_{el}/(\delta\gamma + \gamma_{el}) = \{1 + 4\pi(1-\nu)/[F/(\xi/s)]\}^{-1} \tag{115a}$$

while with eq. (114b) it is

$$G_0^*/G = \gamma_{el}/(\delta\gamma + \gamma_{el})_0 = \left\{1 + 4\pi(1-\nu)/(\mathrm{d}F/\mathrm{d}\xi/s)_{\xi=0}\right\}^{-1} \tag{115b}$$

and at the critical stress

$$G_{crit}^*/G = \left\{1 + \left[4\pi(1-\nu)/F_{max}\right](\xi/s)_{crit}\right\}^{-1}. \tag{115c}$$

Asymptotically, for $s/d_f \Rightarrow \infty$,

$$G_\infty^* \approx G/2.2. \tag{115d}$$

In fig. 68, G_0^*/G and G_{crit}^*/G are shown. Deviations between the two curves indicate changes of the apparent elastic constant from zero applied stress up to the yield stress, i.e. curvature of the quasi-elastic $\tau(\gamma)$ function.

3.14.2. Dislocation cell structures

Matters are rather simpler in cell structures. Namely, at the flow stress, the dislocations bow out reversibly on average through the mean free path of about $l/2 \approx 1/2\sqrt{\rho}$, thereby generating the quasi-elastic strain of

$$\delta\gamma = \rho b l/2 \approx \sqrt{\rho} b/2 \cong \tau/2\alpha G \cong \gamma_{el}. \tag{116}$$

Thus in cell structures the apparent shear modulus is about one half of the actual one and similar to the one for Taylor lattices.

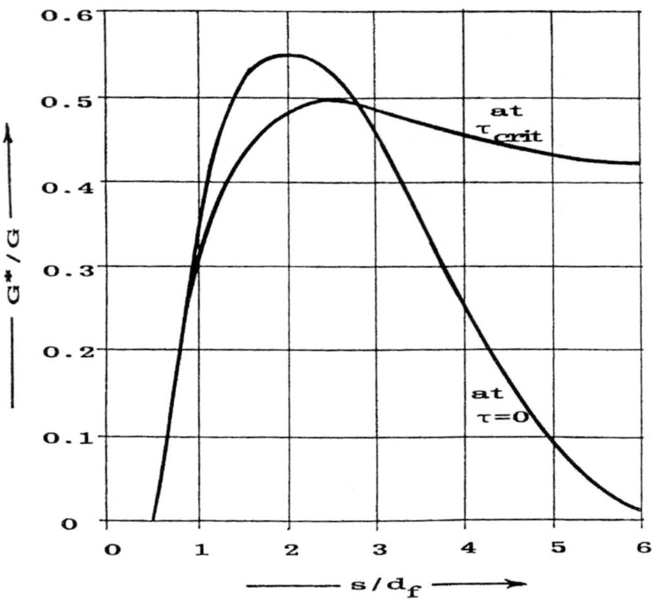

Fig. 68. The apparent relative shear moduli due to anelasticity of Taylor lattices, G^*/G, as a function of s/d_f for low applied stresses (curve "at $\tau = 0$") and at the critical stress (curve "at τ_crit"). As seen, for the most likely case of $s/d_\mathrm{f} = 2\sqrt{3}$ (when A in fig. 7 forms a symmetrical hexagon, see section 1.7.3), G^*/G is near 0.47 at the critical stress, and it is near 0.35 at zero stress.

3.15. Work softening

The major observations regarding work softening have been discussed in section 2.9. In line with these and the results shown in figs 44 and 45, besides dislocation unlocking effects (section 1.8) work-softening takes at least two different forms. Following in outline the more complete exposition of the LEDS theory of work softening presented in [286], these are ascribed to decreases in the dislocation density and to decreases in the friction stress, τ_0, respectively, as follows.

The complete expression for the work-hardening coefficient is, with eq. (40),

$$\mathrm{d}\tau/\mathrm{d}\gamma = \Theta = \mathrm{d}\tau_0/\mathrm{d}\gamma + \mathrm{d}\tau_\mathrm{G}/\mathrm{d}\gamma \approx \mathrm{d}\tau_0/\mathrm{d}\gamma + G\alpha\beta/2g. \tag{117}$$

Normally, but not necessarily, the first term is negligible. Further, g and α are nearly independent of strain so that the strain dependence of $\mathrm{d}\tau_\mathrm{G}/\mathrm{d}\gamma$ essentially mirrors the corresponding gradual decrease of β with strain (section 3.5).

Ordinarily, application of stresses below the previously highest will not affect already established work hardening (compare fig. 66). However, if previous straining produced a dislocation density that is higher than would be evoked by new straining conditions, *and* if the new straining conditions are of a kind to destabilize the existing LEDSs, β may assume negative values, leading to negative Θ values, i.e. work softening. In that case

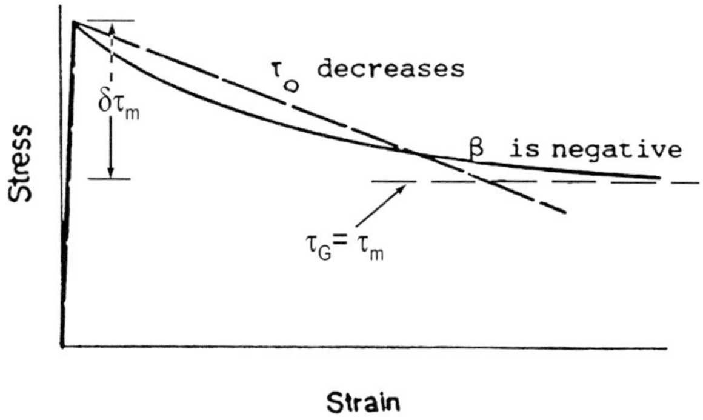

Fig. 69. Work-softening stress–strain curves expected from negative values of the dislocation retention parameter, β (curve "β is negative"), and a friction stress, τ_0, that decreases linearly with strain (curve "τ_0 decreases"), in accordance with eqs (118) to (120) and (121), respectively. Curve "o" in fig. 44 is believed to be an example of transiently negative β values, and figs 45 and 70 examples of decreasing friction stress. (Figure 1 of [307].)

straining depletes rather than increases the dislocation content. An example of this type of work softening is seen in curve 'o' of fig. 44 that, in this particular case, is presumably due to replacement of a Stage II carpet structure formed at low temperature with a Stage III cell structure at the higher deformation temperature.

The simplest and most reasonable dependence of β on flow stress in work softening of this type would be linear, i.e.

$$\beta^* = \beta_0(1 - \tau_G/\tau_m), \tag{118}$$

where τ_m is the flow stress in excess of τ_0 that would have been reached at the applied stress by the otherwise same material had it been strained from an initially lower dislocation density, and β_0 is the correlated β value at τ_m. If so the work-hardening coefficient Θ^* decreases with strain as

$$\Theta^* \approx \Theta_0(1 - \tau_G/\tau_m), \tag{119}$$

so that the flow stress becomes

$$\tau = \tau_0 + \tau_m + \delta\tau_m[\exp(-\gamma\Theta_0/\tau_m)], \tag{120}$$

with $\delta\tau_m$ the initial excess of flow stress above τ_m, as shown in fig. 69 (fig. 1 of [307]).

The characteristic concave shape of the stress–strain curve due to transiently negative β values seen in fig. 69, does not depend on an assumed linear dependence of β on strain but arises whenever the magnitude of β decreases monotonically from its initial largest negative value to zero at some finite, not too large strain increment.

Alternatively, work softening can be caused by a decrease of the friction stress. Most simply there could, again, be a linear dependence on strain, in this case of some excess friction stress, $\delta\tau_0$, above the equilibrium friction stress, τ_0, from an initial value $\delta\tau_{0i}$, i.e.

$$\delta\tau_0 = \delta\tau_{0i}[1 - \gamma/\gamma_f], \tag{121}$$

where γ_f is the strain increment at which the equilibrium friction stress is attained. This relationship yields the curve indicated in fig. 69, provided that τ_G meanwhile remains constant. A deviation from this simple shape will indicate the correspondingly modified strain dependence of the excess friction stress.

A coarse estimate of the critical strain γ_f may be derived from considering that $\delta\tau_0$ will have been eliminated, when the newly generated dislocation loop length (whether or not it may be transiently stored) equals some moderate fraction, C, of the dislocation density ρ_m at the equilibrium stress of τ_m. Herein, clearly C must lie well below unity but can hardly be a very small number. Thus $0.05 < C < 0.3$ may be a reasonable range. With those values and with $b = 3 \times 10^{-10}$ m, $\rho = 4 \times 10^{15}/\text{m}^2$ and $g \approx 25$, one finds γ_f to range between about 1% and 10%.

The technologically most prominent case of this type of work softening is observed in mechanically alloyed metals (MA's) of which fig. 45 gives an example. In it, curves 2 and 3 represent eq. (121) satisfactorily well, albeit the accelerated downturn at the end of such curves is due to necking and fracture. In fact all metals with extremely small grain sizes are prone to the described type of work softening and early fracture. The suspicion that the initially artificially high friction stress in these cases is due to Hall–Petch hardening and that strain somehow lowers the grain boundary resistance against glide dislocation motion does, of course, lie close at hand. Positive evidence to this effect was obtained by a searching analysis of micrographs in conjunction with stress–strain curves of a particular series of high-temperature aluminum MA alloys [307–309] in terms of the theory presented in [286]. In the process, an unexpected additional phenomenon, namely the presence of a 'grain-boundary substance', was discovered, as follows.

The roughly linear, more convex than concave shape of the work-hardening curves in question, e.g., as given in fig. 70, clearly implicates decreasing friction stress as the cause of the observed work softening. In search for the cause of the declining τ_0, a wealth of TEM micrographs were examined. These revealed typical grains ranging between about 0.25 to 0.8 μm diameter. Their shape did not obviously change with straining and neither did the overall appearance of the grain boundaries as shown in figs 71 and 72(a). Before and after straining, the boundaries were decorated with small patches of very fine dislocation grids, e.g., as circled in fig. 71, which are interpreted as epitaxial dislocations at precipitates or dispersoids that anchored the grain boundaries. As the cause of the declining τ_0 one may thus rule out (i) increasing grain size, (ii) grain boundary sliding, and (iii) significant involvement of the precipitates or dispersoids. Rather, the evidence shows that at least the bulk of the deformation is due to ordinary more or less homologous straining of the grains. However, on close examination, there was found one crucial effect: Patches of rather regular, exquisitely fine dislocation networks of average $l \approx 6$ nm spacing in the strained but not the unstrained state, e.g., at right near the top boundary in fig. 72(a).

In terms of the LEDS theory, these networks were interpreted as dislocation rotation boundaries that accommodate a relative misorientation between the adjoining grains above

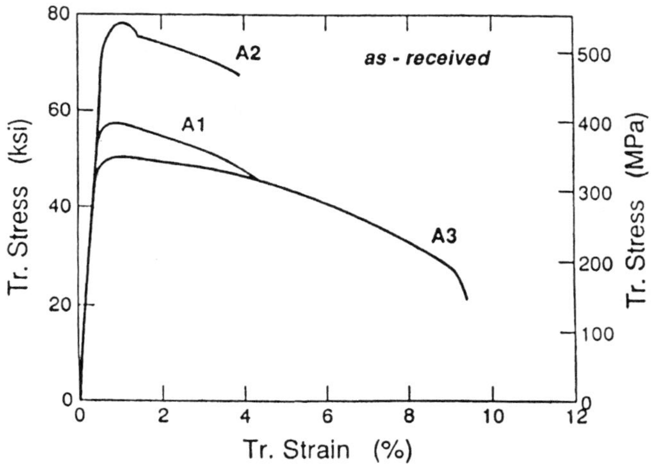

Fig. 70. Room-temperature true tensile stress – tensile strain curves parallel to the extrusion axis of three high-temperature MA aluminum alloys, Al, A2 and A3, in the as-received condition. The major alloying component was 12.1v/o, 10.7v/o and 4.0v/o Al_4C_3, respectively. (Figure 2 of [307].)

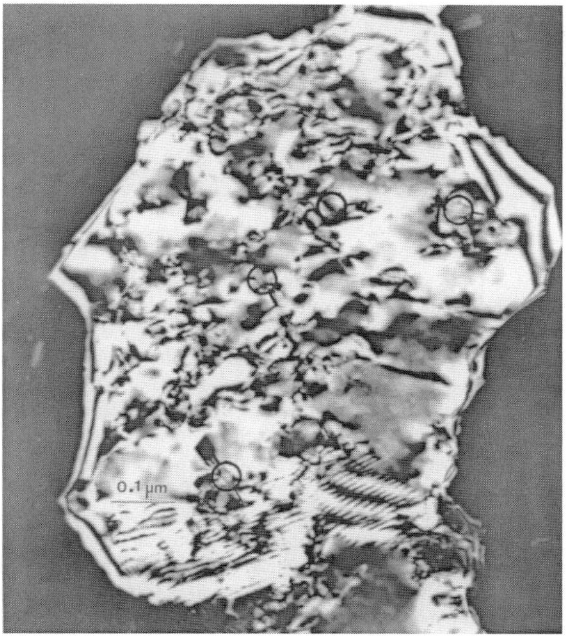

Fig. 71. Plan view of a grain boundary in the MA high-temperature aluminum alloy A3, in the as-received condition. The broad fringes of equal thickness about the periphery indicate how the grain volume tapers through the thickness of the foil. Indistinct irregularities are due to precipitates and/or dispersoids that anchor the boundary. Some small patches of very fine grids of parallel dislocations that have been encircled are interpreted as epitaxial dislocations at such particles. (Figure 3 of [307].)

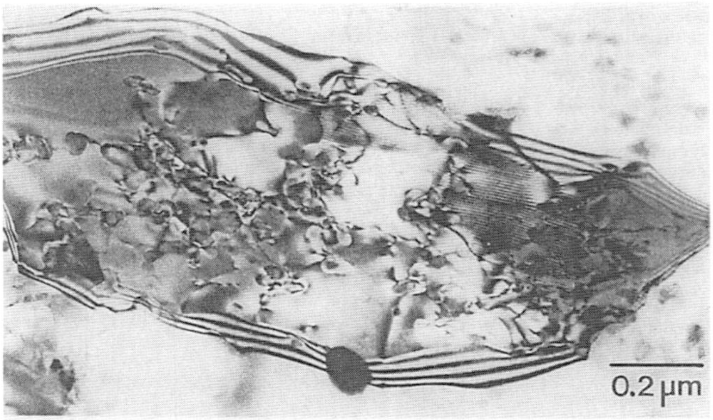

(a)

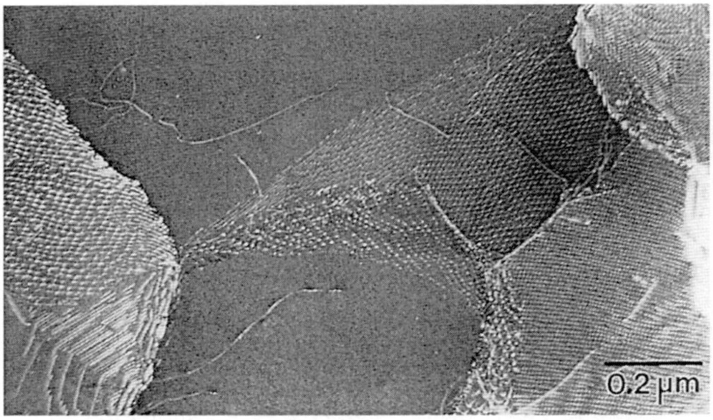

(b)

Fig. 72. (a) As fig. 71 but for a different grain boundary after tensile straining 9.3% to failure. The straining left the grain boundary substantially unchanged, including its anchoring by particles, except for being overlaid with a patch of a very fine dislocation grid, seen right from center near the upper edge of the boundary. Such patches of dislocation grids are found in most if not all boundaries after fracture. They are remnants of rotation boundaries that accommodated the lattice misorientation from one side of the boundaries to the other due to the plastic strain difference in the adjoining grains. Almost certainly, they are formed of glide dislocations that were held up on either side of the grain boundary. These grids persist, instead of being simply annihilated, because of an intervening 'grain boundary substance' (compare fig. 73), in the present case almost certainly due to additives that facilitate the mechanical alloying and ultra-fine grain sizes. (Figure 3 of [309].) (b) Highly regular dislocation networks, that are doubtlessly dislocation rotation boundaries, observed in the same material as in (a) but after 18 h annealing at 550°C. It is unclear whether these networks have any connection with the "patches" such as illustrated in (a), or are simply low-angle sub-boundaries in the now recrystallized material with greatly coarsened particles. Incongruously, this image was used as fig. 7 of [316], in support of SODSs modeling far from thermodynamical equilibrium. (Figure 4 of [309].)

and beyond the overall rotation angles between them determined by the grain boundaries. For the finely spaced patches, with $l \cong 6$ nm and $b \cong 0.3$ nm, the inferred misorientation is $\Delta\phi = b/l \cong 3°$, as would be produced through a shear strain or tensile strain difference

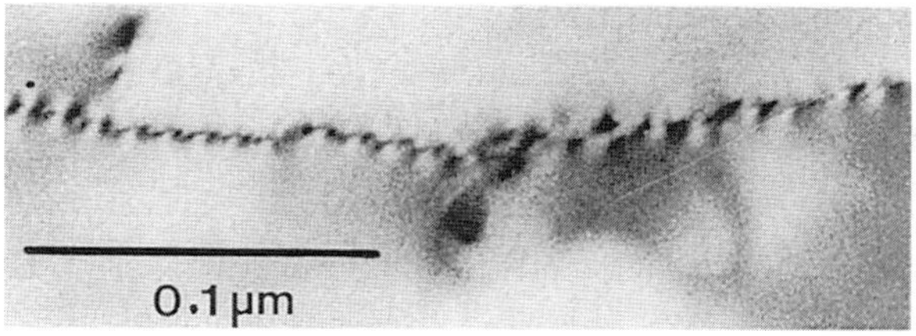

Fig. 73. Cross section of a dislocation 'patch' at a grain boundary in A3 alloy, tensile tested to failure, such as right of centre near the top edge in fig. 72(a). The dislocation diffraction images are interlinked by a barely visible line that is believed to be the image of grain boundary substance. (Figure 5 of [309].)

of $\Delta\gamma \cong 5\%$ and $\Delta\varepsilon = \Delta\gamma/M \cong 2\%$, respectively. Considering that the total tensile strain approached 10%, these numbers suggest that the network patches are due to strain differences from grain to grain built up in the course of tensile straining. Nor is the origin of the dislocations in doubt. They will have arrived as glide dislocations. But, most perplexingly, why did these dislocations not simply annihilate at the grain boundaries? Or in other words, how can there be small-angle dislocation rotation boundaries overlaid on large-angle grain boundaries?

The suggested answer is that the grains are surrounded by an exquisitely thin grain boundary substance that represents a strong, although not impenetrable barrier against glide dislocation passage. Indeed, already the 'father of materials science' Gustav Tammann, proved the existence of 'grain boundary substance' in impure metals and thought that it played a significant role in metal plasticity [310], compare [309]. And on careful search, both in [308] and [309], some persuasive albeit not absolutely certain direct TEM evidence for the boundary substance was found, of which fig. 73 is an example.

After prolonged annealing near the melting temperature, beautifully regular networks were observed in boundaries now free of dispersoids and precipitates and not in the form of patches overlaid on independent grain boundaries (fig. 72(b)). These may well be ordinary low-angle boundaries, without connection to the described patches. Either way, they, and fig. 72(b) in particular, are doubtless equilibrium LEDSs.

With this, the interpretation of the data has become clear: In the face of competing dynamic recrystallization, extremely fine grain sizes are obtained by the addition of insoluble components that form grain-boundary substance. By impeding the annihilation of glide dislocations, this substance causes high Hall–Petch hardening. With straining, the grain-boundary substance initially stretches elastically, but then gives way, thereby causing a precipitous decrease of τ_0, i.e. work softening. This, then, destabilizes the strain distribution, leading to necking and failure.

3.16. Structures formed in constant amplitude fatigue and hysteresis loops

An outline of dislocation structures in fatigue has been presented in section 2.12 where their LEDS nature was documented. The LEDS theory linking those fatigue dislocation structures to mechanical behavior, pertinent to the present section, was developed in [87,

126,311–315] largely in cooperation with C. Laird and partly with P. Charsley [87]. Further, an excellent survey of micrographical evidence on fatigue dislocation structures, mainly in regard to wavy-glide f.c.c. metals but partly also to b.c.c. metals and to f.c.c. wavy glide, was presented by Laird, Charsley and Mughrabi [265]. Again, the evidence that the great majority if not all of the observed structures are LEDSs close to energy minimum, was conceded by the authors to be very powerful. Even so, they expressed some reservation whether all dislocation structures, especially at high amplitudes and low cycles, were necessarily of low-energy type.

The major, perhaps only reason, for this doubt appears not to be any particular dislocation structures but the occasional occurrence of jerky flow, of $\Delta\gamma < 10^{-3}$ amplitude, in hysteresis loops during plus-minus constant strain-amplitude fatigue of copper single crystals (see fig. 19 of [265]). The authors interpreted these strain discontinuities as dislocation avalanches in lieu of microscopically smooth flow, and thus considered them as evidence for deviations from equilibrium. However, the correlated deviations from thermodynamical equilibrium can only be quite minor. E.g., their equivalent temperature change according to eq. (79) falls below 0.1°C. Even so, based on this somewhat tenuous basis, Glazov, Lianes and Laird [316] "*introduce a thermodynamic classification of dislocation structures in fatigued metallic alloys and employ the concept of self-organized dislocation structures* (*"SODS"*). *This classification finds its justification in a fundamental property of strongly non-equilibrium systems – a tendency to display, under certain conditions, highly correlated self-organized behavior* [317]. *It is based on the laws of nonlinear, non-equilibrium thermodynamics and the theory of nonlinear dynamic systems... and we will finally show that SODS should be considered as a general class of dislocation structures that may include LEDS as a particular case.*"

From the theoretician's viewpoint, then, the attraction of the SODS theory is evident, namely that it gives scope for demanding, rigorous mathematical modeling. Alas, the connection to real life dislocation structures is tenuous, to put it conservatively, as in this particular instance clearly shown by the following two points. (1) Almost certainly, the momentary work-softening events that give rise to the strain rate discontinuities and prompted the thought of non-equilibrium dislocation structures in [265] and thence [316], are a Portevin–LeChatelier effect. Specifically the effect is presumably the equivalent of blue-brittleness, wherein the anchoring atmospheres are due to vacancies in the 'mushrooming' syndrome (compare sections 1.4 and 1.8 and fig. 36). Clearly, this phenomenon must be expected to occur and once observed is evidently part and parcel of the LES theory, involving only very minor deviations from equilibrium. (2) The major result of the modeling in [316] is that at small grain sizes, dislocations 'self-organize' to assemble in high concentrations at the grain boundaries (fig. 7 (top) of [316]), – leaving open the question of why, driven by their image forces, the dislocations do not simply annihilate at the boundaries (see section 3.15 above)? Anyway, in support of this result, Glazov et al. [316] reproduce fig. 72(b) as their fig. 7 (bottom). Yet, as explained in section 3.15 above, in fact the structures in fig. 72(b) are the result of prolonged annealing near the melting temperature and doubtlessly are LEDS dislocation rotation boundaries. Hence the networks recorded in fig. 72(b) are certainly no examples of dislocation self-organization far from thermodynamical equilibrium but are as close to equilibrium as any dislocation structures one is liable to find anywhere.

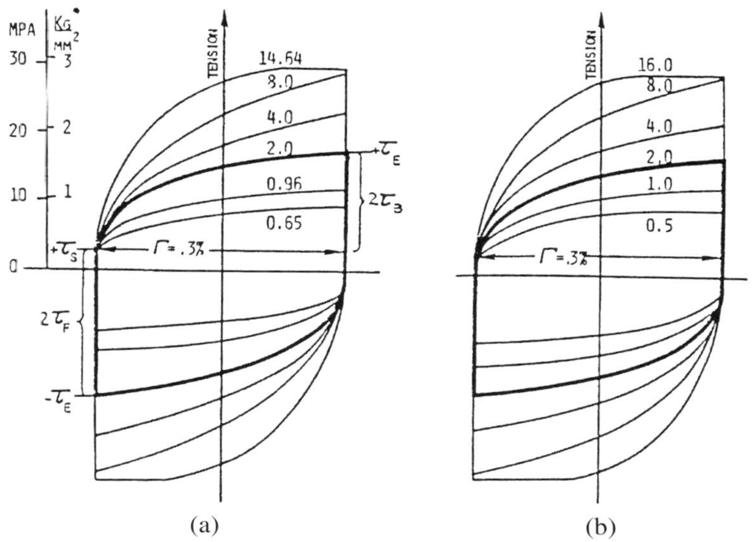

Fig. 74. Fatigue hysteresis loops of Cu crystals at different cumulative strains, $\gamma_{\mathrm{pl,cum}}$, indicated by the corresponding numbers in the upper right quadrant. At left, measured by Mughrabi [320], and at right theoretically derived from the experimental $\gamma_{\mathrm{pl,cum}} = 2$ loop, by scaling both stress and strain axis in proportion with $(\gamma_{\mathrm{pl,cum}})^{1/2}$, in accordance with the LEDS theory. Also indicated is the method of determining the back stress (τ_B) and friction stress (τ_F) according to Cottrell ([52], pp. 111–114). (Figure 1 of [318], see also [126] and [319].)

The already mentioned connections between the fatigue-induced LEDSs and the observed mechanical behavior, derived in [87,126,311–315], are rather varied and intricate. They have been briefly summarized in [318] and together constitute a remarkably closely reasoned system to which no objections known to this writer have as yet been raised. However, the intricacies of the arguments and deductions go beyond the scope of the present paper and only one simple result will therefore be mentioned here. Namely, low-amplitude, constant strain plus/minus fatigue, amounts to cycling in Stage I and yields the corresponding LEDSs, culminating with Taylor lattices in the form of somewhat regularly arranged parallel lancet-shaped loop patches as in fig. 16. At some critical dislocation density these first transform into the ladder structure (fig. 50) and, with the onset of double glide, into cells (figs 53 and 54). The gradual accumulation of the parallel primary edge dislocations of which the loop patches and ladder structure are composed occurs through statistical mutual $^{\perp}{}_{\top}$ trapping, and the dislocation density in the loop patches, ρ_{LP}, rises in proportion with the number of cycles, N, and hence the cumulative strain, $\gamma_{\mathrm{pl,cum}}$.

In the course of fatiguing the loop patches, alias Taylor lattices, 'flip' between neighboring equilibrium positions, and also dislocations cycle in the channels between them, as well as later between the rungs of the ladder structure and/or in the cells. As a result, the same relative positions among the dislocations will be achieved at a resolved shear stress that is proportional to the root of the dislocation density in the loop patches, $\sqrt{\rho_{\mathrm{LP}}}$. However, the shear strain at the point of flipping will also be proportional to $\sqrt{\rho_{\mathrm{LP}}}$, which in turn is proportional to the root of the cumulative strain, $(\gamma_{\mathrm{pl,cum}})^{1/2}$, i.e. the sum of the magnitudes of the plastic strain amplitudes in the course of the preceding cycling. It follows that hysteresis loops should be derivable from each other by simply transforming both the stress and strain axis in proportion to $(\gamma_{\mathrm{pl,cum}})^{1/2}$. This construction has been made in fig. 74 extracted from [318] (see also [126,319]), comparing a set of hysteresis loops between $\gamma_{\mathrm{pl,cum}} = 0.5$ and 16, with measured hysteresis loops by Mughrabi [320]. As seen, the agreement is excellent.

3.17. Outlook

This paper has presented a wide-ranging overview of the LEDS theory, now seen to be but a special case of the more general LES theory. It represents a seamless explanation of all experimentally observed phenomena, including the correlation between dislocation structures and mechanical strength and the shape of the stress–strain curve. While still much needs to be done, especially in gathering systematic data on Taylor lattices, the nature and size of their domains as a function of stress, and similarly on the Stage II carpet structure and its transformation into a Stage III cell structure, there is no indication anywhere of a conflict between observations and the LEDS theory. One may therefore accept its basics with confidence.

With this the way has been opened to begin the systematic development of theoretically based constitutive equations. These would be of great benefit to industry. In fact, throughout, constitutive equations have been the ultimate goal that has motivated the bulk of research on dislocation-based plasticity, if not perhaps consciously in academia then certainly in industry. What is needed now is a two-pronged approach: (1) The determination of functions and parameters that more securely than hitherto tie the mechanical properties to the respective microstructures, – including of course alloying composition and distribution of phases, as also twins if any, but most importantly dislocation structures. (2) The determination of functions and parameters that will relate the microstructure to the thermo-mechanical history.

No doubts there is still a long road ahead before we shall be in the possession of reliable constitutive equations. However, the major obstacle to making secure progress is now removed, namely the previously pervasive uncertainty about how dislocation structures arise in principle, and in which way they control the mechanical properties. The LEDS theory provides the necessary paradigm, and the LES hypothesis will no doubt be an invaluable guide. Let us begin!

Acknowledgements

The LEDS theory has been developed through several decades, beginning with this writer's PhD research in the 1940's. Over the years, the research has been partially supported from different sources, mostly by the Office of Naval Research, Arlington, VA, and currently by the National Science Foundation, Materials Division, grant DMR-98-1476 (Dr. Bruce Mac Donald). This aid is gratefully acknowledged. Much more important yet has been the steadfast encouragement and support of my husband, H.G.F. Wilsdorf († April 17, 2000), and more recently of my friend and colleague E.A. Starke, Jr. Also, I have greatly profited from research cooperation with others, first and foremost my husband, as well as numerous colleagues and graduate students, as documented in common research papers listed among the references. To all of these I owe a great debt of gratitude. Finally, I am grateful to colleagues for permission to reproduce a number of figures from the previous literature and to my current coworkers, Drs. Matthew S. Bednar and Jon T. Moore, for invaluable aid with the technical production of figures.

References

[1] D. Kuhlmann-Wilsdorf and K.I. Winey, J. Appl. Phys. 85 (1999) 6393.
[2] D. Kuhlmann-Wilsdorf, Acta Metall. Mater. 47 (1999) 1697.
[3] D. Kuhlmann-Wilsdorf, Metall. Mater. Trans. (1999), in press.
[4] B.S. Pinheiro, K.I. Winey, D.A. Hajduk and S.M. Gruner, Macromolecules 29 (1996) 1482.
[5] D.L. Polis and K.I. Winey, Macromolecules 29 (1996) 8180.
[6] D.L. Polis and K.I. Winey, Macromolecules 31 (1998) 3617.
[7] P.S. Pinheiro and K.I. Winey, Macromolecules 31 (1998) 4447.
[8] C. Laird, Z. Wang, B.-T. Ma and H.-F. Chai, Mater. Sci. Engng. A 113 (1989) 245.
[9] L. Llanes and C. Laird, Mater. Sci. Engng. A 161 (1993) 1.
[10] F.R.N. Nabarro and H.L. de Villiers, The Physics of Creep (Taylor and Francis, London, 1995).
[11] H.-J. Kestenbach, W. Krause and T.L. da Silveira, Acta Metall. 26 (1978) 661.
[12] E.C. Aifantis, J. Eng. Mater. Technol. 106 (1984) 326.
[13] E.C. Aifantis, Mater Sci. Engng. 81 (1986) 563.
[14] E.C. Aifantis, in: Patterns, Defects and Microstructures in Non-Equilibrium Systems, NATO ASI Series (Martinus Nijhoff Publ., Dordrecht, Netherlands, 1987) p. 221.
[15] D. Walgraef and E.C. Aifantis, Int. J. Eng. Sci. 23 (1985) 1351, 1359, 1365.
[16] D. Walgraef (ed.), Patterns, Defects and Microstructures in Non-Equilibrium Systems, NATO ASI Series (Martinus Nijhoff Publ., Dordrecht, Netherlands, 1987).
[17] D.L. Holt, J. Appl. Phys. 41 (1970) 3197.
[18] M.R. Staker and D.L. Holt, Acta Mater. 20 (1972) 569.
[19] L.P. Kubin, in: Materials Science and Technology: A Comprehensive Treatment, Vol. 6, eds R.W. Cahn, P. Haasen and F.L. Kramer (VCH Weinheim, 1993) p. 138.
[20] L.P. Kubin, in: Stability of Materials, eds A. Gonis, P.E.A. Turchi and J. Kudrnovsky (Plenum Press, New York, 1995) pp. 99–135.
[21] B. Devincre and L.P. Kubin, Mater. Sci. Engng., Special issue of ICSMA Conf. in Prague, Aug. 1997 A234 (1997) 8.
[22] D. Kuhlmann-Wilsdorf, Materials Research Innovations 1 (1998) 265.
[23] D. Kuhlmann-Wilsdorf, Philos. Mag. A 79 (1999) 955.
[24] D. Kuhlmann and G. Masing, Z. Metallkunde 39 (1948) 361.
[25] D. Kuhlmann, G. Masing and J. Raffelsieper, Z. Metallkunde 40 (1949) 241.
[26] D. Kuhlmann, Proc. Phys. Soc. A 64 (1951) 140.
[27] N.F. Mott, Phil. Mag. 43 (1952) 1151.
[28] N.F. Mott, Phil. Mag. 44 (1953) 742.
[29] A. Seeger, J. Diehl, S. Mader and M. Rebstock, Phil. Mag. 2 (1957) 323.
[30] A. Seeger, Kristallplastizität, in: Handb. der Physik, Vol. VII/2 ed. S. Flügge (Springer, Berlin, 1958).
[31] H.G.F. Wilsdorf, D. Kuhlmann-Wilsdorf and J.H. van der Merwe, Phil. Mag. 43 (1952) 632.
[32] D. Kuhlmann-Wilsdorf, Phys. Stat. Sol. A 149 (1995) 225.
[33] D. Kuhlmann-Wilsdorf, Scripta Materialia 34 (1996) 641.
[34] D. Kuhlmann-Wilsdorf, Scripta Materialia 36 (1997) 173.

[35] M.N. Bassim, W.A. Jesser, D. Kuhlmann-Wilsdorf and H.G.F. Wilsdorf eds, Low-Energy Dislocation Structures, Proc. Intl. Conf. on Low-Energy Dislocation Structures, UVA, School of Engr. and Appl. Science, Charlottesville, VA, Aug. 10–14, 1986 (Elsevier Sequoia, Lausanne, 1986); see also Mater. Sci. Engng. 81 (1986) 1–574, and 86 (1987) 19–92.
[36] D. Kuhlmann-Wilsdorf, Phys. Stat. Sol. A 104 (1987) 121.
[37] C.G. Darwin, Phil. Mag. 27 (1914) 315, 675.
[38] J. Moore and D. Kuhlmann-Wilsdorf, J. Appl. Phys. 41 (1970) 4411–4417.
[39] J.T. Moore and D. Kuhlmann-Wilsdorf, J. Appl. Phys. 42 (1971) 953, 3717, 3726.
[40] J.T. Moore and D. Kuhlmann-Wilsdorf, in: Grain Boundaries and Interfaces, Proc. of Intl. Conf. on Structure and Properties of Grain Boundary Interfaces, Yorktown Heights, 1971, eds P. Chaudhari and J.W. Matthews (North-Holland, 1972) pp. 456–478. See also Surface Science (Special Issue) 31 (1972) 456.
[41] W.A. Jesser and D. Kuhlmann-Wilsdorf, Mater. Sci. Engng. 51 (1981) 31.
[42] M.N. Bassim and D. Kuhlmann-Wilsdorf, Cryst. Latt. Defects 4 (1973) 9.
[43] M.N. Bassim and D. Kuhlmann-Wilsdorf, Phys. Stat. Sol. A 15 (1973) 725.
[44] M.N. Bassim and D. Kuhlmann-Wilsdorf, Phys. Stat. Sol. A 16 (1973) 241.
[45] M.N. Bassim and D. Kuhlmann-Wilsdorf, Phys. Stat. Sol. A 17 (1973) 281.
[46] M.N. Bassim and D. Kuhlmann-Wilsdorf, Phys. Stat. Sol. A 17 (1973) 379.
[47] M.N. Bassim and D. Kuhlmann-Wilsdorf, Phys. Stat. Sol. A 19 (1973) 335.
[48] W. Scoble and S. Weissmann, Cryst. Latt. Defects 4 (1973) 123.
[49] F.R.N. Nabarro, Dislocations (Oxford University Press, Oxford, 1987); reprinted at (Dover Publications, New York, 1987); see also Adv. Phys. 1 (1952) 269.
[50] J. Weertman and J.A. Weertman, Elementary Dislocation Theory (Oxford University Press, New York, 1992).
[51] M.A. Myers and K.K. Chawla, Mechanical Behavior of Materials (Prentice-Hall, Upper Saddle River, NJ, 1998).
[52] A.H. Cottrell, Dislocations and Plastic Flow in Crystals (McGraw-Hill, New York, 1953).
[53] R.E. Reed-Hill and R. Abbaschian, Physical Metallurgy Principles (PWS-Kent Publ. Co, Boston, 1993).
[54] G.I. Taylor, Proc. Roy. Soc. London 145 (1934) 362, 388.
[55] D. Kuhlmann-Wilsdorf, in: Physics of Strength and Plasticity, ed. A.S. Argon (MIT Press, 1969) p. 29.
[56] A.H. Cottrell, Dislocations and Plastic Flow in Crystals (McGraw-Hill, New York, 1953).
[57] M. Wilkens, Acta Metall. 15 (1967) 1412.
[58] M. Wilkens, Acta Metall. 17 (1969) 1155.
[59] U.F. Kocks and R.O. Scattergood, Acta Metall. 17 (1967) 1161.
[60] J.T. Moore and D. Kuhlmann-Wilsdorf, Cryst. Latt. Defects 1 (1970) 201.
[61] J.P. Hirth, J. Appl. Phys. 32 (1961) 700.
[62] D. Kuhlmann-Wilsdorf, Acta Metall. 14 (1966) 439.
[63] H.G.F. Wilsdorf and D. Kuhlmann-Wilsdorf, Phys. Rev. Lett. 4 (1959) 170.
[64] D. Kuhlmann-Wilsdorf and H.G.F. Wilsdorf, Acta Metall. 10 (1962) 584.
[65] D. Kuhlmann-Wilsdorf, R. Maddin and H.G.F. Wilsdorf, in: Strengthening Mechanisms in Solids, ed. J.J. Harwood (ASM, Metals Park, Ohio, 1962) p. 137.
[66] D. Kuhlmann-Wilsdorf and H.G.F. Wilsdorf, in: Electron Microscopy and Strength of Crystals, eds G. Thomas and J. Washburn (Interscience, NY, 1963) p. 575.
[67] R.H. Geiss, D. Kuhlmann-Wilsdorf, W.J. Tropf and H.G.F. Wilsdorf, 31st. Ann. Electron Microscopy Soc. Amer., ed. C.J. Arceneaux (New Orleans, LA, 1973).
[68] J.E. Bailey and P.B. Hirsch, Phil. Mag. 5 (1960) 485.
[69] H.G.F. Wilsdorf and J. Schmitz, J. Appl. Phys. 33 (1962) 1750.
[70] A. Seeger, H. Kronmüller, S. Mader and H. Träuble, Phil. Mag. 6 (1961) 639.
[71] A. Seeger, S. Mader and H. Kronmüller, in: Electron Microscopy and Strength of Crystals, eds G. Thomas and J. Washburn (Interscience, NY, 1963) p. 665.
[72] P.B. Hirsch and T.E. Mitchell, Canad. J. Phys. 45 (1967) 663.
[73] H.J. Seemann and F. Stavenow, Z. Metallkunde 52 (1961) 667.
[74] L. Delehouzee and A. Deruyttere, Acta Metall. 15 (1967) 727.

[75] G. Thomas and J. Nutting, J. Inst. Metals 57 (1956) 1.
[76] D.J. Lloyd and D. Kenney, Metall. Trans. A 13 (1982) 1445.
[77] B. Bay, N. Hansen, D.A. Hughes and D. Kuhlmann-Wilsdorf, Acta Metall. Mater. 40 (1992) 205.
[78] D.A. Hughes and Y.L. Liu, in: Hot Deformation of Aluminum Alloys, eds T.G. Langdon, H.D. Merchant, J.G. Morris and M.A. Zaidi (The Minerals, Metals and Materials Soc., 1991) p. 21.
[79] D.A. Hughes, Acta Metall. Mater. 41 (1993) 1421.
[80] D. Kuhlmann-Wilsdorf, H.G.F. Wilsdorf and J.A. Wert, Scripta Metall. Mater. 31 (1994) 729.
[81] B.C. Ellwood, J. Inst. Metals 80 (1951) 605.
[82] K. Sezaki and D. Kuhlmann-Wilsdorf, Acta Metall. 14 (1966) 1131.
[83] J.A. Sigler, PhD thesis, Materials Science (University of Virginia, Charlottesville, 1967).
[84] H.G.F. Wilsdorf, Am. Soc. Test. Mater. Spec. Techn. Publ. 245 (1958) 43.
[85] H.G.F. Wilsdorf, D. Kuhlmann-Wilsdorf, J. Nuclear Materials 5 (1962) 178.
[86] P. Charsley, Mater. Sci. Engng. 41 (1981) 181.
[87] P. Charsley and D. Kuhlmann-Wilsdorf, Phil. Mag. A 44 (1981) 1351.
[88] J.I. Dickson, L. Handfield and G. L'Espérance, Mater. Sci. Engng. 81 (1986) 477–492.
[89] J.I. Dickson, J. Boutin and G. L'Espérance, Acta Metall. 34 (1986) 1505–1514.
[90] J.I. Dickson, Hong Bande and G. L'Espérance, Mater. Sci. Engng. A 101 (1988) 75.
[91] S. Turenne, G. L'Espérance and J.I. Dickson, Acta Metall. 36 (1988) 459.
[92] G.I. Taylor, J. Inst. Metals London 62 (1938) 307.
[93] F.C. Frank, in: Pittsburgh Symp. on the Plastic Deformation of Crystalline Solids (Rept. NAVEXOS-P-834, Off. of Naval Res., Washington, DC, 1950) p. 150.
[94] S. Amelinckx and W. Dekeyser, Solid State Physics, Advances in Research and Applications 8 (1959) 325–499.
[95] W.T. Read and W. Shockley, Imperfections in Nearly Perfect Crystals (Wiley, New York, 1952) Chapt. 13.
[96] D. Kuhlmann-Wilsdorf, J. Appl. Phys. 33 (1962) 648–654.
[97] C.T. Young, T.J. Headley and J.L. Lytton, Mater. Sci. Engng. 81 (1986) 391–407.
[98] D. Kuhlmann-Wilsdorf and E. Aernoudt, J. Appl. Phys. 54 (1983) 184–192.
[99] F.C. Frank, in: Bristol Conference on Defects in Crystalline Solids (The Physical Soc., London, 1955) p. 159.
[100] D. Kuhlmann-Wilsdorf, Workhardening, eds J.P. Hirth and J. Weertman (Gordon and Breach, NY, 1968) pp. 97–132.
[101] G. Langford and M. Cohen, Metall. Trans. A 6 (1975) 901.
[102] D. Kuhlmann-Wilsdorf, Phys. Stat. Sol. A 149 (1995) 225–241.
[103] H. Mughrabi, Phil. Mag. 23 (1971) 869, 897.
[104] H. Mughrabi, Acta Metall. 31 (1983) 1367.
[105] Q. Liu and N. Hansen, Proc. Roy. Soc. London A 54 (1998) 2555–2591.
[106] Q. Liu, D. Juul Jensen and N. Hansen, Acta Mater. 46 (1998) 5819–5853.
[107] D. Juul Jensen, M.T. Lyttle and N. Hansen, in: Hot Deformation of Aluminum Alloys II, eds T.R. Bieler, L.A. Lalli and S.R. MacEwen (The Minerals, Metals and Materials Soc., 1998) pp. 9–21.
[108] Q. Liu, C. Maurice, J. Driver and N. Hansen, Metall. Mater. Trans. 29 A (1998) 2333.
[109] D.A. Hughes, D.C. Chrzan, Q. Liu and N. Hansen, Phys. Rev. Lett. 81 (1998) 4664.
[110] D. Kuhlmann-Wilsdorf, Trans. Met. Soc. AIME 224 (1962) 1047.
[111] W. Bollmann, Crystal Defects and Crystalline Interfaces (Springer, Berlin, 1970).
[112] W. Bollmann, Crystal Lattices, Interfaces, Matrices (W. Bollmann, Geneva, 1982).
[113] C. Solenthaler and W. Bollmann, Mater. Sci. Eng. 81 (1986) 35–49.
[114] W. Bollmann, Mater. Sci. Eng. A 131 (1989) 29.
[115] C. Solenthaler, Mater. Sci. Eng. A 113 (1989) 95–119.
[116] K. Shin and A.H. King, Mater. Sci. Eng. A 113 (1986) 121.
[117] M.N. Bassim, W.A. Jesser, D. Kuhlmann-Wilsdorf and H.G.F. Wilsdorf (eds), Low-Energy Dislocation Structures, Proc. Intl. Conf. on Low-Energy Dislocation Structures, UVA, School of Engr. and Appl. Science, Charlottesville, VA, Aug. 10–14, 1986 (Elsevier Sequoia, Lausanne, 1986); see also Mater. Sci. Eng. 81 (1986) 1–574 and 86 (1987) 19-92.
[118] J.B. Bilde-Sørensen, Mater. Sci. Eng. 81 (1986) 211.

[119] K. Higashida, J. Takamura and N. Narita, Mater. Sci. Eng. 81 (1986) 239.
[120] W.T. Read, Dislocations in Crystals (McGraw-Hill, New York, 1953).
[121] J.C.M. Li, Acta Metall. 8 (1960) 563.
[122] J.C.M. Li, Acta Metall. 8 (1960) 296.
[123] P.J. Jackson and F.R.N. Nabarro, Mater. Sci. Eng. 86 (1987) 29–34.
[124] G. Masing, Z. Physik 124 (1947) 586.
[125] F.R.N. Nabarro, Adv. Phys. 1 (1952) 269, 327/328.
[126] D. Kuhlmann-Wilsdorf, Mater. Sci. Eng. 39 (1979) 231–245.
[127] P. Neumann, Mater. Sci. Eng. 81 (1986) 465–475.
[128] U. Essmann, Phys. Stat. Sol. 12 (1965) 723.
[129] M. Wilkens, Canad. J. Phys. 45 (1967) 567.
[130] P. Gassenmeier and M. Wilkens, Phys. Stat. Sol. 30 (1968) 833.
[131] C.S. Pande and P.M. Hazzledine, Phil. Mag. 24 (1971) 1393.
[132] D. Kuhlmann-Wilsdorf and N. Comins, Mat. Sci. Eng. 60 (1983) 7–24.
[133] G.I. Taylor, Trans. Faraday Soc. 24 (1928) 121.
[134] K. Yamaguchi, Sci. Pap. Inst. Phys. Chem. Res. (Jpn.) 11 (1929) 151.
[135] E. Schmid and W. Boas, Plasticity of Crystals (Chapman and Hall, London, 1950); transl. from original 1935 German edition.
[136] R.A. Foxall, M.S. Duesbery and P.B. Hirsch, Canad. J. Phys. 45 (1967) 903.
[137] D.M. Moon and W.H. Robinson, Canad. J. Phys. 45 (1967) 1017.
[138] J.W. Steeds, Proc. Roy. Soc. London A 292 (1966) 343.
[139] J.W. Mitchell, J.C. Chevrier, B.J. Hockey and J.P. Monaghan, Canad. J. Phys. 45 (1967) 453.
[140] B.J. Hockey and J.W. Mitchell, Phil. Mag. 26 (1972) 409.
[141] R.B. Schwarz and J.W. Mitchell, Phys. Rev. B 9 (1974) 3292.
[142] D.A. Taliaferro, L.F. Henry and J.W. Mitchell, J. Appl. Phys. 45 (1974) 519.
[143] D.A. Taliaferro and J.W. Mitchell, J. Appl. Phys. 45 (1974) 523.
[144] W.E. Nixon, M.H. Massey and J.W. Mitchell, Acta Metall. 27 (1979) 943.
[145] J.W. Mitchell, Phys. Stat. Sol. A 135 (1993) 455.
[146] F.C. Frank, in: Rep. Conf. Strength of Solids, Bristol, 1947 (Physical Soc., London, 1948) p. 46.
[147] P. Müllner, C. Solenthaler, P. Uggowitzer and M.O. Speidel, Mater. Sci. Eng. A 164 (1963) 164.
[148] H. Neuhäuser, in: Dislocations in Solids, Applications and Recent Advances, Vol. 6, ed. F.R.N. Nabarro (North-Holland, Amsterdam, 1983) pp. 319–440.
[149] D. Kuhlmann-Wilsdorf, Mater. Sci. Eng. A 113 (1989) 1–41.
[150] R.L. Patterson and H.G.F. Wilsdorf, in: Fracture, Vol. 1, ed. H. Liebowitz (Academic Press, NY, 1969) Chpt. 8, p. 183.
[151] A.H. Cottrell, in: Rept. of a Conf. on Strength of Solids, H.H. Wills Lab. Bristol 1947 (The Phys. Soc., London, 1948) p. 30.
[152] A.H. Cottrell and B.A. Bilby, Proc. Phys. Soc. London A 62 (1949) 49.
[153] A.H. Cottrell, Dislocations and Plastic Flow in Crystals (Oxford University Press, London, 1953).
[154] F.R.N. Nabarro, in: Rept. of a Conf. on Strength of Solids, H.H. Wills Lab. Bristol 1947 (The Phys. Soc., London, 1948) p. 38.
[155] W. Cochardt, G. Schöck and H. Wiedersich, Acta Metall. 3 (1955) 533.
[156] H. Suzuki, Sci. Rep. Tohoku Univ. Res. Inst. 44 (1952) 455.
[157] G.R. Piercy, R.W. Cahn and A.H. Cottrell, Acta Metall. 3 (1955) 331.
[158] H.B. Aaron and H.K. Birnbaum, Acta Metall. 13 (1965) 205.
[159] H. Kimura and R. Maddin, Quench Hardening in Metals, in: Defects in Crystalline Solids, eds S. Amelinckx, R. Gevers and J. Nihoul (North-Holland, Amsterdam, 1971).
[160] A.E.H. Love, A Treatise on the Mathematical Theory of Elasticity (3rd. ed.) (University Press, Cambridge, 1920).
[161] C.F. Elam, Proc. Roy. Soc. 115 (1927) 148.
[162] M. Masima and G. Sachs, Z. Physik 50 (1928) 161.
[163] T.E. Mitchell, in: Progress in Applied Materials Research, Vol. 6, eds E.G. Stanford, J.H. Pearson and W.J. McGonnagle (Gordon and Breach, 1964) pp. 117–237.

[164] N. Hansen and D. Kuhlmann-Wilsdorf, Mater. Sci. Eng. 81 (1986) 141.
[165] C.Y. Barlow, B. Bay and N. Hansen, Phil. Mag. 51 (1985) 253.
[166] B. Bay, N. Hansen and D. Kuhlmann-Wilsdorf, Mater. Sci. Eng. A 113 (1989) 385.
[167] H. Mughrabi, Acta Metall. 31 (1983) 1367.
[168] T. Ungar, H. Mughrabi and M. Wilkens, Proc. 5th. Int. Symp. on Metall, and Mater. Sci., Risø 1984, eds N. Hessel Andersen, M. Eldrup, N. Hansen, D. Juul Jensen, T. Leffers, H. Lilholt, O.B. Pedersen and B.N. Singh (Natl. Lab. Roskilde, Denmark, 1984) p. 539.
[169] H. Mughrabi, T. Ungar, W. Kienle and M. Wilkens, Phil. Mag. 53 (1986) 793.
[170] H. Mughrabi, T. Ungar and M. Wilkens, Z. Metallkunde 77 (1986) 571.
[171] T. Ungar, H. Mughrabi, D. Rönnpagel and M. Wilkens, Acta Metall. 32 (1986) 333.
[172] T. Ungar, H. Mughrabi, M. Wilkens and A. Hilscher, Phils. Mag. A 64 (1991) 495.
[173] G.Y. Chin and W.L. Mammel, Trans. TMS-AIME 245 (1969) 1211.
[174] J.F.W. Bishop and R. Hill, Phil. Mag. 42 (1951) 414, 1298.
[175] J. Meissner, Z. Metallkunde 50 (1959) 207.
[176] A. Seeger, Kristallplastizität, Handb. der Physik, Vol. VII/2, ed. S. Flügge (Springer, Berlin, 1958).
[177] P. Haasen, Phil. Mag. 3 (1958) 384.
[178] D. Kuhlmann-Wilsdorf and N. Hansen, Scripta Metall. Mater. 25 (1991) 1557.
[179] F. Pfaff, Z. Metallkunde 53 (1962) 411.
[180] P. Haasen, in: Physical Chemistry, Vol. 10 (Academic Press, NY, 1970).
[181] R. Karnop and G. Sachs, Z. Physik 41 (1927) 116.
[182] U.F. Kocks, Acta Metall. 6 (1958) 85.
[183] A.W. Thompson and M.I. Baskes, Phil. Mag. 28 (1973) 301.
[184] A.W. Thompson, in: Work Hardening in Tension and Fatigue, ed. A.W. Thompson (AIME, New York, NY, 1977) p. 89.
[185] D. Kuhlmann-Wilsdorf and H. Wilsdorf, Acta Metall. 1 (1953) 394.
[186] H. Mecking and K. Lücke, Acta Metall. 17 (1969) 279.
[187] H. Mecking and K. Lücke, Z. Metallkunde 60 (1969) 185.
[188] H. Mecking, in: Work Hardening in Tension and Fatigue, ed. A.W. Thompson (AIME, New York, NY, 1977) p. 67.
[189] M. Biermann, Thesis (Technical Unversity of Aachen, Germany, 1975).
[190] D. Kuhlmann-Wilsdorf, Phys. Stat. Sol. A 149 (1995) 131.
[191] E. Voce, J. Inst. Metals 74 (1948) 537.
[192] E. Voce, Metallurgica 51 (1955) 219.
[193] U. Kocks, J. Eng. Mater. Tech. (ASME) 98H (1976) 76.
[194] J. Gil Sevillano and E. Aernoudt, Mater. Sci. Engng. 86 (1987) 35.
[195] O. Richmond, Oral communication (1995).
[196] J. Gil Sevillano, P. VanHoutte and E. Aernoudt, Progr. Mater. Sci. 25 (1980) 69.
[197] D. Rollet, U.F. Kocks and R.D. Doherty, in: Formability and Metallurgical Structure, eds A.K. Sachdev and J.D. Embury (TMS Warrendale, PA, 1987) p. 211.
[198] E. Aernoudt, J. Gil Sevillano and P. van Houtte, in: Proc. 8th. Int. Symp. on Metallurgy and Materials Science, Risø 1987, eds S.I. Andersen, J.B. Bilde-Sørensen, N. Hansen, T. Leffers, H. Lilholt, O.B. Pedersen and B. Ralph (Natl. Lab. Roskilde, Denmark, 1987) p. 1.
[199] H. Mecking and Y. Estrin, ibid. p. 123.
[200] D. Kuhlmann-Wilsdorf, Metall. Trans. 1 (1970) 3173.
[201] D. Kuhlmann-Wilsdorf and N. Hansen, Metall. Trans. A 20 (1989) 2393.
[202] J.L. Sturges, B. Parsons and B.N. Cole, in: Mechanical Properties at High Rates of Strain, ed. J. Harding) The Institute of Physics, London, 1980) p. 35.
[203] J. Gil Sevillano, PhD thesis (Katholieke Universiteit, Leuven, Belgium, 1974).
[204] O.K. Bowen, J.W. Christian and G. Taylor, Canad. J. Phys. 45 (1967) 903.
[205] A.S. Keh and S. Weissmann, in: Electron Microscopy and Strength of Crystals, eds G. Thomas and J. Washburn (Interscience, NY, 1963) p. 231.
[206] S. Mader, in: Electron Microscopy and Strength of Crystals, eds G. Thomas and J. Washburn (Interscience, NY, 1963) p. 183.

[207] H. Mughrabi, Phil. Mag. 23 (1971) 869, 897.
[208] P.R. Swann, in: Electron Microscopy and Strength of Crystals, eds G. Thomas and J. Washburn (Interscience, NY) p. 131.
[209] E.E. Laufer and W.N. Roberts, Phil. Mag. 14 (1966) 65.
[210] D. Kuhlmann-Wilsdorf, E.E. Laufer and H. Nine, J. Appl. Phys. 38 (1967) 896.
[211] S.V. Raj and G.M. Pharr, Mater. Sci. Eng. 81 (1986) 217.
[212] G. Langford and M. Cohen, Trans. ASM 62 (1969) 623.
[213] D. Kuhlmann-Wilsdorf and N. Hansen, Scripta Metall. Mater. 25 (1991) 1557–1562.
[214] L. Titchener and M.B. Bever, Progress in Metal Physics 7 (1958) 247.
[215] D. Kuhlmann-Wilsdorf, Scripta Metall. 4 (1970) 893.
[216] S.S. Kulkarni, Grain Size and Texture Control in Al–Cu and Al–Cu–Si Alloys (M.S. thesis, Materials Science and Engineering, University of Virginia, Charlottesville, VA, Dec. 1997).
[217] S.S. Kulkarni, E.A. Starke, Jr. and D. Kuhlmann-Wilsdorf, Acta Metall. Mater. 46 (1998) 5283.
[218] D. Kuhlmann-Wilsdorf, S.S. Kulkarni, J.T. Moore and E.A. Starke, Metall. Mater. Trans., 1999, in press.
[219] H. Wilsdorf and D. Kuhlmann-Wilsdorf, Naturwissenschaften. 38 (1951) 502.
[220] H. Wilsdorf and D. Kuhlmann-Wilsdorf, Z. ang. Physik 4 (1952) 361–370.
[221] H. Wilsdorf and D. Kuhlmann-Wilsdorf, Z. ang. Physik 4 (1952) 409–418.
[222] H. Wilsdorf and D. Kuhlmann-Wilsdorf, Z. ang. Physik 4 (1952) 418–424.
[223] R. Maddin, C.H. Mathewson and W.R. Hibbard, Trans. AIME 175 (1948) 86, 355.
[224] R. Maddin, C.H. Mathewson and W.R. Hibbard, Metals Trans. 185 (1949) 527.
[225] S. Mader, Z. Physik 149 (1957) 73.
[226] S. Mader and A. Seeger, Acta Metall. 8 (1960) 513.
[227] R. Ebeling, Z. Metallkunde 61 (1970) 42.
[228] W. Boas and E. Schmid, Z. Physik 61 (1930) 767.
[229] G. Masing and M. Polanyi, Erg. Exakt. Naturwiss. 2 (1933) 177.
[230] D. Kuhlmann, Z. Metallkunde 41 (1950) 129.
[231] L.C. Tsien and Y.S. Chow, Proc. Roy. Soc. London 163 (1937) 19.
[232] T.H. Blewitt, R.R. Coltman and J.K. Redman, Rept. Conf. Defects in Cryst. Solids (Physical Soc. London, 1955) p. 369.
[233] H. Rebstock, Z. Metallkunde 48 (1957) 206.
[234] H. Kronmüller, Z. Physik 154 (1959) 574.
[235] U. Essmann, S. Mader and A. Seeger, Z. Metallkunde 52 (1961) 443.
[236] S. Mader, A. Seeger and H.M. Thieringer, J. Appl. Phys. 34 (1963) 3376.
[237] H.G.F. Wilsdorf and J.T. Fourie, Acta Metall. 4 (1956) 271.
[238] J.T. Fourie and H.G.F. Wilsdorf, Acta Metall. 7 (1959) 339.
[239] C.Y. Barlow, B. Bay and N. Hansen, Phil. Mag. 51 (1985) 253.
[240] N. Hansen and D. Kuhlmann-Wilsdorf, Mater. Sci. Eng. 81 (1986) 141.
[241] R.J. Stokes and A.H. Cottrell, Acta Metall. 2 (1954) 341.
[242] A.H. Cottrell and R.J. Stokes, Proc. Roy. Soc. London A 233 (1955) 17.
[243] P.B. Hirsch and D.H. Warrington, Phil. Mag. 6 (1961) 735.
[244] H.G.F. Wilsdorf and D. Kuhlmann-Wilsdorf, Mater. Sci. Engng. A 64 (1993) 1.
[245] H. Westengen, Ardal-Sundal Verk, Sundalsöre, Norway (1982), unpublished report.
[246] U.F. Kocks, H.S. Chen, D.A. Rigney and R.J. Schaefer, in: Workhardening, eds J.P. Hirth and J. Weertman (Gordon and Breach, NY, 1968) p. 151.
[247] D. Kuhlmann and G. Masing, Z. Metallkunde 39 (1948) 361.
[248] P. Haasen and G. Leibfried, Z. Metallkunde 43 (1952) 317.
[249] R. Becker, Physikal. Z. 26 (1925) 919.
[250] R. Becker, Z. Physik 33 (1925) 185.
[251] R. Becker, Z. Techn. Physik 7 (1926) 547.
[252] R. Becker and W. Boas, Metallwirtschaft 8 (1929) 317.
[253] R. Becker and E. Orowan, Z. Physik 79 (1932) 566.
[254] D. Kuhlmann-Wilsdorf, Metall. Trans. A 16 (1985) 2091.
[255] T. Hasegawa and U. Kocks, Acta Metall. 27 (1979) 1705.

[256] E. Nes, Acta Metall. Mater. 43 (1955) 2189.
[257] E.H. Dix, W.A. Andersen and M. Byron Shumaker, Corrosion 15 (1959) 19.
[258] J.T. Michalak and H.W. Paxton, Trans. AIME 221 (1961) 850.
[259] R.E. Sanders, S.F. Baumann and H.C. Stumpf, in: Aluminum Alloys, Their Physical and Mechanical Properties, Proc. Intl. Conf. UVA, Charlottesville, eds E.A. Starke and R.E. Sanders (1986); see also J.E. Hatch, Aluminum, Properties and Physical Metallurgy (Am. Soc. Metals, Metals Park, OH, 1984).
[260] B. Bay and N. Hansen, in: Annealing Processes – Recovery, Recrystallization and Grain Growth, eds N. Hansen, D. Juul Jensen, T. Leffers and B. Ralph (Risø Natl. Lab., Roskilde, Denmark, 1986) p. 215.
[261] T. Furu, R. Örsund and E. Nes, Acta Metall. 43 (1995) 2209.
[262] D. Kuhlmann, Z. Physik 124 (1947) 468.
[263] D. Kuhlmann, G. Masing and J. Raffelsieper, Z. Metallkunde 40 (1949) 241.
[264] A.T. Winter, Phil. Mag. 37 (1978) 457.
[265] C. Laird, P. Charsley and H. Mughrabi, Mater. Sci. Engng. 81 (1986) 433.
[266] P. Charsley, Mater. Sci. Engng. 47 (1981) 181.
[267] Z. Wang, C. Laird and H.-F. Chai, Mater. Sci. Eng. A 113 (1989) 245.
[268] T.K. Lepistö, V.T. Kuokkala and P.O. Kettunen, Mater. Sci. Eng. 81 (1986) 457.
[269] F.D. Rosi, Trans AIME 200 (1954) 1009.
[270] P.J. Jackson and D. Kuhlmann-Wilsdorf, Scripta Metall. 16 (1982) 105.
[271] T.E. Mitchell and R.L. Smialek, in: Workhardening, eds J.P. Hirth and J. Weertman (Gordon and Breach, NY, 1968) p. 365.
[272] F.R.N. Nabarro and D. Kuhlmann-Wilsdorf, Scripta Mater. 35 (1996) 1331.
[273] D. Kuhlmann-Wilsdorf, Mater. Sci. Engng. 86 (1987) 53.
[274] D. Kuhlmann-Wilsdorf, Acta Metall. 37 (1989) 3217.
[275] J.B. Bilde-Sørensen, Acta Metall. 21 (1973) 1495.
[276] P. Lin, S.S. Lee and A.J. Ardell, Acta Metall. 37 (1989) 739.
[277] D. Kuhlmann-Wilsdorf, in: Work Hardening in Tension and Fatigue, ed. A.W. Thompson (AIME, NY, 1977) pp. 1–43.
[278] L.E. Murr and D. Kuhlmann-Wilsdorf, Acta Metall. 26 (1978) 847.
[279] T. Suzuki and T. Ishi, in: Physics of Strength and Plasticity, ed. A.S. Argon (MIT Press, 1969) p. 159.
[280] E.O. Hall, Proc. Phys. Soc. London 64 (1951) 747.
[281] N.J. Petch, J. Iron Steel Inst. 174 (1953) 25.
[282] R. Armstrong, I. Codd, R.M. Douthwaite and N.J. Petch, Phil. Mag. 7 (1962) 45.
[283] M.J. Marcinkowski and R.M. Fisher, Trans. AIME 233 (1965) 293.
[284] T.L. Johnston and C.E. Feltner, Metall. Trans. 1 (1970) 1161.
[285] H.J. Rack and M. Cohen, Mater. Sci. Eng. 6 (1970) 320.
[286] D. Kuhlmann-Wilsdorf and H.G.F. Wilsdorf, Phys. Stat. Sol. B 172 (1992) 235.
[287] D.A. Korzekwa, D. Matlock and G. Kraus, Metall. Trans. A 15 (1984) 1221.
[288] M.C. Mataya, M.J. Carr and G. Kraus, Metall. Trans. A 15 (1984) 347.
[289] D.Z. Yang, E.L. Brown, D.K. Matlock and G. Kraus, Metall. Trans. A 16 (1985) 1385.
[290] M. Dollar, I.M. Bernstein and A.W. Thompson, Acta Metall. 36 (1988) 311.
[291] J. Cuddy and M.N. Bassim, Mater. Sci. Eng. A 113 (1989) 421.
[292] M.N. Bassim and R.J. Klaassen, Mater. Sci. Engng. 81 (1986) 163.
[293] M.F. Ashby, in: Physics of Strength and Plasticity, ed. A.S. Argon (MIT Press, 1969) p. 113.
[294] M.F. Ashby, Phil. Mag. 21 (1970) 399.
[295] M.F. Ashby, in: Strengthening Mechanisms in Crystals, eds A. Kelly and R.B. Nicholson (Elsevier, Amsterdam, 1971) p. 137.
[296] P.B. Hirsch and F.J. Humphreys, in: Physics of Strength and Plasticity, ed. A.S. Argon (MIT Press, 1969) p. 189.
[297] U.F. Kocks, Phil. Mag. 13 (1966) 541.
[298] U.F. Kocks, Acta Metall. 14 (1966) 1629.
[299] U.F. Kocks, Metall. Trans. A 16 (1985) 2109.
[300] A.J. Ardell, Metall. Trans. A 16 (1985) 2131.
[301] L.M. Brown and R.K. Ham, in: Strengthening Mechanisms in Crystals, eds A. Kelly and R.B. Nicholson (Elsevier, Amsterdam, 1971) p. 12.

[302] V. Gerold, in: Dislocations in Solids, Vol. 4, ed. F.R.N. Nabarro (North-Holland Publ., Amsterdam, 1979) p. 221.
[303] E. Hornbogen, in: 3rd. Int. Conf. on Strength of Metals and Alloys 2 (1973) 108.
[304] F.J. Humphreys, in: Dislocations and Properties of Real Materials (The Inst. of Metals, London, 1985) p. 175.
[305] A. Kelly and R.B. Nicholson, Progress in Materials Science 10 (1963) 149.
[306] T.J. Koppenaal and D. Kuhlmann-Wilsdorf, Appl. Phys. Lett. 4 (1964) 59.
[307] H.G.F. Wilsdorf and D. Kuhlmann-Wilsdorf, in: Modelling of Plastic Deformation and Its Engineering Applications, eds S.I. Andersen, J.B. Bilde-Sørensen, N. Hansen, D. Juul Jensen, T. Leffers, H. Lilholt, T. Lorentzen, O.B. Pedersen and B. Ralph (Risø National Laboratory, Roskilde, Denmark, 1992) p. 511.
[308] H.G.F. Wilsdorf and D. Kuhlmann-Wilsdorf, Mater. Sci. Eng. A 164 (1993) 1.
[309] D. Kuhlmann-Wilsdorf and H.G.F. Wilsdorf, Z. Metallkunde 84 (1993) 278.
[310] G. Tammann, Lehrbuch der Metallographie (L. Voss Verlag, Leipzig, Germany); translated as "A Textbook of Metallography", by R.S. Dean and L.G. Swenson (The Chemical Catalog Company, New York, 1925).
[311] D. Kuhlmann-Wilsdorf and C. Laird, Mater. Sci. Eng. 27 (1977) 137.
[312] D. Kuhlmann-Wilsdorf and C. Laird, Mater. Sci. Eng. 37 (1979) 11.
[313] D. Kuhlmann-Wilsdorf, Mater. Sci. Eng. 39 (1979) 127.
[314] D. Kuhlmann-Wilsdorf and C. Laird, Mater. Sci. Eng. 46 (1980) 209.
[315] C. Laird, J.M. Finney and D. Kuhlmann-Wilsdorf, Mater. Sci. Eng. 50 (1981) 127.
[316] M. Glazov, L.M. Lianes and C. Laird, Phys. Stat. Sol. A 149 (1995) 297.
[317] G. Nicolis and I. Prigogine, Exploring Complexity (W.H. Freemen and Co., New York, 1989).
[318] D. Kuhlmann-Wilsdorf, Strength of Metals and Alloys, Proc. ICSMA 5, Aachen, Germany, Aug. 27–31, 1979, eds P. Haasen, V. Gerold and G. Kostorz (Pergamon, NY) pp. 1081–1087.
[319] D. Kuhlmann-Wilsdorf, in: Modeling the Deformation of Crystalline Solids, eds T.C. Lowe, A.D. Rollett, P.S. Follansbee and G.S. Daehn (IMS, Warrendale, PA, 1991) pp. 105–124.
[320] H. Mughrabi, Mater. Sci. Eng. 33 (1978) 207.

CHAPTER 60

Long-Range Internal Stresses in Deformed Single-Phase Materials: The Composite Model and its Consequences

H. MUGHRABI

Institut für Werkstoffwissenschaften
Universität Erlangen-Nürnberg
Martensstr. 5, 91058 Erlangen
Fed. Rep. Germany

and

T. UNGÁR

Department of General Physics
Eötvös University Budapest
H-1518, P.O.B. 32, Budapest
Hungary

Dislocations in Solids
Edited by F. R. N. Nabarro and M. S. Duesbery

Contents

1. Introduction

1.1. A survey of historical developments

Long-range internal stresses have played a significant role in models of the flow stress and, in particular, in Stage II work hardening in face-centred cubic (f.c.c.) crystals during the past 50 years [1–7]. In the fifties, and later, dislocation pile-ups [1–3], also referred to as superdislocations [4,5,7], were considered to be the classical source of long-range internal stresses. The existence of dislocation pile-ups giving rise to long-range internal stresses had been inferred, mainly by the Stuttgart school, from slip-line studies [5,8], magnetic measurements [7] and theoretical considerations [6,9]. The concepts of long-range internal stresses caused by dislocation pile-ups was heavily disputed when the first direct observations of dislocation distributions in deformed f.c.c. single crystals by transmission electron microscopy (TEM) in the early sixties failed to detect dislocation pile-ups in single crystals of metals like copper but, instead, revealed much more complex, tangled and layered dislocation arrangements [10–13]. Although the validity of TEM observations was under discussion at that time, since dislocation rearrangements were expected to occur upon unloading and during subsequent preparation of thin foils, these observations cast doubt on the existence of dislocation pile-ups and internal stresses [14,15]. Among those who did not dispute the existence of long-range internal stresses categorically, the arguments were put forward that, even if long-range internal stresses existed, they would not govern the flow stress [16].

The existence of long-range internal stresses in copper single crystals deformed into Stage II of the work-hardening curve was proved beyond reasonable doubt by Essmann's direct TEM observations of the dislocation arrangement which had been pinned in the unloaded state by fast neutron irradiation [10,11]. In these studies, Essmann observed strongly curved dislocation segments in the absence of an external stress and concluded that these curved dislocations represented local stress probes, giving evidence of the existence of long-range internal stresses of the order of the applied stress. Since classical dislocation pile-ups were rarely observed, Essmann [10,11] and others [6,13,17,18] concluded that the internal stresses were caused by so-called "disguised" pile-ups which were thought to be located mainly in the layered grids [10,11] or sheets [13], consisting of a network of primary and secondary dislocations lying roughly parallel to the primary glide plane. These layered networks were assumed to have originated from dislocation pile-ups that had been (partially) relaxed by secondary glide and thus converted into grids or sheets.

This picture was essentially confirmed and substantiated in subsequent detailed TEM studies on copper single crystals that had been deformed into Stage II and that had then been neutron-irradiated under load at liquid helium temperature (in order to suppress creep) so that the dislocations were pinned in the stress-applied state [19–21]. In these studies, unrelaxed dislocation pile-ups were also observed, though not in sufficient number and strength to confirm the long-range stress theory of Seeger and co-workers [4,5,7] satisfactorily [20,21].

With these developments, a state of partial understanding and ongoing controversy was reached by the end of the sixties and the beginning of the seventies. Some of the main contributions of that time are the original papers [4–7,14,16,22–24] and a number of more or less elaborate reviews by the main experts of their time [15,17,18,21,25,26]. We shall come back to some of the details of competing flow-stress theories in a later section. Here, however, we wish to remark that, in retrospect, it can be stated that progress in the basic understanding had come to a standstill at that time, mainly because a handful of reputed scientists who were at the same time strong personalities took extreme views by advocating one single flow-stress-controlling mechanism which they subsequently defended uncompromisingly. This statement applies in particular to the propagators of the long-range stress theory [4,5,7,17,18], the forest theory [16,26,27], the meshlength theory [14,24] and the statistical hardening theory [23,28]. In a way, Hirsch's theory of Stage II work hardening and flow stress [6,25] and also, to some extent, Seeger's and Wilkens' [17] and Seeger's [18] most recent reviews seem to have been the only attempts to combine and reconcile a number of different mechanisms in one model.

In the early seventies, the analysis of TEM observations of curved free dislocation segments in fatigued copper single crystals in which the dislocations had been pinned under load by fast neutron irradiation provided unambiguous evidence of the existence of substantial long-range internal stresses in dislocation arrangements that consisted predominantly of di- and multipolar edge dislocation clusters [29,30]. These findings of long-range internal stresses in well-screened dislocation distributions that had hitherto been considered almost free of internal stresses were most remarkable and puzzling. In the work of one of the present authors, these observations led to a new and rather simple approach in the late seventies and early eighties. This approach is based on the observation that deformation-induced dislocations are generally distributed heterogeneously in the form of tangles, walls, grids and, frequently, cell structures. It led to a straightforward understanding of deformation-induced long-range internal stresses that did not require the existence of dislocation pile-ups. In these "composite" models which were developed mainly (and independently) by one of the authors [31–33] and by Pedersen et al. [34, 35], the view is taken that real dislocation distributions consist in a first approximation of dislocation-rich hard regions (such as cell walls) which are separated by dislocation-poor soft regions (such as the cell interiors) and that deformed materials containing such dislocation distributions therefore respond to mechanical deformation like a composite. Then, the requirement of compatibility of deformation of soft and hard regions leads more or less automatically to the build-up of long-range internal stresses which cause a redistribution of the applied stress on a local scale in such a manner that the soft and the hard regions can yield plastically simultaneously in a compatible fashion. This is achieved by forward stresses aiding the applied stress in the hard regions and by back stresses opposing the applied stress in the soft regions. The wavelength of these internal stresses corresponds to the more or less periodic spacings of the heterogeneous dislocation distribution.

1.2. Scope of this survey

The purpose of this contribution is to review the composite model of plastic deformation of materials containing a heterogeneous dislocation distribution in its original form and to discuss some more recent extensions and applications of the model. Since the composite model has been worked out primarily as a model of the flow stress, the discussion will focus mainly on the flow stress. At the same time, however, important conclusions regarding the processes of plastic deformation will be drawn, based on the insight gained through the composite model and on the results of microstructural studies.

The scope of this survey will briefly be as follows. First, the main features of the work-hardening stages of f.c.c. crystals will be recalled in order to provide a framework for the subsequent treatment of the flow stress and the plastic deformation mechanisms in terms of the composite model. Then, a definition of (long-range) internal stresses will be given in order to avoid any misunderstandings. This will be followed by a review of the composite models for single slip and multiple slip deformation and their ranges of applicability. In particular, the flow-stress law which is basic to the composite model will be compared with other flow-stress models. The deficiencies of these models will be pointed out. Then, it will be shown that, in spite of the fact that deformation-induced long-range internal stresses are an inseparable part of the composite model, they cause only a modest increase of the energy of the dislocation distribution. Further valuable information is gained by comparing, for a given total dislocation density, in a simple model the flow stress of a one-dimensional wall structure with that of a homogeneous distribution of dislocations. With an extension of this model to include finite strain gradients, it will be shown that some fundamental relations of strain-gradient plasticity can be derived for the deformation of materials containing a heterogeneous dislocation distribution. After a critical assessment of the progress achieved with the composite model, its limitations and deficiencies and possible refinements, some applications of the composite model to more complex forms of loading such as cyclic deformation and deformations involving transients after a strain-path change or a stress-change in a creep test will be summarized. In the final section, recent experimental work on long-range internal stresses in materials deformed to large strains will be reviewed in the light of the composite model and its necessary adjustments for this particular case.

2. *Brief review of work-hardening stages*

The theories of work hardening that were developed after G.I. Taylor's pioneering paper in 1934 [36] originated mainly from the studies in the fifties and sixties of the three-stage work-hardening curves of tensile-deformed f.c.c. crystals that were orientated for single slip. Figure 1(a) shows schematically the three-stage work-hardening curve in a plot of resolved shear stress τ versus plastic shear strain γ_{pl}. In tensile tests, the shear strains till fracture are limited by the occurrence of necking. It is for this reason that a fourth work-hardening stage following Stage III was not observed in the earlier studies. Stage IV is found at large shear strains (in excess of 100%) that are achieved in particular in torsional tests [37–43]. The work-hardening behaviour is perhaps displayed best in a plot of the normalized work-hardening coefficient $\theta/G \approx \frac{1}{G} \cdot \frac{\mathrm{d}\tau}{\mathrm{d}\gamma_{\mathrm{pl}}}$ (G: shear modulus)

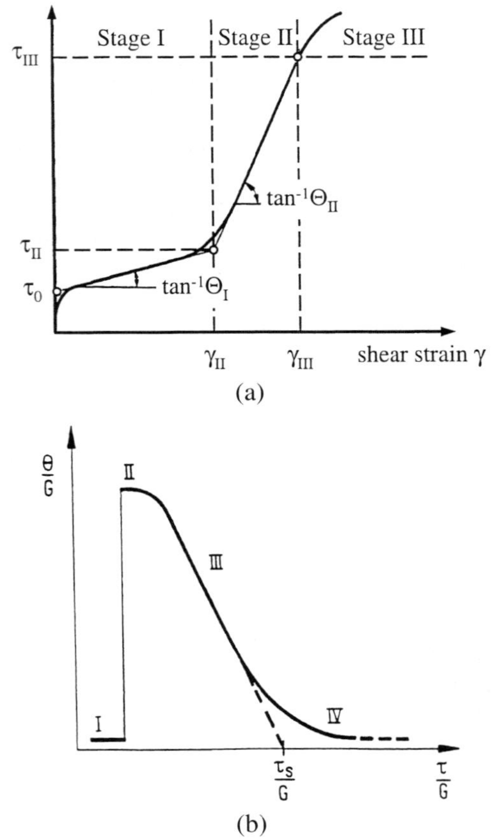

Fig. 1. Schematic review of work hardening stages. (a) three-stage work hardening (stress–strain) curves of tensile-deformed f.c.c. single crystals orientated for single slip; (b) plot of normalized work hardening rate θ against normalized shear flow stress.

versus the normalized shear flow stress τ/G [40–45], as shown schematically in fig. 1(b). The main features are a low (linear) work-hardening rate $\theta_I \approx G/3000$ in Stage I followed by a transition to a larger work-hardening rate $\theta_{II} \approx G/300$ in Stage II and a continuous decrease of θ after the transition from Stage II to Stage III. The extrapolation of Stage III work hardening in fig. 1(b) to a vanishing work hardening rate θ implies the approach of the flow stress to a saturation value, characterized by a (hypothetical) saturation flow stress τ_S. In reality, true saturation is not attained. Instead, the transition to Stage IV work hardening with a low linear work-hardening rate $\theta_{IV} = 0.05–0.1\tau_S$ [40–43] occurs.

In terms of dislocation mechanisms, the stages of work hardening can be characterized crudely as follows. In Stage I, primary dislocation glide over long paths is dominant until, due to the rotation of the stress axis and the build-up of internal stresses as secondary glide systems become activated, the dislocation glide paths are reduced, thus leading to the transition to Stage II work hardening. During Stage II work hardening, the dislocation glide paths become increasingly shorter. Stage III sets in, when processes of dynamic recovery, caused mainly by (correlated) large-scale cross slip and climb of groups of dislocations, lead to a continuous decrease of the work hardening rate [4,5,15]. According to several authors, the mechanisms of Stage IV work hardening are believed to be related to the different local dynamic (recovery) processes occurring in the dislocation cell walls and in the cell interiors [46–50].

The picture outlined here in schematic form applies to single crystals orientated for single slip. In polycrystals, deformation begins in multiple slip in many grains so that macroscopic Stage I behaviour is not observed. It is therefore common practice to describe polycrystal work hardening solely in terms of work hardening Stages II, III and IV.

All theories of work hardening start out from two basic relations, namely $\tau = \tau(\rho)$, where ρ is the (total) dislocation density, and $\rho = \rho(\gamma_{\mathrm{pl}})$. The work-hardening rate θ is then obtained as $\theta = \frac{\mathrm{d}\tau}{\mathrm{d}\rho} \cdot \frac{\mathrm{d}\rho}{\mathrm{d}\gamma_{\mathrm{pl}}}$. In the present paper, we shall be mainly concerned with the flow stress relation $\tau = \tau(\rho)$ and the dislocation models that are commonly envisaged in the derivation of the flow stress relation.

3. *Definition of long-range internal stresses, local plastic yielding*

According to Seeger [51], the shear flow stress τ of a material deforming by a combination of (long-range) elastic interaction of gliding dislocations and of thermally activated dislocation glide consists of two components:

$$\tau = \tau_G + \tau^*(\dot{\gamma}_{\mathrm{pl}}, T). \tag{1}$$

The athermal flow-stress component τ_G arises from the elastic interactions of the dislocations and depends on temperature only via the (weak) temperature dependence of the shear modulus G. In addition, there is the thermal or effective stress component τ^* which is required for the dislocations in order to overcome thermally activatable short-range obstacles and which therefore depends explicitly on both temperature T and shear strain rate $\dot{\gamma}_{\mathrm{pl}}$.

Frequently, the athermal flow-stress component τ_G is (misleadingly) referred to as the internal stress, denoted by the symbol τ_{i}, compare, e.g., [52,53]. While it is true that τ_G frequently is related to (elastic) internal stresses, τ_G itself is not an internal stress in the correct definition. Internal (or residual) stresses are defined as spatially varying stresses existing in a body in the absence of any external stresses (and/or moments). In other words, an internal stress $\tau_{\mathrm{i}}(x)$, varying, for the sake of simplicity, as a function of only one space coordinate x, is correctly defined as an athermal stress whose spatial average vanishes (Albenga's rule):

$$\langle \tau_{\mathrm{i}}(x) \rangle = \frac{\int \tau_{\mathrm{i}}(x)\,\mathrm{d}x}{\int \mathrm{d}x} = 0. \tag{2}$$

On the other hand, when an external shear stress τ is applied, the *locally acting* (*athermal*) *shear stress* τ_{loc} is given by the simple superposition of τ_i and τ as

$$\tau_{loc}(x) = \tau + \tau_i(x). \tag{3}$$

It is therefore clear that the locally acting stress varies spatially and that its spatial average, according to eqs (2) and (3), is identical to the external stress, i.e.

$$\langle \tau_{loc}(x) \rangle = \tau. \tag{4}$$

Since τ_{loc} varies spatially, plastic yielding must be considered *locally*, and the criterion for local plastic yielding is that τ_{loc} must exceed the local flow stress. In this case and quite generally, the local (flow) stress τ_{loc} would of course comprise an athermal and a thermal component. However, in the present study, the main interest lies in the elastic internal stresses which are athermal. Hence, the following considerations will largely be confined to athermal local stresses.

4. *The basic ideas and equations of the composite model*

4.1. The composite model for single slip: experimental evidence and basic equations

The composite model was first formulated for the (cyclic) flow stress of so-called persistent slip bands (PSBs) in cyclically deformed f.c.c. crystals [31–33]. Already in the first publication [31], the general applicability in all cases in which dislocations are distributed heterogeneously, e.g., in a cell structure, was pointed out. PSBs are thin lamellae (thickness: ca. 1 μm) parallel to the primary glide plane in which cyclic straining (which occurs mainly by single slip) becomes localized in the approach to cyclic saturation. The importance of PSBs lies in the fact that they give rise to the evolution of a progressing surface roughness at the sites of their emergence and that fatigue cracks initiate in the notch-peak topography of this surface roughness [30,54,55]. The dislocation arrangement of PSBs is more regular and hence simpler than that observed after unidirectional deformation. In copper, cyclically deformed at room temperature, it consists of rather regularly spaced dense di-/multipolar primary edge dislocation walls of non-negligible thickness (0.1 μm), separated by so-called channels with a width of ca. 1.4 μm in which glide occurs mainly by edge segments bowing out of the walls and traversing the channels and by the to-and-fro glide of bowed-out screw dislocations along the channels [30–33, 55]. In transmission electron microscopy (TEM), the PSBs appear as the so-called "ladder structure" in sections perpendicular to the primary glide plane and parallel to the primary Burgers vector $\mathbf{b}_p$, compare fig. 2(a). In sections parallel to the primary glide plane, the curved screw dislocations and the edge dislocation segments bowing out of the walls can be seen clearly (fig. 2(b)), provided precautions are taken to stabilize the dislocations in the stress-applied state. This was done by low-temperature neutron irradiation in the cases under discussion here [29–33].

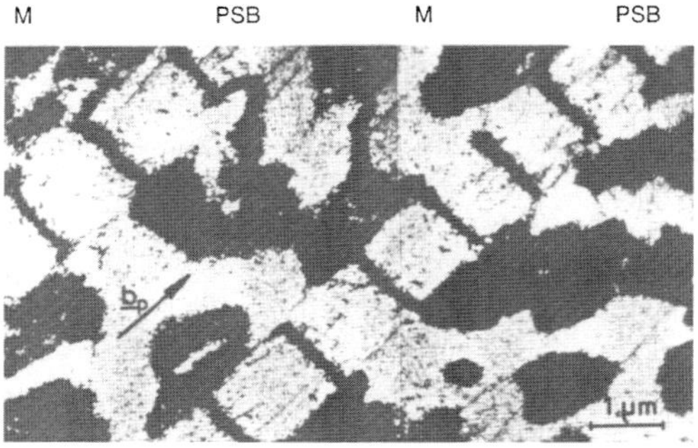

(a)

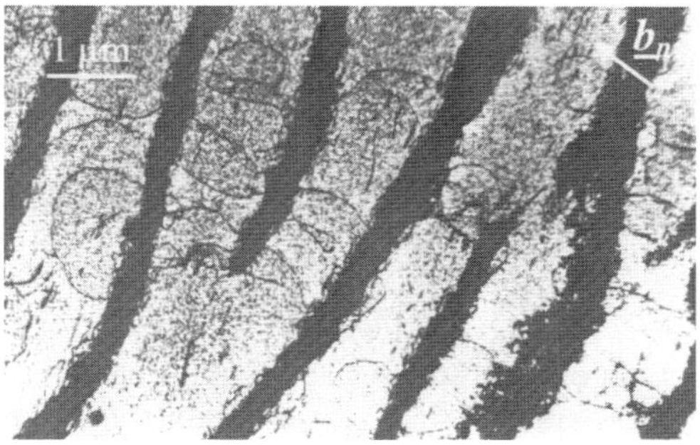

(b)

Fig. 2. TEM views of dislocation distribution in persistent slip bands (PSBs): (a) PSB "ladder" structure in $(1\bar{2}1)$-section perpendicular to the primary glide plane (111) and containing the primary Burgers vector $\mathbf{b}_\mathrm{p} = \frac{1}{2}[\bar{1}01]$; (b) section parallel to primary glide plane (111) showing dislocation walls and dislocation activity in "channels" between walls. Dislocations pinned in the stress-applied state [30,55].

The crucial observation that led to the formulation of the composite model was that the curvatures of the dislocations in the channels, compare fig. 2(b), varied spatially in a characteristic manner, indicating that the locally acting stresses in the channels increased significantly from the centre of the channels to the walls. Based on the fact that the curved dislocations acted as local stress probes, this finding was confirmed quantitatively by evaluating the local radii of curvature of a larger number of curved edge and screw dislocation segments lying at different locations in the channels according to the well-known relation

$$\tau_\mathrm{loc} = \frac{T_\mathrm{i}}{b \cdot r_\mathrm{i}}. \tag{5}$$

Here T_i is the dislocation line tension, r_i the radius of curvature of dislocations of character i and b the modulus of the Burgers vector. Using the experimentally established values of the line tensions of 1 nN and 2.5 nN [21,29] for edge and screw segments,[1]

[1] A detailed account of the determination of the dislocation line tensions by a self-consistent analysis of the spatial distribution of dislocation curvatures in deformed copper crystals has been published only recently [56].

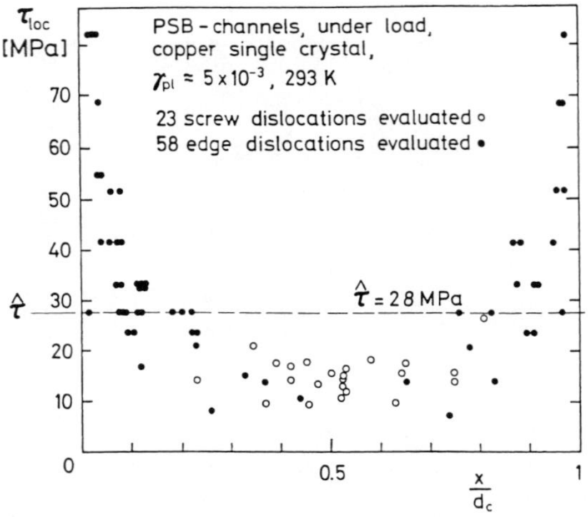

Fig. 3. Distribution of local stresses across channels of PSBs from wall to wall, as evaluated from radii of curvature of free edge and screw dislocation segments [32,33]. x/d_c: distance from wall, normalized with respect to channel width d_c.

respectively, the distribution of local stresses across the channels of PSBs shown in fig. 3 was obtained for the stress-applied state, corresponding to an external stress $\tau_{PSB} =$ 28 MPa [32,33]. The important things to note are that, in the centre of the channels, the local stress is only about half as large as the applied stress, whereas in the regions closer to the walls the local stresses increase sharply, reaching local values almost three times as large as the applied stress. This experimental finding of a significant redistribution of the macroscopic stress on a local scale provided clear evidence of long-range internal stresses which obviously oppose (as back stresses) the applied stress in the centre of the channels and aid it (as forward stresses) in the regions near at and in the walls. The puzzling question was how such long-range internal stresses could arise in a dislocation distribution consisting primarily of dislocation di- and multipoles [29–33].

The answer is provided by the composite model [31–33]. The basic idea is that a crystal containing a heterogeneous dislocation distribution responds to an external stress like a composite, since its local flow stress varies spatially, being higher than the applied stress in regions of high local dislocation density such as the dislocation walls and lower in regions of low dislocation density such as the cell interiors (channels). In the simplest possible formulation of the composite model, a crystal containing a heterogeneous distribution of dislocations is represented by a composite consisting of only two components, namely the hard dislocation-rich and the soft dislocation-poor regions. When deformed, this composite is assumed to be strained compatibly with the elements being deformed in parallel. In this iso-strain approach, applied to the case of single-slip deformation (in PSBs), the total shear strain γ_t is assumed to be constant, thus being equal in the dislocation walls and in the channels (of the PSBs) so that

$$\gamma_t = \gamma_{el,w} + \gamma_{pl,w} = \gamma_{el,c} + \gamma_{pl,c}, \qquad (6)$$

where $\gamma_{el,w}$ and $\gamma_{el,c}$ are the elastic shear strains and $\gamma_{pl,w}$ and $\gamma_{pl,c}$ the plastic shear strains in the walls and in the channels, respectively. If the area (volume) fractions of walls and channels are denoted by f_w and f_c, respectively, then

$$f_c + f_w = 1. \tag{7}$$

The (athermal) macroscopic flow stress τ is then given by a rule of mixtures which represents a spatial average of the local flow stresses τ_c and τ_w of the channels and the walls, respectively, as follows

$$\tau = f_c\tau_c + f_w\tau_w. \tag{8}$$

From eqs (6), (7) and (8), and assuming an elastic accommodation factor Γ according to Eshelby [57] for the elastic accommodation of the matrix (channels) around the hard inclusions (walls), we obtain in a formal manner without specifying the actual dislocation mechanisms:

$$\tau_w = \tau + \Delta\tau_w, \tag{9}$$

$$\tau_c = \tau + \Delta\tau_c \tag{10}$$

with

$$\Delta\tau_w = +G\Gamma f_c(\gamma_{pl,c} - \gamma_{pl,w}) \tag{11}$$

and

$$\Delta\tau_c = -G\Gamma f_w(\gamma_{pl,c} - \gamma_{pl,w}). \tag{12}$$

These relations reflect the fact that the compatible (iso-strain) deformation of the hard walls and the soft channels unavoidably leads to a redistribution of the macroscopic stress on a local scale in such a manner that the local stress is enhanced in the walls by a positive internal stress $\Delta\tau_w$ and reduced in the channels by a negative internal stress $\Delta\tau_c$. These internal stresses are directly proportional to the plastic strain mismatch $(\gamma_{pl,c} - \gamma_{pl,w})$ between channels and walls or, equivalently, to the negative elastic strain mismatch $-(\gamma_{el,c} - \gamma_{el,w})$. In addition, the spatial average of the internal stresses must vanish according to eq. (2). Hence, the equilibrium condition reads:

$$f_c\Delta\tau_c + f_w\Delta\tau_w = 0. \tag{13}$$

The redistribution of the macroscopic stress τ on a local scale in the manner described above permits the simultaneous compatible deformation of hard and soft regions by enhancing the local stress accordingly in the hard regions of high local flow stress while reducing it locally in the regions of low local flow stress.

The envisaged (microscopic) *local* deformation characteristics and their relation to the *macroscopic* deformation behaviour can be visualized in a “Gedankenexperiment”

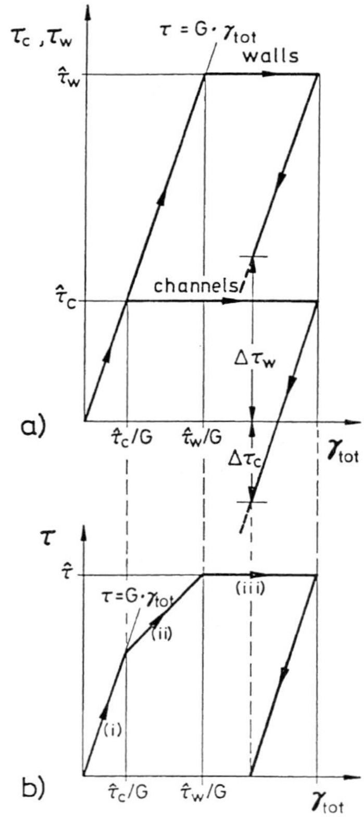

Fig. 4. Local microscopic and global macroscopic yielding in composite model: (a) hypothetical microscopic local stress-strain curves of "hard" dislocation walls and "soft" channels; (b) hypothetical macroscopic stress–strain curve. After [32,33].

according to the original publications [32,33], as illustrated in figs 4(a) and 4(b). These figures refer to a composite made up of a hypothetical dislocation microstructure inherited from a deformation to a given total shear strain in the dislocation-rich walls and in the dislocation-poor channels, respectively. Figure 4(a) shows in a highly idealized manner the hypothetical microscopic local stress–strain curves which the hard dislocation walls and the soft channels would exhibit, if they were now strained again individually. In both cases, ideal elastic–plastic behaviour is assumed. In order to distinguish the current stresses τ_w and τ_c from the respective flow stresses, the latter are denoted by $\hat{\tau}_w$ and $\hat{\tau}_c$, respectively.[2] These idealized stress–strain curves consist of elastic regimes up to the elastic shear strains, $\hat{\tau}_w/G$ and $\hat{\tau}_c/G$, followed by subsequent plastic yielding regimes, respectively. In order to relate this local behaviour to the macroscopic behaviour, we must now consider the

[2] In the following, the symbols for the local flow stresses $\hat{\tau}_c$, $\hat{\tau}_w$ and for the macroscopic flow stress $\hat{\tau}$ will be used only, when the (lower) current values of these stresses are also considered in the same context. Otherwise, the flow stresses will be denoted by τ_c, τ_w and τ, as before.

yielding of the composite consisting of a "sandwich" structure of the bonded components. Assuming that walls and channels are sheared in parallel to equal total shear strains (iso-strain), and that, here only for the sake of simplicity, the volume fractions of walls and channels are the same, we arrive at the hypothetical macroscopic stress–strain curve for the composite, as shown schematically in fig. 4(b). We can distinguish three regimes:

(i) common *elastic* yielding of channels and walls until the channels begin to yield plastically,
(ii) plastic yielding of channels, while the walls are still strained elastically (*microyielding*),
(iii) fully plastic yielding of channels and walls, once the walls yield plastically (*macroyielding*).

If now the composite is unloaded after deformation to a given total shear strain, the important role of the (deformation-induced) long-range internal stresses as the clue to the compatible deformation of the soft and hard components becomes apparent immediately. By definition, unloading along the elastic line in fig. 4(b) is terminated when the external macroscopic stress has been reduced to zero. At the same time, the components are also unloaded, as shown in fig. 4(a). However, the important point to note is that when the external stress becomes zero, an internal positive (forward) shear stress $\Delta\tau_{\mathrm{w}}$ and a corresponding internal negative (back) shear stress $\Delta\tau_{\mathrm{c}}$ remain "frozen in" in the dislocation walls and in the channels, respectively. The relative magnitudes of $\Delta\tau_{\mathrm{c}}$ and $\Delta\tau_{\mathrm{w}}$ follow from the equilibrium condition (13).

If the composite were now loaded again in the original (forward) sense, regime (ii) would not be observed. Instead, the composite would exhibit an elastic regime up to the macroscopic shear flow stress $\hat{\tau}$, followed by macroyielding. While the yielding behaviour discussed would apply in almost all details to the initial yielding behaviour of a real composite, it represents a hypothetical situation for a material which is single-phase in its initial state and then, during plastic deformation, becomes a composite-like material in an evolutionary manner, as the dislocation density increases and the dislocation distribution becomes more and more heterogeneous. Nonetheless, the basic features of fig. 4(a) are hidden in the material and reveal themselves when the material is deformed in the reverse sense. In this case, all three regimes are observed and provide a natural explanation [31,32, 34,58] for the well-known Bauschinger effect [59]. The latter is characterized by reverse yielding at a lower stress which, in our model, would be related to the beginning of reverse microyielding, i.e. beginning of plastic back flow in the softer components, viz. the channels. During reverse microyielding, the deformation-induced long-range internal stresses that were built up during forward stressing are reduced to zero and then built up again in the opposite sense, while the soft components (channels) undergo plastic back-flow. Reverse macroyielding sets in, when the internal stress level has become large enough to permit reverse plastic yielding of the hard components (walls). We shall return to the Bauschinger effect in section 5.4.1.

The intriguing question which remains to be discussed is how, namely by which microstructural dislocation processes, the long-range internal stresses are built up. Referring to the single-slip shear deformation of the PSB wall structure, fig. 5 illustrates schematically the spatial distributions of the dislocations (figs 5(a) and 5(b)) and of

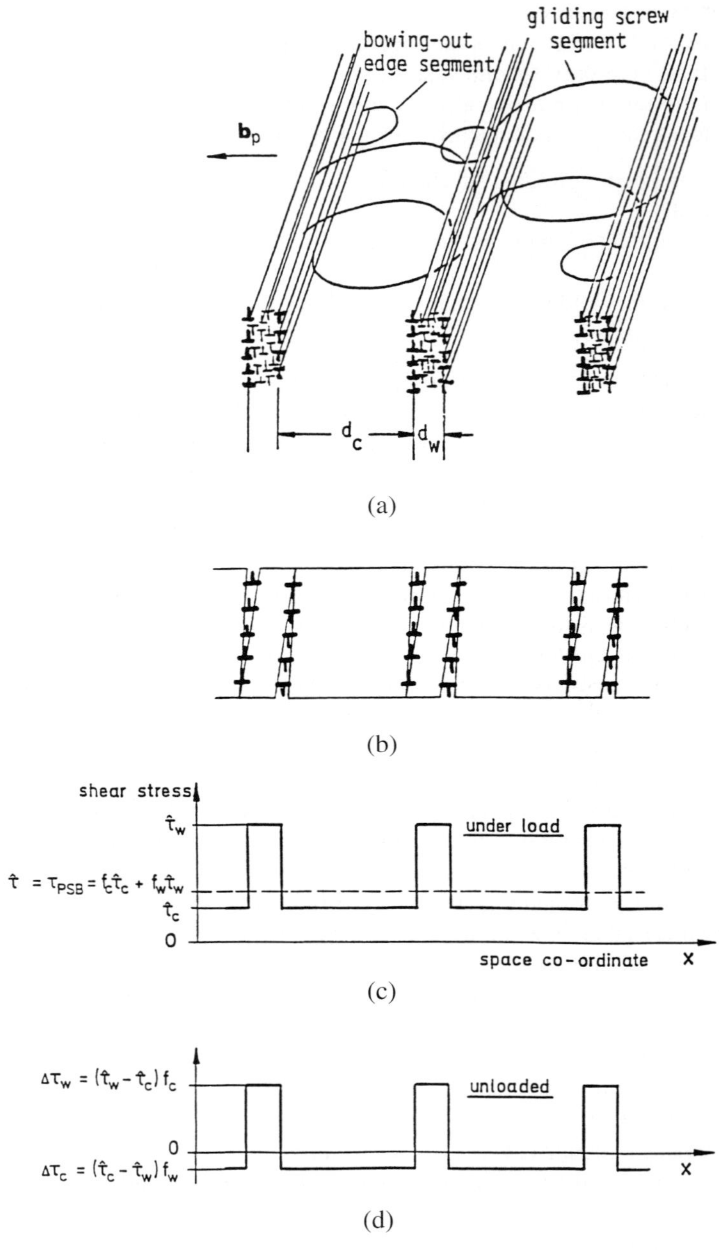

Fig. 5. Schematic illustration of dislocation distribution and corresponding spatial variation of the local (flow) stress in the PSB wall structure: (a) schematic view of dislocations in the walls and channels of a PSB; (b) illustration of accommodation of plastic strain mismatch by interfacial dislocations; (c) spatial variation of local flow stress in stress-applied state; (d) spatial variation of local internal stress after unloading. After [31–33].

the corresponding local flow stresses (fig. 5(c)) and residual internal stresses (fig. 5(d)). Moreover, fig. 5(b) illustrates how the elastic–plastic strain mismatch between the walls and the channels is accommodated by interfacial "misfit" dislocations at the wall-channel interfaces. These interface dislocations correspond to the polarized distribution of the edge dislocation segments which are drawn out by the gliding screw segments and which are constrained in their motion by the walls. We note here that Kocks et al. [60] proposed a similar picture of polarized dislocation walls with (interface) dislocations of opposite sign on either side of the walls. The work of these authors was based on some TEM work on deformed and subsequently annealed aluminium single crystals [60,61], aiming at an explanation of how misorientations and recrystallization nuclei can form. In our hypothetical stress–strain curve (fig. 4(b)), these interface dislocations accumulate in regime (ii). A simple consideration [33] shows that the number n of (edge) interface dislocations per unit length of interface (perpendicular to the glide plane) that is required in order to ensure compatible plastic deformation of soft and hard region is such that the related plastic extra strain nb matches precisely the elastic strain mismatch, i.e.

$$nb = (\gamma_{\mathrm{el,w}} - \gamma_{\mathrm{el,c}}) = \frac{\tau_{\mathrm{w}} - \tau_{\mathrm{c}}}{G}. \tag{14}$$

Inserting typical numbers for the local flow stresses τ_{w} and τ_{c}, namely $\tau_{\mathrm{w}} \approx 90$ MPa, $\tau_{\mathrm{c}} \approx 15$ MPa, compare [32,33], we find $n \approx 7 \times 10^6$ m^{-1}. This value corresponds to a local dislocation density of 10^{14} m^{-2} in the walls or, averaged over the wall spacing, an average dislocation density of $\approx 10^{13}$ m^{-2}. Compared to the local dislocation density ρ_{w} in the walls of some 10^{15} m^{-2} [32,33], the density of interface dislocations is rather low. Nevertheless, these interface dislocations obviously ensure compatibility of deformation by acting as inserted wedges of material at the interfaces between the soft channels and the hard walls, compare fig. 5(b). Moreover, they are closely related to the sources of the deformation-induced long-range internal stresses observed experimentally. A more detailed assessment of the magnitude of the internal stresses and the local flow stresses in the channels and in the walls, combined with other arguments based on TEM observations and, in particular, on the role of the bowed-out screw segments [62] linking the edge segments and on an analysis of the hysteresis loop shape [32,33,58,62], lead to the conclusion that, for the wall structure of the PSBs, the Eshelby accommodation factor Γ introduced in eqs (11) and (12) must be close to unity. In other words, the magnitude of the deformation-induced long-range internal stresses is not or at best only weakly reduced by elastic accommodation.

In the present simple form, the composite model provides a highly idealized profile of the local flow stress distribution (fig. 5(c)) and the residual internal stress distribution (fig. 5(d)) with infinitely steep local flow stress and local elastic/plastic strain gradients at the wall-channel interfaces. It should be noted that, in fig. 5(c), as previously in fig. 4, the local flow stresses of the walls and the channels are again denoted by $\hat{\tau}_{\mathrm{w}}$ and $\hat{\tau}_{\mathrm{c}}$, respectively, in order to distinguish them from the current values τ_{w} and τ_{c}. A more realistic distribution of the local flow stress (with finite gradients) across the channel width which compares very satisfactorily with the experimental data (fig. 3) can be computed in a simple model which takes into account the constraints of the glide of the bowed-out

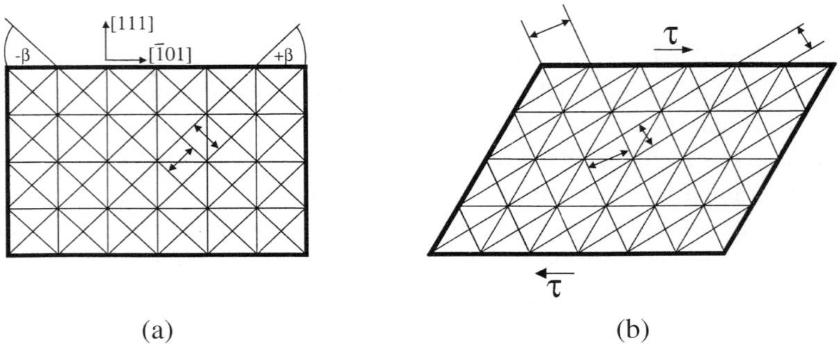

Fig. 6. Schematic illustration of lattice parameter changes during simple elastic shear straining of f.c.c. crystal in the direction [$\bar{1}$01] along the plane (111). View on (1$\bar{2}$1)-plane: (a) before straining; (b) after straining. Note the dependence of lattice spacing changes on the inclination of the planes.

screw dislocation segments in the channels [62], compare the TEM micrograph shown in fig. 2(b). In section 4.6, an extension of the simple composite model will be presented in which the effects of finite strain gradients arising from the geometrical constraints of dislocation bowing and/or from continuous variations of the dislocation density will be considered.

We now discuss briefly further experimental evidence relating to the existence of deformation-induced long-range internal stresses. The most direct indications of internal elastic strains are considered to be local changes in lattice parameter $\Delta d/d$ which should be measurable (under favourable conditions) by diffraction experiments. This topic will be dealt with in more detail in section 4.2, where it will be shown that X-ray diffraction can provide direct and unambiguous evidence of deformation-induced long-range internal stresses in dislocation cell structures produced by (symmetrical) multiple slip. In particular, one finds that the intensity profiles of X-ray diffraction become asymmetric and are in fact composed of mutually displaced symmetric peaks related to the diffraction from regions of tensile and compressive internal stresses, respectively. Hence, it seemed surprising that, in earlier X-ray diffraction studies on monotonically and cyclically [63] strained copper single crystals orientated for single slip, such evidence had not been noticed. In these studies, slight unsystematic asymmetries of {002} X-ray intensity profiles had been observed occasionally and had been considered insignificant. Now, with a better understanding of the expected geometry of the internal strains (and stresses), the conditions under which the X-ray intensity profiles should exhibit the asymmetry effects most convincingly can be specified in more detail. This is illustrated in fig. 6. Here, we consider the single-slip (elastic) shearing of a volume element of a f.c.c. crystal in a direction parallel to the glide plane (111) and along the direction [$\bar{1}$01] of the primary Burgers vector (fig. 6(a)). The view is along the [1$\bar{2}$1] direction. It is obvious that the spacings of lattice planes containing the [1$\bar{2}$1] direction will be affected most severely by the shear strains considered and that there will be no effect for lattice planes parallel to the glide plane (111) or to the plane (1$\bar{2}$1). In particular, a pair of crystallographically equivalent planes inclined at angles $\pm\beta$ to the (111) plane will suffer changes of lattice parameter of opposite sign (fig. 6(b)). Hence, diffraction from such a pair of planes is considered optimally suited in order to

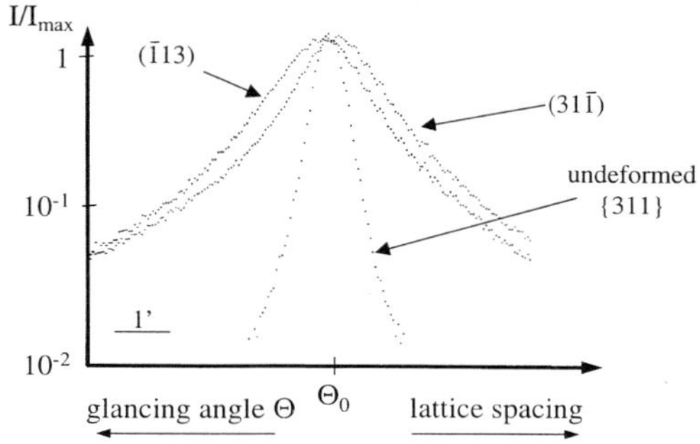

Fig. 7. X-ray reflection profiles. Comparison of the ($\bar{1}13$) and ($31\bar{1}$) normalized intensity profiles I/I_{max} vs. θ (I_{max}: maximum intensity, θ: glancing angle, θ_0: Bragg angle) of an undeformed copper single crystal and of a cyclically deformed crystal (see text). It should be noted that I/I_{max} is plotted on a logarithmic scale.

detect any asymmetry of the intensity profile of X-ray reflections. For the case under consideration, the planes ($\bar{1}13$) and ($31\bar{1}$) fulfill these requirements.

In fig. 7, the X-ray reflection profiles obtained for an undeformed copper single crystal and for a cyclically deformed crystal, using the appropriate {113} reflections, are shown in plots of normalized intensity I/I_{max} versus the diffraction angle θ [64]. The deformed copper crystal had been cycled into saturation at a plastic resolved shear strain amplitude of 7.5×10^{-3} at which the microstructure consists almost exclusively of the PSB wall structure. The specimen had then been unloaded from the peak tensile stress (end of a quarter cycle) in order to obtain the maximum internal stress effect. In addition to the expected broadening of the diffraction peaks, it will be noticed that the ($31\bar{1}$) and ($\bar{1}13$) peaks have become slightly asymmetric and that their maxima have been displaced sideways slightly in opposite directions with respect to the position of the intensity profile of the undeformed specimen. The maxima of these asymmetric diffraction profiles correspond roughly to the maxima of the symmetric sub-profiles associated with the diffraction from the volume regions of the channels (between the dislocation walls) which constitute almost 90% of the total volume. A more detailed evaluation of the asymmetric diffraction profiles yields appropriate values for the deformation-induced internal shear stresses $\Delta\tau_c$ and $\Delta\tau_w$ in the channels and in the walls, respectively, as $\Delta\tau_c \approx -6$ MPa and $\Delta\tau_w \approx +54$ MPa [64]. These values compare very favourably with the values obtained earlier from TEM of dislocation curvatures in the unloaded state, namely $\Delta\tau_c \approx -7$ MPa and $\Delta\tau_w \approx +55$ MPa [62].

4.2. The composite model for symmetrical multiple slip: experimental evidence and basic equations

The composite model for multiple-slip deformation corresponds closely to that for single-slip deformation. It was first formulated [33,65,66] in connection with X-ray diffraction line-broadening studies of tensile-deformed [001]-orientated copper single crystals [65–67]. These X-ray diffraction measurements were performed with a high-resolution double-crystal diffractometer with negligible instrumental line broadening on {002} sections of

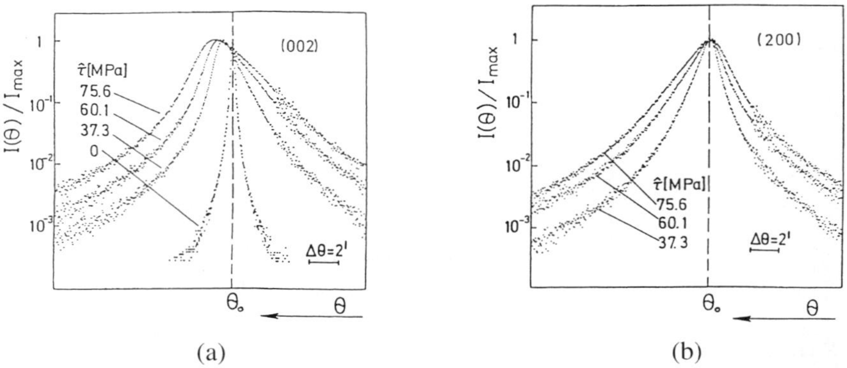

Fig. 8. Normalized X-ray diffraction intensity distributions of [001]-orientated copper single crystals after tensile deformation to different resolved shear stresses: (a) "axial" case, (002) reflections; (b) "side" case, (002) or (020) reflections. It should be noted that the stresses indicated on the left refer also to the outermost, middle and innermost flanks on the right sides of the profiles. After [65,66].

crystals deformed to resolved shear stresses of $\tau = 26.2, 37.3, 60.1$ and 75.6 MPa. The experimentally obtained diffraction intensity distributions $I(\theta)$, where θ is the diffraction angle, are shown for the (002) and the (200) or, equivalently, the (020) Bragg reflections, referred to as the "axial" and "side" case, respectively, in figs 8(a) and 8(b). The important observation to note is that the intensity distributions reveal, in addition to an expected broadening that increases with increasing deformation, an asymmetry which also increases with increasing deformation. Moreover, the asymmetry changes "sign", when comparing the axial and side cases, in the sense that, whereas in the former case the intensity decays more rapidly on the left side (larger θ), it decays more rapidly on the right side in the latter case.

The TEM study of the dislocation distribution revealed an irregular dislocation cell structure with cells which became more and more elongated parallel to the [001] stress axis with increasing deformation. Examples of the TEM micrographs are shown in figs 9(a) and 9(b).

Based on these experimental observations, it was rather straightforward to propose that the response of the elongated dislocation cell structure to an external stress could be viewed in terms of a simple composite model. Here, it is assumed that the soft cell interiors and the hard cell walls are loaded in parallel under the action of the applied axial stress σ so that a redistribution of the stress occurs on a local scale, as described earlier for the case of single slip. Then, the (athermal) macroscopic stress σ will be given by a rule of mixtures, in analogy to eq. (8), as follows:

$$\sigma = f_c\sigma_c + f_W\sigma_W. \tag{15}$$

Here σ_c and σ_W represent the local (flow) stresses in the cell interiors and cell walls, respectively. The axial stresses σ, σ_c and σ_W are converted to the corresponding resolved shear stress τ, τ_c and τ_W by multiplication with the Schmid factor ϕ for dislocation glide on any of the eight equivalent most highly stressed slip systems $\{111\}\langle\bar{1}01\rangle$. For [001]-orientated crystals, $\phi = 0.408$ [65,66]. The redistribution of the axial stress according

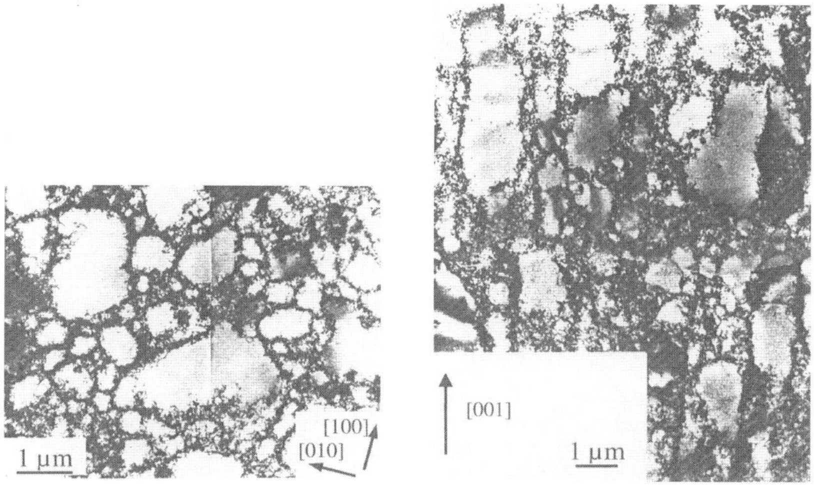

(a) (b)

Fig. 9. TEM micrographs of dislocation cell structure in [001]-orientated copper single crystal deformed in tension to a resolved shear stress of 75.6 MPa: (a) section perpendicular to tensile axis [001]; (b) section parallel to tensile axis [001]. After [66].

to eq. (15) implies, as described earlier for single slip, that long-range internal stresses $\Delta\sigma_c$ and $\Delta\sigma_w$ are induced by the deformation in the cell interiors and in the cell walls, respectively, so that

$$\sigma_w = \sigma + \Delta\sigma_w \tag{16}$$

and

$$\sigma_c = \sigma + \Delta\sigma_c, \tag{17}$$

in analogy to eqs (9) and (10). Furthermore, in analogy to eqs (11) and (12):

$$\Delta\sigma_w = +E\Gamma f_c(\varepsilon_{pl,c} - \varepsilon_{pl,w}) \tag{18}$$

and

$$\Delta\sigma_c = -E\Gamma f_w(\varepsilon_{pl,c} - \varepsilon_{pl,w}). \tag{19}$$

Here, E is Young's modulus (in the [001] direction) and $(\varepsilon_{pl,c} - \varepsilon_{pl,w})$ represents the axial plastic strain mismatch between the axial plastic strains $\varepsilon_{pl,c}$ and $\varepsilon_{pl,w}$ in the cell interiors and in the walls, respectively. In addition, the equilibrium condition, in analogy to eq. (13), must hold:

$$f_c\Delta\sigma_c + f_w\Delta\sigma_w = 0. \tag{20}$$

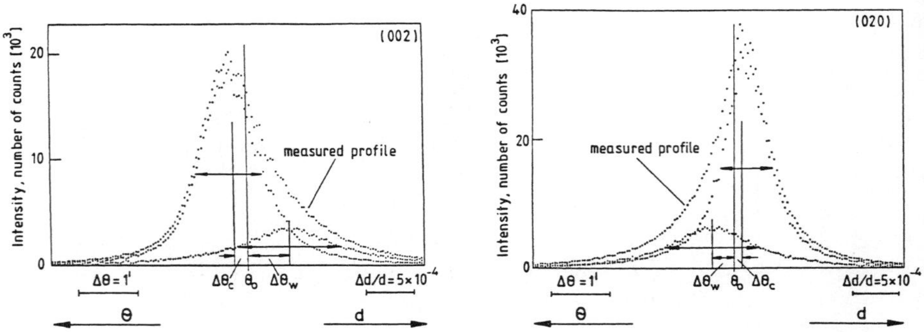

(a) (b)

Fig. 10. Example of decomposition of asymmetric X-ray intensity profiles into two symmetric subprofiles for an [001]-orientated copper single crystal that was deformed in tension to a resolved shear stress of 37.3 MPa: (a) axial case, (002) reflection; (b) side case, (020) reflection. After [66].

The evolution of deformation-induced long-range internal stresses ($\Delta\tau_{\mathrm{w}}$ and $\Delta\tau_{\mathrm{c}}$) even in the case of cell formation during multiple slip deformation is in good accord with some theoretical work by Devincre and Kubin [68]. In a discrete discolation modelling approach, these authors mapped the distribution of local stresses and showed convincingly that, once a heterogeneous dislocation distribution evolves during multiple slip deformation, deformation-induced long-range internal forward and back stresses develop in very much the same manner as described by the composite model.

With this background, we can now return to the interpretation of the asymmetric X-ray diffraction intensity distributions (fig. 8). We put forward the hypothesis that each asymmetric X-ray intensity profile represents, in a first approximation, a superposition of two symmetric intensity profiles which correspond to the cell interior and cell wall regions, respectively, and which are slightly displaced from the origin along the θ-axis because of the elastic strains associated with the internal stresses $\Delta\sigma_{\mathrm{c}}$ and $\Delta\sigma_{\mathrm{w}}$. Figure 10 shows, as an example, that the asymmetric axial (002) and the asymmetric side (020) X-ray intensity profiles can be split quite accurately into two symmetric sub-profiles. The larger sub-profile is assigned to the cell interior regions and the smaller sub-profile to the cell walls. It should be noted that the relative integral intensities of the subprofiles, i.e. the areas under the intensity distributions, correspond to the volume fractions of the cell interiors and cell walls, respectively, and are found to be in good correspondence with direct observations by TEM [65,66]. The displacements of the subprofiles along the abscissa from the value θ_{o} of the profile of the undeformed crystal are denoted by $\Delta\theta_{\mathrm{c}}$ and $\Delta\theta_{\mathrm{w}}$, respectively. These values are easily converted to corresponding relative changes of lattice parameter $\Delta d/d$, according to the scales indicated on the abscissa of figs 8(a) and 8(b). Insertion of the relative lattice parameter changes $(\Delta d/d)_{\mathrm{c}}$ and $(\Delta d/d)_{\mathrm{w}}$ for the "axial" case in the [001]-direction for the cell interiors and cell walls, respectively, into Hooke's law then leads in a straightforward manner to the axial internal stresses $\Delta\sigma_{\mathrm{c}}$ and $\Delta\sigma_{\mathrm{w}}$.

Regarding the absolute values obtained for the long-range internal stresses and for the volume fractions f_{w} (and f_{c}) for different degrees of deformation, the reader is referred to the original publications [33,65,69] and an Erratum [67]. As an example, we merely state the numbers for the data belonging to fig. 8(a): $\tau = 37.3$ MPa, $\tau_c = 34.1$ MPa, $\tau_w = 51.0$ MPa, i.e. $\Delta\tau_c = -3.2$ MPa and $\Delta\tau_w = +6.9$ MPa and $f_w = 0.20$ and for the most strongly deformed specimen: $\tau = 75.6$ MPa, $\tau_c = 67.0$ MPa, $\tau_w = 101.3$ MPa, $\Delta\tau_c = -8.6$ MPa and $\Delta\tau_w = +25.7$ MPa and $f_w = 0.29$. In closing this section, we remark that the values $(\Delta d/d)_c$ and $(\Delta d/d)_w$ obtained for the side case in the transverse direction [010] (or [100]) represent essentially unconstrained transverse strains which correspond in a simple manner to the longitudinal strains, multiplied with Poisson's number ν, under the assumption of a unaxial stress state in the [001] direction.

Another important point with regard to the consistency between the composite model and the experimental observations is the microstructural interpretation of the local flow stresses τ_c and τ_w. Clearly, the simplest and most convenient possible explanation would be to assume that Taylor-type [36] flow-stress relations of the form

$$\tau_c = \alpha_c G b \sqrt{\rho_c} \tag{21}$$

and

$$\tau_w = \alpha_w G b \sqrt{\rho_w} \tag{22}$$

hold locally in the cell interiors and in the cell walls with local dislocation densities ρ_c and ρ_w, respectively [31–33,65,66] with appropriate geometric constants α_c and α_w. If we assume that symmetrical multiple slip occurs in a similar fashion in the cell walls and in the cell interiors, then $\alpha_c \approx \alpha_w = \alpha$, and we can take $\alpha \approx 0.4$ for a Hirsch–Saada process [22], as estimated in [69]. Comparison with TEM observations poses no problems, as far as ρ_w is concerned. However, with respect to ρ_c, it is evident in all published works [66,70] that the local density ρ_c observed in the cell interiors is lower than the ρ_c-values one would expect on the basis of eq. (21). This has been emphasized by Neuhaus and Schwink [71] as a criticism of the composite model. One reason for the discrepancy lies clearly in the fact that, unless the dislocations are pinned in the stress-applied state [19–21], there will be a loss in particular of the free glide dislocations in the cell interiors during unloading and subsequent preparation of thin foils for TEM. It is not clear whether this is the main (and only) reason for the discrepancy. Even if one assumes that the dislocation densities in the cell interiors, deducible from Göttler's TEM observations [70], are too small by a factor of 2, then one can explain at best ca. 40% of the experimental values obtained for τ_c by eq. (21). On the other hand, Pedersen [35] has proposed a model in which a bowing stress of the glide dislocations rather than the local dislocation density controls the flow stress in the soft regions. More recently, Zaiser and Hähner [72] have argued that a bowing stress which is related to the cell size and the latter's variation are important in determining the local flow stress in the cell interiors.

As a last point in this section, a simple model of symmetrical multiple slip [33,66] will be reviewed, This model explains in a simple manner how the observed internal stresses arise during deformation and why the X-ray line profiles exhibit asymmetries of opposite "sign" in the axial case of the (002) and in the side case of the (200)/(020) reflections, respectively.

Figure 11 represents schematically the elongated dislocation cells observed. As is well known, there are 8 equivalently stressed slip systems in [001]-orientated f.c.c.

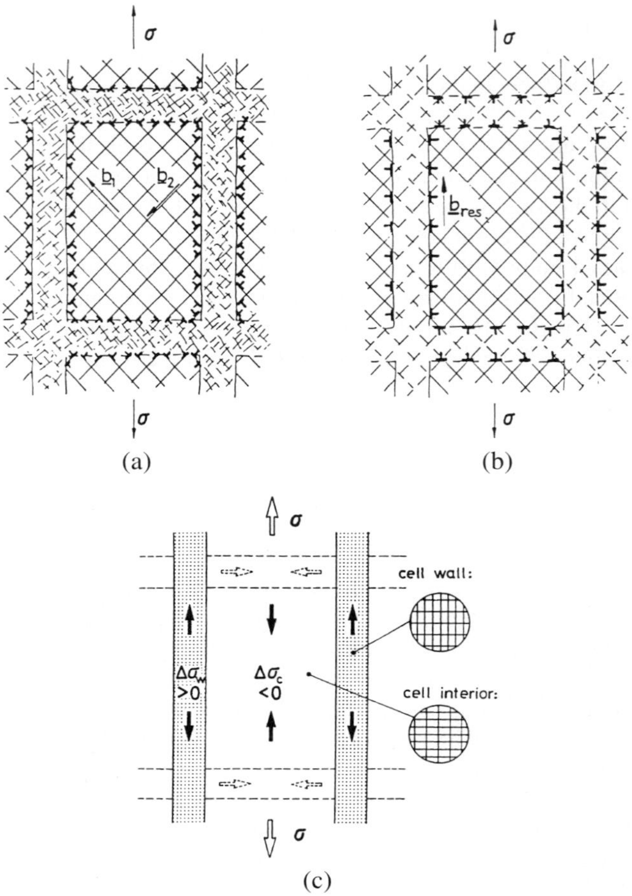

Fig. 11. Schematic illustration of symmetric dislocation glide in the elongated cell structure of the [001]-orientated single crystal and of the resulting lattice distortions: (a) deposition of interface dislocations of two symmetric slip systems at wall/cell interior interfaces; (b) as (a), but with pairs of interface dislocations replaced by one resultant interface dislocation; (c) tetragonal distortions in cell interiors and in cell walls due to interfacial dislocations, as expected after unloading. After [33,66].

single crystals. In fig. 11(a) the slip traces of only two of these slip systems are shown schematically. Some of the dislocations gliding in the soft cell interiors are held up at the wall/cell interfaces. The net effect of these (edge) dislocations at the interfaces is easily seen when one replaces each pair of interface dislocations vectorially by a resultant interface dislocation whose Burgers vector $\mathbf{b}_{res}$ now lies parallel to the wall/cell interface, as shown in fig. 11(b). Regarding only the cell walls parallel to the stress axis, it is obvious that, since the extra half plane of the interface dislocations lies perpendicular to the stress axis in the cell interior, the interface dislocations introduce an axial back stress in the cell interiors and a corresponding axial forward stress in the cell walls. In analogy to eq. (14), one finds that the compatibility of deformation requires that the line density n of resultant interface dislocations adjusts itself so that the related extra strain nb_{res} matches precisely the (tensile) elastic strain mismatch, i.e.

$$nb_{\mathrm{res}} = \varepsilon_{\mathrm{el,w}} - \varepsilon_{\mathrm{el,c}} = \frac{\sigma_{\mathrm{w}} - \sigma_{\mathrm{c}}}{E}. \qquad (23)$$

Turning briefly to the dislocation walls lying perpendicular to the stress axis, we merely note that there the interface dislocations introduce transverse tensile and compressive stresses in the cell walls and in the cell interiors, respectively, and that the overall resulting stresses are such that the cell interiors and both types of cell walls yield plastically compatibly. For a more detailed discussion, the reader is referred to the original paper [66].

The expected axial (and transverse) internal stresses $\Delta\tau_{\mathrm{w}}$ and $\Delta\sigma_{\mathrm{c}}$ are illustrated schematically in fig. 11(c) which shows, as indicated in the inserts, that (in the unloaded state!) the lattice is distorted tetragonally, both in the cell interiors and in the cell walls, with the important difference that, while the cell wall regions are elongated parallel to the stress axis, the cell interior regions are compressed. This is important with respect to the observed asymmetries of the X-ray intensity profiles. When the X-rays are diffracted from the (horizontal) (001) planes in the axial case of the top face (002) reflection, the lattice plane spacings are larger in the cell walls and smaller in the cell interiors than those of an undeformed crystal. On the other hand, for the (200)/(020) side case reflection, the situation is reversed. Now the X-rays "see" the larger lattice plane spacings in the cell interiors and the smaller lattice plane spacings in the cell walls. Thus, the dislocation glide model shown in fig. 11 explains readily the asymmetric shapes of the X-ray intensity profiles and the characteristic differences between X-ray diffraction from the top and the side face, respectively. Regarding the quantitative results, errors are of course introduced by the assumption of elongated cells of quadratic cross section and by neglect of the effects of the cell walls which lie perpendicular to the stress axis. If these walls were considered, the X-ray intensity profiles would have to be split into three and not into two sub-profiles. However, as discussed in a little more detail in [66], the fact that the two subprofiles obtained in the present approximation are found to be very nearly symmetrical indicates that the error made in neglecting the effects of the perpendicular cell walls is small.[3]

When the dislocation model illustrated in fig. 11 is worked out in detail, the Eshelby accommodation factor Γ in eqs (18) and (19) follows as $\Gamma = 0.5$ [66]. However, in this derivation only the contributions of the interfacial edge dislocations to the internal stress were taken into account. It should, however, be noted that, for the dislocation glide geometry under consideration, also interfacial screw dislocations are deposited at the periphery of those (vertical) dislocation walls which lie perpendicular to the walls at which the edge dislocations are deposited. In fig. 11, these interfacial screw dislocations would lie along the traces of the slip planes (i.e. parallel to $\mathbf{b}_1$ and $\mathbf{b}_2$) in a plane

[3]Following the work described in [66], Gromá et al. [73] and Ungár et al. [74] have developed a generalized theory of asymmetric X-ray line broadening originating from a heterogeneous locally polarized dislocation distribution. When applied, this very general theory yields results which are in very good accord with the results obtained in terms of the much simpler two-component model. In addition, a number of other quantities of interest are obtained.

parallel to the plane of the figure. As mentioned in a note-added-in proof of [66], Wilkens (unpublished) has estimated that these interfacial screw dislocations make approximately the same contribution to the internal stress as the interfacial edge dislocations. If this is taken into account, it is found that $\Gamma \approx 1$. More recently, Raabe [75] has performed a detailed three-dimensional numerical simulation of the internal stresses induced by the interfacial edge and screw dislocations. He concluded that the contribution of the screw dislocations amounts to about $(1-\nu) \approx 0.67$ times that of the edge dislocations. With these values, the Eshelby factor would be found to be $\Gamma \approx 0.83$.

4.3. Some conclusions derived from a simple (one-dimensional) dislocation wall model

It has been common practice in the past to express the macroscopic (athermal) shear flow stress in terms of the mean dislocation density ρ in a Taylor-type relationship [15,36]

$$\tau_{\text{hom}} = \alpha_{\text{hom}} G b \sqrt{\rho}. \tag{24}$$

Here, the symbol τ_{hom} has been used for the flow stress and α_{hom} for the appropriate geometric constant in order to indicate that, implicitly or explicitly, the dislocations are assumed to be distributed homogeneously rather than heterogeneously. For the case of multiple slip, we can assume that α_{hom} corresponds to α which is about 0.4 [69]. In the composite model described earlier, the heterogeneity of the dislocation distribution was explicitly taken into account, motivated by the experimental observations outlined in sections 1.1, 4.1 and 4.2 and led to a rule of mixtures, based on the definition of *local* flow stresses (eqs (8), (15)). In this terminology, the macroscopic flow stress can be expressed alternatively and equivalently via the rule of mixtures (eqs (8) or (15)) or in local terms referring to the local flow stress and the local internal stress either in the hard or in the soft regions (eqs (9) and (10) or (16) and (17)).

While it is true that eq. (24) has been used extensively and successfully in the evaluation of experimental data in the past, there can be no doubt that it cannot be strictly valid for heterogeneously distributed dislocation distributions. It is therefore desirable to estimate the magnitude of the deviation between eq. (24) and the flow-stress law of the composite model [76]. For this purpose, we express the mean total dislocation density also by a rule of mixtures

$$\rho = f_{\text{w}} \rho_{\text{w}} + f_{\text{c}} \rho_{\text{c}}, \tag{25}$$

where the small contribution of the interface dislocations has been neglected. Next, we introduce the symbol τ_{het} for the flow stress for a heterogeneous dislocation distribution in terms of the composite model and relate it, in analogy to eq. (24), to the total dislocation density ρ as follows

$$\tau_{\text{het}} = \alpha_{\text{het}} G b \sqrt{\rho}. \tag{26}$$

The introduction of the geometric constant $\alpha_{\rm het}$ emphasizes that this is an effective constant, referring to a heterogeneous dislocation distribution. The constant $\alpha_{\rm het}$ is thus not related in a straight-forward manner to an elementary dislocation mechanism, in contrast to the constant α used earlier for multiple slip, where we assumed that $\alpha = \alpha_{\rm c} \approx \alpha_{\rm w}$. Strictly speaking, eq. (26) expresses more correctly than eq. (24) the real situation, since the dislocation distribution is generally heterogeneous.

With eqs (7), (8), (21), (22), (24) and (25), we obtain an interesting result by comparing the squares $\tau^2_{\rm hom}$ and $\tau^2_{\rm het}$:

$$\tau^2_{\rm hom} - \tau^2_{\rm het} = f_{\rm w} f_{\rm c} (\alpha G b)^2 \left(\sqrt{\rho_{\rm w}} - \sqrt{\rho_{\rm c}}\right)^2 > 0. \tag{27}$$

This result implies that, for a given total (mean) dislocation density ρ, the flow stress will always be smaller, when the dislocations are distributed heterogeneously than when they are distributed homogeneously. A similar result has been obtained by Kocks for the case of point obstacle distributions [77].

Since $\tau_{\rm het} < \tau_{\rm hom}$, we can conclude from a comparison between eqs (24) and (26) that $\alpha_{\rm het} < \alpha_{\rm hom}$. For a particularly simple case of multiple slip in a one-dimensional dislocation wall structure, the author [76] has shown that

$$\tau_{\rm het} = 2\alpha \sqrt{f_{\rm c} f_{\rm w}}\, G b \sqrt{\rho}. \tag{28}$$

By comparison with eq. (26), an effective value $\alpha_{\rm het}$ is obtained as

$$\alpha_{\rm het} = 2\alpha \sqrt{f_{\rm c} f_{\rm w}}. \tag{29}$$

Since $\sqrt{f_{\rm c} f_{\rm w}}$ is always smaller than 0.5, $\alpha_{\rm het} < \alpha$. For typical values $f_{\rm w} = 0.2$ and $f_{\rm c} = 0.8$ [66,70] and $\alpha \approx 0.4$ [69], a value $\alpha_{\rm het} \approx 0.3$ results which is in good accord with experimentally observed values [65,70]. In other words, $\alpha_{\rm het}$ should be regarded as an effective value that would be obtained from experiment under the assumption of eq. (26). The constant $\alpha_{\rm het}$ can be related to the physically meaningful constant α which describes a particular type of dislocation interaction only with the aid of a suitable dislocation model, compare [76]. While the difference between $\alpha_{\rm het}$ and α may be small in numbers, it is fundamental.

4.4. Energy of polarized heterogeneous dislocation distributions containing long-range internal stresses

The heterogeneous dislocation distributions that were discussed in the previous sections can be considered as typical polarized dislocation arrangements of deformed crystals in the stress-applied state. The energy stored in such dislocation arrangements is one particular property of interest. In the heterogeneous dislocation distributions containing long-range internal stresses, the stored energy consists of the strain energy of all dislocations, supplemented by that of the long-range internal strains. The first contribution can generally be written as, compare Nabarro [78]:

$$E_{\mathrm{stor}}^{\mathrm{disl}} = \frac{Gb^2}{4\pi k}\rho \ln \frac{R_{\mathrm{e}}}{r_{\mathrm{o}}}, \tag{30}$$

where $k = (1 - \nu)$ for edge and $k = 1$ for screw dislocations, ν is Poisson's ratio and R_{e} and r_{o} are an appropriate effective outer and the inner cut-off radius, respectively. The second contribution, due mainly to the long-range internal strains (or stresses) resulting from the interface dislocations (if one ignores the short-wavelength oscillations) then can conveniently be written as [58,65,76]:

$$E_{\mathrm{stor}}^{\mathrm{l.r.}} = f_{\mathrm{w}} \frac{\Delta\sigma_{\mathrm{w}}^2}{2E} + (1 - f_{\mathrm{w}})\frac{\Delta\sigma_{\mathrm{c}}^2}{2E}. \tag{31}$$

The really interesting question now is whether this second contribution is a small or a large fraction of the total stored energy $E_{\mathrm{stor}}^{\mathrm{het}} = E_{\mathrm{stor}}^{\mathrm{disl}} + E_{\mathrm{stor}}^{\mathrm{l.r.}}$ for a heterogeneous dislocation distribution. It was already concluded in the early work on deformed [001]-orientated copper single crystals [65] that $E_{\mathrm{disl}}^{\mathrm{l.r.}}$ amounts to only a few per cent of the experimentally measured values [79]. Subsequently, in the one-dimensional wall model [76] referred to earlier, it was confirmed theoretically (compare fig. 5 of [76]) that the contribution of the long-range internal strains to the total stored energy is marginal, e.g., about 6% for a typical value of $f_{\mathrm{w}} \approx 0.3$. This result is remarkable in so far that it shows that a dislocation arrangement containing long-range internal stresses is *not a priori* a "high-energy" dislocation structure of excessively high energy, as one might think in the case of dislocation pile-ups, but still a structure of reasonably low energy without, of course, being equivalent to a low-energy dislocation structure (LEDS), as advocated by Kuhlmann-Wilsdorf [80]. Rather, it might be appropriate to speak of a medium-energy dislocation structure (MEDS).

Another interesting point concerns the elastic strain energies, excluding $E_{\mathrm{stor}}^{\mathrm{disl}}$, involved during straining and after unloading of a specimen containing a heterogeneous dislocation distribution. This topic was discussed for the case of a specimen strained up to a flow stress τ_{het} and with reference to the straining of another specimen containing a homogeneous dislocation distribution of the same mean dislocation density at a flow stress τ_{hom} [58,76, 81] and is illustrated in the form of stress–strain diagrams in fig. 12, from [76]. First of all, we note that, for the same mean dislocation density ρ at the stress levels τ_{hom} (fig. 12(a)) and τ_{het} (fig. 12(b)), respectively, $\tau_{\mathrm{hom}} > \tau_{\mathrm{het}}$, as shown earlier (eq. (27)). Hence, it follows for the corresponding recoverable strain energy densities upon unloading that $\tau_{\mathrm{hom}}^2/2G > \tau_{\mathrm{het}}^2/2G$. However, in the case of the heterogeneous dislocation distribution, the contribution $E_{\mathrm{stor}}^{\mathrm{l.r.}}$ remains stored in the specimen after unloading in the form of the strain energies arising from the long-range internal stresses, compare eq. (31) and fig. 12(c), which shows the loading/unloading for the "microscopic" stress–strain curves for the wall and the cell interior regions, respectively. Here, for the sake of simplicity, equal volume fractions f_{w} and f_{c} were assumed. The shaded triangles indicate the strain energy densities in the wall and the cell interior regions, respectively. As discussed in detail in the one-dimensional wall model in [76], it can be readily shown for a heterogeneous dislocation

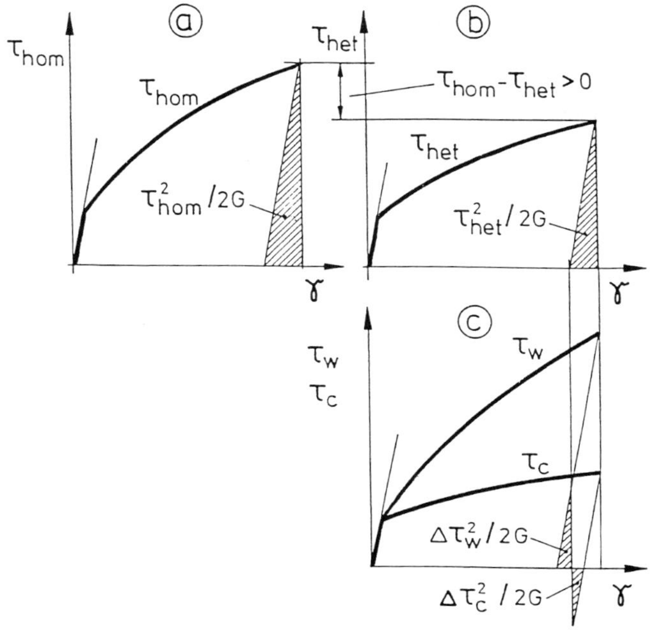

Fig. 12. Stress–strain behaviour and elastic strain energies, from [76]: (a) macroscopic stress–strain behaviour for homogeneous dislocation distribution (identical to microscopic behaviour); (b) macroscopic stress–strain behaviour for heterogeneous dislocation distribution; (c) microscopic stress–strain behaviour corresponding to local behaviour in walls and cell interiors of a heterogeneous dislocation distribution.

distribution without further assumptions that the strain energy density in the stress-applied state (indicated by the symbol σ) is given by

$$E^{\text{het}}_{\text{stor},\sigma} = \left(f_{\text{w}}\tau_{\text{w}}^2 + f_{\text{c}}\tau_{\text{c}}^2\right)/2G \tag{32}$$

and that this is equivalent to

$$E^{\text{het}}_{\text{stor},\sigma} = \left(\tau_{\text{het}}^2 + f_{\text{w}}\Delta\tau_{\text{w}}^2 + f_{\text{c}}\Delta\tau_{\text{c}}^2\right)/2G \tag{33}$$

and that

$$E^{\text{het}}_{\text{stor},\sigma} = E^{\text{hom}}_{\text{stor},\sigma} = \tau_{\text{hom}}^2/2G. \tag{34}$$

Thus, the main difference between the cases of a heterogeneous and a homogeneous dislocation distribution lies in the difference between the recoverable strain energy densities $\tau_{\text{het}}^2/2G$ and $\tau_{\text{hom}}^2/2G$. This difference is found to be identical to the "extra" stored energy density (eq. (31)) which is retained after unloading only in the case of a heterogeneous dislocation distribution but not in the case of a homogeneous dislocation distribution.

4.5. Comparison with other flow-stress models

The original composite model is essentially a model of the athermal component of the flow stress, based on the specific local dislocation mechanisms which control plastic flow by the glide of dislocations through a heterogeneous distribution of dislocations. In order to appreciate its specific features, advantages and deficiencies compared to other models of the flow stress, it is expedient to first summarize and discuss the main characteristics of the other flow-stress models most frequently considered. Then, the progress made with the composite model will become apparent in the light of the limitations of the other models.

4.5.1. Summary and criticism of the main earlier models of the flow stress

When discussing the earlier flow-stress models, it should be kept in mind that they were propagated at a time 20 to 40 years ago, when the main issues under discussion concerned mainly the athermal component of the flow stress and the (Stage II) work hardening of f.c.c. single crystals in single slip [1–28]. Since that time, only a few really new studies were performed, and the state of affairs in the last 10 to 20 years is perhaps reflected best in the more recent publications by Basinski and Basinski [26], Mecking and Kocks [82], Kocks [28], Nabarro [83], Mughrabi [58], Brown [84,85], Gil Sevillano [86] and Kuhlmann-Wilsdorf [87]. For our purpose in this section, it suffices to recapitulate and summarize the main models of the flow stress by the groups of Seeger [3–5,9,18], Saada [22], Basinski [16,26,27], Kuhlmann-Wilsdorf [14,24] and Hirsch [6,25] (and their co-workers). Most other flow-stress models such as that used in the statistical theory of Kocks [23] belong essentially to one of the models listed above.

In the long-range stress theory of Seeger and co-workers [3–5,9,18], the long-range internal stress of primary dislocation pile-ups consisting of n dislocations is assumed to govern the flow stress. (The number n used here should be distinguished from the symbol n in eqs (14) and (23).) In this model, the flow stress τ is described by a relation similar to eq. (24), with a stress enhancement factor $\sqrt{n}$ and with the total dislocation density ρ replaced by the density ρ_{p} of primary dislocations:

$$\tau = \alpha G b \sqrt{n \rho_{\mathrm{p}}}. \tag{35}$$

It should be noted that we shall not discuss the constant α in detail in the present context and shall therefore use the same symbol here and in the following, noting at the same time that the constant α will, of course, vary somewhat from model to model.

The forest hardening theories [16,22,23,26,27] all ascribe the flow stress to the glide of the primary glide dislocations through a "forest" of secondary dislocation "trees" threading the primary glide plane. Therefore, leaving aside refinements, compare Basinski [16,26, 27], the flow stress is given by

$$\tau = \alpha G b \sqrt{\rho_{\mathrm{f}}}, \tag{36}$$

where ρ_{f} is essentially the (mean) density of secondary dislocations. Saada was the first to compute in detail the interaction of glide and forest dislocations and the flow stress resulting from the unzipping of dislocation junctions [22], according to a proposal made earlier by Hirsch [88]. More detailed computations of this Hirsch–Saada process were later performed by Schoeck and co-workers [69]. The statistical theory of Kocks [23] which deals with details of dislocation glide through a random array of point obstacles essentially uses also a flow-stress law of the forest type.

In her "meshlength" theory, Kuhlmann-Wilsdorf [14,24,80,87] starts out from a cell structure in Stage II and postulates that the flow stress is governed by the bowing stress required to bow out segments beyond their critical radius. Thus, the flow stress is given by an equation of the form

$$\tau \approx \frac{Gb}{l_s}, \tag{37}$$

where the "source length" l_s is about three times larger than the average link length and is proportional to $1/\sqrt{\rho}$. Thus, eq. (37) is essentially only a modification of the original Taylor equation (24).

In contrast to the above three flow-stress laws which ascribe the flow stress to *one single* dislocation mechanism which differs from model to model, Hirsch and Mitchell [6] and Hirsch [25] make an Ansatz which is essentially a compromise between all competing mechanisms:

$$\tau = \tau_o + \frac{Gb}{l} + \tau_{for} + \tau_i. \tag{38}$$

Thus, besides a friction stress τ_o, a bowing stress and a forest term τ_{for}, Hirsch and Mitchell employ additively a long-range internal stress contribution τ_i. In a way, eq. (38) reconciles all the other flow stress models but leaves open the question how to weight the different contributions. An important point made by Hirsch and Mitchell is that eq. (38) essentially defines a *local* flow stress, since all terms vary locally because of the heterogeneity of the dislocation distribution.

Today, the inadequacies of the models described above can be summarized as follows:

(1) While the long-range stress theory correctly assumed the existence of long-range internal stresses, the assertion derived from slip-line studies [4,8] that dislocation pile-ups form and act as the only sources of the long-range internal stresses was not quite correct. Nonetheless, in single slip, the dislocation grids [13] which contain about equal amounts of primary and secondary dislocations can be considered as "converted" or "disguised" pile-ups [6,11,17,18,25]. For these reasons, the flow-stress equation (35) is questionable. Moreover, Basinski [16] correctly pointed our that if long-range internal stresses hinder dislocation glide in one region, then there must of necessity be long-range internal stresses of the other sign in other regions which support dislocation glide. In retrospect, the natural consequence of this consideration is that, in this case, the flow stress should be considered as a weighted average of the local stresses necessary for local deformation, very much in the spirit of the composite model described earlier. Furthermore, the flow-stress equation (35) which corresponds to that in the original paper by Seeger et al. in 1957 before the advent of TEM [4] ignores the interaction with secondary dislocations. In later

papers, after TEM had revealed the real, more complex dislocation arrangements, Seeger and Wilkens [17] and Seeger [18] took a much broader view.

(2) The flow-stress equation (36) of the forest theory assumes a homogeneous distribution of forest dislocations, contrary to experimental evidence, compare [21]. It is well known that, in work-hardening Stage II, by far the majority of the secondary dislocations are located in the dense (multipolar) bundles and braids (and in the grids) and only a very small fraction in the larger regions of lower dislocation density in which the dislocation glide processes spread out over larger areas. In a way, this criticism is similar to the statement that there is no single global value of the forest dislocation density that controls the overall flow stress. Instead, eq. (36) would be more reasonable, if applied only locally. In addition, the forest theories ignore completely the interaction with the primary dislocations. The latter interaction is precisely most intense in the same regions, in which most forest dislocations are located, because primary and secondary dislocations tend to occur together in tangles, braids and grids [11,21].

(3) In the meshlength theory, the heterogeneity of the dislocation distribution is taken into account explicitly. The main aim, however, is only to obtain "realistic" segment lengths, while the consequences of the heterogeneous dislocation distribution with respect to the development of deformation-induced long-range internal stresses are ignored completely. Long-range internal stresses are only addressed superficially in a tenor that has changed over the decades [14,24,80,87]. In a way, the source length l_s is adjustable through the variation of the parameters of the theory. It is doubtful whether the source length has any pertinent physical meaning other than being an "effective" quantity with which the overall flow stress can be calculated with eq. (39). In reality, the "source length" l_s is longer than the shorter segments in the dislocation-dense regions and shorter than the longer ones in the dislocation-poor regions. In other words, while eq. (39) may well be suitable to describe dislocation bowing *locally* in the soft and/or the hard regions, it is considered unrealistic to describe the overall flow stress by one single source length.

(4) A common weak point of all theories except that of Hirsch and Mitchell [6,25] is that they attempt to explain the flow stress by one single mechanism, using only one value (out of a whole spectrum!) of a particular microstructural parameter such as the dislocation density or dislocation source length.

(5) None of the above flow-stress theories, aside from the exceptions detailed in the following, has built-in the evolution of deformation-induced long-range internal stresses with a back stress, as is typical of theories of kinematic hardening. They can therefore only describe isotropic hardening satisfactorily and must confine themselves to verbal descriptions of the *kinematic* aspects related to the Bauschinger effect which is observed during reverse loading, compare section 5.4.1. There are two exceptions. In Seeger's original theory with dislocation pile-ups (which was commented upon before), a Bauschinger effect was actually predicted to occur during reverse deformation [5]. In a similar manner, in the case of the statistical theory of Kocks [23], the "piling-up" of dislocations against the hard regions would indeed lead to a Bauschinger effect upon reversing the sense of deformation. All so-called "one-parameter" models such as those of Mecking and Kocks [44,82]

cannot describe the behaviour upon reverse loading or for other abrupt changes of the deformation path. This has led Estrin [89] to extend the earlier work to a two-parameter model.

4.5.2. Critical assessment of progress achieved by the composite model

In the composite model, the very complex phenomenon of plastic deformation has been simplified crudely and radically by considering only one type of soft and one type of hard region rather than, more realistically, the whole spectrum of regions of different local flow stresses, as in several other more demanding theories. Thus, a big sacrifice has been made in terms of accuracy and closeness to reality in detail, compared to the more sophisticated spectrum theories [90,91] which are closely related to the composite model in spirit. However, this deficiency is, perhaps, more than compensated by the simplicity of the model which is not only easily tractable at a modest mathematical level but provides, at the same time, a hitherto lacking clear insight into some of the most basic features of the complexity of dislocation glide.

The idea to consider the deformed material as a composite consisting only of two components was stimulated by the TEM observations of the dislocation distribution. The latter clearly exhibited the distinct differences between the dislocation-dense cell walls and the dislocation-poor cell interiors, suggesting that the deformed material consisted *locally* of soft cell interiors and much harder cell walls. The first notion of such a concept goes back to work of Cullity [92,93] who came to the same qualitative conclusion solely on the basis of X-ray studies of residual stresses in deformed silicon iron and nickel. In 1970, Takeuchi essentially formulated a first composite model quantitatively in order to model the unidirectional deformation of b.c.c. single crystals in which a layered dislocation cell structure develops [94]. In this model, the concept of deformation-induced internal stresses in heterogeneous dislocation structures was already introduced. This early work of the sixties and early seventies was overlooked when the present composite model was developed independently in the later seventies and eighties.

The present composite model really is a two-component Masing-type model [95,96], the main difference being that Masing considered a polycrystal and assumed that every crystallite has a different yield strength, depending on its orientation. Thus, the Masing model and the composite model are similar in a formal sense. In detail, however, they differ fundamentally, since, in the composite model, the local differences in yield strength arise from local differences in the dislocation distribution/density in a single crystal or crystallite and not from different crystallite orientations as in the Masing model. The latter model, in its original formulation, would not be able to explain the Bauschinger effect in single crystals, contrary to observation. While the Masing model was only derived graphically (both for a larger number of components [95] and for only two components [96]) and never formulated mathematically, the mathematical formulation of the composite model follows closely that commonly used in the description of composite materials. Perhaps most important, it must be emphasized that in all Masing-type models, and hence also in the composite model, long-range internal stresses are an essential integral part of the model. Thus the composite model automatically takes into account *kinematic* hardening. It is interesting to note that the composite model and the closely related spectrum theories of Holste and Burmeister [90] and Polák and Klesnil [91] were inspired by work on cyclic deformation which, of necessity, involves kinematic hardening, as reflected in the Bauschinger effect. Also, in cyclically deformed materials, the heterogeneity of the dislocation distribution is even more marked than in unidirectionally deformed materials and in fact suggests strongly the use of a two-component model in a first approximation, as outlined in sections 1.1 and 4.1, compare also [31–33]. These facts qualify the composite model as the natural approach to describe cyclic deformation in the spirit of the original Masing model that was, after all, devised primarily to describe reverse yielding and the Bauschinger effect, compare section 5.3.2.

The central concept of the composite model lies in the definition of *local* (microscopic) flow stresses (eqs (9), (10), (16), (17), (21), (22)) on a *mesoscopic* scale and in the determination of the macroscopic flow stress by a rule of mixtures (eqs (8), (15)) or by a modified formulation of eq. (24) in terms of the model (eq. (15)). Thus, the macroscopic flow stress is described in terms of an appropriate combination of the local flow-stress contributions in the soft and in the hard regions. For a particular dislocation distribution, the local flow-stress laws would have to be formulated in terms of the different known mechanisms. The latter are reflected in the familiar flow-stress equations (35) to (38), the important point being, of course, that these equations should now be applied only locally. For specific dislocation arrangements, this may well pose some problems, compare [81], and local flow stress equations such as eqs (21) and (22) represent only one of several possibilities. Still, there can be no doubt that this description, though perhaps still very crude, is more appropriate for a heterogeneous dislocation distribution than the original Taylor formula (24) or any of the essentially related formulations of the forest theory [16, 22,23,26,27] or the meshlength model [14,24,80,87]. Nonetheless, it cannot be disputed that the rather inappropriate Taylor formula has been applied seemingly quite successfully by many authors to describe experimental observations. This should, however, not be considered as a proof of the rigorous validity of the equation. The reasons are as follows.

Agreement with the Taylor equation (24) is believed to be more or less fortuitous due to the not very well known fact that the flow stress defined by eq. (24) is not very sensitive to changes in dislocation density, if the arrangement changes at the same time, as is usually the case. This is borne out most clearly under conditions of (seemingly) steady state or saturation as in cyclic deformation [30,63,97–99]. There, it has been noted repeatedly that, after saturation of the stress level, the dislocation arrangement (and the general defect population) continue to develop as evidenced by TEM [98,99], by electrical resistivity [97], magnetic property [98] and by X-ray [63] measurements. It is also noted that the screening of the dislocation stress fields through the formation of increasingly sharper subgrain walls, with continuously increasing misorientations, progresses persistently as the dislocation density continues to increase, whereas the stress level almost does not change. These observations show clearly that the flow stress is rather insensitive to changes in the dislocation density. The reason is quite simple: With increasing deformation, there is almost always an increase in the dislocation density which is generally accompanied by a change of the dislocation arrangement towards one of reduced internal stresses. In terms of the Taylor equation (24), this implies that, as the dislocation density ρ increases, the geometrical parameter α decreases in such a manner that the resulting change of τ is frequently only marginal. With respect to unidirectional deformation, we note that the dislocation arrangement changes not only from one work-hardening stage to the next but also continuously within each work hardening stage [11,13,21], implying that the principle of similitude [14,24] which is (perhaps) unduly popular because of its simplicity and usefulness in simple considerations is not really well obeyed [21], when one considers more specific details of the dislocation distribution. Thus, the assumption of a constant value of the constant α over a larger range of deformation is simply unjustified.

Since the composite model starts by assuming the existence of a particular experimentally observed heterogeneous dislocation distribution, it cannot provide direct information on the physical causes of so-called dislocation patterning, compare, for example, the reviews by Kubin [100] and Aifantis [101]. On the other hand, the composite model describes precisely the most important consequences that result from dislocation patterning with respect to plastic flow. Therefore, in retrospect, it could well be that the reverse route is feasible in the sense that the assessment of the characteristic features of the composite model could perhaps provide a means to trace back to the physical origin of dislocation patterning.

4.6. Strain gradients and the role of the geometrically necessary dislocations

The elastic/plastic strain mismatch between the soft and the hard regions is an inherent feature of the composite model. In the simple formulation presented in sections 4.1 and 4.2, sharp interfaces were assumed between soft and hard regions. This implies the existence of infinitely large elastic/plastic strain gradients at the interfaces. While it is clear that finite strain gradients would be more realistic, the model has the merit of simplicity. We shall therefore dwell briefly on this simple picture before considering the more realistic case of finite strain gradients.

The interfacial dislocations illustrated schematically in figs 5(a) and 5(b) and in figs 11(b) and 11(c) for the cases of single and multiple slip, respectively, on the one hand, serve to accommodate the elastic/plastic strain mismatches and, at the same time, in the present simple picture, give rise to the deformation-induced long-range internal stresses which are necessary in order to redistribute the stress on a local scale so as to provide precisely the local stresses which ensure the simultaneous compatible deformation of the soft and the hard regions. In Ashby's terminology, introduced in his work on "plastically non-homogeneous" materials, the interfacial dislocations are called geometrically necessary dislocations in contrast to all the other (redundant) dislocations, referred to as statistically stored dislocations [102]. For the case of sharp interfaces considered so far, the line density of the interfacial dislocations can be stated simply and quantitatively in terms of the difference between either the local elastic/plastic strains or the local flow stresses of the soft and the hard regions, respectively, compare eqs (14) and (23).

In the last 15 years or so, effects of strain gradients have received increasing attention, compare, for example, [102–106]. Hence, the role of the strain gradients in the composite model and of the geometrically necessary dislocations associated with them deserve a more detailed discussion. Here, our main interest lies in the questions whether and how the density of the geometrically necessary dislocations contributes to the overall flow stress and whether the flow-stress equation contains a so-called "internal length scale". The latter has been introduced in a somewhat vague relation to microstructural dimensions by several authors, compare [103–106]. These questions have been addressed recently by one of the authors [107–109]. In the following, only the main conclusions derived in these studies will be presented.

As a starting point, we repeat Ashby's relationship between the (local) density of geometrically necessary dislocations, ρ_G, and the gradient of the plastic shear strain γ_{pl} for the one-dimensional case with dislocation glide in the x-direction:

$$\rho_G = \frac{1}{b} \cdot \frac{\partial \gamma_{pl}}{\partial x}. \tag{39}$$

Ashby argued that both the density of statistically stored dislocations, ρ_S, and ρ_G contribute to the flow stress τ and proposed as the simplest dimensionally correct flow-stress law the relation

$$\tau = CGb\sqrt{\langle\rho_S\rangle + \langle\rho_G\rangle}. \tag{40}$$

In this relationship, $\langle\rho_S\rangle$ and $\langle\rho_G\rangle$ represent the spatially averaged values of the densities of the statistically stored and the geometrically necessary dislocations, respectively, and C is a constant (equal to about 0.3) which is similar to the constants α used earlier in eqs (21), (22) and (24). In fact, since the sum of $\langle\rho_S\rangle$ and $\langle\rho_G\rangle$ is nothing else but the mean total dislocation density ρ in eq. (24), eq. (40) is essentially identical to eq. (24). Thus, the question whether eq. (40) is an appropriate description of the flow stress τ reduces in some essential parts to the discussion given earlier in section 4.3. There, it was concluded that while a flow-stress equation such as eq. (40) (or eq. (24)) may give a satisfactory description of experimental data, it is not strictly correct, since it does not consider explicitly the heterogeneity of the dislocation distribution. If the latter is considered as in the composite model, the flow stress is described by the rule of mixtures as in eqs (8) or (15). In this formulation, the density of the geometrically necessary interface dislocations does not enter the flow-stress equation. Instead, only the *local* densities of statistically stored dislocations, i.e., ρ_c and ρ_w in the cell interiors and in the cell walls, respectively, compare eqs (21) and (22), appear in the flow-stress equation.

Now, extensions of the composite model which allow for more gradual finite elastic/plastic strain gradients will be considered. First, we shall deal with the simple case of a continuous spatial variation of the local density $\rho_S(x)$ of statistically stored dislocations and a corresponding variation of the local flow stress, $\tau_{loc}(x)$, in one dimension. In this case, it is assumed that $\tau_{loc}(x)$ and $\rho_S(x)$ are related through a Taylor-type relationship as $\tau_{loc}(x) = \alpha Gb\sqrt{\rho_S(x)}$. Then, as discussed in [107,108], an alternative description of the local density of geometrically necessary dislocations can be obtained in the framework of the composite model. It is easily shown via eq. (39) and by expressing the plastic shear strain gradient as the negative elastic shear strain gradient that

$$\rho_G(x) = -\frac{1}{bG} \cdot \frac{d\tau_{loc}(x)}{dx} \tag{41}$$

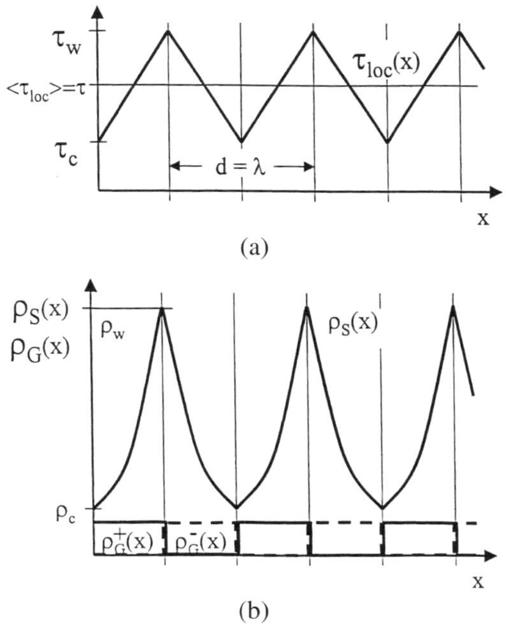

Fig. 13. Schematic illustration of a hypothetical one-dimensional periodic distribution of (a) the local flow stress $\tau_{\mathrm{loc}}(x)$ and (b) the corresponding local densities of the statistically stored and the geometrically necessary dislocations, $\rho_{\mathrm{S}}(x)$ and $\rho_{\mathrm{G}}(x)$, respectively. After [107,108].

or, equivalently,

$$\rho_{\mathrm{G}}(x) = -\frac{\alpha}{2\sqrt{\rho_{\mathrm{S}}}} \cdot \frac{\mathrm{d}\rho_{\mathrm{S}}(x)}{\mathrm{d}x}. \tag{42}$$

In other words, $\rho_{\mathrm{G}}(x)$ can then be expressed alternatively in terms of either the local flow stress gradient or the gradient of the density of the statistically stored dislocations. It is interesting to evaluate the flow-stress law (40) for a hypothetical periodic one-dimensional distribution of the local flow stress and the local density of statistically stored dislocations with linear gradients of alternating sign of the local flow stress $\tau_{\mathrm{loc}}(x)$, as illustrated in figs 13(a) and 13(b). In passing, we note the following. While the hypothetical periodic distribution of the local flow stress and the local density of statistically stored dislocations illustrated in fig. 13 could be interpreted as a very crude model of a one-dimensional periodic wall structure, we emphasize that our sole motivation for the use of this admittedly unrealistic model is to use it as a vehicle in order to explore the effects of the strain gradients on a composite-type flow-stress model.

We now make use of the definitions of $\rho_{\mathrm{G}}(x)$ in terms of eqs (41) and (42), and then obtain the corresponding distribution of the density of geometrically necessary dislocations with constant values for dislocations of positive and negative sign ($\rho_{\mathrm{G}}^{+}(x)$, $\rho_{\mathrm{G}}^{-}(x)$) for the constant positive and negative gradients of the local flow stress. Insertion of the appropriate expressions for $\langle \rho_S(x) \rangle$ and $\langle \rho_G(x) \rangle$ into eq. (40) leads to a lengthy expression for the overall flow stress which contains the "wavelength" λ of the periodic flow stress respective dislocation density distribution as an "internal length scale". When comparing this result with that obtained on the basis of a treatment along the lines of the composite model for the same distribution, it is noted that, in the latter case, the result does not contain an internal length scale [107,108]. Another distinct difference between the two approaches, again referring to fig. 13, concerns the question whether the geometrically necessary dislocations contribute to the overall flow stress. Application of the concepts of the composite model to the hypothetical model illustrated in fig. 13 leads to a simple expression for the overall flow stress which does not contain the density of the geometrically necessary dislocations. Hence, in the hypothetical model discussed (fig. 13), the latter do not contribute to the overall flow stress as was already the case in the original composite model with infinite strain gradients. The main role of the geometrically necessary dislocations is, precisely as before, to ensure compatibility of deformation and to provide the long-range internal stress distribution that redistributes the stress on a local scale, as described earlier. This conclusion applies strictly, within the framework of the simple composite model, in the case of single slip but only in a very good approximation in the case of multiple slip [107–109], as will be explained later.

As discussed in [107,108], finite strain gradients can arise not only from local variations of the density of the statistically stored dislocations but also from details of dislocation glide in the cell interiors such as the constrained glide geometry of bowing-out dislocations and the degree of planarity of dislocation glide. When the simple single-slip two-component composite model of PSBs is modified to include the first effect, a more realistic internal stress distribution than in fig. 5 is obtained in very good agreement with the experimental observation shown in fig. 3 [62,107]. Furthermore, the flow-stress equation then contains the channel width d_c as an internal length scale [107].

This can be understood on the basis of our discussion in section 2.1 (fig. 2(b)). There, it was shown that the multiplication of the dislocations in the channels of the PSBs occurs by the bowing-out of edge dislocation segments from the walls across the channels. Thus, the "glide zones" are bounded by two bowed-out screw dislocation segments which move away from each other and two edge dislocation segments deposited at the wall-channel interfaces, as shown schematically for a number of glide zones in fig. 14. In this figure the slipped-off areas (glide zones) which originated from the bowed-out edge segments are separated by unslipped areas (hatched) in which only elastic shearing has occurred. Thus, elastic/plastic strain mismatches develop between the glide zones and the neighbouring areas. Since the slip path of the bowed-out screw segments is largest in the centre of the glide zones and smallest at their periphery, it is easily seen that, averaged over many glide zones, an elastic (plastic) shear strain gradient exists across the channels in such a manner that the plastic (elastic) shear strain decreases (increases) from the centre of the channels to the walls. The internal stress distribution that arises from this glide configuration can be computed quite easily with simple assumptions [62], and one obtains in closed form an expression for the local flow-stress profile across the channels which can be fitted very satisfactorily to the experimental data of fig. 3 with the density of the screw dislocations, $\rho_{\odot}$, as the only open parameter. Figure 15 shows the experimental data of fig. 3 and computed curves of the flow-stress profile for three different values of the density of screw

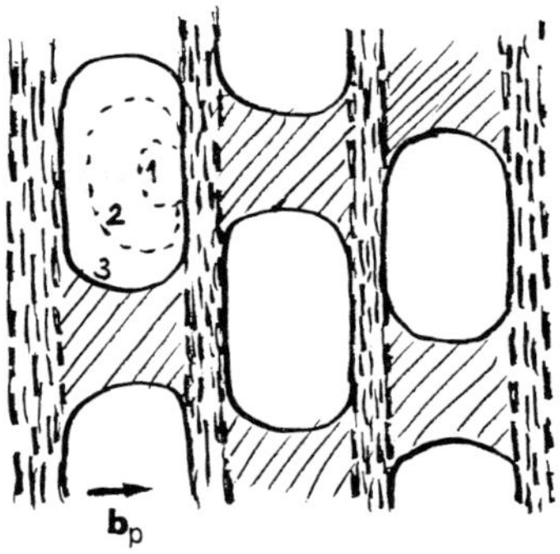

Fig. 14. Schematic illustration of sheared-off "glide zones" and unslipped regions (hatched) in the channels of PSBs. After [62,107].

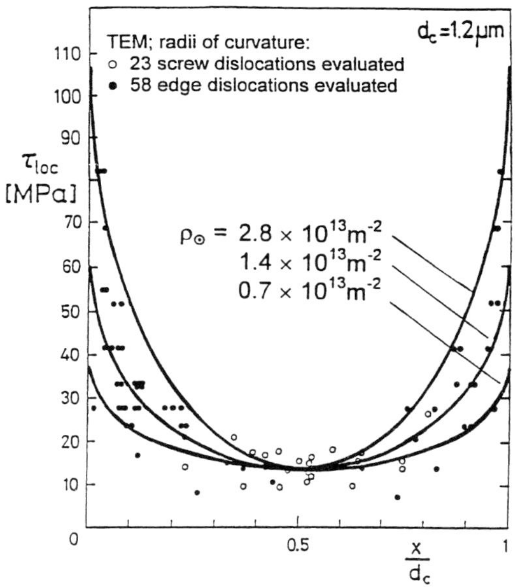

Fig. 15. Computed local flow-stress profiles for three values of the density $\rho_\odot$ of screw dislocations and experimentally measured values of the local stress in the channels (under load) between the multipolar walls of the PSB structure (x: space coordinate across channel). After [62].

dislocations. An excellent fit is obtained for values of $\rho_\odot$ between 10^{13} and 2×10^{13} m^{-2} which agree very well with experimental TEM observations [32,55]. The spatially averaged (athermal) flow stress of the PSBs, i.e. τ_{PSB}, calculated according to the composite model, is obtained as

$$\tau_{\mathrm{PSB}} = \hat{\tau}_c(0) + \frac{Gb\rho_\odot d_c}{2\kappa}\left(1 - \frac{\pi f_c}{4}\right). \tag{43}$$

Here $\hat{\tau}_c(0)$ denotes the flow stress of the channels in the centre of the channels ($x = 0$), d_c the channel width, f_c the volume fraction of the channels (typically, $f_c \approx 0.8$–0.9) and κ the axial ratio of the elliptically bowed-out screw segments (typically $\kappa \approx 2$). As shown in [32], $\hat{\tau}_c(0)$ can be written in good approximation as either the critical screw dislocation dipole passing stress $Gb/4\pi y_{\odot}$, where $y_{\odot}$ is the annihilation distance of screw dislocations (in Cu, $y_{\odot} \approx 50$ nm at room temperature [55]), or as the Orowan bowing stress $1.5Gb/d_c$. Inspection of eq. (43) shows that if $\hat{\tau}_c(0)$ is written as the critical screw dislocation dipole passing stress, then τ_{PSB} contains two characteristic microstructural lengths, namely the channel width d_c and the screw dislocation annihilation distance $y_{\odot}$. On the other hand, if $\hat{\tau}(0)$ is represented by the Orowan bowing stress, then only d_c will appear as a microstructural dimension in τ_{PSB}, namely once in the denominator of the first term and once more in the numerator of the second term.

The important conclusion is that an internal length scale appears only, when details of dislocation glide *in* the cell interior are taken into account. In a similar vein, it can be argued that if dislocation glide is planar, i.e. when the interfacial dislocations will not consist of single layers of individual dislocations but of groups of dislocations "piling-up" at the interfaces and extending into the cell interiors, compare also [110], then the strain gradients will be more gradual, and the cell diameter d_c will appear in the flow-stress equation as an internal length scale.

In passing, we note that, with a more realistic internal stress profile with finite gradients extending into the cell interiors, compare section 4.6, the volume fractions of hard and soft material should be re-interpreted by assigning to the hard component also those peripheral regions of the cell interiors in which forward internal stresses are induced, compare also section 5.4.3.

While most of the arguments presented above for single slip deformation remain valid in the case of multiple slip, one additional feature relating to the role of the interfacial geometrically necessary dislocations in the case of multiple slip must be pointed out. In this case, the interfacial dislocation layer will in general consist of a cross grid of at least two families of dislocations of different Burgers vectors, compare fig. 11(a). Thus, the interfacial dislocations of one glide system will act like trees of a dislocation forest for glide dislocations of the other glide system, as discussed in [107–109] and as pointed out independently by Weertman and Weertman [111]. Hence, there will be an enhancement of the overall flow stress due to the cutting of interfacial "forest" dislocations by the glide dislocations. In this case, the density of the geometrically necessary dislocations will contribute to the *local* flow stress so that the following relationship can be assumed [107]:

$$\tau_{loc}(x) = \alpha Gb\sqrt{\rho_S(x) + \beta\rho_G(x)}. \tag{44}$$

Here, β is a constant ($0 < \beta < 1$) that takes into account the geometrical details of the interaction of glide and forest dislocations. It should be noted that Zaiser had first proposed a similar relationship which was, however, not based on the cutting of interfacial geometrically necessary dislocations but on mesoscopical considerations of statistical fluctuations of the local stresses [112].

For a given dislocation arrangement, the overall flow stress can be evaluated via eq. (44) as $\tau = \langle\tau_{loc}(x)\rangle$. This was explored in detail for multiple slip on 4 glide systems in an idealized cell structure similar to that shown in fig. 11 [109]. It could be shown that, for all practical purposes, the contribution of the interaction of glide and interfacial geometrically necessary "forest" dislocations to the overall flow stress is negligible.

5. *Possible refinements and extensions of the composite model*

5.1. Some general limitations of the original composite model and of other models

In its original form, the composite model has a number of limitations which can be summarized as follows. The composite model is essentially a model of the flow stress of deformed crystals in which a heterogeneous distribution of the dislocations has developed. It is based on the concept of deformation-induced internal stresses and contains, in particular, a "built-in" back stress in the soft regions. The latter is a necessary ingredient of all models that are able to describe *kinematic hardening*, compare section 5.4.1. Since the model incorporates neither an evolutionary development of the heterogeneous dislocation distribution with proceeding deformation nor the operation of time-dependent processes, it lacks essential aspects of work hardening, dislocation kinetics and dynamics. The model considers only two "phases", namely a soft and a hard component, in a very crude approximation of the complex spectrum of dislocation segment lengths, obstacles and other features of the dislocation arrangement.[4] While the local yield stresses of the two components can be modelled globally in terms of discrete dislocation mechanisms, the composite approach must, of necessity, miss important details of the real complex dislocation distribution. One critical aspect concerns the definition of the volume fractions of the soft and the hard regions. In the case of low-temperature deformation, the dislocation cell walls are rather disordered and have a finite measurable thickness. This is especially true for low-temperature cyclic deformation. Hence, it is not surprising that the inspiration to formulate the composite model came from the study of dislocation patterns in fatigued crystals, as described in section 1.1. On the other hand, as described in section 5.2 and in more detail in section 5.4.3, the composite model has also been applied to high-temperature creep, although in this case the cell walls are now rather orderly subboundaries or almost perfect planar low-angle boundaries with no measurable thickness. Thus, the definition of a volume fraction of the walls becomes problematic. As a last point, we mention that the original composite model is not applicable to large-strain deformation, since it does not include the development of increasing misorientations across the dislocation cell walls, as evidenced clearly in experiment, compare, for example, [43,66,114,115]. We shall come back to this point in section 6.

In a way, the most attractive feature of the composite approach is its simplicity, admittedly at the expense of the accurate consideration of details which may be important. We shall in the following mention one specific example to illustrate this point. While the composite model deals with the deformation-induced internal stresses in an easily tractable manner, it is, however, very global. Thus, it does not consider several important details

[4]As shown by Zaiser [113], the composite approach can be generalized to deal with an arbitrary distribution of local dislocation densities just like the spectrum theories [90,91] treat a spectrum of local flow stresses.

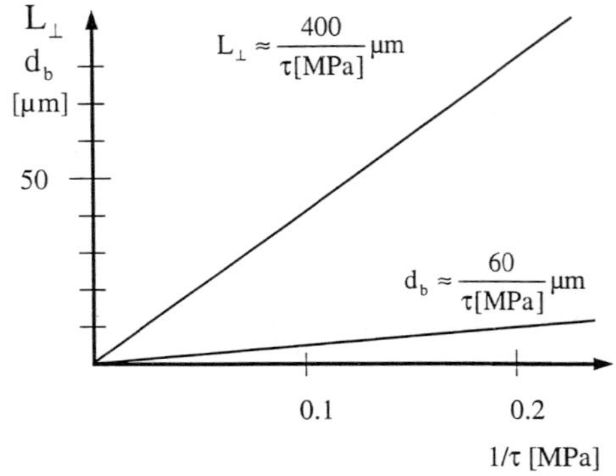

Fig. 16. Dependences on the shear flow stress τ of the slip paths $L_\perp$ of edge dislocations and the mean spacings d_b of dislocation bundles of predominantly edge character in Stage II of deformed copper single crystals. After [21,118].

such as the question of plastic relaxation by secondary slip of internal stresses (provoked by primary slip) or to what extent elastic accommodation, corresponding to an Eshelby factor $\Gamma < 1$, will reduce the internal stresses appreciably for certain geometries of the dislocation arrangement. As an example, we mention the question of plastic relaxation of internal stresses in slip bands which has been dealt with in considerable detail first by Hirsch and Mitchell [6,25], then by Jackson, making use of the Eshelby formalism [116, 117], and, more recently, by Brown [84,85]. On the other hand, these models lack the simplicity of the composite model which is easily applicable not only to unidirectional but also to other, more complex loading modes, compare section 5.4.

While there are other models which deal explicitly with one or some of the above-mentioned specific aspects of crystal plasticity, as illustrated above, there exists in fact no single model which addresses al these aspects satisfactorily in a self-contained manner. Moreover, many other models share shortcomings of the composite model. We shall here mention one shortcoming common to all existing models. In his TEM study of Stage II work hardening of copper single crystals deformed in single slip [20], one of the authors [21,118] noted that the slip paths $L_\perp$ of (edge) dislocations, as deduced from slip-line studies [8], was larger by almost one order of magnitude than the mean spacings d_b of the dislocation bundles or braids. Making use of the "experimental" relations between $L_\perp$ and d_b and the shear flow stress τ given in [21], one obtains the picture shown in fig. 16. Here, the fact was taken into account that, because of the well known soft-surface effect, first discovered by Fourie [119], the slip line paths are shorter in the bulk than near the surface. Nonetheless, the fact remains that, in the bulk, $L_\perp \approx 6\text{–}7 d_\mathrm{b}$ which illustrates clearly that the dislocation bundles only act as "temporary" obstacles [20,21,118] but are essentially transparent to dislocations [20,21], in contrast to the dislocation grids or sheets in which primary dislocation glide is effectively stopped by intense secondary slip and which can hence be considered more as "permanent" obstacles. This situation is illustrated further in fig. 17, taken from [118], which illustrates that the actual slip path of dislocations is really an "effective" slip path L_eff which corresponds to the sum of partial slip paths L_i extending from bundle to bundle. It was concluded from TEM observations

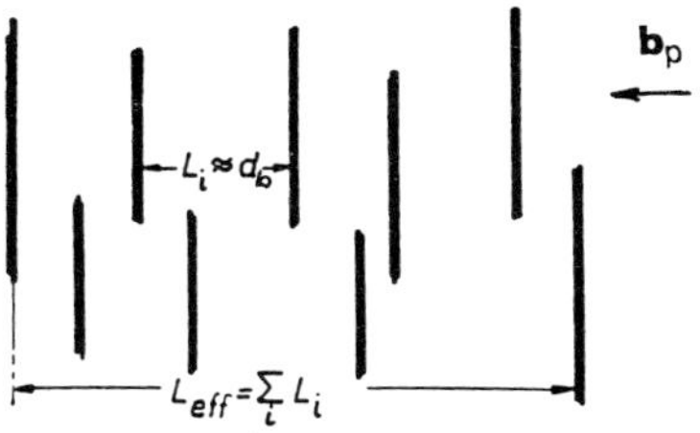

Fig. 17. Schematic illustration of view on the primary glide plane with dislocation bundles of predominantly edge character, showing that the actually observed slip path is an effective slip path L_{eff}, composed of the sum of "partial" slip paths L_i from bundle to bundle. After [118].

that the development of a "glide zone" extending over the length L_{eff} probably starts from a localized dislocation "source region" [20,21] and spreads in a burst- and relay-like fashion, as suggested first by Mader's replica studies of active glide zones [8]. It is quite clear that the penetration of the dislocation bundles which contain not only primary but also secondary dislocations must involve thermally activated overcoming of the "forest" obstacles in the bundles. This dynamic aspect of dislocation propagation is not described in any of the work-hardening models discussed earlier. It is interesting to note in passing that this phenomenon of correlated long-range dislocation propagation is not unique to single slip but occurs obviously also in multiple slip, as emphasized by Göttler [70] in his study of deformed [001]-oriented copper single crystals. Returning to single slip in Stage II, mention should be made of the work of Himstedt who adopted the picture described above in order to model transient effects observed in Stage II during strain-rate changes and stress relaxation tests [120] in terms of correlated dislocation glide processes through several bundles. He arrived at a thermal rate equation which permitted a satisfactory description of his experimental observations. More recently, Thomson and Levine [121] have treated the process of dislocation propagation as the percolation of mobile dislocations through the "blocking walls of a partially ordered structure" in an attempt to derive the foundation of the stress–strain law. Clearly, any satisfactory treatment of the fundamental mechanism of correlated dislocation propagation through a glide zone must go far beyond the simplistic concepts of all existing models.

In spite of its limitations, the composite model has been widely accepted and applied successfully to the interpretation of a number of different cases of plastic deformation such as cyclic deformation, strain path changes, high-temperature creep, including transient deformation behaviour and large-strain deformation (Stage IV). Examples of these applications are discussed in sections 5.4 and 6.

5.2. Criticisms of the composite model

On the whole, the basic ingredients of the two-component composite model have been accepted widely and have in fact served as the basis and starting point for later and sometimes more refined models [46–48,127–129]. There have, however, also been occasional criticisms of the composite model, most notably by Kuhlmann-Wilsdorf [87] and Kassner [122–125]. The criticism by Kuhlmann-Wilsdorf is natural, since her flow stress (and work hardening) models [14,24,80,87] differ in fundamental respects from the composite model. In particular, they do not address *local* variations of the flow stress and do not explain the occurrence of deformation-induced long-range internal stresses, as they are observed. Hence, there is not much point in going into further details beyond those already stated in section 4.5.1 and the general comments made in section 5.1. The doubts expressed by Kassner [122–125] are of a different nature. This author has undertaken a number of different experimental TEM studies (including convergent beam electron diffraction, CBED) with the aim of detecting deformation-induced long-range internal stresses in cyclically deformed aluminium [122,123] and copper [124] single crystals and in creep-deformed aluminium single crystals [125]. In a critical discussion of these studies, the most important points are that all work was done on unloaded specimens and on thin foils, i.e. none on bulk material. Furthermore, no precautions such as pinning of dislocations were taken to avoid loss and rearrangement of dislocations which are known to lead to a relaxation of internal stresses during preparation of thin foils in addition to the relaxation occurring during unloading, compare [126]. Other points of criticism will not be addressed here. Kassner has not ruled out the existence of long-range internal stresses but has concluded that, if present, they must be small. It is the authors' belief and conviction that Kassner's studies, though careful, were not performed under favourable circumstances, as far as the chances to detect deformation-induced long-range internal stresses are concerned.

5.3. Recent and potential refinements and extensions

It is straightforward to extend the composite model for unidirectional deformation semi-quantitatively to strain reversal and to cyclic deformation [32,33,58]. This important application of the composite model is the subject of section 5.4.1. Another (semi-quantitative) extension, treated in section 5.4.3, concerns high-temperature creep, where the extended composite model has been shown to be a powerful tool in the modelling of steady-state and transient creep deformation. In section 6 which is devoted to large-strain deformation, the composite model of dislocation glide is modified conceptually so as to include also the effects of increasing misorientations, in good qualitative correspondence with experimental observations. More details of these examples of modified and extended formulations of the original composite model will be found in the later sections. Here, we wish to refer, in particular, to extensions of the model concerning work hardening, the kinetics and dynamics and the rate dependence of plastic flow and the more recent statistical treatments, including also stochastic models of deformed crystals containing fractal dislocation arrangements.

The first treatments or formulations of models of the composite type that included time-dependent (thermally activated) deformation were those of Nix and Ilschner [127], Nix et al. [128], Prinz and Argon [46], Nabarro [129] for the case of unidirectional deformation and the models of Burmeister and Holste [130] and Polák and Klesnil [91] for cyclic deformation. Quite recently, Roters et al. [131] published a quite detailed three-parameter model for monotonic deformation. It is interesting to note that the first group of authors all considered the simplified "two-component" model, whereas the second group, dealing with cyclic deformation, treated the more complex case of a spectrum of mesoscopic elements in an attempt to describe the continuous changes occurring during a closed cycle. Nix and Ilschner [127] essentially adopted the two-component model developed for low-temperature cyclic deformation and applied it to high-temperature creep in spite of the fact that the dislocation distributions are not at all comparable and that the definition of the volume fraction of the very thin subboundaries in high-temperature creep is problematic, as indicated in section 5.1.

Prinz and Argon [46] and Nix et al. [128] were interested mainly in modelling the later stages of work hardening, taking into account also processes of dynamic recovery. It is interesting to note that the dislocation model used by Nix et al. [128] is identical to that derived from the studies of the deformation processes in the ladder structure of PSBs, as was illustrated in fig. 5(a). Remembering how complex the dislocation distribution actually is in Stage II and also in the later work hardening stages, this is, of course, a gross simplification, which avoids dealing with the actual more complicated dislocation distribution consisting of dislocation bundles, walls and grids/sheets. In the model of Prinz and Argon [46], the dislocation distribution is not specified in so much detail and is simply assumed to be cellular. Nonetheless, in both models, evolutionary laws of dislocation multiplication and recovery in the soft and hard regions are probably good global descriptions, and the results obtained reflect the most important features of rate-dependent deformation.

The aim of Nabarro's extensions and modifications of the composite model for multiple slip also concerns work hardening and dynamic recovery [129], the focus differs, however, from that of the previously mentioned approaches of Nix and co-workers [127,128] and Prinz and Argon [46]. In this paper, Nabarro points out that the composite model for multiple slip can be considered as a complementary model to that of Hart and co-workers [132,133] which is essentially also a two-parameter model. Nabarro's modification of the composite model concerns primarily the introduction of a detailed concept of dynamic recovery in terms of specific dislocation annihilation mechanisms. Nabarro then comes to the conclusion that the two apparently very different models are in fact quite similar and that the so-called hardness parameter σ^* and the back stress σ_a in the model of Hart and co-workers are related, in a crude sense, to the stresses τ_c and $\Delta\tau_c$ of the composite model. An important difference remains, since in Hart's analysis the strain rate $\dot{\varepsilon}$ plays an important role, while the composite model, in its original form, essentially considers only the athermal component of the flow stress.

In a series of papers, Hähner and Zaiser [134–136] investigated the fractal nature of the cellular dislocation structures [134] observed earlier by the authors [65,66]. They then discussed the consequences with respect to dislocation patterning and the composite model and proposed alternative new statistical formulations of the flow stress [135] and the dislocation dynamics and work hardening [136]. An interesting aspect of the latter work, based on a stochastic modelling approach, is that heterogeneous dislocation distributions can be treated as "effective single-phase" structures characterized by distribution functions of the length scales and corresponding configurational averages of the cell structure. This new description includes the simple composite model as a particular case and allows a more general approach to dislocation patterning and dynamics.

None of the models mentioned so far considers the misorientations that develop progressively during large-strain deformation. This aspect has been addressed by Argon and Haasen [48] in an interesting model of Stage IV work hardening, compare section 6. Here we shall discuss briefly only one particular part of their model in which the authors propose an extension and a modification of the interpretation of asymmetric X-ray line broadening given in section 4.2. In their dislocation model, Argon and Haasen discuss two contributions of the cell interiors, saying that one half will experience an axial back stress, while the other half is under a (smaller) axial forward stress. They then construct an asymmetric X-ray intensity peak composed of three sub-peaks. The present authors have some doubt whether this model which was devised primarily for Stage IV deformation with marked and progressively increasing misorientations is applicable to the X-ray studies presented in section 4.2. One assumption of the model is that, in spite of overall multiple slip, single slip prevails locally, inducing lattice misorientations at the walls. In the work on tensile-deformed [001]-orientated copper crystals [65–67] which did not concern Stage IV deformation, lattice misorientations were not a dominant feature, and there was no evidence of local single slip. Perhaps more important is the fact that while the semi-quantitatively constructed asymmetric X-ray diffraction peak looks indeed similar to the asymmetric peaks discussed earlier (figs 8(a) and 10(a)), the residual elastic strains predicted by the model are unrealistically high, amounting to values up to about $+2\%$ in the walls and -0.5% in the cells. Such large elastic strains are not reasonable and would cause plastic relaxation. Therefore, in spite of the many interesting features of the model of Argon and Haasen, it is doubtful whether it is suitable to provide an improved understanding of asymmetric X-ray line broadening.

5.4. Application to different types of loading

5.4.1. Cyclic deformation

The composite model, being of the Masing type [95,96], compare section 4.5.2, contains the basic ingredients of kinematic hardening as a consequence of the deformation-induced long-range internal stresses. In its original form, the Masing model referred essentially to the yielding behaviour of an aggregate of crystallites of different yield stresses upon reverse yielding. It provided the first basically correct interpretation of the so-called Bauschinger effect [59], compare fig. 18, in terms of deformation-induced internal stresses. The most important feature of the Bauschinger effect is the lowering of the yield stress upon stress reversal, shown for the sequence tension–compression in fig. 18. More precisely, one should speak of a lowering of the microyield stress. Another feature frequently observed is the so-called permanent softening which refers to a (permanent) lowering of the macroyield stress, when the level of the extrapolated curve of the first loading is compared with that of the subsequent reversed loading. The Bauschinger effect and permanent softening are two important features of cyclic deformation which govern largely the shape of the hysteresis curves. Both effects follow in a straightforward manner from the composite model. Pedersen et al. [34] have analyzed the Bauschinger effect in copper single crystals oriented for single slip in terms of a composite model on the basis of the dislocation microstructures observed by TEM.

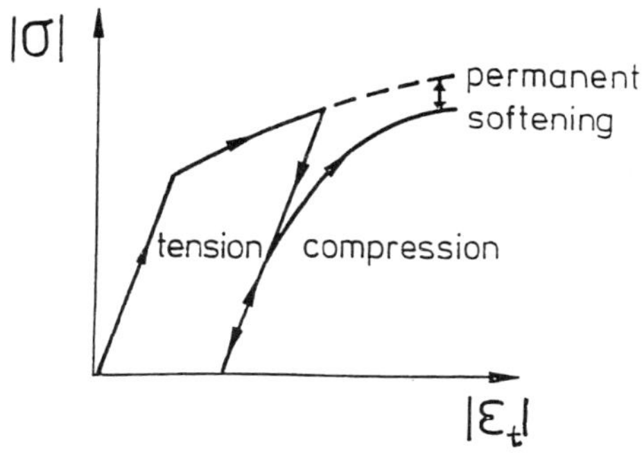

Fig. 18. Schematic illustration of the Bauschinger effect with permanent softening for tension–compression deformation.

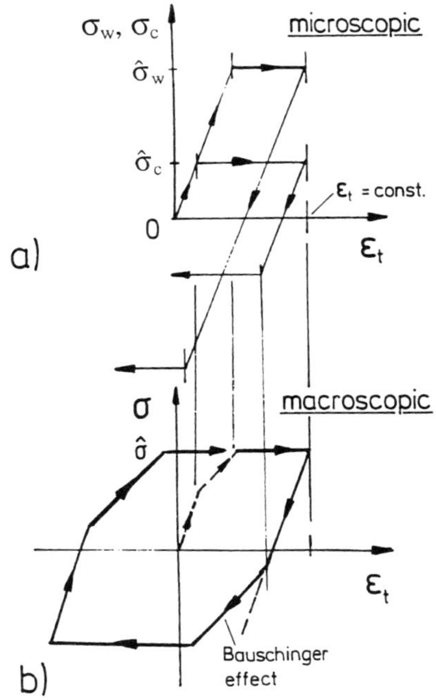

Fig. 19. Composite model of cyclic deformation: (a) microscopic yielding behaviour of the two components (walls, cell interiors), strained in parallel up to a given value of the total strain ε_t and subsequently unloaded and strained in the reserved direction; (b) corresponding macroscopic yielding behaviour with Bauschinger effect in a closed tension–compression cycle. After [58].

We now consider an extension of the original composite model, derived for the description of plastic flow in PSBs [31–33], for the case of reverse yielding, i.e. for cyclic deformation. In fig. 19 the monotonic and reverse microscopic yielding (fig. 19(a)) and macroscopic yielding (fig. 19(b)) curves of a two-component composite [58] are shown in plots of the microscopic axial stresses σ_c and σ_w of the soft cell interiors and hard walls, respectively, and the macroscopic axial stress σ versus the total axial strain ε_t. The monotonic loading and subsequent unloading behaviour is similar to that shown in fig. 4. After unloading at a given total strain, long-range internal forward and back stresses remain in the hard and soft components, respectively. During reverse loading, the long-range internal stresses decrease, as the soft component yields before the hard component (fig. 19(a)). The two reverse loading curves intersect after the soft component has already yielded plastically, while the hard component is still being strained elastically. At that point, the internal stresses would be precisely zero, because the elastic/plastic strains in both components would now be identical, and upon unloading from compression, both unloading curves would follow along the same elastic line. During reverse straining beyond the point of intersection, long-range internal stresses, opposite in sign compared to those induced during forward straining, would build up and increase in magnitude up to the point at which the hard component finally yields.

The macroscopic yielding behaviour follows from the appropriate superposition of the microscopic stress–strain curves, and a closed polygon-type hysteresis loop which exhibits a well-defined Bauschinger effect is obtained, after completing the cycle. In subsequent cycles, the hysteresis curve would repeat itself. Reverse (micro)yielding occurs precisely at that total strain at which the soft component begins to yield in compression, corresponding to a macroscopic stress level that is lower than at the end of the forward straining. Reverse microyielding ends when the hard component begins to yield in compression. The hysteresis curve (polygon) so obtained exhibits so-called "Masing behaviour" in the sense that the deformation curve after stress reversal follows from that of the first forward straining by multiplication by a factor of two. In the case of Masing behaviour, the ascending (descending) branches of a set of hysteresis loops of different amplitudes follow a common curve if the tips of the hysteresis curves are brought into coincidence at the minimum (maximum) peak stresses [137].

In the picture of the composite model, an interesting distinction can be made between cyclic micro- and cyclic macroplasticity, in analogy to monotonic micro- and macro-yielding, compare section 4.1, as illustrated schematically in fig. 20(a) and fig. 20(b), respectively. In the case of cyclic microyielding, the total strain amplitude is so small that plastic yielding occurs only in the soft but not in the hard component, and a narrow pointed hysteresis loop is obtained which exhibits a permanent softening $\Delta\sigma_p$, as indicated in fig. 20(a). The permanent softening can easily be expressed [58] as

$$\Delta\sigma_p = 2|\Delta\sigma_c| = 2\Gamma E f_w(\varepsilon_{pl,c} - \varepsilon_{pl,w}). \tag{45}$$

It should be noted, however, that, in the simple two-component model discussed here, there will only be a finite permanent softening in the cyclic microyield regime, i.e. as long as $\varepsilon_{pl,w} = 0$, while $\varepsilon_{pl,c} \neq 0$. The concept of permanent softening has proved very useful in the analysis of the (cyclic) plasticity of dispersion-hardened copper single crystals by Brown and Stobbs [138]. Once the strain is so large that the hard walls yield, the regime of (cyclic) macroplasticity begins, the hysteresis curve becomes more rectangular, and no permanent softening will occur during reverse loading, as indicated in fig. 20(b). It is important to note that the absence of permanent softening, as in the case of the hysteresis

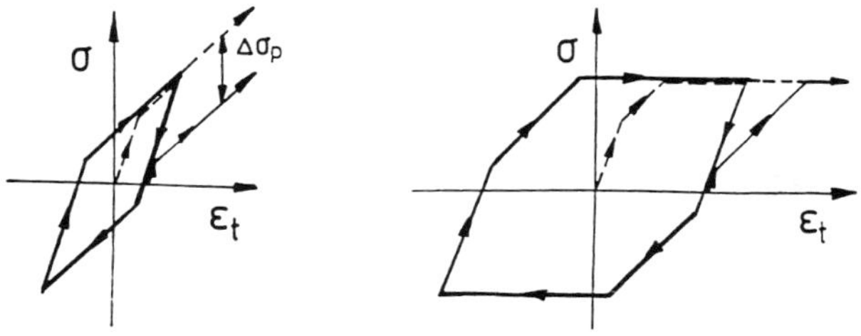

Fig. 20. Schematic illustration of cyclic micro- and cyclic macroyielding in terms of the composite model: (a) cyclic microyielding hysteresis curve, characterized by plastic yielding of the soft and elastic straining of the hard component, with permanent softening $\Delta\sigma_p$; (b) cyclic macroyielding hysteresis curve, characterized by plastic yielding of the soft and the hard components. After [58].

curve of PSBs, compare [31–33,58], is *not* a valid criterion that long-range internal stresses are absent but implies simply that macroyielding has occurred. A more detailed discussion of cyclic deformation in the framework of the simple two-component model can be found in [31–33,58].

In the form described above, the composite model of cyclic deformation is not confined to any particular dislocation mechanisms. In specific cases, details of the dislocation mechanisms in cyclic deformation can be implemented, as has been discussed earlier rather globally for the cases of unidirectional deformation in single slip (section 4.1) and in multiple slip (section 4.2). This has been done in some detail for the case of cyclic deformation of the matrix and of the PSBs of f.c.c. single crystals with the aim to describe the respective hysteresis loops [31,32] and cyclic yield stresses [32,62], and will not be repeated here. More recently, Sedláček [139,140] has extended and generalized the simple dislocation concepts of the original composite model for cyclic deformation in the PSBs and in the matrix of f.c.c. crystals by an analytical formulation of the motion of a continuous flexible screw dislocation bowing out between patches of edge dislocation multipoles, as originally proposed by Kratochvíl and Saxlová [141].

The simple composite concept outlined above has been applied successfully by several researchers in different studies of (cyclic) deformation and of the Bauschinger effect. Thus, Christodoulou et al. [142] have concluded that their observations of transient reversed plastic flow in copper polycrystals cannot be explained by a single structure parameter (such as $\hat{\tau}$) but that a two-component model with three microstructural parameters ($\hat{\tau}_c$, $\hat{\tau}_w$ and f_c or f_w) can provide at least a qualitative description of transient reverse plastic flow. Chicois et al. [143], in a study of cyclic deformation of aluminium polycrystals in which the ultrasonic attenuation was measured during the cycles, found it necessary to invoke the composite principles of compatibility of deformation and of deformation-induced long-range internal stresses in their interpretation of their observations.

Biermann et al. [144] studied the local variation of the long-range internal stresses within a closed cycle during the cyclic deformation of copper polycrystals by performing high-resolution X-ray diffraction experiments on individual surface grains of specimens that had been unloaded from different stress/strain levels within the cycle. By analyzing the asymmetrically broadened X-ray intensity profiles, they could show how the long-range

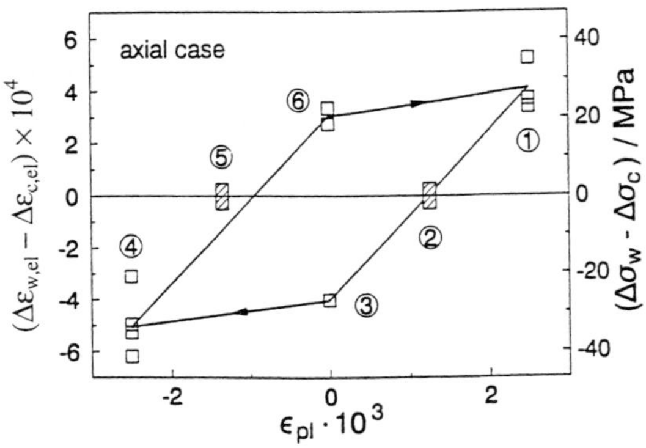

Fig. 21. Difference of local internal strains $(\Delta\varepsilon_{w,el} - \Delta\varepsilon_{c,el})$ and stresses $(\Delta\sigma_w - \Delta\sigma_c)$, after unloading as a function of the plastic strain. The numbers indicate the 6 samples that were unloaded at different plastic strains. The solid line and the arrows indicate the course of the mechanical hysteresis loop which is approximated by a parallelogram. After [144].

internal stresses build up during loading and how they decrease and build up again with the opposite sign during reverse loading. Figure 21 summarizes some of their results. In this diagram, the difference $(\Delta\sigma_w - \Delta\sigma_c)$ between the forward and the back stresses (which is obtained via Hooke's law from the difference $(\Delta\varepsilon_{w,el} - \Delta\varepsilon_{c,el})$ of the corresponding local elastic internal strains) is taken as a measure of the overall strength of the deformation-induced long-range internal stresses. In a detailed analysis of the X-ray results and the shape of the hysteresis loop [90,91,145,146], the changes of the elastic stored energy during a cycle could be evaluated and split into three contributions due to mean stresses (I), the heterogeneous dislocation cell structure (II) and the dislocations that shuttle to-and-fro in the cell interiors (III). In a plot of the stored energy E^{het}_{stored} versus the plastic strain ε_{pl}, butterfly-shaped closed curves are obtained for contributions II and III, as shown in fig. 22. The mean stresses arise from the grain-grain interaction and shift the butterfly-curves upward. These results confirm in substantially more detail the results of earlier work which the Dresden group obtained by magnetic coercive force measurements [147] and by X-ray diffraction [148] on cyclically deformed nickel polycrystals, and thus substantiate the predictions of the composite model.

In a quite different recent study, Feaugas [149] used the Bauschinger effect as a means to probe the internal back stress of unidirectionally deformed polycrystals of the stainless steel AISI 316L with the aim of separating the inter- and the intra-granular components. Feaugas could show convincingly that the intergranular internal stresses are dominant only at small plastic strains ($< 1.5\%$), whereas at larger plastic strains the intragranular contribution to the internal stresses due to the heterogeneous dislocation distribution becomes more important.

While the two-component composite model has been applied, at least semi-quantitatively, with success in the examples discussed above, its limitations are obvious, and more rigorous multi-component models have been developed by Holste and Burmeister [90], Burmeister and Holste [150], Polák and Klesnil [91], Polák et al. [147,151] and Christ [146]. In contrast to the earlier, essentially athermal, Masing-type models of cyclic

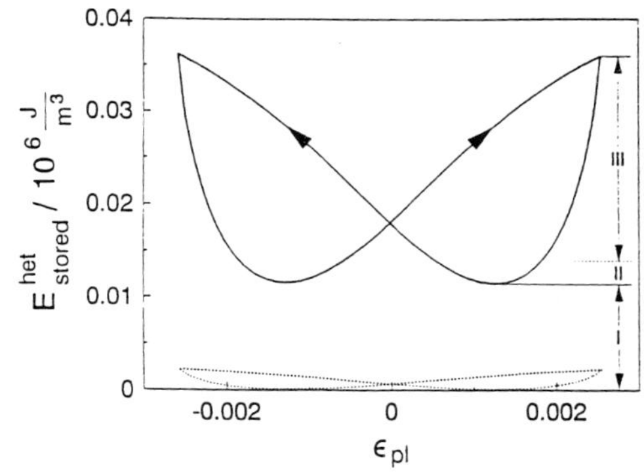

Fig. 22. The elastic stored energy for a heterogenous dislocation microstructure, E^{het}_{stored}, as a function of plastic strain ε_{pl} calculated by the analysis of the hysteresis loop (solid line), composed of three contributions. See text for details. From [144].

deformation, some of these multi-component models [150,151] include also the effects of time-dependent thermally activated plasticity. Thus, they are able to deal with transient phenomena such as stress relaxation and changes of activation area during a deformation cycle.

Because of their complexity, compared to the more easily tractable two-component models, the multi-component models have so far been applied in only a few cases. In the following, we shall discuss as an example a study by Christ [152] which shows how the distribution function of local yield stresses can be obtained from an analysis of the shape of a hysteresis loop and how this result can then be used to model the shapes of other hysteresis loops of different amplitudes, for example in a programmed or random loading test. The multi-component models yield the following relation between the probability density function f_P of the local microscopic yield stresses and the curvature of the hysteresis loop

$$f_P\left(\frac{\varepsilon_r E}{2}\right) = -\frac{2}{E^2}\frac{d^2\sigma_r}{d\varepsilon_r^2}. \qquad (46)$$

In this relation, the stresses and strains are represented in the form of so-called "relative" stresses σ_r and strains ε_r, respectively, which are defined relative to the last load reversal point, i.e. they represent the difference between the maximum (or minimum) and the current values. Figure 23(a) shows a distribution function (in discretized form, with 30 elements) obtained by Christ [152] in this manner from the analysis of a single branch of a hysteresis loop of a copper polycrystal deformed cyclically, under steady-state (saturation) conditions in a so-called incremental step test. In this test, the specimen is subjected to a programmed loading consisting of successive "strain blocks" of typically some ten cycles with linearly increasing and subsequently decreasing strain amplitudes. Thus, the specimen is subjected to a spectrum of amplitudes in a defined sequence. Using the discretized distribution function, Christ was able to compute the entire stress-strain path for this incremental step test. Figure 23(b) shows the corresponding sequence of hysteresis

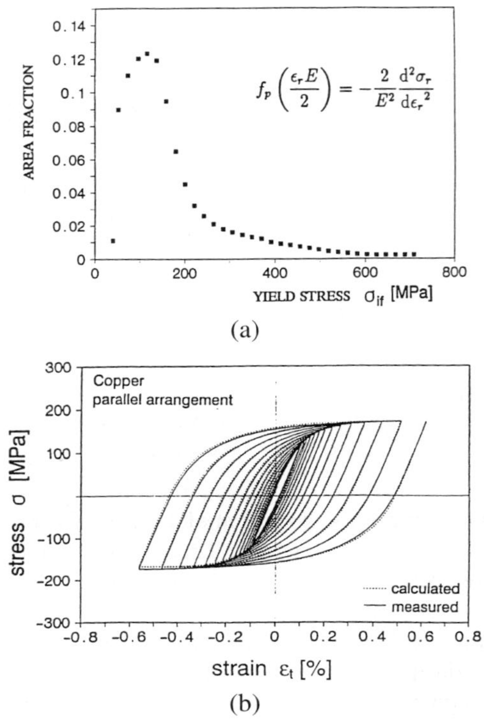

Fig. 23. Modelling of hysteresis curves of copper polycrystal in incremental step test in a multi-component composite model: (a) discrete distribution function of the area fraction versus the flow stress; (b) comparison of calculated and measured stress-strain paths of a half strain block (see text). After [152].

loops of a copper polycrystal, measured during half a strain block of an incremental step test. The computed curves which were obtained, using the results of fig. 23(a), are also shown (dotted curves), and the agreement is found to be excellent. Quite generally, good agreement can be expected when the hysteresis loops exhibit Masing behaviour, which is not always obeyed but which is well fulfilled in the present case [137,152]. These results can be regarded as independent indirect evidence for the correctness of the composite concept.

5.4.2. Other strain-path changes

In cyclic deformation, the sense of the applied stress is reversed but the axis of orientation is retained. Thus, cyclic deformation represents one special case of a strain-path change. More generally, strain-path changes of interest involve a change of deformation axis, as in the classical work of Jackson and Basinski [153], who tensile-strained specimens of different axial orientations cut out of copper single crystals that had been pre-deformed in tension. They thus probed the latent hardening of the slip systems that operated during the first deformation and observed a transient reduced work hardening related to localized softening in coarse slip bands, indicative of a microstructural instability that occurred during the second straining. Subsequent work on polycrystals revealed similar effects.

Here, we shall confine ourselves to a brief discussion of a number of studies that were performed by Wilson and Bate and their associates. These studies concerned not only the Bauschinger effect in unaxial straining of steels [154,155] and two-phase aluminium alloys [156] but also strain-path changes in different sheet materials (aluminium, copper, α-brass and different steels) which had been pre-deformed and from which tensile test specimens for a subsequent deformation had been cut under different angles [157–159]. In their work, the authors emphasized the important roles of the directionality of the deformation-induced long-range internal stresses in heterogeneous dislocation structures, as explained in terms of the composite model. In particular, they explained the microstructural instability observed after a severe strain-path change in terms of the disruption of the dislocation walls as a consequence of the high stresses that arise when geometric accommodation (through the Eshelby factor) is reduced after a strain-path change. In other words, these observations of the mechanical response and the microstructural changes observed after strain-path changes can be explained, at least qualitatively, or semi-quantitatively, in the spirit of the composite model.

5.4.3. High-temperature creep

In high-temperature creep studies, investigations of the internal stresses have frequently played an important role. In particular, residual back stresses were invoked in the interpretation of so-called dip (stress reduction) tests, compare, e.g., [160–162]. The dip test is performed in a creep test by observing the time-dependent forward or reverse deformation occurring after a partial stress reduction. It is then argued that the stress level at which no forward or reverse plastic strain occurs (during a finite time) balances precisely an "internal stress". Referring to our earlier definition of internal stresses, compare section 3, this stress should be called more correctly an athermal stress. In terms of the composite model, this athermal stress should actually be the athermal stress component τ_G or σ_G of the flow stress. It would be most desirable to verify this interpretation directly, e.g., by combining the results of dip tests with those of X-ray diffraction (in order to obtain $\Delta\sigma_w$, $\Delta\sigma_c$ and f_w) and by checking whether the stress determined in the dip test conforms to the stress defined by eqs (15), (16) and (17).

Direct evidence of deformation-induced long-range internal stresses was obtained by Morris and Martin [163] who studied the dislocation distribution in a creep-deformed age-hardenable Al–Zn alloy in which the dislocations had been pinned under load. They concluded, on the basis of their TEM observations of curved dislocation segments, that significant deformation-induced long-range internal stresses existed which varied in space. The discussion of the origin of these internal stresses in terms of the heterogeneity of the dislocation microstructure which is usually a subgrain structure at high temperatures is not as straightforward as in the case of the dislocation cell structure which is observed after deformation at lower temperatures. In the latter case, the dislocation walls are more voluminous than the thin subgrain boundaries observed at higher temperatures. They can hence be characterized more easily by their volume fraction f_w. Thus, a description of high-temperature creep in terms of the composite model is not unproblematic, as already mentioned in sections 5.1 and 5.2. Nevertheless, the composite model has been used and verified with good success in particular by Blum and his research associates in numerous studies of high-temperature creep [163–166].

In a detailed microstructural study of copper polycrystals deformed in compression creep tests at constant stress in the temperature range between 298 K and 633 K, Straub et al. [163] succeeded in characterizing the internal stresses in dislocation microstructures, ranging from dislocation cells formed at the lower temperatures to subgrains formed at the higher temperatures. Using high-resolution X-ray diffraction, they obtained characteristic asymmetrically broadened X-ray diffraction peaks which could be evaluated as described in section 4.2 and yielded consistent data on internal forward and back stresses. It is interesting to note that the volume fractions of the hard component were found to decrease from about 0.2 at a deformation temperature of 298 K to about 0.1 at a deformation temperature of 633 K. Here, it should be noted that these volume fractions were assigned not only to the dislocation walls/subboundaries but, more realistically, to all material with internal forward stresses comprising also peripheral regions of the cell interiors [167]. With this re-definition of the volume fraction of the hard regions, the problems mentioned earlier, compare section 5.1 and 5.2, in defining a volume fraction for thin planar subboundaries are at least partially overcome. Straub et al. supplemented the X-ray data by direct measurements of the local lattice parameters by TEM/CBED and concluded that the local stresses σ_w in the cell walls (at the lower temperatures) corresponded to about 1.3 σ in the transition to subgrain boundaries (formed at the higher temperatures).

Blum and co-workers [164–166] have extended the composite model so as to be able to model the creep behaviour at constant stress both in steady-state creep and in transient creep (during a stress change) in terms of the strain rate. The basic equations can be summarized as follows. Starting from eq. (6), formulated in terms of the corresponding axial strains ε_t, $\varepsilon_{el,w}$, $\varepsilon_{pl,w}$, $\varepsilon_{el,c}$ and $\varepsilon_{pl,c}$, and replacing the elastic strains by $\varepsilon_{el} = \sigma/E$, $\varepsilon_{el,w} = \sigma_w/E$ and $\varepsilon_{el,c} = \sigma_c/E$, respectively, we obtain

$$\varepsilon_t = \frac{\sigma_c}{E} + \varepsilon_{pl,c} = \frac{\sigma_w}{E} + \varepsilon_{pl,w}. \tag{47}$$

From eqs (15) and (47) one obtains after some manipulations,

$$\varepsilon_t = \frac{\sigma}{E} + f_w \varepsilon_{pl,w} + f_c \varepsilon_{pl,c}. \tag{48}$$

Keeping in mind that $f_w + f_c = 1$, compare eq. (7), the total strain rate $\dot{\varepsilon}_t$ follows as

$$\dot{\varepsilon}_t = \frac{\dot{\sigma}}{E} + f_w \dot{\varepsilon}_{pl,w} + f_c \dot{\varepsilon}_{pl,c} + \dot{f}_w (\varepsilon_{pl,w} - \varepsilon_{pl,c}). \tag{49}$$

We find for the variation of the back stress $\Delta\sigma_c$

$$\frac{d(\Delta\sigma_c)}{dt} = \dot{\sigma}_c - \dot{\sigma} = f_w E\left(\dot{\varepsilon}_{pl,c} - \dot{\varepsilon}_{pl,w}\right) + \dot{f}_w (\sigma_w - \sigma_c). \tag{50}$$

The appearance of the change $\dot{f}_w$ complicates matters, since $\dot{f}_w$ is not obtained easily. In the approximation $\dot{f}_w \approx 0$, the last terms of eqs (49) and (50) can be neglected and

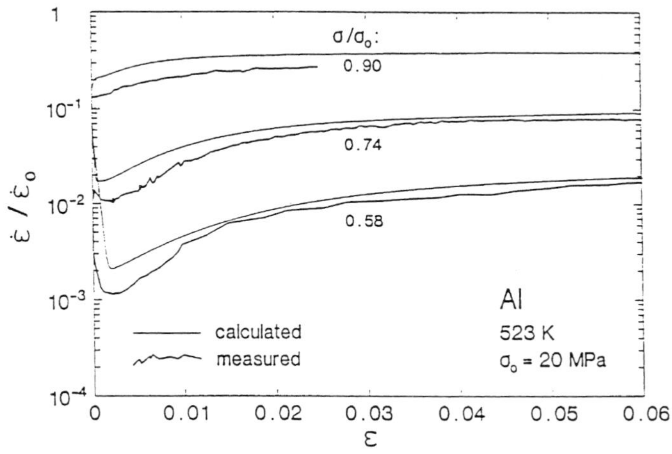

Fig. 24. Measured and calculated creep rates $\dot{\varepsilon}$ (normalized by the initial creep rate $\dot{\varepsilon}_0$) after stress reduction from σ_0 to σ as function of strain ε. Here, the creep rates and creep strains are plastic strain rates and plastic strains, respectively. From [165].

the situation is simplified considerably. For the case of steady-state creep deformation, i.e. $\dot{\sigma} = \dot{\sigma}_c = 0$, $\dot{f}_w = 0$, one obtains from eq. (47)

$$\dot{\varepsilon}_t = \dot{\varepsilon}_{pl,w} = \dot{\varepsilon}_{pl,c}. \tag{51}$$

Blum and co-workers [164–166] have used the above concepts, supplemented by a contribution $\dot{\varepsilon}_m$ of migrating subgrain boundaries to the overall strain rate and by appropriate kinetic laws, and applied them in particular to different cases of transient creep deformation following a stress change which, in a way, corresponds to a change of the loading mode. The behaviour modelled was found to mimic closely the actual material response. Figure 24 illustrates some results obtained on aluminium. These results underline the potential of the composite model in providing a sound basis for the description of the high-temperature creep behaviour.

To conclude this section, it is noted that Orlová [168] has recently pointed out that the well-known relation between the stress σ and the subgrain size d_{sg} observed at high temperatures

$$\sigma = \frac{KGb}{d_{sg}} \tag{52}$$

can be rationalized in terms of an extension of the composite model. In her model, Orlová considers details of the subgrain boundary structure such as the dislocation mesh size and the misorientation angle. The results of Orlová's study of the subgrain size are complementary to those of the one-dimensional model of a dislocation cell wall structure described in section 4.3, which also leads to the relationship (52) for the dislocation cell wall spacing under certain assumptions.

6. Application of composite model to large-strain deformation

6.1. General introductory remarks

In many technical forming operations such as wire drawing or cold and hot rolling, large strains in excess of one are attained. These strains are much larger than typical fracture strains in the range of 0.5 which are encountered in tensile deformation. Therefore, it appears quite natural that studies on large deformations have become a separate chapter in the studies of plastic deformation and why, on the other hand, the importance of Stage IV deformation has only been recognized more recently. Stage IV was first noted explicitly in torsion experiments in the sixties by Stüwe [37] and Kovács [38]. In those days, work hardening was usually analyzed by plotting the stress or shear stress as a function of the square root of strain since, at that time, current work-hardening laws predicted parabolic work hardening [169]. Stüwe and Kovács, independent of each other, realized that the parabolic work-hardening law ceases to hold around shear strains of $\gamma \sim 1$. They found that work hardening increases further up to extremely large strains (up to almost $\gamma \sim 10$) but at a definitely lower rate than below $\gamma \sim 1$ [37,38]. Kocks et al. [170] and Mecking [45] were the first to suggest plotting the work hardening rate θ versus the stress. In such plots, the different stages of work hardening are clearly separated from each other, compare fig. 1(b). Stage IV appears in this plot either as a more or less horizontal curve (line) after Stage III [171]. After the pioneering works of Stüwe and Kovács, Stage IV was studied comprehensively, both experimentally [40–43,172–175] and theoretically [46–50, 176–180]. The experiments have shown that Stage IV can be observed in most metals and alloys, irrespective of whether hardening occurs by dislocation accumulation and cell formation [5–9] or by planar glide without marked dislocation clustering [49]. Most of the experimental investigations aim to study the evolution of the microstructure by TEM, see, for example, [172–175]. There is general agreement that the dislocation cell walls gradually become narrower and that misorientations between adjacent dislocation cells increase with deformation. There is also evidence from electrical resistivity and X-ray line broadening that the dislocation density increases steadily well into Stage IV and even beyond that [173,174].

Haasen [47] developed a Stage IV model based on the composite model of dislocation cells. Essential aspects of the model are the different work-hardening rates in the cell wall and cell interior regions and also the residual internal stresses. Kuhlmann-Wilsdorf [176] and Kuhlmann-Wilsdorf and Hansen [177] attempted to extend the concept of low-energy dislocation structures (LEDS) to account for Stage IV. The composite model of dislocation cell structures has been the basis of a number of further Stage IV models. Argon and Haasen [48] proposed that, when a dislocation traverses a cell wall separating two misoriented cell interior regions, there is a virtual shear left behind in the cell wall which produces internal stresses. Zehetbauer also developed a model based on the composite model of dislocation cells [49]. He assumed that edge and screw dislocations accumulate mainly in the cell wall and cell interior regions, respectively. He attributed different dynamics to edge and screw dislocations with the progression of strain and obtained good agreement with experimental stress–strain curves. Based on experimental data on the evolution of the volume fractions of cell wall regions and long-range internal stresses in

Stage IV [174], Estrin et al. [50] have suggested a Stage IV model in which the dynamics of dislocation storage and annihilation is combined with the decrease of the volume fraction of cell walls and the non-vanishing nature of long-range internal stresses. This summary shows that, even at very large strains, long-range internal stresses remain an important ingredient of the heterogeneous dislocation microstructure. In the following, the results of some rather recent studies, employing high-resolution X-ray diffraction, will be presented, and it will be shown that long-range internal stresses are a useful concept to reveal and describe important (dynamic) properties of the evolving dislocation microstructure. It therefore follows that a comprehensive Stage IV model has to account for these residual internal stresses.

6.2. Experimental verification of long-range internal stresses at large strains

6.2.1. *High-resolution diffraction experiments*

Long-range internal stresses at large strains were determined by high-resolution diffraction techniques in a way similar to that described earlier for smaller tensile strains, compare section 4.2. Characteristically asymmetric X-ray diffraction profiles of the (002) Bragg reflections were obtained and decomposed into two symmetric sub-profiles. Here, the *single crystal* notation is used for the diffracting planes for the following reason. The specimens used were coarsed-grained polycrystals, permitting the diffraction experiments to be performed on single grains whose [001] directions lay close to the surface normal of the specimen. The shifts between the sub-profiles can be expressed in terms of the quantity $|\Delta\sigma_{\mathrm{w}} - \Delta\sigma_{\mathrm{c}}|$ which is a measure of the magnitude of the long-range internal stresses or, in some cases, in terms of the corresponding magnitude of the residual internal shear stresses $|\Delta\tau_{\mathrm{w}} - \Delta\tau_{\mathrm{c}}|$. Here, it is noted that, for all residual long-range internal stress measurements, the {200} type Bragg reflections were used. The simple reason for this is that in copper the $\{h00\}$ type reflections reveal the largest peak broadening [178,181]. These experimental conditions have the consequence that normal and shear deformations or normal and shear stresses are related by the Schmid factor corresponding to the [001] direction: $\varepsilon = 0.408\gamma$, $\sigma = 0.408\tau$ and $\Delta\sigma = 0.408\Delta\tau$, respectively [66]. The X-ray diffraction experiments were carried out either in a special high-resolution double-crystal diffractometer with negligible instrumental broadening [174,178] or at the Austrian small-angle scattering (SAXS) beamline of the ELETTRA synchrotron in Trieste [179]. The first experiments will be referred to as "home", the second ones as "synchrotron" experiments. In the former experiments, individual grains could be studied, in the latter even spatial variations within individual grains.

6.2.2. *Long-range internal stresses in cold-rolled copper, home experiments*

Here, results obtained on coarse-grained copper specimens that were cold-rolled at room temperature to resolved shear strain values γ between 0.2 and 7 in steps of $\Delta\gamma = 0.25$ [174,178] will be presented. Typical profiles of the (002) reflections corresponding to Stage III and Stage IV are shown in fig. 25. (In the present case, (002) corresponds to the diffraction vector perpendicular to the flat surface of the cold-rolled specimen.) The figure shows that the profiles are markedly asymmetric, the tails on the larger angle side decay

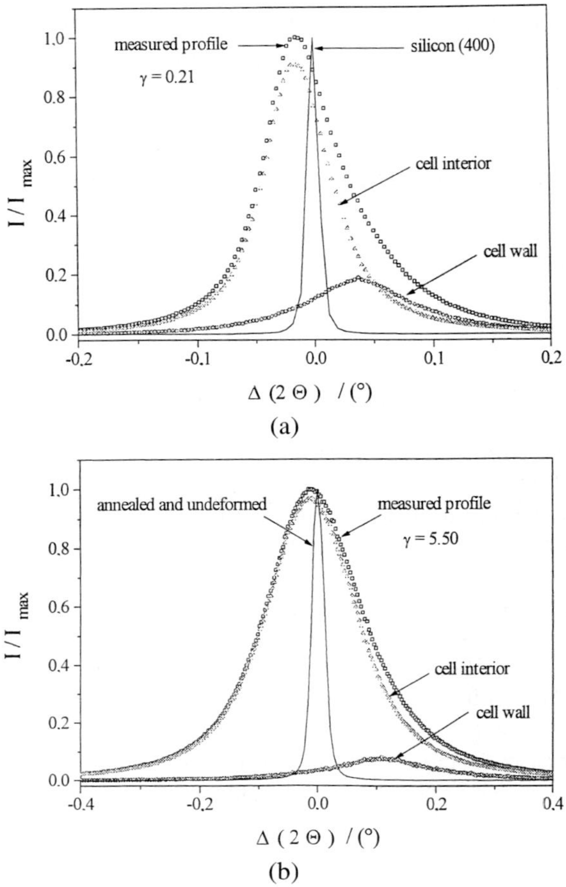

Fig. 25. Typical line profiles of the (002) Bragg reflections of polycrystalline copper cold-rolled to $\gamma = 0.21$ (a) and $\gamma = 5.50$ (b). The subprofiles correspond to the cell interior and cell wall materials, respectively. For reference, two profiles corresponding to a perfect silicon crystal (in (a)) and to an undeformed and annealed Cu specimen (in (b)) are also given in the figure. The centers of gravity of the measured and reference profiles are positioned at zero. From [174].

more slowly than on the smaller angle side. The profiles measured on the narrow edges of the specimen either in the rolling or in the transverse directions revealed the opposite sense of asymmetry. From these observations the following qualitative conclusions were drawn [174]: (i) the reversal of the asymmetry observed on the flat side or on the narrow side in the rolling direction of the specimens indicates that genuine residual internal stresses were induced by plastic deformation, (ii) the sense of the asymmetry shown in the figure is in accordance with long-range internal stresses, as expected after compression, and (iii) the long-range internal stresses are present up to large strains well into Stage IV. The magnitudes of the long-range internal stresses as a function of γ are shown in fig. 26. Each data point in the figure corresponds to a measurement on a different grain in the

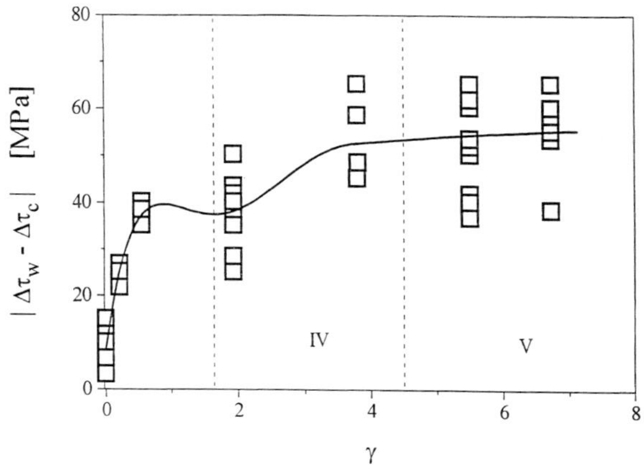

Fig. 26. The magnitude of the residual internal stresses $|\Delta\tau_W - \Delta\tau_c|$ as a function of deformation in polycrystalline cold-rolled copper. Each value in the figure corresponds to a measurement on a different grain in the specimens. The dotted lines indicate the different deformation stages. From [174].

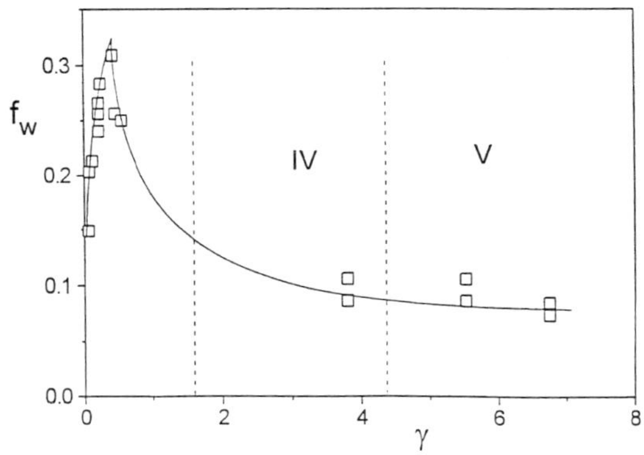

Fig. 27. The volume fraction of the cell wall material as a function of deformation. Each value in the figure corresponds to a measurement on a different grain in the specimens. From [174]. The first four data points in the ascending branch at the beginning of Stage III correspond to measurements on tensile-deformed [001]-oriented copper single crystals [65,66].

specimen. The experimental errors were smaller than the scattering of the internal stress values at the different strain values. This kind of scattering is due to the fluctuating nature of the microstructure from grain to grain. The most striking effect in the evolution of $|\Delta\sigma_W - \Delta\sigma_c|$ versus γ is that, after a fast increase in Stage III, it comes to a standstill around the transition between Stage III and IV. In Stage IV and onwards, it increases again. A similar but not so explicit behaviour has been observed for the "formal" dislocation density ρ^* [181] (which will be explained after this paragraph) as a function of γ [174]. Finally, the volume fraction of the cell wall material, i.e. f_W, plotted against γ, is shown in fig. 27. After a strong increase at the beginning of Stage III it reaches a maximum around

$f_W \cong 0.3$ and thereafter decreases strongly with increasing γ. Well into Stage IV and subsequently, it reaches a steady state around $f_W \cong 0.1$. It is noted that the data related to the beginning of Stage III are those measured on [001]-oriented copper single crystals [66], compare section 4.2. The behaviour of all three parameters, i.e. $|\Delta\sigma_W - \Delta\sigma_c|$, ρ^* and f_W as functions of γ is important in order to understand the evolution of the microstructure and the deformation mechanisms in Stage IV, as will be explained in section 6.3.

The meaning of the term "formal dislocation density" ρ^* is as follows. The contrast of dislocations, as it is well known in TEM studies, depends strongly on the relative orientations between the Burgers vector **b** and the line vector **l** of the dislocations and the diffraction vector **g** [17]. The dislocation density obtained from a broadened diffraction peak profile is denoted by ρ^* and is related to the true dislocation density ρ as:

$$\rho^* = \rho\left(\pi g^2 b^2 C\right)/2, \tag{53}$$

where C are the dislocation contrast factors [174,181].

6.2.3. *The local variation of long-range internal stresses from grain boundary to grain boundary in cold-rolled copper, synchrotron experiments*

In this section, the results of synchrotron microbeam experiments of high spatial resolution [179] which were carried out in order to understand the evolution of the microstructure at large strains will be reported. The {002} diffraction profiles were measured from grain boundary to grain boundary on the same specimens described in the previous section. The X-ray beam was reduced by a pinhole to about 50 μm, and the specimen was moved in front of the beam on an x–y stage. The internal stresses, the formal dislocation density and the dislocation arrangement parameter, $|\Delta\sigma_W - \Delta\sigma_c|$, ρ^* and $M = R_e\sqrt{\rho}$ (where R_e is the effective outer cut off radius of dislocations [181] and ρ is the true dislocation density) are shown for $\varepsilon = 0.086$ ($\gamma = 0.21$) and $\varepsilon = 0.79$ ($\gamma = 1.94$) in fig. 28. At the beginning of Stage III ($\varepsilon = 0.086$, fig. 28(a)), all three parameters are larger in the grain boundary region than in the grain interior indicating that the grain boundary region is in an advanced stage of work hardening as compared to the bulk of the material. In the transition range between Stages III and IV, the three parameters reveal strong fluctuations as a function of position. These fluctuations are larger than the experimental errors. The formal dislocation density shows a strong increase, where $|\Delta\sigma_W - \Delta\sigma_c|$ has a minimum. With further increase of the strain ε, the sense of the fluctuation is reversed: at some positions of the grains a minimum in ρ^* is accompanied by a maximum of $|\Delta\sigma_W - \Delta\sigma_c|$, see, e.g., fig. 3(d) in [179]. These fluctuations will be discussed together with the standstill of $|\Delta\sigma_W - \Delta\sigma_c|$ that was observed as a function of γ in the home experiments.

6.2.4. *Long-range internal stresses in torsionally deformed iron*

Long-range internal stresses at large strains are a general feature which is independent of the mode of deformation. This was illustrated by torsion experiments on pure iron carried out at room temperature at strains up to $\gamma = 9$. The long-range internal stresses $|\tau_W - \tau_c| = |\Delta\tau_W - \Delta\tau_c|$, obtained by the X-ray home experiments, are shown as a function of γ in fig. 29 [180]. After a strong increase in Stage III, the internal stresses decrease to

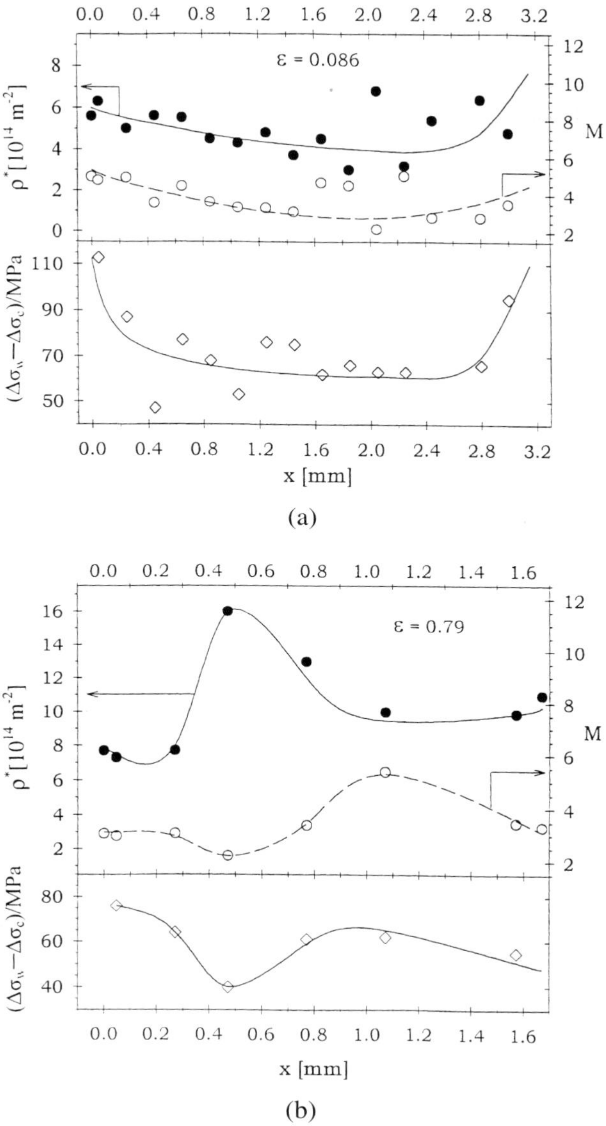

Fig. 28. (a) The formal dislocation density ρ^*, the long range internal stress $|\Delta\sigma_W - \Delta\sigma_C|$, and the dislocation arrangement parameter M, in a grain boundary to grain boundary scan in the specimen cold-rolled to $\varepsilon = 0.086$. (b) The same experiment as shown before in the case of the specimen cold rolled to $\varepsilon = 0.79$. Note the strong fluctuations occurring *in parallel* for all three parameters. From [179].

a relatively low level in Stage IV. The formal dislocation density shows a similar standstill around the transition from Stage III into Stage IV as in copper, see, e.g., fig. 2 in [180]. The results of the home experiments were also confirmed by synchrotron experiments.

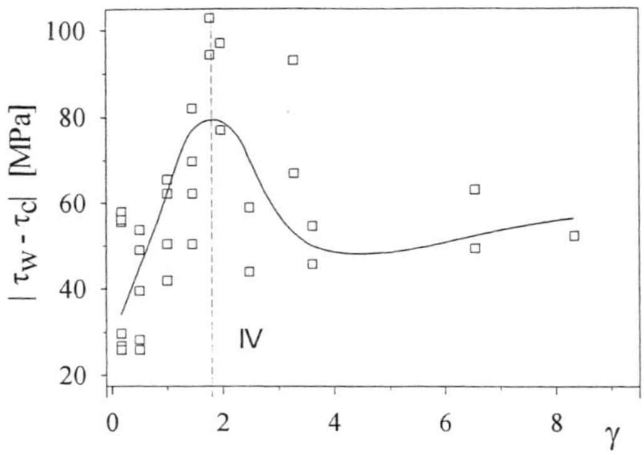

Fig. 29. The magnitude of the residual internal stresses $|\Delta\tau_W - \Delta\tau_C|$ as a function of deformation in polycrystalline iron deformed in torsion. Each value in the figure corresponds to a measurement on a different grain in the specimens. The dashed line indicates the onset of Stage IV. From [180].

6.3. Microstructure, long-range internal stresses and deformation mechanisms at large strains: extension of composite model to include misorientations

In materials characterized by wavy glide, relatively thick dislocation cell walls develop in Stage III. These walls contain dislocation dipoles and multipoles and enclose cell interior regions with considerably smaller dislocation densities, as has been described in detail earlier. Two important features should be stressed here: (i) at the beginning of Stage III, the misorientation between adjacent cells is relatively small and (ii) there is a tendency to cell block (CB) formation [115,182]. The term cell block means that a relatively large number of neighbouring dislocation cells are bounded by cell wall regions separating larger regions with higher misorientations. The typical size of CBs in Stage III is about ten times the dislocation cell diameter [66,182]. In Stage IV, the cell walls become thin, containing dipolar/multipolar dislocations with extremely high densities similar to subgrain boundaries, usually separating increasingly misoriented dislocation-free regions. The cells often form elongated, lamellar structures [183,184], for which the term "lamellar boundaries" (LB), introduced by Hughes and Hansen [183], is used. The LBs frequently lie parallel to the rolling direction or to the shear stress in cold rolled or torsionally deformed specimens, respectively [183,184].

The standstill in the evolution of $|\Delta\sigma_W - \Delta\sigma_C|$ as a function of γ (see figs 26 and 29) can be explained as follows. During Stage III the number of slip systems tends to increase and reaches high values due to latent hardening and the fast increase of dislocation density [28]. The large flux of dislocations towards the thick cell walls in the multiple slip mode produces the polarized dislocation structure shown schematically in figs 11(a) and 11(b). In the following, this type of polarized dislocation structure will be denoted as "polarized dipolar walls": PDW [178].

In Stage IV, the global dislocation density increases steadily in accordance with the extremely large densities within the thin cell walls. Adjacent cells are often alternately tilted, indicating that they have a net polarization similar to that in small-angle grain boundaries. Assuming multiple slip, a simple mechanism has been suggested which

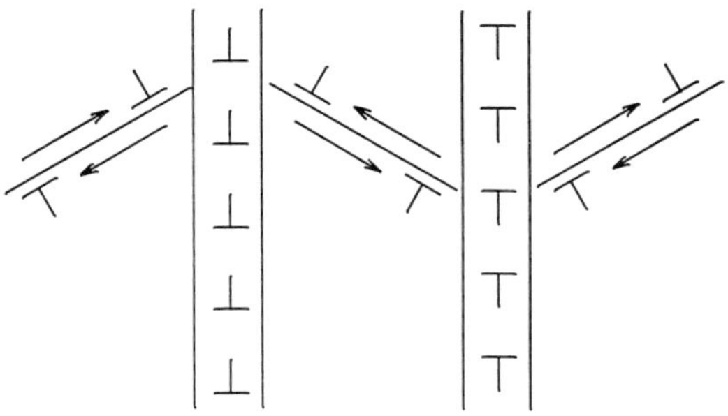

Fig. 30. Schematic drawing of the dislocations deposited inside thin cell walls in Stage IV. The small-angle grain-boundary type of dislocations inside the cell walls are resultant dislocations produced by incoming dislocations along different slip systems on the two opposite sides of a cell wall. This type of polarized dislocation microstructure has been denoted as "polarized dipolar walls": PDWs. From [178].

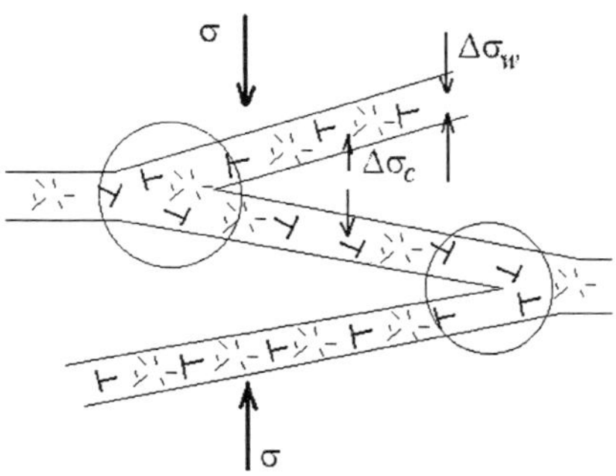

Fig. 31. Schematic picture of the PTW structure. The cell walls (which are lamellar boundaries (LBs) or tilt boundaries) are running almost perpendicular to the applied stress σ. The cell walls form junctions in a random manner, these are indicated by circles, which hold the structure together in the direction of the applied stress.

produces a polarized dislocation arrangement [178]. When dislocations arrive on two opposite sides of a thin cell wall along two different slip systems, they can combine into resultant dislocations which have the small-angle grain boundary configuration, as shown schematically in fig. 30. The inclined lines in the figure indicate traces of different slip planes on the different sides of cell walls. This kind of cell wall has been denoted as "polarized tilt walls": PTW. It is important to realize that the PTWs are *not* simple small-angle grain boundaries. Besides the resultant polarized dislocations, which cause the strong tilts between adjacent cells, (i) they contain a large proportion of redundant dislocations and (ii) they are far from perfect small-angle grain boundaries. In fact they are often curved and almost never run from grain boundary to grain boundary, but merge and bifurcate in a random manner [182,183]. These features are crucial to understanding how this dislocation structure can produce long-range internal stresses. A schematic picture of the PTW structure is shown in fig. 31. The junctions of the lamellar boundaries (LBs) are marked with circles. The applied stress and the residual long-range internal stresses in the cell wall and cell interior regions are denoted by σ, $\Delta\sigma_{\mathrm{w}}$ and $\Delta\sigma_{\mathrm{c}}$, respectively. The

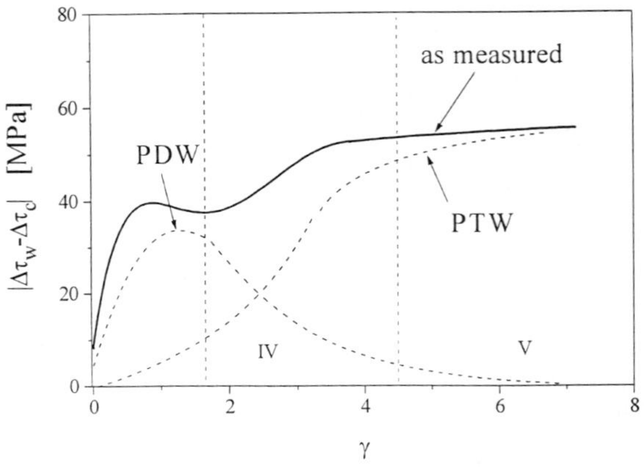

Fig. 32. The magnitude of the residual internal stresses as a function of deformation in cold-rolled polycrystalline copper. The solid line is the same as the one in fig. 26. The dashed lines represent the partial contributions of the PDW and PTW structures to the total values of the long-range internal stresses in accordance with eq. (55). The vertical dashed lines indicate the boundaries of Stages IV and V, respectively. From [178].

junction regions ensure that the soft and hard regions in the whole structure are connected in the direction of the applied stress enabling the evolution of the compressive and tensile internal stresses in the cell wall and cell interior regions after unloading.

The introduction of the terms PDW and PTW was motivated by the necessity to make a clear distinction between the Burgers vector distributions in the cell walls typical in Stage III and Stage IV [178]. In both cases the distribution of the Burgers vectors consists of a random and an oriented part. In the PDWs the Burgers vectors in the oriented part are lying within the cell walls and form the dipolar structure within the cell walls themselves, as shown in fig. 11. In the PTWs the Burgers vectors in the oriented part are perpendicular to the cell walls and the dipolar structure is formed in pairs of the adjacent cell walls as shown in fig. 30. The term PTW refers to the dislocation structure typical of subgrain boundaries.

At the beginning of deformation, the rapid increase of $|\Delta\sigma_W - \Delta\sigma_c|$ and ρ^* with increasing γ is in accordance with (i) the high rate of work hardening in Stage III, θ_{III}, and (ii) the stress-intense dislocation arrangement of the PDW structure. With proceeding deformation: (a) the accumulation of dislocations slows down due to increasing annihilation and (b) the PDW structure starts to become unstable, since the long-range internal stresses grow to extremely high values. Parallel to this, the following changes occur in the microstructure: (1) the CBs shrink and produce growing misorientations between more and more adjacent dislocation cells, (2) the initially thick cell walls become thinner and thinner, transforming into LBs and finally (3) the at first overwhelmingly PDW structure is replaced by the PTW structure. Well into Stage IV and subsequently $|\Delta\sigma_W - \Delta\sigma_c|$ and ρ^* increase at a low rate in accordance with the low work-hardening rate. The strong increase of ρ^* in Stage IV (as it can be seen in fig. 2 in [174]) is in good agreement with the PTW structure of the lamellar boundaries (LBs). The Burgers vectors of the dislocations producing the tilt of adjacent subgrains are aligned parallel to the diffraction vector of the (002) reflection and thus give rise to a large value of the formal dislocation density [178,179]. A semi-quantitative model of the evolution of the long-range internal stresses with deformation can be formulated as follows. Assume that the volume fractions of PDW and PTW dislocation structure are α and $(1-\alpha)$, respectively. Then $|\Delta\sigma_{\mathrm{w}}-\Delta\sigma_{\mathrm{c}}|$ can be represented by:

$$|\Delta\sigma_{\mathrm{w}}-\Delta\sigma_{\mathrm{c}}| = \alpha|\Delta\sigma_{\mathrm{w}}-\Delta\sigma_{\mathrm{c}}|_{\mathrm{PDW}} + (1-\alpha)|\Delta\sigma_{\mathrm{w}}-\Delta\sigma_{\mathrm{c}}|_{\mathrm{PTW}}, \tag{54}$$

where $|\Delta\sigma_{\mathrm{w}}-\Delta\sigma_{\mathrm{c}}|_{\mathrm{PDW}}$ and $|\Delta\sigma_{\mathrm{w}}-\Delta\sigma_{\mathrm{c}}|_{\mathrm{PTW}}$ are the magnitudes of the residual long-range internal stresses in the PDW and PTW structures, respectively. Assuming that α is a simple sigmoidal function of γ, the variation of $|\Delta\sigma_{\mathrm{w}}-\Delta\sigma_{\mathrm{c}}|$ as a function of γ can be given as in fig. 32, where the dotted lines represent the partial contributions of the PDW and PTW structures to the long-range internal stresses. This approach thus permits a semi-quantitative decription of the experimentally observed evolution of the long-range internal stresses (fig. 26) up to Stage V in terms of an extended two-component composite model.

7. *Summary and closing remarks*

In this article, the experimental observations that proved directly the existence of deformation-induced long-range internal stresses in deformed metals and that lead to the formulation of the so-called composite model of crystal plasticity are reviewed, and the consequences are discussed. The composite model takes into account the fact that dislocations are usually distributed heterogeneously, e.g., in a cell structure. The model therefore considers the *local* spatial variation of the dislocation density and of the flow stress. In its simplest form, the model treats a deformed crystal as a two-component material consisting of hard cell walls of high local dislocation density, separated by soft cell interiors of low local dislocation density. The most important consequence of the composite model is that the development of deformation-induced long-range internal stresses (whose magnitude and wavelength correspond closely to observation) follows in a straightforward manner as a result of the compatibility requirements, when the soft and the hard regions are deformed in parallel. Thus, the model predicts the evolution of an internal back stress in the soft regions and an internal forward stress in the hard regions without requiring the existence of classical dislocation pile-ups as sources of the internal stresses. Rather, the internal stresses arise from the so-called interfacial dislocations that accommodate the elastic–plastic strain mismatch and from the geometrical constraints of dislocation glide. The overall macroscopic flow stress follows as a weighted spatial average of the *local* flow stresses in the soft and hard regions (rule of mixtures). Because of the "built-in" back stress, the composite model can describe *kinematic* hardening and cyclic yielding with a Bauschinger effect after stress reversal. The model provides a simple natural picture of both unidirectional and cyclic micro- and macroyielding.

The composite model is first discussed in detail for single slip and then for multiple slip. The flow-stress law is compared critically with different flow-stress models that, in general, consider only spatially averaged mean dislocation densities and, in most cases, only one single governing dislocation mechanism. It is concluded that such flow stress laws are inadequate for a sound description. In spite of the extra elastic stored energy resulting from the internal stresses, the overall elastic stored energy is nonetheless, contrary to a widespread belief, enhanced only very moderately. In an extended formulation of the composite model, the strain gradients between the hard and the soft regions and the role of the interfacial geometrically necessary dislocations are taken into account more explicitly. In particular, it is shown that it is generally not justified to describe the macroscopic flow stress by a Taylor-type law in which the mean densities of the geometrically necessary and the statistically stored dislocations are simply superimposed.

After a critical discussion of the limitations of the composite model in its original form, some recent refinements and extensions are reviewed, including models which address also kinetic aspects and multi-component models in which a spectrum of local flow stresses is considered. It is then shown that extended variants of the composite model have been applied successfully to different types of loading such as cyclic deformation, strain-path changes and high-temperature creep and that the composite principle allows also the satisfactory description of deformation transients. In the final part of this review of deformation-induced long-range internal stresses in terms of the composite model, an account of recent studies is given, showing that long-range internal stresses persist up to very large strains in cold-rolled and torsionally deformed metals. The experimental observations can be explained satisfactorily by extending the composite model to take into account that, at larger strains, the nature of the dislocation cell walls changes gradually, as the latter become thinner and as larger and larger misorientations develop in the adjacent regions.

In conclusion, it can be stated that the composite model has been successful in reconciling earlier models of crystal plasticity and in going beyond these models in terms of a better justification of the underlying principles, of a more convincing interpretation of experimental observations and in providing a basis for a simpler and, at the same time, significantly broader approach to problems of crystal plasticity.

Acknowledgements

The authors acknowledge gratefully the inspiring cooperation with Dr. Manfred Wilkens and Dr. Uwe Essmann in the seventies and eighties, and the fruitful interaction over the years with Professor Alfred Seeger at the Max-Planck-Institut für Metallforschung in Stuttgart and the contributions by their co-authors in former publications. Sincere thanks are extended to Professor Michael Zehetbauer for his substantial contributions to section 6 as a co-author in all related publications. The encouragement of Professor Frank Nabarro, University of the Witwatersrand, Johannesburg, to write this contribution and his many helpful remarks on the manuscript are deeply appreciated. The authors thank sincerely Professor Wolfgang Blum for critical comments on the manuscript and Mrs. Waltraud Kränzlein and, in particular, Dr. Heinz Werner Höppel for their able support during the preparation of the manuscript.

Note-added-in-proof

Originally, the composite model was developed for and applied to deformed single-phase materials, based on the hypothesis that deformed single-phase materials behave mechanically like two-phase materials, because they consist of hard dislocation-rich and soft dislocation-poor regions. Only in later years were deformed real two-phase alloys, in particular monocrystalline nickel-base superalloys, studied extensively in one of the authors' (H.M.) research group. The results obtained conformed in detail to the expectations according to the composite model with respect to the deformation-induced internal stresses, as witnessed by asymmetric X-ray line broadening, e.g. [185,186], and convergent beam electron diffraction, e.g. [187,188]. While these results on real two-phase materials were not too surprising, the strong similarity between these findings and the results obtained earlier on deformed single-phase materials provides, in retrospect, comforting confidence in the correctness of the composite concept developed originally for deformed single-phase materials.

References

[1] N.F. Mott, Phil. Mag. 43 (1952) 1151.
[2] J. Friedel, Phil. Mag. 46 (1955) 1169.
[3] A. Seeger, in: Dislocations and Mechanical Properties of Crystals, eds J.C. Fisher, W.G. Johnston, R. Thomson and T. Vreeland, Jr. (Wiley, 1957) p. 243.
[4] A. Seeger, J. Diehl, S. Mader and H. Rebstock, Phil. Mag. 2 (1957) 323.
[5] A. Seeger, S. Mader and H. Kronmüller, in: Electron Microscopy and Strength of Crystals, eds G. Thomas and J. Washburn (Interscience Publishers, 1963) p. 665.
[6] P.B. Hirsch and T.E. Mitchell, Can. J. Phys. 45 (1967) 663.
[7] H. Kronmüller, Can. J. Phys. 45 (1967) 631.
[8] S. Mader, Z. Phys. 149 (1957) 73.
[9] H. Kronmüller and A. Seeger, J. Phys. Chem. Solids 18 (1961) 93.
[10] U. Essmann, Phys. Stat. Sol. 3 (1963) 932.
[11] U. Essmann, Phys. Stat. Sol. 12 (1965) 723.
[12] Z.S. Basinski, Discuss. Faraday Soc. 38 (1964) 93.
[13] J.W. Steeds, Proc. Roy. Soc. A 292 (1966) 343.
[14] D. Kuhlmann-Wilsdorf, Trans. Metall. Soc. AIME 224 (1962) 1047.
[15] F.R.N. Nabarro, Z.S. Basinski and D.B. Holt, Adv. Phys. 13 (1964) 193.
[16] Z.S. Basinski, Phil. Mag. 4 (1959) 393.
[17] A. Seeger and M. Wilkens, in: Realstruktur und Eigenschaften von Reinststoffen, Vol. III, ed. E. Rexer (Akademie-Verlag, Berlin, 1967) p. 29.
[18] A. Seeger, in: Work Hardening, eds J.P. Hirth and J. Weertman (Gordon and Breach, New York, 1968) p. 27.
[19] H. Mughrabi, Phil. Mag. 18 (1968) 1211.
[20] H. Mughrabi, Phil. Mag. 23 (1971) 897.
[21] H. Mughrabi, in: Constitutive Equations in Plasticity, ed. A.S. Argon (MIT Press, Cambridge, MA, 1975) p. 199.
[22] G. Saada, Acta Metall. 9 (1961) 166.
[23] U.F. Kocks, Phil. Mag. 13 (1966) 541.
[24] D. Kuhlmann-Wilsdorf, in: Work Hardening, eds J.P. Hirth and J. Weertman (Gordon and Breach, New York, 1968) p. 97.
[25] P.B. Hirsch, in: Defects, Vol. 2, ed. P.B. Hirsch (Cambridge University Press, 1975) p. 189.
[26] S.J. Basinski and Z.S. Basinski, in: Dislocations in Solids, Vol. 4, ed. F.R.N. Nabarro (North-Holland Publishing, 1979) p. 261.
[27] Z.S. Basinski, Scripta Metall. 8 (1974) 1301.
[28] U.F. Kocks, in: Dislocations and Properties of Real Materials (The Institute of Metals, London, 1985) p. 125.

[29] H. Mughrabi, in: Proc. ICSMA 3 (Cambridge, 1973) p. 407.
[30] J.C. Grosskreutz and H. Mughrabi, in: Constitutive Equantions in Plasticity, ed. A.S. Argon (MIT Press, Cambridge, MA, 1975) p. 251.
[31] H. Mughrabi, in: Proc. ICSMA 5, Vol. 3, eds P. Haasen, V. Gerold and G. Kostorz (Pergamon Press, 1980) p. 1615.
[32] H. Mughrabi, in: Continuum Models of Discrete Systems 4 (CMDS 4), eds O. Brulin and R.K.T. Hsieh (North Holland, 1981) p. 241.
[33] H. Mughrabi, Acta Metall. 31 (1983) 1367.
[34] O.B. Pedersen, L.M. Brown and W.M. Stobbs, Acta Metall. 29 (1981) 1843.
[35] O.B. Pedersen, Acta Metall. 35 (1987) 2567.
[36] G.I. Taylor, Proc. Roy. Soc. A 145 (1934) 362.
[37] H.P. Stüwe, Z. Metallkd. 56 (1965) 633.
[38] I. Kovács, Acta Metall. 15 (1967) 1731.
[39] J. Gil Sevillano, P. van Houtte and E. Aernoudt, Proc. Mater. Sci. 25 (1981) 69.
[40] E. Aernoudt, J. Sevillano and P. van Houtte, in: Constitutive Relations and their Physical Basis, Proc. 8th Risø International Symposium on Metallurgy and Materials Science, eds S.I. Andersen, J.B. Bilde-Sørensen, N. Hansen, T. Leffers, H. Lilholdt, O.B. Pedersen and B. Ralph (Risø National Laboratory, 1987) p. 1.
[41] H. Mecking and Y. Estrin, in: Constitutive Relations and their Physical Basis, Proc. 8th Risø International Symposium on Metallurgy and Materials Science, eds S.I. Andersen, J.B. Bilde-Sørensen, N. Hansen, T. Leffers, H. Lilholdt, O.B. Pedersen and B. Ralph (Risø National Laboratory, 1987) p. 123.
[42] J.M.G. Alberdi, Ph. D. Thesis (University of Navarra, Spain, 1984).
[43] A.D. Rollett, U.F. Kocks, J.D. Embury, M.G. Stout and R.D. Doherty, in: Proc. 8th Int. Conf. on Strength of Metals and Alloys (ICSMA 8), Vol. 1, eds P.O. Kettunen, T.K. Lepistö and M.E. Lehtonen (Pergamon Press, Oxford, 1988) p. 433.
[44] U.F. Kocks, J. Eng. Mater. Technol. 98 (1976) 76.
[45] H. Mecking, in: Work Hardening in Tension and Fatigue, ed. A.W. Thompson (The Metallurgical Society of AIME, 1977) p. 67.
[46] F. Prinz and A.S. Argon, Acta Metall. 32 (1984) 1021.
[47] P. Haasen, J. Physique 50 (1989) 2445.
[48] A.S. Argon and P. Haasen, Acta Metall. Mater. 41 (1993) 3289.
[49] M. Zehetbauer, Acta Metall. Mater. 41 (1993) 589.
[50] Y. Estrin, A. Molinari, L.S. Tóth and Y. Bréchet, Acta Mater. 46 (1998) 5509.
[51] A. Seeger, Phil. Mag. 45 (1954) 771.
[52] J.C.M. Li, Canad. J. Phys. 45 (1967) 493.
[53] W. Spitzig and A.S. Keh, Metall. Trans. 1 (1970) 3325.
[54] N. Thompson, N.J. Wadsworth and N. Louat, Phil. Mag. 1 (1956) 113.
[55] H. Mughrabi, F. Ackermann and K. Herz, in: Fatigue Mechanisms, ASTM STP 675, ed. J.T. Fong (American Society for Testing and Materials, 1979) p. 69.
[56] H. Mughrabi, in: Proc. Int. Conf. DISLOCATIONS 2000, Mater. Sci. Eng. A 309–310 (2001) 237.
[57] J.D. Eshelby, Proc. Roy. Soc. London A 241 (1957) 376.
[58] H. Mughrabi, Rev. Phys. Appl. 23 (1988) 367.
[59] J. Bauschinger, Mitteilungen aus dem Mechanisch-technischen Laboratorium der königlichen Technischen Hochschule in München 13 (1886) 1.
[60] U.F. Kocks, T. Hasegawa and R.O. Scattergood, Scripta Metall. 14 (1980) 449.
[61] T. Hasegawa and U.F. Kocks, Acta Metall. 27 (1979) 1705.
[62] H. Mughrabi, Phys. Stat. Sol. A 104 (1987) 107.
[63] M. Wilkens, K. Herz and H. Mughrabi, Z. Metallkd. 71 (1980) 376.
[64] H. Mughrabi and T. Ungár, unpublished (1981).
[65] T. Ungár, H. Mughrabi, D. Rönnpagel and M. Wilkens, Acta Metall. 32 (1984) 333.
[66] H. Mughrabi, T. Ungár, W. Kienle and M. Wilkens, Phil. Mag. 53 (1986) 793.
[67] T. Ungár, H. Mughrabi, M. Wilkens and A. Hilscher, Phil. Mag. 64 (1991) 495.
[68] B. Devincre and L.P. Kubin, Mater. Sci. Eng. A 234–236 (1997) 8.

[69] W. Püschl, R. Frydman and G. Schoeck, Phys. Stat. Sol. A 74 (1982) 211.
[70] E. Göttler, Phil. Mag. 28 (1973) 1057.
[71] R. Neuhaus and C. Schwink, Phil. Mag. A 65 (1992) 1463.
[72] M. Zaiser and P. Haehner, Mater. Sci. Eng. A 270 (1999) 299.
[73] I. Gromá, T. Ungár and M. Wilkens, J. Appl. Cryst. 21 (1988) 47.
[74] T. Ungár, I. Gromá and M. Wilkens, J. Appl. Cryst. 22 (1989) 26.
[75] D. Raabe, Phil. Mag. A 73 (1996) 1363.
[76] H. Mughrabi, Mater. Sci. Eng. 85 (1987) 15.
[77] U.F. Kocks, in: Physics of Strength and Plasticity, ed. A.S. Argon (MIT Press, Cambridge, MA, 1975) p. 143.
[78] F.R.N. Nabarro, The Theory of Crystal Dislocations (Oxford University Press, 1967).
[79] D. Rönnpagel and C. Schwink, Acta Metall. 26 (1978) 319.
[80] D. Kuhlmann-Wilsdorf, Mater. Sci. Eng. 86 (1987) 53.
[81] H. Mughrabi, South African J. Phys. 9 (1986) 62.
[82] H. Mecking and U.F. Kocks, Acta Metall. 29 (1981) 1865.
[83] F.R.N. Nabarro, in: Proc. ICSMA 7, Vol. 3, eds H.J. McQueen, J.P. Bailon, J.I. Dickson, J.J. Jonas and M.G. Akben (Pergamon Press, 1985) p. 1667.
[84] L.M. Brown, Metall. Trans A 22A (1991) 1693.
[85] L.M. Brown, in: Understanding Materials, ed. C.J. Humphreys (Institute of Materials, London, 2002) p. 135.
[86] J. Gil Sevillano in: Plastic Deformation and Fracture of Materials, Vol. 6 of R.W. Cahn, P. Haasen and E.J. Kramer (eds), Materials Science and Technology, VCH, 1993, ed. H. Mughrabi, p. 19.
[87] D. Kuhlmann-Wilsdorf, Mat. Res. Innovat. 1 (1998) 265.
[88] P.B. Hirsch, in: Internal Stresses and Fatigue in Metals, eds G.M. Rassweiler and W.L. Grube (Elsevier, 1959) p. 139.
[89] Y. Estrin, in: Unified Constitutive Laws of Plastic Deformation, eds A.S. Krausz and K. Krausz (Academic Press, 1996) p. 69.
[90] C. Holste and H.J. Burmeister, Phys. Stat. Sol. A 57 (1980) 269.
[91] J. Polák and M. Klesnil, Fat. Eng. Mat. Struct. 5 (1982) 19.
[92] B.D. Cullity, Trans. Metall. Soc. AIME 227 (1963) 356.
[93] B.D. Cullity and O.P. Puri, Trans. Metall. Soc. AIME 222 (1963) 359.
[94] T. Takeuchi, J. Phys. Soc. Japan 28 (1970) 955.
[95] G. Masing, Wissenschaftl. Veröffentl. aus dem Siemens-Konzern 3 (1923) 231.
[96] G. Masing, Z. Techn. Physik 6 (1925) 73.
[97] J. Polák, Czech J. Phys. 19 (1969) 315.
[98] H. Mughrabi, R. Kütterer, K. Lubitz and H. Kronmüller, Phys. Stat. Sol. A 38 (1976) 261.
[99] H. Mughrabi, K. Herz and X. Stark, Int. J. Fract. 17 (1981) 193.
[100] L.P. Kubin, in: Materials Science and Technology, Plastic Deformation and Fracture of Materials, ed. H. Mughrabi, Vol. 6, eds R.W. Cahn, P. Haasen and E.J. Kramer (VCH, 1993) p. 137.
[101] E.C. Aifantis, Int. J. Eng. Sci. 33 (1995) 2161.
[102] M.F. Ashby, Phil. Mag. 21 (1970) 399.
[103] N.A. Fleck, G.M. Muller, M.F. Ashby and J.W. Hutchinson, Acta Metall. 42 (1994) 475.
[104] J. Ning and E.C. Aifantis, in: Unified Constitutive Laws of Plastic Deformation, eds A.S. Krausz and K. Krausz (Academic Press, 1996) p. 319.
[105] W.D. Nix and H. Gao, J. Mech. Phys. Sol. 46 (1998) 441.
[106] H.M. Zbib, in: Material Instabilities in Solids, eds R. de Borst and E. van der Giessen (Wiley, 1998) p. 473.
[107] H. Mughrabi, in: Deformation-Induced Microstructures: Analysis and Relation to Properties, Proc. 20th Risø Int. Symp. on Materials Science, eds J.B. Bilde-Sørensen et al. (Risø Nat. Lab., Roskilde, Denmark, 1999) p. 417.
[108] H. Mughrabi, Mater Sci. Eng. A, Special Volume, Proc. TMS Symposium in Honor of Professor Ali Argon (San Diego, CA, March 1999), Mater. Sci. Eng. A 317 (2001) 171.
[109] H. Mughrabi, in: Proc. ICSMA 12, Asilomar, CA, Aug. 2000, Mater. Sci. Eng. A 319–321 (2001) 139.

[110] O.B. Pedersen and L.M. Brown, in: Deformation of Multi-Phase and Particle Containing Materials, Proc. 4th Risø Int. Symp. on Metallurgy and Materials Science, eds J.B. Bilde-Sørenson et al. (Risø Nat. Lab., Roskilde, Denmark, 1983) p. 83.
[111] J.R. Weertmann and J. Weertman, in: Deformation-Induced Microstructures: Analysis and Relation to Properties, Proc. 20th Risø Int. Symp. on Materials Science, eds J.B. Bilde-Sørensen et al. (Risø Nat. Lab., Roskilde, Denmark, 1999) p. 201.
[112] M. Zaiser, personal communication (1999).
[113] M. Zaiser, Mater. Sci. Eng. A 249 (1998) 145.
[114] A.D. Rollett, U.F. Kocks and R.D. Doherty, in: Formability and Metallurgical Structure, eds A.K. Sachdev and J.D. Embury (TMS, Warrendale, PA, USA, 1987) p. 211.
[115] M. Wilkens, T. Ungár and H. Mughrabi, Phys. Stat. Sol. A 104 (1987) 157.
[116] P.J. Jackson, Acta Metall. 33 (1985) 449.
[117] P.J. Jackson, Progr. Mater. Sci. 29 (1985) 139.
[118] H. Mughrabi, Phil. Mag. 23 (1971) 931.
[119] J.T. Fourie, Phil. Mag. 17 (1968) 148.
[120] N. Himstedt, Acta Metall. 26 (1978) 351.
[121] R. Thomson and L.E. Levine, Phys. Rev. Lett. 81 (1998) 3884.
[122] M.E. Kassner, M.A. Wall and M.A. Delos-Reyes, Metall. and Mater. Trans. 28A (1997) 595.
[123] M.E. Kassner and M.A. Wall, Metall. and Mater. Trans. 30A (1999) 777.
[124] M.E. Kassner, M.-T. Pérez-Prado, K.S. Vecchio and M.A. Wall, Acta Mater. 48 (2000) 4247.
[125] M.E. Kassner, M.-T. Pérez-Prado and K.S. Vecchio, in: Proc. 12th Int. Conf. on the Strength of Materials (ICMA 12), Mater. Sci. Eng. A 319–321 (2001) 730.
[126] A. Borbély, W. Blum and T. Ungár, Mater. Sci. Eng. A 276 (2000) 186.
[127] W.D. Nix and B. Ilschner, in: Proc. 5th Int. Conf. on Strength of Metals and Alloys (ICSMA 5), Vol. 3, eds P. Haasen, V. Gerold and G. Kostorz (Pergamon Press, 1980) p. 1503.
[128] W.D. Nix, J.C. Gibeling and D.A. Hughes, Metall. Trans. 16A (1985) 2215.
[129] F.R.N. Nabarro, Acta Metall. 37 (1989) 1521.
[130] H.J. Burmeister and C. Holste, Phys. Stat. Sol. A 64 (1981) 611.
[131] F. Roters, D. Raabe and G. Gottstein, Acta. Mater. 48 (2000) 4181.
[132] E.W. Hart, Acta Metall. 18 (1970) 599.
[133] E.W. Hart, C.Y. Li, H. Yamada and G.L. Wire, in: Constitutive Equations in Plasticity, ed. A.S. Argon (MIT Press, Cambridge, MA, 1975) p. 149.
[134] M. Zaiser, K. Bay and P. Hähner, Acta Mater. 47 (1999) 2463.
[135] M. Zaiser and P. Hähner, Mater. Sci. Eng. A 270 (1999) 299.
[136] P. Hähner and M. Zaiser, Mater. Sci. Eng. A 272 (1999) 443.
[137] H.J. Christ and H. Mughrabi, Fatigue Fract. Engng. Mater. Struct. 19 (1996) 335.
[138] L.M. Brown and W.M. Stobbs, Phil. Mag. 23 (1971) 1185.
[139] R. Sedláček, Key Engineering Materials 97–98 (1994) 497.
[140] R. Sedláček, Mater. Sci. Eng. A 234–236 (1997) 643.
[141] J. Kratochvíl and M. Saxlová, Scripta Metall. Mater. 26 (1992) 113.
[142] N. Christodoulou, O.T. Woo and S.R. MacEwen, Acta Metall. 34 (1986) 1553.
[143] J. Chicois, R. Fougères, G. Guichon, A. Hamel and A. Vincent, Acta Metall. 34 (1986) 2157.
[144] H. Biermann, T. Ungár, T. Pfannenmüller, G. Hoffmann, A. Borbély and H. Mughrabi, Acta Metall. Mater. 41 (1993) 2743.
[145] J. Polák, M. Klesnil and J. Helešic, Fat. Eng. Mat. Struct. 5 (1982) 33.
[146] H.J. Christ, Wechselverformung von Metallen (Springer, Berlin, 1991) p. 83.
[147] H.D. Schumann, Phys. Stat. Sol. A 18 (1973) K 27.
[148] K.-H. Weile and G.K. Schmidt, Phys. Stat. Sol. A 31 (1975) K155.
[149] X. Feaugas, Acta Mater. 47 (1999) 3617.
[150] H.J. Burmeister and C. Holste, Phys. Stat. Sol. A 64 (1981) 611.
[151] J. Polák, M. Klesnil and J. Helešic, Fat. Eng. Mat. Struct. 5 (1982) 45.
[152] H.J. Christ, in: Proc. FATIGUE '93, Vol. 1, eds J.-P. Bailon and J.J. Dickson (EMAS Ltd., UK, 1993) p. 115.

[153] P.J. Jackson and Z.S. Basinski, Canad. J. Phys. 45 (1967) 707.
[154] P.S. Bate and D.V. Wilson, Acta Metall. 34 (1986) 1097.
[155] D.V. Wilson and P.S. Bate, Acta Metall. 34 (1986) 1107.
[156] P. Bate, W.T. Robeerts and D.V. Wilson, Acta Metall. 30 (1982) 725.
[157] M. Zandrahimi, S. Platias, D. Price, D. Barrett, P.S. Bate, W.T. Roberts and D.V. Wilson, Metall. Trans. 20A (1989) 2471.
[158] P.S. Bate, Metall. Trans. 24A (1993) 2679.
[159] D.V. Wilson and P.S. Bate, Acta Metall. Mater. 42 (1994) 1099.
[160] G.B. Gibbs, Phil. Mag. A 13 (1966) 317.
[161] C.N. Ahlquist and W.D. Nix, Acta Metall. 19 (1971) 373.
[162] W. Blum and A. Finkel, Acta Metall. 30 (1982) 1705.
[163] S. Straub, W. Blum, H.J. Maier, T. Ungár, A. Borbély and H. Renner, Acta Mater. 44 (1996) 4337.
[164] W. Blum, A. Rosen, A. Cegielska and J.L. Martin, Acta Metall. 37 (1989) 2439.
[165] S. Vogler and W. Blum, in: Creep and Fracture of Engineering Materials and Structures, eds B. Wilshire and R.W. Evans (The Institute of Metals, London, 1990) p. 65.
[166] M. Meier, Q. Zhu and W. Blum, Z. Metallkde. 84 (1993) 263.
[167] R. Sedláček, S. Straub, A. Borbély, T. Ungár and W. Blum, Key Engineering Materials 97–98 (1994) 461.
[168] A. Orlová, Mater. Sci. Eng. A 297 (2001) 281.
[169] A. Seeger, in: Handbuch der Physik, Vol. VII/2, ed. S. Fluegge (Springer, 1958) p. 1.
[170] U.F. Kocks, M.G. Stout and A.D. Rollett, in: Proc. 8th Int. Conf. on Strength of Metals and Alloys, ICSMA 8, eds P.O. Kettunen, T.K. Lepistö and M.E. Lehtonen (Pergamon, 1988) p. 25.
[171] J. Gil Sevillano and E. Aernoudt, Mater. Sci. Eng. 86 (1987) 35.
[172] D.A. Hughes, J.C. Gibeling and W. Nix, in: Proc. 7th Int. Conf. on Strength of Metals and Alloys, ICSMA 7, Montreal, Canada, eds H.J. McQueen, J.-P. Bailon, J.I. Dickson, J.J. Jonas and M.G. Akben (Pergamon, Oxford, 1985) p. 51.
[173] M. Zehetbauer and V. Seumer, Acta Metall. Mater. 41 (1993) 577.
[174] M. Müller, M. Zehetbauer, A. Borbély and T. Ungár, Scripta Metall. Mater. 35 (1996) 1461.
[175] A.D. Rollett, PhD Thesis (Drexel University, Philadelphia, USA, 1988).
[176] D. Kuhlmann-Wilsdorf, Mater. Sci. Eng. A 113 (1989) 1.
[177] D. Kuhlmann-Wilsdorf and N. Hansen, Metall. Trans. 20A (1989) 2393.
[178] T. Ungár and M. Zehetbauer, Scripta Mater. 35 (1996) 1467.
[179] M. Zehetbauer, T. Ungár, R. Kral, A. Borbély, E. Schafler, B. Ortner, H. Amenitsch and S. Bernsdorff, Acta Mater. 47 (1999) 1053.
[180] E. Schafler, M. Zehetbauer, A. Borbély and T. Ungár, Mat. Sci. Eng. A 234–236 (1997) 445.
[181] M. Wilkens, Phys. Stat. Sol. A 2 (1970) 359.
[182] D.A. Hughes, in: Proc. 16th Risø Int. Symp. Mater. Sci.: Microstructural Crystallographic Aspects of Recrystallisation, eds N. Hansen, D. Juul Jensen, Y.L. Liu and B. Ralph (Risø Nat. Lab., Roskilde, Denmark, 1995) p. 63.
[183] D.A. Hughes and N. Hansen, Metall. Trans. 24A (1993) 2021.
[184] T. Ungár, L.S. Tóth, J. Illy and I. Kovács, Acta Metall. 34 (1986) 1257.
[185] H.-A. Kuhn, H. Biermann, T. Ungar and H. Mughrabi, Acta Metall. Mater. 39 (1991) 2783.
[186] H. Mughrabi, H. Biermann and T. Ungár, in: Superalloys, eds S.D. Antolovich, R.W. Stusrud, R.A. MacKay, D.L. Anton, T. Khan, R.D. Kissinger and D.L. Klarstrom (The Minerals, Metals & Materials Society, 1992) p. 599.
[187] R.R. Keller, H.J. Maier, H. Renner and H. Mughrabi, Scripta Metall. Mater. 27 (1992) 1167 and Erratum in Scripta Metall. Mater. 28 (1993) 661.
[188] Z. Lu, F. Pyczak, H. Biermann and H. Mughrabi, Phil. Mag. A 82 (2002) 1219.

CHAPTER 61

Work Hardening of Face Centred Cubic Crystals. Dislocations Intersection and Cross Slip

GEORGES SAADA and PATRICK VEYSSIÈRE

Laboratoire d'étude des microstructures
CNRS-ONERA BP 72
F92322 Châtillon cedex
France

Dislocations in Solids
Edited by F. R. N. Nabarro and M. S. Duesbery

Contents

1. *Introduction. Scope of the paper*

1.1. General introduction

Plastic deformation generates irreversible microstructural changes which, in most cases, increase the resistance of crystalline solids to subsequent deformation, a phenomenon known as *work hardening*. In the simple case of uniaxial testing at a fixed temperature, the effect is reflected either by a positive slope of the stress–strain curve (strain-hardening rate) or by an increase in yield stress upon reloading (strengthening). The two main goals of a theory of work hardening are to predict the evolution of the microstructure and to relate this to the mechanical behaviour of a given class of crystals.

From now on, we restrict the following review to pure crystals deformed under uniaxial conditions at temperatures low enough for long-distance diffusion phenomena to be ignored. In these conditions, grain boundaries are immobile and dislocation motion proceeds by glide. The change in the microstructure consists mostly in modifications via conservative processes of the fine structure, content and organisation of dislocations. Mutual immobilisation as well as self-locking result in *dislocation storage* that generally increases the crystal resistance to subsequent deformation. Annihilation between dislocations of opposite sign results in *dislocation recovery* that softens the crystal. In the same vein, dislocation elimination at free surfaces may greatly contribute to recovery, especially in single crystals. Cross slip is a conservative process known to allow for dislocation annihilation but which, in some systems such as ordered alloys, is responsible for dislocation locking. Non-conservative processes such as the spontaneous annihilation of edge dipoles by fast climb are commonly invoked to contribute to recovery. We believe that this process needs direct confirmation before it can be fully accepted. Although generated during deformation, sometimes in great quantities, point defects do not clearly influence mechanical behaviour.

1.2. The microstructure. Multiscale patterning

During plastic flow, the microstructure is progressively transformed under the combined effect of the applied and internal stresses. Transmission electron microscopy (TEM) observations reveal that the distribution of dislocations may assume more or less well defined patterns. To quote Basinski [1] "*A whole glossary of qualitative terms such as carpets, walls sausages, strings, braids, chaff, . . . has grown up, and there is no guarantee that different authors apply the same definition to a given term*". In fact, patterns had rather be defined by well-chosen structural elements clearly characterised by specific length scales. In effect, dislocation organisation involves phenomena that cover a wide range of length scales from subnanometric, the size of atoms, to millimetric, the size of samples.

Theoretical calculations of internal stresses have reached high degrees of precision and sophistication for any dislocation distribution [2]. Closed rigorous analytical expressions are indeed available in every case of interest, allowing one to calculate the force acting on a dislocation segment or on a dislocation loop, whatever its shape. However, predicting the macroscopic mechanical behaviour of crystals with sufficient accuracy remains a major issue in the theory of plasticity, beyond reach in the largest majority of cases.

- Due to the large amount of elastic singularities (the dislocation density is of the order of $10^{12} - 10^{15}$ m^{-1}) and due to the long-range nature of stress fields, the calculation of the internal stress on every dislocation segment in a crystal of 1 cm^3 lies beyond the present capabilities of the fastest computers (see section 3.4).
- The maximum of the forces engendered during dislocation intersection occurs when dislocations are so close that the elastic approximation fails. This is for instance the case of jog creation.
- Since microstructural evolution is in general controlled by various processes, dislocation dynamics is characterised by various time scales.
- Most important is the fact that dislocation dynamics under stress critically depends on dislocation core structure, and that implies a description at atomic level.

A dislocation is a rather complex object, be it perfect or dissociated. This is especially true at the atomic scale as reflected by the fact that the core structure may strongly depend on dislocation character. In the same vein, the overall mobility of a dislocation is in some cases determined by properties of much smaller features, such as atomic-sized kinks. Similarly, jogs, which are generated during deformation every time two dislocations intersect, are quite involved objects especially when dislocations are dissociated. In some cases jogs oppose dislocation motion through a trailing force that depends not only on their density but also on their fine structure. On the other hand, jogs may help recovery by providing constrictions that in turn encourage cross-slip nucleation.

In general not accessible to direct observations, our knowledge of the fine structure of a defect is often rather uncertain and this structure is quite frequently described by several alternative hypotheses. At present, *ab initio* atomic scale calculations consume too much computer time in order to be really helpful in deciding which hypothesis accounts best for a given situation. Atomic simulations using phenomenological interatomic potentials that correctly represent selected properties – elastic modulii, phase stability, point defect energy... – are being more and more widely employed. These simulations are, however, applied to situations such as properties taking place at dislocation cores, that may be extraneous to those to which potentials were fitted. Then the issue turns out to be that of setting up appropriate validation procedures for the potentials.

In brief, available information on the processes that control dislocation mobility is rather poor while our current knowledge on long-range interactions is, despite its sophistication, partly fruitless because of still undersized computing power and, to some extent, for lack of appropriate methods, but this is gradually improving.

1.3. Average plastic strain rate, and strain hardening

Current mechanical tests measure average quantities. The stress–strain curve is a plot of the applied stress σ versus the *average* total strain ε for a fixed value of the imposed *average* strain rate $\dot{\varepsilon}$.

The average plastic strain rate is expressed as [3]:

$$\dot{\varepsilon}_{\mathrm{p}} = K\rho_{\mathrm{m}} b v, \tag{1}$$

where ρ_{m}, b, v are the average density of mobile dislocations, the length of the Burgers vector and the average dislocation velocity, respectively. K is a number that depends on the situation under consideration. An accurate evaluation of all the quantities entering formula (1) from an analysis at the relevant levels is a formidable task which one is commonly confronted with in every theory of work hardening.

Let $\varepsilon_{\mathrm{el}}$ and ε_{p}, be the elastic strain and the plastic strain averaged on the volume of a sample deformed in tension (or compression), respectively. Then:

$$\varepsilon = \varepsilon_{\mathrm{el}} + \varepsilon_{\mathrm{p}}, \tag{2a}$$

$$\delta\sigma = M\delta\varepsilon_{\mathrm{el}} = M(\delta\varepsilon - \delta\varepsilon_{\mathrm{p}}), \tag{2b}$$

where M is of the order of the Young modulus. By dividing by δt:

$$\dot{\sigma} = M(\dot{\varepsilon} - \dot{\varepsilon}_{\mathrm{p}}), \tag{3a}$$

$$\theta \equiv \frac{\mathrm{d}\sigma}{\mathrm{d}\varepsilon} = M\left(1 - \frac{\dot{\varepsilon}_{\mathrm{p}}}{\dot{\varepsilon}}\right). \tag{3b}$$

θ is the strain-hardening rate (SHR).

Except in very special cases, crystals exhibit a SHR far smaller than $10^{-2}M$ which, in view of eq. (3b), implies that $0.99\dot{\varepsilon} < \dot{\varepsilon}_{\mathrm{p}} < \dot{\varepsilon}$. Measuring the SHR to within an accuracy of 10% then allows the plastic strain rate to be estimated within a precision of 10^{-3}. Interestingly, in order to predict the SHR to within an accuracy of 10% by means of eq. (3b), one should be able to estimate the average plastic strain rate to within an accuracy of 10^{-3} which, on account of σ, ε, and $\dot{\varepsilon}$ being averaged quantities, is not accessible.

In conclusion for this section, a complete and quantitatively accurate theory of work hardening is inherently beyond reach. One may instead more realistically focus analyses to processes that occur at the various scales and to estimating their relevance in predicting the macroscopic behaviour.

1.4. Mobility-controlled and obstacle-controlled processes

1.4.1. General definition

At least in pure materials, the length scales that account for the characteristics of the rate-controlling obstacles may vary by orders of magnitude; this is even true for the time scales that govern dislocation mobility. In case of alloys, distances between alloying elements and/or size of precipitates naturally add to the relevant length scales.

In order to discriminate between mobility- and obstacle-controlled situations, one usually considers the very approximate picture wherein dislocation dynamics combines viscous motion during t_f the free-flight time, and immobilisation at obstacles during a waiting time t_w. Whenever $t_w \gg t_f$, deformation is regarded as obstacle-controlled, whereas $t_w \ll t_f$, defines a mobility-controlled process.

Rather than a rigid classification, the above distinction in terms of mobility- and obstacle-controlled processes must be regarded as a useful method to analyse phenomena that occur at very different time and length scales. Genuine processes sometimes combine both effects with great subtlety as illustrated in what follows on pure face-centred cubic (f.c.c.) metals.

1.4.2. Obstacle-controlled processes

(1) When mobility effects are negligible, the plastic strain rate is of the order of the imposed strain rate as long as the applied stress exceeds the stress σ_{obs} that a given obstacle distribution opposes to dislocation propagation. This is associated with a sharp elastic–plastic transition. Assuming a reasonably homogeneous distribution of obstacles, σ_{obs} scales with the square root of the obstacle real density ρ. Then the flow stress σ is given by

$$\sigma \approx \alpha \mu b \sqrt{\rho}, \tag{4a}$$

where μ is the shear modulus, α is a numerical coefficient of the order of unity for strong obstacles, and smaller than 1 for weak obstacles [4,5].

For pure crystals, a reasonable approximation for the obstacle density is given by that of intersected dislocations, the forest, which typically varies between $10^{10}\ \mathrm{m}^{-2}$ and $10^{15}\ \mathrm{m}^{-2}$ for a well-annealed and a heavily deformed crystal, respectively.

The average distance l between obstacles scales with $\rho^{-1/2}$. Eq. (4a) becomes:

$$\sigma \approx \frac{\alpha \mu b}{l}. \tag{4b}$$

For typical values (i.e. $\alpha = 0.3$, $b = 0.25$ nm), formulae (4) show that the flow stress varies from about $10^{-5}\mu$ for a well-annealed crystal to about $2.10^{-3}\mu$ for a heavily deformed crystal [1].

(2) The resistance opposed by the microstructure to the moving dislocations combines two kinds of interactions that result in a moderate, negative temperature dependence of the flow stress [6] (section 3):

- Elastic interactions whose thermal dependence is that of the elastic modulus.
- Non-elastic interactions, such as the creation of a jog upon dislocation intersection, which are thermally activated. Their contribution to the flow stress is at most of the same level as those of elastic interactions.

(3) The strain rate sensitivity $S = \mathrm{d}\sigma/\mathrm{d}\ln\dot{\varepsilon}$ is expected to be modest in comparison with σ and this is essentially for two reasons:

- The inertia of a moving dislocation is small.
- The activation area for obstacle thwarting (see section 3) is of the order of the average distance l between obstacles multiplied by a characteristic obstacle width w, of the order of a few Burgers vectors. A rough estimate for S is given by (see section 3.2.3):

$$S \approx \frac{kT}{bw}\sqrt{\rho} \approx \frac{kT}{\alpha \mu b^2 w}\sigma \approx 1.5.10^{-5} T\sigma \tag{5}$$

which is small relative to the flow stress.

(4) From eqs (3b) and (4), one gets:

$$\theta \equiv \frac{\mathrm{d}\sigma}{\mathrm{d}\varepsilon} \approx \frac{\alpha \mu b \, \mathrm{d}\rho}{2\sqrt{\rho}\,\mathrm{d}\varepsilon}. \tag{6a}$$

Then, combining eqs (4a) and (6a) yields:

$$\theta\sigma \approx \frac{(\alpha \mu b)^2 \, \mathrm{d}\rho}{2\,\mathrm{d}\varepsilon}. \tag{6b}$$

By plotting the product $\theta\sigma$ versus stress or strain, one may obtain an estimate of the variation of the dislocation density [7]. In case of a constant SHR, integration of eq. (6a) gives:

$$\sqrt{\rho} = \frac{\theta}{\alpha \mu b}\varepsilon + \text{const.}, \tag{7}$$

which, for a deformation of 100% would give $\rho \approx 2.10^{15}\ \mathrm{m}^{-2}$.

(5) A simple reasoning based on the analysis of the junction unzipping mechanism (section 3.2) shows that the atomic concentration c of point defects created during plastic deformation may be expressed as [8,9]:

$$c \approx \frac{C}{\mu}\int \sigma \, \mathrm{d}\varepsilon, \quad C \approx 0.5. \tag{8}$$

This formula has been checked experimentally. It shows that a rather large atomic concentration of point defects, of the order of 10^{-4}, is created after some strain but, as mentioned above, these have non-perceivable influence on mechanical properties.

1.4.3. Mobility-controlled processes

In pure crystals, plastic flow is controlled by dislocation mobility whenever interaction with the lattice is large. Lattice resistance to dislocation motion depends in general quite strongly on dislocation character and, in some cases, on dislocation orientation with respect to certain crystallographic orientations. Friction is significant only for specific dislocation orientations, denoted hard orientations, where dislocation motion is strongly temperature- and stress-dependent. Well known examples are dislocations along $\langle 110 \rangle$ in diamond cubic semi-conductors below approximately $0.6T_M$ (T_M is the melting temperature) [10], screw dislocations along $\langle 111 \rangle$ in body centred cubic (b.c.c.) metals below room temperature [11], screw dislocations along $\langle 110 \rangle$ in $L1_2$ alloys such as Ni_3Al-based alloys, below 600–1000 K [12].

The as-grown microstructure comprises randomly oriented dislocations that can be moved under a certain range of applied stress. This is a regime of microplasticity where the plastic strain rate is relatively small and multiplication is modest. At a given stress, the population of mobile dislocations is progressively exhausted by annihilation and/or by locking in the hard orientations. Then, the stress must be increased in order to satisfy the condition of a constant total strain rate, and deformation must make use of some of the dislocations of the as-grown microstructure that have remained immobile so far. The macro-yield stress σ_M, is attained when the multiplication rate compensates for the exhaustion rate.

The following properties characterise mobility-controlled plastic flow.

- In the microplastic domain the slope of the stress–strain curve is large (formula (3b)).
- The elastic–plastic transition is smooth.
- After loading a crystal beyond the macro-yield stress and unloading it, most of the dislocation lines are aligned in the hard orientations. Consistently, the microplastic stage disappears upon reloading.
- The microyield behaviour resembles that of obstacle-controlled plastic flow, in particular with moderate strain-rate sensitivity (SRS) and thermal dependence.
- In a number of cases, such as in diamond cubic semi-conductors or in b.c.c. metals at low temperature, dislocation locking is athermal (or almost athermal), while unlocking is markedly thermally activated. The macroyield stress exhibits a negative thermal dependence. The low temperature macroyield stress is much larger than that inferred from obstacle density. The SRS is large and the SHR may be quite large too and temperature dependent.
- In some cases though, the locking process is thermally activated while unlocking is almost athermal. The macroyield stress exhibits a positive thermal dependence. In the domain of temperature where it peaks, the macroyield stress is much larger than what obstacle density alone would contribute to. Properties accompanying a yield stress anomaly may, however, differ from one alloy to the other. In Ni_3Al-based alloys, for instance, the SRS is small while the SHR is very large, up to $\mu/10$, and temperature dependent (it peaks at a temperature that differs from that of the yield stress for a given load orientation). The yield stress is orientation dependent and it exhibits a tension-compression asymmetry.

1.5. Scope of the paper

The mechanical properties of f.c.c. single crystals have been the object of intensive investigations and of controversial discussions [1,13–17]. A fair account of them all would take much longer than the space allowed to the present contribution. Therefore we shall concentrate on selected experimental facts (excluding twinning) emphasizing those that seem to have been overlooked so far. The theoretical analysis of work hardening will be restricted to certain recent developments.

F.c.c. metals are ductile at all temperatures, deforming essentially by dislocations with $\frac{1}{2}\langle 110\rangle$ Burgers vector in $\{111\}$ planes. Since dissociation takes place in the glide plane, f.c.c. metals obey the Schmid law. Lattice friction is modest and dislocation mobility is large. It is thus rather tempting to regard f.c.c. metals as a clear-cut example of obstacle-controlled systems. A complication arises though from dislocation cross slip which is quite common in f.c.c. metals. Cross slip influences deformation under several respects:

- Cross slip is thermally activated, manifesting a strong interaction with the lattice.
- It is by double cross slip that glide bands are widened and dislocations, to some extent, multiplied [18,19].
- Cross slip hardens the crystal by increasing the forest density.
- It is simultaneously a recovery process, actually the only effective recovery process at the temperatures considered here.

Hence, although f.c.c. metals appear at first as an ideal system to analyse internal stresses from dislocation interactions, there are reasons to believe that mobility-controlled phenomena will combine their effects with those of obstacle-controlled processes.

2. *The mechanical behaviour of f.c.c. metals single crystals. Experimental aspects*

2.1. Introduction

It is noteworthy that, to a very good approximation, relations (4) hold over a wide range of length scales, practically, from 10 nm, to 10 μm, even in the presence of rather heterogeneous dislocation distributions. The SHR increases with increasing strain up to a value θ_m, then decreases to eventually more or less stabilise. In terms of dislocation processes, this indicates that storage prevails over recovery at the beginning of plastic flow and that recovery becomes gradually important until it almost compensates for storage. Since crystal plasticity depends markedly on the nature of the processes activated in the very early stages of plastic flow, most of section 2 will, for clarity, address properties of work hardening in single crystals oriented for single slip.

2.2. Single crystals oriented for single glide

2.2.1. *General features of the stress–strain curve*

The stress–strain curve of a single crystal oriented for single slip exhibits a sharp elastic–plastic transition. At the onset of plastic flow, deformation proceeds mostly, but *not only*, by slip on the primary system. It is therefore relevant to plot the shear stress versus the shear strain $\tau(\gamma)$ (fig. 1), and to define the SHR from the slope θ_p of this curve:

$$\theta_p = \frac{d\tau}{d\gamma} = f_p^2 \frac{d\sigma}{d\varepsilon} = f_p^2 \theta, \tag{9}$$

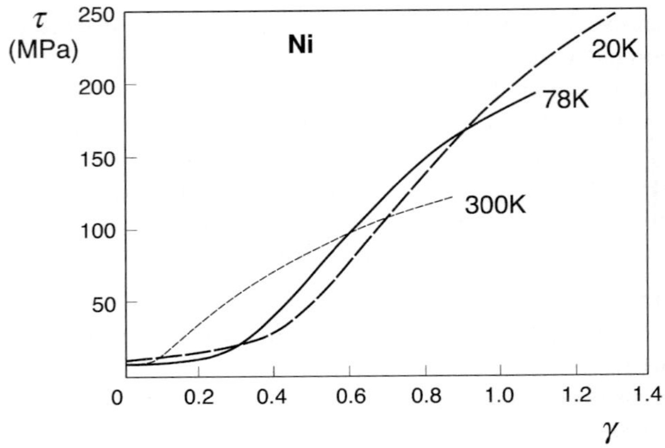

Fig. 1. Shear stress *versus* shear strain for a nickel single crystal deformed initially in single glide at various temperatures. The extent of the plateau-like behaviour at low strain is significantly reduced at 300 K. After [20].

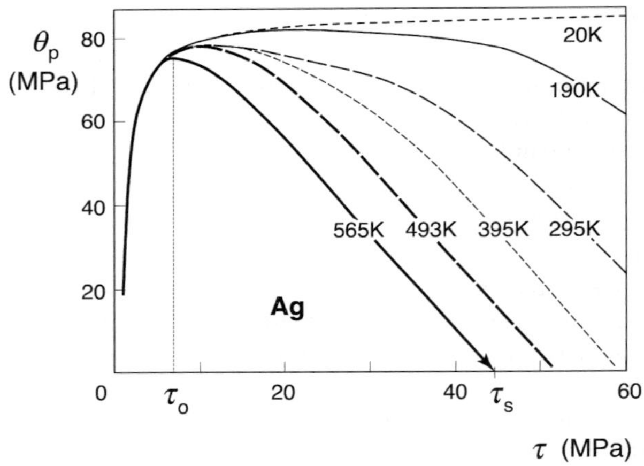

Fig. 2. Dependence of the strain-hardening rate (SHR) θ_p of a silver single crystal at various temperature on the resolved shear stress τ. Notice that the maximum of the SHR is almost temperature independent whereas the peak widens with increasing temperature. Domain 2b is not observed in this case (see fig. 15(a)). After [7].

where f_p is the Schmid factor in the primary plane. Formulae (4)–(7) are easily transposed by substituting τ, γ and θ_p for σ, ε and θ, respectively.

Figures 1 and 2 are representative of the behaviour of f.c.c. single crystals deformed at low temperature. Figure 2 indeed shows that $\theta_p(\gamma)$ exhibits a maximum, θ_m, which we use to distinguish two domains:

Domain 1: the SHR increases up to a maximum value θ_m. This domain can be in turn divided into two subdomains, 1a and 1b, whose limits are more or less well defined.

Subdomain 1a (the early stage of deformation) is where the SHR increases slowly, whereas it increases more rapidly in subdomain 1b, up to θ_m which marks the end of domain 1.

Domain 2: the SHR decreases steadily from θ_m. This domain can be also divided into two subdomains, 2a and 2b, whose limits are again not precisely defined, roughly corresponding to the fact that the SHR decreases almost linearly (2a), then almost stabilises (2b). At the end of domain 2, the level of SHR more or less compares with that measured at yield at the same temperature.

The reader has certainly noticed that the above classification does not conform to the usual description of the stress–strain curve which, traditionally, is truncated into a succession of stages, each associated with a specific mechanism and SHR. In the usual description indeed, the *easy glide* stage (Stage I) is defined by its moderate SHR. After a more or less rapid transition, Stage I is followed by a stage of large SHR, Stage II, referred to as the *linear hardening* stage. Then Stage III is characterised by a more or less linearly decreasing SHR up to Stage IV, itself associated to a constant SHR in fact rather close to that of Stage I. This is eventually followed by various more or less well-defined stages denoted Va, b....

It may be argued that the difference between the traditional Stages I to V and the present domains 1 and 2, and their subdomains, is more semantic than founded on solid physical principles. We felt it useful though to refer to the $\tau(\gamma)$ curve under a different perspective essentially because of the many controversial debates on workhardening in f.c.c. metals. We believe that stages are not clearly distinct in terms of the underlying physical processes and that instead the various domains of workhardening reflect a smooth, well-defined change in dislocation patterning that takes place as deformation proceeds.

2.2.2. *The early stage of plastic deformation (subdomain 1a)*

The critical resolved shear stress (CRSS) is small, of the order of $10^{-5}\mu$. At the onset of plastic flow θ_p is small, of the order of $3.10^{-4}\mu$. It then increases slowly with deformation, and this is accompanied by a small amount of secondary slip. The upper limit of this early stage domain is not well defined.

The slip-line pattern is essentially made of parallel lines whose length is about the same (of the order of a fraction of millimetre). These lines may reflect:

- elementary steps with a height of the order of the interatomic distance,
- slip lines comprised of elementary steps,
- glide lamellae themselves formed upon congregation of steps and whose height is of the of the order of a few nanometres.

It is quite probable that our knowledge of the fine structure of the slip lines as well as our understanding of the processes occurring at the elastic–plastic transition will further improve in the next few years with the extensive use of atomic force microscopy.

Analysis of etch pits indicates that the primary glide bands may deviate by a couple of degrees from the trace of the glide plane. The density of secondary dislocations is estimated to be one order of magnitude lower than that of the primaries. It is in addition reported that the glide bands become more closely spaced as deformation proceeds.

TEM investigations reveal isolated, long bundles of interlaced edge dislocation dipoles. There is a close correspondence between the bundle distribution and that of the etch pits. Some bundles have their ends all aligned parallel to the intersection of the glide plane by the cross-glide plane, which indicates that *cross-slip is active from the very beginning of plastic flow*. We shall see in the following that the latter remark is supported by several other experimental, numerical and theoretical arguments.

The tensile axis rotates as a result of the predominance of single glide, f_P decreases while f_S, the Schmid factor of the secondary system, increases until $f_S = f_P$, however, with no clear changes in the deformation mode at that point. Slip carries on on the primary slip system, indicating that extensive glide on one system impedes glide on intersecting systems. This phenomenon known as *overshoot*, or *latent hardening*, has been demonstrated by direct experiments [1,21]. A reference microstructure is generated in one large single crystal deformed on a single slip system, and referred to in what follows as primary. Then smaller crystals are cut from the first sample and deformed along selected orientations. What occurs is that the CRSS is unchanged for the crystals that deform on the primary slip system whereas the CRSS of crystals deforming on systems other than primary is increased by 10% to 40%.

2.2.3. The rapid increase of SHR (subdomain 1b)

As deformation proceeds, the SHR increases rapidly up to a about $4.10^{-3}\mu$, and then decreases.

The rapid raise of the SHR is accompanied by increasingly active secondary glide. However:

- Primary glide still predominates.
- The glide lines corresponding to secondary glide are much shorter than those reflecting primary glide.
- Large scale plastic strain heterogeneities take place. Sample surfaces appear organised in domains associated with different slip-line patterns, which means that plastic flow is activated on different glide systems in different regions of the crystal. This observation originates from latent hardening that inhibits glide on intersecting systems in large volumes of crystal. The resulting strain heterogeneities are the source of gradually increasing internal stresses that become eventually strong enough to trigger extensive slip on other systems.
- The increase in dislocation density is attested by etch pits and confirmed by TEM. Dislocations congregate in bands. The density of bands and the density of dislocations in a given band both increase with deformation, and since glide bands are inclined to the glide plane, neighbouring bands tend to impinge which results in a progressively uniform distribution of etch pits.
- TEM investigations indicate that dislocations concentrate in the glide bands and that the density of secondary dislocations grows rapidly. Primary dislocations remain the most profuse of all activated systems and new features such as Lomer–Cottrell locks begin to appear upon reaction between intersecting dislocations.
- Patterning becomes increasingly intricate but, before the end of domain (1b), dislocations start to rearrange into more or less equiaxial cells.

2.2.4. The decrease of SHR (subdomain 2a)
The rearrangement into cells is increasingly pronounced as the SHR decreases. The dislocation density in the walls is much larger than that inside the cells. The slip line pattern attests to large scale cross slip, usually referred to as prominent cross slip.

Cells are randomly misoriented. Misorientation increases slightly up to a value of the order of 0.02 radians.

2.2.5. The stabilisation of the SHR (domain 2b)
SHR stabilisation occurs at a large stress and small SHR, i.e. when plastic flow becomes unstable in tension, generating striction. The stabilisation is more easily observed in twist than in tensile testing. It is not systematically achieved in f.c.c. metals. It has been reported in polycrystals tested at temperatures down to 77 K, but it is not observed in tensile tests on single crystals at liquid helium temperature.

Domain 2b is not as well documented as the preceding domains. It has been argued for a while that the SHR does stabilise at a value comparable to that measured at the onset of plastic flow. More accurate studies have revealed that the SHR may in some cases increase and then decrease. The cell size diminishes with deformation while the misorientation between cells increases. Whenever it is observed at relatively high temperature, it is associated with non compact glide on planes of the types {11x} and {1xx}.

2.2.6. Phenomenological discussion
We rewrite eq. (6b) so as to account for deformation in single slip:

$$\frac{\mathrm{d}\rho}{\mathrm{d}\gamma} = \frac{2\theta_{\mathrm{p}}\tau}{(\alpha\mu b)^2} \tag{10}$$

suggesting that information on the multiplication rate can be derived by plotting $(\theta_{\mathrm{p}}\tau)$ versus τ [7]. Such a plot is exemplified in fig. 3 which, by comparison with fig. 2, indicates that the SHR and the multiplication rate do not peak for the same value of the flow stress.

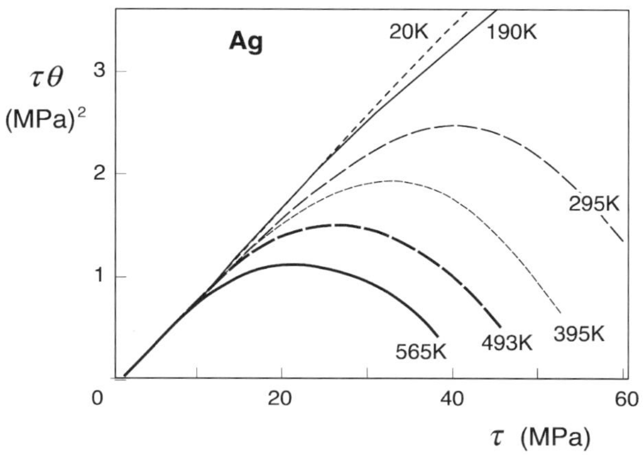

Fig. 3. The $\tau\theta_{\mathrm{p}}$ versus τ diagram that corresponds to fig. 2. After [7].

Assuming dislocations move on average over a distance $d\lambda$,

$$d\gamma = C\rho b\, d\lambda, \tag{11}$$

where C is a geometrical factor of the order of 1. Then from eqs (10) and (11), one gets for $C = 1$:

$$d\lambda = \Lambda \frac{d\rho}{\rho}, \tag{12a}$$

$$\Lambda \approx \frac{(\alpha\mu)^2 b}{2\theta_p \tau}. \tag{12b}$$

Hence, Λ can be regarded as the average distance λ at which the relative creation rate reaches 100%. In evaluating Λ, dislocation elimination at free surfaces has been ignored which is reasonable whenever Λ is smaller than the size of the crystal. Alternatively, Λ provides a typical sample dimension below which storage should remain modest. In subdomain 1a, $\theta_p/\mu \approx 3.10^{-4}$ and $\tau/\mu \approx 10^{-4}$, corresponding to $\Lambda \approx 0.2$ mm. This finding is rather convincingly supported by experiments since no stage of rapid hardening is actually observed for crystals whose lateral dimension is smaller than about 0.2 mm.

Consider now that part of the $\theta(\tau)$ curve where the SHR decreases almost linearly with τ. This behaviour can be expressed as [22]:

$$\frac{d\tau}{d\gamma} \equiv \theta_p = \theta_m \frac{\tau - \tau_s}{\tau_0 - \tau_s}, \tag{13}$$

where τ_0 and τ_s are quantities indicated on fig. 2. By integration, one has:

$$\tau = \tau_s - (\tau_s - \tau_0) \exp - \frac{\theta_m \gamma}{\tau_s - \tau_0}. \tag{14}$$

Then combining eqs (4), (10), and (13) yields the variation of dislocation density:

$$\frac{d\rho}{d\gamma} = \frac{2\theta_p \tau}{(\alpha\mu b)^2} = \frac{2\theta_m \tau}{(\alpha\mu b)^2} \frac{\tau_s - \tau}{\tau_s - \tau_0} \tag{15}$$

or equivalently:

$$\frac{d\rho}{d\gamma} = \frac{2\theta_m}{\tau_s - \tau_0} \left[\sqrt{\rho_s \rho} - \rho\right]. \tag{16}$$

The first term of the right-hand side of the expression in brackets corresponds to dislocation multiplication, the second is the recovery term regardless of the mechanisms involved. By integration of eq. (15) one has:

$$\rho = \left[\frac{\tau_s - (\tau_s - \tau_0) \exp - \left\{\frac{\theta_m \gamma}{\tau_s - \tau_0}\right\}}{\alpha\mu b}\right]^2 \tag{17}$$

that accounts for the occurrence of a saturation of the dislocation density. It is noted that θ_m and τ_0 are almost temperature independent whereas τ_s decreases with temperature.

2.2.7. *Stability of the microstructure with respect to path changes*

Unloading and reloading the crystal at the same temperature results in almost the same value of the CRSS (actually within 1%), which attests to the stability of the microstructure. On the other hand, when a single crystal oriented in single glide (for uniaxial straining) is twisted along its axis, unloaded, then reloaded in tension, one finds that (fig. 4):

(1) Twisting always increases the CRSS.
(2) The SHR is always smaller than that of the crystal directly deformed in tension at the same stress level.
(3) The stress–strain curve exhibits serrations.

Surface observations show that the clustering of slip lines and of slip bands becomes increasingly pronounced as the amount of twist prestrain is increased [23].

During the tensile test, dislocations are forced to propagate through a forest originating from the simultaneous operation of several slip systems and referred to as the *alien* microstructure. As a result the forest density at the onset of plastic flow is large, which is consistent with the increase in CRSS. The decrease of the SHR indicates that the multiplication rate of the forest dislocations is smaller than that observed at the same level of strength in a specimen which has undergone monotonous straining. The twist-induced alien microstructure thus appears to be unstable with respect to tensile deformation. In addition to the intrinsic instability of the microstructure, the Considère criterion for stable deformation imposes that in a tensile test the SHR be larger than the CRSS. An increase

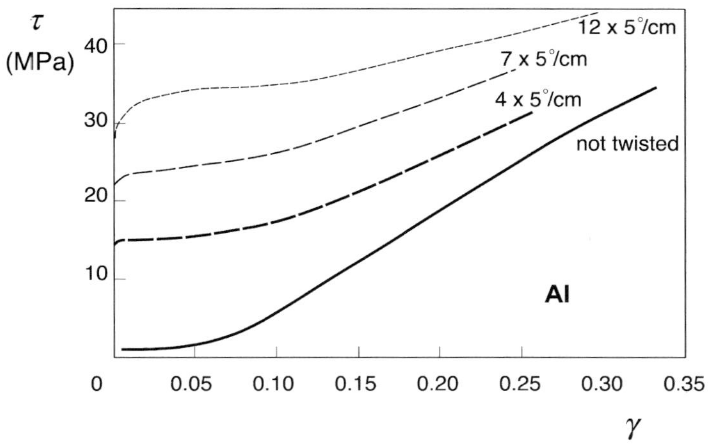

Fig. 4. Effects of changes in strain paths. The "untwisted" curve corresponds to a single crystal of Al well annealed and deformed in tension at room temperature. The other curves represent the stress–strain curve of the same material after various twist deformations. The occurrence of plastic instabilities is represented by dashed lines. After [23].

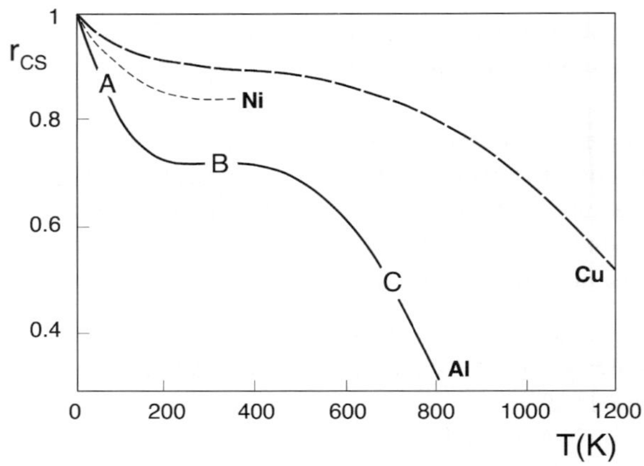

Fig. 5. Temperature dependence of the Cottrell–Stokes ratio (expression (18)) of Al, Cu, Ni showing the three domains A, B, C for Cu and Al. To the authors' knowledge, systematic measurements of the behaviour of Ni at high temperature are not available. After [20,24,25].

of the latter simultaneous to a decrease of the former amplifies the tendency towards instability.

The twist prestrain experiments indicate that, given the test temperature, the thermodynamical state of the crystal is determined both by the yield stress and by the SHR.

2.2.8. *Temperature dependence of the mechanical strength of f.c.c. crystals*

The change in yield stress between two test temperatures arises from two contributions: the irreversible change in the microstructure and the reversible, thermally-activated defeating of obstacles.

The stability of the microstructure with respect to temperature changes is tested by measuring the yield stress (from the unloaded state) of a sample in tension (or in compression) at a temperature $T_1, \tau(T_1)$, then at T_2, and again at T_1, $\tau^*(T_1)$. A stable microstructure is when $\tau^*(T_1) = \tau(T_1)$.

In the cases of f.c.c. metals, the change in yield stress is to a very good approximation reversible over a wide range of temperatures, attesting to the stability of the microstructure controlling sample strength (see section 3.2.3).

The temperature dependence of the flow stress is measured by the Cottrell–Stokes ratio r_{cs}:

$$r_{cs} \equiv \frac{\tau(T)/\mu(T)}{\tau(4.2\text{ K})/\mu(4.2\text{ K})}. \tag{18}$$

The temperature dependencies of the yield stress of Al, Cu, Ni, are displayed in fig. 5. These curves define three domains for the temperature dependence of the yield stress corresponding to the following properties [20,24,25]:

- At low temperature ($T < 100$ K), the flow stress decreases with increasing temperature which determines domain A.

- At intermediate temperature the yield stress is almost temperature independent in the case of Al, Cu, Ni, and this is domain B.
- At higher temperature, the yield stress decreases for all the crystals considered here.
- The microstructure remains stable with respect to changes in temperature even for relatively high temperature and, in certain cases such as Al and Cu, even beyond $0.5T_M$.
- This description does not apply in the case of Ag. Here the yield stress decreases continuously with temperature. This is probably a consequence of the very small stacking fault energy.

Systematic studies in Cu and Al [26,27] have shown that domain C is related to the activation of slip on non-compact planes {100}. Non-compact slip on other systems of the {11x} type has been reported but no systematic analyses are available. There is evidence of non-compact slip in Ni and Au but, again, systematic analyses are missing. As all cross-slip processes, non-compact slip depends on the stacking fault energy.

Another important result is that the Cottrell–Stokes ratio r_{cs} is almost independent of plastic strain up to quite large strains. This is consistent with the assumption that the temperature dependent part and the temperature independent part of the resistance opposed by the microstructure have the same origin, up to rather large strains.

Figure 2 shows that $\theta_p(\tau)$ exhibits a maximum θ_m which is almost temperature independent. The hump become wider with decreasing temperature, while the SHR at the end of plastic flow remains comparable to that measured at yield.

2.2.9. Remark

Figure 6 reveals that the yield stress scales astonishingly well with the square root of the average intersecting dislocation (forest) density, however not with that of the overall dislocation density, over about 5 decades of density. This remains true whether dislocation distribution is homogeneous or heterogeneous, and in spite of the fact that the microstructure evolves significantly from yield up to the end of domain 2. Although accurate measurements, such as in fig. 6, show in fact a slight departure from formula (4), this expression represents reasonably well the experimental situation, with $\alpha \approx 0.3$.

2.3. Deformation of single crystals oriented for multiple glide

The main properties of deformation of single crystals under controlled multiple slip can be summarised as follows [29–31]:

- Deformation is generally heterogeneous including domains where one system develops preferentially. Multiple glide does not take place homogeneously over the sample.
- In some rare cases, particularly at very low temperature, a stage of deformation in *single* glide has been observed. In this case the curve $\theta(\sigma)$ shows a peak similar to that of crystals oriented in single glide. In most cases though the SHR decreases monotonically from the very beginning of plastic flow.
- The glide line pattern is quite coarse, involving profuse cross-slip.
- Unstable (discontinuous) flow leading to multiple necking is observed.

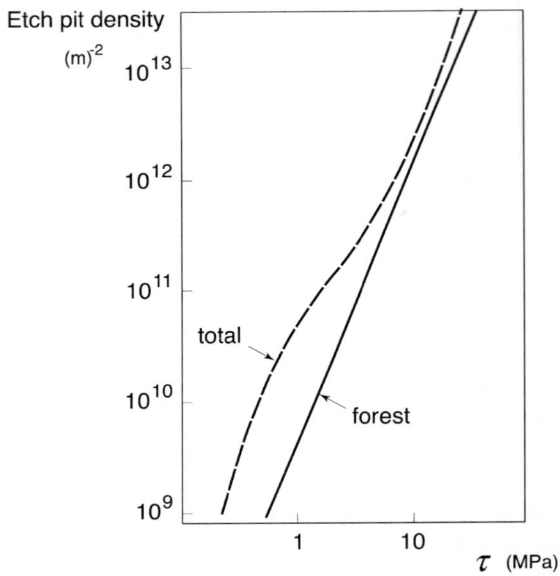

Fig. 6. Simplified representation of the relationship between the flow stress and the dislocation density. The $\rho^{1/2}$ dependence of the flow stress when considering the density of straight line represents the dislocations intersected by the primary slip plane. After [1].

These observations are consistent with properties of deformation in single slip. In general, different glide systems are activated in different regions of the specimen. The activation of one glide system in one given region inhibits the operation of other systems in that region. If, by chance the same glide system is activated from the beginning in most of the sample, it inhibits glide of the others systems in the whole sample, and easy glide takes place.

Whenever, due to the state of stress, a glide system starts to develop in an already deformed region, it does so in an alien dislocation distribution. This is associated with a small local SHR that in turn causes local instabilities.

It is noted that in case of deformation under multiple slip, one has no choice other than plot a true stress–true strain curve $\sigma(\varepsilon)$ instead of the shear-stress–shear-strain curve $\tau(\gamma)$. From formula (9), θ may thus be 10 times larger than θ_p for purely geometrical reasons. Comparing the SHRs under single and multiple slip has not very much physical sense.

2.4. Conclusion

The dominating role of *dislocation intersection* is probably one of the most important experimental results of almost fifty years of experimental studies of *work hardening*.

The role of dislocation intersection is for instance confirmed by noticing (i) the *large* SHR of crystals where dislocations intersections occur, such as f.c.c. crystals, diamond cubic semiconductors deformed at high temperature, b.c.c. metals, in contrast with (ii) the *small* SHR of crystal where dislocations intersections cannot occur such as some h.c.p. crystals like Zn, Cd, Mg, deformed at low temperature.

The definition of the mechanical state of the crystal needs at least the knowledge of *both* the flow stress and the SHR.

The amount of secondary slip during easy glide, and the shortening of the low SHR domain as temperature or specimen size is increased are evidence that cross slip occurs from the beginning of plastic flow and not at a given level of stress (see section 5).

Besides their contribution to the flow stress, jogs should have an effect on dislocation motion or on the widening of slip lines. However, except for the fact that constrictions facilitate cross-slip, this effect has not been shown to be important.

3. Theoretical analysis of plastic flow in single crystals

3.1. Introduction

In the absence of a comprehensive *ab initio* theory for the flow stress we proceed by steps in discussing the several ingredients which, in our opinion, should be accounted for in a model of strain-hardening. In section 3.2 we examine the propagation of one dislocation within a homogeneous forest, under the combined effect of the applied and internal stresses and we incorporate the contribution of core effects. In section 3.3 we address the question of large-scale heterogeneities. Section 3.4 deals with interactions between moving dislocations and with the influence of plastic strain on the evolution of the microstructure.

3.2. Resistance to dislocation glide from a random microstructure

3.2.1. Introduction

In this paragraph, internal stresses are assumed to originate from a roughly homogeneous, random dislocation distribution with a density ρ. The average distance between dislocations is $l \approx \rho^{-1/2}$. Intersecting dislocations react forming junctions. It has been shown that the unzipping of the attractive junctions contributes about 80% of the resistance to plastic flow and this is why we restrict the following to considerations on junction zipping and unzipping [32–34]. The geometry and properties of junctions are addressed in section 3.2.2. A simplified elastic evaluation of the stress necessary to unzip a junction and a phenomenological estimate of the core contribution are given in the next paragraph. The expressions derived are compared with experimental observations in section 3.2.4. Investigations at the atomic level are reviewed in section 3.2.5.

3.2.2. Dislocation junctions in f.c.c. crystals

Because of the importance attributed to junctions, we have fond it insightful to first recall selected properties of dislocation reactions in the f.c.c. structure. For simplicity, dislocation dissociation is ignored.

A dislocation gliding in the plane ABC (fig. 7) and intersecting a dislocation of the plane ACD yields attractive junctions in the form of the following configurations (fig. 8):

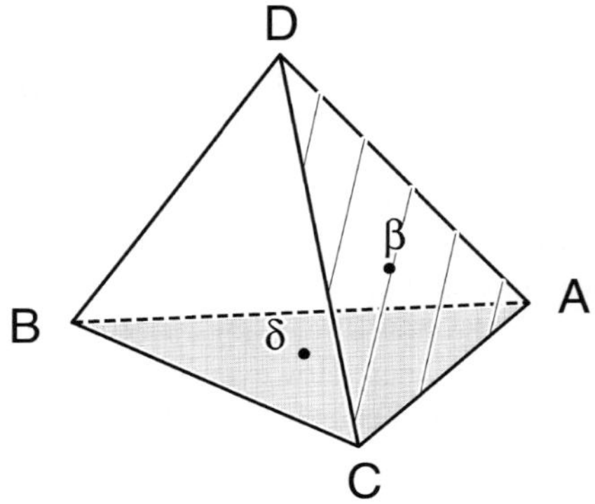

Fig. 7. The Thompson tetrahedron.

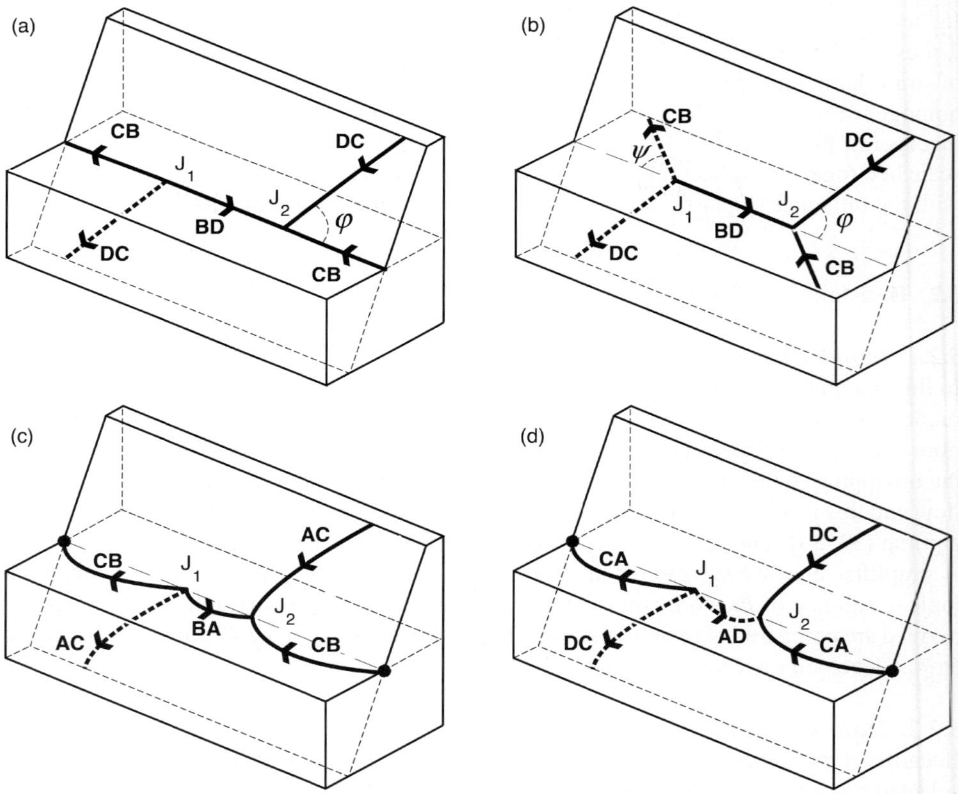

Fig. 8. Attractive junctions: (a) the symmetrical Lomer lock; (b) the asymmetrical Lomer lock; (c) the attractive junction is glissile in the primary plane; (d) the attractive junction is glissile in the cross-slip plane. l is the length of the tree, the dislocation AC in (c), l' is the length of the moving dislocation, CB in (c).

- A Lomer lock

$$\mathbf{BC} + \mathbf{CD} \to \mathbf{BD}, \tag{19a}$$

$$\mathbf{BA} + \mathbf{AD} \to \mathbf{BD}. \tag{19b}$$

- A junction glissile in ABC

$$\mathbf{BC} + \mathbf{CA} \to \mathbf{BA}, \tag{20a}$$

$$\mathbf{BA} + \mathbf{AC} \to \mathbf{BC}. \tag{20b}$$

- A junction glissile in the cross-slip plane ACD

$$\mathbf{DC} + \mathbf{CA} \to \mathbf{DA}, \tag{21a}$$

$$\mathbf{CA} + \mathbf{AD} \to \mathbf{CD}. \tag{21b}$$

Under the effect of the stress applied locally, τ, these junctions can be unzipped from either side. Figure 8(a) shows the simple case of a Lomer lock when $\psi = 0$. Here portion **CB** assumes a bowing angle that will increase with stress while the length of the lock $J_1 J_2$ will decrease. Depending on the relative values of l, l', φ and ψ, unzipping occurs either before the lock is completely destroyed or else when the two nodes J_1 and J_2 are in contact. Figure 8(c) represents the glissile, hence flexible, junction **BA**. A third case of an unzipping is provided by junction **AD** (fig. 8(d)) which, may possibly cross slip. If cross slip is not possible, the situation is similar to that of the above sessile lock.

Upon crossing through one another two dislocations take one jog each. Junctions and jogs have distinct contributions to the mechanical strength and will be examined separately in the following.

3.2.3. Elastic analysis

The elastic theory of dislocations allows for an accurate calculation of the stress field of a dislocation of any shape, and of the effect of stresses at each point of a dislocation line including the contribution of the rest of the line. Calculations can be achieved under elastic anisotropy. The corresponding analytical expressions are, however, very involved and they necessitate computer calculations. A great deal of information is, nevertheless, obtained by using the line tension approximation that has the advantage of providing a vivid picture of the physical processes involved. Here again, many levels of approximation are available. For simplicity, we restrict the following analysis to the character-independent and elastically isotropic line tension approximation, ignoring effects of dislocation curvature. In this case, the line tension Γ is simply written:

$$\Gamma = \beta \mu b^2, \quad 0.5 < \beta < 1. \tag{22}$$

Neglecting core interactions, the resistance of a microstructure consisting of sessile junctions is given by expression:

$$\tau_a = \alpha_a \mu b \sqrt{\rho}, \tag{23a}$$

α_a is temperature independent, hence τ_a/μ is temperature independent.

It is intuitively evident that, for a glissile junction, unzipping occurs for an applied stress smaller than for a sessile junction. The calculation done in the case $l = l',\ \psi = 0$, gives for $\varphi = \pi/3$ a stress reduction by a factor 10, and by a factor of 3 for $\varphi = \pi/4$. Similar considerations apply to a junction that is liable to cross slip, although two complications occur in this case:

(1) One must take the shear stress into account both in the primary and in the cross-slip planes.
(2) Since cross slip is thermally activated, cross slip aided unzipping shows a thermal dependence which we ignore until the next paragraph.

These rather straightforward considerations indicate that the strength of a junction, hence the coefficient α_a in expression (23a), may vary quite significantly according to which category of junctions is actually generated.

In the same vein, the stress necessary to unzip a given junction depends on the position and actual structure of its neighbouring junctions. Hence, difficulties arise in deciding how the averaging is performed over the junctions present. The latter issue is by far the most difficult to solve. Averaging is actually the weakest point of every theoretical evaluation of α.

A further difficulty in evaluating junction resistance to unzipping arises from core interactions (including dissociation effects). As pointed in section 1.2, core effects are effective only when dislocations are in contact. The physical origin of core effects is hard to define with precision, a reasonable approximation consists in attributing it to jog generation, and to identify them as localised pinning centres, whose resistance is generally expressed as:

$$\tau_{th} = \alpha_{th} \frac{\mu b}{l}. \tag{23b}$$

The coefficient α_{th} is an adjustable parameter whose value depends on the strength of the obstacles and which should reflect the thermal dependence of the yield stress. Furthermore α_{th} should increase with decreasing stacking fault energy (see section 3.2.3).

Finally, we tentatively postulate that the flow stress is the sum of the two above contributions [6]:

$$\tau = \tau_a + \tau_{th}. \tag{23c}$$

In what follows, we discuss the effects of various contributions to the flow stress which we confront to this simplified approach. In particular, we show that the form of expressions (23b, c) is *validated* experimentally by the fact that formula (4) is verified over a large temperature range.

It is worth briefly recalling at this stage that in a microstructure made of randomly distributed dislocations and such that the intersection process is unique (one type of obstacle, here of junction), the plastic strain rate under an applied stress τ is expressed very generally as:

$$\dot{\varepsilon}_p = \nu \exp\left[-\frac{\Delta G(\tau, T, y)}{kT}\right], \tag{24}$$

where $\Delta G(\tau, T, y)$ is an activation enthalpy, y is the set of variables defining the microstructure and ν is a microstructure-dependent frequency factor.

Assuming that ν is stress-independent, the SRS is written:

$$S \equiv \frac{\partial \tau}{\partial \ln \dot{\varepsilon}} = -kT\frac{\partial \Delta G}{\partial \tau} = -\frac{kT}{bA} \tag{25}$$

which in turn defines the activation area A, that is, the area swept by the dislocation during the activation process. A is more conveniently expressed as:

$$A = lw, \tag{26}$$

where w reflects the obstacle width, that is, the distance between the positions of stable and unstable equilibrium under stress.

Let ΔG_0 be the activation enthalpy at 0 K and zero stress. The theory of thermally activated glide gives:

$$\alpha_{\text{th}} = \frac{\Delta G_0}{\mu b^2 w} f(T). \tag{27}$$

The exact expression of $f(T)$ depends on the expression of ΔG. For example, if ΔG depends linearly on τ, then $f(T)$ is a linearly decreasing function of T.

Another useful expression is given by:

$$\tau bA = \frac{\tau kT}{S} = \alpha_{\text{th}} \mu b^2 w. \tag{28}$$

3.2.4. Comparison with experiment

Despite its simplicity, the analysis that yielded expressions (23) is consistent with the following experimental results:

- The resistance to plastic flow scales with the square root of the dislocation density and is therefore consistent with formula (4).
- The slight deviation from the simple law which has been mentioned in section 2.2.9 results from the variation of the line tension with dislocation curvature, or equivalently with l. When the latter is taken into account, the agreement is good [35,36].
- With regards to expressions (23), the Cottrell–Stokes ratio defined by formula (18) is written as:

$$r_{\text{cs}} = \frac{\alpha_a + \alpha_{\text{th}}}{\alpha_a}. \tag{29}$$

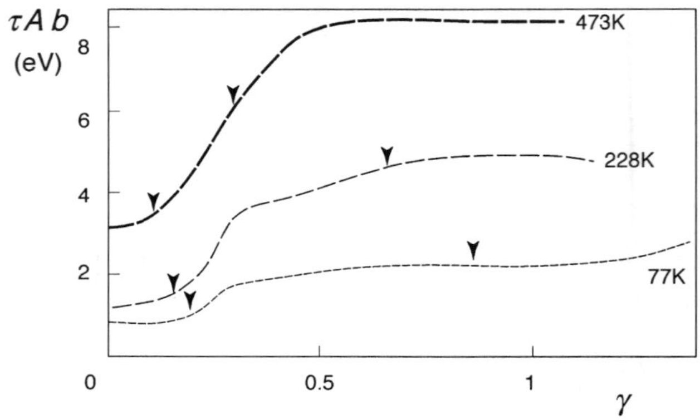

Fig. 9. Plot of τAb versus the shear strain for a Cu crystal at various temperatures. The arrows indicate the limit of deformation stages showing that the absence of a correlation between the changes in τAb and the transition from one stage to the next. After [13].

It is independent of plastic strain and temperature, in agreement with experimental results up to strains well beyond those corresponding to the maximum SHR.

- The theoretical predictions on the stability of junctions is confirmed experimentally by the fact that after unloading the stress at which the crystal yields at T_1 is unchanged to within 1%, attesting to the stability of these junctions (see section 2.2.8).
- Assume now that the temperature is raised just after unloading. Since the interaction producing the junctions is elastic, one expects no change in the microstructure. This is in agreement with the observed flow stress reversibility.

The following discrepancies have however been reported:

- The Cottrell–Stokes ratio depends on deformation at very large strains.
- Formula (28) predicts that the SRS is proportional to the flow stress whenever w is stress independent. Conversely, a departure from linearity reflects a stress dependence of the obstacle width. Experiments show that $S(\tau)$ exhibits a positive curvature which indicates in turn that w is a decreasing function of τ [37,38]. Additional information [13,20] may be gained by plotting the quantity τbA versus γ as is done in fig. 9 that suggests the coexistence of several processes. It is noted that the transitions between two plateaux of τbA are not related with transitions between the various stages of the SHR.
- According to formulae (23) and (24), the temperature dependence of the yield stress scaled by the shear modulus is consistent with two domains A and B. These formulae do not account though for the decrease of the yield stress that defines domain C.

We propose that the latter discrepancy takes its origin in the cross-slip assisted junction unzipping which we have ignored so far. This effect is difficult to evaluate (see section 5). It is likely that it cannot be accounted for by means an additive term similar to expression (23c). Besides one may also have to incorporate cross-slip on non-compact planes which, in the case of Al, and Cu, has been shown to be associated with domain C.

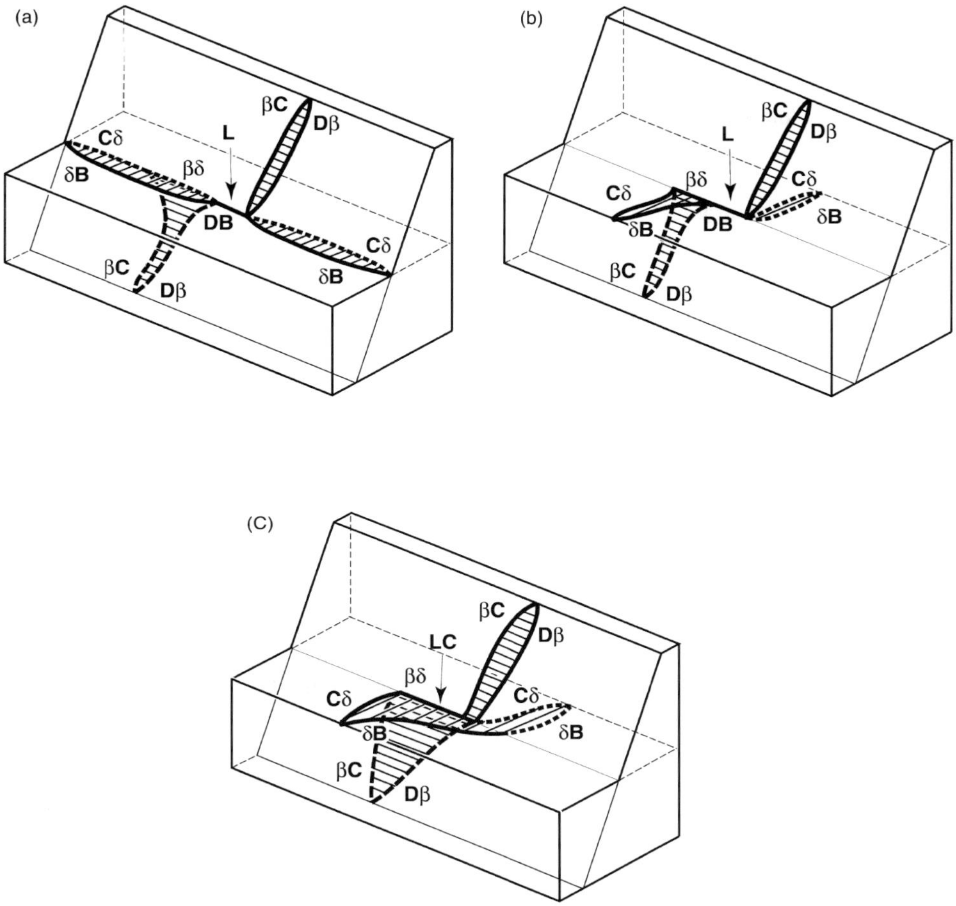

Fig. 10. Simulations of a Lomer–Cottrell lock under no applied stress: (a) $\psi = 0$, intermediate stacking-fault energy. Atomic simulation. The junction (Burgers vector **DB**) is undissociated. After [39]. (b) $\psi = \pi/3$, intermediate stacking-fault energy. Atomic simulation. Ibid. (c) $\psi = \pi/3$, small stacking-fault energy. Elastic simulation. The junction (Burgers vector **DB**) is dissociated, only one of the partials of the primary dislocation has reacted to form the junction [40].

3.2.5. Attractive junctions at the atomic level

Because of computing limitations, numerical simulations of attractive junctions were achieved by combination of atomic simulations in the core region with finite element methods for the surrounding volume. To the authors' knowledge, this quasi-continuum method has so far been used for Lomer–Cottrell barriers in Al for selected configurations [39]. Figures 10(a) and (b) show the result of the simulation for the case $\varphi = \pi/6$, $\psi = 0$, corresponding to fig. 8(a). The structure of the junction appears to be more complex than is inferred from elasticity. This holds true for the unzipping process. However, the breaking stress is not very different from that calculated in the simple elastic approximation. It is thus likely that the details of the unzipping process influence the thermal part of the stress. Interestingly, vacancies created during the intersection process come naturally from the simulation.

Systematic calculations involving various metals and geometrical configurations are still required for the intersection process to be satisfactorily documented and, more importantly, for the stress necessary to unzip the junction to be appropriately estimated. The field is however in rapid progress.

As indicated in section 3.2.3, the stress to unzip a given junction depends also on neighbouring junctions. In the elastic approximation, this complicates an evaluation of the average stress that the microstructure opposes to a mobile dislocation. In the atomic simulation though, the linear dimension of the volume analysed is much smaller than the average distance between neighbouring dislocations. As a consequence the effect of the microstructure is taken into account through boundary conditions that may influence result quite dramatically. This difficulty may be eliminated by means of an elastic treatment down to a critical distance, itself evaluated from atomic simulations whose results serve also to parametrise interactions at close distance [40]. Preliminary analyses achieved with this technique compare quite well to that predicted by atomic simulations. They show the influence of the stacking fault energy on the structure of an attractive junction, as indicated in fig. 10(c). Although detailed processes differ in many respects from the simplified description made in section 3.2.2, the calculated stress is close to the results obtained from eq. (23a).

These calculations reveal that the junction zipping-unzipping process is by far more complex than expected from the early naive description. The preliminary calculations indicate neither a change in the conditions for the stability of the junction, nor a difference in the unzipping stress. These points should certainly be checked by systematic calculations. In any case, the thermal part of the unzipping stress is certainly dependent on the details of the process.

3.2.6. *Conclusion*

The analysis of dislocation junctions has benefitted from dramatic refinements in the recent years. One insight from these achievements is that sophisticated elastic calculations need to be performed in parallel with atomic simulations. The approach should be expanded to a range of selected metals and to attractive configurations other than Lomer and Lomer–Cottrell locks. In particular, it is suggested that attention be paid to the degree of freedom brought about by junction glide in the primary and in the {111} cross-slip planes as well as in non-compact cross-slip planes. One question indeed that seems worth addressing is whether cross slip relaxes the unzipping stress. Besides, the comparison with experiment relies on averaging methods that are questionable even in the simple approximation of a random distribution of obstacles. A more satisfactory approach requires that complementary methods such as mesoscopic simulations be employed in order to test the various factors separately (section 3.4).

Finally, the fact that experiments generally lead to α less than 0.5 indicates that forest dislocations are relatively weak obstacles. They are not bypassed and they may not serve as anchoring points for source operation.

3.3. Effect of long-range stress fields

3.3.1. *Introduction*

For clarity, we have so far regarded the microstructure as formed of a random distribution of dislocations. In particular we have ignored the long-range heterogeneities induced by plastic strain incompatibility. In some regions, long bent dislocations indicate the existence of a stress field whose length scale is much larger than l [41]. Long-range elastic strain fields are also attested by X-ray diffraction [42,43]. The purpose of this paragraph is to examine to which extent the above conclusions are altered by effects of long-range stress fields.

Tractable, analytical expressions for internal stresses are available in some limited cases such as the planar pile up, the ellipsoidal inclusion and the dipolar wall. We restrict the following development to the dipolar wall (section 3.3.2) for its close relationship with experimental situations of interest in plasticity. The general dipolar wall is applied to the two phase model in section 3.3.3. The state of stress originating from an inclined glide band is discussed in section 3.3.4. We conclude in section 3.3.5.

3.3.2. *The dipolar wall*

Consider an infinite solid medium containing a distribution of parallel, identical dislocations, equidistant in the plane P. Formally, this situation may result from a homogeneous plastic deformation restricted to a half space. Let $\boldsymbol{t}$ and $\boldsymbol{b}$ be the common unit vector and the common Burgers vector of the dislocations, respectively. $l = g^{-1}$ is the distance between the dislocations. The average surface dislocation density tensor $\boldsymbol{\alpha}$ is defined as:

$$\alpha_{ij} = g t_i b_j. \tag{30}$$

The condition $\alpha_{1j} = 0$ implies that x_2Ox_3 is the habit plane of the network. In general $\boldsymbol{\alpha}$ has 6 independent components. The stress field $\underline{\sigma}_{\text{net}}$ of the network is the sum of two terms:

$$\underline{\sigma}_{\text{net}} = \underline{\sigma}^{\infty} + \underline{\sigma}_{\text{p}}, \tag{31}$$

$\underline{\sigma}^{\infty}$ is constant in each half space bounded by the network. The difference in stress between each half space is a linear function of $\boldsymbol{\alpha}$. $\underline{\sigma}_{\text{p}}$ is a periodic function of x_2, and x_3 that decreases as $\exp[-2\pi g|x_1|]$ with the distance from P. Hence, the internal stress due to the wall (as for any periodic planar distribution of dislocations) is approximately a constant at a distance from P roughly larger than l [44].

A dipolar wall is defined as a pair of walls forming a slab of thickness $w > l$, and whose long-range stress field σ^{∞} vanishes outside the slab. This occurs, for example, when plastic deformation is confined either inside the slab, or outside the slab (fig. 11).

Consider now an infinite succession of hard slabs periodically distributed inside a soft matrix (fig. 12). Assume now that the matrix undergoes homogeneous deformation while the slabs do not deform. Then the above results combined with Albenga's theorem (that states that the average internal stress is zero) enables one to evaluate the stress at any point

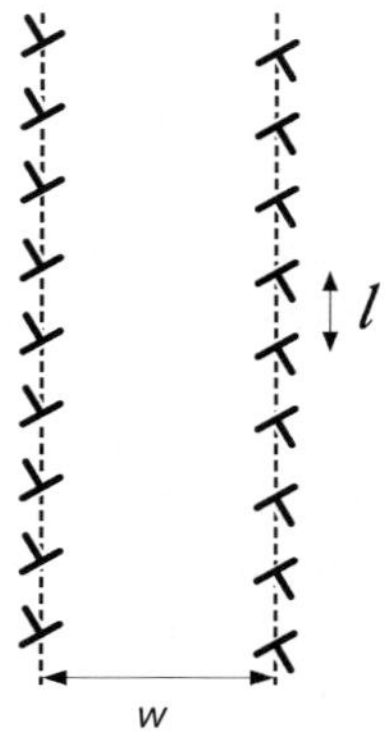

Fig. 11. Dipolar dislocation wall limiting a slab of material with thickness w.

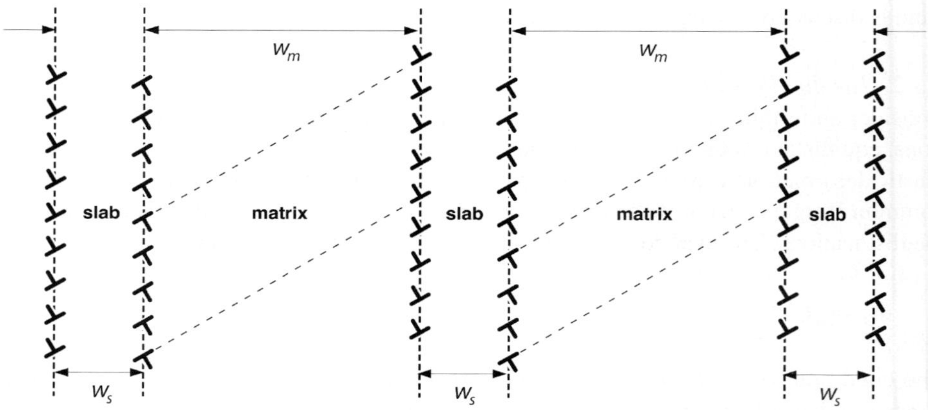

Fig. 12. Periodically distributed slabs. Initially, the slabs are assumed to be less deformable than the matrix. It is the plastic deformation of the latter that generates dipolar walls along the slabs.

of the system. In the case of fig. 12, we have [44]:

$$\gamma_{\mathrm{m}} = \frac{b}{l \cos\varphi}, \tag{32a}$$

$$\Omega = -\frac{b\cos\varphi}{l}, \tag{32b}$$

$$\sigma_{33} = \nu\sigma_{22} = -(1-f)\frac{2\mu}{1-\nu}\frac{b\sin\varphi}{l}, \tag{32c}$$

$$\sigma_{33}^{\mathrm{m}} = \nu\sigma_{22}^{\mathrm{m}} = f\frac{2\mu}{1-\nu}\frac{b\sin\varphi}{l}, \tag{32d}$$

$$f = \frac{w_{\mathrm{s}}}{w_{\mathrm{m}} + w_{\mathrm{s}}}, \tag{32e}$$

where φ is the angle between the glide plane and the walls, γ_{m} is the plastic shear in the matrix, Ω is the rotation between two successive matrix lamellae when the limits of the slab are crossed from left to right, ν is the Poisson ratio, f is the volume fraction of the hard slabs, σ_{ij}^{m} and σ_{ij} are the stresses outside and inside the slab, respectively.

Interestingly, the elastic shear stress opposes and assists dislocation motion in the matrix and in the slabs, respectively. Very generally:

$$\tau_{\mathrm{m}}^{\mathrm{el}} = -kf\gamma_{\mathrm{m}}, \tag{33a}$$

$$\tau_{\mathrm{w}}^{\mathrm{el}} = k(1-f)\gamma_{\mathrm{m}}, \tag{33b}$$

where k is a numerical coefficient that accounts for geometry and γ_{m} is the average plastic shear of the matrix. Given γ_{m}, the thinner the slabs the larger the stress inside the slab. It is noted that when $\varphi = 0$, the stress vanishes everywhere and the boundaries are low angle tilt boundaries whose rotation given by $\pm b/l$.

3.3.3. *The two phases model*

Consider now the case where the slabs deform beyond some threshold stress. We assume that the thin slabs contain a large density of dislocations ρ_{s}. They oppose a resistance to dislocation crossing which is written:

$$\tau_{\mathrm{s}0} = \alpha\mu b\sqrt{\rho_{\mathrm{s}}}. \tag{34a}$$

The large slabs, referred to in the following as the matrix, contain a small density of dislocations $\rho_{\mathrm{m}} \ll \rho_{\mathrm{s}}$, opposing a resistance:

$$\tau_{\mathrm{m}0} = \alpha\mu b\sqrt{\rho_{\mathrm{m}}}. \tag{34b}$$

We assume in addition that there is no initial long-range stress field and that the thickness D of the soft phase is larger than $\mu b/\tau$, where τ is the applied stress.

The matrix starts deforming when the applied stress reaches the value $\tau_{\mathrm{m}0}$. The slabs cannot deform. If the matrix deforms homogeneously, dipolar walls build up on both sides of the thin slabs. From eqs (33), the corresponding elastic shear stress is $-kf\gamma_{\mathrm{m}}$ in the matrix and opposes dislocation motion inside the matrix. In the slabs, the elastic shear stress amounts to $+k(1-f)\gamma_{\mathrm{m}}$ and assists deformation. From eqs (33) the shear stress τ_{m} (resp. τ_{s}) opposing the deformation in the matrix and in the slabs is:

$$\tau_{\mathrm{m}} = \tau_{\mathrm{m}0} + kf\gamma_{\mathrm{m}}, \tag{35a}$$

$$\tau_{\mathrm{s}} = \tau_{\mathrm{s}0} - k(1-f)\gamma_{\mathrm{m}}. \tag{35b}$$

The matrix deforms alone until $\tau_{\mathrm{m}} = \tau_{\mathrm{s}}$, that corresponds to the strain γ_{m}^{*} given by:

$$k\gamma_{\mathrm{m}}^{*} = \tau_{\mathrm{s}0} - \tau_{\mathrm{m}0}. \tag{36}$$

Then both phases deform simultaneously. The flow stress is written:

$$\tau = \tau_{\mathrm{s}} = \tau_{\mathrm{m}} = f\tau_{\mathrm{s}0} + (1-f)\tau_{\mathrm{m}0}. \tag{37}$$

This formula, which was established in a different way by Mughrabi [43], combined with expressions (34) yields:

$$\tau = \alpha \mu b\left[f\sqrt{\rho_s} + (1-f)\sqrt{\rho_m}\right]. \tag{38}$$

The calculation applies only if the dipole wall does not shrink under the attraction of its two boundaries. The dipole wall is stable provided [44]:

$$l > \frac{l_s}{\alpha} \approx 3l_s. \tag{39}$$

On the other hand, the average dislocation density $\langle\rho\rangle$ is:

$$\langle\rho\rangle = f\rho_s + (1-f)\rho_m. \tag{40}$$

We now define $\bar{\alpha}$ from:

$$\tau = \bar{\alpha}\mu b\sqrt{\langle\rho\rangle}. \tag{41}$$

From eqs (37), (39) and (40):

$$\frac{\bar{\alpha}}{\alpha} = \frac{f\sqrt{q} + 1 - f}{\sqrt{fq + 1 - f}}, \tag{42a}$$

$$\rho_s = q\rho_m; \tag{42b}$$

q measures the *hardness* of the *hard* phase relative to the *soft* phase. It is emphasised that for a given value of f, the ratio $\bar{\alpha}/\alpha$ is equal to 1 for $q = 1$, and decreases steadily as q increases, down to $f^{1/2}$ for $q \gg 1$.

Equations (41), (42) show that:

- $\bar{\alpha}$ depends explicitly on both f and q but is insensitive to the average dislocation density,
- the form of the scaling law (4) is not affected by microstructure heterogeneities,
- neither the average dislocation density nor the dislocation density in the dense regions constitute a relevant measure of the state of the crystal.

Whenever the dislocation distribution is markedly heterogeneous the measured value of $\frac{\tau}{\mu b\sqrt{\rho}}$ is not uniquely related to local dislocation interactions but also to the volume fraction and hardness of the hard phase relative to the soft phase.

By differentiation of expression (4), and since $\bar{\alpha}$ depends on strain, formula (6a) is written:

$$\theta = \theta_h + \mu b\sqrt{\langle\rho\rangle}\frac{d\bar{\alpha}}{d\varepsilon}. \tag{43}$$

Here θ_h is the SHR of a homogeneous microstructure with average density $\langle\rho\rangle$. When the microstructure becomes heterogeneous, $d\bar{\alpha}/d\varepsilon < 0$ (see expression (42a)), the SHR increases less rapidly; it may even decrease while the average dislocation density increases.

3.3.4. The stress inside glide bands

As mentioned in section 2.2.2, the primary glide bands are inclined by about 0.02 radians to the glide planes. From formulae (32), this corresponds to a stress of the order of $0.05\mu\gamma$, large enough to generate secondary glide which in turn inhibits primary glide. A detailed discussion of a model of linear SHR based on this remark, is given in Brown's contribution to this volume.

3.4. Mesoscopic simulations

In order to investigate the plastic flow of an heterogeneous microstructure, mesoscopic simulations offer a promising alternative to statistical methods. Whenever required by computer capability or else by the limits of crystal meshing, mesoscopic simulations make use of simplified rules to represent physical elementary processes including their temperature dependence. A great deal of the method is concerned both with the design and with the validation of such rules that have to be adjusted to dynamical processes [45,46].

The dislocation lines are discretised into assemblies of segments. The interaction stress on a given segment is the sum of the contributions from every other segment. Then, in mimicking the in-plane motion of a segment, advantage is taken of the almost negligible inertia of dislocations (and of the smallness of the friction stress in f.c.c. metals) to relate the dislocation velocity of this segment linearly to the local stress.

Even at this elementary stage of the simulation process, non-trivial difficulties had to be solved which arise because:

- The correlated motion of neighbour segments must be ensured.
- The time and space scales must be chosen realistically. When these are too small, the computation is indeed too time-consuming. When they are too large, transition stages may be lost, hence directing the overall evolution along irrelevant though self-consistent paths.

Up to now the minimum length taken for the elementary segment is of the order of a few nanometers, a constraint that naturally excludes processes like jog formation, jog trailing, cross-slip at least in its critical stages, spontaneous edge dipole annihilation, if any, etc.

Starting from a given dislocation distribution, one may follow quite a number of parameters of the simulated tensile test that would be otherwise extremely difficult to estimate; amongst these the density of mobile dislocations, the number and the nature of the junctions, cross-slip events etc.

The efficiency of numerical simulations has been demonstrated in several situations and, with regard to the present review, in the case of collective effects such as planar glide. Processes like the creation and the unzipping of a junction have been reproduced and adjusted to results obtained by simple-minded elastic calculations (see section 3.2). Mesoscopic simulations have fully confirmed expression (4), including the value of the constant $\alpha = 0.3$, and they support the earlier estimation that attractive junctions contribute about 80% of the flow stress. Cross slip is accounted for via a reduced rule that assumes that a screw segment has a temperature dependent probability to cross slip. Although a little rudimentary, the method shows that cross slip may occur from the earliest stage of plastic flow and that it enhances strain hardening. Work is in progress to incorporate more elaborate elastic evaluations of the relevant parameters.

This method has shown quite useful to test the efficiency of elementary mechanisms. The main limitation is indeed computation time. At present, the maximum size of the sample is restricted to a cube with edges 15 μm long, that is, edges one order of magnitude smaller than the storage length Λ defined by formula (12). Besides, the total deformation achieved so far is of the order of 1 to 2%.

4. *Deformation of polycrystals*

4.1. General introduction

The yield stress of polycrystals varies with temperature in the same way as that of single crystals and the Cottrell–Stokes ratio (eq. (18)) has the same properties in both cases. A slight temperature increase changes the value of the maximum SHR, decreases the extent of the plateau and accelerates the decrease of SHR after the plateau. Both dislocation storage and recovery are influenced by grain-size dependent interactions.

4.2. Compatibility constraints

4.2.1. General introduction

Except in very special crystallographic orientation relationships between neighbouring grains, dislocations can neither by-pass nor cross through grain boundaries (GBs). Instead, they must accumulate in the GB vicinity as deformation proceeds. Dislocation aggregation is a source both of internal stresses and of rotations. In order to analyse in which way the presence of GB's modifies the mechanical behaviour solids, let us assume first that plastic deformation is homogeneous in each grain.

Giving each grain an arbitrary deformation generates voids and matter in excess (fig. 13(b)). Plastic deformation is said to be incompatible. Nevertheless, plastic incompatibility may be elastically accommodated which results in internal stresses throughout

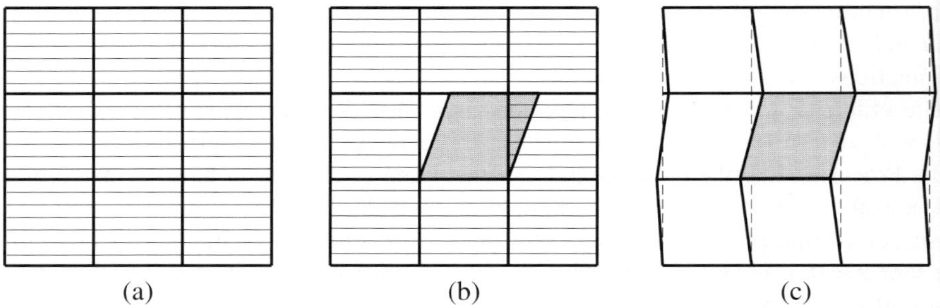

Fig. 13. (a) Stress-free undeformed polycrystal; (b) incompatibility resulting from homogeneous plastic straining of a single grain in the polycrystal; (c) elastic accommodation of the plastic incompatibility.

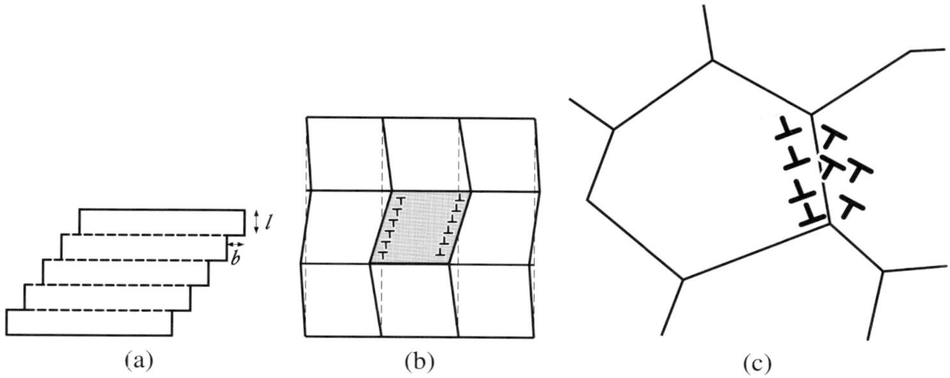

Fig. 14. (a) Homogeneous plastic shear of an isolated grain; (b) homogeneous plastic shear of a single grain embedded in a polycrystal. Note the relation with fig. 13(c). (c) When each grain undergoes an arbitrary homogeneous deformation, a network of dislocations is created at the grain boundaries.

the crystal (fig. 13(c)). Compatible plastic deformation requires that all grains undergo the same plastic deformation. The von Mises theorem shows that then the activity of 5 independent glide systems is required and this is satisfied in f.c.c. metals at all temperatures.

These rather general statements are now analysed at the scale of individual dislocations. Homogeneous plastic deformation of a single crystal results from dislocations gliding in parallel equidistant planes. Figure 14(a) represents a simple situation giving rise to the plastic shear $\gamma_p = b/l$. In this case, all the dislocations are eliminated at the free surfaces after deformation and the crystal is unstressed. When, on the other hand, the grain is embedded in a polycrystal, dislocations accumulate at the GB. These interfacial dislocations generate an internal stress field and a rotation field (fig. 14(b)). When the plastic deformation is the same for all grains, the stress field exerted by the interfacial dislocations tends to zero at a distance l from the boundary. There are no equivalent constraints on the rotation field. Finally, when the average plastic strain of each grain is arbitrarily fixed, the interface dislocations generate both a rotation field and a long range stress field (fig. 14(c)).

Consider the situation where the polycrystalline sample has undergone a homogeneous plastic deformation and assume that one particular grain G is given an excess plastic deformation $\delta\gamma_p$, and that this excess deformation is accommodated *elastically*. In the case of an ellipsoidal grain, the resulting shear stress opposing further plastic strain in G is written [47]:

$$\tau = -\beta\mu\delta\gamma_p, \tag{44a}$$

$$\beta \approx 0.5. \tag{44b}$$

The expression of the stress in the neighbouring grains is not straightforward but its order of magnitude close to the boundary is approximated by eqs (44). Hence, a difference in plastic strain $\delta\gamma_p$ of about 10^{-3} between a given grain and the surrounding matrix inhibits forward plastic flow in the grain. On the other hand, if the polycrystal satisfies the von Mises conditions, then an excess strain $\delta\gamma_p$ of the order of 10^{-3} is sufficient to relax stresses in the matrix plastically.

This forms the basis of Taylor's model [48].

4.2.2. *Taylor's model*

Taylor's model imposes complete compatibility between the various grains which, in every grain, is expressed as:

$$\boldsymbol{\varepsilon} = \boldsymbol{E}, \tag{45}$$

where $\boldsymbol{\varepsilon}$ is the plastic strain tensor in the grain, and $\boldsymbol{E}$ is the macroscopic deformation imposed to the sample. Formula (45) has no general solutions for a crystal deforming with less than 5 independent glide systems. For a crystal with exactly 5 independent glide systems, formula (45) completely determines the contribution of each slip system in each grain. This is not true though in the case of f.c.c. metals which possess 8 independent glide systems. In order to predict which systems are active, Taylor assumes that the critical shear resistance τ_R is the same for all the systems and that the applied axial stress σ necessary to activate a given system a, satisfies:

$$\tau_R = f^a \sigma \tag{46}$$

where f^a is the Schmid factor of system a. The total yield stress is calculated as the average of the yield stresses of the grains. It then depends on texture and is generally expressed as:

$$\sigma = m\tau_R, \tag{47}$$

where m is known as the Taylor factor. In a fully random texture, m amounts to 3.06. From eq. (9), the SHR is rewritten as:

$$\frac{d\sigma}{d\varepsilon} = m^2 \frac{d\tau}{d\gamma}. \tag{48}$$

Hence, the SHR of a polycrystal is about 10 times larger than for a single crystal of the same element deformed in single glide. These results are further discussed in section 4.3.

4.2.3. *Discussion*

The Taylor model suffers from several difficulties.

(1) Homogeneous strain is assumed in each grain whereas observations show that homogeneous multiple glide does not occur. Under such conditions, it is unlikely that the choice of the active glide systems depends on any averaged estimate.
(2) The Taylor model does not refer to a scaling length which is, for instance, in contradiction with the observed grain size dependence of the yield stress (see next section).

(3) The model assumes full plastic deformation compatibility. This is certainly an excessively strong constraint. Accordingly, the model has been further improved by taking incompatibility stresses into account. Each grain is assumed to be embedded in an *average* matrix made of all the other grains. This matrix is determined by averaging the properties of all grains self-consistently. The formalism of this class of models is, however, too complex to be even outlined here. It must be noticed that they still suffer criticisms (1) and (2).

(4) The Taylor and self-consistent models do not take into account the time dependence of plastic flow. This has been improved by expressing the plastic shear rate of a glide system as a function of the total shear stress on this system. The approximation has the advantage to give a physical criterion for the amount of deformation of the various glide systems, but it still suffers from difficulties (1) and (2).

4.3. The yield stress

In a polycrystal deformed in tension, plastic flow starts in those grains whose Schmid factor is the highest, which in turn explains why the elastic–plastic transition is not sharp. What is termed yield stress σ_Y in a polycrystal is the 0.2% proof stress, i.e. the flow stress for a plastic strain conventionally taken as 0.2%. Experiments show that σ_Y varies with the reciprocal of the square root of the grain size as:

$$\sigma_Y = \sigma_0 + K d^{-1/2}. \tag{49}$$

Both σ_0 and K depend on the nature of the crystal and on texture. This equation, known as Hall–Petch law [49,50] represents the situation satisfactorily for grain sizes ranging from about 1 μm up to 1 mm. Departures from the Hall–Petch law have been observed in very fine-grained polycrystals [51], as well as in thin films [52].

Early theories of the yield stress made the assumption that slip started in some well-oriented grains and was blocked under stress at GB's under the form of planar dislocation pile-ups. The force $b\tau_1$ on the leading dislocation can be conveniently written in terms of the size D of the pile up as:

$$b\tau_1 = F\frac{(\tau - \tau_0)^2 D}{\mu}, \tag{50}$$

where τ_0 is a resistance opposed to each dislocation of the pile-up, F is a geometrical constant, τ the applied stress. Formula (49) is then derived by assuming that the size of the pile up is equal to the diameter of the grain and by identifying τ_1 with the stress beyond which the GB supposedly yields. This is not quite satisfactory since, as emphasised earlier, the first dislocation cannot in general cross through the grain boundary. One may argue, instead, that the stress τ_{eff} at a distance d ahead of the spearhead is written:

$$\tau_{\text{eff}} = C(\tau - \tau_0)\sqrt{\frac{D}{d}}. \tag{51}$$

Then the yield stress would represent the stress necessary to activate slip in the adjacent grain at a given distance ahead of the pile-up. Formula (51) is valid only very close to the spearhead. The distance d may be identified with that to activate the motion of dislocations either in the next grain, or in the grain boundary [53]. Finally, pile-ups should tend to be relaxed by intragranular deformation or else by cross slip, in particular those whose length approaches 1 mm, where the Hall–Petch law is still valid.

In view of the above inconsistencies, it has been alternatively considered that the yield stress of a polycrystal corresponds to the stress at which the sample has already undergone a permanent deformation ε that corresponds to an average plastic shear γ of each grain given by $\gamma = m\varepsilon \approx 3\varepsilon$. It is then supposed that loops homogeneously nucleated in the grain expand until they reach the grain boundary at which they are blocked. For a grain of diameter D, this is achieved by a number $n \approx 3\gamma D/b$ of dislocation loops that represent a total dislocation length:

$$\lambda \approx n\pi D = \frac{3\pi D^2\varepsilon}{b}. \tag{52}$$

This is the basis of some models of the Hall–Petch law [54] (see remarks 1 and 2, below). Observations, however, reveal a network intimately entangled in the interior of the grains. In order to take these into account, we make the assumption that the additional length λ' of dislocations created inside the grains during deformation somehow scales with λ ($\lambda' = a\lambda$). Then the dislocation density ρ in the grain is

$$\rho \approx \frac{6a\lambda}{\pi D^3} = \frac{18a\varepsilon}{bD}, \tag{53}$$

where a is an unknown constant that relates the dislocation length in the grain to that at the interface. Under the simplifying assumption that dislocations responsible for the deformation of a given grain are more or less homogeneously distributed in this grain, one can see that the combination of formulae (4) and (53) yields the Hall–Petch size dependence of the 0.2% flow stress

$$\sigma = \sigma_0 + 3\alpha\mu\sqrt{\frac{2a\varepsilon b}{D}} \tag{54a}$$

with

$$\varepsilon = 2 \cdot 10^{-3}. \tag{54b}$$

REMARK 1. Dividing λ (formula (52)) by the volume of the grain, and applying formula (4), actually yields the Hall–Petch law, but the derivation is inconsistent since then the internal stress field due to the interface dislocations is independent of grain size (formulae (32)).

REMARK 2. Formulae (54) rely on the undemonstrated assumption that intragranular dislocation storage depends linearly on interfacial storage. This assumption is certainly not correct for very small-grained crystals. Departure from Hall–Petch law is consistent with this remark.

REMARK 3. At least part of the stress due to the GB dislocations in a given grain should be relaxed by the GB dislocations of the neighbouring grain, but this modifies only the term σ_0 in formula (54a).

REMARK 4. In general, grains have a polyhedral rather than ellipsoidal shape which in turn generates stress concentrations at edges. Depending on the symmetry of the crystal, the stress close to an edge scales either with $\ln(r/D)$ or with $(r/D)^{-n}$, where r, D and n are the distance to the edge, the size of the grain and an exponent smaller than 1, respectively [55]. It is thus plausible that, in some cases, grain boundaries act as dislocations sources. Similar stress concentrations may also occur at GB ledges.

REMARK 5. The situation is probably more involved than the simple description given above. There are, indeed experimental evidences of single crystals deformed in multislip that exhibit a yield stress larger than for large-grained polycrystals. We come back to this point in the next paragraph.

4.4. Work hardening

Typical stress dependencies of θ and $\sigma\theta$ in twisted polycrystals are shown figs 15(a) and (b). The rapid initial decrease of $\theta(\sigma)$ expresses the fact that the number of grains participating in deformation increases gradually as the threshold stress for this grain is attained. The curve exhibits a plateau for a SHR of the order of 10 times the maximum SHR for a single crystal deformed in single slip. Subsequently, $\theta(\sigma)$ decreases almost linearly to more or less stabilise at a very small level. This behaviour compares reasonably well with that of a $\langle 100\rangle$- or $\langle 111\rangle$-oriented single crystals deformed in multislip from the beginning of deformation.

The following puzzling facts should be noticed though:

- The SHR of large grained samples may be very small at the onset of the macroplastic stage, and then increase rapidly as for single crystals oriented in single glide [57,58]. This is inconsistent with the operation of multiple glide.
- Formulae (54) apply at any strain. They predict that the coefficient K of eq. (49) should increase as $\varepsilon^{1/2}$ which is in reasonably good agreement with experiments. However, they also predict that stress–strain curve of crystals of various grain size should not present any cross over which is not always supported by experiments [59]. In the light of the previous paragraph, this indicates that the coefficient a might vary with the amount of plastic strain.
- Systematic experiments conducted on Cu deformed at 4.2 K show that large-grained crystals exhibit a behaviour definitely different from that of small-grained crystals [29] (fig. 16). Up to a grain size of approximately 0.1 mm, the SHR of polycrystals is larger

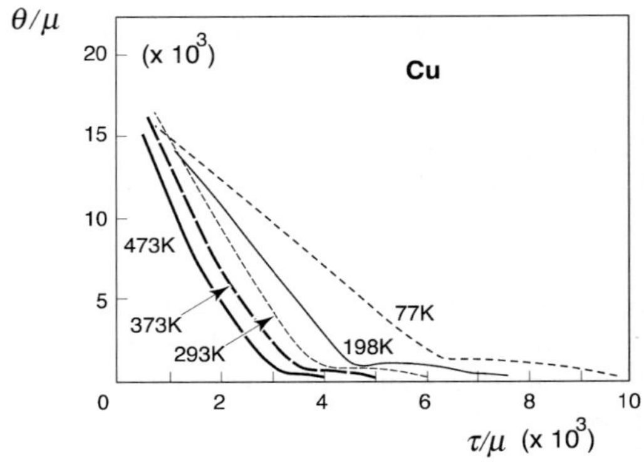

(a)

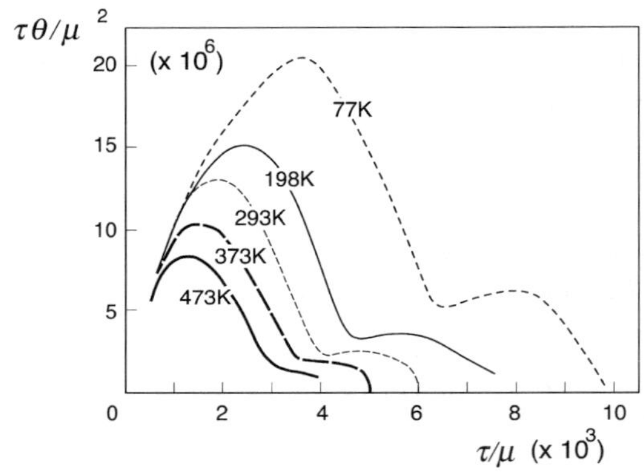

(b)

Fig. 15. (a) Stress dependence of the strain-hardening rate in polycrystalline Cu deformed at various temperatures. The SHR decreases from an ill-defined upper value which is almost temperature independent. Notice the strong effect of temperature on the rate at which the SHR decreases. (b) Corresponding $\sigma\theta(\sigma)$ diagram. After [56].

than that of a $\langle 100 \rangle$-oriented single crystal. However, this is no longer true for grain sizes larger than 0.4 mm. Interestingly, the 0.1–0.4 mm range corresponds reasonably well to the storage length Λ defined by formula (12), thence suggesting that for crystal sizes smaller (or larger) than Λ, storage at the GB (or in the grain) predominates.

- Large-grain polycrystals may have a smaller SHR than $\langle 100 \rangle$ oriented single crystals. One possible reason may be that some grains in large-grain polycrystals may deform in single glide.
- Additional information in fig. 16 is that both the SHR at the plateau and the extension of the latter depend on grain size. The conclusion is that both the storage and the recovery rate depend on the grain size.

These observations as well as remark 5 in the previous paragraph indicate that the locking of dislocations at the interface influences the evolution of the microstructure and that this depends on the grain size. The origin of the effect is not yet understood.

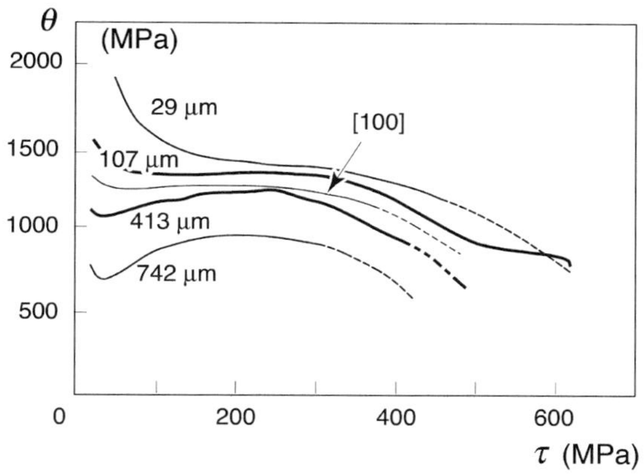

Fig. 16. $\theta(\sigma)$ diagram for a $\langle 110 \rangle$-oriented single crystal compared with those of polycrystals with various grain sizes. Depending on grain size, the SHR of the single crystal may be larger or smaller than that of the polycrystal. After [29].

Finally, localised shear bands in highly deformed pure metals may originate from the fact that new glide systems develop in alien structures after some strain and this may generate instabilities of plastic flow (sections 2.2.7 and 2.3).

5. Cross-slip

5.1. Introduction

The importance of the cross-slip process, both as a potential multiplication mechanism and as the only conceivable low-temperature recovery process, has been emphasised as early as in the fifties. Many microscopic mechanisms have been proposed and discussed. With regard to strain hardening, almost all early analyses on the subject were addressed to interpret the Stage III of the stress–strain curve of f.c.c. single crystals.

In all the models of cross slip, the evaluation of the activation energy relies more or less on the calculation of the energy of a constricted segment of the screw dislocation and of the further development of this segment in the cross-slip plane. The activation energy is estimated as the sum of the constriction energy and of the work to expand the segment in the cross-slip plane. This kind of approach has been in particular applied to the analysis of plastic flow in f.c.c. metals by Escaig [60,61], from a model originally outlined by Friedel [62,63] and whose occurrence has been recently confirmed by atomic simulations [64]. Escaig's analysis has been conducted in the elastic approximation and under several simplifying assumptions. Two interesting aspects were emphasised (i) the necessity to take into account the resolved shear stresses both on the primary and on the cross-slip planes and (ii) the importance of pre-existing constrictions.

5.2. The Friedel process

5.2.1. General introduction

The process of cross slip of a screw dislocation gliding in the primary plane is assumed to occur in three steps:

(1) The screw dislocation takes a constriction associated with an energy E_c.

(2) The constriction extends along the dislocation line, splits and bows out into the cross-slip plane (fig. 17). The configuration comprises two constrictions E and S (E and S stand for edge and screw, respectively) separated by a distance l, and has an energy $E_{dc}(\tau_{cs}, l)$ where τ_{cs} is the resolved shear stress in the cross-slip plane.

(3) The dislocation spreads in the cross-slip plane. For a given value of τ_{cs}, $E_{dc}(\tau_{cs}, l)$ presents a maximum E_{cs}, which is the activation energy for cross slip.

In order to exemplify the difficulties in evaluating the various relevant energies, consider the difference in line energy ΔE between the undissociated form of a screw dislocation and its form dissociated into Shockley partials at equilibrium at a distance w_e. A simple calculation in the isotropic approximation shows that:

$$\Delta E = \frac{\mu b^2}{24\pi}\frac{(2-3\nu)}{(1-\nu)}\left[\frac{(4-3\nu)}{(2-3\nu)}\ln\frac{r_1}{r_c} + \ln\frac{w_e}{e r_c}\right], \tag{55}$$

where r_1, and r_c are the cut-off radii of the perfect and Shockley dislocations. The usual assumption that these radii are proportional to the length of the Burgers vectors and taking $\nu = 1/3$, implies that:

$$\Delta E \approx 0.02\mu b^2 \ln\frac{w_e}{14 r_c}. \tag{56}$$

Hence, in most cases, both the absolute value and the sign of ΔE depend very markedly on the assumptions made for the size of the cut-off radius. This result remains true in elastically anisotropic crystals.

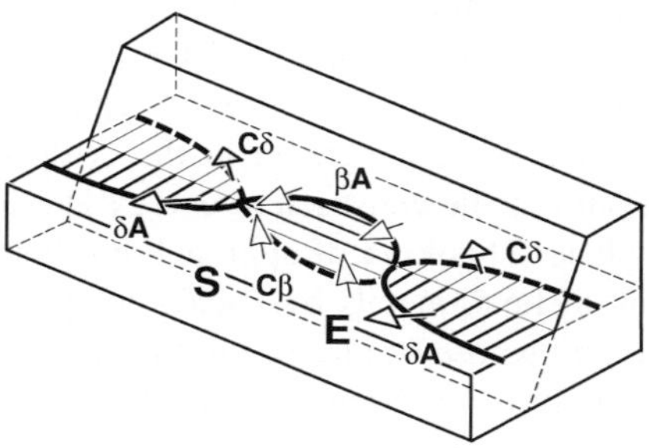

Fig. 17. Step 2 of the Friedel process for cross slip. Notice that the two constrictions, E and S, differ in terms of the characters of the interconnecting dislocation partials.

Under isotropic elasticity and the line tension approximation, the constriction energy E_c is written [65]:

$$E_c = \sqrt{2w_e\gamma\Gamma}\left(1 + \frac{E_l}{\Gamma}\right)w_e I(w_e), \tag{57}$$

$$I(w_e) = \int_{\frac{r_c}{w_e}}^{1} \sqrt{u - \ln u - 1}\, du, \tag{58}$$

where γ, Γ and E_l are the stacking fault energy, the line tension and the line energy of the straight Shockley partials, respectively. This approach takes partly into account the orientation dependence of the line tension. The weakest point of relations (57), (58) is the evaluation of the function I, for this depends critically of the assumption made on the cut-off radius r_c whenever the splitting width is smaller than $30b$, that is, for all cases of interest.

A second source of uncertainty comes from the variety of possible constriction configurations [66]. Besides, the double constriction may nucleate on a pre-existing constriction on a jog which lower its energy.

The calculation has been further improved, as in the case of junction unzipping, via a direct estimate of the elastic interaction between dislocation segments and by taking anisotropy into account. One interesting result is that the two constrictions E and S are different. Constriction E is more rounded than the constriction S. This difference stems directly from the orientation dependence of dislocation line tension [67,68].

5.2.2. Atomistic simulations of the cross-slip process

So far, only a few attempts at analysing the cross-slip process in Cu and Ni have been achieved by atomistic simulations [64,69,70]. They confirm the operation of the Friedel process and the existence of the two types of constrictions (E and S). However, numerical results depend strongly on the choice of the potential and on the postulated process (table 1):

1. The energy of the screw constriction (S) is negative and relatively small, while that of the edge (E) constriction is positive and large.
2. The cross-slip energy E_{cs} depends quite markedly on the choice of the potential.
3. The energy E_D of a pair of constrictions varies with their distance x as:

$$E_D = E_{cs}\left[1 - 0.82\exp\left\{-4.88\cosh 4.88\left(1 - \frac{x}{60b}\right)\right\}\right], \tag{59a}$$

$$E_D = E_{cs}\left[0.5 + \frac{1}{\pi}\,\mathrm{Arctan}\,\frac{1.8x}{w_e}\right]. \tag{59b}$$

   Formulae (59a) and (59b) are fitted from numerical values for Cu and Ni, respectively.
4. $E_c' = E_{cs} - E_D(0)$ is an approximate expression for the cross-slip energy on a preexisting constriction.

Table 1
(Energies are in eV)

		S	E	E_{cs}	E_c/E_{cs}	E'_c	E_j	E_k
Cu1	[64]	−1.1	3.8	2.7	0.6	1.1		
Cu2	[69]			3.4			0.87	0.86
Ni1	[70]	−1.45	6.3	4.85	0.5	2.42		
Ni2	[70]			2.35	0.5	1.15		

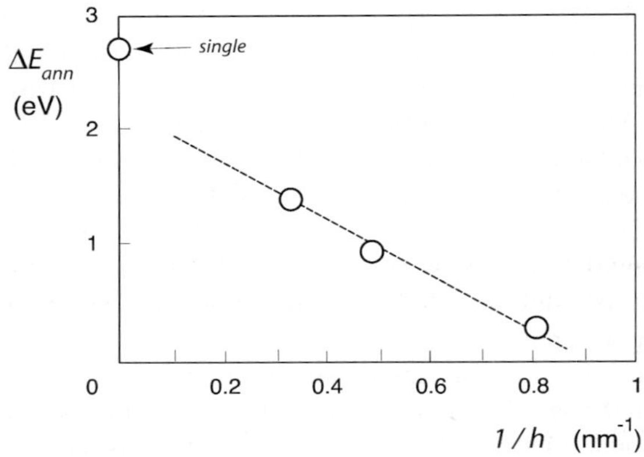

Fig. 18. The activation energy for annihilation of rigid screw dipoles as a function of dipole height. After [69]. The dipole of infinite height is a single dislocation.

5. E_j and E_k are the cross-slip energies on a pre-existing jog. They depend the on jog structure. The pronounced difference between these and E'_c should be noted.

The activation volume, calculated in the case of Ni, is of the order of $20b^3$ for a stress of $10^{-3}\mu$.

The method has been applied to address the issue of the annihilation of a screw dipole comprised of unjogged dislocations in Cu. It is found that rigid screw dipoles located 1.2 nm apart recombine spontaneously. For rigid screw dipoles of height h (expressed in nm), the activation energy per unit length for annihilation E_{an} expressed in eV/b is fitted by:

$$E_{an} = 0.13 - \frac{0.61}{h}. \tag{60}$$

For flexible dislocations, the process depends on line tension, here both on the height and on the length of the dipole. For a given height, the energy increases with the length, and saturates for a length of order 4 to 6 times this height, implying that simulations should be conducted on sufficiently long dislocations. The annihilation energy depends on h as indicated in fig. 18. An activation volume of the order of that found for Ni has been inferred from these results.

5.3. Discussion

It is rather clear that, in their present state, calculations based on the elastic approximation are not adapted to a reliable analysis of the cross-slip processes. A tractable and accurate description of these will combine precise atomic simulation estimates for the cut-off radius with rigorous elastic calculations. One condition for this is most evidently to employ several reliable potentials for a given metal, paying particular attention to the fitting of the dislocation width. It seems also insightful to cross-check the results for a large set of f.c.c. metals, keeping in mind that the variety of constrictions is rather large. Since the activation energy for cross slip is lowered by about 40% at a constricted dislocation, another direction is to compare between simulations achieved on unconstricted and constricted dislocations. A systematic analysis of the effect of stress is also essential.

The analysis of dipole annihilation is a good start for the analysis of recovery and it is expected that this will be further developed in the future.

Inherent difficulties will, however, remain in comparing theoretical calculations and experimental measurements. This is because the former are aimed at idealised systems whose relevance to real situations must be validated. The two following examples may illustrate this point of view:

1. *The critical height for spontaneous annihilation of a screw dipole in Cu*:

This distance is all the more important since it is an instrumental parameter in certain mesoscopic simulations. From atomic calculations this height is 1.2 nm in Ni and Cu, whereas no screw dipoles below a height of approximately 50 nm could be evidenced by TEM [71]. In order to explain this discrepancy, it may be argued that real screw dipoles contain jogs which may actually ease significantly the annihilation of a dipole with a relatively large *average* height. Recovery would then depend not only on stress and temperature but also on the fine structure of screws, that is, on strain hence, on mechanical history. On the other hand, one might also argue that more systematic TEM observations are necessary to confirm the value of 50 nm.

2. *The onset of Stage III*:

It has been stated for years that the onset of Stage III, or more precisely the stress τ_{III} corresponding to the maximum of SHR reflects the onset of prominent cross slip. Experiment shows that τ_{III} follows an exponential law:

$$\frac{\tau_{\mathrm{III}}(T)}{\tau_{\mathrm{III}}(0)} = \frac{\mu(T)}{\mu(0)} \exp\left(-\frac{T}{T_{\mathrm{c}}}\right), \tag{61}$$

where T_{c} is a parameter. Assuming that the yield stress at the onset of Stage III is controlled by a single cross-slip mechanism, one may evaluate the activation energy of the latter and its activation volume. In practice, the activation energy measured with this method indeed agrees reasonably well with the calculated value for a preconstricted dislocation but there is a discrepancy by a factor of 10 in the activation volume.

This suggests that the assumption of a single cross-slip process controlling plastic flow at the maximum of SHR is not indisputable.

6. Conclusion

Based on early TEM observations, the forest theory has succeeded in relating the stress opposing dislocation motion to interactions with intersecting dislocations. This interaction defines the average distance l between the latter as a fundamental scaling length. The connection between forest interaction and latent hardening had been established as early as 1959. The latter is responsible for the building of large scale heterogeneities as soon as multiple slip occurs. Observed differences in mechanical behaviour between single crystals oriented in multislip and polycrystals with various grain sizes confirm that phenomena occurring at various scales combine their effects. It is frustrating, however, to realise that the only well identified scaling length is the grain size, a result which has been known for a long time. Local analysis at an intermediate scale (1–10 microns) which has brought some results on the scaling of the distributions of the misorientations between dislocation cells, might provide the missing clues.

What is lacking is information on the dynamics of the three-dimensional evolution of the microstructure, which is controlled for a large part by cross slip. In this case, the elementary processes cannot be observed at the atomic scale and one has to rely on a theoretical description, which is still in progress.

In its evolution, the microstructure combines processes which are active to various extents, from the onset of plastic flow. The stress–strain curve averages their effect in a quite complex way. Observed changes on the SHR are connected neither with a given level of stress nor with a single process.

It must be noticed finally that the SHR is a state variable at least as relevant as the flow stress.

References

[1] S.J. Basinski and S.J. Basinski, Dislocations in Solids, Vol. 4, ed. F.R.N. Nabarro (North-Holland, Amsterdam, 1979) p. 261.
[2] J.P. Hirth and J. Lothe, Krieger (Malabar, Florida, 1992).
[3] F. Seitz and T.A. Read, J. Appl. Phys. 12 (1941) 470.
[4] E. Orowan, Discussion, Symposium on Internal Stresses (Inst. Metals London, 1947) p. 451.
[5] L.M. Brown and R.K. Ham, Strengthening Methods in Crystals, eds A. Kelly and R. Nicholson (Applied Science Publishers Ltd, London, 1971) p. 9.
[6] A. Seeger, Phil. Mag. 45 (1954) 771.
[7] H. Mecking, Workhardening in Tension and Fatigue, ed. A.W. Thompson (TMS-AIME, New York, 1977) p. 67.
[8] G. Saada, Acta Metall. 9 (2) (1961) 166.
[9] G. Saada, Physica 27 (1961) 657.
[10] A. Georges and J. Rabier, Revue de Physique Appliquée 22 (1987) 941.
[11] M.S. Duesberry, Dislocations in Solids, Vol. 8, ed. F.R.N. Nabarro (North-Holland, Amsterdam, 1989) p. 89.
[12] P. Veyssière and G. Saada, Dislocations in Solids, Vol. 10, eds F.R.N. Nabarro and M.S. Duesberry (North-Holland, Amsterdam, 1996) p. 53.
[13] F.R.N. Nabarro, Strength of Metals and Alloys ICSMA 7, Vol. 3, eds H.J. McQueen, J.P. Bailon, J.I. Dickson, J.J. Jonas and M.G. Akben (Pergamon, Oxford, 1986) p. 1667.
[14] F.R.N. Nabarro, Z.S. Basinski and D.B. Holt, Adv. Phys. 13 (1964) 193.

[15] T.E. Mitchell, Prog. Appl. Mater. Res. 6 (1964) 117.
[16] J. Gil Sevillano, Materials Science and Technology, Vol. 6, ed. H. Mughrabi (V.C.H., Cambridge, 1993) p. 19.
[17] L.M. Clarebrough and M.E. Hargreaves, Prog. Met. Phys. 8 (1959) 1.
[18] J.S. Koehler, Phys. Rev. 86 (1952) 52.
[19] W.G. Johnston and J.J. Gilman, J. Appl. Phys. 31 (1960) 632.
[20] P. Haasen, Phil. Mag. 3 (1958) 384.
[21] P. Franciosi, M. Berveiller and A. Zaoui, Acta Metall. 28 (1980) 273.
[22] E. Voce, J. Inst. Met. 74 (1948) 537.
[23] S. Basinski and P.J. Jackson, Phys. Stat. Sol. 9 (1965) 805.
[24] Z.S. Basinski, Phil. Mag. 4 (1959) 393.
[25] P.R. Thornton, T.E. Mitchell and P.B. Hirsch, Phil. Mag. 7 (1962) 337.
[26] M. Carrard and J.L. Martin, Phil. Mag. A 58 (3) (1988) 491.
[27] P. Anongba, J. Bonneville and J.L. Martin, ICSMA 8, eds P.O. Kettunen, T.K. Lepistö and M.E. Lehtonen (Pergamon, Oxford, 1988) p. 265.
[28] A.H. Cottrell and R.J. Stokes, Proc. Roy. Soc. London A 233 (1955) 17.
[29] M. Niewczas and J.D. Embury, The Integration of Material, Process and Product Design, ed. Zabaras (Balkema, Rotterdam, 1999) p. 71.
[30] H. Suzuki, S. Ikeda and S. Takeuchi, J. Phys. Soc. Japan 11 (1956) 82.
[31] W. Hosford, R.L. Fleisher and W.A. Backhofen, Acta Metall. 8 (1960) 187.
[32] G. Saada, Acta Metall. 8 (1960) 200.
[33] G. Saada. Acta Metall. 8 (1960) 841.
[34] G. Saada, Electron Microscopy and Strength of Crystals, eds G. Thomas and J. Washburn (Interscience, New York, 1963) p. 651.
[35] G. Schoeck and R. Friedman, Phys. Stat. Sol. A 53 (1972) 661.
[36] W. Püschl, G. Schoeck and R. Friedman, Phys. Stat. Sol. A 74 (1982) 211.
[37] Z.S. Basinski, Scr. Metall. 8 (1974) 1301.
[38] H. Mecking and U.F. Kocks, Acta Metall. 29 (1981) 334.
[39] D. Rodney and R. Phillips, Phys. Rev. Lett. 82 (8) (1999) 1704.
[40] V.B. Shenoy, R.V. Kutka and R. Phillips, Phys. Rev. Lett. 84 (7) (2000) 1491.
[41] U. Essmann, Phys. Stat. Sol. 12 (1965) 707.
[42] T. Ungar, H. Mughrabi, D. Rönnpagel and M. Wilkens, Acta Metall. 32 (1984) 333.
[43] H. Mughrabi, Acta Metall. 31 (1983) 1367.
[44] G. Saada, Solid State Phenomena, Vol. 59–60, eds N. Clément and J. Douin (Scitec Publications Ltd, Zürich, 1998) p. 77.
[45] B. Devincre and L.P. Kubin, Modelling Simul. Mater. Sci. Eng. 2 (1994) 559.
[46] L.P. Kubin and B. Devincre, Deformation-induced Microstructures: Analysis and Relation to Properties, eds J.B. Bilde-Sørensen, J.V. Carstensen, N. Hansen, D. Juul Jensen, T. Leffers, W. Pantleon, O.B. Petersen and G. Winther (Risoe National Laboratory, Roskilde, Denmark, 1999) p. 61.
[47] J.D. Eshelby, Proc. Roy. Soc. A 241 (1957) 376.
[48] G.I. Taylor, J. Inst. Metall. 62 (1938) 307.
[49] E.O. Hall, Proc. Phys. Soc. London B 64 (1951) 747.
[50] N.J. Petch, J. Iron Steel Inst. 174 (1953) 25.
[51] C.S. Pande, R.A. Masamura and R.W. Armstrong, Nano-Structured Materials 2 (1993) 323.
[52] A. Misra, M. Verdier, Y.C. Lu, H. Kung, T.E. Mitchell, M. Nastasi and J.D. Embury, Scripta Mater. 39 (1998) 555.
[53] J.C.M. Li and G.C.T. Liu, Phil. Mag. 38 (1967) 1059.
[54] M.F. Ashby, Phil. Mag. 21 (1970) 399.
[55] G. Saada, J. Phys. France 50 (1989) 2505.
[56] J.M. Alberdi Garitaonandia, PhD thesis (Faculty of Sciences, University of Navarra, San Sebastian, Spain, 1984).
[57] H. Suzuki, S. Ikeda and S. Takeuchi, J. Phys Soc. Japan 11 (1956) 382.
[58] W.F. Hosford, R.L. Fleisher and W.A. Backhofen, Acta Metall. 8 (1960) 187.

[59] A. Lasalmonie and J.L. Strudel, Journal Mater. Sci. 21 (1986) 1837.
[60] B. Escaig, J. Phys. 29 (1968) 255.
[61] J. Bonneville, B. Escaig and J.L. Martin, Acta Metall. 36 (8) (1988) 1989.
[62] J. Friedel, Les Dislocations (Gauthier-Villars, Paris, 1956).
[63] J. Friedel, Dislocations (Pergamon, London, 1964).
[64] T. Rasmussen, Multiscale Phenomena in Plasticity, eds J. Lépinoux, D. Mazière, V. Pontikis and G. Saada, (Nato Science Series E 367, Kluwer, Dordrecht, 2000) p. 281.
[65] A.N. Stroh, Proc. Phys. Soc. B 67 (1954) 427.
[66] G. Saada, Mater. Sci. Eng. A 137 (1991) 177.
[67] M.S. Duesbery, N.P. Louat and K. Sadananda, Acta Metall. Mater. 40 (1992) 149.
[68] M.S. Duesbery, Modelling Simul. Mater. Sci. Eng. 6 (1998) 35.
[69] T. Rasmussen, T. Vegge, T. Leffers, O.B. Petersen and K.W. Jacobsen, Phil. Mag. A 80 (2000) 1273.
[70] S. Rao, T. Partharasathy and C. Woodward, Phil. Mag. A 79 (1999) 1167.
[71] U. Essmann and H. Mughrabi, Phil. Mag. 40 (1979) 731.

CHAPTER 62

Work Hardening in Some Ordered Intermetallic Compounds

B. VIGUIER

Centre Interuniversitaire de Recherche et d'Ingénierie des Matériaux
UMR CNRS 5085, Institut National Polytechnique de Toulouse
31077 Toulouse cedex 4
France

J.L. MARTIN

Ecole Polytechnique Fédérale de Lausanne
IPMC-FSB
1015 Lausanne
Suisse

and

J. BONNEVILLE

Université de Poitiers, LMP-UMR 6630 CNRS
B.P. 179, 86960 Futuroscope cedex
France

Dislocations in Solids
Edited by F. R. N. Nabarro and M. S. Duesbery

Contents

1. Introduction

Several ordered intermetallic compounds exhibit a positive temperature dependence (TDFS) of their flow strength, which make them very attractive for both research investigations and industrial applications. During approximately the last forty years, a considerable effort has been devoted to the characterisation and the understanding of such strength anomalies with temperature, leading to a tremendous amount of publications for a variety of ordered intermetallic alloys; the most studied being certainly the $L1_2$ intermetallic Ni_3Al. Excellent reviews have been dedicated to the subject with regards to both their potential applications and degree of understanding of the fundamental mechanisms controlling their strength and plastic properties. One of the more complete and recent review on the latter subject can be found in the volume 10 of '*Dislocations in Solids*' [1], which consists in several chapters dealing with mechanical properties and modelling of numerous intermetallic alloys of different structures, investigated by various experimental techniques.

In addition to the well known strength anomalies, peculiar behaviours of the related work-hardening rates as a function of temperature have also been reported in the literature, but to a lesser extent. This is at variance to pure metals, such as, for instance, the face centred cubic (f.c.c.) metals, for which an abundant literature concerning work-hardening processes is available (see, for instance, [2]). A reason for that may be that the proposed models are essentially focused at explaining the flow stress anomaly without provisions for other physical parameters such as the work-hardening rate and/or the strain-rate sensitivity of the stress. Flow stress anomalies do not all originate from the same dislocation mechanism and, depending on the ordered phase considered, more or less understanding of the anomaly phenomena has been achieved. For instance, in the case of the Ni_3Al intermetallic compound and its alloys, it is now well admitted that the flow stress anomaly is a direct consequence of a cross-slip mechanism from the primary octahedral plane to the cube cross-slip plane. Depending on the dislocation dynamics considered in the models, this cross-slip process leads to a decrease in the mobility and/or in the density of the dislocations which control plastic flow. Since this cross-slip mechanism is thermally activated, with increasing temperatures a decrease in either dislocation mobility and/or mobile density, or both is therefore expected, yielding the flow stress anomaly. Therefore, the question which is now addressed concerns the predictions of such models on the work-hardening rate (WHR), i.e., dislocation storage and recovery, and eventually on the strain-rate sensitivity of the stress (SRSS), which characterises dislocation mobility. Such an achievement has not yet been completed for all materials exhibiting a positive TDFS, and we shall therefore restrict our investigations to the $L1_2$ structure, for which Ni_3Al and Cu_3Au will be considered, and to the case of γ-TiAl alloys of the $L1_0$ structure. We shall see that even for these well documented intermetallics the experimental data concerning the WHR are rather limited.

The first part of this contribution presents the experimental characterisations of the WHR found in the literature for single and polycrystals of Ni_3Al, Cu_3Au and γ-TiAl. Since the experimental results are usually very limited and sometimes controversial, special attention has been paid to extracting the major trends from the data. The second part concerns investigations on the dislocation mobility and mobile dislocation density (i.e., in a manner correlated to dislocation exhaustion) in the intermetallics Ni_3Al and γ-TiAl. Experimental results obtained by various techniques, such as etch pit experiments, repeated stress relaxations, successive creep tests and mechanical spectroscopy, are presented and compared. Section 4 presents an attempt to correlate the internal stress of the intermetallic compounds Ni_3Al and γ-TiAl with dislocation densities that are measured by transmission electron microscopy (TEM), using a modified Taylor relationship. Then, complex multistep deformation experiments which have been designed for investigating the reversible and irreversible parts of the flow stress (i.e., influence of pre-straining and Cottrell–Stokes type experiments), are presented and critically examined in part 5. Part 6 is aimed at discussing the origin of the flow stress anomaly in Ni_3Al in connection with the work-hardening behaviour. Some guide lines are given for interpreting the WHR in this structure. Finally, the case of γ-TiAl is briefly approached.

2. *Experimental measurements of the work-hardening rates*

As a general remark, it must be emphasised that the experimental values of the work-hardening rates (WHR), which are available in the literature for various intermetallic compounds are rather scarce and unfortunately given at different strain levels. In addition, several definitions of WHR have been used, leading as well to some difficulties for a direct comparison. Therefore, we shall consider, for presenting a given feature of the WHR, the study which illustrates the considered subject most clearly; of course, some attention will be also paid to similar results established by other authors.

In this section, we shall present the WHR of intermetallic compounds of the $L1_2$ and the $L1_0$ structures, which exhibit an anomalous behaviour of their flow stress with temperature. We shall begin with the $L1_2$ phase for which two ordered compounds will be considered, i.e. Ni_3Al with various alloying elements and Cu_3Au. The features addressed will essentially concern the hardening stages, together with the associated lattice rotations, and the influence of several parameters such as strain rate, temperature and crystallographic orientation. Particular attention will be also paid to the WHR behaviour with temperature in connection with its strain dependence. The second part of this section will be devoted to the $L1_0$ structure. Experimental results concerning the WHR are rather scarce for this structure. Therefore, we shall only present single-phase γ-TiAl based alloys for which the most complete set of experimental data referring to the WHR is available. Here again the influence of various experimental parameters will be checked.

2.1. Work-hardening rates in $L1_2$ ordered intermetallics

2.1.1. Single crystalline Ni_3Al based alloys

2.1.1.1. Hardening stages. The plastic behaviour of Ni_3Al has been essentially examined close to yielding, that is, at a plastic offset strain usually smaller than or equal to few per cent. Only a few studies have been devoted to investigate the flow strength of this intermetallic over a large deformation range.

Aoki and Izumi [3] have examined the operation of the secondary (conjugate) slip system during tensile deformation of $Ni_{77.6}Al_{22.4}$ single crystals at room temperature (RT) for an initial imposed strain rate of $5.6 \times 10^{-4}\ s^{-1}$. The initial crystallographic orientation of the tensile axis is located in the central part of the [001-011-$\bar{1}$11] standard triangle. They report that the shear-stress/shear-strain curve can be roughly divided into three stages, namely, easy glide, linear hardening and parabolic hardening regions, though the distinction between each stage is apparently not so clear than in usual face centered cubic (f.c.c.) single crystals. In these experiments, the shear strain before fracture was found to be higher than 150%. An overshooting phenomenon was evidenced, that is, although lattice rotations take place, the primary slip system [$\bar{1}$01](111) continues to operate until the Schmid factor ratio (SFR) between primary and secondary slip [011]($\bar{1}\bar{1}$1) systems reaches approximately 1.44, or in other words, the secondary slip system was not active below a SFR of 1.44. At 673 K, the stress–strain curves start directly in Stage III, i.e. with a parabolic WHR, and failure occurs before the secondary octahedral slip system is activated. This corresponds in these experiments to lattice rotations leading to a SFR of 1.1. At high temperatures, 873 K, the specimens break before reaching the symmetry line, i.e. a SFR = 1, and overshooting is not observed. By considering the relation between observed and calculated shear strains, Aoki and Izumi postulate that the contribution of the secondary slip system is small at lower temperature, but increases with increasing temperature and strain. The discrepancy between observed and calculated shear strains is entirely attributed to the activation of the secondary octahedral slip system (the possibility of cube slip is not envisaged by the authors). Based on these results the authors concluded that work-hardening effect is controlled by two superimposed processes which consist of a cross-slip mechanism of the Kear–Wilsdorf type, leading to the positive temperature dependence of the WHR, and another mechanism that only contributes to latent hardening and which exhibits a negative dependence with temperature.

Large deformations and lattice rotations have also been investigated by Kim et al. [4]. Their experiments were conducted in tension on $Ni_{75}Al_{20}Ti_5$ and $Ni_{76}Al_{19}Ti_5$ + B. They observed a step-like increase of the flow stress with increasing deformation, similar to what has already been reported for $Ni_{75}Al_{20}Ti_5$ and Ni_{75} $Al_{20}Ta_5$ (see [5]). This step-like shape of the stress–strain curve is ascribed to the operation of alternate octahedral slip systems, having the maximum Schmid factor value.

2.1.1.2. Influence of the strain rate. Miura et al. [6] have reported stress–strain curves of single crystalline Ni_3(Al,5at%Ti) specimens deformed at various strain rates, varying over two orders of magnitude (i.e., from $1.4 \times 10^{-5}\ s^{-1}$ up to $1.4 \times 10^{-3}\ s^{-1}$). By using their data, we have estimated the corresponding WHRs by measuring on the (σ, ε) curves the slope of the linear portion after yielding. The deduced values are plotted in fig. 1 as a

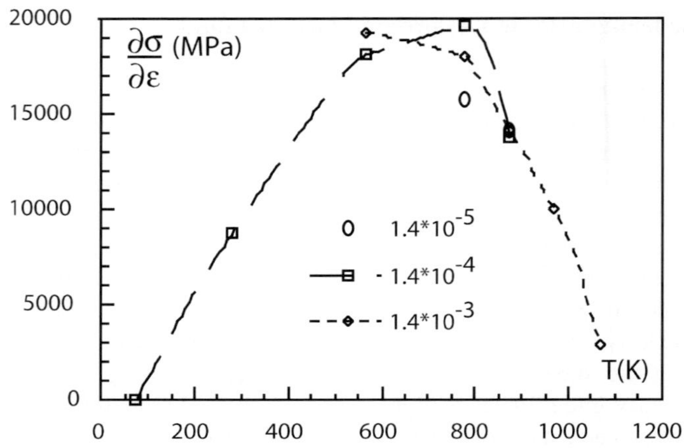

Fig. 1. Work-hardening rate as a function of temperature for two strain rates, i.e., 1.4×10^{-5} s^{-1} and 1.4×10^{-3} s^{-1}, deduced from Miura's stress–strain curves [6]. Ni_3(Al,5at%Ti) single crystals.

function of temperature. For each strain rate, the results do not cover the entire temperature range, but in the temperature domain where they overlap no clear distinction can be made between the different strain rates. This suggests that, as for the yield stress, the WHR is rather strain-rate insensitive.

Support for this conclusion is found in the work of Baluc [7], who has investigated the effect of strain-rate on the WHR in $Ni_{75}Al_{24}Ta_1$. In this study, ⟨123⟩ oriented samples have been deformed in compression at increasing temperatures for two strain rates, namely 7×10^{-5} s^{-1} and 7×10^{-4} s^{-1}. The temperature range was 300–1270 K. Figure 2 shows the WHR measured at 0.2% plastic strain offset. It is seen that, within the experimental scatter, the behaviour of WHR is identical for the two strain-rates with absolute values that are not very far from each other. In this figure, the WHR exhibits for both strain rates two maxima in the anomalous temperature domain. A very sharp peak is first recorded at the temperature of 470 K, followed by a hump with a maximum in the WHR at about 770 K. The presence of these two WHR peaks as a function of temperature results from an artefact, as will be shown below. Above the yield stress peak, the WHR goes to negative values, as reported by Staton-Bevan [8] before rising again at very high temperatures.

In the case of binary stoichiometric $Ni_{75}Al_{25}$ single crystals and for temperatures up to 400 K, it has been reported by Masahiko and Hirano [9] that a downward strain-rate change (from 8.3×10^{-4} to 8.3×10^{-6} s^{-1}) was initially accompanied by a stress drop, which is followed by a gradual return to the stress level extrapolated from the steady state at the previous strain rate as if the strain rate did not change. This indicates that under these deformation conditions the long-term WHR is insensitive to the strain rate. For higher temperatures, the transition temperature being located between 400 and 460 K, the stress again returns after the strain-rate change to the level extrapolated from the previous steady rate similarly to what is observed below 400 K, but the WHR in the subsequent steady state is different from that of the previous values (see fig. 3) indicating in this case a strain-rate dependence of the WHR. A behaviour somewhat similar has also been

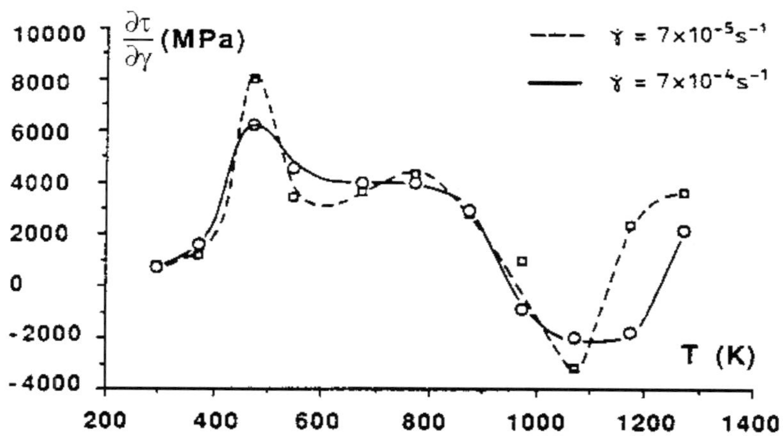

Fig. 2. Work-hardening rate as a function of temperature for two strain rates, i.e., 7×10^{-5} s^{-1} and 7×10^{-4} s^{-1}, $Ni_{75}Al_{24}Ta_1$ single crystals [7].

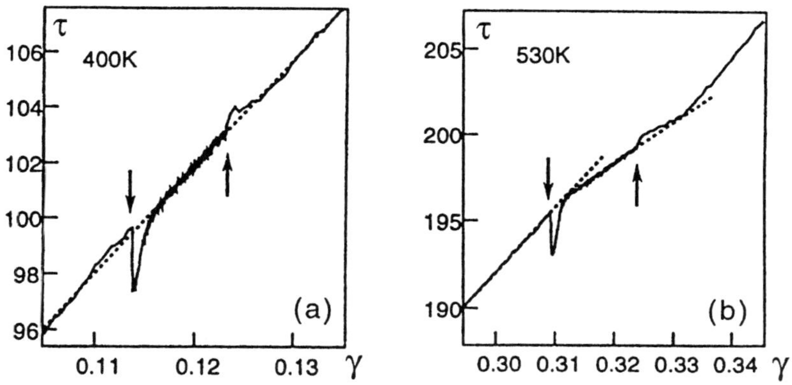

Fig. 3. Portions of stress–strain curves at two different temperatures which illustrate the behaviour of the WHR as a function of strain-rate. The ↓ arrows indicate a downward strain-rate change and the ↑ arrows indicate a upward strain rate jump [9].

observed by Thornton et al. [10], who have divided the flow stress anomaly domain into two types of stress response modes to a strain-rate change, with a transition temperature at approximately 673 K for their deformation conditions (see section 6.2.2.2).

2.1.1.3. Temperature and orientation dependences. Copley and Kear [11] have studied the yield stress and the WHR of Ni_3Al single crystals of off-stoichiometric composition, i.e., $Ni_{77.5}Al_{22.5}$, deformed at a strain rate of 5×10^{-4} s^{-1}. Two crystallographic orientations have been investigated (see fig. 4) in order to discriminate between the proposed work-hardening theories, which have been based on either the production of antiphase boundary defects by the slip process or the pinning of dislocations by a

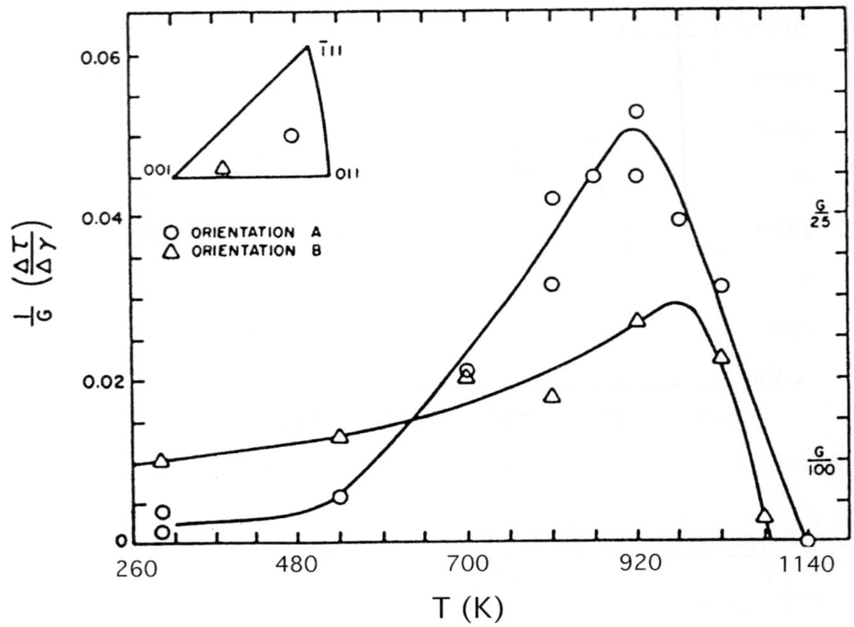

Fig. 4. Normalised work-hardering rate as a function of temperature of binary $Ni_{77.5}Al_{22.5}$ single crystals for two different crystallographic orientations, according to [11] (G is the shear modulus).

cross-slip mechanism. They predict and observe, as required by the cross-slip model for hardening, that crystal orientation A (which favoured the trapping of dislocations) should exhibit a WHR greater than orientation B (favouring cross-slip into $(1\bar{1}1)$). Figure 4 clearly illustrates the generic behaviour of the WHR of Ni_3Al based single crystals with temperature, that is, as does the yield stress, it increases with increasing temperature up to a peak temperature $T_{p\theta}$, above which it finally decreases.

One of the more complete studies concerning the work-hardening rate of Ni_3Al has been reported by Staton-Bevan [8]. The experiments were performed on single crystalline specimens of various compositions: $Ni_{74.4/76}Al_{18.8/20.4}Ti_{5/6.1}$ and for four different crystallographic orientations, namely $\langle 111\rangle$, $\langle 123\rangle$, $\langle 144\rangle$ and $\langle 001\rangle$. A strain rate of approximately 3×10^{-4} s^{-1} was used and the test temperatures were in the range 163 to 1273 K. In this study, it was not possible to clearly identify a Stage I or II on the $\sigma(\varepsilon)$ curves (see fig. 5) and the WHR has been arbitrarily measured at 1% plastic strain offset. The stress–strain curves given in fig. 5, for a $\langle 111\rangle$ orientation, clearly illustrate the difficulty of identifying a plastic offset strain that would lead to an unambiguous definition of the WHR. This point is invoked by Bontemps-Neveu [12] as a possible explanation for the second WH peak observed at high temperature, that is at temperatures higher than $T_{p,\sigma}$. However, this is at variance to what is usually observed by other authors, since for temperatures below $T_{p,\theta}$ the WHR is nearly constant for plastic strains ranging between 1 up to at least 5%.

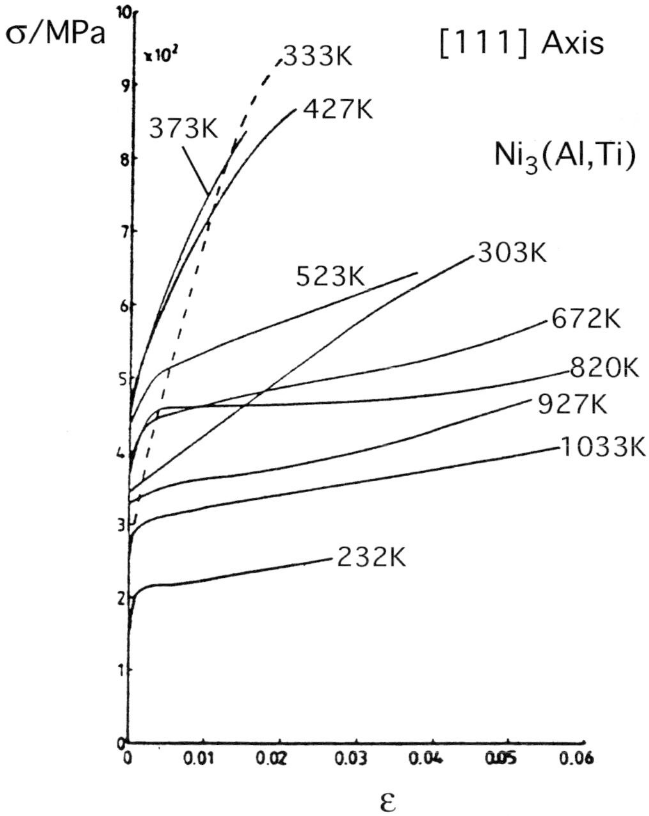

Fig. 5. Stress–strain curves of Ni_3(Al,Ti) single crystals deformed in the ⟨111⟩ orientation [8].

The main features of the orientation dependence of the WHR observed by Staton-Bevan [8] can be summarised as follows (see also fig. 6).

- For all investigated orientations in the anomalous temperature regime of the yield stress the WHR exhibits an anomalous temperature dependence, with one or two peaks depending on the crystallographic orientation. The amplitude of the WHR anomaly is orientation dependent and maximum for the ⟨111⟩ orientation, less prominent for the ⟨123⟩ and ⟨144⟩ orientations and very small for the ⟨001⟩ orientation.
- For the ⟨123⟩ and ⟨144⟩ orientations, the WHR exhibits two peaks in the anomalous regime, similarly to what was observed by Spätig et al. [13] when using specimens successively deformed at increasing temperatures (see section 2.1.1.4).
- The temperatures at which the WHR peaks appear to be different, but the peak temperature for the ⟨001⟩ orientation is not very well defined.
- A striking feature, which is not observed either by Bontemps-Neveu [12] or by Spätig [14] is that, for all orientations and compositions, the WHR falls to an almost zero value at a temperature of nearly 770 K.

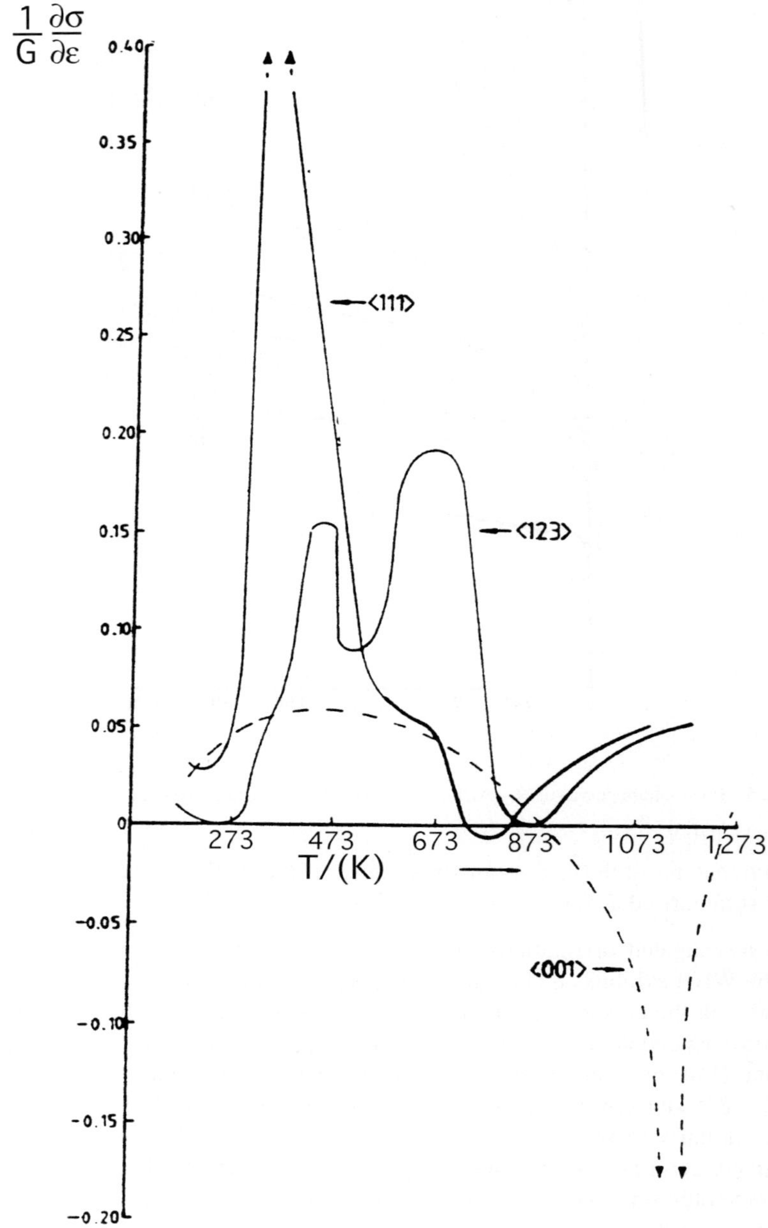

Fig. 6. Work-hardening rate of $Ni_3(Al,Ti)$ single crystals as a function of temperature for various orientations [8].

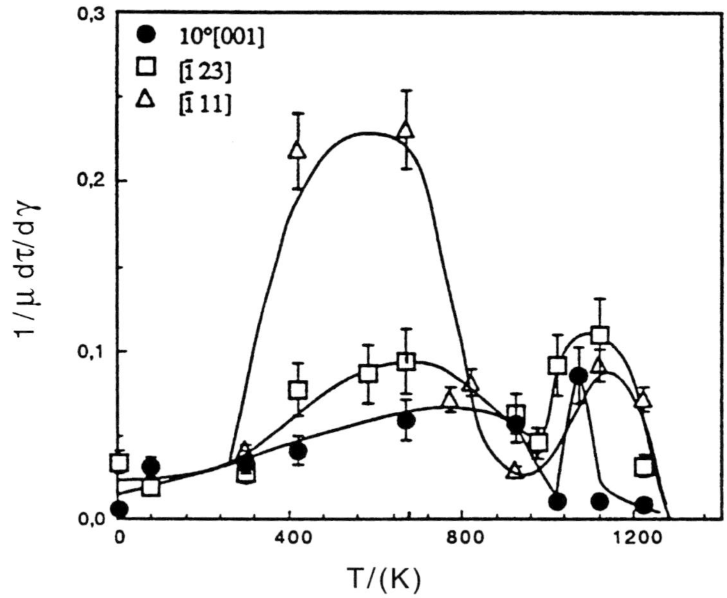

Fig. 7. Orientation and temperature dependences of the work-hardering rate in $Ni_{77.04}Al_{22.6}Hf_{0.26}$ single crystalline specimens [12].

Bontemps-Neveu [12] has measured the WHR for three different orientations in single crystalline specimens having the nominal composition of $Ni_{77}Al_{22.7}Hf_{0.26}$ (see fig. 7) and for two orientations in single crystalline specimens with the nominal composition $Ni_{75.6}Al_{22.7}Hf_{1.6}$. In the temperature range investigated, 4–1200 K, the WHR exhibits for all conditions two peaks: one below $T_{p,\sigma}$ and one above $T_{p,\sigma}$. Therefore, while two peaks are present for the ⟨123⟩ orientation, only one peak is observed in the anomaly domain, since the second WHR peak above $T_{p,\sigma}$ belongs to the region of flow stress decreasing with temperature. Basically for the ⟨111⟩ and near ⟨001⟩ orientations, the results are qualitatively similar to those reported by Staton-Bevan for $Ni_3(Al,Ti)$; the highest WHR is measured for the ⟨111⟩ orientation, which peaks sharply, while only a small increase exists for the near ⟨001⟩ orientation. For the $Ni_{77}Al_{22.7}Hf_{0.26}$ composition, the temperatures at which the WHR peaks in the anomaly domain is orientation dependent and can be classified as follows: $T_{p,\theta}[111] < T_{p,\theta}[\bar{1}23] < T_{p,\theta}[001]$. It must be noted that the peak temperature of the critical resolved shear stress ($T_{p,\tau}$) follows the same orientation dependence. Due to the lack of experimental data, this is less clear for the 1.6%Hf compound. However, when compared with the 0.26% composition, it is observed that, for the two orientations investigated the WHR peak temperatures are shifted to lower temperatures, while $T_{p,\tau}$ are rather similar for both compositions.

2.1.1.4. Influence of testing procedure. Spätig et al. [13,15] have examined the WHR of $Ni_{75}Al_{24}Ta_1$ single crystals by using two experimental procedures. In procedure I, yielding and the related WHR have been investigated by deforming a specimen several times up to

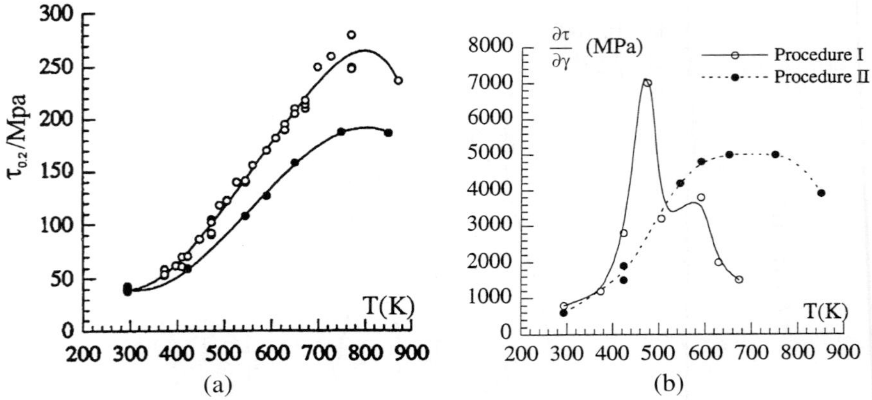

Fig. 8. Characteristic parameters of ⟨123⟩ oriented $Ni_{75}Al_{24}Ta_1$ single crystalline specimens: (a) $\tau_{0.2\%}$ proof stress determined as a function of temperature by using two different deformation procedures (see text) and (b) related WHR, $\theta_{0.2\%}$, according to [13].

slightly above 0.2% plastic strain at increasing temperatures. As emphasised above and in section 5.2.2.1, this procedure is currently used for $L1_2$ crystals. Procedure II has consisted in deforming a virgin specimen at each temperature in the temperature range investigated (300–800 K). The latter procedure is of course considered as more reliable as compared to the former one, in spite of the metallurgical dispersion. For both types of experiments the temperature variations of the $\tau_{0.2\%}$ proof stress, i.e. the conventional shear-stress measured at 0.2% offset strain, are fairly similar, with a $(\tau_{0.2\%}, T)$ slope which is slightly higher in the case of procedure I than for procedure II. In addition, the temperatures at which $\tau_{0.2\%}$ peaks occur are similarly in fair agreement (see fig. 8(a)). However, while the procedure I does not change fundamentally the shape of the $\tau_{0.2\%}(T)$ curve as compared to the one obtained with virgin specimens, drastic consequences on other physical parameters such as the WHR and the strain-rate sensitivity of the flow stress have been evidenced. As shown in fig. 8(b) the WHR measured at $\tau_{0.2\%}$ behaves quite differently according to the investigation procedure. For a specimen deformed by using procedure I, the WHR exhibits sharp variations as a function of temperature, with two peaks at $T = 470$ K and 600 K, while procedure II yields a more monotonic dependence of the WHR with temperature. In this latter case, the WHR increases from RT up to about 650 K and remains constant up to 750 K, that is a temperature lower than that of the peak temperature of the $\tau_{0.2\%}$ stress, above which it finally decreases.

2.1.1.5. The peak temperature ($T_{p,\theta}$) and its strain dependence. Experiments performed on single crystalline specimens, specifically dealing with WHR, have been reported for binary $Ni_{76.6}Al_{23.4}$ [16], $Ni_{74.8}Al_{21.9}Hf_{3.3}$ and $Ni_{75}Al_{24}Ta_1$ [14]. The WHRs measured at 0.2% plastic strain are given in fig. 9. This figure also shows for each composition the temperature at which the $\tau_{0.2\%}$ stress is peaking ($T_{p,\tau}$). The peak temperature of the WHR is located at a temperature which is lower than that of $\tau_{0.2\%}$. Similar observations hold for a variety of $L1_2$ intermetallic compositions (see, for instance, [9,12,17–19]). While it is not

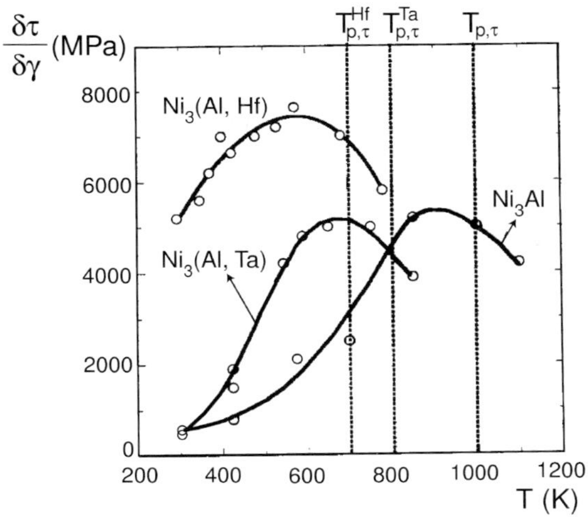

Fig. 9. $\theta_{0.2\%at}$ as a function of temperature in compression tests at constant strain- rate for three different Ni_3Al compounds, according to [14]. Note that for each composition the peak temperature $T_{p,\tau}$ of the corresponding $\tau_{0.2\%}$ proof stress is indicated.

always possible to directly check this occurrence, due to the lack of experimental data in the reported studies, it seems however that this behaviour is a general characteristics of the WHR of the $L1_2$ intermetallic compounds exhibiting a positive temperature dependence of their yield stress, at least when it is measured at and above 0.2% plastic strain. It has been suggested by Saada and Veyssière [20] that $T_{p,\theta}$ roughly corresponds to the temperature at which the $\tau(T)$ curve is inflected.

A different comportment is noted when the WHR is measured at plastic strains lower than about 0.2%, as illustrated in fig. 10 for $Ni_{75}Al_{22.7}Hf_{1.51}B_{0.2}$ [22]. This figure shows the WHR measured at 1.5%, 0.15% and 0.05% plastic strain. At 1.5%, the WHR exhibits the main trends that have already described above. At 0.15%, the WHR increases steadily with increasing temperature up to $T_{p,\sigma}$ and at 0.05% it is almost constant. The WHR drastically increases with decreasing strains, which clearly demonstrates its strain dependence at small strains. Therefore, as emphasised by Veyssière and Saada [23], the relevance of measuring the WHR in a transient domain, where microplasticity is predominant, may be questionable and they proposed that reliable WHR values should not be measured at least below a permanent plastic strain of about 0.5%.

2.1.2. Polycrystalline Ni_3Al based alloys

The WHR of polycrystalline $Ni_{75.9}Al_{24.1}B_{0.095}$ specimens has been studied at large strains under dynamic deformations conditions (1000 to 8000 s^{-1}) by Gray and Embury [21]. They suggest that the hardening response of dynamically deformed Ni_3Al specimens displays some common features with the classical Stages II and III of pure f.c.c. metals. In addition, they show that Ni_3Al exhibits prolonged high WHR, up to very high stress level ($\sim E/80$), attributed by the authors to the suppression of dynamic recovery processes.

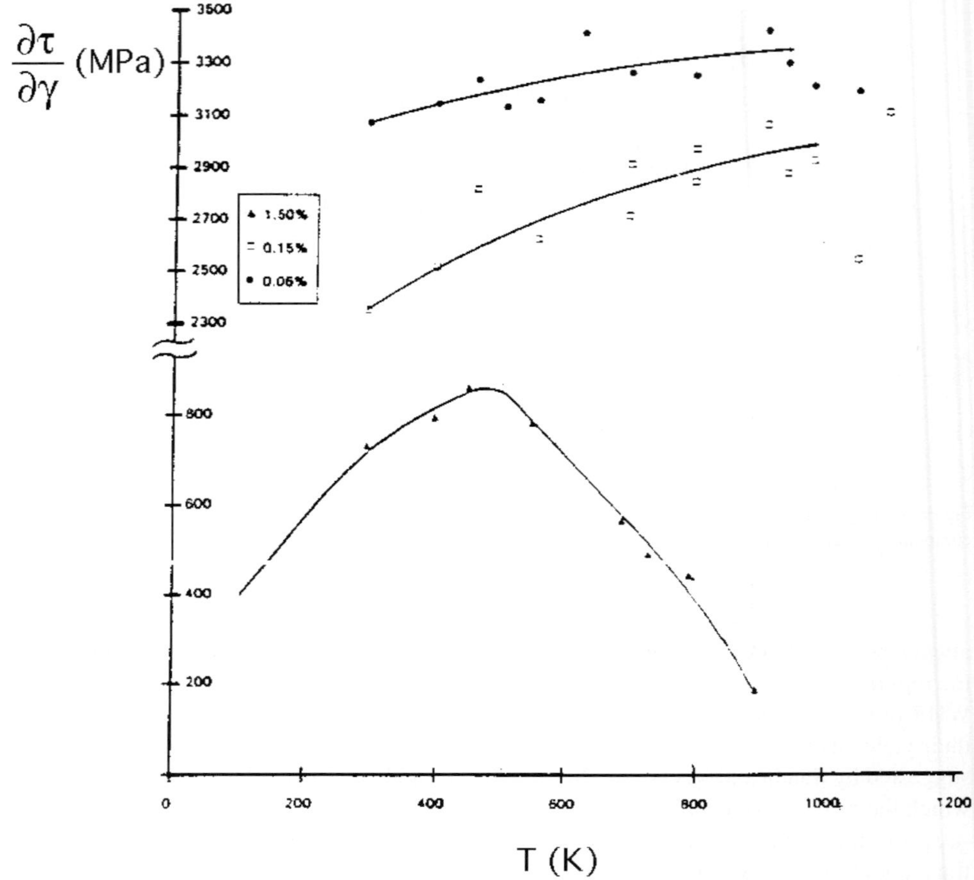

Fig. 10. Temperature dependence of the WHR measured at three different plastic strain levels in $Ni_3(Al,Hf)+B$ single crystals [22].

The existence of an anomalous behaviour of WHR in the binary $Ni_{75}Al_{25}$ polycrystalline alloy with increasing temperature has been evidenced by Thornton et al. [10]. They have reported the flow stress measured at different strain levels for various temperatures (fig. 11). As represented, the gap between two curves is a direct measure of the WHR. At given temperature, one can deduce the variation of the WHR as a function of strain, while the difference between two curves at constant strain with increasing temperature yields the WHR temperature dependence. Therefore, fig. 11 indicates that the WHR increases with increasing temperature, for all strains, up to nearly the peak in flow stress, which is strain independent, and is strongly strain dependent between 10^{-4} and 10^{-5}.

Another study has been performed more recently by Lo Piccolo [24] who has investigated three compositions, namely $Ni_{76}Al_{24}$, $Ni_{75}Al_{25}$ and $Ni_{74}Al_{26}$. For the three alloys, it has been observed that, certainly due to the polycrystalline nature of the materials,

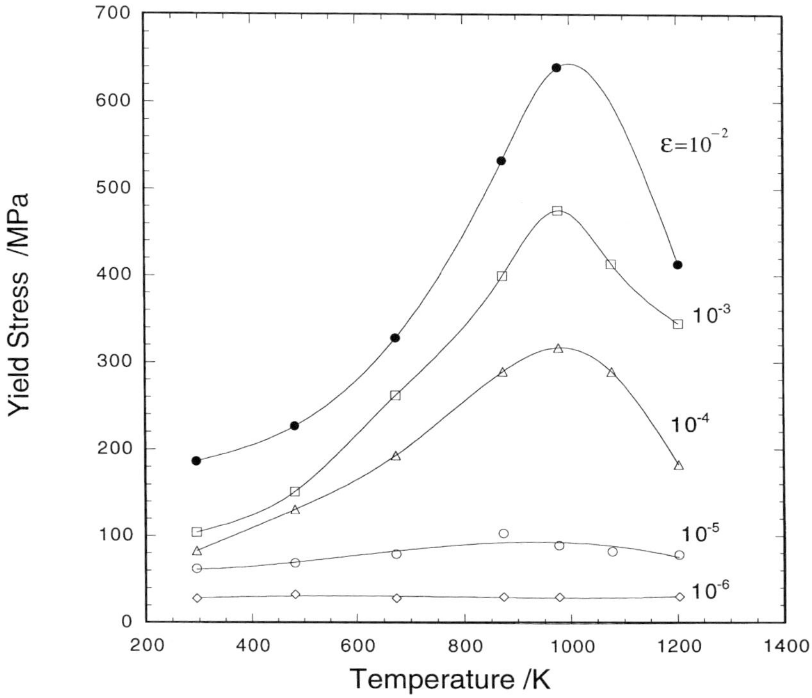

Fig. 11. Temperature dependence of the flow stress at various plastic strain levels of binary $Ni_{75}Al_{25}$ polycrystalline specimens [10].

the transition between the micro and macroplastic domains is very gradual, leading to some uncertainties in the WHR measurements at low strains. Therefore, more representative values of the actual WHR have been measured at 3% plastic offset strain ($\theta_{3\%}$). Figure 12 shows the variation of $\theta_{3\%}$ with temperature for the three compositions. It is observed that, for each composition, the WHR curve exhibits a shallow profile, which does not permit a precise determination of the peak temperature. Nevertheless, since for the three compositions the related flow stresses peak at temperatures, $T_{p,\sigma}$. which are above 800 K, the temperature *maxima* of the WHR, $T_{p,\theta}$, are necessarily at temperatures lower than $T_{p,\sigma}$, i.e., $T_{p,\theta} < T_{p,\sigma}$.

2.1.3. Cu_3Au compound

The WHRs in Cu_3Au and Ni_3Al exhibit similar features as functions of orientation and temperature. The absolute values for both $L1_2$ phases appear to be drastically different, but are not given at homologous strain levels. For comparison, typical WHR values are given in table 1 in units of G, where G is the shear modulus.

Kuramoto and Pope [25] have measured the WHR in Cu_3Au at 30 and 40% plastic strain (referred to as θ_{II} in the following). This study was aimed at clarifying the origin of the WHR, for which two possibilities have been proposed. One of these possibilities is the

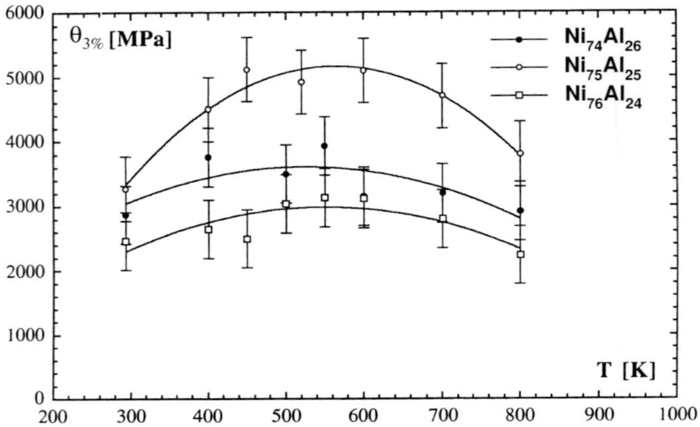

Fig. 12. Temperature dependence of the WHR measured at 3% plastic strain for three different compositions of Ni_3Al polycrystals [24].

Table 1
Maximum work hardening rate values measured for different $L1_2$ compounds

Alloy composition		Strain (%)	Temperature (K) – orientation	Maximun WHR (in units of G)	Reference
Cu_3Au	S	30	300	0.006	Kuramoto and Pope [25]
Ni_3(Al,Ti)	S	1	350 – ⟨111⟩	0.6	Staton-Bevan [8]
Ni_3(Al,Ti)	S	1	673 – ⟨123⟩	0.2	Staton-Bevan [8]
Ni_3(Al,0.25%Hf)	S	1	673 – ⟨123⟩	0.1	Bontemps-Neveu [12]
Ni_3(Al,1.5%Hf)	S	1	653 – ⟨123⟩	0.15	Bontemps-Neveu [12]
Ni_3(Al,3.3%Hf)	S	0.2	560 – ⟨123⟩	0.11	Spätig [14]
Ni_3Al	S	0.2	900 – ⟨123⟩	0.08	Spätig [14]
Ni_3(Al,1%Ta)	S	0.2	700 – ⟨123⟩	0.07	Spätig [14]

cross slip of screw dislocation segments from {111} planes to {010} planes, as suggested by Kear and Wilsdorf [26], while the other one is based on the formation of APB tubes by non-aligned jogs on superdislocations, as proposed by Vidoz and Brown [27]. According to Kuramoto and Pope, it is possible to distinguish between the two possibilities by studying the orientation dependence of the WHR, for the following reasons:

- the cross-slip mechanism is expected to depend essentially on the ratio N of the Schmid factors between the primary octahedral slip plane and the cube cross-slip plane, that is, N is a measure of the promoting stress for cross slip,
- the APB tube mechanism, which involves an interaction between moving dislocations on primary and secondary octahedral slip systems, should depend on the ratio M of the Schmid factors between the primary and secondary octahedral slip planes.

Therefore, by selecting a judicious set of crystallographic orientations, it should be possible to discriminate between the two proposed explanations. Four orientations have been selected in this study (see fig. 13 for orientation B). The stress–strain curves have been converted into shear stress-shear strain curves by using the Schmid factor of the primary

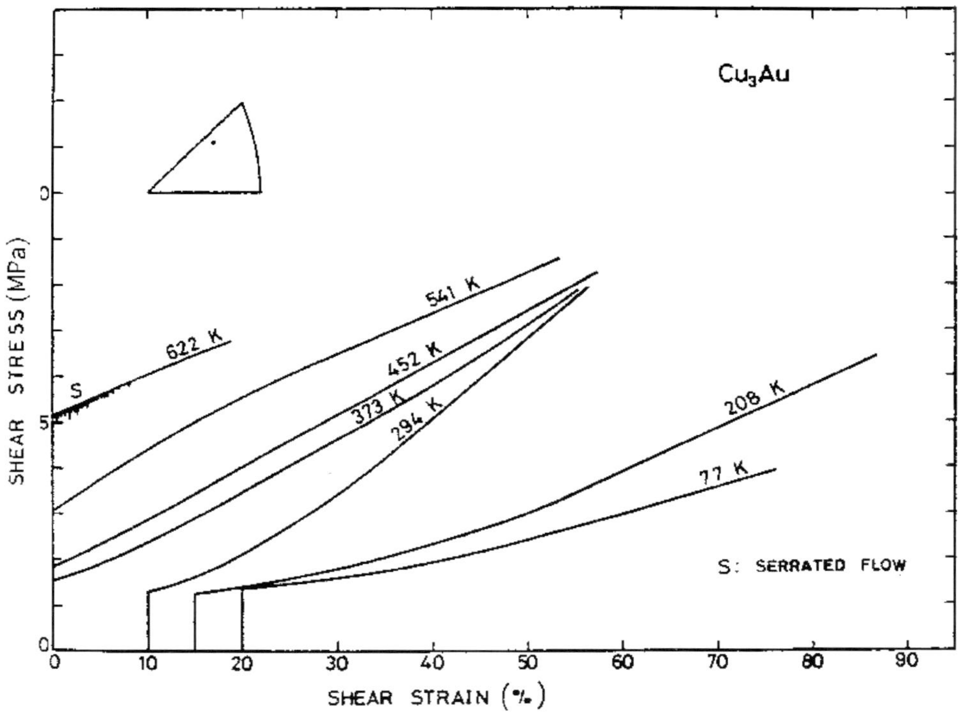

Fig. 13. Typical stress–strain curves of Cu_3Au single crystals showing the usual WHR stages of pure f.c.c. crystals. Note that serrations are observed for the 622 K curve [25].

octahedral plane, without taking into account the lattice rotations. For all orientations at low temperature the stress–strain curves show an easy glide stage and all deformation curves display a stage of linear hardening (Stage II) which, depending on temperature and amount of deformation, is followed by a region of parabolic hardening (fig. 13).

Note that serration at high temperatures (622 K), visible at the beginning of plastic deformation for orientations B and D, have been interpreted as the result of twinning. Twinning has already been observed in fully ordered Cu_3Au by Chakrabortty and Starke [28] at and below RT.

The temperature and orientation dependences of the WHR, θ_{II}, given by the authors of [25] are reported in fig. 14. In this figure, the reported values of θ_{II} have been measured at 30% shear strain for the high temperature range and at about 40% shear strain for the low temperature domain. θ_{II} shows two maxima, separated by a minimum for all investigated orientations. This result is different from the behaviour reported by Davies and Stoloff [29], who did not observe the minimum around 373 K. Because Kuramoto and Pope found that θ_{II} is an increasing function of N, which is not obvious in regards of fig. 14, and while no direct correlation could be made with M, they conclude, as Sastry and Ramaswami [30] did before, that the WHR in Cu_3Au is determined by the formation of sessile screw dislocations by cross slip and the formation of superdislocation edge dipoles. In order to

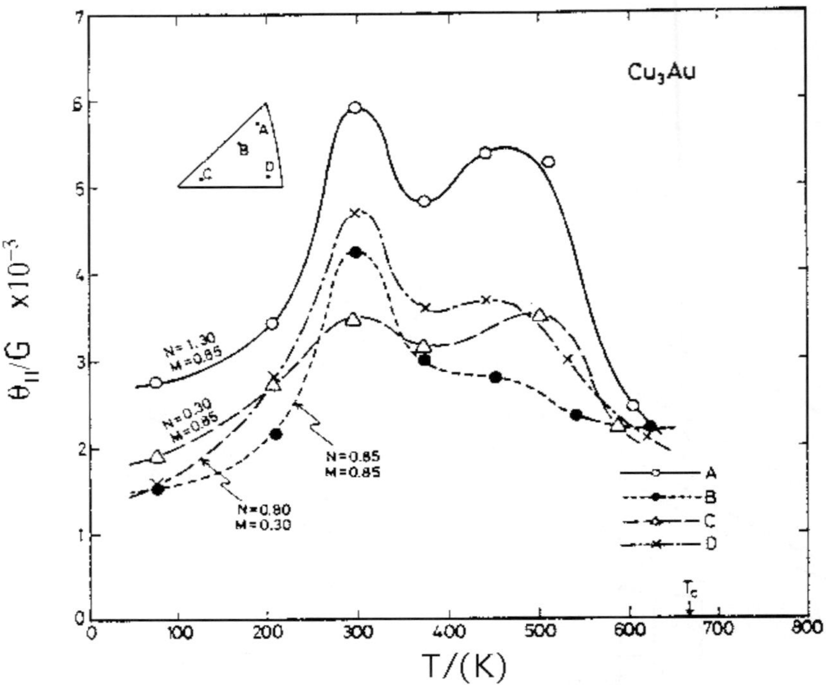

Fig. 14. Temperature dependence of the normalised WHR measured in Stage II, θ_{II}/G, of Cu_3Au single crystals for various orientations [25]. Orientation A, $N = 1.3$; B, $N = 0.85$; C, $N = 0.3$; D, $N = 0.8$.

test this hypothesis, an additional experiment was proposed that consists in testing the effect of secondary dislocations by predeforming a specimen. This predeformation was performed by twisting the crystal around its axis and the new flow stress was compared with that of a virgin sample. The result indicates that this treatment increases the yield stress by about 20%, but does not change the WHR, suggesting that non-aligned jogs formed on primary dislocations can contribute to the increase in the yield stress, but do not increase the WHR. No explanation is proposed for the intermediate minimum, and the decrease of θ_{II} at high temperatures is interpreted in term of a decreasing resistance to dislocation motion on the (010) cube cross-slip plane.

2.2. Work-hardening rates in $L1_0$ ordered intermetallics

2.2.1. Single crystals

This section will be mainly devoted to the intermetallic γ-TiAl of the $L1_0$ phase, for which a few studies on the WHR have been reported in the literature. Single crystals of TiAl are difficult to produce and the orientation dependence of the WHR of this $L1_0$ intermetallic compound is far less documented than that of the $L1_2$ phase. In the first reported study on $Ti_{44}Al_{56}$ single crystals by Kawabata et al. [31], the WHR was

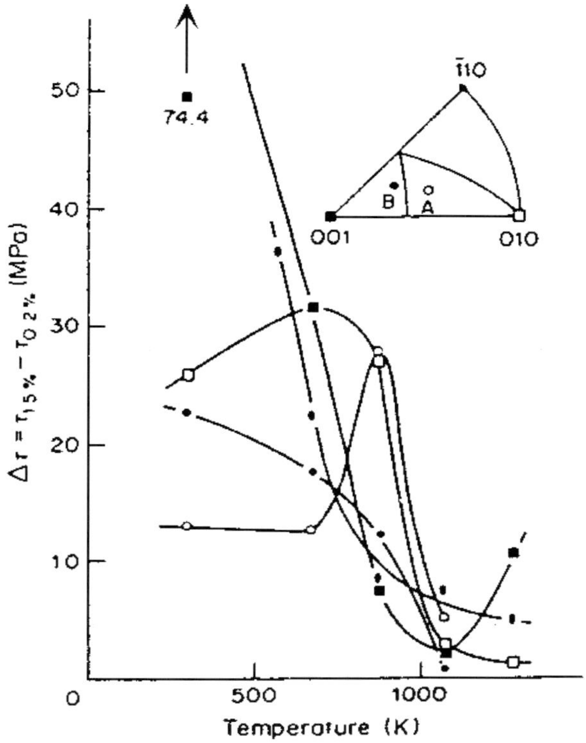

Fig. 15. WHR, i.e., $\tau_{1.5\%} - \tau_{0.2\%}$, for γ-TiAl single crystals of various crystallographic orientations [33]. The symbols used to identify the crystallographic orientation of the reported WHR in the $\Delta\tau - T$ ordinates are given in the standard stereographic projection [31].

defined as the difference between the resolved shear stresses at respectively 1.5% and 0.2% compressive plastic strains, i.e. WHR $= \tau_{1.5\%} - \tau_{0.2\%}$. A more complete work, which includes low-temperature data, has been published elsewhere [32]. However, the WHR, which is defined as $(\sigma_{1.5\%} - \sigma_{0.2\%})/1.3 \times 10^{-2}$, exhibits unexpected values of nearly 60 GPa, which are comparable to the shear modulus. In addition, as seen above, such a definition may lead in some circumstances to an inappropriate representation of the actual WHR and must be cautiously interpreted. Therefore, fig. 15 only reproduces the former values of the WHR for five orientations. It must be noted that, for all orientations tested, the $\tau_{0.2\%}$ proof stress shows a positive temperature dependence with a common peak temperature around 873 K, except for the $[\bar{1}10]$ for which $\tau_{0.2\%}$ peaks at nearly 1073 K. Two cases exist for the WHR, which shows either a positive temperature dependence for the orientations A and [010] or a continuous decrease with increasing temperature for the three other orientations. Based on a Schmid factor analysis, Kawabata and co-workers attribute the anomalous behaviour of the WHR as resulting from the activation of $\frac{a}{2}\langle 110\rangle$ ordinary dislocations as primary glide dislocations. On the contrary, in the orientations where the WHR does not reveal the positive temperature dependence, $a\langle 011\rangle$

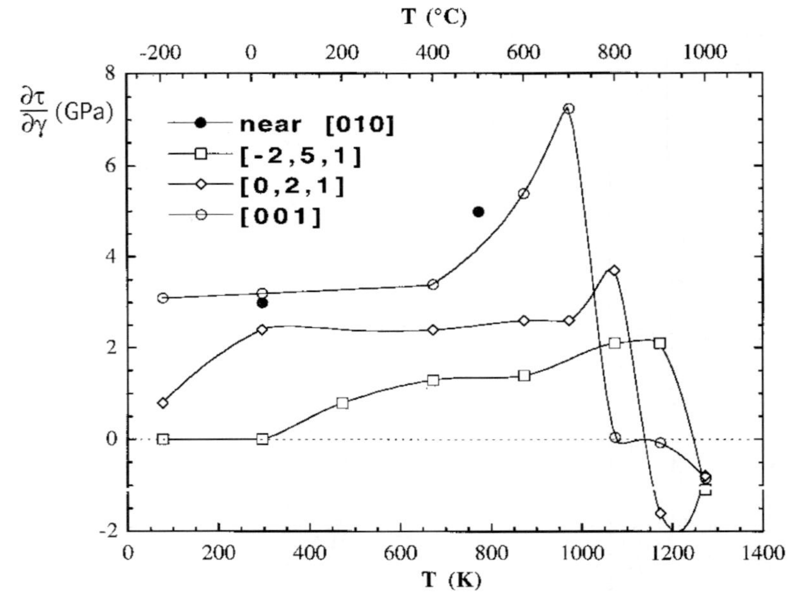

Fig. 16. Temperature dependence of the WHR of γ-TiAl single crystals deduced from stress–strain curves reported in [33] for three different orientations, together with WHR data reported in [34] for the orientation near $\langle 010 \rangle$.

superdislocations should operate. A careful examination of the stress–strain curves in [31] (see fig. 2 of this reference) indicates that, in the domain of the flow stress anomaly, the WHR is rather insensitive to an increase in strain rate of one order of magnitude, starting from the value of $1 \times 10^{-4}\ s^{-1}$. A slight increase in WHR is only detectable at 673 K in the [001] and $[\bar{1}10]$ orientations.

A detailed study on TiAl single crystals of similar composition has been performed more recently by Inui et al. [33]. Seven orientations have been carefully selected within the three standard crystallographic triangles necessary for the $L1_0$ structure, leading to a variety of Schmid factor combinations for ordinary and superdislocations. The operative deformation modes have been systematically identified as functions of crystal orientation and temperature by slip traces and TEM (Transmission Electron Microscope) observations. For all orientations, the yield stresses exhibit within a given temperature range a positive temperature dependence, in agreement with the findings of Kawabata et al. [31]. However, the extent of the orientation range for the operation of ordinary slip is found to be limited to within a few degrees around the [021] orientation. This is in contrast with the assumption of Kawabata et al. [31], who considered that the Schmid law was obeyed in terms of primary Burgers vector. Unfortunately, the WHR has not been systematically investigated in this work [33]. Stress–strain curves obtained at a strain-rate of $2 \times 10^{-4}\ s^{-1}$ have been presented for the [001], [021] and $[\bar{2}51]$ orientations. The authors highlight the following characteristics: (i) a peak in WHR occurs at a temperature 100–200 K lower than that of the peak temperature for the yield stress, (ii) at a given temperature, [001] oriented crystals have the highest WHR, most probably due to the activation of multiple slip. We have used their fig. 4 to estimate the WHR dependences as functions of temperature and orientation. Due to the lack of accuracy in both stress and strain, the WHRs have been estimated from that portion of the stress–strain curve which is linear after yielding. Figure 16 shows the WHR values so obtained, together with those published by Stucke et al. [34] measured for similar experimental conditions but with a different orientation. For all cases, the WHRs are observed to increase with increasing temperature, that is for both ordinary and super-dislocations. This is in contrast with the results of Kawabata et al. where the WHR is a decreasing function of temperature for all orientations in which superdislocations are involved. Note that in the work of Stucke et al. the nature of the dislocations controlling the flow stress has not been determined. Revisiting the results of Kawabata et al. in the light of the Burgers vector identifications done by Inui et al. as a function of specimen orientation would lead one to consider that both super and ordinary dislocations can lead to an anomalous behaviour of the WHR with temperature, but not for all orientations in the case of superdislocations. Although it has been shown that heat treatments may have drastic effects on the yield stress [35], it is hard to believe that the annealing heat treatment performed by Kawabata et al. prior to deformation can account for such a difference. These discrepancies need further clarification.

Jiao et al. [36] have deformed $Ti_{55.5}Al_{44.5}$ single crystals along two crystallographic orientations (see fig. 17). The $\tau_{0.2\%}$ stresses are about the same at RT and do not vary too much up to nearly 673 K, above which they increase with increasing temperature with almost equal slope. They peak at about 1073 K and about 1173 K for C and D orientations, respectively. Therefore, except for the differences in $T_{p,\tau}$, $\tau_{0.2\%}$ shows similar trends for both orientations. The operating systems have been investigated by optical and transmission electron microscopy and it was found that, for the two orientations and below the yield stress peak, the major part of slip was accommodated by the motion of [011] superdislocations gliding on the $(1\bar{1}1)$ planes. However, as recognised by the authors, the corresponding WHRs, $d\tau/d\gamma$, averaged over the plastic strain interval 0.015–0.02 differ considerably from each other (fig. 17). Both show a small peak around 673 K, which is followed in the case of orientation C by a continuous decrease while, for orientation D, it increases again to a maximum at $T_{p,\tau}$. This yields above $T_{p,\tau}$ to very low values of the WHR for orientation C, while for orientation D it remains very high. This latter result also contrasts with the results obtained by Inui et al., who have reported that the peak in WHR was usually at temperatures 100–200 K below $T_{p,\tau}$.

2.2.2. *Polycrystals*

The WHRs have also been examined by a few authors in polycrystalline TiAl-based alloys. For instance, Sriram et al. [37] have investigated binary Ti-50Al and Ti-52Al alloys. They report for both alloys a pronounced WHR (10–15 GPa) which follows with temperature a similar trend to the flow strength. That is, for work hardening values, an initial decrease from low temperature (77 K) up to nearly 600 K, then a continuous increase up to 1073 K and a final decrease up to 1173 K, the maximum temperature investigated in this study. The WHR's at 1% plastic strain are usually smaller than at 0.5%.

Viguier et al. [38] have investigated the WHR of the $Ti_{47}Al_{51}Mn_2$ alloy. A moderate serrated flow accompanies yielding from 120 K up to approximately 700 K, which does

a)

b)

c)

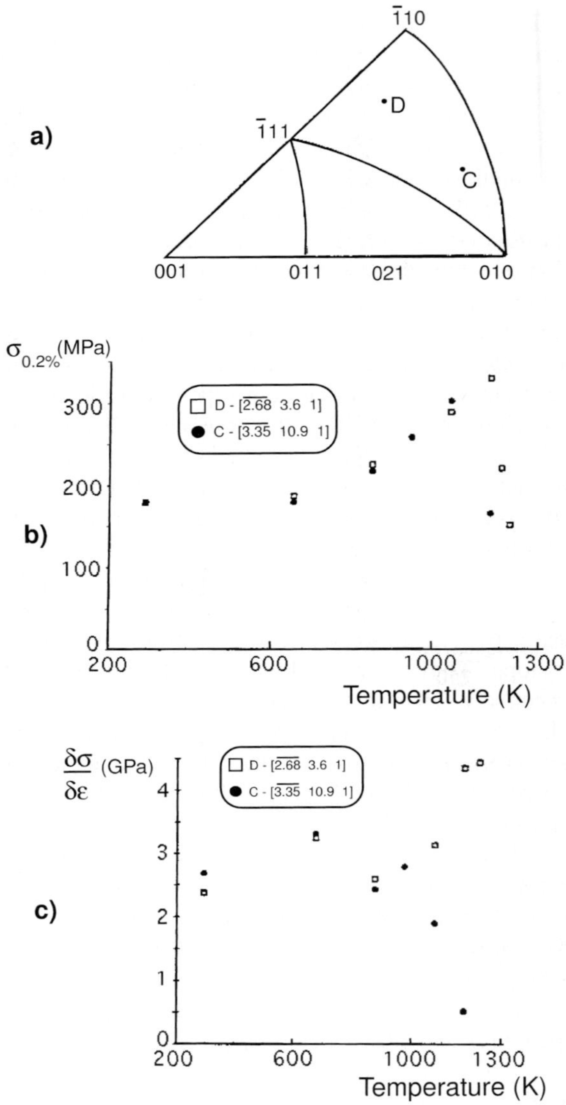

Fig. 17. Data concerning the study of $Ti_{55.5}Al_{44.5}$ single crystals reported in [36]: (a) crystallographic orientations of the specimens tested; (b) temperature dependence of the $\sigma_{0.2\%}$ proof stress for both orientations; (c) corresponding WHR averaged over the plastic strain interval 0.015–0.02.

not permit the determination of the work-hardening rate at the conventional 0.2% plastic offset strain. More valuable WHR values have been measured at 1% plastic strain ($\theta_{1\%}$). The normalised WHR, $\theta_{1\%}/G$ (G is the polycrystalline shear modulus), is shown in fig. 18. $\theta_{1\%}/G$ exhibits with temperature three types of behaviour whose boundaries roughly

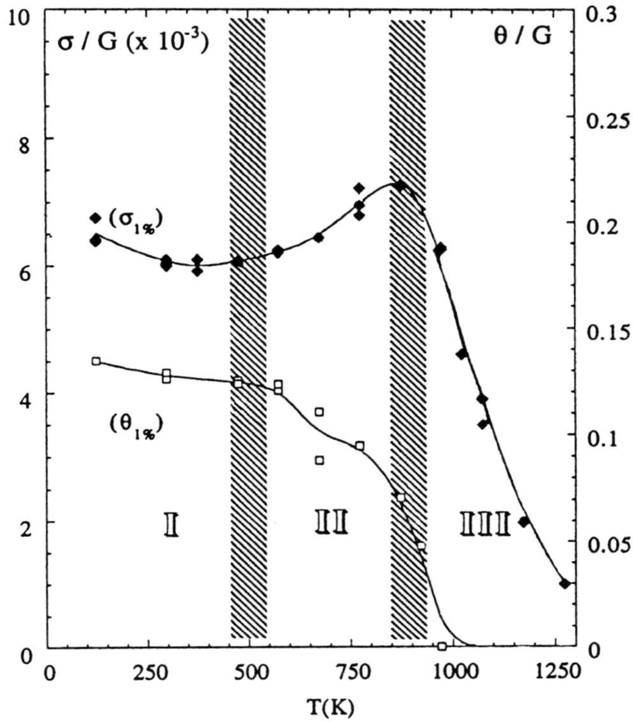

Fig. 18. Temperature dependence of the WHR measured at 1% plastic strain in a polycrystalline γ-TiAlMn alloy, [38].

correspond to those of $\sigma_{0.2\%}/G$. From low temperature up to ≈ 500 K, $\theta_{1\%}/G$ is high (> 0.1) and weakly decreases with increasing temperature. With increasing temperatures, in the second temperature domain, $\theta_{1\%}/G$ always decreases with temperature, but at a rate which is greater than in the previous domain. At the onset of the third domain, $\theta_{1\%}/G$ falls off rapidly to very low values and this regime is characterised by $\theta_{1\%}/G$ values that are almost zero.

The decreasing tendency of θ/G with increasing temperature and its smooth decrease with strain is consistent with the progressive disappearance of the stress anomaly, which is observed when the flow stress is measured at higher plastic strain levels [38].

3. Contribution of mobile dislocation densities and velocities to hardening

The Orowan equation stipulates that the plastic strain-rate $\dot{\gamma}_p$ is proportional to the mobile dislocation density ρ_m and their average velocity v. It is commonly used to correlate mechanical properties and dislocation mechanisms. However, after the early attempts of Gilman and Johnston at measuring values of dislocation velocities in LiF [39], the determination of the respective contributions of v and ρ_m to $\dot{\gamma}_p$ has never been performed convincingly in other types of crystals. In particular, conventional mechanical tests in which $\dot{\gamma}$ is either imposed (constant strain-rate experiments) or measured (creep or relaxation) are useless in this respect. Then the question remains open on whether the crystal contains few dislocations moving at high velocities or many dislocations at low velocities for a given rate of straining. In addition, the separation of v and ρ_m may allow to clarify a point which has been raised [40], concerning strength anomalies: is the latter property the result of an anomalous behaviour with temperature of the mobile dislocation density or velocity or both?

The present section reviews some of the available data from a variety of experimental techniques which partially answer the above question.

Dislocation velocities have been measured mostly using etch-pit experiments in a variety of crystals in the sixties (see, e.g., [39]). Section 3.1.1 reports such recent attempts in Ni_3Al crystals. In section 3.1.2, strain-rate sensitivities and activation volume values are reported for Ni_3Al compounds and γTiAl, together with the related information about deformation mechanisms. These are measured by transient techniques described in the appendix.

Internal friction experiments are a sensitive tool to study dislocation dynamics. Since the sample undergoes very low strains (and stresses) no dislocation multiplication takes place and it is possible in principle to relate the deformation rate to the mobile dislocation densities and velocities. Section 3.2.2 reports some results of mechanical spectroscopy in a $Ni_{75}Al_{24}Ta_1$ single-crystal alloy in the temperature range of the flow stress anomaly.

In contrast to dislocation velocities, mobile dislocation densities are a poorly documented parameter. These can be determined in principle using etch-pit experiments as well, but in most corresponding publications this quantity is ignored. We have developed repeated transient mechanical tests, which aim at separating the respective contributions to $\dot{\gamma}_p$ of ρ_m and v. The procedure followed is recalled in the appendix together with the main assumptions about their interpretation. Results of such tests applied to Ni_3Al and γTiAl are presented in section 3.2.

3.1. Dislocation mobility

3.1.1. Dislocation velocities in Ni_3Al measured from etch-pit experiments

Two groups have measured dislocation velocities as a function of stress and temperature in Ni_3Al compounds (Nadgorny and Iunin [41] and Jiang et al. [42]). Both studies have used the single crystal orientation illustrated on fig. 19 which was deformed in three-point bending (around $[1\bar{1}0]$). Under such conditions, two octahedral planes (111) and $(\bar{1}\bar{1}1)$ respectively can be activated, each one along two different $\langle 110\rangle$ directions shown on the figure with a Schmid factor of 0.408. Glide can also take place along two cube planes ((100) and (010) respectively) each one containing two active $\langle 110\rangle$ slip directions with a Schmid factor of 0.354. For this orientation, the latter planes correspond to both the cube cross-slip planes and the primary cube plane. Etch pits are produced and observed on the (001) face. Under such conditions, octahedral slip traces are along $[1\bar{1}0]$ and cube slip

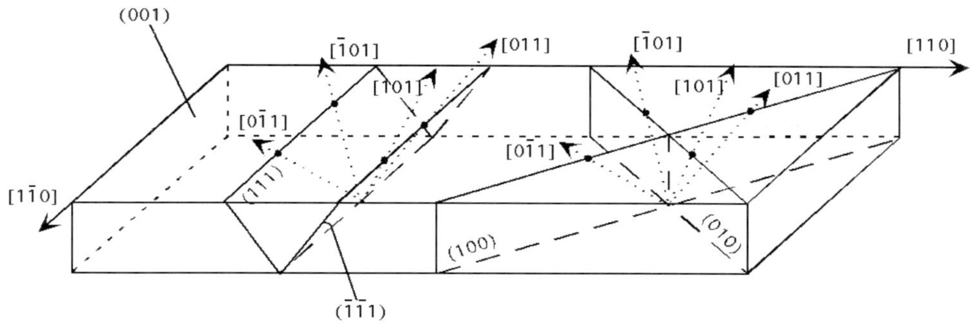

Fig. 19. Single crystal orientation for the 3 point-bending experiments around [1$\bar{1}$0] used by Nadgorny et al. [41] and Jiang et al. [42].

traces along [010] and [100] respectively. Thus the active slip planes can be unambiguously determined.

The procedure consists in annealing the crystal, then introducing fresh dislocations by scratch or Vickers indentation at RT. A first etching is performed to reveal these dislocations. At a given temperature the sample is then loaded during a time interval Δt, then unloaded and cooled down prior to a second etching sequence to reveal the new positions of the dislocations. Δt has to be adjusted for each load and temperature so that the distance d between the initial and final positions of a dislocation can be safely measured using an optical microscope. The velocity is then considered to be:

$$v = d/\Delta t. \tag{1}$$

Since the loading mode induces a stress gradient along the sample, a single experiment provides a range of velocities at different stresses. The velocity is expressed as a function of stress and temperature using the relations:

$$v(\tau, T) = v_0(\tau/\tau_0)^m, \tag{2}$$

where v_0 and τ_0 are constants and m a function of temperature. The exponent m and the activation volume V are connected by:

$$m = \frac{\tau V}{kT}. \tag{3}$$

m is obtained by plotting $\ln v$ as a function of $\ln \tau$ and V by using relation (3).

The materials used together with the experimental conditions are listed in table 2 for both studies. The resolved stresses achieved in compression tests and in bending experiments are compared in fig. 20 for binary compounds. The bending stresses are from Nadgorny et al. [41] in $Ni_{77.1}Al_{22.9}$ while $\tau_{0.2\%}$ has been measured in the same material by Dimiduk [43] and in $Ni_{76.6}Al_{23.4}$ by Spätig [14] who also determined activation volumes through macroscopic relaxation tests. For both compounds the $\tau_{0.2\%}$ data as a function

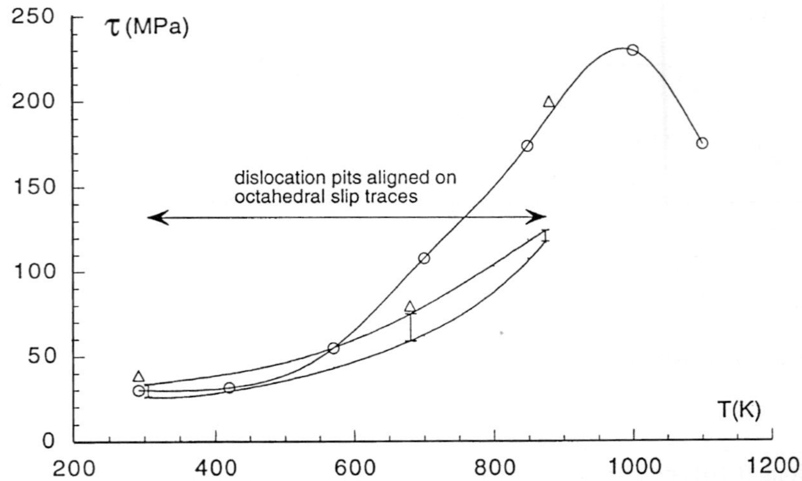

Fig. 20. Comparison of shear stresses achieved at different temperatures in three-point bending experiments (I from Nadgorny et al. [41] and $\tau_{0.2\%}$ in compression tests respectively on $Ni_{76.6}Al_{23.4}$ (○) (from Spätig [14]) and $Ni_{77.1}Al_{22.9}$ (Δ) (from Dimiduk [43]). Stresses are resolved along the primary octahedral slip systems.

Table 2
Comparison of experimental conditions for dislocation velocity measurements using etch-pit observations

Experiment	Nadgorny et al. [41]	Jiang et al. [42]
Material	$Ni_{77.1}Al_{22.9}$	$Ni_{75.1}Al_{21}Ti_{3.9}$
Annealing conditions	120 h 1473 K	168 h 1473 K + 168 h 1523 K
Introduction of fresh dislocations	scratch	Vickers indentation
Deformation test	3 point bending then unloading and fast cooling	3 point bending
Dislocation arrangements during glide	dislocations move in arrays along parallel slip planes	
Comparison of dislocation properties under tension or compression	– differences in dislocation array geometries – velocity data in tension only	differences in dislocation velocities are quantified

of temperature are along the same curve within the experimental error. Figure 20 shows that the bending stresses achieved are lower or equal to $\tau_{0.2\%}$. They correspond to the temperature range of the strength anomaly, i.e. deformation proceeds by octahedral glide, as confirmed by the observed etch-pit arrays [41]. On fig. 21, similar data are represented for a $Ni_{75.1}Al_{21}Ti_{3.9}$ compound. The bending stresses are from Jiang et al. [42] and the $\tau_{0.2\%}$ values are computed from Korner's data [44]. In this case, the bending experiments are performed in a wider range of temperature, i.e. below and above the stress peak temperature ($\sim$ 733 K) respectively. The etch-pit arrays are consistent with the primary

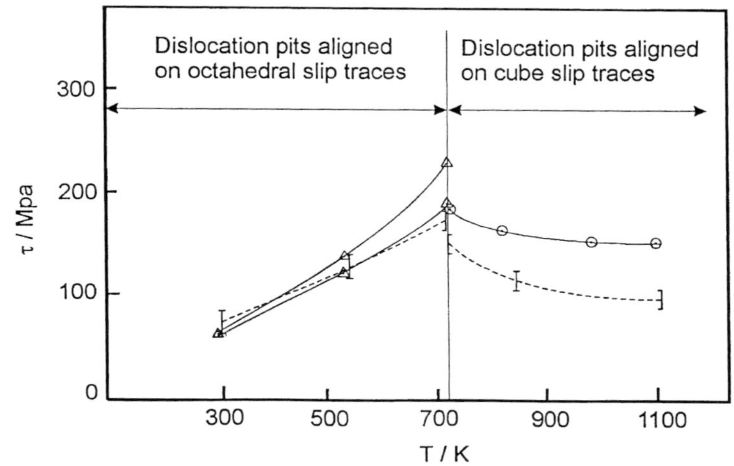

Fig. 21. Same type of plot as fig. 2 for $Ni_{75.1}Al_{21}Ti_{3.9}$. The bending stresses [42] are resolved respectively on the primary octahedral system (Δ) and on the cube cross-slip system (O). I $\tau_{0.2\%}$ computed from Korner data [44] resolved respectively along the primary octahedral systems for $T \leqslant 733$ K and the cube cross-slip systems for $T \geqslant 733$ K.

octahedral slip systems for 295 K $\leqslant T \leqslant$ 733 K and with cube slip for 733 K $\leqslant T \leqslant$ 1133 K. The Burgers vectors of the moving dislocations cannot be determined in such experiments but they are thought to correspond to the less mobile dislocation species [42]. The bending stresses are equal or slightly higher than the resolved stresses that correspond to a 0.2% plastic strain in compression over the temperature range explored. Note that the compression data in figs 20 and 21 refer to tests performed on a virgin sample at each temperature (see section 2.1.1.4).

The velocities measured for both types of compounds are represented in fig. 22 as functions of the resolved shear stress. Average curves only have been reported for the sake of clarity. Nadgorny and Iunin [41] measure the velocities of dislocations that belong to loose arrays around the scratch. Below the peak temperature, the results are represented on fig. 22(a). At each of the temperatures investigated, relation (2) is rather well satisfied. The figure also shows that for a given compound, as the temperature is raised m increases and the velocity at a given stress decreases. The data related to the ternary compound show slightly different curves in tension and compression at a given temperature, which reflects the well known asymmetries on the corresponding stress–strain curves (see, e.g., [45]). Comparing the data for the two compounds at one temperature (e.g., RT) shows that at a given stress dislocations are slower in the ternary alloy, while m is lower. For temperatures equal to or above the peak, data are available only for the ternary compound (fig. 22(b)). Around the peak temperature both the octahedral and cube systems are observed while above the peak only the latter glide systems are present. Again relation (2) is satisfied. m values are found to be lower than those measured for octahedral glide. As the temperature is raised m values are decreasing, unlike below the stress peak temperature,

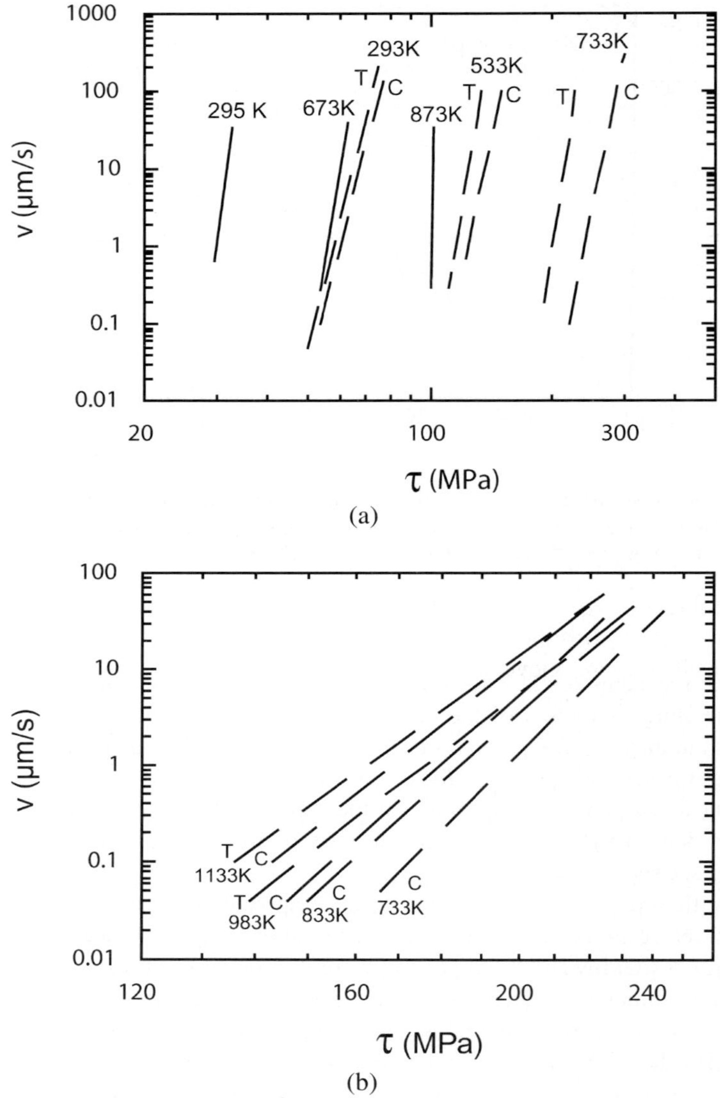

Fig. 22. Dislocation velocities as a function of stress measured using etch-pit experiments: (a) octahedral glide; (b) cube glide; — [41] data in tension, – – – [42]. Temperatures are indicated. T and C refer to samples in tension and compression.

and for a given stress, the velocity increases as in a conventional thermally activated mechanism.

At this stage, it is interesting to compare the activation volumes measured for dislocation velocities in dynamic etch-pit experiments (relation 3) and in compression tests. Typical values are listed in table 3, expressed in b^3 units where b is the Burgers vector of the superpartial dislocation ($b = 0.254$ nm).

For compression tests, a binary compound and an alloy are considered. Inspection of table 3 shows that below the peak temperature, etch-pit experiments yield low values of V, between 80 and $190b^3$, which are comparable in both types of studies. The volumes measured from transient tests under similar conditions are larger. They vary between $1370b^3$ and $1800b^3$ in the binary compound, between $200b^3$ and $700b^3$ in the Hf alloy. The latter volumes are lower in the ternary compound as compared to the binary one, which is not visible in the etch-pit data. However, above the peak temperature, the volumes measured by Jiang et al. [42] are in the range 56–$85b^3$ for the Ti alloy. For the highest temperatures achieved above $T_{p,\tau}$ [14], small values are measured for V, such as $95b^3$ and $30b^3$ in the binary and the ternary compounds respectively.

The following interpretation aims at clarifying the above results on volume values. A number of deformation parameters are different when dynamic etch-pit experiments and compression tests are compared:

- An internal stress around the scratch [41] or the indentation [42] is likely to alter the V values determined by considering the applied stress in relations (2) and (3).
- However, the following interpretation seems more plausible.

In the temperature range below the peak temperature, the formation of Kear–Wilsdorf locks is likely to disturb the dislocation velocity measurement. Indeed, as the sample is stressed, dislocations move and may be locked after a time interval smaller than Δt so that the velocities measured are likely to be underestimated. Under these conditions, relations (2) and (3) show that V is also underestimated. On the contrary, in compression tests, the volumes measured refer to average velocities of continuously moving dislocations. This latter effect seems to explain the differences between the volume data of table 2 below $T_{p,\tau}$ which suggests that dynamic etch-pit experiments do not yield actual dislocation velocities under such conditions. A check of the validity of the above velocity measurements would be to measure v for different time intervals Δt. A correct experiment should yield v values independent of Δt. This is reported in the early papers on velocity measurements in materials obeying conventional thermal activation (Gilman and Johnston [39]). Unfortunately such data are not available for Ni_3Al in the two works considered.

At and above the peak temperature, several microstructural studies report that straight locked screws are not present any more (see, e.g., [46]). Dislocations are observed to bow along cube planes (cross-slip or primary plane). Macroscopic experiments report a stress decreasing with temperature (see, e.g., [14,45,47]). Although the corresponding deformation mechanism is poorly documented, one can expect a viscous motion of the dislocations overcoming a pseudo-Peierls Nabarro friction on the cube plane, which is thermally activated. Under such conditions it is not surprising to measure comparable activation volume values using dynamic etch-pit experiments and constant strain-rate compression tests.

For experiments below the stress-peak temperature, Nadgorny and Iunin [41] concluded that their data evidenced an anomaly of the dislocation mobility. They also stated that multiplication processes were playing an important role in monotonic tests, while they

Table 3
Activation volumes for dislocation velocities

Method	Reference	Material	$T_{p,\tau}$	Activation volumes and temperatures							
				Octahedral slip				Cube slip			
Dynamic etch-pit experiments	Nadgorny et al. [41]	$Ni_{77.1}Al_{22.9}$	~ 1000 K	295 K	673 K	873 K					
				$80b^3$	$90b^3$	$190b^3$					
	Jiang et al. [42]	$Ni_{75.1}Al_{21}Ti_{3.9}$	~ 733 K	293 K	533 K	733 K		733 K	833 K	983 K	1133 K
				$99b^3$	$86b^3$	$84b^3$		$56b^3$	$62b^3$	$73b^3$	$85b^3$
Macroscopic transient tests	Spätig [14]	$Ni_{76.6}Al_{23.4}$	~ 1000 K	300 K	440 K	600 K	670 K	1000 K	1100 K		
	Bonneville et al. [16]			$1370b^3$	$1800b^3$	$1800b^3$	$1480b^3$	$100b^3$	$96b^3$		
	Spätig [14]	$Ni_{74.8}Al_{21.9}Hf_{3.3}$	~ 630 K	300 K	350 K	550 K	600 K	680 K	780 K		
	Bonneville [71]			$200b^3$	$350b^3$	$600b^3$	$700b^3$	$400b^3$	$30b^3$		

were not effective in the dynamic etch-pit experiments. Under the same temperature conditions, Jiang et al. [42] confirm the anomaly of the dislocation mobility. However, as explained above, dislocation mobility on the octahedral planes deserves a deeper analysis.

3.1.2. Activation volumes and strain-rate sensitivities in Ni_3Al and γ TiAl

During a constant strain-rate experiment, transient tests can be performed at a given point of the stress–strain curve (see, e.g., [46,48] for a review on such tests): the strain rate can be altered in a strain-rate jump experiment, the strain (or stress) can be kept constant in a relaxation (or creep) test. The two last types of transients have been extensively used in a variety of Ni_3Al compounds. They provide deformation parameters which are not accessible using monotonic tests.

As recalled in the appendix, the observation of repeated logarithmic relaxation or creep transients allows one to define a microscopic activation volume. In what follows, this parameter will be used to characterize the dislocation mobility mechanisms for given deformation conditions.

3.1.2.1. Preplastic and plastic domains in Ni_3Al single crystals. It has generally been accepted that at stresses lower than $\tau_{0.2\%}$ (the 2×10^{-3} offset stress) deformation corresponds to the rapid motion of edge dislocation segments. For stresses larger than $\tau_{0.2\%}$, flow is associated with the more difficult propagation of screws that yields the anomalous temperature dependence of the stress [49]. However, the validity of using $\tau_{0.2\%}$ as the critical stress for extensive screw dislocation motion in Ni_3Al has been questioned [50,51], with the following arguments: (i) as shown by Thornton et al. [10], the 10^{-4} offset stress exhibits an anomalous behaviour with temperature which cannot be explained by edge propagation, (ii) extensive octahedral glide occurs during primary creep at stresses lower than $\tau_{0.2\%}$ [50], (iii) the stress–strain curves do not exhibit any distinct yield point.

To evidence a possible critical stress between the micro- and the macro-plastic domains, microscopic activation volumes V have been measured as functions of strain by the technique of relaxation series (see the appendix) in $Ni_{74.8}Al_{23.9}Hf_{3.3}$ single crystals oriented along $\langle 123 \rangle$ (single-glide orientation). Three temperatures have been investigated in the anomaly domain of the flow stress, namely 293 K, 423 K and 573 K. At least three samples have been deformed at each temperature and a good reproducibility has been observed comparing the different tests [14,52]. A typical stress–strain curve is represented on fig. 23(a). As emphasized above, it exhibits a gradual transition from the micro- to the macro-plastic domains. V values are also reported as functions of strain on fig. 23(a). The strain dependence of V evidences two distinct domains with a well defined transition strain. It corresponds to a critical stress $\tau_{cr} = 130 \pm 15$ MPa on the stress–strain curve. τ_{cr} can be compared with the mean value of $\tau_{0.2\%} = 140 \pm 6$ MPa measured on the stress–strain curves of the three tests. Accordingly, the $V(\tau)$ curve on fig. 23(b) exhibits a similar behaviour. The variation of V as a function of stress confirms the existence of two stress domains. The same type of observation holds for the two other temperatures which were investigated, i.e. the two strain domains for V were clearly identified and the corresponding τ_{cr} values could be determined. The results are summarized on fig. 24 which represents the τ_{cr} values as a function of $\tau_{0.2\%}$ for the three temperatures. This figure shows a reasonable

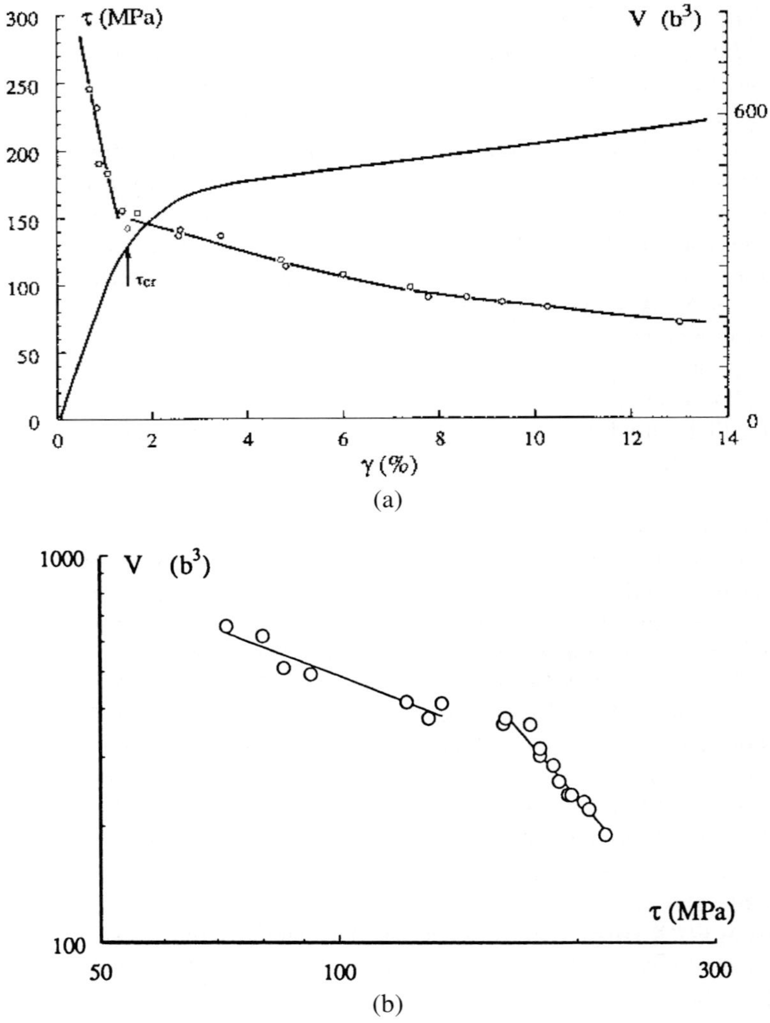

Fig. 23. Mechanical test results of a $Ni_{74.8}Al_{23.9}Hf_{3.3}$ single crystal. ⟨123⟩ orientation. $T = 293$ K. (a) Stress–strain curve and microscopic activation volumes as a function of strain. The determination of τ_{cr} is indicated. (b) V as a function of stress. $\dot{\gamma} = 10^{-4}\ s^{-1}$ [52].

correlation between the conventional $\tau_{0.2\%}$ stress and the critical stress deduced from the strain dependence of the activation volume.

The peculiar strain or stress dependence of V illustrated above suggests that two deformation domains are present along the stress–strain curve. Such a dependence has already been observed in quite different classes of materials [53–55]. Therefore τ_{cr} in the present study can be considered as the critical stress for macroplasticity. The preplastic domain ($\tau < \tau_{cr}$) corresponds to long segments (large V) moving easily under low

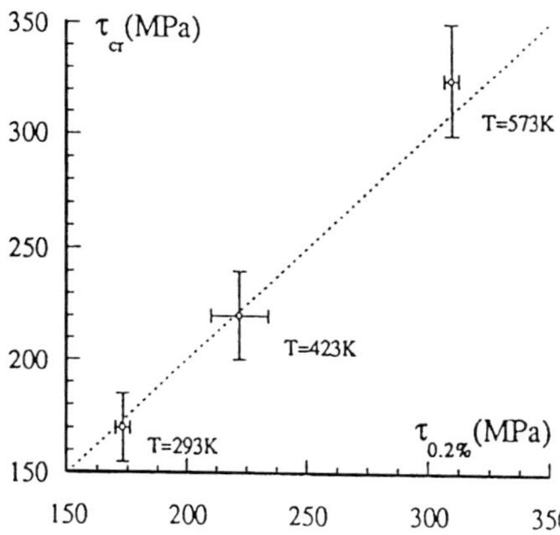

Fig. 24. Transition stress τ_{cr} as a function of $\tau_{0.2\%}$ at three different temperatures. Same material as fig. 23.

stresses, very likely of edge character, the more mobile species. Shorter segments are then activated as τ increases and V decreases. In the macro-plastic domain ($\tau > \tau_{cr}$), large scale dislocation multiplication takes place, as well as their motion over large distances, with the locking of screw segments being responsible for the strength anomaly. The corresponding V values of fig. 23 suggest that the length of the moving dislocation segments decreases slowly as strain (or stress) increases. Consequently, the good agreement established above between the stresses $\tau_{0.2\%}$ and τ_{cr} indicates that $\tau_{0.2\%}$ can be considered as representative of the critical stress for macroscopic deformation. Let us note that the rough description presented here of the micro- and macro-plastic domains has to be refined somewhat. Indeed, it should account for the anomalous behaviour of the 10^{-4} offset stress reported above. Screw dislocation locking is likely to operate already in the preplastic domain as well, but to a lesser extent.

3.1.2.2. Activation volume values in single crystals of Ni_3Al compounds and polycrystals of γ TiAl. Among the parameters measured to understand the mechanical behaviour of intermetallics, the strain rate sensitivity of the flow stress S has received some attention. It is measured by comparing tests performed at different strain rates by, e.g., [6,56,57] or through strain-rate jump experiments by [10,18]. But S is still a poorly documented parameter. The relaxation experiments described in the appendix allow one to measure an apparent activation volume V_r defined by:

$$V_r = kT\,\partial \ln \dot{\gamma}_p / \partial \tau,$$

which is obtained by eliminating dt between the time derivatives of relations (2a) and (3a). Let us note that this volume can also be measured in strain-rate jump experiments. If we adopt the definition [10]:

$$S = (1/T)(\partial \ln \tau / \partial \ln \dot{\gamma})_T,$$

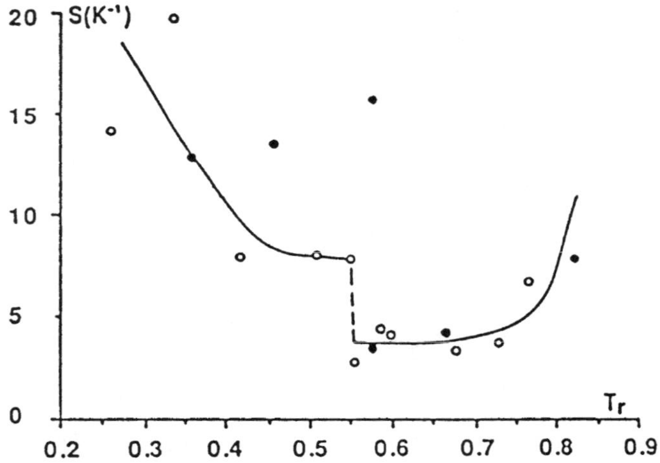

Fig. 25. Strain-rate sensitivity values S as a function of the reduced temperature T_r for: (○) $Ni_3(Al,Cr)$ [10] and (●) $Ni_{75}Al_{24}Ta_1$ [58].

V_r and S are related by:

$$S = k/\tau V_r, \tag{4}$$

assuming that $\dot{\gamma}_p$ and $\dot{\gamma}$ are not too different.

Data about V_r in a $Ni_{75}Al_{24}Ta_1$ single crystal in the ⟨123⟩ orientation performed between 300 and 800 K ($T_{p,\tau} \approx 800$ K) at two strain-rates ($\dot{\gamma} = 7 \times 10^{-5}$ s^{-1} and 7×10^{-4} s^{-1} respectively) have been published [58]. They are compared on fig. 25 with the S measurements of [10] on a $Ni_3(Al,Cr)$ polycrystal, as a function of T_r, the temperature normalized to the stress peak temperature. If both compounds are compared, although there is a large discrepancy in S values at low homologous temperatures, both sets of data exhibit a discontinuity of the $S(T_r)$ variation at a temperature close to $0.55T_{p,\sigma}$. This discontinuity suggests a change in deformation mechanisms at the corresponding temperature. In addition for higher temperatures a fair agreement is found for S values in both compounds. As $T_{p,\sigma}$ is approached, S increases steeply with T, which corresponds to a decrease of V_r down to low values, as expected for the process of high-temperature glide on the cube plane (Peierls–Nabarro mechanism).

In γTiAl polycrystals, activation volumes have been measured through relaxation tests at plastic strains between 0.3 and 0.5%. Some strain-rate jump experiments have also been performed. The variation of the microscopic volume V (expressed in b^3 units, $b^3 = 2.2 \times 10^{-2}$ nm^3 representing the atomic volume) as a function of temperature is plotted in fig. 26. In fact, the signature of a given deformation mechanism is given by the variations of V as a function of stress and the curves of fig. 26 are a consequence of the $\sigma(T)$ dependence. The $V(\sigma)$ curves are represented on fig. 27. Three distinct curves are observed on this figure which correspond to the three different deformation mechanisms described above. The variation $V(\sigma)$ is first studied in regime I, where deformation

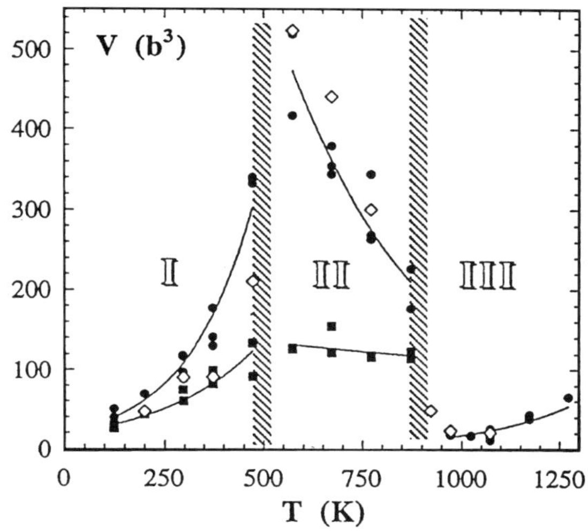

Fig. 26. Activation volumes as a function of temperature in a γTiAl polycrystal, measured by stress relaxations ((●) apparent and (■) microscopic) and strain-rate jumps (◇). $\varepsilon_p = 0.3$ to 0.5%, $\dot{\varepsilon} = 3.3 \times 10^{-5}\ s^{-1}$ [38].

results from the motion of superdislocations including the formation of faulted dipoles. These dislocations are often aligned along particular crystallographic directions as shown by TEM observations [59,60]. This suggests that their mobility is controlled either by a Peierls–Nabarro lattice resistance or a pseudo Peierls–Nabarro friction due to a core extension out of the glide plane. Atomistic computations indicate that both types of friction are effective in TiAl. In particular, the core configuration of $\langle 101]$ superdislocations has been studied by Yamaguchi et al. [61]. It dissociates as follows:

$$[\bar{1}01] \rightarrow 1/6\,[\bar{1}\bar{1}2]_{\text{leading}} + \text{ISF} + 1/6\,[\bar{5}14]_{\text{trailing}}.$$

The computations show that the trailing dislocation core is extended out of the (111) plane. This type of core extension has also been invoked by Viguier and Hemker [62] to explain the formation of faulted dipoles. The stress decrease with temperature in this regime (fig. 18) and the values of V ranging from 25 to $130b^3$ are in agreement with a thermally activated glide process controlled by a pseudo Peierls–Nabarro friction, as shown by Viguier et al. [38]. Such a mechanism has been described by Friedel [63] for screw dislocation glide along the prismatic plane of hexagonal close-packed metals. Such dislocations split along the basal plane. A recombination of the screw over a length l_c is postulated; the screw thus becomes mobile along the prism plane. This mechanism is thought to operate as well for the mobility of the trailing superpartial of the superdislocation above. The model predicts a variation of V as the reciprocal of the effective stress squared. This type of dependence is checked in the present case on fig. 28. The relation $V = \alpha(\sigma - \sigma_\mu)^{-2}$ has been fitted to the experimental curve, yielding $\sigma_\mu = 265$ MPa in fair agreement with the experimental data. Let us note that this mechanism has been experimentally studied in a restricted number of cases: prismatic

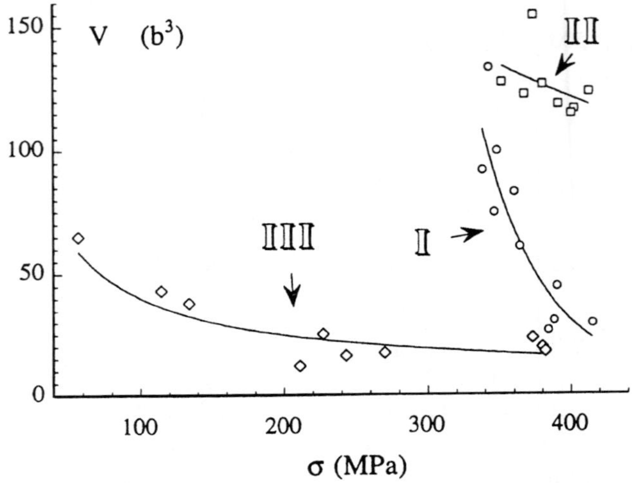

Fig. 27. Same data as fig. 26 as a function of stress (microscopic volumes). The temperature domains are indicated [38].

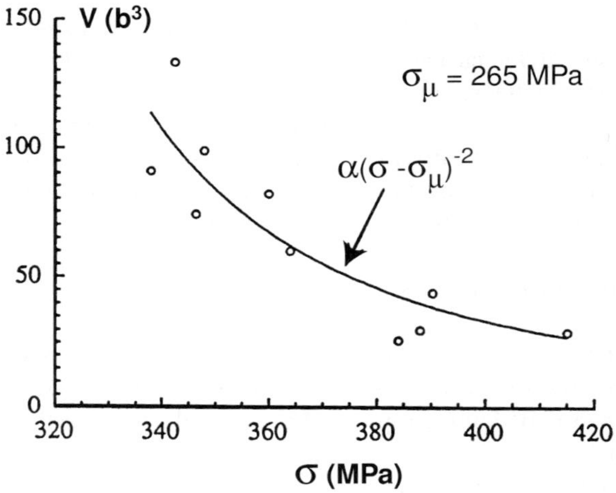

Fig. 28. Stress dependence of the microscopic volume V in temperature regime I. The fit with the model curve is shown (see text). γTiAl polycrystals [38].

glide (i) in Ti by Biget and Saada [64] using conventional mechanical tests and (ii) in Mg by Couret and Caillard [65] through in TEM situ experiments.

In regime II (yield stress anomaly) glide of ordinary screw dislocations is rate controlling. It has been described by a local pinning-unzipping model [66] in agreement with TEM observations. Ordinary dislocations exhibit numerous pinning points aligned along the screw direction (see section 6.3). Unzipping of the dislocation is assumed to occur by the lateral motion of the cusps. The model predicts that a strong exhaustion of mobile dislocations is responsible for the strength anomaly. This agrees with the high dislocation exhaustion rates measured under such conditions (see section 3.2). In such a model, the strength anomaly is of the "mobile dislocation density" type as opposed to a "velocity" type. At fixed temperature and plastic strain rate, the model assumes deformation to occur under constant effective stress. This agrees with experimental values of V which are found to be constant as a function of strain along a stress–strain curve, in the macroplastic domain [38]. V being a function of $\sigma^* = \sigma - \sigma_\mu$, the latter result suggests that this stress component is constant under these conditions.

In temperature regime III, dislocation density measurements [60] indicate that ordinary dislocations dominate. They are observed to bow out of their glide planes, climb processes being involved. The mechanical properties have the following characteristics: $\sigma_{0.2\%}$ is strongly thermally activated, the stress–strain curves do not exhibit any hardening in the macroplastic domain and the apparent activation volume V_a is small ($V_a \sim 15$ to $60b^3$) so that the relaxation curves are better fitted by a power function as compared to a logarithmic function of time. All these features suggest that a recovery process takes place at these temperatures ($T \geqslant T_{p\sigma} \sim 900$ K).

The apparent activation volume values which have been determined [59] allow for a comparison with other works on similar compounds. V_a values have been transformed into S values, using eq. (4) above. These S values are plotted as a function of temperature on fig. 29, together with those of Kawabata et al. [31] on single crystals of various orientations, for comparison. The $S(T)$ curves are fairly similar, which suggests that dislocation mobility mechanisms are the same as functions of temperature in single and polycrystals. More precisely, fig. 29 shows that at low temperature the polycrystal curve fits the $\langle 001 \rangle$ single crystal one. Since, in the latter orientation, ordinary dislocations have a zero Schmid factor, this confirms that deformation proceeds via the motion of superdislocations which appear to be rate controlling in both materials under such conditions. When temperature is raised (regimes II and III), the $S(T)$ curves of fig. 29 indicate that polycrystals behave as single crystals of orientations A and [010] respectively, for which ordinary dislocations have the largest Schmid factor. This confirms that the mobility of the latter dislocations is rate controlling for these deformation conditions. All these considerations are fully supported by the TEM observations [60].

3.2. Exhaustion of mobile dislocations

3.2.1. Direct measurements of exhaustion rates

As explained in the appendix, the methods of repeated transients provide, in addition to the microscopic volume V, an estimation of the mobile dislocation exhaustion parameter $\Delta\rho_m/\rho_{mo}$ during plastic deformation. So far it has been derived from relaxation series. In most results presented below, it expresses the relative decrease of the mobile dislocation density which corresponds to a decrease of one order of magnitude of the initial plastic strain rate (i.e. $q = 10$ in relation (A15) of the appendix), unless otherwise specified. The results reported in what follows indicate that although measured during a transient, this exhaustion parameter correlates well with features of the stress–strain curves. This can be

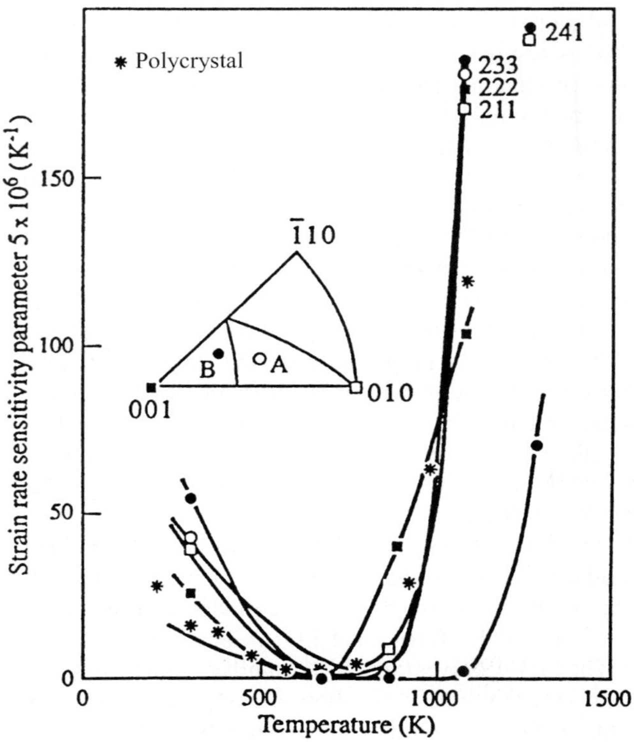

Fig. 29. Strain-rate sensitivity as a function of temperature in γTiAl. Comparison of polycrystals [38] and single crystals [31] of indicated orientations.

explained by considering that during the short transient, the structural state which develops under constant strain-rate conditions is only slightly modified.

Since the parameter $\Delta\rho_m/\rho_{mo}$ has been measured only recently, values are first given for comparison of several types of crystals and straining conditions which are known to correspond to different deformation mechanisms. These values are listed in table 4 where the deformation conditions are recalled together with the work-hardening coefficient θ normalized to the shear modulus G. θ is defined here as the slope of the stress–strain curve.

In table 4, the parameters given for the Ni_3Al single crystals have been measured in the strength anomaly domain at a shear strain-rate of 10^{-4} s^{-1}. The normalized work-hardening coefficient is high for Ni_3Al as compared to other materials in the table. For incrementing strain, stress must be increased to activate shorter segments. The locking process occurs through the cross-slip of the leading superpartial dislocation. The latter mechanism is strongly dependent on the complex stacking fault energy of the compound as established by weak-beam measurements of this parameter by Kruml et al. [67] (see section 6.1.1.1). The corresponding value of $\Delta\rho_m/\rho_{mo}$ in table 4 indicates that more than 80% of the mobile dislocations are exhausted in the conditions specified above. This value

Table 4
Mobile dislocation exhaustion parameter and work-hardening coefficient for different types of crystals and deformation conditions (compression tests)

Material	Ni_3(Al,3at%Hf) $\bar{1}23$ single crystal	Ge 110 single crystal	Cu 110 single crystal
Deformation temperature	293 K	750 K	77 K
θ/G	300×10^{-3}	35×10^{-3}	$3–4 \times 10^{-3}$
$\Delta\rho_m/\rho_{mo}$	0.74–0.87	0.28	0.20–0.25
References	Martin et al. [48]	Charbonnier et al. [69]	Couteau et al. [70]

is also high as compared to the other materials of the table 4. Let us note that the high exhaustion rate of dislocations in Ni_3Al compounds has been evidenced by using different types of experiments, e.g., internal friction tests [68] in $Ni_{75}Al_{24}Ta_1$ single crystals or creep experiments [51].

For the Ge single crystals in symmetrical orientation at 750 K ($\dot{\gamma} = 3.8 \times 10^{-5}$ s^{-1}), the stress–strain curve exhibits an upper yield point which corresponds to dislocation multiplication followed by a lower yield point after which the parameters are measured [69]. Work hardening is thought to occur from dislocation intersections, in the presence of a strong Peierls–Nabarro glide resistance in this covalent material, along 4 distinct slip systems. The number of stored dislocation increases with strain which necessitates higher stresses for deformation to proceed. The normalized work-hardening coefficient is found to be lower than in the intermetallic compounds, as well as $\Delta\rho_m/\rho_{mo}$. Finally, [70], for copper single crystals also in symmetrical orientation, θ/G is measured in Stage II using the same strain rate as in Ge. The relatively low value of this parameter is attributed to dislocation intersections as in the Ge crystal above, but with no lattice resistance to glide in this metal. The parameter $\Delta\rho_m/\rho_{mo}$ is found to be low too, as compared to the other materials. The fair correlation between mobile dislocation exhaustion rates and the normalized work-hardening coefficients emphasizes the validity of the description proposed for transient tests in the appendix. It also gives a common interpretation of hardening in terms of dislocation exhaustion, although the exhaustion mechanisms are of different nature in the examples of table 4.

The exhaustion of mobile dislocations has been studied in more detail in Ni_3Al compounds and in γTiAl. The parameter $\Delta\rho_m/\rho_{mo}$ has been measured along a stress–strain curve in a single crystal of $Ni_{74.8}Al_{23.9}Hf_{3.3}$ in the $\langle 123\rangle$ orientation at RT [48]. The variations of $\Delta\rho_m/\rho_{mo}$ and θ are plotted as a function of stress in fig. 30. It shows that the mobile dislocation exhaustion rate and the work-hardening coefficient decrease simultaneously as stress increases. If θ and $\Delta\rho_m/\rho_{mo}$ are measured at 0.2% plastic strain, their variations as a function of T exhibit again parallel trends as illustrated on fig. 31, for a binary single crystal of the same orientation [71]. The peak temperature for θ (lower than the stress peak temperature) is indicated; it coincides fairly well with the maximum value of $\Delta\rho_m/\rho_{mo}$. The same property is observed for θ and $\Delta\rho_m/\rho_{mo}$ when both parameters are plotted as functions of $\tau_{0.2\%}$ (see [72] for binary single crystals). The discrepancy between the peak temperature for θ (or $\Delta\rho_m/\rho_{mo}$) and that for $\tau_{0.2\%}$ will be discussed in section 6.1.

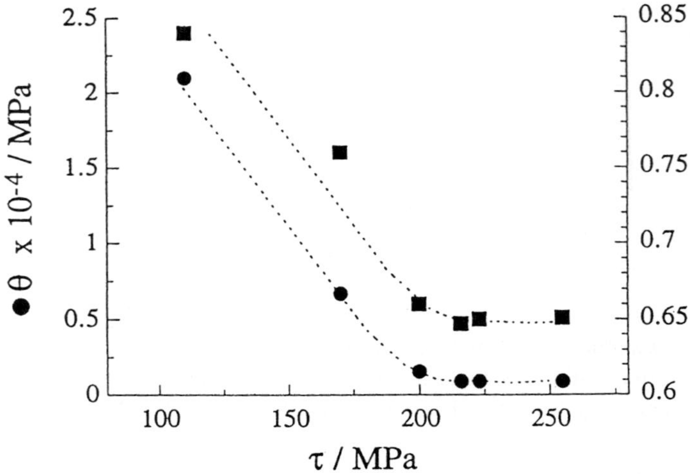

Fig. 30. Work-hardening coefficient and mobile dislocation exhaustion rate $\Delta\rho_m/\rho_{mo}$ as functions of stress along a stress–strain curve. $Ni_{74.8}Al_{23.9}Hf_{3.3}$ in ⟨123⟩ orientation. $T = 293$ K [48].

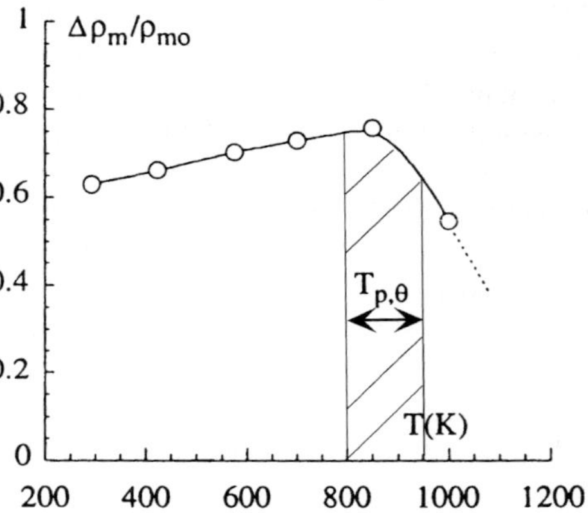

Fig. 31. $\Delta\rho_m/\rho_{mo}$ measured at a 0.2% plastic strain as a function of temperature. $Ni_{76.6}Al_{23.4}$ single crystal. ⟨123⟩ orientation [71].

In polycrystals of γ phase of $Ti_{47}Al_{51}Mn_2$, the mobile dislocation exhaustion rates have also been measured, at stress levels close to the 0.2% proof stress. Figure 32 represents this parameter as a function of temperature [38]. The temperature regimes which correspond to peculiar deformation mechanisms have been detailed by Viguier et al. [59,60]. In temperature regime I (see section 3.1.2.2) $\Delta\rho_m/\rho_{mo}$ increases between 120 K and reaches a maximum value at the onset of regime II ($T \sim 500$ K). In this temperature domain, θ has been shown to decrease with increasing temperature (see section 2.2.1), thus exhibiting a trend opposite to that of $\Delta\rho_m/\rho_{mo}$. This is an exception to the rule of parallel trends illustrated above for other materials. To explain this discrepancy, one can consider that the

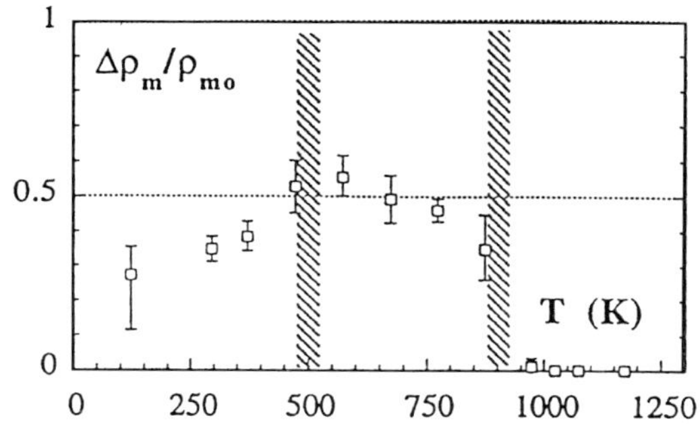

Fig. 32. Exhaustion rate of mobile dislocations as a function of temperature. γTiAl polycrystal. The three temperature domains are indicated [38].

exhaustion of mobile superdislocations occurs through the annihilation of dislocations of opposite signs, the frequency of which increases with temperature. This causes a decrease of θ under the same conditions. In regime II, $\Delta\rho_{\mathrm{m}}/\rho_{\mathrm{mo}}$ slightly decreases with increasing temperatures but keeps high values, while θ is also observed to decrease. At the onset of regime III which is dominated by dislocation climb [73], $\Delta\rho_{\mathrm{m}}/\rho_{\mathrm{mo}}$ as well as the work-hardening coefficient fall off to zero. Therefore in regimes II and III, θ and $\Delta\rho_{\mathrm{m}}/\rho_{\mathrm{mo}}$ exhibit again parallel trends, as generally observed.

To conclude this section on mobile dislocation densities, the above results suggest that in the strength anomaly domain of Ni_3Al single crystalline compounds, as well as for γTiAl polycrystals, high mobile dislocation exhaustion rates are observed in agreement with high values of the corresponding work-hardening coefficients.

3.2.2. *Internal friction experiments in* Ni_3Al

Although mechanical spectroscopy is known as a very sensitive tool for providing information about dislocation dynamics, it has been scarcely used to investigate the rate-controlling mechanisms in Ni_3Al ordered intermetallic alloys. A few studies have been reported in the literature [68,74,75]. These latter experiments have identified a well defined Debye peak which is located at a temperature of nearly 950 K for an excitation frequency of 1 Hz. This internal friction peak has been interpreted as resulting from a short-range diffusion process leading to a stress-induced reorientation of Al–Al dipoles on $\langle 111\rangle$ planes [75]. This interpretation has been confirmed by a detailed study undertaken by Numakura et al. [76], who show that the characteristics of this peak are compatible with those observed by [74] in Ni_3Al polycrystals, stoichiometric and of various compositions. This peak of Zener type divides the damping spectrum in two temperature regimes. At high temperatures, the exponential increase in mechanical loss is related to dislocation motion on the $\{001\}$ cube planes [75], while below 800 K the mechanical loss level is rather low and almost independent of temperature.

In a more interesting analysis, Cheng et al. [68] have shown that in order to observe any dependence of the mechanical loss on the temperature in the anomaly domain of the flow stress, two conditions have to be fulfilled: (i) specimens should be pre-deformed and (ii) the oscillation amplitude must be higher than 10^{-4} in order to induce

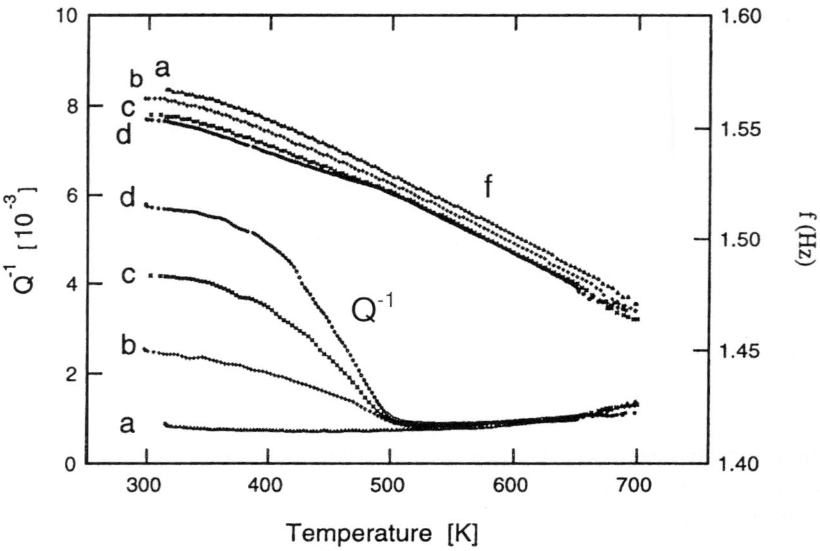

Fig. 33. Temperature dependence of mechanical loss (Q^{-1}) and vibration frequency (f; proportional to the square root of the shear modulus) of $Ni_{75}Al_{24}Ta_1$ single crystals. Strain amplitude 5×10^{-4}, heating rate 2 K mn^{-1}. Specimens are as cast (a) and in situ predeformed by torsion in the pendulum up to 1% (b), 3.4% (c) and 5.6% (d) [68].

a damping process. This is confirmed by the results of fig. 33 showing the mechanical loss over the temperature range 300–700 K for an as-cast specimen and for pre-deformed specimens. These experiments were performed in a torsion pendulum on $Ni_{75}Al_{24}Ta_1$ single crystalline plates along a $\langle 111 \rangle$ torsion axis. The pre-deformation strains were carried out *in situ* in the pendulum at RT and different amounts of plastic strain (ε_p) were imposed, i.e., 1.0%, 3.4% and 5.6%. The relevance of using large strain amplitudes has been presented in [77]. This behaviour in mechanical loss is in good correlation with the results of Thornton et al. [10] showing that the anomalous increase in flow stress with temperature only appears for strain amplitudes higher than about 10^{-4}. Figure 33 clearly shows an increase in the internal friction background (Q^{-1}) in the low-temperature range. In addition, it appears that at RT Q^{-1} strongly depends on ε_p. With increasing temperature, Q^{-1} decreases for all amounts of pre-strain and recovers the value of the as cast specimen at nearly 500 K. It is remarkable that this transition temperature does not depend on the amount of pre-deformation. It is also worth noting that, whatever is the level of pre-deformation, for any further thermal cycles, the Q^{-1} background of pre-deformed specimens behaves like that of the as-cast specimen, which clearly indicates that heating above 500 K eliminates the predeformation effects.

The occurrence of a mechanical damping for an excitation amplitude higher than 1×10^{-4} only has been attributed to the existence of a critical length for the motion of the superkinks, while the increase in mechanical loss induced by plastically pre-deforming the specimens has been interpreted as induced by an increase in the mobile dislocation density [68,78]. The other features of mechanical loss, which are essentially:

- a sharp decrease with increasing temperature,
- a transition temperature at nearly 500 K, above which it is temperature independent,

have been explained by the authors as resulting from a combination of two phenomena [68,78]:

- a pinning effect of the screw dislocation segments via cross-slip from the (111) onto the (010) planes and the subsequent formation of Kear–Wilsdorf locks,
- an exhaustion of the mobile dislocation segments, that is the irreversible glide and/or bowing out process of the superkinks lying on the (111) planes between the Kear–Wilsdorf locks.

4. *Work hardening and total dislocation density*

The aim of this paragraph is to analyse the evolution of the total dislocation density during the deformation of intermetallic compounds with special emphasis on the role of this density on the strain hardening rate.

Most of the TEM observations of dislocation microstructures in intermetallic compounds are more focused on the detailed configuration of individual dislocations than on the global density and relative arrangement. For instance papers reporting global dislocation densities as a function of plastic strain are very scarce. Values are reported by Kruml et al. [79]. In the case of $Ni_{74.8}Al_{21.9}Hf_{3.3}$ single crystals deformed along the $\langle 123 \rangle$ axis at 373 K, the initial dislocation density was estimated to be 10^{12} m^{-2} before straining; after 1% resolved plastic strain the density increased up to 2.5×10^{12} m^{-2} and after 6% plastic strain it reached 17×10^{12} m^{-2}. Dislocation density extrapolated at a given plastic strain level (8%) was reported as a function of the deformation temperature in the same alloy [80]. It is shown that the density is maximum in the temperature range 500–600 K that corresponds to the higher values of the work hardening. Similar observations were also reported in a Ni_3(Al,Hf)B compound [81]. Dislocation densities were also measured for different strain amounts in γ-TiAl polycrystalline specimens [73]. It is shown that the global dislocation density increases with strain within the anomalous domain, while it is nearly constant for straining experiments at temperatures higher than $T_{P\sigma}$.

The role of dislocation storage on the strain hardening of intermetallics is the topic of a controversial debate. It is often argued that work hardening rates as high as $\mu/2$ or $\mu/10$ where μ is the shear modulus (as measured in Ni_3Al based compounds [8,12]) cannot be interpreted by mechanisms based on dislocation interactions such as long-range elastic interactions or forest hardening. Indeed such hardening mechanisms in the case of f.c.c. metals result in a strain hardening rate the magnitude of which is of the order of about $\mu/200$. However at the same time many of the models developed for explaining the yield stress anomaly of intermetallic compounds invoke dislocation interactions and dislocation storage when they tackle the topic of strain hardening or for modelling creep experiments [82]. For instance, Ezz et al. [81] assume that the strain hardening comes from forest hardening, Greenberg and Ivanov [83] established their modelling of the effects of temperature changes considering dislocation interactions with forest dislocations and parallel dislocations. Louchet [84] claims that high dislocation storage is necessary in order to explain the high strain hardening rate.

In the case of a pure forest hardening, as in f.c.c. metals, the flow shear stress is related to the total dislocation density ρ through the Taylor equation [85]:

$$\tau = \alpha\mu b\sqrt{\rho}. \tag{5}$$

For the metals for which there exists a strong friction stress or obstacles hindering dislocation motion independently of the dislocation–dislocation interactions, the Taylor equation is modified in order to include this supplementary stress τ_0:

$$\tau = \tau_0 + \alpha\mu b\sqrt{\rho}. \tag{6}$$

This relation has been applied to the strain hardening of body centered cubic (b.c.c.) [86,87] as well as of hexagonal close packed (h.c.p.) metals [88]. The parameter α refers to specific mechanisms (long range elastic interaction, junction with forest dislocations – attractive or repulsive – shortening of the Frank–Read segment) operating for a given material and deformation condition. Numerical values for α were compiled for metals of f.c.c., b.c.c. and h.c.p. structures [89], which range from 0.05 to 1.3 on the basis of experimental results and from 0.2 up to 1.1 based on theoretical modelling.

Dislocation densities measured on different intermetallic compounds at different strain (and stress) levels [7,74,90–92] were compiled and compared with the respective stress increase, during plastic strain (see table 5). In order to check the validity of eq. (6), one would need measurements of the dislocation density at different strain levels for a given temperature in each material. As mentioned earlier, dislocation density measurements are very scarce and no data concerning the evolution of the density versus the plastic strain have been reported to the authors' knowledge (except for the two points measured in Ni_3AlHf). However, at temperatures for which the dislocation density was known and under the assumption that within the microsplastic domain no large multiplication of dislocation occurs, and knowing the initial dislocation density ρ_0, one knows two points on the stress – density relation which are enough to determine the α coefficient, reducing eq. (6) to:

$$\tau - \tau_{0.2\%} = \alpha\mu b\left(\sqrt{\rho} - \sqrt{\rho_0}\right). \tag{7}$$

Experimental results displayed in table 5 have been used to represent the above relation in fig. 34, together with data for pure Cu that were previously compiled [93,94] and for silver [92] for comparison. For the case of NiAl-based compound, b in eq. (7) corresponds to the total Burgers vector ($b = [101]$), since the stress scales as the total Burgers vector of the dislocation involved. For TiAl, b is taken as the ordinary dislocation Burgers vector ($\frac{1}{2}\langle 110]$) since in the polycrystal that was examined these are the more numerous in the region of the anomaly [60]. From that figure it appears that the points corresponding to intermetallic compounds fit well with the points related to f.c.c. metals; that is the stress increase during plastic deformation obeys pretty well eq. (7) with α ranging from 0.2 to 1. The stress increment is somewhat too high for one point, corresponding to a plastic shear strain of 1% – indicated by the white arrow in fig. 34 – suggesting that this correlation may mainly apply for relatively large plastic strains. This observation implies that dislocation

Table 5

Experimental measurements of the total dislocation density in different intermetallic compounds. The stress and strain are resolved in the primary slip system for single crystals, for polycrystals a Taylor factor of 3 was used

Material	T	$\tau_{0.2\%}$	τ	γ_p	ρ	$\frac{\theta}{\mu} = \frac{1}{\mu}\frac{\tau - \tau_{0.2\%}}{\gamma_p}$	Λ
	(K)	(MPa)	(MPa)	(%)	$(10^{12}\ m^{-2})$	$(\times 10^{-3})$	(μm)
$Ni_{74.8}Al_{24.9}Hf_{3.3}$ [123]	84	121	156	9	16	6	12
[79]	373	195	244	1	2.5	75	13
$\rho_0 \approx 10^{12}\ m^{-2}$	373	195	300	6	17	27	7.4
	423	224	310	4	52	33	1.5
	573	322	410	5	96	27	1.04
	683	395	480	2.3	67	57	0.7
$Ni_{77}Al_{23}$ [123]	373	30	66	9.4	85	5	2.2
[91]	673	110	173	7.4	152	12	0.97
$\rho_0 \approx 10^{12}\ m^{-2}$*							
$Ni_3(Al,Ta)$ [123]	300	37	124	22	28	5	16
[7]	544	126	318	22	72	12	6.3
$\rho_0 \approx 10^{11}\ m^{-2}$	758	231	412	22	160	12	2.8
	1050	174	253	22	35	5	13
$Ti_{47}Al_{51}Mn_2$	300	117	140	3	16	11	6.7
(polycrystal) [73]	673	117	142	3	10	4	15
$\rho_0 \approx 3 10^{11}\ m^{-2}$	773	133	177	15	36	12	10.8
Cu (polycrystal)	300	11	16.3	3	19	4.2	6.2
[94]	300	11	50	24	150	3.9	6.2
$\rho_0 \approx 10^{12}\ m^{-2}$*	300	11	73	51	300	2.9	6.6
	300	11	91	90	570	2.1	6.2
Ag (polycrystal)	300	7	33	33	220	3.4	5.9
[92]	300	7	48	63	520	2.6	4.7
$\rho_0 \approx 10^{12}\ m^{-2}$*	300	7	59	96	680	2.1	5.5

*The initial dislocation density was not reported in the study, the value of $10^{12}\ m^{-2}$ was considered as an upper limit.

interactions play a significant role on the stress increment occurring during plastic strain of intermetallics. It can be seen in table 5 that the points in fig. 34 do correspond to a mean work hardening rate $(\tau - \tau_{0.2\%})/\gamma_p$ ranging from about 2–5 × $10^{-3}\mu$ for copper and silver up to values of about one order of magnitude higher 1–5 × $10^{-2}\mu$ for Ni_3Al based compounds. The question arises of the origin of these differences in the work hardening rates, since the total density of dislocations have somewhat the same magnitude and are related to stress through the same law (eq. (7)). This difference may arise from the fact that the same amount of dislocations is produced during a somewhat lower plastic strain in intermetallics as compared to metals. Alternatively, the mean free path Λ of the dislocations can be estimated using the integral form of the Orowan equation:

$$\gamma_p = \rho b \Lambda. \tag{8}$$

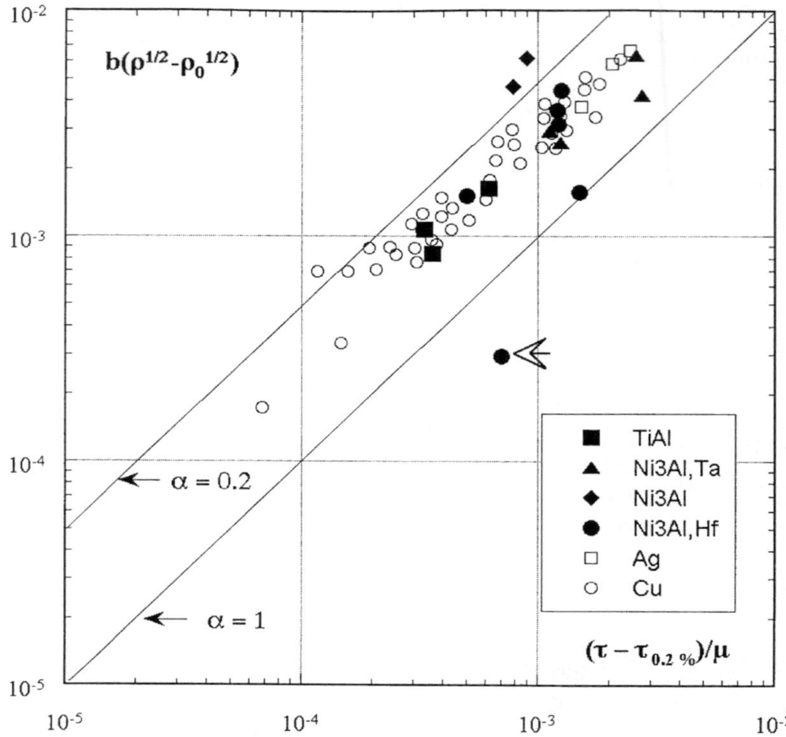

Fig. 34. Dislocation density versus the increase of applied resolved shear stress during plastic deformation of various intermetallic alloys. The two lines drawn correspond to different values of the α parameter in eq. (7) equal to 0.2 and 1 respectively. The density obtained by TEM observations were collected in the literature: TiAl [73], $Ni_{74.3}Al_{24.7}Ta_1$ [7], Ni_3Al [91], $Ni_{74.8}Al_{23.9}Hf_{3.3}$ [79], Ag [92], Cu values compiled in [93] (the TEM values from fig. 1 in this reference were multiplied back by 2 to get the original measured densities).

In Ni_3Al compounds, in the temperature range of the yield stress anomaly, Λ can reach values lower than a micrometer, whereas in f.c.c. metals Λ is nearly constant around 6 micrometers (see table 5), which corroborates the idea of self locking of dislocations. In the same vein, this faster exhaustion of mobile dislocations necessitates further multiplication leading to a higher work hardening rate.

Results of experimental measurements of the dislocation density in intermetallic compounds as a function of strain and temperature that were collected allow us to draw the following conclusions:

- the order of magnitude of the dislocation densities measured in intermetallic compounds is comparable to that measured in metals,
- the dislocation density increases with plastic strain,
- the increase of stress during plastic strain is proportional to the difference of the square roots of the dislocation density, according to a modified Taylor relation (eq. (7)),

– relatively high work hardening rates, after a consequent plastic strain, are associated with a low mean free path of dislocation, i.e. during an imposed strain-rate test, to higher dislocation multiplication and storage rates.

5. Multistep deformation experiments

Temperature change experiments were first performed by Cottrell and Stokes in 1955 [95] and since that time were widely applied to f.c.c. metals. The aim of such experiments is to discriminate in the divergence of the flow stress at a given strain and at two different temperatures the part produced by work-hardening and the part due to the reversible change of flow stress (in other words to test the relative importance of the dislocation mobility versus the dislocation substructure on the flow properties). The flow stress at a given strain and temperature is assumed to be the sum of an "irreversible" stress depending solely on the microstructure (MS – density and distribution of dislocations), which in turn may depend on the temperature at which prestrain is performed and a reversible part which depends essentially on the actual temperature and strain rate:

$$\sigma = \sigma_{\mathrm{i}}(\mathrm{MS}) + \sigma_{\mathrm{r}}(T, \dot{\varepsilon}). \tag{9}$$

Temperature changes, as initiated by Cottrell and Stokes, consist in performing a prestrain at a given temperature (T_{PS}) up to a stress level σ_{PS} then unloading the specimen, changing the temperature and restraining the sample at the new temperature (T) (see fig. 35). The flow stress upon reloading σ_{R} is then compared to σ_{PS} and to the flow stress at the same temperature for a virgin sample σ_{V}. As stated by Cottrell and Stokes

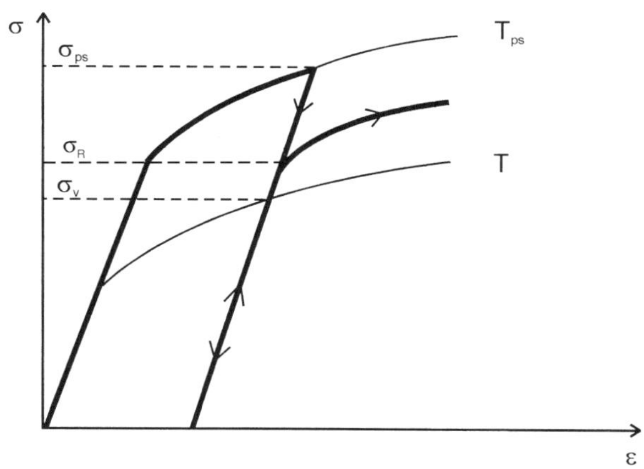

Fig. 35. Schematic representation of the temperature-change experiments as first performed by Cottrell and Stokes [95]. Note that in the temperature range of intermetallic yield stress anomaly, prestrain at higher temperature (T_{PS}) involves higher stress than the reloading at $T < T_{\mathrm{PS}}$. Upon reloading, the difference in stresses $\sigma_{\mathrm{PS}} - \sigma_{\mathrm{R}}$ measures the irreversibility of the stress, while the difference $\sigma_{\mathrm{R}} - \sigma_{\mathrm{V}}$ measures the difference in work hardening rates between T and T_{PS}.

in their original paper it is worth remembering that the stress offset $\sigma_{\mathrm{PS}} - \sigma_{\mathrm{R}}$ measures the reversibility of stress while $\sigma_{\mathrm{R}} - \sigma_{\mathrm{V}}$ measures the difference in the levels of work hardening reached at the two temperatures. For the cases where the flow is solely substructure controlled (i.e. depends mainly on long-range dislocation interactions) the stress upon reloading consequently reflects the prestrain history and one may find: $\sigma_{\mathrm{R}} = \sigma_{\mathrm{PS}}$. In this case, the flow stress is said to be fully irreversible. On the other hand, if the flow is controlled solely by the temperature dependent dislocation velocity the stress upon reloading may correspond to the yield stress at the actual temperature plus the stress due to the work hardening achieved during prestrain. This original analysis implicitly recognises the validity of the usual decomposition of the applied stress σ into two independent terms, namely the so-called internal stress σ_{μ} and the effective stress σ^*. The internal stress arising from long-range interaction with the microstructure essentially depends on strain level and the temperature at which straining was performed. The effective stress is dictated by the overcoming of local obstacles and depends mainly on temperature, strain rate and strain level. This decomposition can be summarised as follows [96]:

$$\sigma = \sigma_{\mu}(\varepsilon) + \sigma^*(T, \dot{\varepsilon}). \tag{10}$$

Most of the studies performed on f.c.c. structure metals have shown that the stress difference obeys the so-called Cottrell–Stokes law, i.e. the ratio $(\sigma_{\mathrm{PS}} - \sigma_{\mathrm{V}})/\sigma_{\mathrm{V}}$ does not depend on strain, provided the temperature jumps are performed within the beginning of the deformation process (i.e. within strain hardening Stages I and II where no recovery process occurs [93]). By identifying the decomposition of the stress as described in eq. (9) to the decomposition of eq. (10), it has been shown that the Cottrell–Stokes (CS) experiment is in many instances equivalent to a strain rate change experiment and the conclusions driven from temperature change could be applied to strain rate change results, at least for those cases where the dislocations have to overcome obstacles of finite range and strength [97].

In the case of intermetallic compounds as a rule not enough data are available in order to check the validity of Cottrell–Stokes law during temperature changes. Ezz and Hirsch [18,19] could however establish that the CS law is obeyed for that part of the stress which is due to the work hardening (τ_{h} in their notation). As exposed in sections 5.1 and 5.3, the interpretation is also much less obvious and there still exist some controversies on the "reversible" character of the flow stress. Furthermore the temperature changes performed on intermetallic compounds do not reflect exactly the original CS experiments (isolated tests against the jumps originally performed at different plastic strain levels). This paragraph also includes a discussion on the successive deformations of the same sample at different temperatures. Thus, apart from testing the pure reversibility of the flow stress, the interest for reviewing the results of CS experiments in intermetallic alloys is two-fold:

(1) determine if the results reported in the literature are affected or not by the mechanical history the specimen overcame, and
(2) check the influence of the microstructure on the flow stress, which would allow one to draw some conclusions on the role of dislocation storage.

5.1. Cottrell–Stokes experiments in TiAl compounds

5.1.1. Polycrystalline specimens

Temperature change experiments have been performed on $Ti_{47}Al_{51}Mn_2$ polycrystals by Lu and Hemker, between 573 K and 823 K (i.e. within the anomaly domain) performing temperature jumps from low temperature prestrain to high temperature and drops from high temperature to low. In both cases, prestrains were performed up to a total strain level of about 1%. The authors demonstrate that in this situation both the flow stress and the work hardening rate appear to be essentially reversible [98], this reversibility being quasi immediate for a temperature rise, while it is delayed (within about 0.5% strain) for the temperature drop. This behaviour remains valid when two temperature changes (either 573 then 823 and finally 573 or 823 then 573 and 823) are performed. The same type of experiment has been performed by us on the same alloy between 773 and 373 K, see fig. 36. The reversibility of the plastic flow is actually observed for the first two temperature changes (up to the dashed line in fig. 36) in full agreement with the results of Lu and Hemker (see fig. 9 in [99]). However for increasing strain levels the behaviour becomes much more irreversible, particularly the curve corresponding to the third loading at 773 K (see the white arrows in fig. 36) is above the monotonic one, and the work hardening rate is higher on the CS curve than on the monotonic one. Lu and Hemker performed

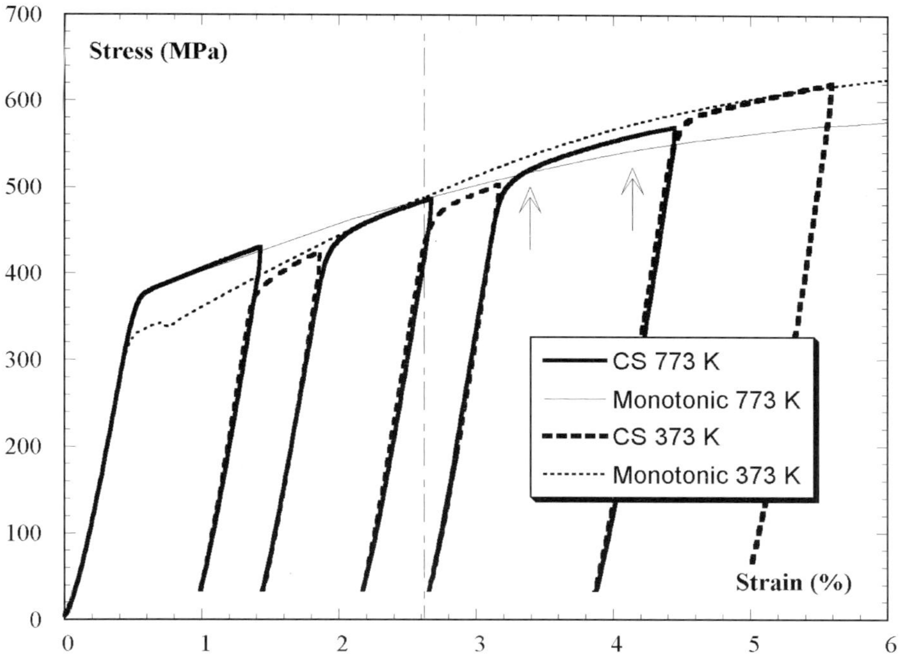

Fig. 36. Cottrell–Stokes experiment in $\gamma Ti_{47}Al_{51}Mn_2$ polycrystals realised between 773 and 373 K. Flow stress seems to be reversible up to the second temperature change (dashed vertical line), while beyond this strain level the third loading at 773 K indicates a clear strengthening as compared to the monotonic curve.

TEM observations after straining at alternate temperatures and showed that the 573 K deformation is associated with the activation of ⟨101] type super dislocations and the concomitant formation of "faulted dipoles" whereas the 823 K deformation involves the activation of ordinary dislocations. Moreover they show that changing the deformation temperature produces a reversible change in the operative deformation system. From their mechanical tests and TEM observations, they deduced that the motion of both ordinary and super dislocations are "fully reversible processes". This conclusion seems to be valid only for moderate strain levels (up to about 2% plastic strain), whereas for higher strains some irreversible character appears. Notably the motion of ordinary dislocations seems to be hindered (the stress and work hardening rate are higher) by the microstructure developed at low temperature.

5.1.2. Single crystals

In the last decades, it has become possible to grow TiAl single crystals more readily, and CS type experiments have been reported for the nominal composition $Ti_{46}Al_{54}$. The first temperature change experiments on TiAl single crystals were performed by Stucke et al. [34]; they deformed single crystals with a load axis close to [010]. This orientation corresponds to the activation of superdislocations, as assessed by maximal Schmid factor calculations, slip line analysis and further direct dislocation observations on single crystals with the same composition [33]. Stucke et al. reported that when strain at RT follows prestrain at 773 K, the ultimate stress during prestrain is retained, indicating a complete irreversibility of the flow stress, fig. 37(a). They also reported that the work hardening rate is completely reversible (i.e. depends only on the deformation temperature and not on the sample history). The same authors also performed some temperature change experiments by deforming up to 0.5% total strain the same specimen at different increasing temperatures (successively 473, 673 and 873 K). In these experiments they noted that the yield stress at a given temperature depends on the prestrain that the sample overcame, and is appreciably higher than for virgin specimens deformed at the same temperature. The same kind of experiments was repeated by Inui et al. [33] in a study of the same alloy. By deforming a crystal along the $[\bar{2}51]$ axis, an orientation which also activates the ⟨101] superdislocations, they showed that after a prestrain at 1173 K the yield stress measured at RT was only slightly higher than when measured on a virgin specimen, fig. 37(b). It is worth noting that in this case the prestrain temperature corresponds to the peak temperature for this orientation, thus to the onset of ordinary dislocation activation. However in this experiment it was again observed that the work hardening rate is only temperature dependent. The above results have also been confirmed by the work of Mahapatra et al. [100] who show that for load axes oriented along the [001] and [011] directions the RT flow stress is irreversible when predeformation is performed at 973 K. For the orientation [011] after prestrain at 1173 K the flow stress at RT is reported to be reversible. This observation was also made more recently by Jiao et al. [101], who show that prestraining $Ti_{44.5}Al_{54.5}$ crystal with a load axis near $[\bar{3}41]$ at 1073 K leads to a partial reversibility of the flow stress at RT. The same alloy was also deformed along $[\bar{1}\bar{1}, 3, 6]$ which activates ordinary dislocation glide; it was observed that if predeformation is done in the anomaly domain (973 K in this case) the RT flow stress retains completely the predeformation stress, i.e. the flow stress of ordinary dislocations is also irreversible. Moreover, one must note that for this experiment

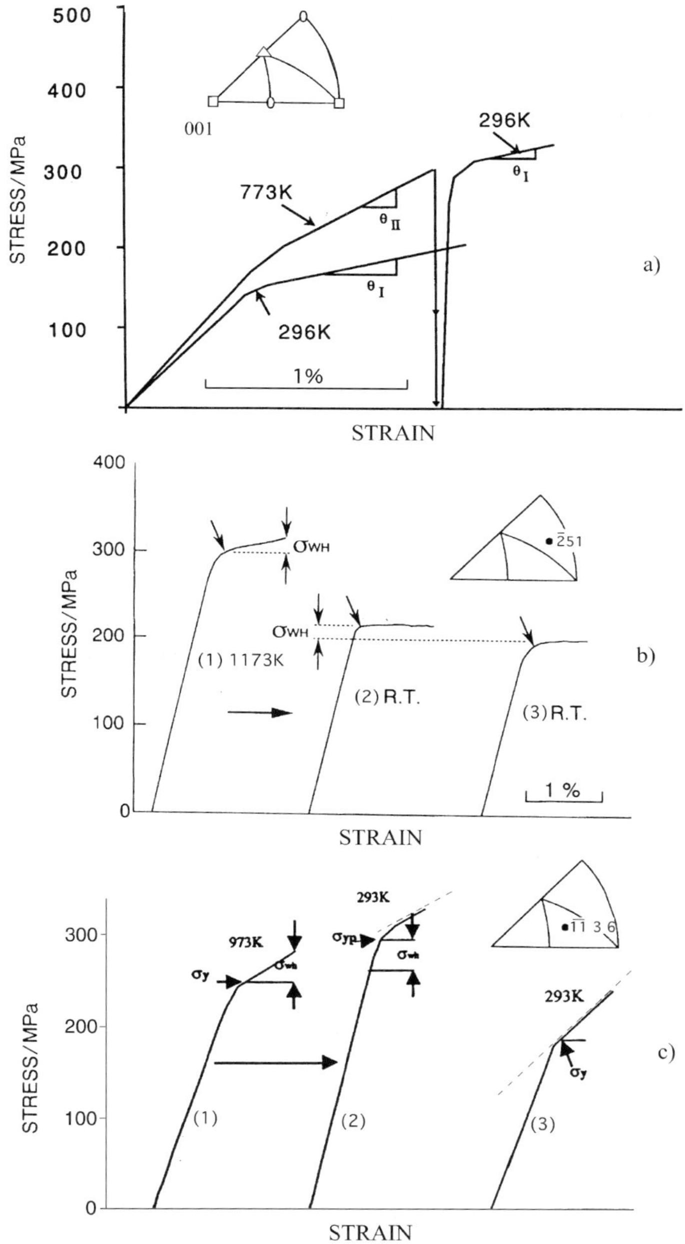

Fig. 37. Typical results of temperature change experiments in TiAl single crystals. When superdislocations are activated the stress is fully irreversible as in (a) (schematics from [34]), except if the prestrain is realised above or at the peak temperature as in (b) (after [33]). In both cases, the work hardening rate is reversible. When ordinary dislocations are activated, both flow stress and work hardening rates are irreversible as in (c) (from [101]).

the work hardening rate was reported to be irreversible, since the work hardening value corresponding to 973K was retained during further strain at RT, fig. 37(c).

5.1.3. Reversibility of the plastic flow in TiAl compound
From the above results it can be stated that the plastic flow of the γ phase TiAl is essentially an irreversible process for both the ordinary and the super dislocation motion. Indeed for large plastic strain in polycrystals and for single crystals, the flow stress after predeformation is retained for both types of glide system. The work hardening rate seems to be reversible for superdislocations, while it is apparently irreversible for the motion of ordinary dislocations.

5.2. Temperature changes in Ni_3Al and related $L1_2$ compounds

Among the numerous mechanical test results published on Ni_3Al type compounds one can find several examples of prestrain experiments performed on alloys of various compositions and with different load axis orientations. These experiments may consist of Cottrell–Stokes like straining processes but also on different types of straining–unloading–reloading procedures. For the sake of clarity, these experiments have been sorted depending on the relation between the prestrain temperature T_{PS} and reload temperature T, leading to three groups: (i) $T = T_{PS}$; (ii) $T > T_{PS}$ and (iii) $T < T_{PS}$.

5.2.1. Prestrain and reload at the same temperature
In this class of two-step experiments the sample is first strained, then unloaded and strained again (without any holding time) at the same temperature. This allows one to test the occurrence of any stress relaxation (or softening) during the unloading process. This kind of experiment has been conducted by Ezz and Hirsch on $Ni_{75}Al_{22.7}Hf_{1.5}B_{0.2}$ crystal for load axis orientations $\langle 123\rangle$, close to $\langle 001\rangle$ and close to $\langle 111\rangle$ at temperatures varying from RT to 800 K [81]. It is shown that in most cases no relaxation of the stress is observed, i.e. the yield stress for the second straining equals the stress at the end of the prestrain (within an experimental scatter of maximum 1.5%). The only exception is observed for two specimens deformed at 300 K and 470 K, respectively, along $\langle 123\rangle$ for which the yield stress (measured at 0.05%) upon reloading was observed to be lower (by 5.5% for RT deformation) than the stress at the end of the prestrain [102]. Note that the same experiment performed on a slightly different alloy but with the same load axis and deformation temperature show absolutely no stress relaxation [103].

The general observation that the flow stress does not relax on unloading and further straining at the same temperature is very important in two ways:

(i) it indicates that the dislocation substructure established during the deformation is stable during unloading and therefore justifies TEM post-mortem observations,
(ii) it shows that there is no recovery of the dislocation substructure for temperatures up to 800 K ($0.8T_{p\sigma}$ for this alloy and load axis orientation).

Within the same categories of experiments, Shi et al. [103], working on $Ni_3(Al,0.25Hf)$, designed a procedure which consists in straining a sample at 300 K, partially unloading it

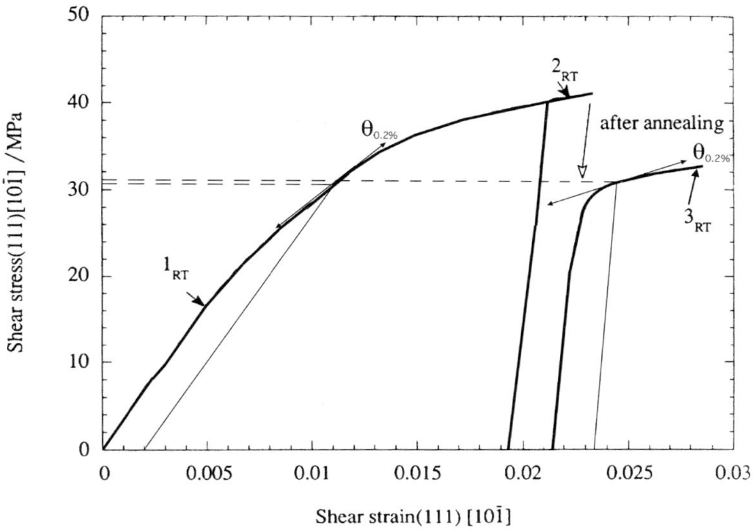

Fig. 38. Unload–reload experiment at RT on a $Ni_{77.05}Al_{22.7}Hf_{0.25}$ single crystal oriented along ⟨123⟩ (after [103]). The first load–reload sequence shows no recovery of the microstructure during unloading. After annealing the same specimen by heating 24 hours at 1173 K in the load machine, the microstructure recovers completely, restoring the original yield stress. However, one may notice the huge difference in work hardening measured at 0.2% offset strain on the two specimens.

and then annealing it in the machine before straining it at the desired temperature, fig. 38. Such a procedure results firstly in perfect alignment of the sample in the loading train and secondly in the exhaustion of the defects responsible for microplasticity. The major incidence on the reloading stress–strain curves is that: (i) there exists a 'pure elastic' part (the slope of which is very close to the Young's modulus as measured for this specific alloy by ultrasonic wave velocity determination [12]) and (ii) the transition between elastic and plastic parts of the curve is much more narrow and well defined than in a virgin sample, evidencing the existence of a critical resolved shear stress as assessed in section 3.1.2.1 using a completely different approach. The conventional "critical" stress measured at 0.2% offset strain is not changed by this procedure. The experiment highlights the necessary care that should be taken in order to measure first deviations from linearity, since they may be related to the microplastic process which involve different mechanisms than those involved in macroscopic flow. It also puts some doubt on the relevance of the work hardening rate measured at quite small plastic strain levels; for instance, $\theta_{0.2\%}$ varies by nearly a factor of two when measured either on a virgin sample (curve 1_{RT} in fig. 38: $\theta_{0.2\%} \approx 1.5$ GPa) or a prestrained and annealed sample (curve 3_{RT} in the same figure: $\theta_{0.2\%} \approx 0.8$ GPa). However, at larger plastic strain fig. 38 suggests that the WHR's become comparable.

5.2.2. Prestrain at lower temperature

5.2.2.1. Successive testing of a single specimen. The type of experiments where $T > T_{PS}$ is of great importance since it includes not only the classic Cottrell–Stokes type experiments but also many experiments for which a single specimen has been deformed over a wide range of temperature. Indeed, since we are concerned with experiments performed within the anomalous yield stress domain, increasing temperature increases flow stress; it was thus suspected that slight predeformation at a temperature T_1 does not influence plastic behaviour at higher temperature (i.e. higher stress levels). Following this assumption it was possible to examine the plastic behaviour over a wide range of temperature using only few specimens. This type of experiments has been first performed in 1967 using one specimen for scanning the temperature range from 300 K to 1200 K [11] and then by Mulford and Pope on $Ni_{75}Al_{22}W_3$ [49], using a unique specimen deformed at different temperatures and doing a prestrain at 77 K before each higher temperature test. By this method they report values for microyield ($\varepsilon_p = 5 \times 10^{-5}$) and macroscopic yield strength ($\varepsilon_p = 10^{-3}$) as functions of the temperature (fig. 39). Their results are compared to the previous data from Thornton et al. [10] in binary Ni_3Al polycrystalline alloy, in this case, the authors were using a unique specimen for each tested temperature (fig. 11). It can be observed that the macroscopic yield stress follows exactly the same tendencies in both cases. The microplastic limit exhibits a positive temperature dependence in Thornton et al. experiments and a near flat or even decreasing trend in the Mulford and Pope work. This suggests that the microyield limit measured in this case characterises more the 77 K predeformation than the actual temperature test, while the macro limit is quite insensitive to the slight predeformation performed at such low temperature and stress. Kuramoto and Pope [104] validated the procedure of using one specimen for different temperatures by comparing the yield stress values obtained by straining a specimen at only one temperature and those arising from the deformation of the same specimen at different temperatures.

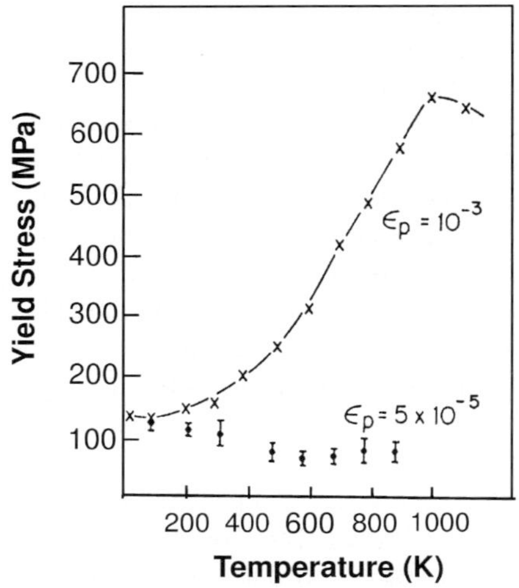

Fig. 39. Flow stress for different plastic strain levels in $Ni_3(Al,W)$, after [49]. The curves were obtained by straining one specimen at different temperatures (the specimen was restrained at 77 K between each deformation), while those in fig. 11 were obtained using one specimen per examined temperature. The comparison reveals that the microstress values are very sensitive to prestrain, contrary to the macroscopic yield stress level.

They showed that subtracting the small stress increase due to work hardening during previous strains from the yield measured on the unique sample gives the same results as to deforming only one sample at a given temperature. The same procedure was then used by Umakoshi, Pope and Vitek [57] for tension–compression tests on $Ni_{75}Al_{20.5}Ta_{4.5}$ compounds. During the following years different workers used the same procedure for testing Ni_3Al compounds [105–107]. However, as presented in section 2.1.1.4, it has been shown [14] that the yield stress values obtained by this procedure may lead to slightly different results than the ones obtained when a single virgin specimen is deformed at a given temperature (see fig. 8(a)). Furthermore the work hardening rates measured in both procedures (either on virgin samples or predeformed ones) lead to results quite different in terms of quantitative values and moreover in terms of temperature dependence. Indeed as exposed in section 2, Spätig observed two peaks for the work hardening rate in $Ni_{75}Al_{24}Ta_1$ deformed successively at increasing temperature, while only one peak could be observed for virgin samples (fig. 8(b)). The same behaviour was also observed by Spätig [14] in the temperature dependence of activation volume that exhibits two discontinuities when measured on a unique sample and only one when measured on virgin samples [14]. Provided a very small amount of plastic strain is imposed, the procedure of deforming a unique specimen at several temperatures thus seems to give reasonable results for the values of yield stress, but this is no more the case for other parameters of the plastic flow such as the work-hardening rate or the strain-rate sensitivity.

5.2.2.2. Temperature changes. The reversibility of the flow stress has also been tested through more conventional temperature change experiments such as the one sketched in fig. 35, i.e. the predeformation is performed over a significant plastic strain and the stress–strain curve upon reloading is directly compared to the curve of a virgin specimen. In that situation one has to distinguish two cases depending on whether the ultimate stress reached during prestrain is higher or not than the "virgin" yield stress at the second deformation temperature (one has to keep in mind that we consider materials exhibiting a yield stress anomaly, so that increasing the temperature increases the yield stress).

In the first case, i.e. where the ultimate prestrain stress is higher than the expected yield stress, most results indicate that the specimen retains almost all its low temperature strength and as a result the positive temperature dependence of the yield stress is hindered by a plateau corresponding to the prestrain stress [10,108]. Figure 40 drawn after [108,109] is a schematic of the variation of the yield stress for a specimen prestrained at low temperature, the slope of the line representing the stress on reload versus the reload temperature $\sigma_R(T)$ may be zero or slightly positive. For the temperature T_c, the stress at the end of the prestrain σ_{ps} corresponds to the stress $\sigma_y(T_c)$ (T_c corresponds to the temperature labelled T_{2c} in [108]).

Above the temperature T_c in fig. 40, one is in the case where the ultimate stress during prestrain is lower than the expected yield stress. In such cases yielding (or at least departure from the elastic slope) occurs at a stress level in between the ultimate tensile stress reached at T_{PS} and the yield stress of a virgin specimen at T. There exists then a transient strain regime, similar to some micro-plastic deformation process, before the stress–strain curve reaches back to the curve of a virgin specimen deformed at T. This is illustrated in fig. 41 taken from [110] where the temperature change experiment was performed between 77 K

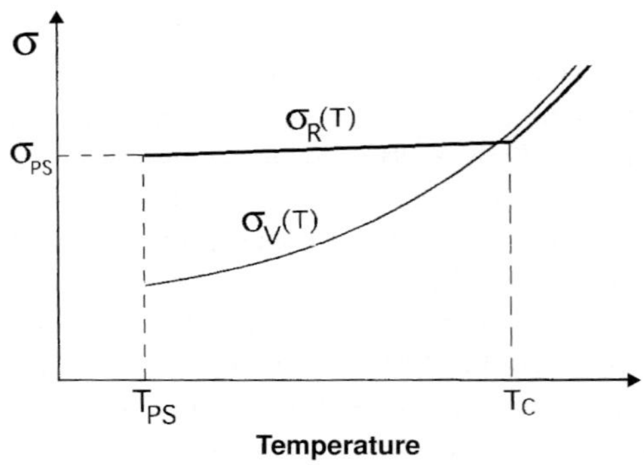

Fig. 40. Schematics of the general behaviour of flow stress for a specimen prestrained up to σ_{PS} at a low temperature T_{PS} and subsequently deformed at higher temperature. $\sigma_R(T)$ and $\sigma_V(T)$ represent the values of flow stress on prestrained and virgin specimens respectively.

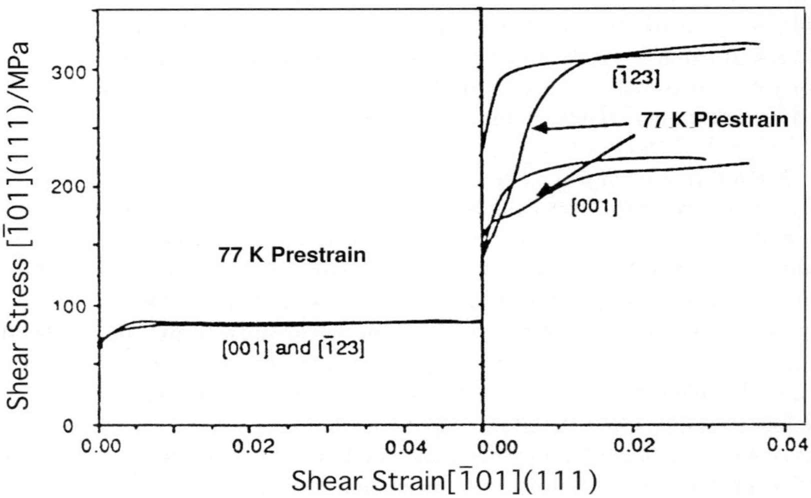

Fig. 41. Cottrell–Stokes type experiment with $T_{PS} < T$ of a $Ni_{76.2}Al_{23.3}Cr_{0.3}B_{0.2}$ specimen deformed along ⟨123⟩ axis. Prestrain at 77 K is followed by further deformation at 823 K. The curves corresponding to a virgin sample are also reported for comparison [101].

and 823 K on a $Ni_{76.2}Al_{23.3}Cr_{0.3}B_{0.2}$ specimen along two load axes: ⟨123⟩ and ⟨001⟩. The transient regime is characterised by a very strong work hardening rate (this regime may be assimilated to a change – a decrease – in the value of apparent elastic slope). Its extent depends on the plastic strain level imposed during the predeformation (the more extended the predeformation, the longer the transient); also, the stress at which this regime starts seems to decrease (toward σ_{PS}) with an increase of the amount of prestrain [103]. After this transient, the stress–strain curve upon reloading at the higher temperature T is pretty similar to the one of a virgin specimen directly deformed at T. The same stress level and almost the same work hardening rate could be measured, see fig. 41. Dowling and Gibala [110] performed TEM examinations of dislocation microstructures along the total strain path at both temperatures for specimens with load axis along ⟨123⟩ direction. After low-temperature prestrain the microstructure is mainly formed of ⟨101⟩ dislocations without any preferential orientation and numerous intrinsically faulted dipoles. After only 0.1% strain at 550°C, they observed the disappearance of faulted dipoles, as well as a tendency for dislocations to align along their screw direction; after 4% strain at 550°C the typical deformation microstructure for this temperature (long screw segments joined by superkinks, no large edge segments, no faulted dipoles) was observed. It is deduced from such experiments that the microstructure introduced at low temperature provides long edge segments that are exhausted during the "microplastic" regime occurring at high temperature.

5.2.3. Prestrain at higher temperature

This deformation path includes two kinds of experiments quite different in nature: (i) "real" Cottrell–Stokes type experiments in which a virgin specimen is first deformed at T_{PS}, then tested at $T < T_{PS}$ (so that the ultimate stress during prestrain is always higher than the yield stress at T); (ii) some recovery experiments during which a specimen is first deformed at a temperature T_1, then heated up to temperature T_2 at which some further deformation is imposed and finally cooled down back to T_1 for measuring the flow stress, this complicated deformation path allowing for the measure of the recovery of deformation microstructure at the temperature T_2.

The recovery tests have been first performed by Thornton et al. [10] who showed that the deformation structure formed at RT (T_1) was stable for intermediate temperatures (T_2) up to 773 K, above which (for $T_2 \geqslant 873$ K) recovery of the RT microstructure occurs, the higher T_2 the more important the recovery. This kind of tests has been repeated more recently by Ezz and Hirsch [108] on $Ni_3(Al,Hf)B$ alloy for different load axis orientations (⟨123⟩ near ⟨011⟩ and near ⟨001⟩). The authors observed that recovery occurs for T_2 larger than about 600 K. This behaviour was also observed by Ezz [18] on Ni_3Ga single crystals for which he showed that recovery occurs for temperatures higher than 503 K. The temperature T_2 at which recovery starts to operate can be normalised by the stress peak temperature for these different authors: (900 K/1200 K = 0.75) for binary Ni_3Al polycrystals [10], (600/1000 = 0.6) in $Ni_3(Al,Hf)B$ [108] and (503/873 = 0.6) in Ni_3Ga [18]. These results contrast with the relaxation testing presented in section 5.2.1 where the sample is prestrained, unloaded, then restrained at the same temperature, for which no relaxation occurs up to $0.8T_{P\sigma}$. This indicates that the microstructure which is set up at high temperature is stable at this temperature (or a lower one) while a microstructure set up at low temperature (RT in the above tests) may strongly relax during strain at higher temperature.

The temperature change experiments with $T < T_{PS}$, such as the one presented in fig. 42, have been more widely performed by different workers. The first CS type experiments in intermetallic structures were reported by Davies and Stoloff in 1965 [111] on Ni_3Al polycrystals, on which prestrains at 773 and 298 K were followed by deformation at 298 and 77 K respectively. In both cases the results showed that the flow stress is *only partially reversible*, that is, the stress upon reloading lies between the ultimate stress reached during

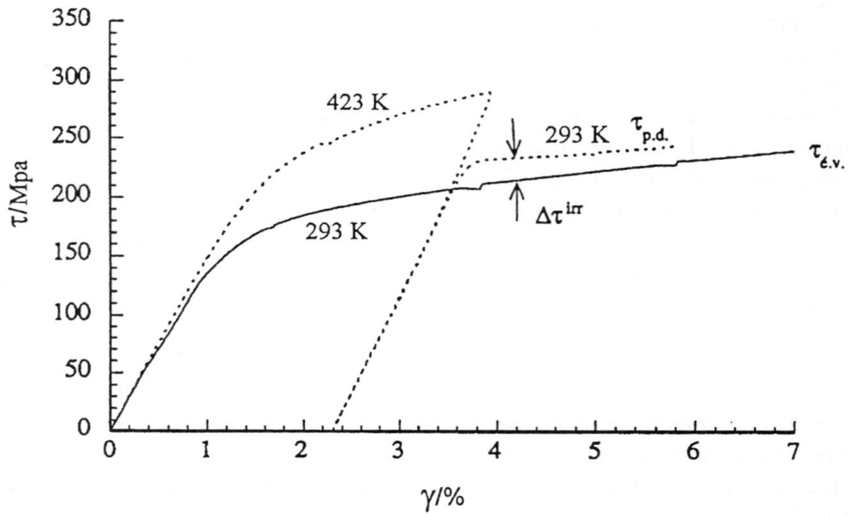

Fig. 42. Cottrell–Stokes type experiment in $Ni_{74.8}Al_{21.9}Hf_{3.3}$ deformed along ⟨123⟩ showing the partial reversibility of the flow stress [14].

prestrain and the monotonic stress for the second deformation. This finding was then supported in single crystals by many researchers for Ni_3Al based compounds [10,12, 103,110,112] as well as for Ni_3Ga [17]. The general finding is that the flow stress upon reloading is somewhat higher than the flow stress at the same temperature as measured on a virgin specimen. It is also observed that this stress increment $\Delta\sigma = \sigma_R - \sigma_V$ increases with the plastic strain reached during prestrain; in the extreme case where the specimen is prestressed below the corresponding yield stress, the mechanical properties upon reloading are not affected at all [113]. Some authors [17,51,114] claimed that $\Delta\sigma$ equals the work hardening stored in the sample during prestrain. However this conclusion may be questionable since it depends strongly on the "exact definition of the flow increase due to work hardening [during the prestrain at high temperature]" [110] which is not precisely defined due to the smooth transition between "elastic" and plastic domains. In a more recent study Veyssière and Saada [23] show that the stress increase $\Delta\sigma$ is actually half of the stress increase caused by work hardening at higher temperature. It is worth noting that in most cases this stress increase upon reloading $\Delta\sigma$ is usually quite small as compared to the stress level obtained at the end of the prestrain σ_{PS} (i.e. the ratio $\Delta\sigma/\sigma_{PS}$ takes values of: 0.05 [14]; 0.09 [114]; 0.15 [23]; 0.15 [103]; 0.09–0.19 [12]; 0.33 [51]).

TEM observations of the microstructure after deformations in two steps have been performed in two cases. For a specimen of $Ni_{75.7}Al_{23.7}Hf_{0.4}B_{0.2}$ deformed along the ⟨123⟩ axis Dowling and Gibala [110] reported, after 2.5% strain at 823 K followed by 4.5% at 77 K, a microstructure essentially representative of the high-temperature deformation (namely the presence of long screw dislocations bowing in (010) plane). Dimiduk and Parthasarasy [114] deformed $Ni_{77.1}Al_{22.9}$ successively at 875 K then at 300 K. They observed that two distinct microstructures coexist, with some regions representative of T_{PS} (long locked screw dislocations) and other areas where both high- and low-temperature microstructures are superimposed (long screw and more isotropic dislocation configurations).

A general finding is also that the work hardening rate is fully reversible upon temperature changes, i.e. the work hardening rate upon reloading at T is the same as the one measured at T on a virgin sample [12,51]. However, a close examination of the CS curves shows that in some cases the work hardening appears to be slightly influenced by the prestrain, either lowered [12,14] or increased [110,112,114].

5.2.4. *Summary of the temperature change experiments in Ni_3Al and related compounds*

The experimental results that have been reviewed in this paragraph may be summarised as follows:

(1) the deformation microstructure which is formed at a given temperature is stable at this temperature (i.e. no relaxation occurs) up to $0.8T_{p,\sigma}$;
(2) there exists a microplastic domain responsible for the smooth transition between the elastic and plastic domains;
(3) the use of a single specimen for studying the mechanical properties over quite a large range of temperatures must be considered with great care, since this procedure reproduces quite well the yield stress values but gives very different results in terms of work-hardening rate and strain-rate sensitivity (activation volumes) as compared to tests performed on virgin specimens;
(4) prestrain at low temperature over a significant strain implies a transient behaviour before reaching the monotonic curve;
(5) a microstructure developed at low temperature may be destabilised by further straining at higher temperature;
(6) the flow stress is partially reversible when predeformation is performed at a temperature higher than the test temperature, the irreversible part of the stress is most often a small fraction of the prestrain stress and increases with plastic strain during the predeformation step. This stress increment is never higher than the work hardening achieved during the prestrain.

These experimental observations have often been interpreted in connection with the dislocation microstructure typical of $L1_2$ compounds. A comprehensive review of such microstructure can be found in [23]. Within the yield-stress anomaly domain, the dislocation landscape is mainly composed of screw segments in form of Kear–Wilsdorf locks connected by mixed character segments, the super-kinks (SK's). During the microsplastic domain, the higher SK will move upon moderate stress, trail screw segment of dislocations and exhaust [103]. This exhaustion may take place either by gliding out of the specimen or by shortening its height – the so called screw dipole hardening sketched in fig. 11(b), (c), (d) of [102]. The shear stress necessary to move a SK scales inversely as its height, so that after strain at a given shear stress τ, SK's larger than the corresponding height h $(= \mu b/\tau)$ are nearly all exhausted. During macroplastic deformation, as new dislocations are created, the number of SK's whose height is smaller than h increases with the total dislocation density. So that after a prestrain experiment, the remaining SK's height distribution is representative of the prestrain temperature and their density depends on the amount of plastic deformation during prestrain.

This description explains quite well the behaviour of $L1_2$ compounds during temperature-change experiments for which $T > T_{PS}$ and corresponding to the case where the ultimate stress during prestrain is lower than the corresponding yield stress at T. Indeed for a significant prestrain at low temperature T_{PS}, a large number of SK are created which are mobile at a stress level lower than the virgin yield stress at T. So that upon reloading at T the first deviation of the stress–strain curve from the elastic slope occurs near the stress able to move these SK, i.e. the ultimate stress of the prestrain τ_U, constituting the transient of point (4) here above. Since the number of SK available in the prestrained specimen increases with the amount of plastic deformation during prestrain, so does the extent of the transient.

For the temperature changes with $T > T_{PS}$ but with an ultimate stress during prestrain higher than the corresponding yield stress at T, the situation does not seem to be as clear. This case was recently considered by Greenberg and co-workers in a series of papers on the interpretation of temperature-change experiments on intermetallic alloys [83,109,115–117]. This analysis extends to more general situations some ideas presented by Ezz and Hirsch [108]. According to Greenberg et al., two parallel and independent *phenomena* may control plastic strain: (i) the activation of dislocation sources related to the dislocation locking probability (denoted τ_F – essentially a temperature-dependant stress) and (ii) the overcoming of dislocation–dislocation interactions ($\tau(\rho) = \kappa\sqrt{\rho}$). The stress which is measured thus corresponds to the greater of these two components ($\tau = \mathrm{Max}\{\tau_F(T), \tau(\rho)\}$). The authors then speculate that upon reloading different situations may occur depending on the slip systems involved during prestrain and reloading. If in both cases the activated slip systems are the same, moving dislocations upon reloading interact with the same neighbours as during prestrain, the stress $\tau(\rho)$ thus predominates and the stress upon reloading equals the ultimate stress during prestrain in agreement with experimental results (fig. 40). However, we have shown in section 4 that the Taylor equation in its original form does not apply directly in the case of intermetallic compounds, so that the assumption $\tau(\rho) = \kappa\sqrt{\rho}$ may not be valid. Also according to this model, in the case of Ni_3Al compounds and for experiments performed with $T_{PS} > T$ (experiment for which a partial reversibility is observed) the slip system activated during the low temperature step should be different from the one active during the prestrain. This was not observed for the set of experiments for which dislocation microstructures were observed by TEM [110,114]. The model also does not seem to apply in TiAl alloys, since for the CS type experiments reported in [99] and in fig. 36, while different slip systems are activated at the respective high and low temperature the CS experiment shows a clear irreversibility of the stress (see fig. 36).

For the CS experiments corresponding to the case where $T_{PS} > T$ (section 5.2.3) for which a partial reversibility is observed (point (6) here above), this approach based on the height distribution of SK's has also been invoked. In this case some authors [102,114] highlight the necessity of some substructure recovery upon changing temperature and/or unloading. Ezz and Hirsch proposed that this recovery is insured by a run back of the edge part of dislocations that form the sources. However this run-back would give a stress relaxation when prestrain and reload are performed at the same temperature, which was only observed by these authors in two cases, while the partial reversibility of the stress is widely recognised. So this explanation of the partial reversibility of the stress is not so convincing. Also based on the microstructural features, and on some inhomogeneity in the spatial distribution of dislocations during prestrain, Shi and co-workers [103] assign the stress increase upon reloading to interactions between the dislocations formed at T and the dead dislocation structure that was established at T_{PS}. This description explains why the stress increases with the amount of prestrain; however for a very large prestrain the reversible part would disappear (since the specimen is fully occupied with dead dislocations) in contrast with the observation that the stress increase is always limited to the stress gained through work hardening.

6. *Discussion*

6.1. The various components of stress (Ni_3Al and γTiAl)

As assessed by the experimental results collected in the literature (see section 4) the total dislocation density increases with the plastic deformation in intermetallic compounds. The magnitude of the total density measured after a few percent of plastic strain is of the order of 10^{13}–10^{14} m^{-2}, similarly to what has been currently measured in metals. We have demonstrated that the stress increase along the stress–strain curve (i.e. the mean work-hardening rate) is related to this increase in total dislocation density according to a modified Taylor equation. However this statement is only valid for quite large plastic deformation, and not at the beginning of the plastic domain (where the work-hardening rate is maximum). This initial hardening is more likely to be due to the exhaustion of superdislocation sources (SK or more generally non-screw segments). In this way such a hardening mechanism is more related to a micro-plasticity phenomenon. This is fairly well supported by the results of Shi et al. [103] reported in section 5.2.1 (see fig. 38). The hardening resulting from source exhaustion is believed to be strongly related to the initial microstructure and not to vary with temperature, while the hardening caused by the storage of dislocations is related to the probability for the arresting of a moving dislocation, thus a strongly temperature dependent phenomenon (see section 6.1.1 for Ni_3Al). This corroborates pretty well the results of Ezz and Hirsch in Ni_3Al presented in fig. 10 in which the WHR measured at very low strain level exhibits high values but with quite a small temperature dependence, while when measured at higher plastic strain levels, the WHR is smaller but varies strongly with deformation temperature.

The question also arises of the relative decomposition of the applied stress τ_{appl} in terms of an effective stress τ^* (necessary to overcome localised obstacles) and an internal stress τ_i (due to the long-range interaction dislocation microstructure). Indeed as shown by the pioneer work of Thornton et al. [10], for sufficiently low strain (i.e. before any work hardening has occurred, that is before any internal stress is established) the yield stress anomaly is not observed. Thus Thornton deduces that the flow stress anomaly is related to "structural conditions" [the increase with T of cube glide] rather than to a single lattice interaction effect. The role of this structural part of the stress on the macroscopic stress measured during experiments has been denied in the last years. For instance, Devincre et al. [118] postulate that "in the anomalous domain of temperature, τ^* is close to τ_{appl}". This is also implicitly done by Caillard et al. [119,120] who compare the theoretical stresses necessary to unlock a Kear–Wilsdorf configuration with the values of the applied shear stress experimentally measured. Also the experimental measure of the effective and internal parts of the stress are still challenging, the values of stress measured from dislocation curvatures during in situ experiments (obviously an effective stress) [121] compare rather well with macroscopic stresses and moreover exhibit the same temperature dependence. This result suggests that the anomaly of the yield stress measured during conventional mechanical testing may be associated with the effective part of the stress. This is also supported by the results of Ezz and Hirsch [19] who decompose the flow stress into two terms: the yield stress (which by itself accounts for the anomalous temperature dependence and which may be assimilated to our effective stress) and some work hardening stress ($\tau_h = \tau - \tau_y$). The stress resulting from the storage of dislocations may thus only play a role in the work hardening rate. Also this internal stress is certainly responsible for the non-reversible part of the stress which is recorded during CS experiments for which $T < T_{PS}$. Such a decomposition would be adequate in γTiAl, where the effective activation volume is measured to be constant along a stress–strain curve in the anomaly domain, indicating that the effective stress is also constant (in this case work hardening is solely ascribed to an increase of the internal stress through the continuous storage of dislocations [38]). However, in Ni_3Al compounds the continuous decrease of the effective activation volume values along the stress–strain curve (see examples in [15,52]) indicates an increase of the effective stress [14].

6.2. The peak temperatures for stress and work-hardening for Ni_3Al compounds

As shown in section 2.1.1.5 for single crystals and section 2.1.2 for polycrystals, these peak temperatures, respectively $T_{p\tau}$ and $T_{p\theta}$ are different, with $T_{p\tau} > T_{p\theta}$. Additional data can be found, e.g., in Ni_3Ga [18,122], in $Ni_{75}Al_{22.7}Hf_{1.51}B_{0.2}$ [19], in $Ni_{77}Al_{22.7}Hf_{0.26}$ and $Ni_{76.1}$ $Al_{22.7}Hf_{1.2}$ [12]. The present state of understanding of the stress-peak temperature is first exposed.

6.2.1. The stress peak temperature

As recalled in a preceding review [123], early studies have shown that it corresponds to the transition between the anomalous regime and the high-temperature deformation mechanism which is thermally activated [124]. Below $T_{p,\tau}$, dislocations gliding on the primary octahedral planes are exhausted via the formation of Kear–Wilsdorf locks as shown by a variety of experimental observations (see, e.g., [23]). This involves a thermally activated cross-slip mechanism of the leading superpartial dislocation. Under such conditions, as the temperature rises, larger stresses are required for deformation to proceed since more and more dislocations are exhausted. Above $T_{p,\tau}$, cube glide is observed.

A few recent observations are now reported which shed some additional light on these various mechanisms and consequently on the peak temperature. They concern three types of Ni_3Al single crystalline compounds deformed along $\langle 123 \rangle$ and polycrystals of binary compounds of various compositions. Since the dislocation core geometry is of key importance for the above mobility processes, fault energy values are first reported for crystals in which the mechanical properties have also been characterized, in view of a tentative comparison.

6.2.1.1. Dislocation core characteristics. The technical problems encountered in measuring fault energies by the weak-beam technique in TEM have been discussed elsewhere (see [67]) and are not exposed here. However, one has to bear in mind that image shift corrections have to be performed by using dedicated simulation programs (see [125]) to deduce the real fault width from dislocation image separation. This is essential to prevent an underestimation of the fault energies. In addition, depending on the core configuration and the diffraction conditions used, the number of intensity peaks on the image can be equal to, smaller or larger than the number of partial dislocations in the core. This has been shown by image simulations and TEM observations [67,126,127]. The determination of fault energies from the measured dissociation widths requires programmes which compute interactions between non co-planar dislocations in the frame of elastic anisotropy [128]. In the case of narrow core extensions, the more sophisticated approach of Schoeck [129], which describes the dislocation core in terms of Peierls–Nabarro dislocations, is necessary. It takes into account the possible influence of the dislocation core energy on the determination of fault energy when the latter is high. In such a situation, various assumptions included in the image simulation programs are at their limit of validity. This is the case for the measurement of the complex stacking-fault energy, which is very high in the Hf-containing compound. Only a lower limit of the stacking fault energy could be provided (Kruml et al. [67]).

Table 6 summarizes the fault energy values, at least the most reliable ones to our knowledge, i.e. those for which image shift corrections have been considered. These include the antiphase boundary (APB) energies on {010} and {111} planes, respectively γ_{010} and γ_{111} and the complex stacking fault (CSF) energy, γ_{CSF}.

Complementary information is available on the APB energies in the Hf compounds as a function of concentration. They are illustrated in fig. 43. Figure 43(a) shows the variation of the APB energies. For the binary compound, the corrected and uncorrected values of γ_{111} were measured by Hemker and Mills [130] in $Ni_{76}Al_{24}$, while the uncorrected

Table 6
Comparison of the fault energies in three types of Ni_3Al compounds

Compounds	γ_{010} (mJ m^{-2})	γ_{111} (mJ m^{-2})	$\gamma_{111}/\gamma_{010}$	γ_{CSF} (mJ m^{-2})	References
$Ni_{74.8}Al_{21.9}Hf_{3.3}$	250 ± 25	300 ± 25	1.28 ± 0.18	$\geqslant 460$	[67]
$Ni_{74.3}Al_{24.7}Ta_1$	200 ± 25	237 ± 30	1.2 ± 0.3		[127]
				352 ± 50	[135]
$Ni_{74}Al_{26}$				277 ± 50	[133]
$Ni_{75}Al_{25}$				236 ± 30	[133]
$Ni_{75.8}Al_{24.2}$			1.34 ± 0.21		[41]
		180 ± 20		206 ± 30	[130]
$Ni_{76}Al_{24}$				206 ± 27	[133]
$Ni_{78}Al_{22}$	104 ± 15	175 ± 15	1.68 ± 0.38	235 ± 40	[136]

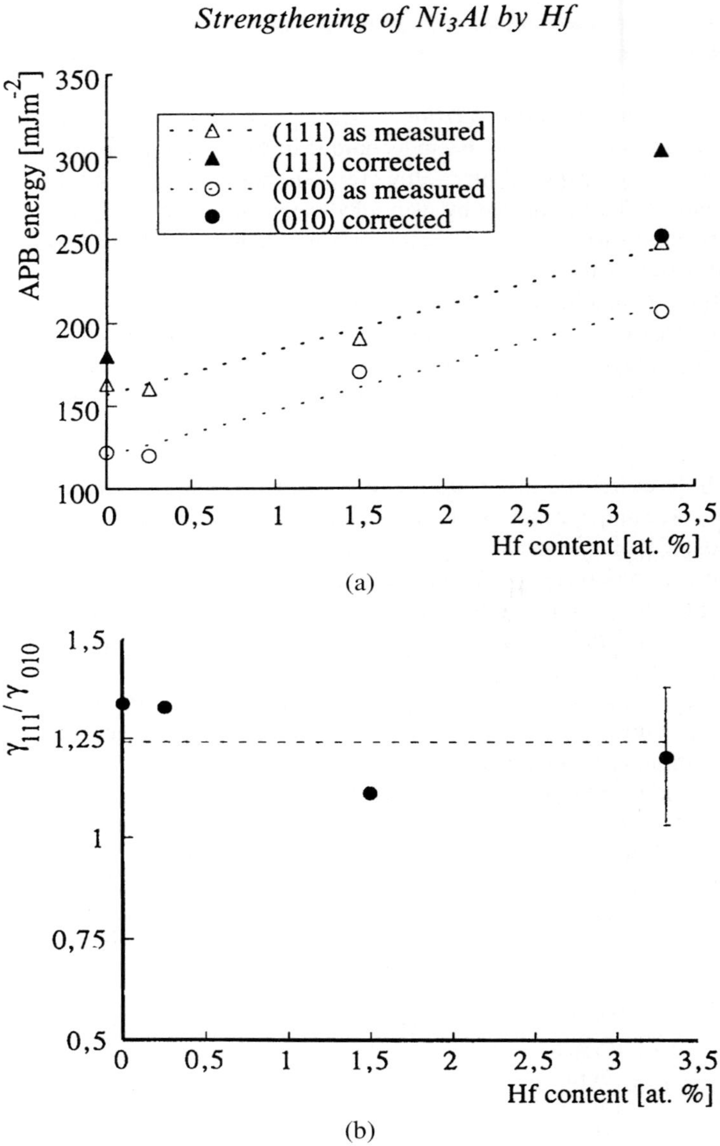

Fig. 43. Compilation of APB energy values on 111 and 010 respectively in $Ni_3(Al,Hf)$ as a function of Hf concentration. See text for references. (a) Absolute values. Open symbols correspond to experimental dislocation distances and filled symbols to the corrected ones (Cufour programme [125]) respectively. (b) APB energy ratio (uncorrected values of energies).

value for γ_{010} is from Dimiduk et al. [131] in $Ni_{77.1}Al_{22.9}$. The "uncorrected values" for 0.2 at% and 1.5 at% Hf respectively are from Bontemps-Neveu [12]. Figure 43(a) shows that the "uncorrected values" of γ_{111} and γ_{010} respectively increase linearly with the Hf concentration. A similar trend is observed for the "corrected values" of γ_{111}. The APB energy ratio $\gamma_{111}/\gamma_{010}$ computed from the "uncorrected values" of fig. 43(a) for lack of better information, is represented on fig. 43(b). Such an APB energy ratio is considered as acceptable since in the 3.3 at% Hf compound a ratio of 1.20 is obtained whether corrected or uncorrected values of the APB energies are used [67]. Figure 43(b) also shows that $\gamma_{111}/\gamma_{010} = 1.20 \pm 0.18$ for the Hf compounds considered. Table 6 shows in addition that only a few groups have attempted a correct measurement of γ_{CSF} in these compounds. In most cases, the latter energy is computed in the frame of classical elasticity theory for the interactions of the corresponding Shockley partials. However, in the case of the 3.3 at% Hf compound, Schoeck's approach is preferred. It provides a significantly higher CSF energy value as compared to the classical treatment which yielded values larger than 350 mJ/m^2 [132]. Similar attempts at measuring γ_{CSF} in binary Ni_3Al compounds can be found in [133].

6.2.1.2. Dislocation core and strength (single crystals). The mechanical properties of the three types of single crystal compounds of table 6 are summarized in fig. 44. It shows the 0.2% offset stress $\tau_{0.2\%}$ as a function of temperature, measured on virgin specimens (section 2.1.1.4) by Spätig [14] and Kruml et al. [132].

A correlation between the APB energy ratio and the flow stress has been already attempted in Hf compounds for various concentrations. Heredia and Pope [134] showed that the critical resolved shear stress (CRSS) on {111} increases with Hf content. They compared a binary compound $Ni_{76.6}Al_{23.4}$ and 3.3 at% Hf alloy. At, e.g., 600 K, the CRSS in the latter compound is 1.5 times higher than in the binary compound. At the same temperature, fig. 44 indicates a factor of 5.3 for $\tau_{0.2\%}$ when both materials are compared (see also [71,132]). The difference between the two results can be explained

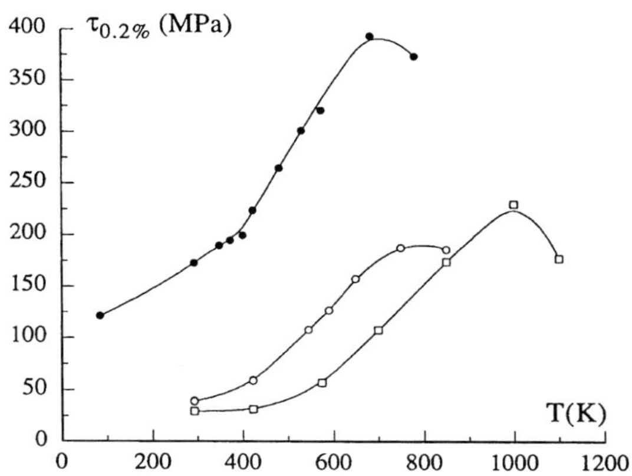

Fig. 44. $\tau_{0.2\%}$ as a function of temperature for three ⟨123⟩ Ni_3Al single crystalline compounds. (□), $Ni_{76.6}Al_{23.4}$; (○), $Ni_{74.3}Al_{24.7}Ta_1$; (●), $Ni_{74.8}Al_{21.9}Hf_{3.3}$ [14].

by different experimental procedures: Heredia and Pope [134] used one sample for all the temperatures investigated unlike Spätig [14]. Spätig's procedure is safer, all samples having the same initial dislocation structure. Therefore, in the present case, the stress is composition dependent, while the APB energy ratio is not (fig. 43(b)).

Therefore, the APB energy ratio cannot be correlated with the strength of the Hf compounds in a simple manner. Along the same line, table 6 shows that the anisotropy ratio is about the same for the Hf and the Ta compounds while fig. 44 indicates that their respective strengths are different in the anomaly regime (300 K $\leqslant T \leqslant$ 700 K). This confirms earlier similar conclusions by Dimiduk et al. [131] who studied a range of binary compounds and ternary ones containing Sn and V additions.

Let us now consider the mechanical test results of fig. 44 and the parameter γ_{CSF} of table 6. This figure shows that over the temperature domain 300 K $\leqslant T \leqslant$ 700 K the three compounds considered exhibit a strength anomaly. In this domain, the binary compound appears to be the weakest and the Hf compound the strongest. Table 6 shows that the Hf compound has the highest CSF energy value while the binary compound has the smallest one. These facts can be understood in terms of the cross-slip mechanism which leads to the formation of Kear–Wilsdorf locks, the frequency of which is controlled by the constriction of the leading superpartial. When two compounds are compared under the same conditions, the latter process operates more easily in the material which exhibits the higher γ_{CSF} value. Consequently, at a given temperature of the strength anomaly domain, dislocation exhaustion is more frequent in the compound of high CSF energy. Therefore, to keep a constant strain-rate, a higher stress is required for deformation to proceed via the motion of shorter dislocation segments which can eventually multiply. Hemker and Mills [130] were the first to mention such a correlation between strength and γ_{CSF} as an interpretation of their experimental results. Their study consisted in comparing two different compounds, namely a binary and a boron-doped Ni_3Al crystal. They measured the CSF energy using the weak-beam technique, as well as $\tau_{0.2\%}$ at one temperature only. The correlation they have proposed is confirmed here over a temperature range of several hundred degrees and for much larger differences in stress.

In addition, for the three compounds of fig. 44, the mobile dislocation exhaustion rates have been measured directly by the technique of repeated relaxations (see the appendix). In the present case, the exhaustion rates $\Delta\rho_m/\rho_{mo}$ are defined as the decrease $\Delta\rho_m$ of the initial dislocation density ρ_{mo} when the deformation rate becomes half its initial value during a relaxation test performed at a 0.2% plastic strain ($q = 2$ in eq. (A15)). At 573 K, the measured values for $\Delta\rho_m/\rho_{mo}$ are 31% for Ni_3Al, 35% for $Ni_3(Al,Ta)$ and 40% for $Ni_3(Al,Hf)$. Therefore, the direct measurement of mobile dislocation exhaustion rates supports the above interpretation of the dependence of the stress on γ_{CSF}.

A direct consequence of the cross-slip mechanism of the APB from {111} to {010} is illustrated in fig. 45. This figure summarizes some data obtained in various TEM observations of superdislocations split along the {111} plane, in an attempt to measure γ_{111}. In the various studies available [128,130,131,135,136] it is remarkable that no data are reported for dislocations approaching screw orientation. In what follows, the parameter α_{min} accounts for the minimum angle, extracted from the previous references, between superdislocations on {111} and the screw orientation. The fact that α_{min} is different from zero is thought to be the consequence of APB cross-slip from {111} on to {010} as the

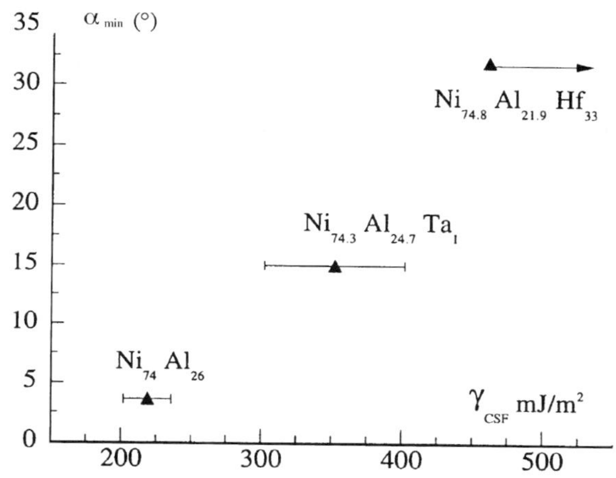

Fig. 45. Compilation of values of parameter α_{min} corresponding to superdislocations on {111} observed in TEM as a function of γ_{CSF} for various Ni_3Al crystals. See text for definition of α_{min} and references.

dislocation line approaches screw orientation. Along this line, α_{min} is plotted as a function of CSF energy in fig. 45. For $Ni_{74.3}Al_{24.7}Ta_1$, α_{min} is 15° [128,135]) while γ_{CSF} has an average value of 350 J m^{-2} (table 6). For $Ni_{74.8}Al_{23.9}Hf_{3.3}$, α_{min} is close to 32° [67] while γ_{CSF} is larger than 460 mJ m^{-2}. Along the same line, Dimiduk et al. [131] report a very low α_{min} value of 3° in $Ni_{74.1}Al_{25.9}$ while γ_{CSF} is close to 219 mJ m^{-2} (table 6). Figure 45 clearly shows that as γ_{CSF} increases, α_{min} increases too. This fully agrees with the cross-slip mechanism quoted above. A few other experiments [130,136] provide data points situated above those of fig. 45. This is likely due to a lack of interest for α_{min} or to stress and temperature conditions less favourable to cross-slip in the corresponding experiments.

6.2.1.3. Dislocation core and the stress peak temperatures (single crystals). Another parameter which is expected to be influenced by γ_{CSF} is the stress peak temperature. It corresponds to the transition between the strength anomaly mechanism and the cube glide process which takes over at high temperatures. The latter process is by far less studied than octahedral glide (see, e.g., [23]) and is therefore a poorly documented one. Dislocation observations by TEM after deformation at temperatures higher than $T_{p\tau}$, reveal rather square dislocation arrays in the active 010 plane. Screws as well as edges appear as quasi-linear along crystallographic directions in Ni_3Al [22], $Ni_3(Al,Ti)$ [137], $Ni_{74.8}Al_{21.9}Hf_{3.3}$ [132] and Ni_3Ga [138]. This indicates that both types of dislocations undergo a Peierls–Nabarro friction along this plane. This interpretation is in agreement with the rapid stress decrease above $T_{p\tau}$ as illustrated by fig. 44 for the three compounds in question, as well as with the small values measured for the microscopic activation volumes (in table 3 see the volumes for $Ni_{76.6}Al_{23.4}$ and $Ni_{74.8}Al_{21.9}Hf_{3.3}$ [14,16]). For the screw, the friction is due to a non-planar (therefore sessile) core configuration: the APB lies on {010} while each superpartial is dissociated along an intersecting octahedral plane. For the edge, dissociation into a super Lomer–Cottrell [135] or a double Lomer–Cottrell [139] lock have been established well above the peak. For the screws to glide along the cube plane, the core has to recombine by stress assisted thermal activation. Under such conditions, the glide friction depends directly on γ_{CSF}, i.e. is expected to be small for large CSF energies (a smaller constriction is necessary for cube glide to proceed). In this case, for given temperature and strain-rate, a lower stress is required. Therefore, the stress-temperature curve is shifted towards the left. For the same crystal, the stress-temperature curve in the strength anomaly domain is also shifted towards lower temperatures and so consequently is $T_{p\tau}$. Figure 44 agrees fairly well with this description: the Hf compound which exhibits the highest CSF energy (see table 6) has the lowest $T_{p\tau}$ value. The binary alloy which has the lowest CSF energy also has the highest stress peak temperature. To conclude, the frequency of splitting of the core out of the cube plane accounts fairly well for the variation with the CSF energy of the yield stress at one temperature of the anomaly domain as well as the magnitude of the stress peak temperature.

Nevertheless, the above description is only quantitative, although it accounts for some trends which are experimentally observed. In particular it does not account for some features of fig. 44: if the binary and the tantalum compounds are compared, as γ_{CSF} increases by a factor of 2, the stress at, e.g., 600 K is doubled. At the same temperature, the yield stress of the Hf compound is about 2.5 time larger than in the Ta alloy, while the ratio of γ_{CSF} values is 1.3 only. It is likely that other parameters and effects not considered so far have some influence. One can think of:

(a) the values of the elastic constants of each compound and their variation with temperature,
(b) the evolution of γ_{CSF} with temperature,
(c) the fact that the strengths of the three compounds are compared at one temperature for which the stress levels are very different,
(d) other aspects of solid-solution strengthening such as interaction of moving dislocations and solute atoms and solute segregation at the faults which also interfere.

Parameters and effects (a) to (c) influence the core extension of the dislocation, therefore their cross-slip frequency. The processes recalled in (d) influence the mobility of the dislocation, superimposed on the Peierls–Nabarro friction described above. Quantitative information is lacking about the effects listed, but they are expected to play a role.

6.2.1.4. Comparison of single and polycrystals. The case of polycrystals has been studied [24,133] using binary compounds of various compositions, namely $Ni_{74}Al_{26}$, $Ni_{75}Al_{25}$ and $Ni_{76}Al_{24}$. The dislocation core characteristics can be found in table 6 while the variation of yield stress ($\sigma_{0.2\%}$) with temperature is illustrated in fig. 46. The general trend of the curves of fig. 46 is similar to the one of single crystals in fig. 44. However, to be valid in the case of polycrystals, the comparison of the strengths requires a constant average grain size. This parameter is smaller in $Ni_{74}Al_{26}$ (0.5 mm) as compared with the two other crystals (0.8 mm). This causes an additional increase of yield stress for the Al-rich compound. Indeed, it has been estimated in independent measurements [140] that the yield stress decreases by 25% when the average grain size changes from 0.5 mm to 1 mm. Applying a similar grain size correction to $Ni_{74}Al_{26}$ lowers the corresponding curve on fig. 46, but this compound still exhibits the highest strength over the temperature range studied. At RT, $\sigma_{0.2\%}$ is about the same for the stoichiometric and the nickel-rich alloy and

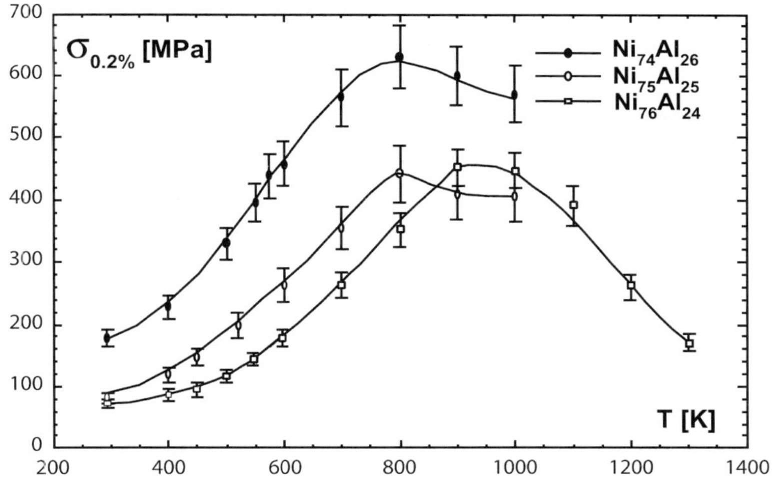

Fig. 46. Variation of $\tau_{0.2\%}$ as a function of temperature for Ni_3Al and related polycrystals. From [24,133]. Arrows indicate the stress peak temperatures.

higher for the aluminium-rich alloy. At a given temperature of the strength anomaly regime, as the aluminium content increases, the yield stress rises. The peak temperatures are the same, within the experimental uncertainty for the stoichiometric and the aluminium-rich compound (close to 800 K) while the peak temperature for the nickel-rich compound is shifted to the right (950 K). Similar measurements in polycrystals of compositions ranging between $Ni_{76.5}Al_{23.5}$ and $Ni_{73.5}Al_{26.5}$ confirm the above trend of variation of the yield stress as a function of composition in the strength anomaly domain [141]. To summarize, as the aluminium content increases, the $\sigma_{0.2}(T)$ curve is shifted towards higher stresses in the latter temperature range and the CSF energy increases (table 6). The correlation between the strength and the complex stacking-fault energy is the same as for the single crystalline ternary alloys which have been considered above. Therefore, in the present case, the same interpretation in terms of the cross-slip mechanism seems to hold, which leads to mobile dislocation exhaustion by Kear–Wilsdorf lock formation.

6.2.2. The work-hardening peak temperature

According to the preceding experimental review, reliable data about the work-hardening rate θ and in particular the corresponding peak temperature are those determined using a virgin sample at each temperature (section 2.1.1.4). The anomalous behaviour of the work-hardening rate and the discrepancy between the two peak temperatures $T_{p,\tau}$ and $T_{p,\theta}$ have stimulated research only recently.

6.2.2.1. General considerations. The curves $\theta(T)$ corresponding to single and polycrystals respectively appear to be different. For the former, a rather well defined maximum is observed (fig. 9), while for polycrystals a shallow maximum is present (fig. 12). This can be understood in terms of athermal processes taking place in the polycrystal (e.g., dislo-

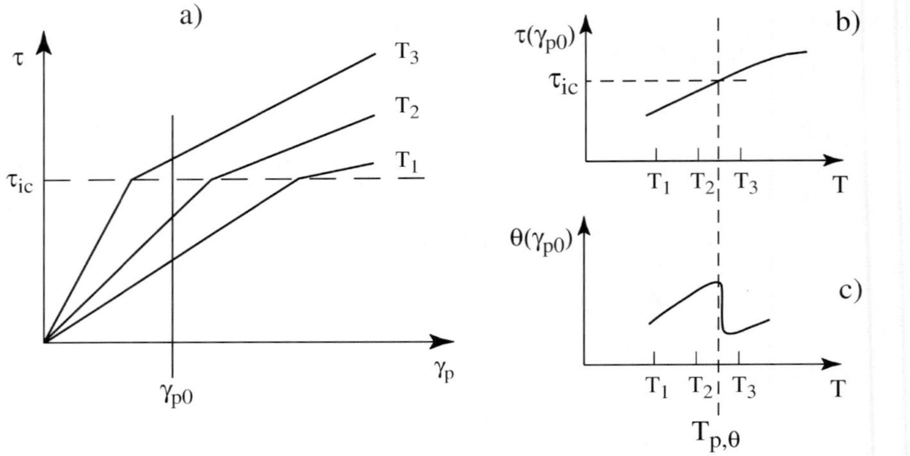

Fig. 47. Schematic representation of the presence of two different work-hardening mechanisms depending on stress and temperature: (a) stress-plastic strain curves; (b) stress at given plastic strain (γ_{po}) as a function of temperature (τ_{ic} is the transition stress between the two mechanisms); (c) work-hardening coefficient (at γ_{po}) as a function of temperature. $T_{p\theta}$ is the temperature that corresponds to the peak in work hardening (D. Caillard unpublished work).

cations piling up at grain boundaries) which are expected to smear out the maximum of work-hardening rate as a function of temperature.

The interpretation of the existence of a peak in work-hardening rate is illustrated on the schematics of fig. 47. It shows the correlation between the evolution of the stress–strain curve with temperature and the variation of the work-hardening rate with the latter parameter. On fig. 47(a), for the sake of simplification, the stress–(plastic) strain curve consists of two linear portions, with a weaker slope at high stresses, which correspond respectively to two distinct hardening mechanisms. For a given plastic strain γ_{po}, the positive variation of stress with temperature is represented on fig. 47(b), while the curve $\theta(T)$ is illustrated on fig. 47(c). The possible hardening mechanisms operating at low and high stresses respectively at given temperature or at low and high temperature respectively for given plastic strain are discussed in the following section.

6.2.2.2. Various interpretations of the peak temperature for work hardening. In the early and very enlightening work of Thornton et al. [10], the existence of $T_{p\theta}$ is not explicitly formulated. However, the schematics of their fig. 3(b) representing the increasing part of $\sigma(T)$ for Ni_3Al polycrystals exhibits a corresponding inflexion point. It divides the temperature regime of positive variation of the flow stress into two domains (labeled I and II respectively by the authors). In domains I and II the response of the sample to strain-rate jumps is observed to be different and two types of creep transients are observed: in domain I a rapid decrease of the creep rate is observed, contrary to domain II. Two different glide mechanisms are then postulated by the authors, implemented by microstructural observations: mobile dislocation exhaustion in domain I (exhaustion hardening) and a "debris mechanism" in domain II attributed to the increase with temperature of the activity of the cube cross-slip system. Much later, dislocation activity on the latter system has been evidenced by TEM (see, e.g., Douin et al. [142] and Kruml et al. [132] at temperatures lower than the stress peak temperature.

Another reason for the existence of $T_{p\theta}$ is the presence of APB tubes at the lower end of the anomalous temperature regime. This type of defect forms at the gliding superdislocation and therefore slows down its motion. They have been observed in a variety of Ni_3Al compounds (see, e.g., [143,144]. As the temperature rises, these tubes are observed to annihilate, presumably by diffusion, well below $T_{p\tau}$. A subsequent increase of the superpartial velocities is thus expected with a direct influence on the WHR.

The third type of interpretation is more quantitatively developed and is described in the following section.

6.2.2.3. Kear–Wilsdorf lock stability and work-hardening rate. The work-hardening mechanisms which correspond to the strength anomaly domain consist of mobile dislocation exhaustion by Kear–Wilsdorf lock formation (see sections 3.2 and 6.2.1.2). Consequently the stability of these locks against stress has been questioned. The early motivation for such estimations was to estimate whether these locks can resist the peak stress.

Two types of locks have been experimentally observed, i.e. complete and incomplete ones. TEM observations (in situ and post mortem) have reported for a long time complete Kear–Wilsdorf locks in the strength anomaly domain. Such locks are the ones which were considered in section 6.2.1.1 for the determination of the core characteristics of the dislocations. Their formation is responsible for the exhaustion of dislocations gliding along the octahedral planes for $T < T_{p\tau}$. The formation (and destruction) of incomplete Kear–Wilsdorf locks is responsible for a spectacular dislocation mobility processes called "APB jumps". It has been seen during in situ observations in TEM [121,145]: rectilinear screw dislocations which are dissociated approximately along their octahedral plane exhibit jerky glide; they jump over distances which scale exactly with their dissociation width, the trailing superpartial stopping exactly at the previous position of the leading one.

The detailed mechanism is described by Caillard [120]. The two types of locks are formed through different processes: for the complete one, the leading superpartial recombines over a certain length and cross slips on the cubic plane over a larger distance before it dissociates again on an intersecting octahedral plane. In this case, the activation energy refers to the recombination and bulging in the cube plane. On the contrary, the incomplete lock forms by cross slip of the leading superpartial over some interatomic distances on the cube plane. Then it cross-slips back onto the octahedral plane, resulting in the formation of kinks on {010} (simple macrokinks [23]). In this case, the activation energy corresponds to the kink pair formation.

In the APB jump process, another type of macrokink is formed (elementary macrokinks) along the octahedral plane, the height of which is equal to the APB width on {111} [120] Such elementary macrokinks have been observed by TEM to be dissociated along the {111} plane in various Ni_3Al compounds [17,23].

Estimates of the stability of these locks with respect to stress are as follows [146,147]. An incomplete lock is represented in the schematic of fig. 48. Under an applied stress the leading superpartial is pushed forward along {111} while the trailing one can be locked

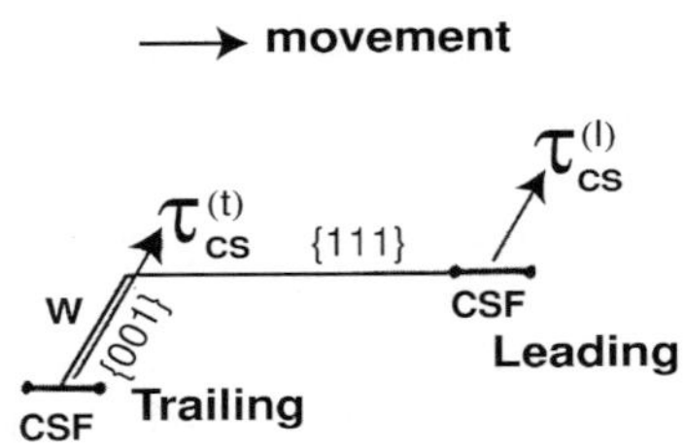

Fig. 48. Schematic of an incomplete Kear–Wilsdorf lock seen end on. The leading and trailing superpartials are represented. W is the extension of the APB on {001}. τ_{cs} is the shear stress acting on the superpartials along {001}, the superscripts t and l referring to trailing and leading dislocations respectively [120].

with respect to {001} glide. The stresses acting on the superpartials along the cube plane are respectively τ_{cs}^{t} and τ_{cs}^{l}.

These stresses have been computed [148] (for infinitely long dislocations) taking into account elastic interactions between superpartials in the frame of anisotropic elasticity and the surface tension of the APB on {001}. Unlocking takes place when the trailing superpartial yields, i.e. τ_{cs}^{t} overcomes the friction stress τ_{cs}^{min} on the cube plane. However, the leading superpartial can also glide along {001} under the stress τ_{cs}^{l} which tends to lock the superdislocations further. Therefore, for a successful unlocking event, cross-slip on {001} of the trailing superpartial must be faster than for the leading one. This condition can be expressed by a second threshold stress τ_{ul} [148]. It has been shown by Caillard and Paidar that τ_{ul} is in most cases the relevant threshold stress. It has been computed [146–148], and the following estimations have been found. For sake of simplification, the locks are separated into two extreme types with the corresponding critical stresses for yielding:

(i) Incomplete locks with a short cross-slip distance on {001} ($w \sim b$)

$$\tau_1 = \frac{\gamma_{010}}{b}\left[1 - \frac{\gamma_{010}}{\gamma_{111}}\frac{1+2/A}{\sqrt{3}}\right], \tag{11}$$

where A is the elastic anisotropy ratio and b the Burgers vector of the superpartial dislocation. τ_1 corresponds to the stress necessary to start an APB jump by unlocking of the configuration of fig. 48.

(ii) Complete locks which correspond to a larger cross-slip distance on {001}

$$\tau_2 = \frac{\gamma_{010}}{b}\left[1 - \frac{\gamma_{010}}{\gamma_{111}}\frac{1}{\sqrt{3}}\right]. \tag{12}$$

Since A is close to 3, and using the APB energy values of table 6, the above estimations of τ_1 and τ_2 confirm that incomplete locks are easier to destroy than complete ones. The estimation of τ_1 agrees with some observations of APB jumps or elementary macrokinks: APB jumps have been observed in Ni_3(Al,1.5at%Hf) at 150°C and Ni_3(Al,0.25at%Hf) at 400°C in samples which had been predeformed at a stress close to τ_1. Conversely, in the second material, deformed at 20°C under a stress smaller than τ_1, the number of observed "elementary macrokinks" was not abnormally large [120].

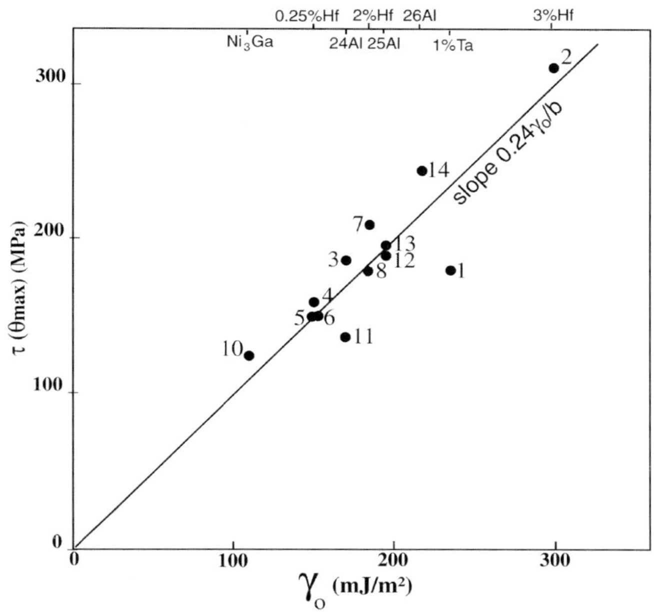

Fig. 49. Stress corresponding to the maximum of work hardening rate $\tau(\theta_{max})$ as a function the APB energy on 111 (γ_{111}). The straight line corresponds to relation (11) (see text). Data points refer to the materials indicated at the top of the figure. Points 1–3 are from [7,14,24,132], points 4–8 are from [12], point 10 from [122], point 24 from [24].

It has then been proposed by Caillard that the stress corresponding to the peak in work hardening, $\tau(\theta_{max})$, is the unlocking stress of incomplete locks. Referring again to fig. 47, the two mechanisms proposed for work hardening would be: (i) dislocation exhaustion by lock formation at low strain or temperature, (ii) dislocation exhaustion combined with the yielding of incomplete locks when the stress becomes large enough, at high stresses or temperatures.

The comparison between τ_l and $\tau(\theta_{max})$ can be seen on fig. 49 which refers to the values estimated for τ_l. In relation (11), the anisotropy ratio A is close to 3.3, and the ratio of the APB energies is between 0.77 and 0.84 for values of γ_{010} and γ_{111} listed in table 6, found in a previous review [23] or a compilation by Caillard and Molénat [119]. An average value of 0.8 is considered for the APB energy ratio, so that relation (11) yields a linear variation of τ_l as a function of γ_{111} with a slope close to $0.26/b$. Figure 49 is a plot of $\tau(\theta_{max})$ as a function of γ_{111} for a range of $L1_2$ compounds for which these two quantities are available. An average straight line is drawn, the slope of which is found to be $0.24/b$. This is considered by the author to be close enough to the previous estimate for τ_l and supports the idea that $\tau(\theta_{max})$ corresponds to the yielding of incomplete locks in the $L1_2$ compounds. Let us note that according to the considerations of section 4, a local stress should be considered instead of the applied stress. However, for lack of a more unifying theory, this description of the existence of $T_{p,\theta}$ is acceptable.

Concerning the estimate of τ_2 in relation (12), Caillard mentions some results about $\theta(T)$ curves which exhibit two peaks (see [18] for Ni_3Ga). The stress corresponding to the second peak compares rather well with the τ_2 value computed in this compound. This could be an example of a second softening mechanism taking place at higher temperature and corresponding to the yielding of complete locks.

6.3. The strength anomaly in γTiAl

The anomalous behaviour of polycrystalline single-phase γTiAl has been modelled on the basis of both mechanical experiments and dislocation observation in TEM [73]. It was shown that the yield stress anomaly observed in the intermediate temperature regime (domain II) is associated with the activation of ordinary dislocations [60]. While their core structure is shown to be compact [149,150], their morphology evolves with temperature, exhibiting pinning points aligned along the screw direction, the density of which increases with the deformation temperature in the anomaly domain [59,151,152] (note that the pinning point density decreases back at a temperature lower than $T_{\mathrm{p}\sigma}$). The origin of these pinning points is still the subject of controversies regarding their nature (intrinsic [59,151] or extrinsic [153,154]), as well as the detailed mechanism that leads to their formation in the intrinsic case. Whatever the details of their formation, it is generally believed that these anchoring points play a significant role in the particular mechanical behaviour of γTiAl [59,82,151]. However some recent results of stress measurements during in-situ experiments seem to indicate that the pinning points are weak obstacles as compared to the friction stress [155]. Theoretical modelling of the dynamic behaviour of ordinary dislocations and thus of the mechanical properties of TiAl proposed so far are based either on a single dislocation mechanism (either jog dragging or dipole trailing [151]), or on the evolution of the mobile dislocation density (Local Pinning Unzipping – LPU – [66, 156]). These different models have been compared through numerical calculations [157]; it was shown that only the LPU model can account for the temperature increase of the yield stress, while the jog dragging mechanism may explain the stress drop above $T_{\mathrm{p}\sigma}$. The main characteristics of the LPU model reside in the fact that the dislocation has a finite probability to be permanently locked during its motion. This locking probability is related to the statistical distribution of the length of dislocation segments between the pinning points. This type of approach highlights the role played by the longest segment along the dislocation (and not the mean segment length). An equivalent approach has been applied to $L1_2$ compounds, by considering the height of superkinks on superdislocations [84]. In TiAl, for an imposed strain rate type of experiment, the exhaustion of mobile dislocations has to be compensated by some multiplication. Some dynamic equilibrium thus occurs between the essentially temperature-dependent exhaustion and the stress-dependent multiplication processes leading to an implicit relation between flow stress and temperature. Since the locking probability increases with temperature, this leads to the yield stress anomaly. Exhaustion of mobile dislocations (see section 3.2.1) also provides quite a high storage of dislocations which may explain the work- hardening rate in agreement with the modified Taylor equation (see section 4). Numerical calculations have been performed [73,158] on these bases and showed that it is possible with one set of parameters to simulate the most striking features of the mechanical behaviour (yield stress anomaly, stress–strain curve, i.e. correct WHR, strain-rate jump response, stress relaxation ...). The same approach, focussed on the behaviour of the mobile dislocations as a whole, and not on the velocity of a single dislocation, has been extended to model the behaviour of $L1_2$ based compounds [39,84,159].

Acknowledgements

The authors would like to express their sincere thanks to Mmes. Lovato and Bettinger for typing the manuscript, to MM Proietti and Halter and Dr. Conforto for preparing the figures. They would like to acknowledge fruitful discussions with Dr. Kruml as well as the financial support of the Swiss National Science Foundation for the research they have performed that is quoted in this chapter. They are grateful to Pr. Sir P.B. Hirsch for useful comments.

Appendix. Characterization of thermally activated dislocation mechanisms using transient tests

The transient tests presented here have been developed by Bonneville et al. [160], Spätig [14] and Lo Piccolo [24]. They have been partially described in several publications, their interpretation evolving as a function of time. The development below is a synthetic summary of the present state of the methods.

The Orowan equation has been already recalled together with the definition of the parameters involved:

$$\dot{\gamma}_{\mathrm{p}} = \rho_{\mathrm{m}} v b. \tag{A1}$$

A constant strain-rate test neither allows one to determine how many dislocations are moving in the solid and at which velocity, nor yields information about dislocation multiplication and storage. However, transient mechanical tests can bring some answer to these questions.

These can be performed as follows: a monotonic deformation test at given strain rate and temperature is interrupted at a selected stress while the specimen is allowed to relax (or to creep). The idea of performing a transient is not new. In a relaxation test (see a review by Dotsenko [161]), the total strain is kept constant and the stress is recorded as a function of time. For a transient creep experiment, the load is kept constant, strain being measured as a function of time. Provided the transient is not too long, the corresponding dislocation mechanisms are not too different from those operating at constant strain rate, and useful information can therefore be gained about them.

With the aim of separating the respective contributions to the plastic strain rate of mobile dislocation densities and velocities, repeated transient mechanical tests have been developed. They include repeated relaxation tests and repeated creep experiments. The principles of such transients are exposed, together with the interpretation of the behaviour of the material and the related assumptions, with emphasis on the information they provide.

Single transients

The interpretation of single transients is first recalled, to introduce relations necessary for the description of repeated transients. Logarithmic transients are considered, i.e. stress for relaxation or strain for creep, is a logarithmic function of time. This is usually the case at low enough temperature, although a non-logarithmic behaviour has been observed occasionally (see, e.g., Bonneville et al. [58] in $Ni_3(Al,Ta)$ over a limited temperature range). As a rule, resolved shear stresses and strains, τ and γ respectively, will be considered. However, relations similar to those below can apply to true stresses and strains (respectively σ and ε). The subscripts or superscripts r and c refer respectively to relaxation and creep conditions. A complete description of relaxation tests can be found in [46] and of creep tests in [162,163]. Additional considerations have been developed in [49,164].

For logarithmic relaxations, the stress decreases by an amount $\Delta\tau$ as a function of time t according to the relation (see [165]):

$$\Delta\tau = -(kT/V_r)\ln(1 + t/c_r) \tag{A2a}$$

while for creep the plastic strain increase is:

$$\Delta\gamma_p = (kT/MV_c)\ln(1 + t/c_c), \tag{A2b}$$

where k is the Boltzmann constant, T the absolute temperature, M the elastic modulus of the specimen-machine assembly, V_r and V_c have the dimension of a volume, c_r and c_c are time constants. A fit of relations (A2) with the corresponding transient curve yields V_r and c_r for relaxation, V_c and c_c for creep. The time derivative of relation (A2) provides the plastic strain rate for relaxation and creep respectively:

$$\dot\gamma_p = (kT/MV_r)\big[1/(c_r + t)\big], \tag{A3a}$$

$$\dot\gamma_p = (kT/MV_c)\big[1/(c_c + t)\big]. \tag{A3b}$$

Relation (A3a) is obtained by considering the specimen-machine assembly equation:

$$\gamma = \tau/M + \gamma_p$$

and its time derivative under relaxation conditions:

$$\dot\gamma_p = -\dot\tau/M.$$

There are two ways of interpretation of relations (A2), either using Hart's equation as suggested by Saada et al. [165] or in the frame of the theory of thermally activated deformation which is considered below.

Some assumptions are made during the short transients:

(i) The applied stress τ can be decomposed into an internal stress τ_μ and an effective stress τ^*: $\tau = \tau_\mu + \tau^*$ which yields for relaxation conditions: $\Delta\tau_\mu = K^{\mathrm{r}}\Delta\gamma_{\mathrm{p}}$, where ΔX ($X = \tau^*$, τ_μ, τ or γ_{p}) is the variation of parameter X during the transient, K^{r} and K^{c} are work-hardening coefficients during the transients. The machine equation under relaxation conditions gives $\Delta\gamma_{\mathrm{p}} = -\Delta\tau/M$, whence:

$$\Delta\tau^* = \left(1 + K^{\mathrm{r}}/M\right)\Delta\tau. \tag{A4a}$$

For creep:

$$-\Delta\tau^* = \Delta\tau_\mu = K^{\mathrm{c}}\Delta\gamma_{\mathrm{p}}. \tag{A4b}$$

(ii) The dislocation velocity is thermally activated:

$$v = v_1 \exp\left[-(\Delta G_0 - \tau^* V)/kT\right], \tag{A5}$$

where the activation energy depends on the effective stress with value ΔG_0 when $\tau^* = 0$. V is the activation volume of the dislocation velocity. When the microstructure changes, V is different from V_{r} and V_{c} in relations (A2), and therefore cannot be determined via a single transient experiment. Generally, V depends on τ^* but it is considered as constant during the short relaxation or creep test. During the transient, the velocity is:

$$v = v_0 \exp\left(V\Delta\tau^*/kT\right) \tag{A6}$$

from relations (A4) and (A5) with v_0 being the velocity at the onset of the transient.

(iii) The mobile dislocation density is a power function of the velocity [166]:

$$\rho_{\mathrm{m}}/\rho_{\mathrm{mo}} = (v/v_0)^{\beta}, \tag{A7}$$

where ρ_{mo} is the mobile density at the onset of the transient. The variation of ρ_{m} during the transient has to be taken into account. Indeed, considering ρ_{m} constant yields K_{r} values larger than the work-hardening coefficient along the stress–strain curve which is very unlikely [48,167]. The variation of ρ_{m} during stress relaxation had been already emphasised [168].

Comparing the experimental strain rate (A3) and the value obtained via the Orowan relation (A1), using (A6) and (A7) yields:

$$(kT/MV_{\mathrm{r}})\left[1/(c_{\mathrm{r}} + t)\right] = v_0\rho_{\mathrm{mo}} b \exp\left[(V/kT)(1 + \beta_{\mathrm{r}})\left(1 + K^{\mathrm{r}}/M\right)\Delta\tau\right],$$
$$(kT/MV_{\mathrm{c}})\left[1/(c_{\mathrm{c}} + t)\right] = v_0\rho_{\mathrm{mo}} b \exp\left[-(V/kT)(1 + \beta_{\mathrm{c}})K^{\mathrm{c}}\Delta\gamma_{\mathrm{p}}\right]$$

for relaxation and creep respectively.

The above equation for relaxation can be transformed by expressing $\Delta\tau$ via relation (A2a) and $v_0\rho_{mo}$ using (A1) and (A3a) at $t = 0$. This yields a relation between V_r and V:

$$V_r = \Omega_r V \quad \text{with } \Omega_r = (1 + \beta_r)(1 + K_r/M). \tag{A8a}$$

The above equation for creep can be transformed similarly: $\Delta\gamma_p$ is expressed via relation (A2b) whence:

$$V_c = \Omega_c' V \quad \text{with } \Omega_c' = (1 + \beta_c)\big(K^c/M\big). \tag{A8b}$$

V_r and V_c are obtained from a single transient (relations (A2)), while repeated transients are necessary to obtain Ω_r or Ω_c' (i.e. V, using relations (A8)) as shown below.

If Ω_r is known, relation (A8a) shows that information can be gained about the structural parameters β_r and K_r. The same remark holds in creep for β_c and K_c.

Repeated transients

The technique of repeated relaxations, although proposed long ago [169], has not been used extensively (see, e.g., [170,171]). A new procedure has been defined [52] (see fig. A1), in which the first relaxation starts at stress τ_M over a time interval Δt, with a corresponding stress decrease $\Delta\tau_1$. The specimen is reloaded again up to τ_M, then allowed to relax by an amount $\Delta\tau_2$ during an equal Δt, etc. At low enough temperatures and for positive work-hardening coefficients, the following features are observed: (i) $\Delta\tau_j$ decreases as j, the relaxation number in the series, increases, (ii) when the successive relaxations are analyzed in terms of relation (A2a), V_r is constant while the time constant $c_{r,j}$ depends on j.

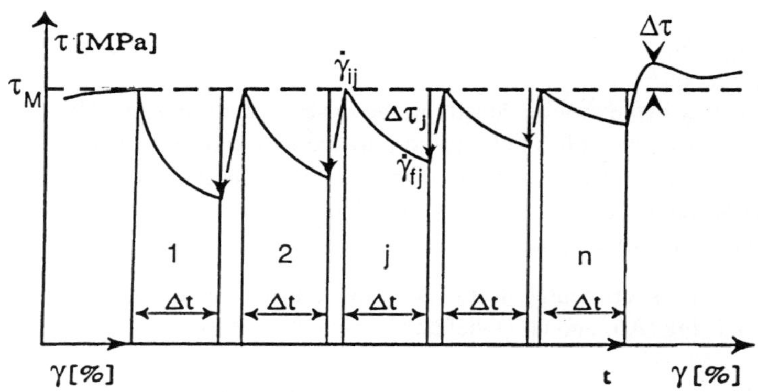

Fig. A1. The procedure of successive relaxations with constant duration. Definition of the parameters used. The relaxation number in the series j is indicated. τ_M is the stress at the onset of each relaxation, $\Delta\tau_j$ the amount of stress relaxation for transient no. j which starts and ends with plastic strain rates $\dot\gamma_{i,j}$ and $\dot\gamma_{f,j}$ respectively. $\Delta\tau$ is the stress increment observed on reloading after the transient.

The slowing down of the relaxations with j is due to hardening, which reduces τ^* despite a constant applied stress τ_M at the onset of each relaxation. This hardening is the result of dislocation exhaustion along a transient, so that the initial mobile dislocation density is lower as j increases.

Three methods are available to determine the volume V. For various parameters below, the subscripts i and f will be used which refer respectively to the onset and the end of a relaxation test in the series. The quasi-elastic reloading conditions ensure that no structural changes take place, i.e. ρ_m and τ_μ are the same at the end of relaxation no. j and at the onset of relaxation no. $(j+1)$. Consequently, relations (A1) and (A6) yield:

$$\dot{\gamma}_{i,j+1}/\dot{\gamma}_{f,j} = v_{i,j+1}/v_{f,j} = \exp(-V\Delta\tau_j/kT). \tag{A9}$$

Relation (A9) provides a first method of determining V, by measuring the $\dot{\gamma}$ values as proportional to the slope of the relaxation curves. A second method consists in combining (A2) and (A9). This yields:

$$V = (kT/\Delta\tau_j)\ln\left(\frac{c_{r,j} + \Delta t}{c_{r,j+1}}\right). \tag{A10}$$

In this case, V is computed from the time constants of (A2a) for two successive relaxations in the series.

The comparison between successive relaxations can be further developed, thus introducing a third method based on the estimation of Ω_r (relation (A8a)). To express Ω_r as a function of experimental data, the ratio $\dot{\gamma}_{i,j+1}/\dot{\gamma}_{i,j}$ is computed following two different ways. In the first one, the strain-rates are compared at the onset and the end of test no. j, using (A3a) for $t = 0$ and $t = \Delta t$ respectively. This yields

$$\dot{\gamma}_{ij}/\dot{\gamma}_{fj} = (c_{r,j} + \Delta t)/c_{r,j} = \exp(-V_r\Delta\tau_j/kT),$$

where relation (A2a) has been used to compute the ratio of the time constants. Combining this expression with (A9) gives:

$$\dot{\gamma}_{i,j+1}/\dot{\gamma}_{i,j} = \exp\big[(V_r - V)\Delta\tau_j/kT\big], \tag{A11}$$

which provides a recurrent relation between $\dot{\gamma}_{in}$ and $\dot{\gamma}_i$ (for a series of n relaxations)

$$\dot{\gamma}_{i,n}/\dot{\gamma}_{i,1} = \exp\left[(V_r - V)\sum_{1}^{n-1}(\Delta\tau_j/kT)\right].$$

The ratio can be replaced by $c_{r,1}/c_{r,n}$ according to (A3a). The ratio of time constants is evaluated via (A2a) for $t = \Delta t$, which yields:

$$\dot{\gamma}_{i,n}/\dot{\gamma}_{i,1} = \exp\big[(V_r\Delta\tau_n/kT) - 1\big]/\exp\big[(V_r\Delta\tau_n/kT) - 1\big].$$

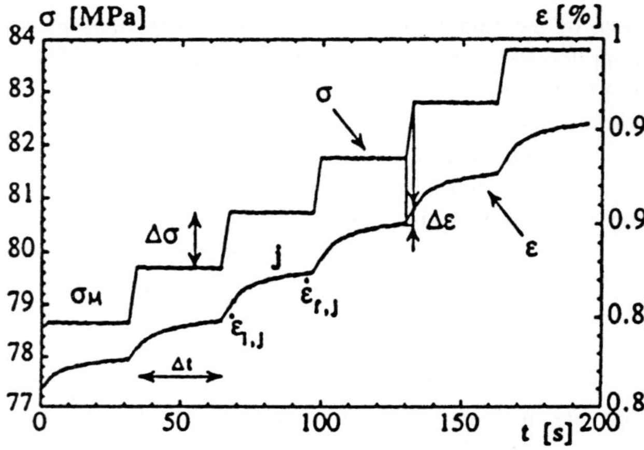

Fig. A2. Example of successive creep experiment in a $Ni_{75}Al_{25}$ polycrystal at 300 K [24]. The imposed stress σ and the measured strain ε are represented as functions of time. σ_M is the transition stress between constant strain-rate and the transient, $\Delta\sigma$ and $\Delta\varepsilon$ respectively the stress and strain increments between each creep period, j the creep number in the series, Δt the duration.

Eliminating $\dot{\gamma}_{i,n}/\dot{\gamma}_i$, between the two preceding relations and replacing V by $V_r\Omega_r$ (relation (A8a)) gives:

$$\Omega_r^{-1} = 1 - \left(kT\Big/\left(V_r\sum_1^{n-1}\Delta\tau_j\right)\right)\ln\left[\exp(-V_r\Delta\tau_n/kT) - 1\right]/\left[\exp(-V_r\Delta\tau_1/kT) - 1\right]. \tag{A12}$$

Relation (A12) allows one to determine Ω_r as a function of experimental data such as V_r measured along the first relaxation and the $\Delta\tau_j$'s. Then V is obtained via (A8a).

A creep experiment, similar to the repeated relaxation test, was proposed [172] and applied successfully to γTiAl polycrystals at 300 K [160]. However, because of very low creep strains, it could not be used in Cu or Ni_3Al single crystals. Therefore the repeated creep test is now performed by the following procedure (see fig. A2): As the constant strain-rate test is interrupted, the specimen is allowed to creep during Δt (total creep strain $\Delta\gamma_1$). It is then quasi-elastically reloaded by a small amount $\Delta\tau$, then allowed to creep during Δt, under stress $\tau_M + \Delta\tau$ (strain $\Delta\gamma_2$), etc. The interpretation of the specimen behaviour is very close to the one used above for successive relaxations. Similarly, three methods are available to determine V. In the present case, a relation equivalent to (A9) is:

$$\dot{\gamma}_{i,j+1}/\dot{\gamma}_{f,j} = v_{i,j+1}/v_{f,j} = \exp(V\Delta\tau/kT), \tag{A13}$$

where $\Delta\gamma$ is the strain increment which corresponds to $\Delta\tau$ during reloading. Relation (A13) indicates in particular that the strain-rate or velocity ratio in question is independent of j, if the dislocation mobility mechanism (i.e. V in relation (A5)) is the same.

Relation (A13) provides a first method of determining V, by measuring the $\dot{\gamma}$ values as the slope of the creep curves, and knowing $\Delta\tau$. A second method provides a relation equivalent to (A10) following the same procedure as for relaxations, $\Delta\tau_j$ being replaced by $\Delta\tau$ in the case of creep:

$$V = (kT/\Delta\tau)\ln\left(\frac{c_{c,j}+\Delta t}{c_{c,j+1}}\cdot\frac{V_{c,j}}{V_{c,j+1}}\right). \tag{A14}$$

(A14) allows one to compute V using the fitted parameters of (A2) along two successive creep tests in the series (usually the first and second ones).

The comparison between two successive creep tests can be pursued to estimate Ω'_c (relation (A8b)). As above, the ratio $\dot{\gamma}_{i,j+1}/\dot{\gamma}_{i,j}$ is evaluated, using two different procedures, which yields:

$$\dot{\gamma}_{i,j+1}/\dot{\gamma}_i = \exp\big((MV_{c,j}/kT)(\Delta\gamma/\Omega'_{c,j} - \Delta\gamma_j)\big),$$

which is equivalent to (A11).

The second way of computing the above strain-rate ratio consists in using (A3b):

$$\dot{\gamma}_{i,j+1}/\dot{\gamma}_{i,j} = V_{c,j}\cdot c_{c,j}/(V_{c,j+1}\cdot c_{c,j+1}).$$

The ratio of the time constants is calculated by considering relation (A2b) for $t=\Delta t$ for tests nos. j and $j+1$ respectively, whence:

$$\dot{\gamma}_{i,j+1}/\dot{\gamma}_{i,j} = (V_{c,j}/V_{c,j+1})\big[\exp(MV_{c,j+1}\Delta\gamma_{j+1}/kT) - 1\big] \\ /\big[\exp(MV_{c,j}\Delta\gamma_j/kT) - 1\big].$$

By eliminating $\dot{\gamma}_{i,j+1}/\dot{\gamma}_{i,j}$ between this last relation and the first one, an expression is obtained which contains $\Omega'_{c,j}$ and various parameters such as the V_c's and the $\Delta\gamma$'s, measured along two successive tests respectively nos. j and $j+1$ (in practice nos. 1 and 2):

$$(V_{c,j}/V_{c,j+1})\big[\exp(MV_{c,j+1}\Delta\gamma_{j+1}/kT) - 1\big]/\big[\exp(MV_{c,j}\Delta\gamma_j/kT - 1\big] \\ = \exp\big((MV_{c,j}/kT)(\Delta\gamma/\Omega'_{c,j} - \Delta\gamma_j)\big).$$

Although this expression is complicated, the V_c's and the $\Delta\gamma$'s are determined rather safely, so that this latter method has been used most of the time. $\Omega'_{c,j}$ being known as well as $V_{c,j}$, V is computed by relation (A8b).

Assessments of the above transient description

A number of experimental facts support the assumptions made for the above transient interpretation. First, the continuity of the plastic strain rate at the end of the monotonic test and at the onset of the transient is expected [165]. Indeed, the average mobile dislocation density and velocity should not change abruptly at the transition, as a rule. Various experiments performed on $Ni_{76}Al_{24}$ and $Ni_{75}Al_{25}$ polycrystals at 300 K and two applied strain rates provide data which confirm this prediction [24,165,173]. Second, in a different type of experiment, the activation volume V of the dislocation velocity is measured along a stress–strain curve using alternately creep and relaxation series. A unique curve of V decreasing monotonically as a function of strain is obtained (see, e.g., [163] for a $Ni_{75}Al_{25}$ polycrystal tested at 300 K). This can be interpreted as the same mobility dislocation mechanisms operating during constant strain-rate deformation, independently of the transient technique used to characterize them.

Therefore, the relaxation and creep transients just described allow one to safely determine the microscopic volume V of dislocation velocity. In addition they also yield useful information about dislocation exhaustion rates $\Delta\rho_{\mathrm{m}}/\rho_{\mathrm{mo}}$. $\Delta\rho_{\mathrm{m}}$ is the decrease of the mobile dislocation density which corresponds to a decrease of the initial deformation rate by a factor q (q is an integer) or a transient duration equal to $(q-1)c$ – relation (A3a). This exhaustion rate is estimated as follows in the case of relaxation. A combination of (A6) and (A7) gives $\rho_{\mathrm{m}}/\rho_{\mathrm{mo}}$ as a function of $\Delta\tau^*$ or $\Delta\tau$ (using (A4a)):

$$\rho_{\mathrm{m}}/\rho_{\mathrm{mo}} = \exp\bigl(\beta_{\mathrm{m}} V \Delta\tau^*/kT\bigr) = \exp\bigl[\beta_{\mathrm{r}} V (1 + K_{\mathrm{r}}/M)\Delta\tau/kT\bigr].$$

$\rho_{\mathrm{m}}/\rho_{\mathrm{mo}}$ can be expressed as a function of time by eliminating $\Delta\tau$ between this latter relation and (A2a):

$$\rho_{\mathrm{m}}/\rho_{\mathrm{mo}} = (1 + t/c_{\mathrm{r}})^{-\beta_{\mathrm{r}}/(1+\beta_{\mathrm{r}})},$$

where relation (A8a) has been used. For the transient duration in question, the mobile dislocation exhaustion parameter is found to be:

$$\Delta\rho_{\mathrm{m}}/\rho_{\mathrm{mo}} = 1 - (1/q)^{\beta_{\mathrm{r}}/(1+\beta_{\mathrm{r}})}. \tag{A15}$$

Ω_{r} being known from repeated relaxations with relation (A12), β can be obtained from (A8a) under two extreme assumptions: K_{r} is close to θ, the work-hardening coefficient along the stress–strain curve, and $K_{\mathrm{r}} = 0$. This accounts for the error bar in the values of $\Delta\rho_{\mathrm{m}}/\rho_{\mathrm{mo}}$. An exhaustion parameter can also be obtained using successive creep experiments [24].

References

[1] F.R.N. Nabarro and M.S. Duesbery, Dislocations in Solids 10 (Elsevier, 1996).
[2] F.R.N. Nabarro, Acta Metall. 37 (1989) 1521.
[3] K. Aoki and O. Izumi, Trans. Jpn Inst. Metals 19 (1978) 145.
[4] M.S. Kim, S. Hanada, S. Watanabe and O. Izumi, Trans. Jpn Inst. Metals 29 (1988) 790.
[5] M.S. Kim, S. Hanada, S. Watanabe and O. Izumi, Acta Metall. 36 (1988) 2615.
[6] S. Miura, S. Ochia, Y. Oya, Y. Mishima and T. Suzuki, in: High Temperature Ordered Intermetallic Alloys III 133, eds C.T. Liu, A.I. Taub, N.S. Stoloff and C.C. Koch (Materials Research Society, 1989) p. 341.
[7] N.L. Baluc, PhD thesis (Ecole Polytechnique Fédérale Lausanne, 1990).

[8] A.E. Staton-Bevan, Phil. Mag. A 47 (1983) 939.
[9] M. Masahiko and T. Hirano, Phil. Mag. Lett. 75 (1997) 143.
[10] P.H. Thornton, R.G. Davies and T.L. Johnston, Metall. Trans. 1 (1970) 207.
[11] S.M. Copley and B.H. Kear, Trans. Metall. Soc. AIME 239 (1967) 977.
[12] C. Bontemps-Neveu, PhD thesis (Université de Paris-Sud Orsay, Paris, 1991).
[13] P. Spätig, J. Bonneville and J.-L. Martin, in: Proc. of the 4th European Conf. on Advanced Materials and Processes (Euromat, 1995) p. 97.
[14] P. Spätig, PhD thesis (Ecole Polytechnique Fédérale de Lausanne, 1995).
[15] P. Spätig, J. Bonneville and J.L. Martin, in: High Temperature Ordered Intermetallic Alloys, eds I. Baker, R. Dariola, J.D. Whitttenberger and M.H. Yoo (Material, Research Society, Pittsburgh, 1995) p. 713.
[16] J. Bonneville, J.L. Martin, P. Spätig, B. Viguier and B. Matterstock, in: High Temperature Ordered Intermetallic Alloys VII 460, eds C.C. Koch, C.T. Liu, N.S. Stoloff and A. Wanner (Materials Research Society, Pittsburgh, 1997) p. 419.
[17] A. Couret, Y.Q. Sun and P.B. Hirsch, Phil. Mag. A 67 (1993) 29.
[18] S.S. Ezz, Acta Mater. 44 (1996) 4395.
[19] S.S. Ezz and P.B. Hirsch, Phil. Mag. A 69 (1994) 105.
[20] G. Saada and P. Veyssière, in: Int. Symp. on Struct. Intermetallics (Champion, USA, 1993) p. 379.
[21] G.T. Gray and J.D. Embury, in: High Temperature Ordered Intermetallic Alloys V 288, eds I. Baker, R. Dariola, J.D. Whitttenberger and M.H. Yoo (Materials Research Society, Pittsburgh, 1993) p. 585.
[22] P. Veyssière, Phil. Mag. A 50 (1984) 189.
[23] P. Veyssière and G. Saada, in: Dislocations in Solids, Vol. 10, eds F.R.N. Nabarro and M.S. Duesbery (North-Holland, Amstedam, 1996) p. 253.
[24] B. Lo Piccolo, PhD thesis (Ecole Polytechnique Fédérale Lausanne, 1999).
[25] E. Kuramoto and D.P. Pope, Phil. Mag. 34 (1976) 593.
[26] B.H. Kear and H.G.F. Wilsdorf, Trans. Met. Soc. AIME 224 (1962) 382.
[27] A.E. Vidoz and L.M. Brown, Phil. Mag. 7 (1962) 1167.
[28] S.B. Chakrabortty and E.A. Starke, Acta Metall. 23 (1975) 63.
[29] R.G. Davies and N.S. Stoloff, Phil. Mag. 12 (1965) 297.
[30] S.M. Sastry and B. Ramaswami, Phil. Mag. 32 (1975) 801.
[31] T. Kawabata, T. Kanai and O. Izumi, Acta Metall. 33 (1985) 1355.
[32] T. Kawabata and O. Izumi, in: High Temperature Aluminides and Intermetallics, eds S.H. Whang, C.T. Liu, D.P. Pope and J.O. Stiegler (The Minerals, Metals & Materials Society, 1990) p. 403.
[33] H. Inui, M. Matsumuro, D.H. Wu and M. Yamaguchi, Phil. Mag. A 75 (1997) 395.
[34] M.A. Stucke, D.M. Dimiduk and P.M. Hazzledine, in: High Temperature Ordered Intermetallic Alloys V 288, eds I. Baker, R. Dariola, J.D. Whitttenberger and M.H. Yoo (Materials Research Society, Pittsburgh, 1993) p. 471.
[35] F. Grégori, P. Penhoud and P. Veyssière, Phil. Mag. A (2000) in press.
[36] S. Jiao, N. Bird, P.B. Hirsch and G. Taylor, Phil. Mag. A 78 (1998) 777.
[37] S. Sriram, V.K. Vasudevan and D.M. Dimiduk, Mat. Sci. Eng. A 192/193 (1995) 217.
[38] B. Viguier, J. Bonneville and J.L. Martin, Acta Metall. 44 (1996) 4403.
[39] J.J. Gilman and W.G. Johnston, J. Appl. Phys. 31 (1960) 687.
[40] F. Louchet, Phil. Mag. A 72 (1995) 905.
[41] E.M. Nadgorny and Y.L. Iunin, in: High Temperature Ordered Intermetallic Alloys VI 364, eds J.A. Horton, I. Baker, S. Hanada, R.D. Noebe and D.S. Schwartz (Materials Research Society, Pittsburgh, 1995) p. 707.
[42] C.B. Jiang, S. Patu, Q.Z. Lei and C.X. Shi, Phil. Mag. Lett. 78 (1998) 1.
[43] D.M. Dimiduk, PhD thesis (Carnegie Mellon University, 1989).
[44] A. Korner, Phil. Mag. A 63 (1991) 407.
[45] S.S. Ezz, D.P. Pope and V. Paidar, Acta Metall. 30 (1982) 921.
[46] J. Bonneville and J.L. Martin, in: Multiscale Phenomena in Plasticity Nato Science Series, Vol. 367, eds J. Lepinoux, D. Mazière, V. Pontikis and G. Saada (Kluwer Academic Publishers, London, 2000) p. 57.
[47] T. Kruml, B. Viguier, J. Bonneville, P. Spätig and J.L. Martin, in: High Temperature Ordered Intermetallic Alloys VII 460, eds C.C. Koch, C.T. Liu, N.S. Stoloff and A. Wanner (Materials Research Society, Pittsburgh, 1997) p. 529.

[48] J.L. Martin, B. Matterstock, P. Spätig and J. Bonneville, in: Proceedings of the 20th Risø Int. Symp. on Mater. Sci., eds J.B. Bilde-Sørensen, J.V. Cartensen, N. Hansen, D.J. Jensen, T. Leffers, W. Pantleon, O.B. Pedersen and G. Winther (Risø Nat. Lab., Roskilde, Denmark, 1999) p. 103.
[49] R.A. Mulford and D.P. Pope, Acta Metall. 21 (1973) 1375.
[50] K.J. Hemker, M.J. Mills, K.R. Forbes, D.D. Sternbergh and W.D. Nix, in: Strength of Metals and Alloys, Vol. 1, eds D.G. Brandon, R. Chaim and A. Rosen (Freund Publishing, London, 1991) p. 271.
[51] K.J. Hemker, M.J. Mills and W.D. Nix, J. Material Res. 7 (1992) 2059.
[52] P. Spätig, J. Bonneville and J.L. Martin, Mater. Sci. Eng. A 167 (1993) 73.
[53] B. Escaig, J. Physique 35 C7 (1974) 151.
[54] J.L. Farvacque, J. Crampon, J.C. Doukhan and B. Escaig, Phys. Stat. Sol. 14 (1972) 623.
[55] J.D. Meakin, Canad. J. Phys. 45 (1967) 1124.
[56] T. Takasugi, S. Watanabp, D. Izumi and N.K. Fat-Halla, Acta. Metall. 37 (1989) 3425.
[57] Y. Umakoshi, D.P. Pope and V. Vitek, Acta Metall. 32 (1984) 449.
[58] J. Bonneville and J.L. Martin, in: High Temperature Orded Intermetallic Alloys IV 213, eds L.A. Johnson, D.P. Pope and J.O. Stiegler (Materials Research Society, Pittsburgh, 1991) p. 629.
[59] B. Viguier, K. Hemker, J. Bonneville, F. Louchet and J.L. Martin, Phil. Mag. A 71 (1995) 1295.
[60] B. Viguier, M. Cieslar, H.J. Hemker and J.L. Martin, in: High Temperature Ordered Intermetallic Alloys VI 364, eds J.A. Horton, I. Baker, S. Hanada, R.D. Noebe and D.S. Schwartz (Materials Research Society, Pittsburgh, 1995) p. 653.
[61] M. Yamaguchi, Y. Umakoshi and T. Yamane, in: Dislocations in Solids, eds H. Suzuki, T. Ninomiya, K. Sumino and S. Takuchi (University of Tokyo Press, 1985) p. 77.
[62] B. Viguier and K.J. Hemker, Phil. Mag. A 73 (1996) 575.
[63] J. Friedel, in: Internal Stresses and Fatigue in Metals (Elsevier, Amsterdam, 1959) p. 220.
[64] M.P. Biget and G. Saada, Phil. Mag. A 59 (1989) 747.
[65] A. Couret and D. Caillard, Acta Metall. 33 (1985) 1455.
[66] F. Louchet and B. Viguier, Phil. Mag. A 71 (1995) 1313.
[67] T. Kruml, J.L. Martin and J. Bonneville, Phil. Mag. A 80 (2000) 1545.
[68] B. Cheng, E. Carreno-Morelli, N. Baluc, J. Bonneville and R. Schaller, Phil. Mag. A 79 (1999) 2227.
[69] C. Charbonnier, T. Kruml and J.L. Martin, MRS Fall Meeting (2000) accepted. To appear in the Proceedings 2001.
[70] O. Couteau, C. Charbonnier, T. Kruml and J.L. Martin, MRS Fall Meeting (2000) accepted. To appear in the Proceedings 2001.
[71] J. Bonneville, T. Kruml, J.L. Martin, B. Matterstock, B. Viguier and P. Spätig, Mater. Sci. Eng. A 234–236 (1997) 770.
[72] B. Matterstock, J.L. Martin, J. Bonneville, T. Kruml and P. Spätig, Mater. Sci. Eng. A 239–240 (1997) 169.
[73] B. Viguier, PhD thesis (Ecole Polytechnique Fédérale Lausanne, 1995).
[74] P. Gadaud and K. Chakib, Mater. Sci. Forum 119–121 (1993) 397.
[75] A. Mourisco, N. Baluc, J. Bonneville and R. Schaller, Mater. Sci. Eng. A 239–240 (1997) 281.
[76] H. Numakura, N. Kurita, M. Koiwa and P. Gadaux, Phil. Mag. A 79 (1999) 943.
[77] E. Carreno-Morelli, B.L. Cheng, M. Demura, R. Schaller and N. Baluc, Mat. Res. Symp. 578 (2000) in press.
[78] B. Cheng, E. Carreno-Morelli, N. Baluc, J. Bonneville and R. Schaller, in: Mater. Res. Soc. Proc. 552, eds E.P. George, M.J. Mills and M. Yamaguchi (1998) p. KK5.38.1.
[79] T. Kruml, V. Paidar and J.L. Martin, Intermetallics 8 (2000) 729.
[80] B. Matterstock, J.L. Martin, J. Bonneville and T. Kruml, in: High Temperature Ordered Intermetallic Alloys VIII 552, eds E.P. George, M.J. Mills and M. Yamaguchi (Materials Research Society, Pittsburgh, 1999) p. KK5.17.1.
[81] S.S. Ezz, Y.Q. Sun and P.B. Hirsch, Mater. Sci. Eng. A 192/193 (1995) 45.
[82] G.B. Viswanathan, V.K. Vasudevan and M.J. Mills, Acta Metall. 47 (1999) 1399.
[83] B.A. Greenberg and M.A. Ivanov, The Physics of Metals and Metallography 79 (1995) 373.
[84] F. Louchet, J. Phys. Paris III 5 (1995) 1803.
[85] G.I. Taylor, Proc. Roy. Soc. A 145 (1934) 362.

[86] A. Orlova, Acta Metall. Mater. 39 (1991) 2805.
[87] W. Püschl, Phil. Mag. Lett. 80 (2000) 199.
[88] I. Hesse, Phys. Stat. Solid. 21 (1967) 495.
[89] F.F. Lavrentev, Mater. Sci. Eng. 46 (1980) 191.
[90] B.L. Cheng, E. Carreno-Morelli, N. Baluc, J. Bonneville and R. Schaller, Phil. Mag. A 79 (1999) 2227.
[91] B.M. Grabner, H.P. Karnthaler and C. Rentenberger, in: Proc of EUREM-11, Dublin 1996 (CESM, European Society of Microscopy, Brussels, 1998) p. 357.
[92] J.E. Bailey and P.B. Hirsch, Phil. Mag. 5 (1960) 485.
[93] H. Mecking and U.F. Kocks, Acta Metall. 29 (1981) 1865.
[94] J.E. Bailey, Phil. Mag. 8 (1963) 223.
[95] A.H. Cottrell and R.J. Stokes, Proc. Roy. Soc. A 233 (1955) 17.
[96] A. Seeger, J. Diehl, S. Mader and H. Rebstock, Phil. Mag. 2 (1957) 323.
[97] F.R.N. Nabarro, Acta Metall. Mater. 38 (1990) 161.
[98] M. Lu, PhD thesis (The Johns Hopkins University, Baltimore, 1997).
[99] M. Lu and K.J. Hemker, Phil. Mag. A 77 (1997) 325.
[100] R. Mahapatra, Y.T. Chou and D.P. Pope, Mater. Sci. Eng. A 239–240 (1997) 97.
[101] S. Jiao, N. Bird, P.B. Hirsch and G. Taylor, in: High Temperature Ordered Intermetallic Alloys VIII 552, eds E.P. George, M.J. Mills and M. Yamaguchi (Materials Research Society, Pittsburgh, 1999) p. KK8.11.1.
[102] S.S. Ezz and P.B. Hirsch, Phil. Mag. A 72 (1995) 383.
[103] X. Shi, G. Saada and P. Veyssière, Phil. Mag. A 73 (1996) 1419.
[104] E. Kuramoto and D.P. Pope, Phil. Mag. 33 (1976) 675.
[105] C. Bontemps, P. Veyssière and G. Saada, Scr. Metall. Mater. 24 (1990) 2385.
[106] H.F. Heredia, PhD thesis (University of Pennsylvania, 1990).
[107] S.S. Ezz, D.P. Pope and V. Vitek, Acta Metall. 35 (1987) 1879.
[108] S.S. Ezz and P.B. Hirsch, Phil. Mag. A 73 (1996) 1069.
[109] B.A. Greenberg and M.A. Ivanov, Mater. Sci. Eng. A 239–240 (1997) 813.
[110] W.E. Dowling and R. Gibala, in: High-Temperature Ordered Intermetallic Alloys III 133, eds C.T. Liu, A.I. Taub, N.S. Stoloff and C.C. Koch (Materials Research Society, 1989) p. 209.
[111] R.G. Davies and N.S. Stoloff, Trans. Met. Soc. AIME 233 (1965) 714.
[112] D.M. Dimiduk, J. Physique III 1 (1991) 1025.
[113] M.H. Yoo and C.T. Liu, J. Mater. Res. 3 (1988) 845.
[114] D.M. Dimiduk and T.A. Parthasarathy, Phil. Mag. Lett. 71 (1995) 21.
[115] B.A. Greenberg and M.A. Ivanov, in: Gamma Titanium Aluminides, eds Y.W. Kim, R. Wagner and M. Yamaguchi (The Minerals, Metals and Materials Society, 1995) p. 299.
[116] B.A. Greenberg, M.A. Ivanov, T.O. Barabash and A.G. Blokhin, The Physics of Metals and Metallography 81 (1996) 374.
[117] B.A. Greenberg, M.A. Ivanov, T.O. Barabash and A.G. Blokhin, The Physics of Metals and Metallography 81 (1996) 381.
[118] B. Devincre, P. Veyssière and G. Saada, Phil. Mag. A 79 (1999) 1609.
[119] D. Caillard and G. Molenat, in: Deformation Induced Microstructures: Analysis and Relation to Properties, 20th Risø Int. Symp. on Mater. Sci., eds J.V.C.J.B. Bilde-Sørensen, N. Hansen, D. Juul Jensen, T. Leffers, W. Pantleon, O.B. Pedersen and G. Winther (Risø Nat. Lab., Roskilde, Denmark, 1999) p. 1.
[120] D. Caillard, Mater. Sci. Eng. A (2001) submitted.
[121] G. Molenat and D. Caillard, Phil. Mag. A 69 (1994) 939.
[122] K. Suzuki, M. Ichiara and S. Takeuchi, Acta. Metall. 27 (1979) 193.
[123] D. Caillard and A. Couret, in: Dislocations in Solids, Vol. 10, eds F.R.N. Nabarro and M.S. Duesbery (North-Holland, Amstedam, 1996) p. 69.
[124] S. Takeuchi and E. Kuramoto, Acta Metall. 21 (1973) 415.
[125] R. Schaüblin and P. Stadelmann, Mater. Sci. Eng. A 164 (1993) 373.
[126] K.J. Hemker, Phil. Mag. A 76 (1997) 241.
[127] N. Baluc, R. Schaüblin and K. Hemker, Phil. Mag. Lett. 64 (1991) 327.
[128] N.L. Baluc, H.P. Karnthaler and M.J. Mills, Phil. Mag. A 64 (1991) 137.

[129] G. Schoeck, Phil. Mag. Lett. 75 (1997) 7.
[130] K.J. Hemker and M.J. Mills, Phil. Mag. A 68 (1993) 305.
[131] D.M. Dimiduk, A.W. Thompson and J.C. Williams, Phil. Mag. A 67 (1993) 675.
[132] T. Kruml, J.L. Martin, B. Viguier, J. Bonneville and P. Spätig, Mater. Sci. Eng. A 239–240 (1997) 174.
[133] B. Matterstock, E. Conforto, T. Kruml, J. Bonneville and J.L. Martin, in: High Temperature Ordered Intermetallic Alloys VIII 552, eds E.P. George, M.J. Mills and M. Yamaguchi (Materials Research Society, 1999) p. KK10.3.1.
[134] F.E. Heredia and D.P. Pope, J. Phys. 1 (1991) 1055.
[135] N. Baluc and R. Schaüblin, Phil. Mag. A 74 (1996) 113.
[136] H.P. Karnthaler, E.T. Mühlbacher and C. Reutenberger, Acta Mater. 44 (1996) 547.
[137] A. Korner, Phil. Mag. Lett. 60 (1989) 103.
[138] Y.Q. Sun, P.M. Hazzeldine, M.A. Crimp and A. Couret, Phil. Mag. A 64 (1991) 371.
[139] G. Molénat and D. Caillard, Phil. Mag. A 65 (1992) 1327.
[140] J.A. Lopez and G.F. Hancock, Phys. Stat. Sol. A 2 (1970) 469.
[141] O. Noguchi, Y. Oya and T. Suzuki, Metall. Trans. A 12 (1981) 1647.
[142] J. Douin, P. Veyssière and P. Beauchamp, Phil. Mag. A 54 (1986) 375.
[143] J. Bonneville, N. Baluc and J.L. Martin, in: 6th International Symposium on Intermetallic Compounds (JIMIS-6), ed. O. Izumi (The Japan Institute of Metals, 1991) p. 323.
[144] X. Shi, G. Saada and P. Veyssière, Phil. Mag. A 73 (1995) 1159.
[145] G. Molénat and D. Caillard, Phil. Mag. A 64 (1991) 1291.
[146] G. Saada and P. Veyssière, Phil. Mag. A 66 (1992) 1081.
[147] C.T. Chou and P.B. Hirsch, Phil. Mag. A 68 (1993) 1097.
[148] D. Caillard and V. Paidar, Acta Mater. 44 (1996) 2759.
[149] K.J. Hemker, B. Viguier and M.J. Mills, Mater. Sci. Eng. A 164 (1993) 391.
[150] J.P. Simmons, M.J. Mills and S.I. Rao, in: High Temperature Ordered Intermetallic Alloys VI 364, eds J.A. Horton, I. Baker, S. Hanada, R.D. Noebe and D.S. Schwartz (Materials Research Society, Pittsburgh, 1995) p. 137.
[151] S. Sriram, D.M. Dimiduk, P. Hazzledine and V.K. Vasudevan, Phil. Mag. A 76 (1997) 965.
[152] F. Gregori and P. Veyssière, in: Gamma Titanium Aluminides, eds Y.W. Kim, D.M. Dimiduk and M.H. Loretto (The Minerals, Metals and Materials Society, 1999) p. 75.
[153] S. Zghal, A. Menand and A. Couret, Acta Mater. 46 (1998) 5899.
[154] U. Messerchmidt, M. Bartsch, M. Haussler, M. Aindow, R. Hattenhauer and I. Jones, in: High Temperature Ordered Intermetallic Alloys VI 364, eds J.A. Horton, I. Baker, S. Hanada, R.D. Noebe and D.S. Schwartz (Materials Research Society, Pittsburgh, 1995) p. 47.
[155] A. Couret, Phil. Mag. A 79 (1999) 1977.
[156] F. Louchet and B. Viguier, Scripta Metall. 31 (1994) 369.
[157] F. Louchet and B. Viguier, Phil. Mag. A 80 (2000) 765.
[158] F. Louchet and B. Viguier, in: High Temperature Ordered Intermetallic Alloys VI 364, eds J.A. Horton, I. Baker, S. Hanada, R.D. Noebe and D.S. Schwartz (Materials Research Society, Pittsburgh, 1995) p. 163.
[159] F. Louchet, Phil. Mag. A 77 (1998) 761.
[160] J. Bonneville, B. Viguier and P. Spätig, Scripta Mater. 36 (1997) 275.
[161] V.I. Dotsenko, Phys. Stat. Sol. B 93 (1979) 11.
[162] B. Lo Piccolo, J.L. Martin and J. Bonneville, in: Proceedings of COM 2000 (Ottawa, 2000) under print.
[163] J.L. Martin, B. Lo Piccolo, T. Kruml and J. Bonneville, Mater. Sci. Eng. (2001) accepted for publication.
[164] F. Guiu and P.L. Pratt, Phys. Stat. Sol. 6 (1964) 111.
[165] G. Saada, J. Bonneville and P. Spätig, Mater. Sci. Eng. A 234–236 (1997) 263.
[166] B. Lo Piccolo, P. Spätig, T. Kruml, J.L. Martin and J. Bonneville, Mater. Sci. Eng. A (2001) accepted for publication.
[167] J. Bonneville, P. Spätig and J.L. Martin, in: High Temperature Ordered Intermetallic Alloys VI 364, eds J.A. Horton, I. Baker, S. Hanada, R.D. Noebe and D.S. Schwartz (Materials Research Society, Pittsburgh, 1995) p. 369.
[168] R.W. Rohde and T.V. Nordstrom, Mater. Sci. Eng. 12 (1973) 179.
[169] G. Sargent, G. Jones and H. Conrad, Scripta Metall. 3 (1969) 481.

[170] L.P. Kubin, Phil. Mag. 30 (1974) 705.
[171] P. François, D. Melot, J.M. Lefebvre and B. Escaig, J. Mater. Sci. 27 (1992) 2173.
[172] A. Orlova, J. Bonneville and P. Spätig, Mater. Sci. Eng. A 191 (1995) 85.
[173] B. Matterstock, G. Saada, J. Bonneville and J.L. Martin, in: High Temperature Ordered Intermetallic Alloys VIII 552, eds E.P. George, M.J. Mills and M. Yamaguchi (Materials Research Society, Pittsburgh, 1999) p. KK9.12.1.
[174] S.S. Ezz, Y.Q. Sun and P.B. Hirsch, in: High Temperature Ordered Intermetallic Alloys VI 364, eds J.A. Horton, I. Baker, S. Hanada, R.D. Noebe and D.S. Schwartz (Materials Research Society, Pittsburgh, 1995) p. 695.

CHAPTER 63

Dislocations and High-Temperature Plastic Deformation of Superalloy Single Crystals

T.M. Pollock

Department of Materials Science and Engineering, The University of Michigan
2300 Hayward, Ann Arbor
MI 48109, USA

and

R.D. Field

Los Alamos National Laboratory
MS G770, Los Alamos
NM 87545, USA

Dislocations in Solids
Edited by F. R. N. Nabarro and M. S. Duesbery

Contents

1. Introduction

Since the pioneering papers of Taylor and Orowan [1,2], innumerable studies on the physics of plastic deformation and mechanical behavior of metallic materials have been conducted. These studies, many of which have been reviewed and discussed in past volumes of the present series, have focused to a large degree on plastic deformation in single phase systems, and in many cases on high purity materials. The present paper concentrates on a number of aspects of plasticity in two-phase nickel-base superalloys. In their single crystal form, superalloys are composed of a continuous disordered fcc nickel matrix reinforced with ordered Ni_3Al ($L1_2$) precipitates. In many respects, these materials serve as "model" two-phase systems. First, they are easily grown in single crystal form, so deformation can be studied quantitatively in terms of the resolved shear stresses acting on particular slip systems. Second, a high degree of control can be exerted on the morphological development of the two-phase microstructure, through adjustments in the compositions of the constituent phases, permitting various aspects of strengthening to be studied.

Nickel-base superalloys are also commercially important materials, due to their application in numerous components of aircraft engines, including turbine blades where they are utilized in single crystal form. Thus, a great deal of research has been conducted on fairly complex alloys, and some of the deformation phenomena to be discussed are only observed in these multicomponent alloys. For this reason, a brief background on the constitution and microstructure of nickel-base superalloys, including commercial compositions, is given. To quantify deformation phenomena to the greatest degree possible, the focus here is on deformation of single-crystal superalloys, rather than on polycrystalline forms of these materials. Of greatest interest is deformation at temperatures above half of the melting temperature, where a wide variety of dislocation glide, climb and precipitate shearing phenomena are operative. To highlight the constraints imposed by the presence of a high volume fraction of ordered precipitates, the high-temperature deformation characteristics of the individual phases in their bulk form will first be reviewed. Following this, deformation in the two-phase materials will be discussed and contrasted with the deformation processes that occur in the corresponding bulk, single phase materials. Aspects of the two-phase deformation problem that will be highlighted include the influence of misfit stresses, the difficulty and inhomogeneity of deformation in highly constrained, small volumes of material, the development of interfacial dislocation networks, precipitate shearing and plasticity-induced microstructural instabilities.

2. Constitution and microstructure of nickel-base superalloys

Ni-base superalloys consist of two major phases: an f.c.c. matrix (γ) and an $L1_2$ precipitate (γ'). A comparison of the crystal structures of the two phases is shown in fig. 1. As can

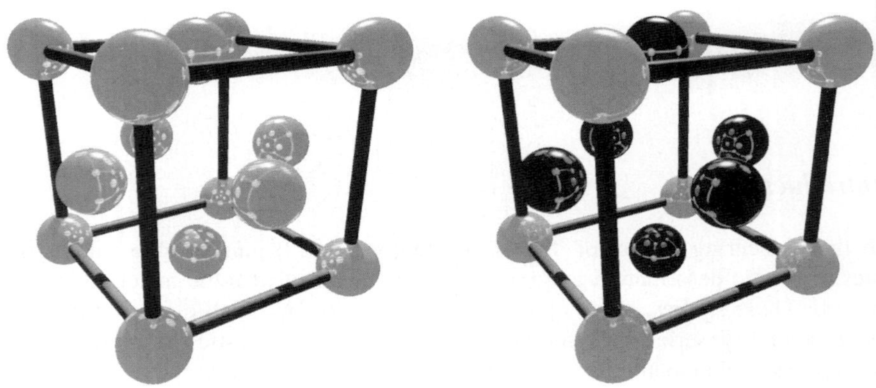

(a) (b)

Fig. 1. The f.c.c. γ matrix phase (a) and the $L1_2$ γ' precipitate (b) crystal structures.

be readily seen in this figure, the $L1_2$ phase is an ordered variation of the f.c.c. structure, based on Ni_3Al, with Al atoms in the cube corners and Ni atoms in the face-centered sites. This ordering changes the lattice from face centered to primitive, so that the shortest lattice translation vector changes from $\frac{a}{2}\langle 110\rangle$ to $a\langle 100\rangle$. This has obvious consequences for the dislocation substructure and deformation modes, which will be discussed in more detail in section 4.

The lattice mismatch between the two phases is usually quite small ($< 1\%$). In the as-heat-treated condition, the phases are therefore fully coherent, with a cube-on-cube orientation relationship. Commercial single crystal alloys generally contain from 60 to 75% of the γ' phase in the as-heat-treated condition. These precipitates typically assume the form of cuboidal particles, 0.2 to 0.5 μm in cube edge length, with the sides parallel to {100} planes, as shown in fig. 2. The precipitates are often aligned along $\langle 100\rangle$ directions. The degree of alignment can be controlled to a certain extent through heat treatment and, as discussed in sections 4 and 5, this alignment is believed to be important during high-temperature creep, particularly with respect to directional coarsening.

The use of the term "single crystal" in the context of these materials has been questioned for two reasons. First, the microstructure actually consists of two phases: the γ matrix and γ' precipitates. As stated above, these phases are generally completely coherent in the as-heat-treated condition and, since the γ' phase is simply an ordered version of the matrix, atoms from both phases can be referred to a common lattice. Second, low-angle grain boundaries, with misorientation angles below 12°, are commonly present in commercial castings due to the growth of misoriented dendrites. Additionally, lower-angle boundaries with misorientations of less than 1° are often present in interdendritic regions. These low-angle boundaries are difficult to eliminate from the castings and do not result in significant reduction in properties. They do, however, contribute to the deformation process, serving as dislocation sources during the early stages of creep (see section 4). Thus, single-crystal Ni base alloys are best thought of as "engineering single crystals".

Polycrystalline (conventionally cast) nickel-base superalloys may contain several additional phases. Most prevalent of these are the carbides, consisting of primary MC

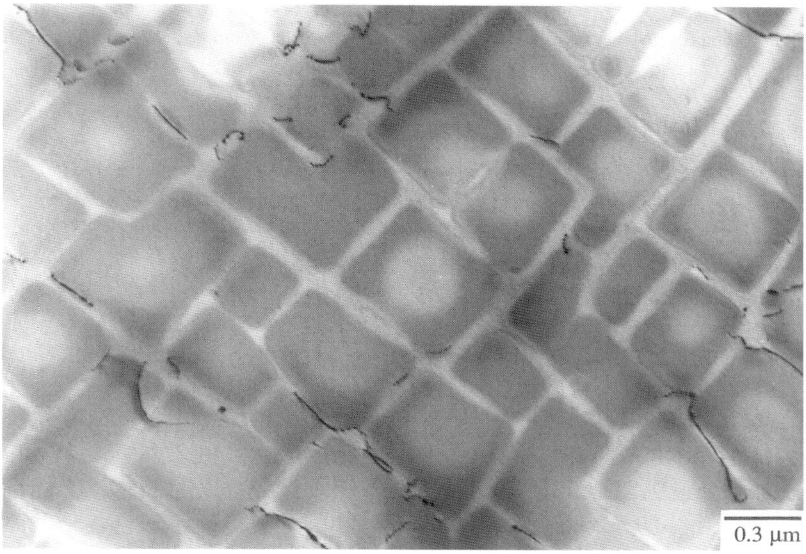

Fig. 2. TEM micrograph of superalloy single crystal CMSX-3 in the solution treated and aged condition, containing approximately 70 vol% cube-shaped, coherent γ' precipitates.

particles (where M represents a variety of metallic elements), which are stable up to the melting point of the alloy, and more complex carbides, such as M_6C and $M_{23}C_6$, that form during intermediate temperature heat treatment. The MC carbides contain γ' forming elements such as Ti, Ta, and Nb. These particles partially dissolve during heat treatment, in favor of the complex carbides, which are rich in Cr, W, and Mo. This release of γ' forming elements at the grain boundaries associated with the carbide reactions results in individual carbides surrounded by a sheath of γ', a microstructure which is thought to enhance the damage resistance of the grain boundary [3]. In addition, B additions to the alloys result in the formation of a small number of borides, primarily M_5B_3 and M_3B_2. The major function of these particles is to act as a source for B in the grain boundaries, the presence of which is also believed to contribute to grain boundary strength [4].

Single-crystal alloys have several advantages over conventionally cast alloys, or even directionally solidified (DS) alloys. The most obvious of these is the fact that the creep damage initiation sites in the microstructure (i.e. the grain boundaries) have been totally eliminated. In addition, the absence of grain boundaries precludes the necessity for compromises in alloy design required to strengthen grain boundaries, as discussed above. For example, B is a particularly potent grain boundary strengthener. However, it is also a strong melting point depressant in Ni base alloys, lowering significantly the incipient melting temperature of the alloy. As the volume fraction of γ' is increased, the solution temperature is naturally also increased. By raising the incipient melting point, the solution temperature can also be raised while maintaining an acceptable "heat treatment window" between these two temperatures. Thus, the reduction or elimination of B in the alloy allows increased volume fraction of γ' and therefore higher strength.

Table 1
Nominal compositions of several commercial Ni base single crystal alloys (compositions in wt%, with the balance of Ni)

Alloy	Co	Cr	Mo	W	Ta	Al	Ti	Re	Other
René N4 [5]	7.5	9.8	1.5	6.0	4.8	4.2	3.5	–	0.5 Nb, 0.15 Hf
PWA1480 [6]	5.0	10.0	–	4.0	12.0	5.0	1.5	–	–
CMSX-2 [7]	4.5	8.0	0.5	8.0	5.8	5.6	1.0	–	–
CMSX-3 [8]	4.6	8.0	0.5	8.0	5.8	5.6	1.0	–	0.1 Hf
NASAIR100 [9]	–	9.5	1.0	10.0	3.2	5.5	1.2	–	–
AM1 [10]	6.5	7.8	2.0	5.5	8.0	5.3	1.2	–	–
SRR99 [11]	5.0	8.5	–	9.5	2.8	5.6	2.2	–	–
SC16 [12]	–	16.4	2.8	0.1	3.6	3.5	3.5	–	–
AF56 [13]	8.0	12.0	2.0	4.0	5.0	3.4	4.2	–	–
René N5 [14]	7.5	7.0	1.5	5.0	6.5	6.2	–	3.0	0.15 Hf
PWA1484 [15]	10.0	5.0	2.0	6.0	8.7	5.6	–	3.0	0.1 Hf
CMSX-4 [16]	9.2	6.6	0.6	5.9	6.2	5.6	0.9	3.3	–
René N6 [17]	12.5	4.2	1.4	6.0	7.2	5.75	–	5.4	–
CMSX-10 [18]	3.0	2.0	0.4	5.0	8.0	5.7	0.2	6.0	0.1 Nb
Alloy SX-1 [19]	12.5	4.5	–	5.8	7.0	6.0	–	6.3	–

Table 2
Compositions of the constituent phases in several multicomponent single-crystal alloys (atomic %)

Alloy	Co	Cr	Mo	W	Ta	Al	Ti	Re	Ni
CMSX-2-γ [20]	7.4	21.6	0.5	3.0	0.04	4.4	0.5	–	62.4
CMSX-2-γ' [20]	3.2	2.3	0.2	2.3	3.1	16.7	1.6	–	70.6
NASAIR100-γ [21]	–	28.6	0.9	2.0	–	5.3	–	–	63.0
NASAIR100-γ' [21]	–	2.4	0.5	4.4	2.0	17.4	2.1	–	70.8
Alloy SX-1 γ [22]	21.2	11.9	–	2.4	0.45	6.4	–	5.7	51.9
Alloy SX-1 γ' [22]	8.7	2.1	–	1.5	2.1	16.3	–	0.4	68.8

Table 1 lists the compositions of a number of commercial single crystal alloys. Chemically, these alloys are complicated, since the compositions have been selected on the basis of a balanced set of physical and mechanical properties, including creep, fatigue, oxidation and corrosion. Alloying elements may partition to either the γ or γ' phase or both, and plastic deformation must be studied in terms of the compositions of the individual phases. Table 2 lists the compositions of the constituent γ and γ' phases for several alloys in table 1. As can been seen in table 2, Ti and Ta strongly partition to the γ' phase, while Re, Cr, and Co strongly partition to the matrix phase. Other elements reside substitutionally in both the γ and γ', providing the opportunity for strengthening of both phases.

Most of the alloys listed toward the top of table 1 are early single-crystal compositions that were among the first introduced into commercial aircraft applications [5,6]. Several of the alloys near the bottom of the list, which contain higher levels of refractory alloying elements such as Ta, W and Re, have recently been transitioned into commercial applications [17,18]. As the understanding of alloying and single crystal solidification has improved over the past two decades, the maximum temperatures to which these materials are exposed in service have risen to values in excess of 90% of their melting temperature. Thus, the mechanisms of high-temperature strengthening in these alloys warrant close examination for their possible application to other alloy systems.

3. High-temperature deformation of bulk single phase nickel and Ni_3Al

3.1. Creep deformation in single phase nickel alloys

A number of studies have been conducted on the creep behavior of pure nickel single crystals and polycrystals at temperatures above 600°C ($T/T_m > 0.5$) [23–31]. The characteristics of creep in nickel are similar to those of other pure f.c.c. metals, displaying a normal three stage creep curve, with a primary transient of decreasing creep rates with time, followed by steady-state deformation and a tertiary transient of accelerating creep rates, eventually leading to fracture. Steady-state creep rate can be described by a power law relationship, with the expression of Dorn and co-workers [32] most frequently used:

$$\dot{\varepsilon} = A\left(\frac{\chi}{\mu b}\right) D \frac{\mu b}{kT}\left(\frac{\sigma}{\mu}\right)^n, \tag{1}$$

where σ is the applied stress, A is a material constant, μ is the shear modulus, k is Boltzmann's constant, T is the temperature, D is the self diffusion coefficient, χ is the stacking fault energy, b the Burgers vector and n is the creep stress exponent. For pure nickel, creep stress exponents in the range of 4–6 have been reported, typical of other f.c.c. metals [33]. Activation energies for creep are reported to be in the range of 269 kJ/mol to 280 kJ/mol [23–29], and are within the range of the activation energies reported for self diffusion [34]. This is indicative of the importance of dislocation climb to the creep process for the pure material. It has also been shown [31] that primary creep data at different temperatures can be normalized with a parameter, θ, that includes the activation energy for creep:

$$\theta = t(T)\exp\left\{-\frac{Q}{RT}\right\}, \tag{2}$$

where $t(T)$ is the time elapsed in the primary transient, R is the gas constant and the other terms have already been defined. This suggests that the same kinetic recovery processes that govern steady-state creep are also operative during primary creep.

During creep, slip occurs on $\frac{a}{2}\langle 110\rangle\{111\}$ type systems [32,33], as for other f.c.c. metals. Starting from an initially low dislocation density, there is substantial hardening and substructure development during primary creep. As steady state is approached, well defined subgrain structures are present. The dimensions of the subgrains present at steady state have been observed to scale inversely as the level of the applied stress [35]. Although a number of theories of creep deformation which account for this "composite" structure have been proposed [36,37], they will not be reviewed here, since they are not directly relevant to the two-phase deformation problem, due to the qualitatively different substructures that develop during creep, as discussed further in section 5.

In the two-phase superalloy system, a number of elements are typically added to improve high-temperature mechanical or environmental properties. Among these, W, Cr, Co and Re have a relatively high solubility in the nickel matrix. Therefore, it is worth while to consider their potential effect on the creep behavior of nickel. The most detailed study is

Table 3
Activation energies for diffusion of various elements in nickel.

Element	Activation energy (kJ/mol)	Reference
Ni (self diffusion)	279	Hoffman et al. [34]
Ti	257	Swalin and Martin [40]
W	321	Swalin and Martin [40]
Mo	288	Swalin et al. [41]
Co	286	Ruder and Birchenall [42]
Cr	272–290	Monma et al. [43]
Al	268	Swalin and Martin [41]

that of Johnson, Barrett and Nix [38]. In polycrystalline alloys containing up to 6 wt% W in solution, minimum creep rates decreased with increasing levels of W. The addition of tungsten was shown to influence creep through increases in the activation energy for creep (through changes in the effective diffusion coefficient) as well through the reduction in stacking-fault energy in the following manner:

$$\dot{\varepsilon} = A\left(\frac{\chi}{\mu b}\right)^{4.2} D_{\text{eff}}\left(\frac{\sigma}{E}\right)^{4.8}, \tag{3}$$

where E is the Young's modulus, and the effective diffusion coefficient, D_{eff}, was defined as:

$$D_{\text{eff}} = \frac{D_{\text{Ni}} D_{\text{W}}}{X_{\text{Ni}} D_{\text{W}} + X_{\text{W}} D_{\text{Ni}}}, \tag{4}$$

where $D_{\text{Ni}}, D_{\text{W}}, X_{\text{Ni}}$ and X_{W} are the diffusion coefficients and atomic fractions of Ni and W in the alloy, respectively. A good correlation between this effective diffusion coefficient and activation energies for creep was demonstrated. Creep deformation has also been studied in a series of single-phase Ni–Cr alloys containing up to 30 at% Cr [39]. Although large additions of Cr also lower the creep rate of nickel, Cr is a far less potent alloying element, compared to W. Table 3 shows activation energies for diffusion of several elements in nickel. Unfortunately, there are several common elements contained in superalloys, such as Ta, for which there are no diffusion data.

The reduction of stacking-fault energy may influence the ease of cross slip as well as dislocation climb. However, since activation energies for creep show a strong correlation with the activation energies for diffusion, climb of extended dislocations is apparently the rate-limiting recovery mechanism. An analysis of the climb velocity, v_{c}, for extended jogged dislocations [44], suggests the following dependence on stacking fault energy:

$$v_{\text{c}} = \frac{A\sigma\Omega D}{bkT} c_{\text{j}} \left(\frac{\chi}{\mu b}\right)^{2}, \tag{5}$$

where A is a material constant of the order of 1000, Ω is the atomic volume, c_{j} is the concentration of jogs and the other constants have been defined earlier. Note that the power law exponent for the stacking-fault energy is higher in the experimental studies, compared to the theoretical analysis; the reasons for this are not well understood. One possibility is that there is some parallel contribution of cross slip to the recovery process, which tends to be very jerky in nature, with collapses of tangles following the diffusion-induced removal of dipolar barriers in subgrain walls [45]. Nevertheless, alloying additions to nickel that change the stacking-fault energy are likely to significantly influence the creep process. Several elements, including Cr, Co, Mo, Ti and W, have been reported to reduce the stacking-fault energy of nickel [46–48].

Reviewing the available data, it appears that W would provide the most potent strengthening of Ni in creep, through its influence on both diffusion and stacking-fault energy. Recent measurements of interdiffusion of Re and Ni reveal diffusivities that are a factor of 5 to 6 times lower, compared to the diffusion of W in Ni [49]. Thus, Re and W in solution are both expected to be very effective at inhibiting diffusion-controlled processes during creep. It is worth noting that, in the case of the W additions, no changes in the power law exponents occur with the solute addition.

One more recent investigation of single-phase nickel alloys aimed to examine the creep properties for compositions similar to that of the matrix phase of superalloys. In this study by Nathal et al. [50] on a single-phase γ alloy with composition Ni–4.4Al–26.5Cr–3.3W–0.6Mo–18.3Co (at%), the minimum creep rates and stress exponents of the multicomponent alloy were very similar to those of the Ni–2wt%W alloy of Johnson, Barrett and Nix [38]. Interestingly, large variations in Co, from 0 to 18.3 at% did not significantly influence minimum creep rates in the multicomponent alloys [50]. This would again suggest that W is much more potent with regard to creep strengthening, compared to Cr and Co. In the alloys of Nathal et al. [50], the creep stress exponents were $n = 4.2$, slightly lower than the exponents in the binary Ni–W alloys. The activation energy for creep was 355 kJ/mol, somewhat higher than that observed in the Ni–6wt%W alloy [38].

3.2. Creep in Ni_3Al alloys

The characteristic feature of deformation in Ni_3Al that has received the greatest attention is the anomalous increase in flow stress with temperature, typically in the range of 500°C to 700°C. This behavior is also apparent in two-phase single crystals superalloys, fig. 3, when they are strained at relatively high rates in the range of 10^{-3} to 10^{-4} s^{-1}. This phenomenon has been the subject of many detailed electron microscopy studies and dislocation models for the increasing flow stress effect have been under discussion for the previous two decades. While this aspect of deformation is beyond the scope of the present paper, it has been treated in detail in several excellent reviews [51–54]. With regard to the anomalous rise in flow stress, the primary observation of importance is that the motion of $a\langle 101\rangle$ screw superdislocations gliding on $\{111\}$ octahedral planes is inhibited by cross slip of segments onto $\{100\}$ planes in the temperature range near the peak stress.

There is no corresponding anomalously high creep strength in the temperature range of the anomalous increase in flow stress. Minimum creep rates increase with temperature, even with relatively high applied stresses of 745 MPa in the temperature range of 600°C to 650°C in [001] oriented Ni_3Al containing small amounts of Hf and B [55]. The creep curve in the temperature range of the anomalous flow stress has been characterized [55,

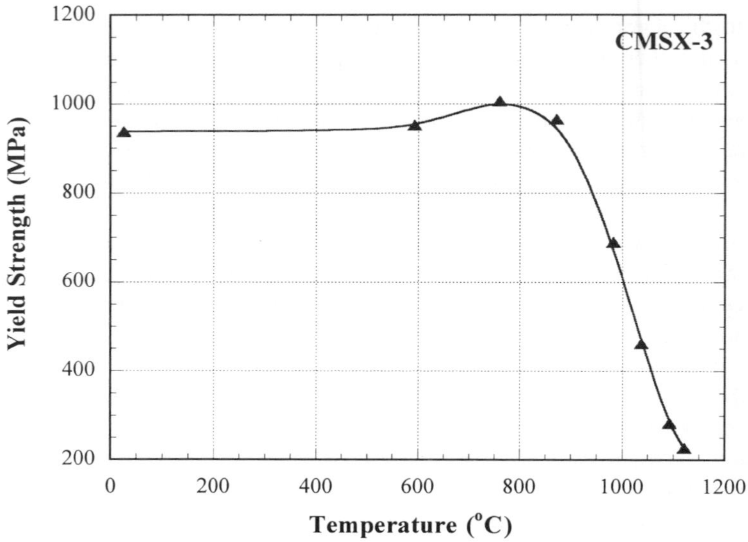

Fig. 3. Yield stress as a function of temperature for CMSX-3, showing a slight increase with temperature in the vicinity of 600°C to 700°C.

56] as displaying a brief primary creep transient, immediately followed by an "inverse creep" response, consisting of a continuously increasing creep rate to failure, with no steady state. The primary transient was attributed to exhaustion of the dislocations gliding on primary octahedral planes, leaving primary ⟨101⟩ screws that are locked by the Kear–Wilsdorf mechanism [57]. Subsequent creep was then attributed to glide of dislocations against a high lattice resistance on the (010) cube cross slip plane, rather than on the most highly stressed cube plane or the octahedral plane. Deformation by cube slip was confirmed by surface slip traces [55] and is apparently responsible for the continuously increasing "inverse creep" behavior. Detailed TEM investigations of dislocations on {001} planes have also revealed Lomer–Cottrell configurations which can be unlocked by interactions with secondary slip systems by forming mobile $\boldsymbol{b} = \langle 100\rangle$ segments [58,59]. Crystals tested in [001] orientations have no resolved shear stress for slip on cube planes and therefore have higher creep resistances compared to other orientations in this temperature range.

In the temperature range of 900°C to 1100°C the creep behavior of Ni_3Al is somewhat different from that observed at lower temperatures. In binary nickel-rich [001] crystals, Shah [60] measured a creep stress exponent of $n = 3.5$, characteristic of mobility-controlled creep. However, no electron microscopy studies were conducted. For soft-oriented (non-⟨001⟩) single crystals containing small amounts of Hf and B, Hemker and Nix [61] observed limited primary creep and a true steady state with $n = 4.3$. Microscopy revealed subgrain formation during creep and this would seem to indicate substructure rather than mobility controlled creep, as at lower temperatures. However, the creep experiments of Nathal et al. [50] in single crystals of Ni_3Al containing significant amounts of Cr, Ti, Ta, W and Co showed lower stress exponents, indicative of solute drag controlled creep, along with isolated dislocations and the absence of a tendency for subgrain

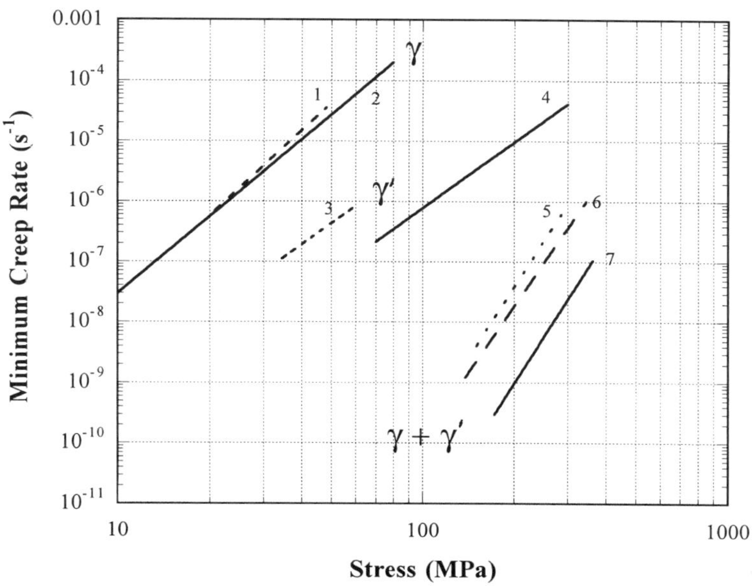

Fig. 4. Comparison of the minimum creep rates of f.c.c. nickel alloys and Ni_3Al type alloys to those of single-crystal superalloys. Curves are labeled by number as follows: 1 = Ni–2W (wt%) [38], 2 = Ni–26.5Cr–4.4Al–3.3W–0.6Mo–18.3Co (at%) [50], 3 = Ni–12.4Al (wt%) [60], 4 = Ni–2.2Cr–18.8Al–2.4Ti–1.7Ta–1.5W–5.7Co (at%) [50], 5 = NASAIR 100 [64], 6 = René N4, 7 = Alloy SX-1.

formation. In these experiments the activation energy for creep was 380 kJ/mol. Studies on polycrystalline Ni_3Al containing Fe have also shown stress exponents of less than 3.5 [62,63]. There are apparently some complexities associated with alloy composition, stoichiometry and/or crystal orientation that influence creep in this temperature range that are not yet fully understood. However, as will be apparent in subsequent sections, with few exceptions the creep behavior of the two-phase material is not related to the deformation phenomena observed in single-phase Ni_3Al due to its isolation as small, initially dislocation-free precipitates.

It is worth while to compare the minimum creep rates of f.c.c. nickel alloys and Ni_3Al type alloys to those of single-crystal superalloys, fig. 4. From this figure it is clear that the creep resistance of the ordered Ni_3Al phase is higher than that of the disordered nickel alloys. Table 4 also lists reported values of the activation energies for creep in a variety of two-phase superalloys containing volume fractions of γ' of 0.5 or higher. While the activation energies for creep in the single-phase nickel alloys, both f.c.c. and Ni_3Al type, are in the expected range for self diffusion, the two-phase materials clearly have activation energies much higher than those that could be expected from self diffusion. Finally, from fig. 4 it is clear that the two-phase superalloys have a significantly higher creep resistance than their constituents. Thus, it is essential to understand the additional strengthening mechanisms associated with the fine-scale two-phase mixture. The following sections consider a number of aspects of the two-phase material, starting with the initial mechanical

Table 4
Activation energies for creep and creep stress exponents for superalloy single crystals containing high volume fractions of γ' precipitates

Alloy	Q_{creep} (kJ/mol)	Stress exponent, n	Reference
NASAIR 100	345	7.2–7.6	[64]
SC Mar-M-200	627	7.7	[65]
Ni–7Ta–6Cr–5Al–2Mo	565	11	[66]
SC IN-100	–	11.3	[67]
Ni–5.8Al–14.6Mo–6.2Ta	475	8.7–10.9	[68]
CMSX-3	495	7.1	[69]
CMSX-4	–	8.8–13.5	[70]
René N6	403	6.1–10.8	[17]

state of the γ–γ' microstructure, and following with detailed observations on deformation mechanisms across a range of temperature and stress. The observations reported draw primarily from the past and present work of the co-authors on a subset of alloys from table 1 as well as several model ternary alloys. However, many of the phenomena discussed are fairly generally observed over a wide range of alloy composition, and similar observations by other research groups will be noted.

4. *The initial structure of superalloy single crystals*

Several aspects of the initial structure of the single crystal can significantly affect its high-temperature deformation behavior. First is the initial distribution of dislocations in the crystal, which is sensitive to the single-crystal solidification process as well as to subsequent solution and aging heat treatments. Additional microstructural parameters of importance include the size, shape, and distribution of the precipitates as well as the γ–γ' lattice misfit.

Superalloy single crystals are typically produced via investment casting techniques, where a pre-alloyed charge is melted, poured into a ceramic mold and withdrawn from the hot zone of a furnace at rates in the range of 0.14–3.5 mm/s. Given the thermal gradients typically present in this directional solidification technique, the solidification is dendritic in nature. Due to the high degree of alloying and the associated solidification partition coefficients, there is pronounced segregation in the as-grown crystals, fig. 5. For these reasons, the initial dislocation densities present in superalloy single crystals may be somewhat higher than for pure materials grown by other melt growth techniques (such as Czochralski-grown Si) and dependent on the scale of the dendritic structure.

Because of the association of the grown-in dislocations with the dendritic network, the initial distribution of dislocations is somewhat inhomogeneous, and associated with the interdendritic and/or eutectic regions of the crystal where the final liquid freezes. Subsequent solution heat treatments just below the incipient melting temperature are typically used to homogenize the crystals to some degree and to permit re-precipitation of the γ' phase. These heat treatments presumably result in some recovery of grown-in networks, but the networks nevertheless persist in interdendritic regions. Figure 6 is an optical micrograph that shows enhanced etching of the low-angle boundaries in the interdendritic regions. Figure 7 is a bright field TEM micrograph of an interdendritic

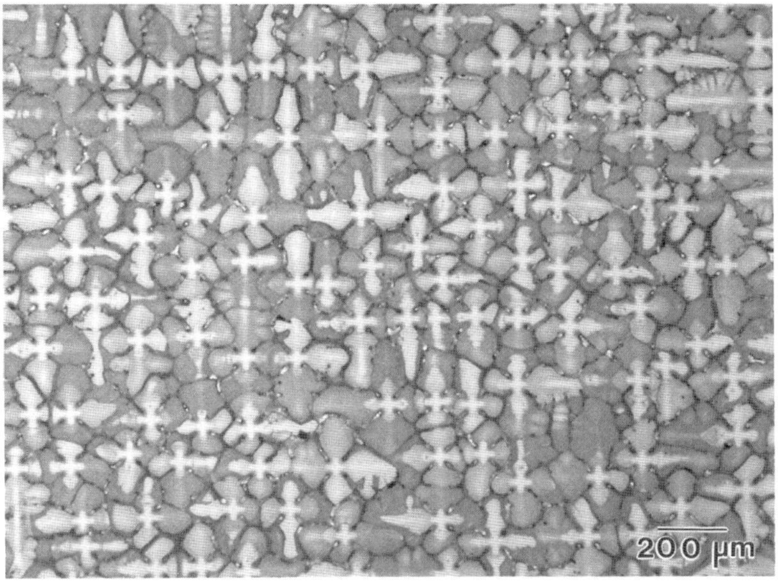

Fig. 5. Dendritic segregation in an as-grown CMSX-3 crystal, viewed normal to the ⟨001⟩ solidification direction. The four-fold character of the dendrites is apparent.

grown-in network. The grown-in networks consist of loose arrays of $\frac{a}{2}\langle 110\rangle$ dislocations that reside in the f.c.c. matrix. These networks are widely spaced (infrequently encountered in TEM foils), and it has been estimated [69] that the initial dislocation density in single-crystal superalloys is approximately $10^8/\mathrm{cm}^2$.

As mentioned in section 2, the precipitates are cuboidal, with {100} faces. They are completely coherent and possess a cube-on-cube orientation relationship with the matrix. There is typically a small degree of lattice misfit between the precipitate and matrix, which changes with temperature due to the differing thermal expansion coefficients of the γ and γ' phases. As the misfitting precipitates coarsen during aging, their shape and distribution typically evolve from spheres to weakly aligned cubes to strongly aligned cubes [71–76].

Several techniques have been used to measure lattice misfit in superalloys, including X-ray diffraction [77–83], neutron diffraction [84], synchrotron radiation [85], High Order Laue Zone (HOLZ) lines (a TEM technique) [82,86–88] and interfacial dislocation networks [82,89,90]. Techniques which measure the unconstrained mismatch at high temperature (i.e. the creep temperature) are the most meaningful. Figure 8 shows a high-temperature X-ray scan of the (400) γ and γ' reflections for a quaternary Ni–14.8Al–1.9Ti–1.9Ta (at%) single crystal, designated PMA alloy. Prior to hot-stage measurements the sample was subjected to an extended anneal at 1100°C for 50 hrs in order to measure the unconstrained misfit. From the separation of the γ and γ' peaks a misfit of $+0.83\% \pm 0.01\%$ can be measured for a temperature of 1000°C. Figure 9 is a weak-beam transmission electron micrograph of the interfacial dislocation structure after annealing at

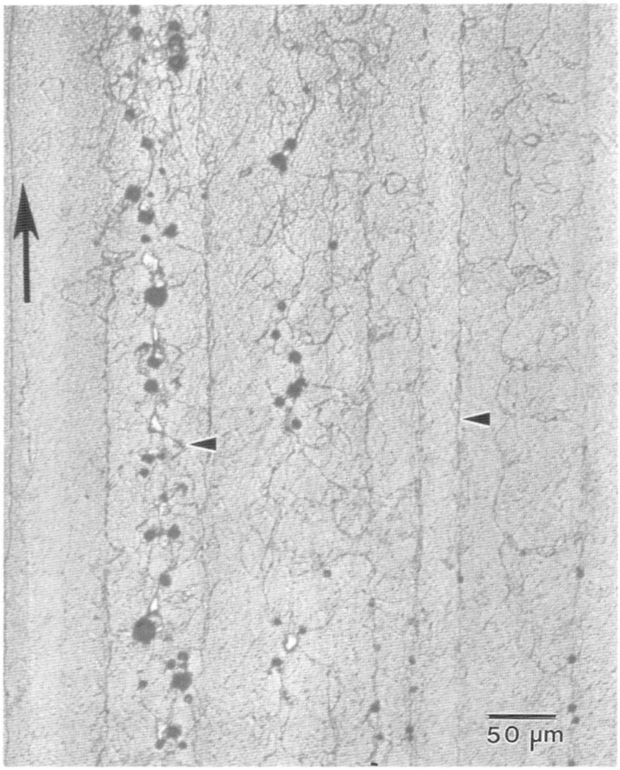

Fig. 6. Optical micrograph showing enhanced etching associated with low-angle boundaries (indicated by small arrows) in the interdendritic regions of a CMSX-3 single crystal. The section is parallel to the ⟨001⟩ growth direction, marked by the large arrow.

982°C for 100 hrs. Analysis of dislocation spacings in the interfacial dislocation network structure suggests a misfit of $0.63 \pm 0.4\%$, in reasonable agreement with X-ray analysis. Due to the fact that the misfit stresses in the matrix which drive dislocations into the interface decrease with time as the interfacial nets develop, measurements of interfacial dislocations spacings may predict misfits that are as much as 10–20% lower than X-ray analysis, depending on the volume fraction of the precipitates. Since rearrangements of dislocations within these nets is generally very sluggish, as will be discussed further in section 5.2, it should be noted that misfit determined on the basis of interfacial dislocation spacings is representative of the misfit at the annealing temperature, rather than the misfit at the temperature where the spacings are measured.

It is worth mentioning that there are a number of other potential sources of error in the measurement of lattice misfit either by X-ray analysis or by interfacial dislocation network measurements. With regard to X-ray experiments, extended annealing prior to X-ray scanning is required to permit relaxation of misfit strains in the matrix. Measurements on microstructures where the precipitates remain in their coherent state may be in significant

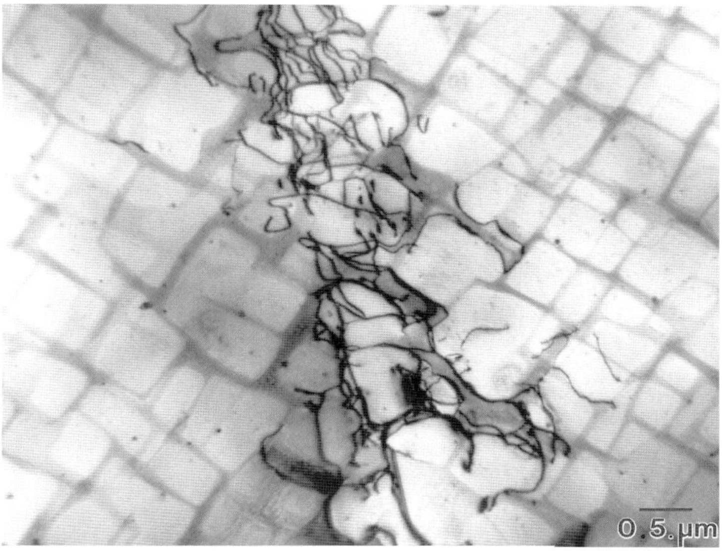

Fig. 7. Bright field TEM micrograph of an interdendritic grown-in dislocation network [69].

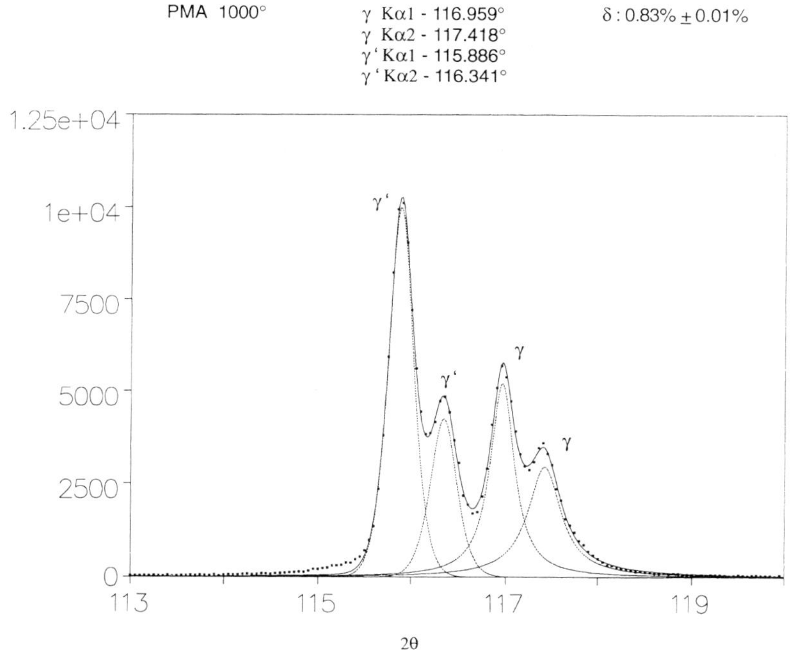

Fig. 8. High-temperature X-ray scan of the (400) γ and γ' reflections for a quaternary Ni–14.8Al–1.9Ti–1.9Ta (at%) single crystal, designated PMA alloy.

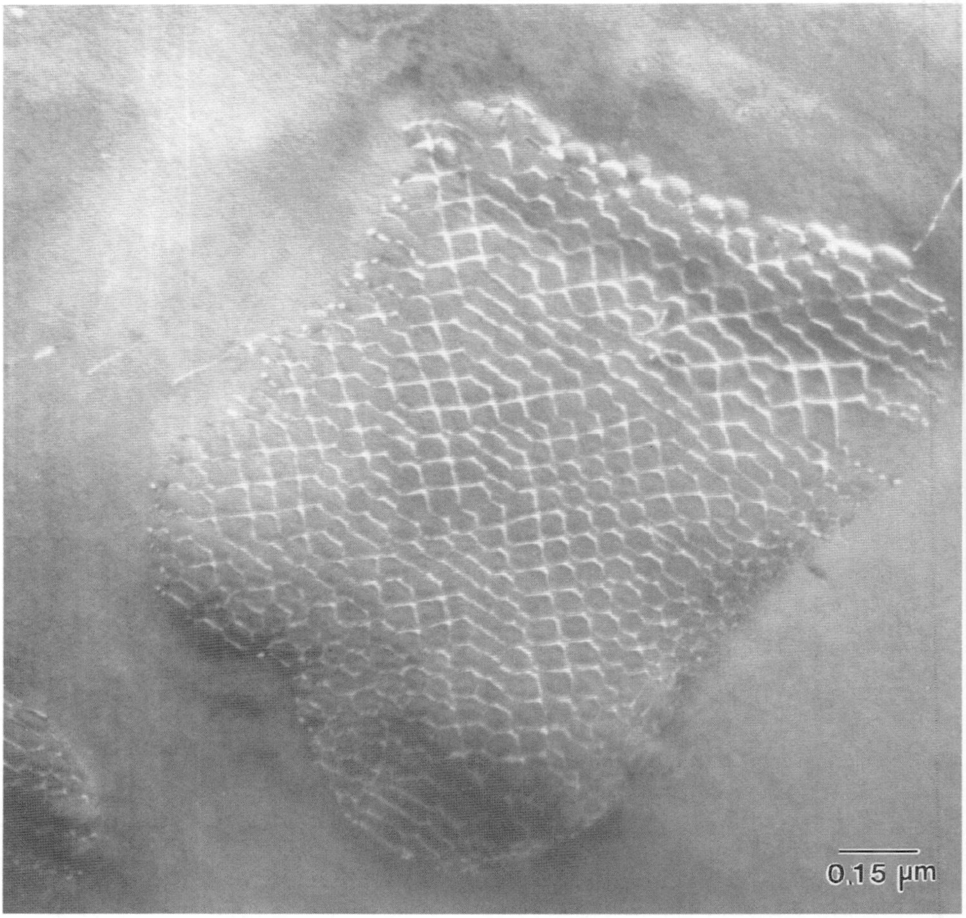

Fig. 9. Weak-beam transmission electron micrograph of the interfacial dislocation structure in Alloy PMA after annealing at 982°C for 100 hrs.

error due to peak broadening and/or shifting. This is particularly true for the matrix phase, which is more highly strained in the coherent state. Also, if samples are cooled slowly from an annealing temperature prior to making lower-temperature measurements, fine-scale γ' particles that precipitate due to the high supersaturation of elements in the matrix at lower temperatures also influence misfit measurements [83]. Since interfacial dislocation networks develop very inhomogeneously, as discussed in detail in section 5.3, annealing for insufficient lengths of time will not permit complete development of interfacial nets and will lead to an under estimation of misfit from dislocation spacing measurements.

Table 5 shows lattice mismatch values for several alloys as measured by dislocation spacings of interfacial networks in specimens which were equilibrated at the temperatures

Table 5
Lattice mismatch measurements from interfacial dislocation spacings for various alloys

Alloy	Temperature (°C)	% mismatch	Std deviation
René N4	1080	(−) 0.25	0.023
René N5	1080	(−) 0.26	0.029
PWA 1484	1080	(−) 0.22	0.028
PMA	982	(+) 0.63	0.021

Table 6
Finite element analysis of thermal misfit stresses in the CMSX-3 γ–γ' microstructure. The Mises equivalent stress, σ_e, and the individual stress components, σ_1, σ_2 and σ_3 at selected nodes in the co-ordinate system shown in fig. 10 are listed [92]

Temperature (°C)	Node (location)	σ_1 (MPa)	σ_2 (MPa)	σ_3 (MPa)	σ_e (MPa)
850	32 (γ)	−433	50	−373	456
850	148 (γ)	50	−433	−373	456
850	44 (γ')	54	52	111	58
900	32 (γ)	−447	35	−384	454
900	148 (γ)	5	−470	−405	450
900	44 (γ')	26	28	95	75
950	32 (γ)	−454	35	−388	461
950	148 (γ)	3	−478	−412	456
950	44 (γ')	24	26	94	77
1050	32 (γ)	−461	35	−388	467
1050	148 (γ)	−2	−488	−422	461
1050	44 (γ')	19	22	92	82

specified. The positive/negative sense of the mismatch was determined by X-ray measurements. Note that the commercial alloys all have similar negative misfits of the order of −0.25%. As discussed above, the PMA alloy was specifically designed to have a large positive mismatch.

At these levels of misfit, there are substantial stresses in the coherent microstructure. Two- and three-dimensional finite element analysis has been used by a number of groups to analyze these stresses [91–96]. For cuboidal precipitates, the misfit induces a state of stress in the matrix channels which is nearly biaxial, with significant compressive stresses in the plane of the precipitate faces for a negative misfit. Table 6 shows the magnitudes of individual stress components that develop due to misfit for a material with a misfit of −0.3% at 850°C for several locations in the mesh shown in fig. 10, as determined by generalized plane strain analysis [91,92]. Figure 11 shows the distribution of Mises equivalent stress in the matrix and precipitate phases due to this misfit. For even this low level of γ–γ' misfit, the Mises equivalent stresses in the matrix are very high, of the same order of magnitude as applied stresses that cause significant creep deformation. The stresses in the precipitate are much lower and nearly hydrostatic in nature. Additionally, table 6 shows the variations in misfit stresses with temperature. Note that over the temperature range of 850°C to 1050°C these stresses are fairly constant, due to the compensating effect of the increasing misfit and the decreasing elastic constants with temperature.

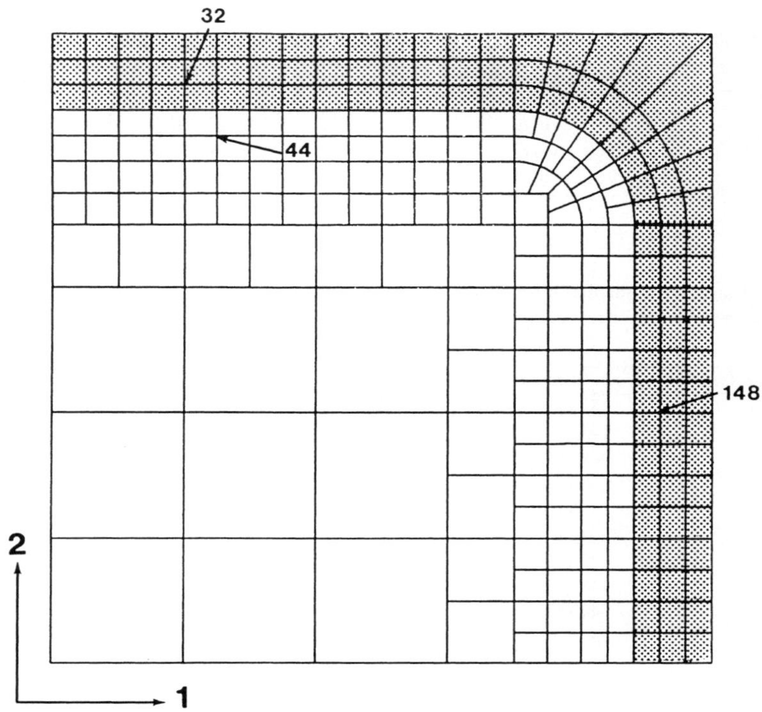

Fig. 10. Finite element mesh utilized for modeling the internal stress state of superalloy single crystals [69]. Due to symmetry, the mesh includes one quarter of an individual precipitate and the surrounding matrix material. A reference co-ordinate system is shown; the model is subjected to uniaxial loading along the '2' direction.

5. *Creep deformation in γ–γ' superalloy single crystals*

A wide variety of creep deformation behavior is observed in single-crystal superalloys over a range of temperatures from 760°C to 1100°C. For example, at lower temperatures, the creep curve typically exhibits a brief primary transient where the strain rate decays rapidly, followed by a gradual increase in strain rate with no actual steady-state regime [65]. Conversely, at temperatures above approximately 1000°C a steady-state creep rate is often observed [64]. Additionally, over the range of temperatures to be discussed, incubation periods of unmeasurably low creep rate prior to the onset of primary creep may also be present at low levels of applied stress [20,65,69]. The stress dependence of the minimum creep rate and activation energies for creep for a number of superalloy single crystal alloys were previously shown in fig. 4 and table 6. Compared to their single-phase counterparts, superalloy single crystals typically have the following characteristics:

(I) Creep rates that are slower by an order of magnitude or more for a fixed temperature and stress.
(II) Relatively low magnitudes of instantaneous straining and primary creep strain, sometimes accompanied by an incubation period prior to primary creep.

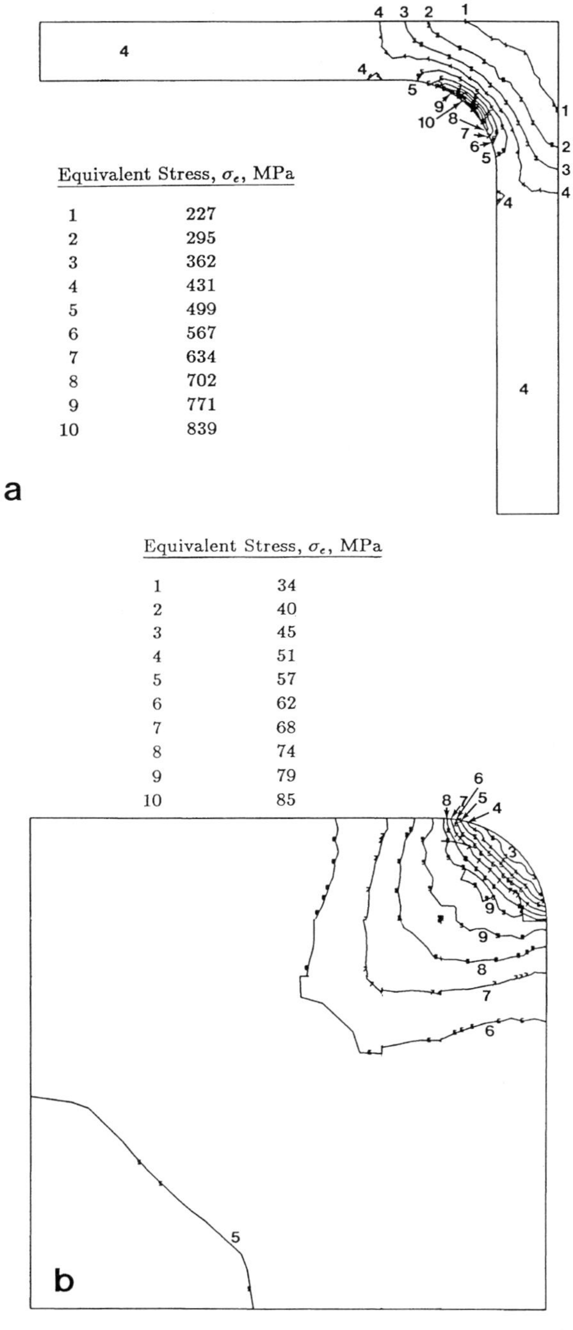

Fig. 11. The distribution of Mises equivalent stresses associated with the initial lattice misfit of −0.3% [69].

(III) Very high creep stress exponents and activation energies for creep.
(IV) Lack of a well-defined steady-state creep behavior over a large range of temperature and stress.

In this section we discuss mechanisms responsible for these characteristics of creep in superalloy single crystals. A wide variety of deformation phenomena are observed over broad ranges of temperature and stress of interest for these alloys. For this reason it is convenient to divide the discussion on creep deformation into three temperature regimes along with the most prominent feature of deformation in each regime:

(I) Low temperatures $T/T_m < 0.6$; due to the accompanying high stresses, shearing of the Ni_3Al precipitates is common.
(II) Intermediate temperatures, $T/T_m \approx 0.6$–0.7; deformation is predominantly confined to the matrix phase.
(III) High temperatures, $T/T_m \geqslant 0.7$; deformation is accompanied by stress-induced changes in precipitate morphology.

In the following sections each of these temperature regimes is discussed with regard to operative deformation mechanisms. The role of alloying and an overall assessment of the sources of creep resistance in the two-phase material are given. Important differences in the deformation characteristics of the two-phase material compared to its single-phase constituents will be highlighted. In this work we focus primarily on crystals stressed along the [001] direction. We also limit ourselves to deformation in materials that contain relatively high volume fractions of the precipitate phase. The analysis of strengthening for high volume fractions of precipitates compared to relatively low volume fractions is qualitatively different, since in the former case deformation is confined to physically small and relatively isolated regions, while for low volume fractions of the second phase substructure development in the matrix phase is similar to that observed in the single-phase f.c.c. materials. Also, at lower volume fractions, Orowan looping of individual particles is common, while the formation of loops around single precipitates due to dislocation motion through a high volume fraction array of precipitates very rarely occurs, due to the constrained geometry of the matrix channels and the associated glide phenomena.

5.1. Low temperature creep

Probably the most interesting aspect of low-temperature creep in superalloy single crystals is the activation of $\langle 112 \rangle$ slip during primary creep. This phenomenon was first discovered by Kear et al. in the late 1960's and the dislocation mechanisms were described in several papers in the late 1960's and early 1970's [65,97–101]. These authors provided a model for $\langle 112 \rangle$ slip along with detailed experimental evidence. Numerous papers have been written subsequently concerning modifications to this model [102–110] and its implications for creep behavior, particularly with respect to deviations from the [001] stress axis [111–115], effects of γ' size [113–118] and composition effects [102,114,116,119,120].

Although a detailed analysis of the dislocation reactions in the original model and proposed modifications to that model are beyond the scope of this paper, a brief description

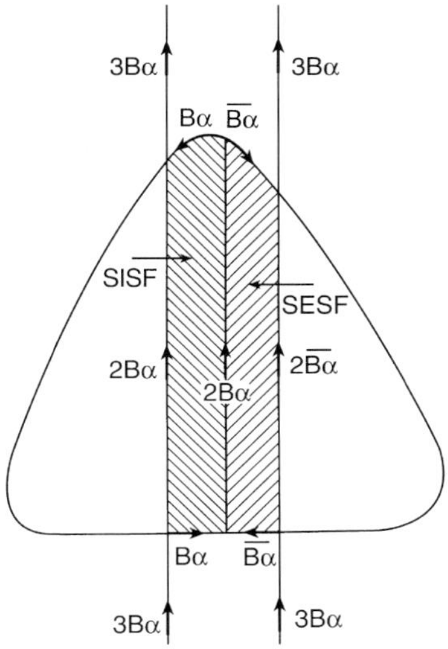

Fig. 12. A schematic representation of the dislocation interactions and faults within an individual precipitate for $\langle 112 \rangle$ slip occurring on a $\{111\}$ plane within the precipitate.

is warranted. In the original model due to Kear et al., reactions between dissociated Shockley partials of matrix $\frac{1}{2}\langle 110 \rangle$ dislocations occur at the γ/γ' interface. An example is:

$$\frac{1}{2}[011] + \frac{1}{2}[\bar{1}01] \rightarrow \frac{1}{3}[\bar{1}12] + \frac{1}{6}[\bar{1}12].$$

The $\frac{1}{3}[\bar{1}12]$ dislocation then enters the precipitate, leaving a superlattice intrinsic stacking fault (SISF) in its wake, while the $\frac{1}{6}[\bar{1}12]$ dislocation is left at the interface. When subsequent $\frac{1}{2}\langle 110 \rangle$ dislocations with the same Burgers vectors arrive at the interface, additional reactions occur which result in a total slip vector of $[\bar{1}12]$, eliminating all of the intermediate stacking faults. A schematic representation of the final dislocation structure is shown in fig. 12. The structure consists of three identical $\frac{1}{3}\langle 112 \rangle$ dislocations within the precipitate and two $\frac{1}{2}\langle 112 \rangle$ dislocations within the matrix, leaving $\frac{1}{6}\langle 112 \rangle$ Shockleys temporarily at the interface.

Modifications to this model are mainly concerned with where the reactions occur. One alternative [102] involves a single $\frac{1}{2}\langle 110 \rangle$ dislocation dissociated at the γ/γ' interface via the following scheme:

$$\frac{1}{2}[011] \rightarrow \frac{1}{3}[\bar{1}12] + \frac{1}{6}[21\bar{1}].$$

The $\frac{1}{3}[\bar{1}12]$ dislocation then enters the precipitate, leaving a SISF in its wake, while the $\frac{1}{6}[21\bar{1}]$ is left at the interface. The latter dislocation can then interact with a subsequent

$\frac{1}{2}[\bar{1}01]$, ultimately resulting in the same structure as in the original model. Although this mechanism results in an initial increase in line energy, which must be overcome by high stress, it results in an attractive force between the $\frac{1}{6}[21\bar{1}]$ interfacial dislocation and the subsequent $\frac{1}{2}[\bar{1}01]$ arriving from the matrix. In a variation of this model, an intermediate step occurs in which the entire $\frac{1}{2}[011]$ enters the precipitate, leaving an APB which is then eliminated by the nucleation of the $\frac{1}{6}[21\bar{1}]$ as in [103]. Evidence for both of these mechanisms has been observed in superalloys [103–105]. One implication of these modifications to the model of Kear et al. is that the $\frac{1}{2}\langle 110\rangle$ dislocations remain as such in the matrix, rather than reacting to form $\frac{1}{2}\langle 112\rangle$'s, at least in the initial stages.

It should be noted that deformation via the $\langle 110\rangle$ slip vector can also involve shearing of the precipitates by $\frac{1}{3}\langle 112\rangle$ dislocations (e.g., $[011] \rightarrow \frac{1}{3}[121] + \frac{1}{3}[\bar{1}12]$), so that the presence of stacking faults in the precipitates does not necessarily indicate a $\langle 112\rangle$ total slip vector. If multiple $\frac{1}{2}\langle 110\rangle$ dislocations with the same Burgers vector interact at the interface, a $\langle 110\rangle$ slip vector results. If two different Burgers vectors interact (e.g., $[011] + [\bar{1}01] \rightarrow [\bar{1}12]$), the dislocation complex associated with the $\langle 112\rangle$ slip vector can form. Thus, differentiating between $\langle 110\rangle$ and $\langle 112\rangle$ slip via TEM can only be accomplished by detailed analysis of the dislocation complexes, not simply observation of stacking faults within the γ'. Measurements of rotations during creep, particularly after the primary transient, can be used to determine which slip vector dominates.

Deformation by $\langle 112\rangle$ slip is characterized by an extended primary transient with high levels of accumulated strain, resulting from the planar nature of this slip system. Since each $\langle 112\rangle$ slip vector is mobile on only one $\{111\}$ plane, cross slip is not possible. Thus, extended planar slip is observed until a duplex slip orientation is reached. Once two slip systems are activated, interactions between them shuts down $\langle 112\rangle$ slip and the $\langle 110\rangle$ slip systems are reestablished during secondary creep [112]. These interactions have been observed to result in structures consisting of stacking faults on two planes joined by a $\frac{1}{3}\langle 110\rangle$ stair rod dislocation [102]. Evidence for another locking mechanism, analogous to Kear–Wilsdorf locks for $\langle 110\rangle$ slip, has also been observed for $\frac{1}{3}\langle 112\rangle$ dislocations within the γ' [106,108–110,121].

The $\langle 001\rangle$ and $\langle 111\rangle$ orientations are stable multiple-slip orientations and therefore display lower primary creep and dramatically improved rupture life compared to the $\langle 110\rangle$, which is far from a stable duplex-slip orientation and thus undergoes large rotations and extended primary creep [111,112]. The effects of orientation near [001] (of primary interest for turbine blade applications) on rotations during creep are shown in fig. 13. For $\langle 112\rangle$ slip, higher primary creep strains are expected for orientations deviated along the $[001]/[\bar{1}11]$ boundary, since these orientations are far from the duplex-slip line located along the [001]/[011] boundary [111,112]. This high primary creep is associated with large reductions in rupture life. Also shown in fig. 13 are the expected rotations associated with $\langle 112\rangle$ versus $\langle 110\rangle$ slip.

Recent experiments on the effects of orientation on the activation of $\langle 112\rangle$ slip, and associated variations in creep behavior, have led to new insights on the interactions between $\langle 110\rangle$ and $\langle 112\rangle$ slip systems [122,123]. It is noted by the authors that the end of primary creep does not necessarily require rotation to a duplex $\langle 112\rangle$ slip orientation. Since $\langle 112\rangle$ slip requires reactions between $\frac{1}{2}\langle 110\rangle$ dislocations, these systems are active

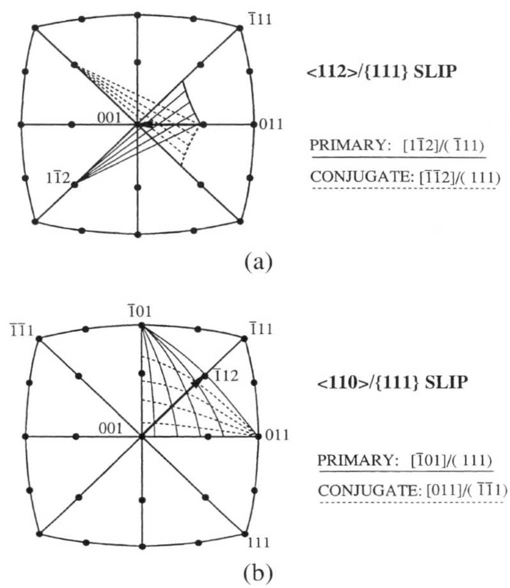

Fig. 13. Stereographic triangles showing expected rotations for tensile specimen orientations near [001]: (a) ⟨112⟩ slip vector; (b) ⟨110⟩ slip vector. The thick solid arrow indicates the boundary at which duplex slip occurs and the direction of rotation expected during duplex slip.

simultaneously. The authors propose that a "window of opportunity" exists for ⟨112⟩ slip, during which sufficient numbers of $\frac{1}{2}\langle 110\rangle$ dislocations are present to create the ⟨112⟩ dislocations, but the density of these $\frac{1}{2}\langle 110\rangle$ dislocations in the matrix channels and on the γ/γ' interfaces is not high enough to prevent precipitate cutting via ⟨112⟩ slip. Thus, continued action of ⟨110⟩ slip in the matrix, not necessarily interactions between multiple ⟨112⟩ systems, can provide the mechanism to shut down ⟨112⟩ slip. The same authors also note that the most active ⟨110⟩ slip systems for orientations near [001] along the [001]/[011] boundary combine to form ⟨110⟩ Burgers vectors rather than ⟨112⟩ and it may be this effect, not the duplex nature of ⟨112⟩ slip, which is responsible for limited activity of ⟨112⟩ slip, and consequent low primary creep strains, for these orientations. Similar observations have been made for orientations far from [001] [115].

A study of this effect has been performed on single-crystal alloys René N4 and René N5. Specimens with various misorientations with respect to [001] were interrupted after primary creep at 760°C and were subjected to TEM dislocation analysis as well as X-ray Laue measurements to determine rotations during creep. Results from the rotation measurements are shown in fig. 14. The N5 specimens display rotations consistent with those expected for ⟨112⟩ slip (fig. 13(a)). However, the N4 specimens display seemingly random rotations, inconsistent with extensive ⟨112⟩ slip. N4 specimens were also found to have very little orientation dependence of rupture life, while N5 specimens displayed a 10 to 15 × spread in rupture life as a function of orientation.

Results from TEM investigations are presented in figs 15–17. Specimens with three orientations were examined by TEM: exactly along [001], 15° away from [001] toward

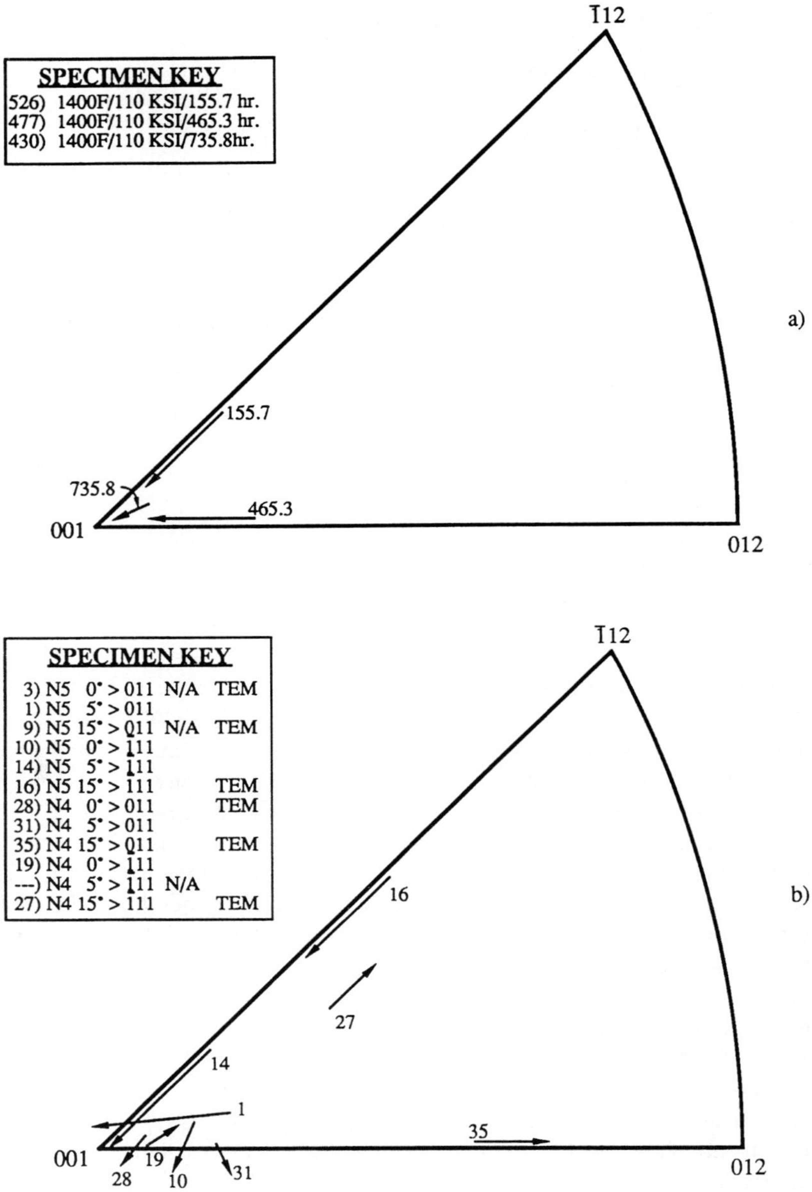

Fig. 14. Observed rotations during tensile creep testing at 760°C: (a) René N5 specimens tested to failure; (b) René N5 and René N4 specimens tested through primary creep. N/A in the "specimen key" indicates that rotation data were not available; specimens which were investigated using TEM are indicated.

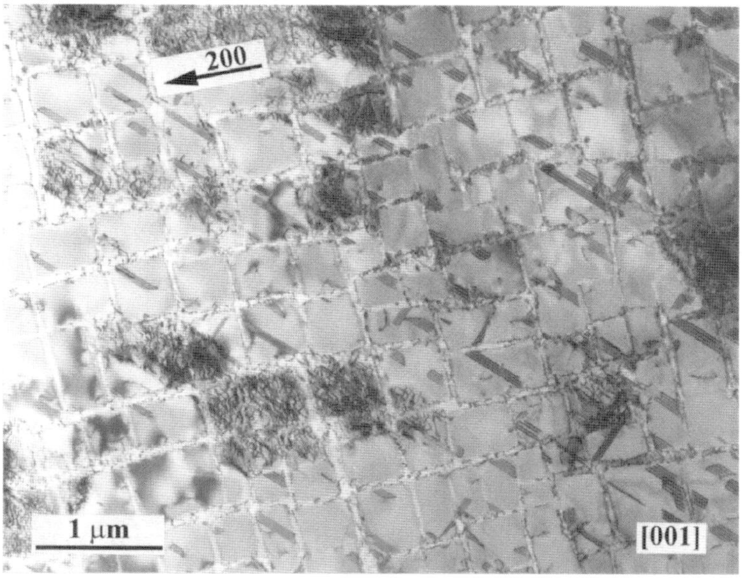

(a)

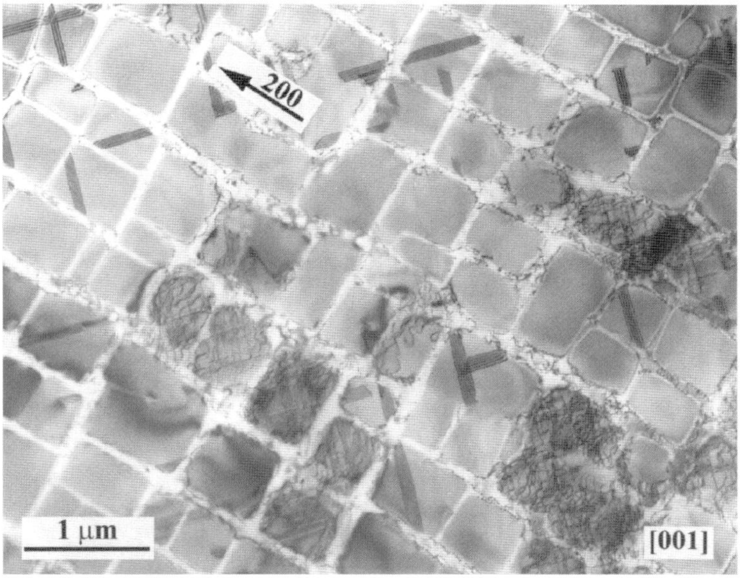

(b)

Fig. 15. TEM micrographs of interrupted creep specimens oriented along the [001] direction: (a) René N5; (b) René N4.

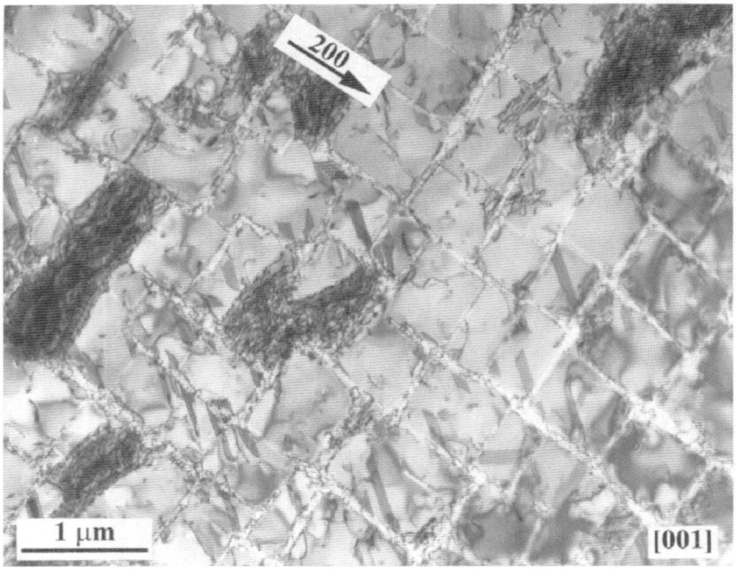

(a)

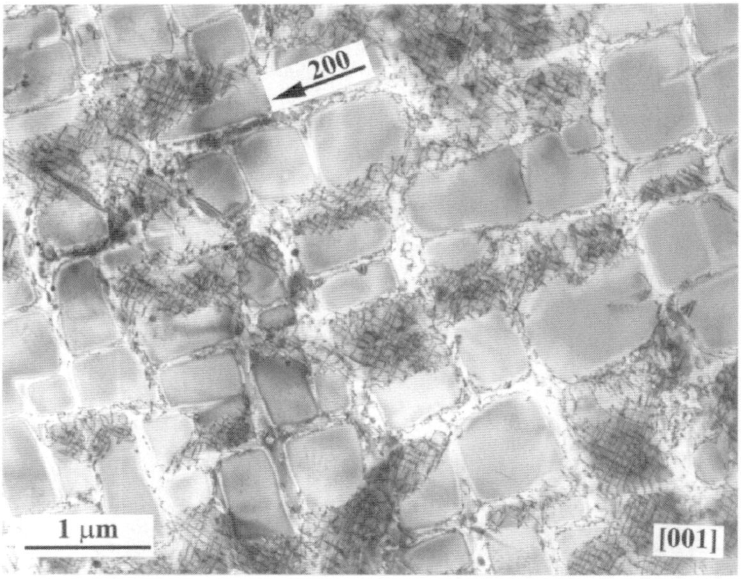

(b)

Fig. 16. TEM micrographs of interrupted creep specimens oriented 15° away from [001] toward [011]: (a) René N5; (b) René N4.

Fig. 17. TEM micrographs of interrupted creep specimens oriented 15° away from [001] toward [$\bar{1}$11]: (a) René N5; (b) René N4.

[011] ([001]/[011]) and 15° away from [001] toward [$\bar{1}$11] ([001]/[$\bar{1}$11]). Although detailed analysis of the dislocation reactions at the γ–γ' interfaces within the specimens was not performed, differences in the general dislocation distributions were found between the N4 and N5 specimens. Specimens of both alloys with the [001] orientation contain some stacking faults in the γ', suggesting the possibility of $\langle 112 \rangle$ slip. This orientation equally stresses four different $\langle 112 \rangle$ slip systems, and at least two different orientations of stacking faults are observed for both alloys, indicating that at least two different slip vectors have been activated. For the [001]/[011] orientation, numerous stacking faults are observed in the N5 specimen, but very few in N4. This orientation would be expected to produce duplex $\langle 112 \rangle$ slip, and stacking faults on at least two different planes are observed in the N5 specimen. For the [001]/[$\bar{1}$11] orientation, most of the stacking faults appear to be on a single {111} plane, consistent with planar slip of a single $\langle 112 \rangle$ slip vector. Again, the N5 specimen displays higher densities of stacking faults, all on a single {111} plane, while the N4 specimen still shows stacking faults on two different {111} planes. The observation that stacking faults are more prevalent and more prone to be on a single {111} plane in N5 compared to N4 is consistent with the rotation and creep data which indicate that N5 is more prone to planar $\langle 112 \rangle$ slip effects than N4.

The reason for this difference in the two alloys is unclear. One possible explanation is that the presence of Re lowers the stacking-fault energy of the matrix, making the formation of the $\langle 112 \rangle$ dislocation complex at the γ–γ' interfaces easier (there is no appreciable Re in the γ' phase). Studies by other investigators have shown that alloying additions which lower the stacking-fault energy of the matrix promote $\langle 112 \rangle$ slip [100, 116,119]. Heat treatments which result in fine γ' in the matrix, thereby reducing the lengths of $\langle 112 \rangle$ dislocations in the γ phase, have also been shown to promote $\langle 112 \rangle$ slip [114]. Alternatively, it has been suggested that alloying additions in the γ' phase, particularly Ti, can significantly lower the superlattice intrinsic stacking-fault (SISF) and superlattice extrinsic stacking-fault (SESF) energies, and thus influence the propensity for $\langle 112 \rangle$ slip [120]. On this basis, the presence of Ti in the N4 and its absence in N5 would be expected to promote $\langle 112 \rangle$ slip in the former, by decreasing stacking-fault energy in the γ'. The opposite is observed experimentally, i.e. $\langle 112 \rangle$ slip is more prevalent in N5. Thus, it appears to be the stacking-fault energy in the matrix which is dominant. It is also interesting to note that a $\langle 110 \rangle$ slip vector has been reported for directionally solidified René 80 crept at 790°C [106]. Although this alloy does contain slightly higher levels of Co than N4 and N5, it does not contain Re, so that the stacking-fault energy of the matrix might be expected to be higher than that in N5. René 80 also contains high levels of Ti and thus would be expected to have low stacking-fault energy in the γ'. This reported $\langle 110 \rangle$ slip vector for low-temperature creep in René 80 again supports the hypothesis that it is the stacking-fault energy in the matrix which determines the propensity for $\langle 112 \rangle$ slip, rather than the superlattice stacking fault energy (SISF/SESF) in the γ'.

It is interesting to consider why it is the matrix stacking-fault energy which determines the slip vector. As discussed above, precipitate shearing by $\frac{1}{3}\langle 112 \rangle$ dislocations can occur for both $\langle 110 \rangle$ and $\langle 112 \rangle$ slip, although it is the $\langle 112 \rangle$ slip vector which results in the high anisotropy in low-temperature creep and rupture data. If the matrix dislocations transform to $\frac{1}{2}\langle 112 \rangle$, as in the model of Kear et al. (and recently confirmed experimentally [123]), $\langle 112 \rangle$ slip is greatly facilitated, since these dislocations are then "locked in" to the $\langle 112 \rangle$

slip configuration. If the matrix dislocations remain as $\frac{1}{2}\langle 110\rangle$'s, they can move more independently and precipitate shearing can take place by either the $\langle 110\rangle$ or $\langle 112\rangle$ slip vector, depending on the reactions which take place at the interface. Whether the matrix dislocations exist as $\frac{1}{2}\langle 112\rangle$ or $\frac{1}{2}\langle 110\rangle$ is dependent on the stacking-fault energy of the matrix, since the $\frac{1}{2}\langle 112\rangle$ dislocations need to be widely dissociated in the matrix to lower their line energy. Thus, which total slip vector dominates is expected to depend on the nature of the matrix dislocations, which is in turn dependent on the stacking-fault energy of the matrix. It also should be noted that Re greatly increases the strength of the matrix, which is expected to promote precipitate shearing in general. Although γ' cutting can occur for either a $\langle 110\rangle$ or $\langle 112\rangle$ total slip vector, $\langle 112\rangle$ slip is associated exclusively with cutting, while $\langle 110\rangle$ slip is associated both with cutting and bypass mechanisms. Therefore, alloying elements which promote cutting would be expected to increase the propensity for $\langle 112\rangle$ slip.

The $\langle 112\rangle$ slip phenomenon, which can dramatically lower rupture life and increase primary creep at 760°C, might be considered a serious problem for engineering components. However, the extended primary creep and reduction in rupture properties associated with this slip mechanism appear to be restricted to uniaxial stress states. Notched creep specimens of René N5 display very little orientation dependence of creep behavior and do not undergo high primary creep strains, even for the most unfavorable orientations, due to the activation of multiple slip systems by the triaxial state of stress at the notch. The effects of $\langle 112\rangle$ slip also diminish with decreasing stress, since precipitate cutting in general is reduced [101]. The anisotropy of rupture life was observed to be reduced at lower stresses in the René N4/N5 study, consistent with reduced precipitate shearing (via $\langle 112\rangle$ slip) at lower stresses.

5.2. Intermediate-temperature creep

At higher temperatures and correspondingly lower stresses, precipitate shearing becomes more difficult, resulting in confinement of deformation to a large degree to the f.c.c. matrix phase. At intermediate temperatures and stresses, three-stage creep curves are typically observed. However, the creep rates usually exhibit a minimum just beyond the primary creep stage and then begin to accelerate without a well-defined steady state. Additionally, incubation periods, during which no macroscopically measurable straining occurs, are often present prior to the primary creep transient [20,65,69,124–126]. The evolution of dislocation substructure for each stage of creep in this temperature range is discussed in the following sections.

5.2.1. The incubation period

The incubation process has been studied in detail in CMSX-3 single crystals stressed along the $\langle 001\rangle$ growth direction [69,91]. Figure 18 shows a TEM micrograph of the substructure during the incubation period for deformation at 825°C and 450 MPa. In the upper left corner of this micrograph is a "grown-in" network of dislocations. During the incubation period, this network serves as a source for $\frac{a}{2}\langle 110\rangle$ dislocations which glide through the narrow matrix channels, leaving segments of mixed or pure screw

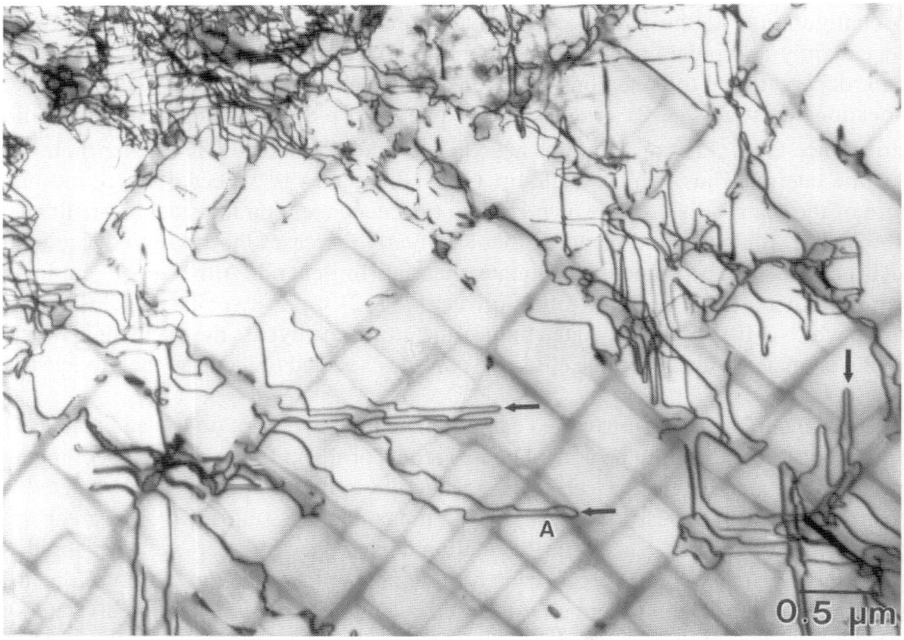

Fig. 18. TEM micrograph of the substructure during the incubation period for deformation at 825°C and 450 MPa [69]. Grown-in network at upper left serves as a source for gliding dislocations in the early stages of creep deformation.

character on the γ–γ' interfaces. The leading segments, indicated by arrows in fig. 18, are screw in character. The prominence of leading screw segments is due to the need for extensive cross slip for an individual loop to expand over any significant distance along the {111} slip planes, which intersect {100} precipitate-matrix interfaces. As the loops expand through the narrow matrix channels, segments of mixed or edge character are held at the interface (partially compensating misfit between the precipitate and matrix), while the screw segments undergo cross slip and continue to glide. Even though the matrix is highly alloyed, presumably resulting in a low stacking-fault energy, cross slip occurs relatively frequently, as apparent from the 90° configurations along individual loops, marked by "arrow" in fig. 19. As noted in section 5, Orowan looping of individual precipitates by a single dislocation is rarely observed, due to the complex cross-slip patterns required to accomplish this.

As shown in figs 18 and 19, dislocations are required to bow through a small radius to glide through the narrow matrix channels. The resolved shear stress, τ_{OR}, required for this is:

$$\tau_{\mathrm{OR}} = \sqrt{\frac{2}{3}}\frac{\mu b}{h}, \tag{6}$$

where μ is the shear modulus, b is the Burgers vector and h is the width of the channel. For CMSX-3, typical values of these material properties at 850°C are $\mu = 48.2$ GPa,

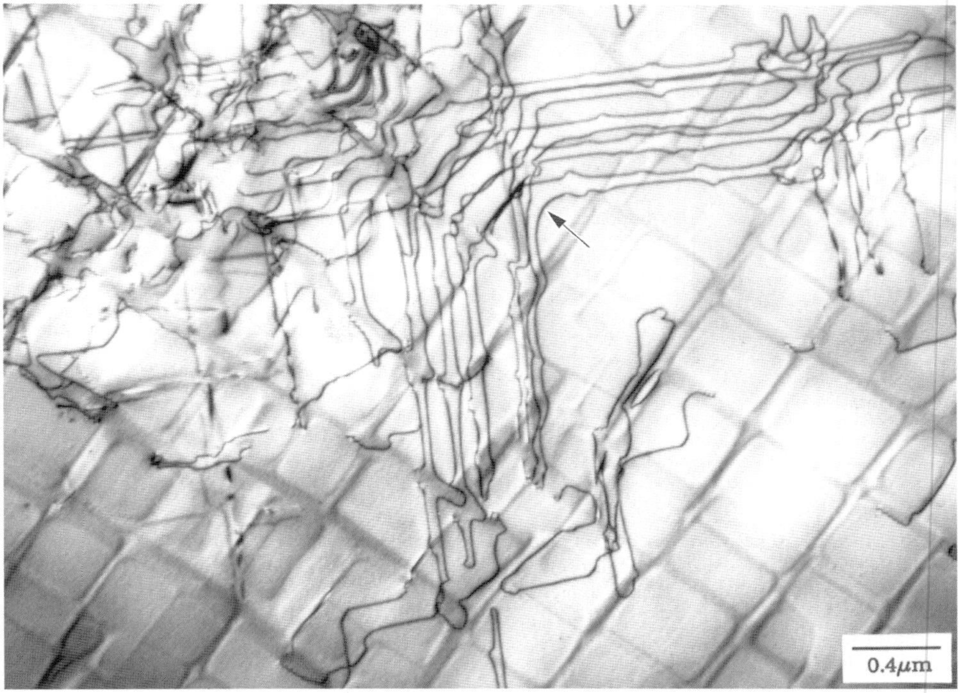

Fig. 19. Dislocation cross slip in the matrix during the early stages of creep, characterized by the 90° configurations along individual loops, marked by “arrow” [69].

$b = 0.254$ nm, and $h = 60$ nm [69]. This results in an Orowan stress for bowing of 166 MPa, which corresponds to a uniaxial applied stress of 408 MPa. Thus, this resistance accounts for a large fraction, but not all, of the creep resistance of the two-phase material.

An interesting feature of deformation during the incubation is the preferential glide of dislocations through horizontal channels which lie normal to the axis of the applied stress (the plane of the micrographs in figs 18 and 19). This occurs due to the fact that the misfit stresses in the two sets of channels, the horizontal channels which lie normal to the axis of the applied stress and the vertical channels which are parallel to the applied stress, are unbalanced by the application of an external stress, fig. 20. Finite element analysis has been utilized to determine the magnitude and distribution of these stresses in the initial stages of creep [69,91]. For an applied tensile stress of 450 MPa and a misfit of $\delta = -0.3\%$, the initial resolved shear stresses on the $\langle 110\rangle\{111\}$ systems in the horizontal channels are approximately $\tau = 380$ MPa on average, while in the vertical channels they are a factor of 2.4 lower at $\tau = 160$ MPa. For this reason glide in the horizontal channels is strongly favored in the early stages of creep, while glide into vertical channels which are only slightly wider than average would also be favored, figs 18, 19. It is worth noting that dislocations that are pushed into the vertical channels will initially serve to aggravate the misfit, if they remain in their glide positions [127,128]. As a result, in the early stages

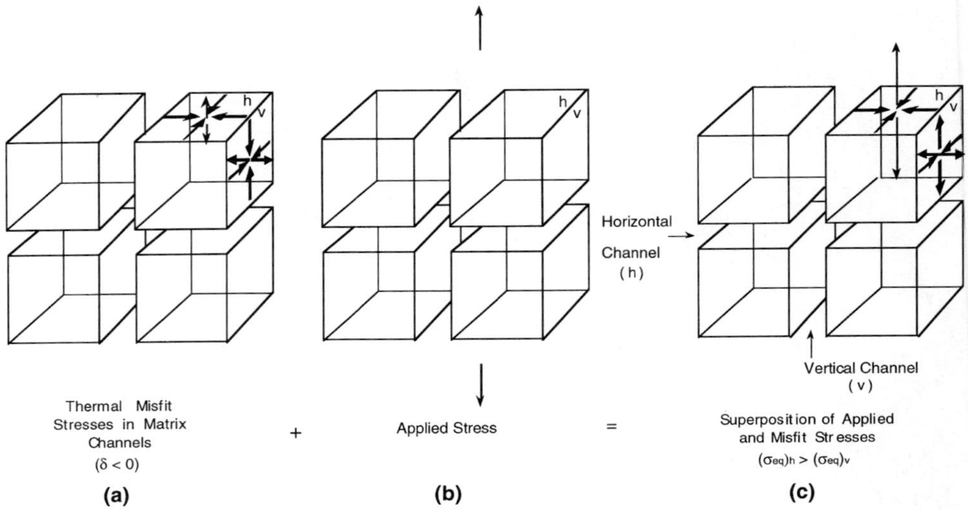

Fig. 20. Schematic of microstructure showing that stresses in horizontal channels, which lie normal to the axis of the applied stress, and in the vertical channels which are parallel to the applied stress, are unbalanced by the application of an external stress, leading to preferential glide in the horizontal channels in the early stages of creep.

of creep there is considerable inhomogeneity in the development of the substructure due to two factors: the non-uniform distribution of the "grown-in" networks and the misfit-induced differences in the local stresses in the horizontal channels compared to the vertical.

Another interesting feature of the incubation process shown in fig. 18 is the leading screw segments that are in the process of moving sluggishly through the channels, rather than always being held up at γ–γ' interfaces. This suggests a significant solid solution resistance to their motion. From the average glide distances of dislocations during the incubation, a solid solution resistance of $\tau_{SS} = 17.5$ MPa has been estimated for the matrix of CMSX-3 [69]. Comparing the overall creep properties of the alloys in table 4, it is apparent that refractory solutes that partition strongly to the matrix are the most effective strengtheners. This includes W and Re, with atomic radii of 0.141 and 0.138 nm, respectively, relative to nickel, with a radius of 0.125 nm.

The observation of incubation periods suggests that creep deformation does not proceed until this rather difficult percolation of dislocations through the matrix channels is complete for some cross section of the sample. Thus, in terms of resisting the onset of deformation, it is clearly most desirable to have a high volume fraction of small, closely spaced cuboidal precipitates. Most commercial single-crystal alloys, which presumably have been optimized with attention to maximizing creep resistance, contain between 60 and 70 vol% cuboidal precipitates with edge lengths between 0.35 μm and 0.5 μm. Nathal [129] and Khan [130] have shown creep properties to be at their maximum for precipitate sizes in this range for several alloys. In systems with some degree of γ–γ' misfit, as precipitates decrease in size below 0.3 μm, they typically undergo a transition in shape from cuboidal to spherical (as the interfacial energy becomes dominant). With this shape transition, the

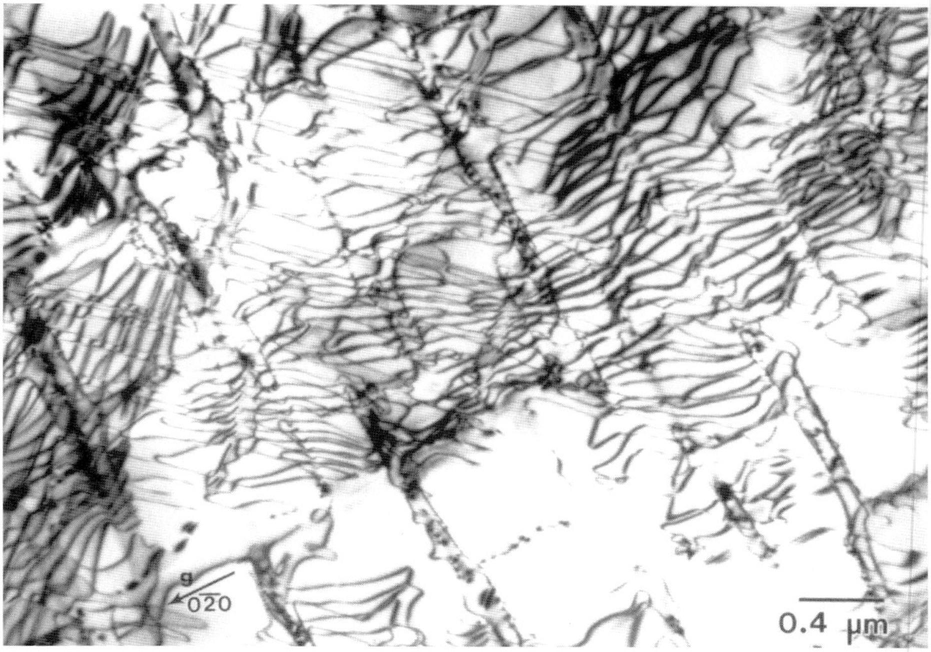

Fig. 21. Bright-field TEM micrograph of substructure during primary creep. The dislocation density in the matrix channels continues to increase until all channels are filled. Foil normal is parallel to the axis of the [001] applied stress.

constraint of the planar γ–γ' interfaces is substantially reduced, apparently resulting in a reduction in creep resistance. As the cube-shaped precipitates coarsen beyond 0.5 μm, the Orowan stress is reduced, again reducing the creep resistance of the material. Thus, an optimum at intermediate sizes where precipitates are cuboidal, but relatively small, is expected.

5.2.2. Primary and secondary creep

Beyond the incubation period and into primary creep, the dislocation density in the matrix channels continues to increase, until all channels are filled, as shown in fig. 21 for CMSX-3. In the early stages of primary creep, regions containing dislocations predominantly of a single Burgers vector are present, due to the limited numbers of widely spaced dislocation sources. Because glide occurs on $\{111\}$ planes, the segments deposited at the interfaces are aligned at 45° angles to the cube faces [69,91,131,132]. If reactions between dislocations with multiple $\frac{a}{2}\langle 110\rangle$ Burgers vectors do not occur in the early stages of creep, climb along the interfaces results in segments oriented parallel to the cube edges, fig. 22, resulting in a net decrease in line length and a more favorable alignment for misfit relief. The early stages of this climb process are apparent in regions where cross slip has occurred, indicated by the arrow in fig. 19.

It is interesting that the magnitude of the strain accumulated in the primary creep transient is often consistent with the magnitude of straining required to relieve the

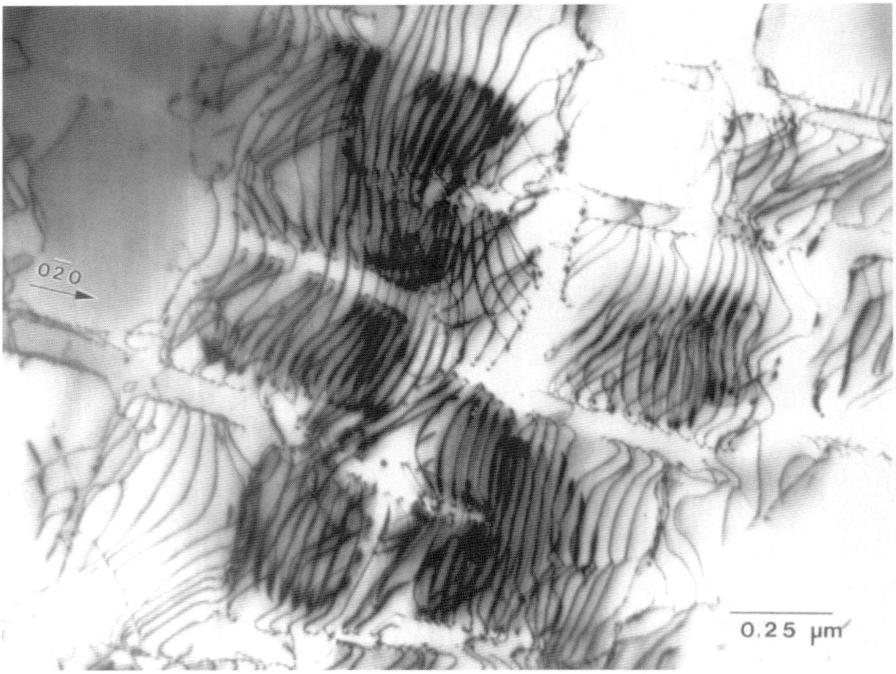

Fig. 22. Dislocation climb along γ–γ' interfaces during primary creep of CMSX-3 at 850°C. Climb is apparent due to the alignment of dislocations parallel to the cube edges in regions populated preferentially with one set of $\frac{a}{2}\langle 110\rangle$ dislocations, prior to network formation.

initially high stresses in the channels, due in large part to the misfit between the matrix and precipitate. Finite element analyses for CMSX-3, $\delta = -0.3\%$, show that the Mises equivalent stresses in the horizontal matrix channels decay from their initial value of 1030 MPa to the applied stress of 552 MPa during the primary transient, with a strain accumulation of 7×10^{-4} [92]. This strain is of the same order as the primary creep strain that is measured experimentally. Thus the decay in the creep rate is due to the reduction of the initially high stresses in the matrix channels, rather than hardening processes that occur during primary creep of single-phase nickel alloys. Because of the preferential glide in the horizontal matrix channels, higher densities of dislocations than required for relief of misfit have been observed in the later stages of creep in this temperature range, with an overall higher dislocation density in the horizontal channels, compared to the vertical [133].

As the end of the primary creep transient is approached, dislocations with different Burgers vectors on the stressed slip systems penetrate into individual channels and react. These reactions, which are considered in more detail in the next section, result in the development of a three-dimensional nodal network that fills the matrix channels and is associated in large part with the γ–γ' interface, fig. 23. As will be discussed in section 5.3, these networks have been analyzed in detail in a number of different alloys and consist of Burgers vectors from the primary stressed slip systems as well as a number of reaction products.

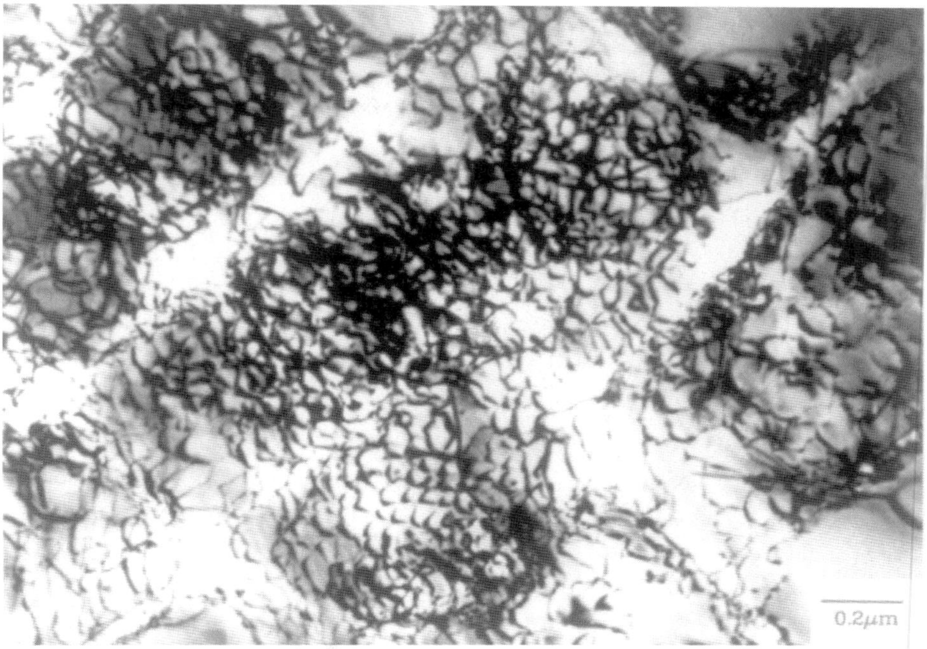

Fig. 23. Development of a three-dimensional nodal network during creep of CMSX-3 at 850°C that fills the matrix channels and is associated in large part with the γ–γ' interface [69].

The three-dimensional nodal networks are extremely stable during creep at intermediate temperatures. Figure 24(a) shows a matrix network of dislocations in a CMSX-3 sample that was pre-crept to the "steady state" regime at 850°C. This foil was annealed in-situ in a TEM heating stage. Only very local rearrangements of dislocations that were not well knit into the networks were observed during in-situ annealing at 845°C for periods of up to 45 minutes, fig. 24(b). Diffusion-controlled static recovery processes within the network are apparently very sluggish at temperatures below 850°C. At a higher temperature of 882°C, recovery in the foil became more rapid, with a measurable reduction in local dislocation density, fig. 25. Creep experiments interrupted by static annealing under no load show very limited reloading transients in this temperature range [69], again apparently due to the stability of the three-dimensional networks.

It is worth while to consider the approximate local resistance, τ_{DIS} offered by these matrix dislocation networks to further glide through the channels. This is given approximately by:

$$\tau_{\mathrm{DIS}} = \alpha \mu b \sqrt{\rho}, \tag{7}$$

where ρ is the density of dislocations in the matrix channel, α is a material constant and other terms have already been defined. Since the dislocations are confined to the matrix

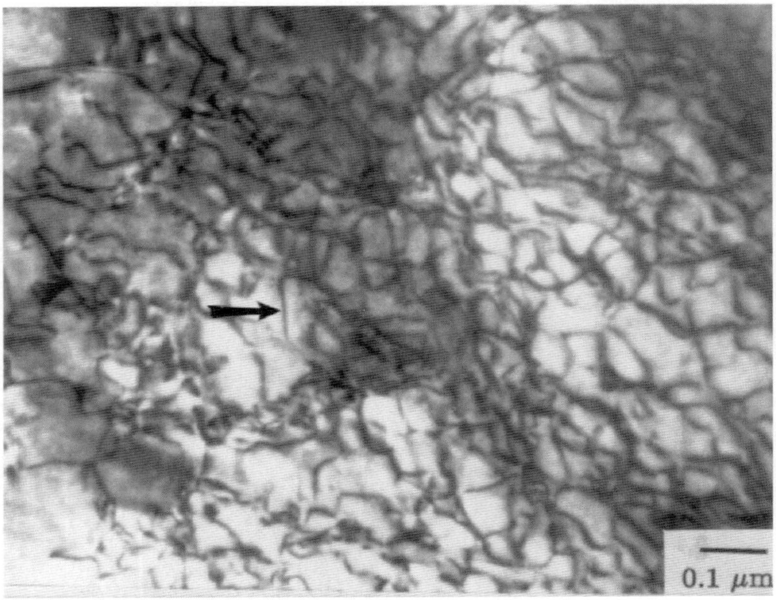

(a)

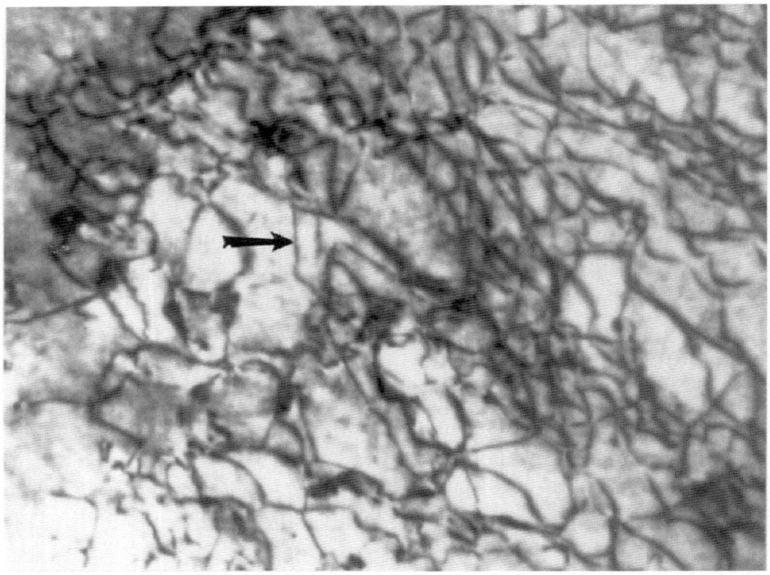

(b)

Fig. 24. Matrix network of dislocations in a CMSX-3 sample that was pre-crept to the "steady state" regime at 850°C and annealed in-situ in a TEM heating stage at 845°C. The starting structure is shown in (a), while (b) shows the structure after 17 minutes of annealing. Arrows mark reference points in the two micrographs. The networks are very resistant to static recovery at this temperature [92].

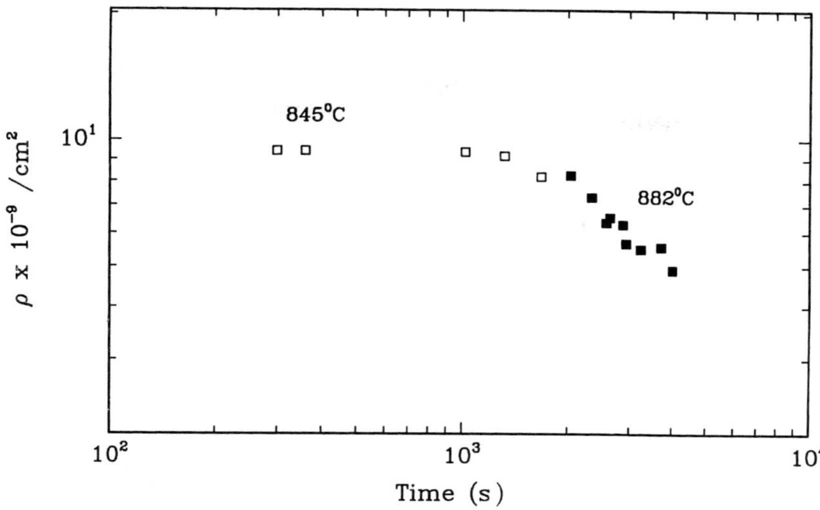

Fig. 25. Dislocation densities as a function of time and temperature measured in in-situ TEM annealing experiments for the material shown in fig. 24 [92].

channels, using fig. 25 we obtain a matrix dislocation density, ρ_m of

$$\rho_m = \frac{\rho_{tot}}{1-c}, \tag{8}$$

where c is the volume fraction of the precipitates. Using $c = 0.7$, $\rho_{tot} = 9 \times 10^9/\text{cm}^2$ (fig. 25), $b = 0.254$ nm, $\mu = 48$ GPa and $\alpha = 0.1$, then $\tau_{DIS} = 21$ MPa. Relating this to a Mises equivalent stress:

$$\sigma_{eq} = \sqrt{3}\,\tau_{DIS} = 36 \text{ MPa}. \tag{9}$$

It is interesting that this is similar to the Mises equivalent stresses in the matrix channels predicted by the finite element analyses of the creep problem, fig. 26, where the creep properties of solid-solution strengthened nickel in its bulk form are used as input for the constitutive creep response of the matrix. However, compared to the contribution of the Orowan resistance, eq. (6), this resistance is minor.

Significant diffusion-induced changes in the character of the networks established during creep at 850°C and 552 MPa have also been observed upon annealing (after removal of the applied stress) for periods of 750 hours at 850°C. After these extended anneals the dislocations remaining are primarily at the interfaces, serving to relieve the misfit stresses. The mechanisms by which these complex networks coarsen are presently not well understood. At higher temperatures, of the order of 1000°C, the networks formed during creep are again associated almost exclusively with the interfaces, as discussed further in section 5.3. These observations taken collectively suggest that dynamic recovery is an

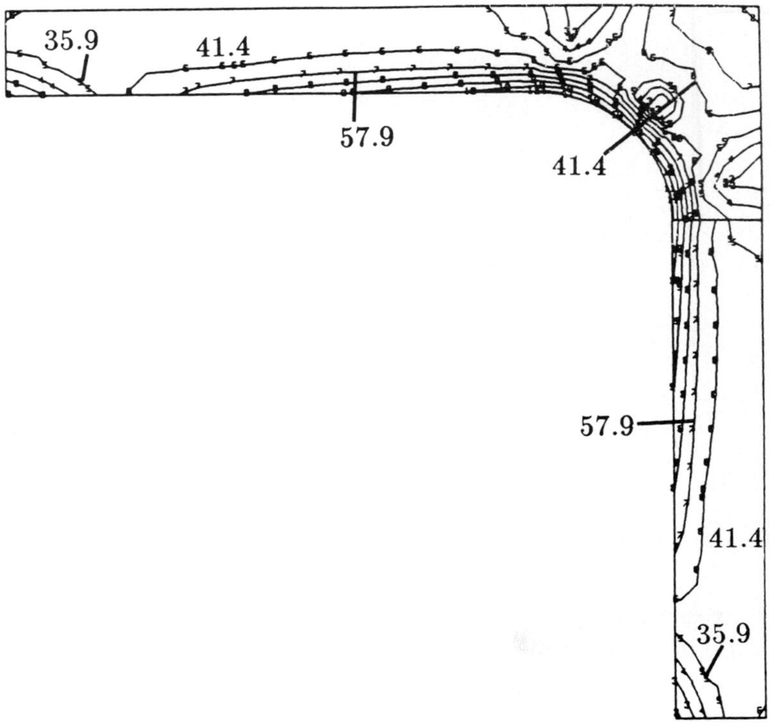

Fig. 26. Mises equivalent stresses (in units of MPa) in the matrix channels predicted by the finite element analyses of the creep problem [69].

essential component of the creep process at lower temperatures, while static recovery may be of greater importance at higher temperatures.

5.2.3. The γ' precipitates: shear resistance and mechanical constraint

An important feature of creep deformation in this intermediate temperature regime is the resistance of the γ' particles to shearing by dislocations entering from the highly deformed matrix. Evidence of dislocations shearing the γ' is only apparent in the later stages of creep, after strains of the order 2–3% are accumulated [69], fig. 27. Again, finite element analysis provides some insight to this behavior. In this generalized plane strain model [69,91,92] the precipitates remain elastic and coherent (no interfacial sliding) during deformation. As a result of these constraints, the stresses within the precipitate continue to rise over the typical time period utilized during creep testing, as the matrix accumulates strain by power law creep at the rate characteristic of the bulk matrix material, fig. 28. For this reason, a true steady state should not typically be expected in these materials. Clearly, the stresses rise to levels high enough to permit shearing by dislocations at some point during creep. Since there are no sources of dislocations within the precipitates, matrix dislocations must be pushed into the precipitates. This is a complex process, given the presence of the interfacial

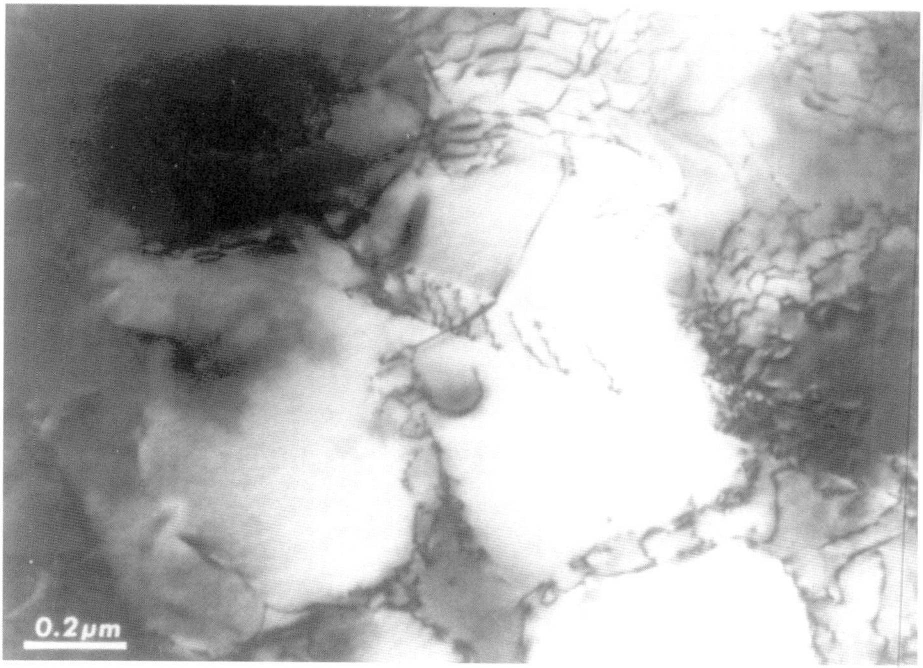

Fig. 27. Alloy CMSX-3 subjected to deformation at 850°C and 552 MPa. Dislocations shearing the γ', only in the later stages of creep, after strains of the order of 2–3% are accumulated [69].

networks, the ordered nature of the precipitates and the difficulty of capturing the early stages of the shearing process by transmission electron microscopy in this temperature range.

The precipitate shearing process at intermediate temperatures is clearly much different from that observed at 760°C. Stacking faults are not typically observed and shearing occurs due to the penetration of pairs of $\frac{a}{2}\langle 110\rangle$ matrix dislocations coupled by an APB [69, 134]. The shearing of ordered precipitates by superlattice dislocations under conditions where relatively low volume fractions of precipitates are present has been considered in detail, as reviewed by Nembach and Neite [135]. However it is worth noting that the shearing of Ni_3Al precipitates under conditions where a high volume fraction of these precipitates with cuboidal morphology are present is qualitatively different from the low volume fraction situation for two reasons. First, the precipitates are enclosed by a "cage" of misfit dislocations by the time the shearing process begins, so shearing is likely to involve some combination of a dislocation gliding up to the interface along with extraction of a second $\frac{a}{2}\langle 110\rangle$ dislocation from the network. Secondly, with cuboidal precipitates, the dislocations enter from a flat interface, rather than first partially looping the precipitate, so an increase in line energy is required for initial penetration. A simple analysis of shearing that considers these factors follows.

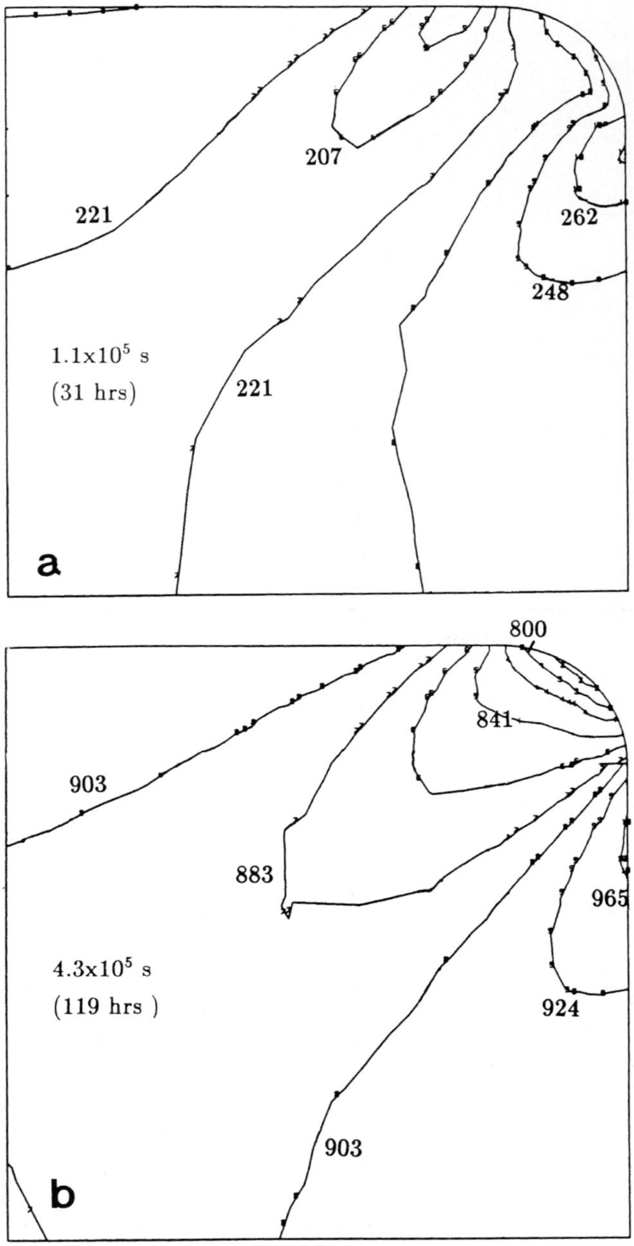

Fig. 28. Finite element calculations of Mises equivalent stresses (in units of MPa) in the precipitate at 31 h (a) and 119 h (b). Stresses continue to rise as strain accumulates in the matrix, eventually resulting in precipitate shearing [69].

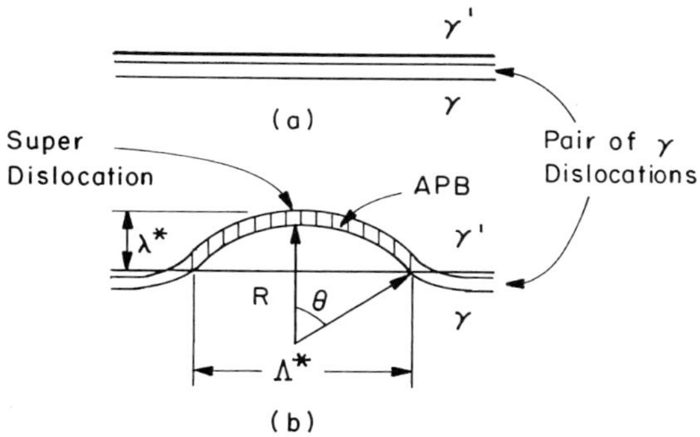

Fig. 29. A model for the penetration of pairs of $\frac{a}{2}\langle 110\rangle$ dislocations into the ordered precipitate.

We consider a pair of dislocations being pushed into the ordered γ' from the matrix under the influence of an applied shear stress. Assuming an activated configuration shown in fig. 29(a), an energy of ΔG is required:

$$\Delta G = E_{\rm sd}(2R\theta) - E_{\rm 2d}(L) - 2\tau b R^2\left(\theta - \frac{\sin 2\theta}{2}\right), \tag{10}$$

where R and θ relate to the geometry of the bowing dislocations. $E_{\rm 2d}$ and $E_{\rm sd}$ are the energies per unit length, respectively, of the superdislocation in the γ' and the two ordinary dislocations pushed against the interface before entering the γ':

$$E_{\rm 2d} = 2E_{\rm o} + I_{\rm o-o},$$

$$E_{\rm sd} = 2E_{\rm s} + I_{\rm s-s} + \Delta\chi_{\rm APB},$$

where:

$$E_{\rm o} = 2E_{\rm p} + \delta_{\rm SF}\chi_{\rm SF},$$
$$E_{\rm s} = 2E_{\rm sp} + \delta_{\rm CF}\chi_{\rm CF}, \tag{11}$$

with $E_{\rm o}$ and $E_{\rm s}$ being the energy per unit length of the ordinary $\frac{a}{2}\langle 110\rangle$ dislocations and the corresponding superpartials, respectively, in the matrix and precipitate phases. The interaction energies between these dislocations in the matrix and precipitate, respectively, are $I_{\rm o-o}$ and $I_{\rm s-s}$. The ordinaries are further dissociated in the matrix (each partial with an energy $E_{\rm p}$ per unit length), and also within the precipitate (partials with energy $E_{\rm sp}$), with stacking and complex fault energies of $\chi_{\rm SF}$ and $\chi_{\rm CF}$, and fault widths of $\delta_{\rm SF}$ and $\delta_{\rm CF}$, respectively. Finally it is apparent that there is an additional contribution to the energy of the pair of dislocations once they glide into the precipitate, due to the creation of an anti-phase boundary of energy $\chi_{\rm APB}$ with a separation of Δ between dislocations in the pair.

In this calculation we assume an APB energy of 110 $\rm mJ/m^2$ ($\Delta = 6.9$ nm), and a stacking-fault energy of 90 $\rm mJ/m^2$ ($\delta = 1.5$ nm), $\mu = 48.2$ GPa, $\nu = 0.33$ and $b =$

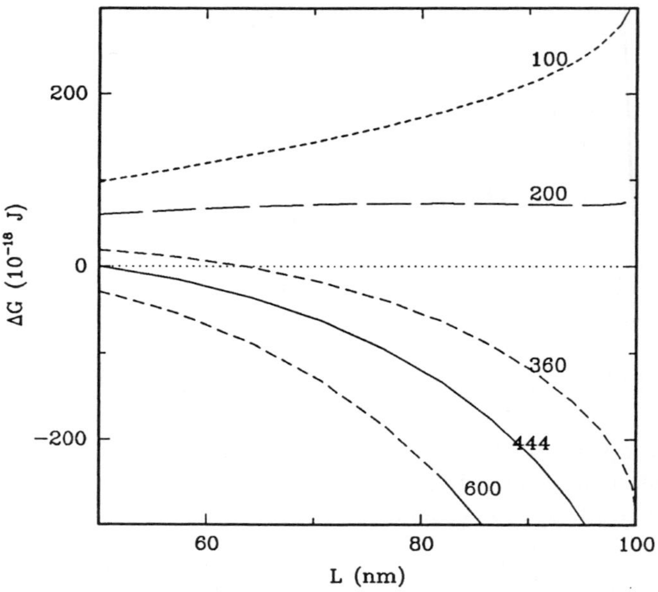

Fig. 30. Free energy required for precipitate shearing as a function of segment length along the interface. The results are plotted for a range of τ (MPa), with $R = 50$ nm.

0.254 nm. (In the absence of conclusive experimental data, this calculation assumes that the complex fault energy in the precipitate and the stacking-fault energy of the matrix are of the same magnitude; however, the results of the calculation would not be substantially influenced by a somewhat higher complex fault energy.) Two of the three geometrical bowing parameters, R, L and θ are independent, so the free energy can be minimized with respect to R with a fixed L, or with a fixed R with varying initial length of the segment at the interface, L. The length of the initial segment along the interface L, is likely to be set by the spacing of the interfacial dislocations. Additionally, for a given L, there is a minimum R, below which only one of the two dislocations have penetrated the precipitate.

The mechanical threshold stress, τ_s, the stress at which the precipitates are sheared with no thermal assistance, can be determined by evaluation of eq. (10) to determine the stress level at which there is no energy barrier. The energy required to push the pair of dislocations into the γ' as a function of the initial length of the segment at the interface, L, is shown for $R = 50$ nm in fig. 30. As can be seen from fig. 30, τ_s is around 444 MPa. This would correspond to a uniaxial stress of 1088 MPa, which is comparable to the yield strengths of these materials at room temperature, fig. 3. The variation in the athermal shearing stress as a function of R is shown in fig. 31. Complete penetration at lower stresses for a larger R would require higher L, and thus larger spacings in the interfacial network. A similar calculation for a single dislocation would show that higher stresses are required due to the continuous creation of APB. Referring back to the finite-element plots of fig. 26, the equivalent stresses in the γ' precipitate on the flat portion of the interface just below the corner reach 1000 MPa after 4.3×10^5 seconds. This is

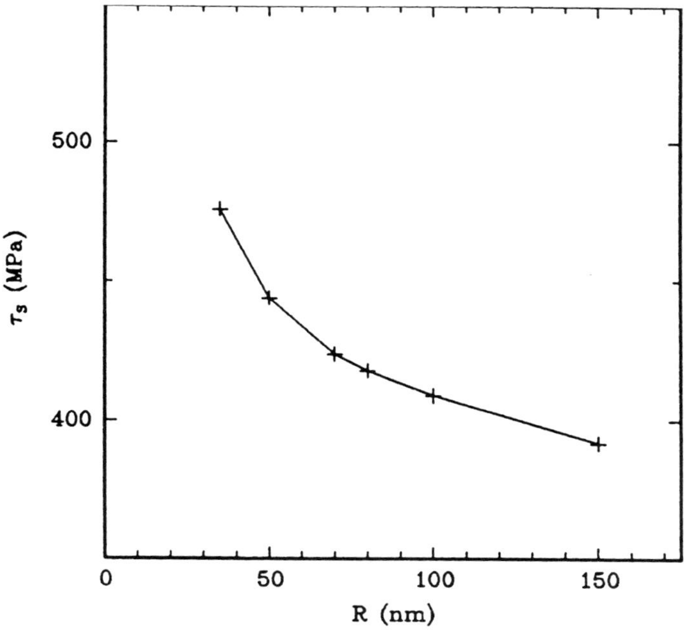

Fig. 31. The variation in the athermal shearing stress as a function of dislocation bowout radius.

consistent with the microscopy observations that show dislocations shearing the γ' only in the later stages of creep, and may in part contribute to the long, gradual tertiary transient characteristic of creep deformation in this temperature range. However, damage in the form of microcracks that develop in interdendritic regions in the later stages of creep also contribute to the acceleration, and the individual contributions of these mechanisms have not been considered in any detail.

Another consequence of the presence of planar precipitate/matrix interfaces is also highlighted by the finite element analyses [69]. Because of the constraint of the coherent interfaces, no sliding between the phases occurs, resulting in the build-up of large pressure gradients in the matrix channels, fig. 32. In fig. 32, for an imposed creep rate of 2.5×10^{-8}/s, the peak mean normal stress at the centre of the horizontal channels is 414 MPa. Note that in this stage of creep, the Mises equivalent stresses in the matrix on average are about 40 MPa, consistent with the resistance offered by the dislocation network, from eq. (7). Experimentally, for this creep rate of 2.5×10^{-8}/s, an applied stress of 552 MPa is sustained in CMSX-3. Thus, it is apparent that the constraint of the coherent interfaces can result in a significant increase in the load-carrying capacity of the two-phase material.

This constraint effect is analogous to the "friction hill" problem in forging [136], where friction at the interface between a rigid die and a thin disk of deforming material elevates the stress P required for deformation of a material with a uniaxial flow stress of σ_0. For complete sticking at the interface (analogous to maintaining coherency) the average

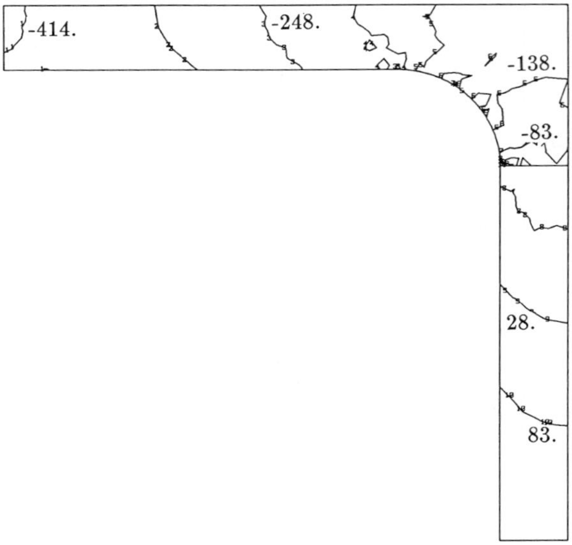

Fig. 32. Finite-element calculations of pressure distributions in the matrix following creep deformation [69].

increase in the deformation resistance of the material is dependent on the aspect ratio of the disk (R/t, radius/thickness):

$$\left(\frac{P}{\sigma_{\mathrm{o}}}\right)_{\mathrm{avg}} = 1 + \frac{2k}{3}\left(\frac{R}{t}\right), \tag{12}$$

where k is a constant equal to 0.577. In the γ–γ' deformation problem, it is necessary to account for both the horizontal and the vertical matrix channels by superposition [69]. Given a cube edge length L and a channel height h, the total increase in the deformation resistance is approximately:

$$\left(\frac{P}{\sigma_{\mathrm{o}}}\right)_{\mathrm{tot}} \approx 2\left[1 + \frac{2k}{3}\left(\frac{L}{h\sqrt{\pi}}\right)\right]. \tag{13}$$

For a superalloy with a high volume fraction of γ', typical dimensions of L and h are 0.45 μm and 60 nm, respectively. This would suggest that the creep strength of the two-phase material could be higher than that of the matrix by a factor of 5.3, in the limit of no slippage along the γ–γ' interface. However, as seen in fig. 22, it is apparent that dislocation climb along the interface occurs, reducing the effective frictional constraint and the magnitude of $(P/\sigma_{\mathrm{o}})_{\mathrm{tot}}$.

It follows then that the total creep resistance for the superalloy in this intermediate temperature range has contributions from the Orowan resistance, the solid solution and dislocation resistances and the flow constraint of the coherent interfaces:

$$\frac{\dot{\varepsilon}}{\dot{\varepsilon}_o} = \left[\sigma_{app} - \left\{\zeta\left(\frac{L}{h}\right)\left[\sigma_{OR} + \sigma_{SS}(T) + \sigma_{DIS}(T)\right]\right\}\right]^n, \tag{14}$$

where $\dot{\varepsilon}_o$ is a material constant and $\zeta(L/h)$ is a constraint factor that ranges approximately between 1 and 2, depending on the dimensions of the matrix channels (and thus the volume fraction of the precipitates) as well as the shape of the precipitates, and accounts for the frictional resistance of the interface. The terms σ_{OR}, σ_{SS} and σ_{DIS} are the contributions of the Orowan, solid solution and dislocation resistances, respectively. The dislocation and solid solution resistances are expected to be temperature sensitive, with a characteristic activation energy that correlates with static and/or dynamic recovery processes and accounts for their temperature dependence.

5.3. High-temperature creep

High-temperature deformation in single-crystal superalloys is dominated by two prominent features: (I) the rapid development of equilibrium interfacial dislocation networks [82,133, 137–139], and (II) directional coarsening or "rafting" of the precipitates [140–146]. As in the intermediate temperature range, during the early stages of creep, deformation is somewhat inhomogeneous due to limited dislocation sources, fig. 18. However, due to more rapid diffusion, as dislocations with Burgers vectors of all the stressed slip systems percolate through the structure, nodal networks that compensate misfit form rapidly. The kinematic details of their formation are important, as the directional coarsening of the precipitates is influenced by the local state of the misfit. Network formation is discussed in detail in the following section. Following this, a number of aspects of the early stages of creep deformation in several different alloys are discussed.

Figure 33 summarizes the major features of the directional coarsening process as it is influenced by lattice misfit in two ternary Ni–Al–Mo alloys. The alloy compositions and lattice misfit at the temperature of stressing are given in table 7. For the alloy with negative misfit at elevated temperatures (Alloy R3), under uniaxial loading the initially cuboidal precipitates evolve to a plate structure, with the broad faces of the plates oriented normal to the axis of the applied tensile stress or parallel to the axis of the applied compressive stress. Fourier transforms of the image in the plane parallel to the axis of the applied stress clearly show a preferential alignment. Conversely, for the positive misfit material (Alloy R1), the plates align parallel to the tension axis and perpendicular to the compression axis. It is apparent that the superposition of the misfit and applied stresses has a pronounced influence on the evolution of microstructure during creep. For this reason, both the initial elastic state of stress and any local plastic deformation and network formation will influence the development of the rafted structure; experiments on this aspect of high-temperature deformation will be discussed. Finally, new observations on the influence of rafting on high-temperature creep properties will be presented.

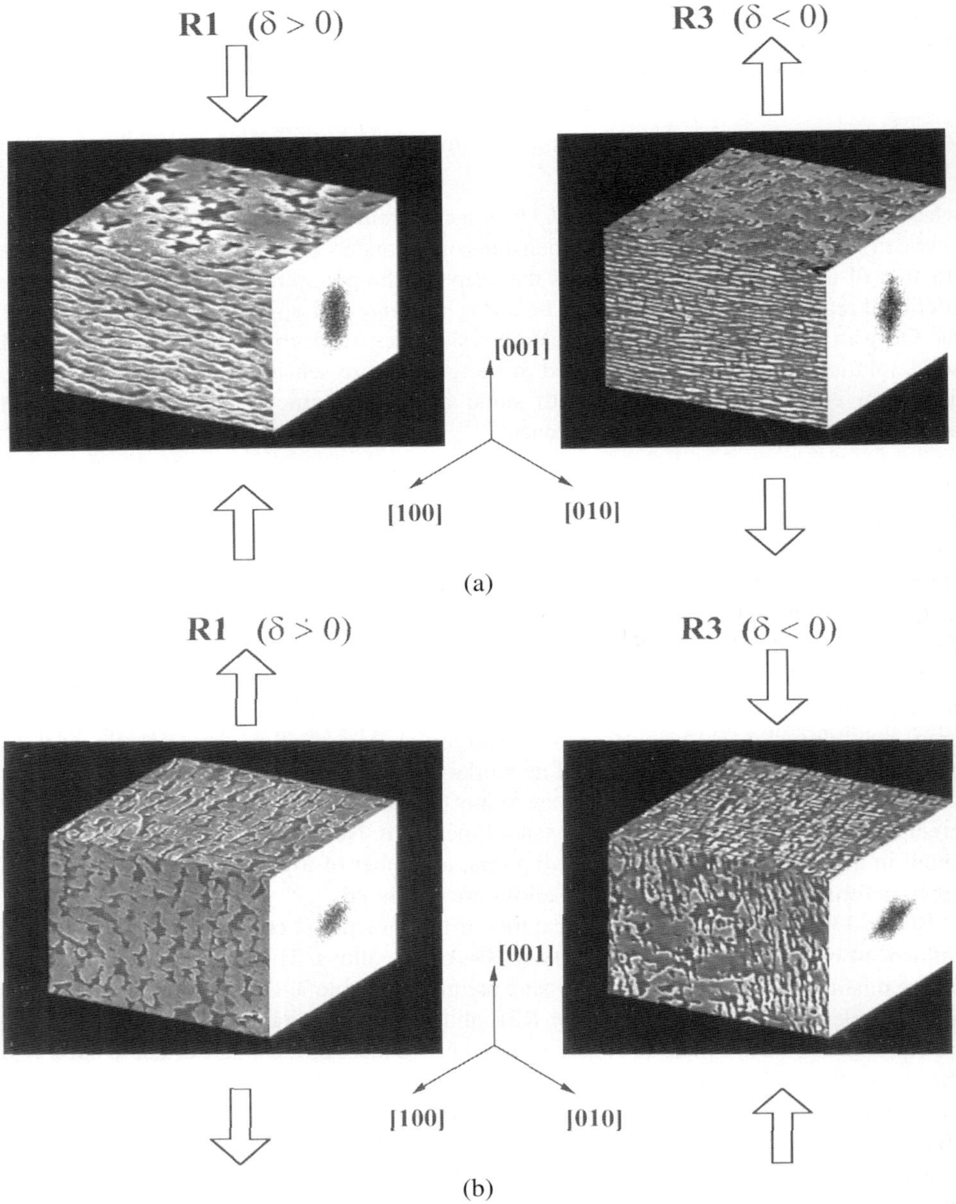

Fig. 33. Directional coarsening in ternary Ni–Al–Mo alloys, resulting in the formation of plates of the precipitate phase. Plate orientation is sensitive to the positive/negative sense of the lattice misfit, δ, and the tensile/compressive nature of the applied stress. Alloy R1 has a positive misfit, while Alloy R3 has a negative misfit (table 7). The [010] faces show a Fourier transform of the image of the [100] faces, demonstrating the directionality of the rafting process in tension vs. compression as a function of misfit.

Table 7

Ternary Ni–A1–Mo alloys utilized for directional coarsening studies. Courtesy of M. Fahrmann

Alloy designation	Composition (wt%)	Misfit at 850°C
R1	Ni–9.6Al–0.9Mo	+0.4%
R3	Ni–8.8A1–13.0 Mo	−0.45%

5.3.1. Network formation

During high-temperature exposure, the lattice misfit between the γ and γ' phases can be relieved by the formation of networks of edge dislocations at the interfaces between the two phases. For a negative lattice mismatch, slip-deposited dislocations, resulting from deformation in the horizontal matrix channels as discussed in section 5.2, are arranged such that their extra half planes are in the phase with the smaller lattice parameter (i.e. the γ'), thus accommodating the lattice parameter mismatch between the phases. Larger mismatch alloys require more closely spaced dislocations to relieve the mismatch, and measurement of the dislocation spacing in the nets can yield a measure of the lattice mismatch [82], as discussed in section 4.

The formation of interfacial mismatch-accommodating dislocation networks occurs concurrently with rafting. As discussed in section 5.3.2, there is increasing evidence that these interfacial dislocations, whether in equilibrium mismatch configuration or in the less efficient slip orientations, provide the driving force for rafting during creep by relieving coherency stresses in the horizontal channels, thus stabilizing the horizontal γ/γ' interfaces with respect to the vertical interfaces. In order to understand high-temperature creep, the mechanisms for the formation of the mismatch networks and the dislocation knitting and unknitting processes necessary for continued deformation in microstructures with the fully developed networks need to be considered.

The nature of the interfacial nets is well established [82,133,137–139]. The configuration of the equilibrium mismatch-accommodating network consists of pure edge dislocations, representing all six of the $\frac{1}{2}\langle 110\rangle$ Burgers vectors in the f.c.c. lattice. This includes Burgers vectors which have no resolved shear stress for the applied stress (e.g., $\frac{1}{2}[110]$ for a [001] tensile stress). Since the dislocations within the nets originate from slipping matrix dislocations captured on the interface, these non-slip dislocations must result from interactions between slip-deposited dislocations. Also, the line directions of slip-deposited dislocations are not in the proper orientation to efficiently relieve the mismatch. As dislocations move from "slip" to "mismatch" orientations, there is a reduction in misfit energy as well as a considerable decrease in dislocation line energy, due to the more efficient accommodation of misfit. The line directions of the dislocation segments at the interfaces give an indication of how well established the mismatch accommodating networks have become in the specimen. The "slip" and "mismatch" orientations for dislocations slipping on {111} planes in a [001] oriented specimen are given in table 8.

A mechanism has been proposed for the reorientation of Orowan loops around the γ' particles to yield mismatch orientated line segments [147]. In this model, the loops must rotate by 90° along the γ/γ' interface, by a combination of glide and climb. While this model is appropriate for isolated γ' particles, it does not apply to high γ' volume fraction modern single crystal alloys in which dislocations do not tend to form Orowan

Table 8
Comparison of slip-deposited and mismatch-accommodating dislocation line directions on (001) γ/γ' interfaces for a [001] oriented creep specimen

Burgers vector	Slip/mismatch plane*	Line direction
011	$1\bar{1}1$	$\bar{1}\bar{1}0$
011	$\bar{1}\bar{1}1$	$\bar{1}10$
011	001_m	100
$01\bar{1}$	$\bar{1}11$	$\bar{1}\bar{1}0$
$01\bar{1}$	111	$\bar{1}10$
$01\bar{1}$	001_m	100
101	$\bar{1}11$	110
101	$\bar{1}\bar{1}1$	$\bar{1}10$
101	001_m	010
$10\bar{1}$	111	$\bar{1}10$
$10\bar{1}$	$1\bar{1}1$	110
$10\bar{1}$	001_m	010

*The subscript "m" indicates a mismatch orientation.

loops around individual γ' particles, but travel long distances through the γ channels, as discussed in section 5.2.

A more appropriate model for alloys containing high volume fractions of precipitates has been proposed in which the networks form by nodal reactions between slip-deposited dislocations followed by localized rearrangements [138]. An example of a process by which the equilibrium mismatch network configuration can be developed by reactions between slip-oriented dislocations is shown in fig. 34. In this example, two perpendicular sets of $\frac{1}{2}[0\bar{1}1]$ dislocations, fig. 34(a), one resulting from slip on (111) and the other from slip on $(\bar{1}11)$, react to form a single set with an average line direction of [100], as shown in fig. 34(b). In this manner, the "mismatch" configuration is achieved without long-range climb and/or glide. The addition of $\frac{1}{2}[10\bar{1}]$ dislocations, resulting from slip on the $(1\bar{1}1)$ plane, and subsequent reactions result in a net containing $\frac{1}{2}[0\bar{1}1]$, $\frac{1}{2}[1\bar{1}0]$, and $\frac{1}{2}[10\bar{1}]$ dislocations, all oriented in optimum mismatch accommodating configurations, figs 34(c) and (d). The $\frac{1}{2}[101]$ dislocations, resulting from slip on the $(\bar{1}\bar{1}1)$ plane may also be added and knitted into the net, as shown in figs 34(e) and (f). Subsequent reactions of the type: $\frac{1}{2}[101] - \frac{1}{2}[0\bar{1}1] \to \frac{1}{2}[110]$ and $\frac{1}{2}[101] - \frac{1}{2}[1\bar{1}0] \to \frac{1}{2}[011]$ result in the net shown in fig. 34(g), the equilibrium mismatch configuration. Alternatively, the $\frac{1}{2}[101]$ and $\frac{1}{2}[10\bar{1}]$ dislocations can combine to form the [100] Burgers vector, as shown in fig. 34(h). Further dissociation of the [100] segments once again yields the standard (001) net. The formation of the mismatch nets may also be aided by the extensive cross-slip known to occur in these alloys. The "zig-zag" morphology of fig. 34(a) and (b), for example, could result from cross-slipping of a single set of $\frac{1}{2}[0\bar{1}1]$ dislocations. This is depicted schematically in fig. 35. Here, cross-slipping dislocations with several different Burgers vectors are deposited on the γ/γ' interface. Diffusional rounding of the corners at the cross-slip locations then occurs, along with climb of dislocation segments along the interface and reactions between these segments to form the mismatch network. It is worth noting that at these high temperatures multiple types of $\frac{1}{2}\langle 110\rangle$ dislocations are typically

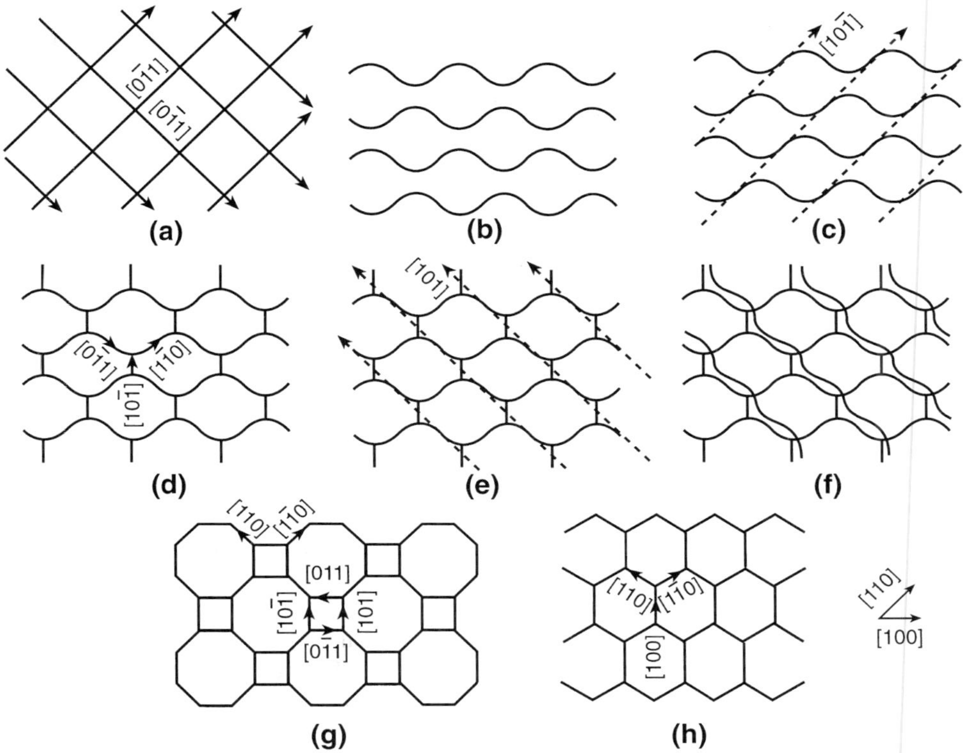

Fig. 34. Schematic representation of dislocation reactions leading to the formation of mismatch-accommodating nets on a (001) γ/γ' interface during creep. Line directions and Burgers vectors are marked on the figures. See text for details [138].

available, regardless of orientation or imposed stress state; recent experiments on crystals loaded in pure shear along $\langle 100 \rangle$ directions revealed interfacial network formation in a manner similar to outlined above [148,149], even though the slip systems required for their formation have largely different Schmid factors.

Many permutations of these basic themes are possible, all of which rely only on local dislocation rearrangements within the nets, without the necessity of long-range climb and/or glide around the interfaces. Of course, the actual sequence of dislocation knitting into the nets will vary within the interface according to the local dislocation content, resulting in many different intermediate configurations within a given net. This is particularly true for creep specimens, in which dislocations continue to enter and leave the nets, so that the steady-state configuration will deviate from that of the unstressed condition. Also, the scale of these interactions will not necessarily be that of the equilibrium mismatch spacing, particularly early in creep, when the dislocation densities in the interfaces are low. At this stage, longer climb/glide distances will be required, as governed by the actual dislocation spacing in the interface.

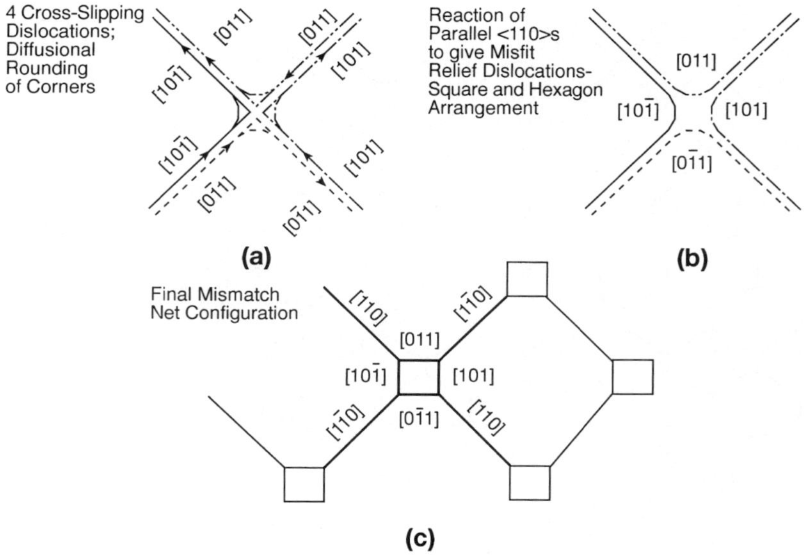

Fig. 35. Schematic representation of an alternative mechanism for mismatch net formation on γ/γ' interfaces. Corners on cross-slipped dislocations are rounded through diffusional processes, followed by nodal reactions similar to those described in fig. 34 [138].

Experimental evidence which supports this model has been reported [138]. One example is shown in fig. 36, which shows a transverse cross-section of a PWA 1484 specimen tested to 400 hours ($\sim$10% life) at 982°C. The interfacial dislocation nets in this specimen are in distorted mismatch configurations in some areas; however, a large number of the interfacial dislocations are in slip orientations as seen in the upper left corner of the micrograph. Several slip oriented dislocations are seen entering into a distorted mismatch network, demonstrating the transition from slip-deposited dislocation structures to mismatch-accommodating networks. For example, many of the $\boldsymbol{b} = \frac{1}{2}[10\bar{1}]$ dislocations in the lower right region of the micrograph (and depicted schematically in fig. 35(c)) are in slip orientation outside of the mismatch net and curve into the mismatch orientation as they enter the net.

An example of diffusional rounding at a cross-slip site is shown in fig. 37. This René N5 specimen was tested to 0.4% creep strain (235 hours, $\sim$10% expected life) at 982°C and also contains partially developed dislocation networks at the γ/γ' interfaces. The dislocation at the center of this micrograph is a $\boldsymbol{b} = [011]$ with segments in both mismatch (m) and slip (s) orientations. The mismatch-oriented segment has formed by the mechanism depicted in fig. 35.

An example of a fully developed network is shown in fig. 38. Note the combination of square and hexagon shaped elements (compare to the schematic in fig. 34(g)), made up of all of the $\frac{1}{2}\langle 110\rangle$ Burgers vectors in the f.c.c. structure. A few $\langle 100\rangle$ dislocations are also present in this network, as depicted in the schematic of fig. 34(h). Although the spacing of the networks is highly variable in this specimen, it is interesting to compare measurements

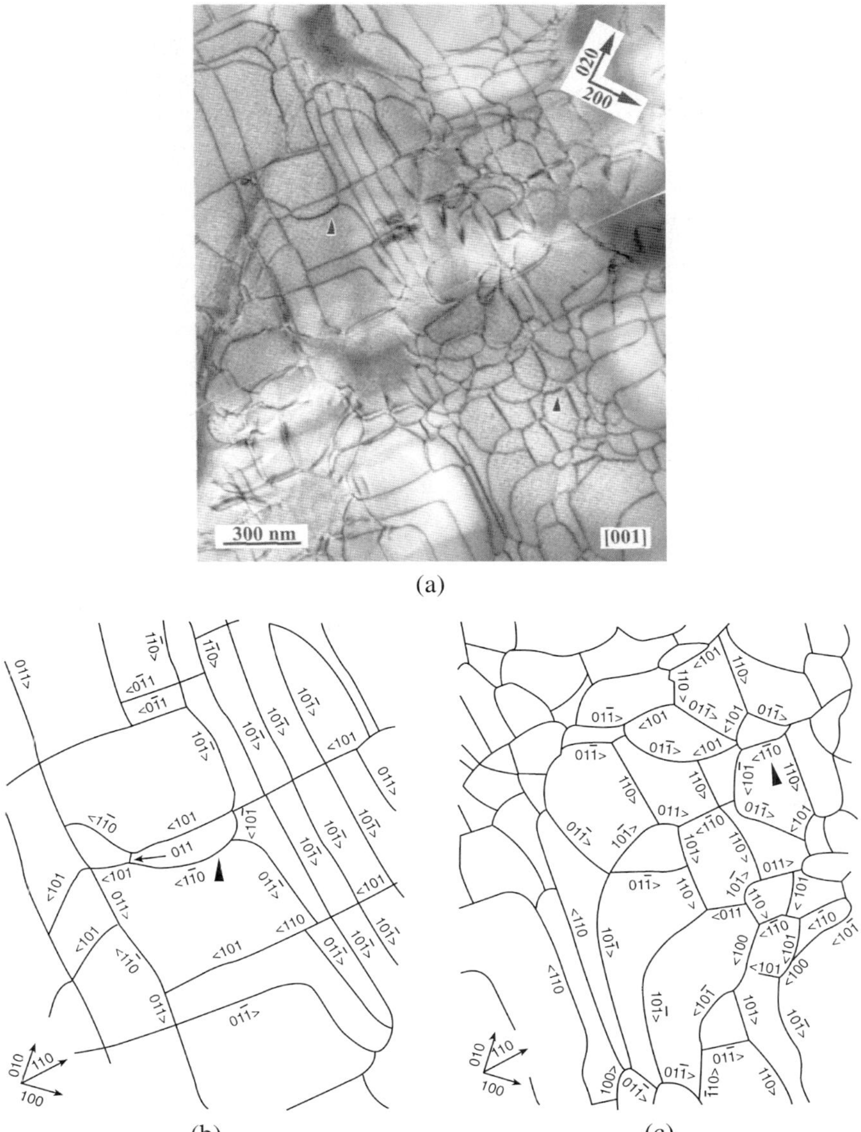

Fig. 36. TEM micrograph (a) and accompanying schematics ((b) and (c)) of interrupted PWA 1484 creep specimen tested to 400 hours (∼10% expected life) at 982°C showing dislocation networks on the γ/γ' interfaces (overlay of two micrographs to show all of the dislocations within the net, some double images of dislocations appear due to tilt between images). Most of the dislocations in the upper left are slip oriented (schematic in (b)), while the dislocations in the lower right hand corner of the micrograph are forming a distorted mismatch network (schematic in (c)). The arrows provide common reference points in the micrograph and schematics. All $\langle 110\rangle$ Burgers vectors have a magnitude of $\frac{1}{2}\langle 110\rangle$ (transverse cross-section) [138].

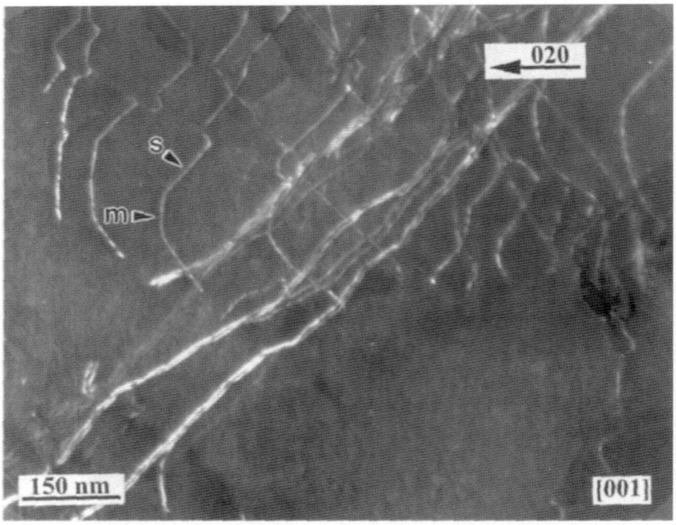

Fig. 37. Weak-beam TEM micrograph of an interrupted René N5 creep specimen tested to 0.4% creep strain (235 hours, ~10% expected life) at 982°C, showing incomplete mismatch networks with some slip oriented dislocations within the nets (example shown with arrows is a $\boldsymbol{b} = [011]$ dislocation in both mismatch (m) and slip (s) orientations) as well as paired $\boldsymbol{b} = \frac{1}{2}[011]$ dislocations in slip orientation (line direction $= [1\bar{1}0]$) within the γ matrix (transverse cross-section) [138].

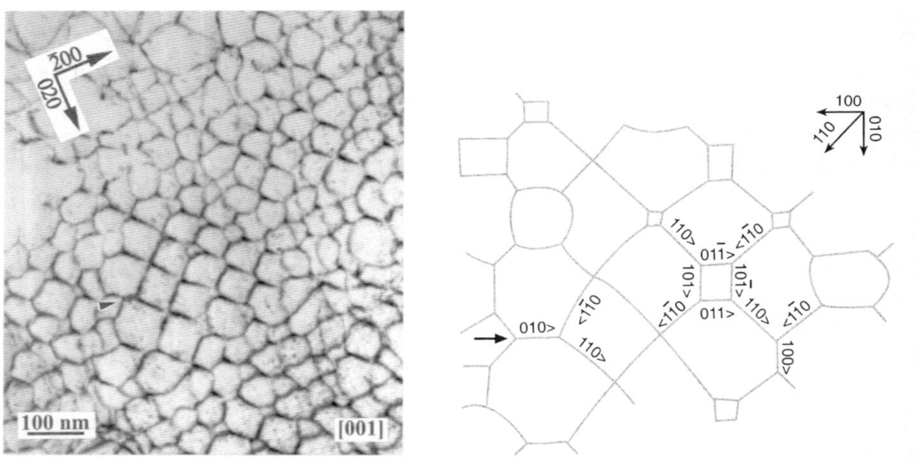

(a) (b)

Fig. 38. TEM micrograph (a) and accompanying schematic (b) of an interrupted René N5 creep specimen tested to 0.5% creep strain (20 hours, ~ 5% expected life) at 1093°C, showing dislocation networks on the γ/γ' interfaces. Arrows in (a) and (b) show equivalent positions (overlay of two micrographs to show all of the dislocations within the net, some double images of dislocations appear due to tilt between images) (transverse cross-section) [138].

from this area to those from samples aged at the same temperature under no stress. In the unstressed condition, the spacing is approximately 98 nm, yielding a lattice mismatch of 0.26%. In fig. 38, the spacing is approximately 60 nm. The decreased spacing reflects extra dislocations at the interface due to the external stress. An increasing density of dislocations at the interface at higher creep strains has also been observed in several other alloys [131, 150].

5.3.2. Deformation processes at high temperatures

In discussing high-temperature deformation processes, a distinction must be made between the early stages of creep, before the development of rafted γ' with fully developed interfacial dislocation networks, and later stages when the rafted microstructure necessitates γ' cutting through well developed interfacial networks.

Similar to the process discussed in section 5.2, during the early stages of high-temperature creep the γ' is cuboidal and deformation occurs by cross slip assisted motion of loops through the matrix channels, fig. 18. Again, because of the limited sources for dislocations, in the early stages of creep, deformation is inhomogeneous, and often dominated by dislocations of a single Burgers vector. Also similar to the intermediate-temperature behavior, incubation periods prior to primary creep may be present. During this stage, the creep rate of the specimen is highly dependent on the magnitude of the misfit and the creep resistance of the matrix material.

At these high temperatures, other matrix deformation processes are occasionally observed. Figure 39 shows transverse cross-section TEM micrographs of a specimen in the early stages of creep ($\sim$ 10% of expected rupture life) at 982°C. The stress axis is [001]. This is an experimental alloy which has very high Co content. In fig. 40, a schematic diagram is provided, showing the results of a Burgers vector analysis. All of the dislocations are in the matrix or on the γ–γ' interfaces. Slip oriented $\frac{1}{2}[101]$ and $\frac{1}{2}[011]$ dislocations are present, as well as a stacking fault on a $(11\bar{1})$ plane. In fig. 41, a longitudinal cross section of the same specimen is shown, revealing stacking faults bridging the matrix channels.

A possible mechanism for the formation of these stacking faults is as follows. The $\frac{1}{2}[011]$ and $\frac{1}{2}[101]$ dislocations interact to form a $\frac{1}{2}[1\bar{1}0]$ segment. This segment is in edge orientation, ideal for mismatch accommodation, but with no resolved shear stress with respect to the [001] applied stress. The constituent Shockley partials of this dislocation on the $(11\bar{1})$ plane ($\frac{1}{6}[1\bar{2}\bar{1}]$ and $\frac{1}{6}[2\bar{1}1]$) do have resolved shear stresses in equal and opposite directions with respect to the applied stress and can dissociate by slip. This process is aided by a low stacking-fault energy in the γ, expected for this high Co alloy, particularly if elemental partitioning to the fault is considered. This mechanism could explain the high number of faulted dislocations in the matrix of this specimen. Stacking faults bridging the matrix channels serve as effective barriers to dislocation motion. It should also be noted that the low stacking-fault energy of this alloy retards cross slip, necessary for the dislocation percolation process described above. This alloy displays extremely sluggish rafting and excellent creep resistance at 982°C, as expected from the observed dislocation structures. This example illustrates the importance of stacking-fault energy in the creep behavior of superalloys.

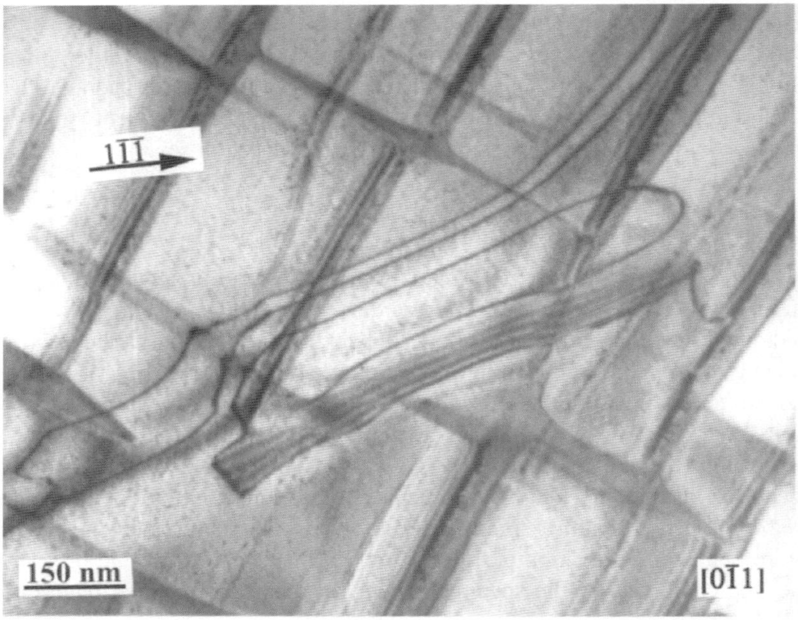

(a)

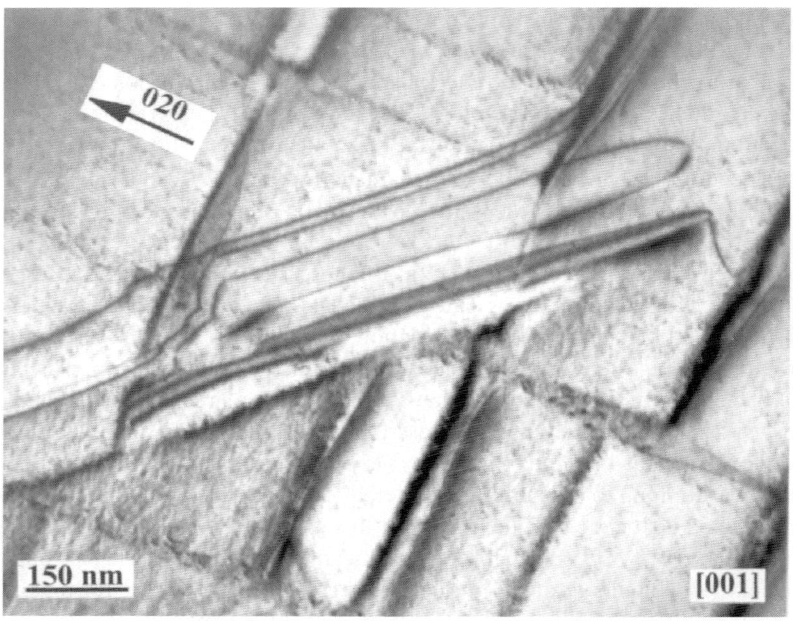

(b)

Fig. 39. Bright-field TEM micrographs showing dislocations in a transverse cross-section of a low stacking-fault energy alloy tested to 0.3% creep strain at 982°C (237 hours, ~10% of expected life).

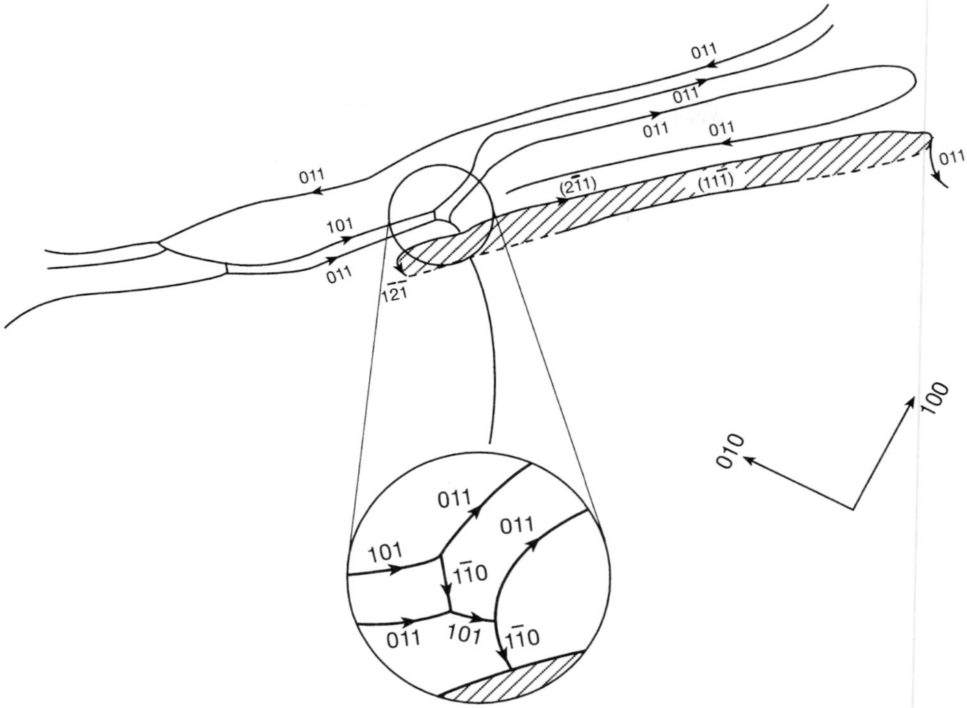

Fig. 40. Schematic diagram showing the results of a Burgers vector analysis on the dislocations shown in fig. 39.

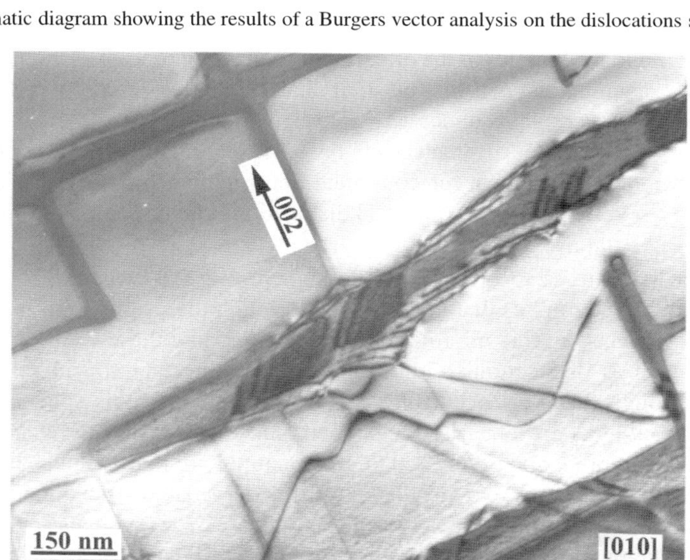

Fig. 41. Bright-field TEM micrographs showing stacking faults in the matrix channels of the specimen shown in fig. 39 (longitudinal cross-section).

Concurrent with the initial plastic deformation processes is the directional coarsening of the γ' precipitates. In high volume-fraction γ' single crystals, plastic deformation has a very strong influence on the directional coarsening process. This is apparent because of the inhomogeneity of the plastic deformation process. Two regions from a single creep specimen tested at 1050°C and 50 MPa are shown in the bright field TEM micrographs in fig. 42. The horizontal matrix channels are contained in the plane of the foils and the applied tension axis is normal to the plane of the micrograph. In the region of the sample where the dislocation density in the horizontal matrix channels is high, the rafting process is complete. However, in the region of the sample where the density of dislocations is low, matrix channels are still visible.

Referring again to table 6, it is apparent that an applied stress of 50 MPa superimposed on the misfit stress in CMSX-3 would result in resolved shear stresses that exceed the Orowan stress in the horizontal channels, but not in the vertical channels. This deformation-induced inhomogeneity can also be detected by X-ray measurements of lattice parameters in the horizontal and vertical channels following straining at 1050°C [151–153]. The importance of the early stages of the plastic straining process is highlighted by the experiments by Véron et al. [154] and Fahrmann et al. [155,156], where negative misfit single crystals were subjected to brief straining in tension at intermediate temperatures to place dislocations in the horizontal matrix channels without any changes in the precipitate morphology. In subsequent static annealing experiments *under no external stress*, directional coarsening was still observed in a manner consistent with the application of a tension stress. Using Small Angle X-ray Scattering, Fahrmann et al. [156] also showed that prestraining a sample of the Ni–Al–Mo alloy shown in fig. 34(b) in tension caused rafting under an applied compressive stress to occur initially in the opposite direction to that shown in fig. 34(b). These experiments collectively suggest that plasticity alters both the driving force and the kinetics of the directional coarsening process. A number of models of directional coarsening that consider plasticity have recently been proposed [92,154,155, 157–159], and this area remains under active investigation.

After rafting is complete, the matrix channels no longer provide a continuous path for dislocation motion necessary for continued deformation, forcing shearing of the γ' phase. In fig. 37 several pairs of $\boldsymbol{b} = \frac{1}{2}[011]$ dislocations can be seen, with line directions approximately parallel to $[\bar{1}10]$ (thus lying on $(\bar{1}\bar{1}1)$ slip planes). Reverse $\boldsymbol{g}$ experiments showed these dislocations to have the same sign of Burgers vector. The fact that the dislocations are paired implies that they have cut through the γ' precipitates, even though they are in the γ phase in this micrograph. Dislocations like these were common in this specimen, as well as dislocations within the γ', indicating that considerable cutting of dislocations through the γ' is occurring. The γ' has rafted in this specimen; however, the interfacial networks are not fully developed, and many dislocations are still in slip orientations. The combination of a rafted microstructure with poorly developed interfacial nets may be responsible for the extensive γ' cutting observed in this specimen. It is difficult for dislocations to loop around the rafts, and the networks, which can act as barriers to dislocation movement through the γ', are not yet fully established.

In another area of the same specimen, several faulted dislocations can be seen within the γ' (fig. 43). In one instance (center of the micrograph), a $\boldsymbol{b} = \frac{1}{3}[11\bar{2}]$ dipole can be seen emanating from $\boldsymbol{b} = [10\bar{1}]$ dislocation bowed out on a (111) slip plane. A more detailed

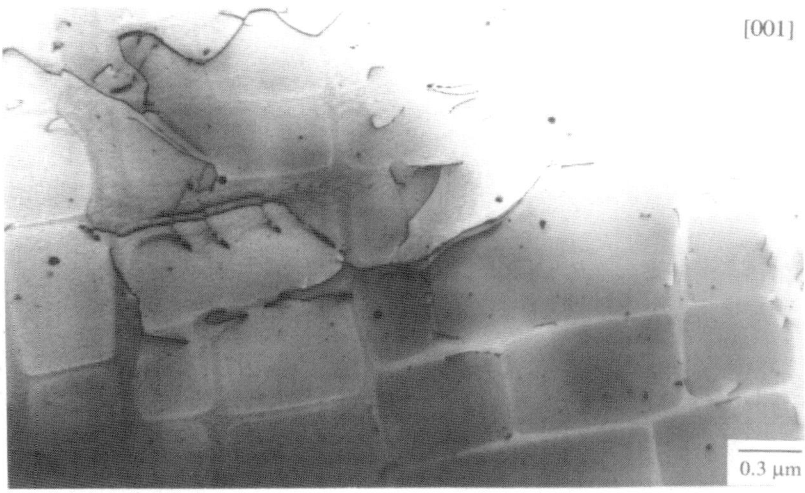

(a)

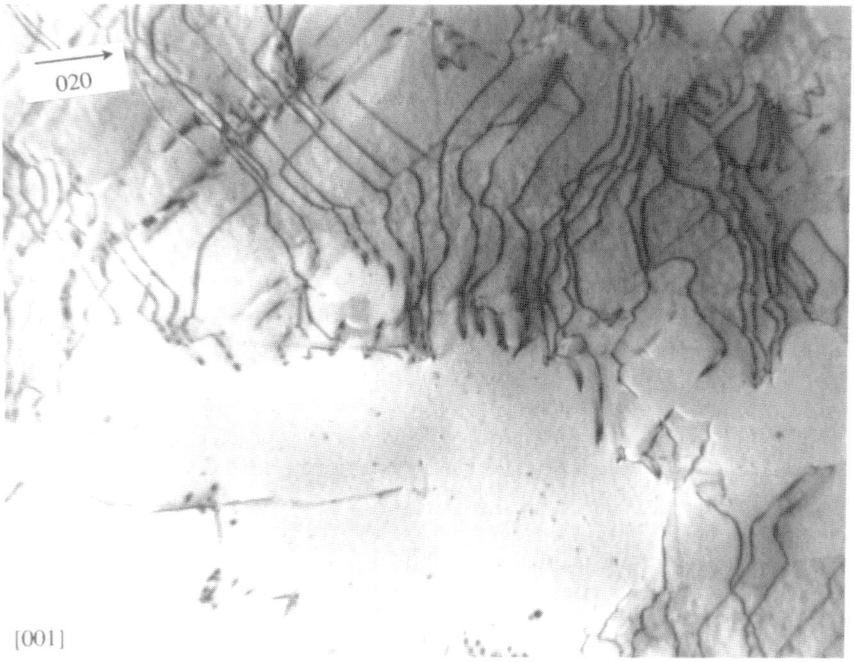

(b)

Fig. 42. TEM micrographs showing two regions from a single CMSX-3 creep specimen tested at 1050°C and 50 MPa. Rafts are developed to a greater degree in regions where plastic deformation in the horizontal matrix channels has occurred. Foil normal is [001] [92].

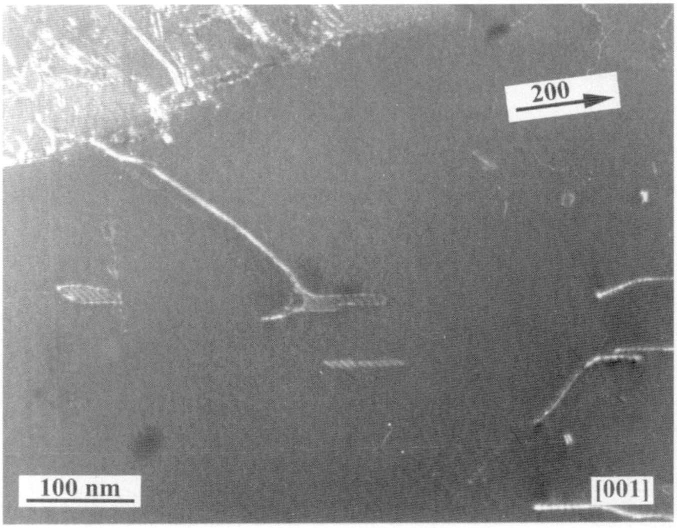

Fig. 43. Weak-beam TEM micrograph of a René N5 creep specimen tested to 0.4% creep strain at 982°C (235 hours, ~10% of expected life) showing dislocations within the γ'. The dislocation marked in the center of the micrograph is a $\boldsymbol{b} = [10\bar{1}]$ [dislocation bowed out on a (111) plane in the γ' and with an extended dissociated region ($[10\bar{1}] \rightarrow \frac{1}{3}[11\bar{2}] + \frac{1}{3}[2\bar{1}\bar{1}]$) (longitudinal cross-section)].

analysis of a similar dipole is shown in fig. 44. This is a René N4 specimen tested at 982°C for 30 hours (~ 40% expected life) at a relatively high stress, so that well developed networks were not able to form. An image simulation program [160] was used to identify this feature as a $\frac{1}{3}[112]$ dipole with a [011] line direction (lying on a $(11\bar{1})$ plane). The implications of these features are unclear, but they probably served as pinning points for $\boldsymbol{b} = \langle 110 \rangle$ dislocations in the γ' phase, as evidenced by the bowing of the dislocation in fig. 43. Similar dislocation structures have been observed previously within γ' precipitates and have been associated with dislocation pinning [161].

Once the mismatch accommodating interfacial networks are fully established, the cutting mechanism changes. The dislocation density in both phases is low, with dislocations mainly confined to the interfacial networks. Some paired $\frac{1}{2}\langle 110 \rangle$ dislocations are observed in the γ'. An example is given in fig. 45. This is a René N5 specimen, tested at 982°C for 150 hours (~ 40% expected life).

In fig. 46, a stereo pair is presented showing another example of paired $\frac{1}{2}\langle 110 \rangle$ dislocations within the γ' phase. This specimen was first given a "pre-raft" treatment at 1093°C for 20 hours (~ 5% of expected life) in order to establish a rafted structure with fully developed interfacial networks. It was then subjected to a 982°C/235 h exposure (~ 10% expected life). One of the paired dislocations (arrow, lower left) can clearly be seen as originating from two dislocations within the network. Two other $\frac{1}{2}\langle 110 \rangle$ dislocations (arrows, upper right) can be seen coming out of the network into the matrix phase.

Another example of dislocations in the γ' originating from the interfacial network is shown in fig. 47. This René N5 specimen was interrupted after creep at 1093°C for 40 hours

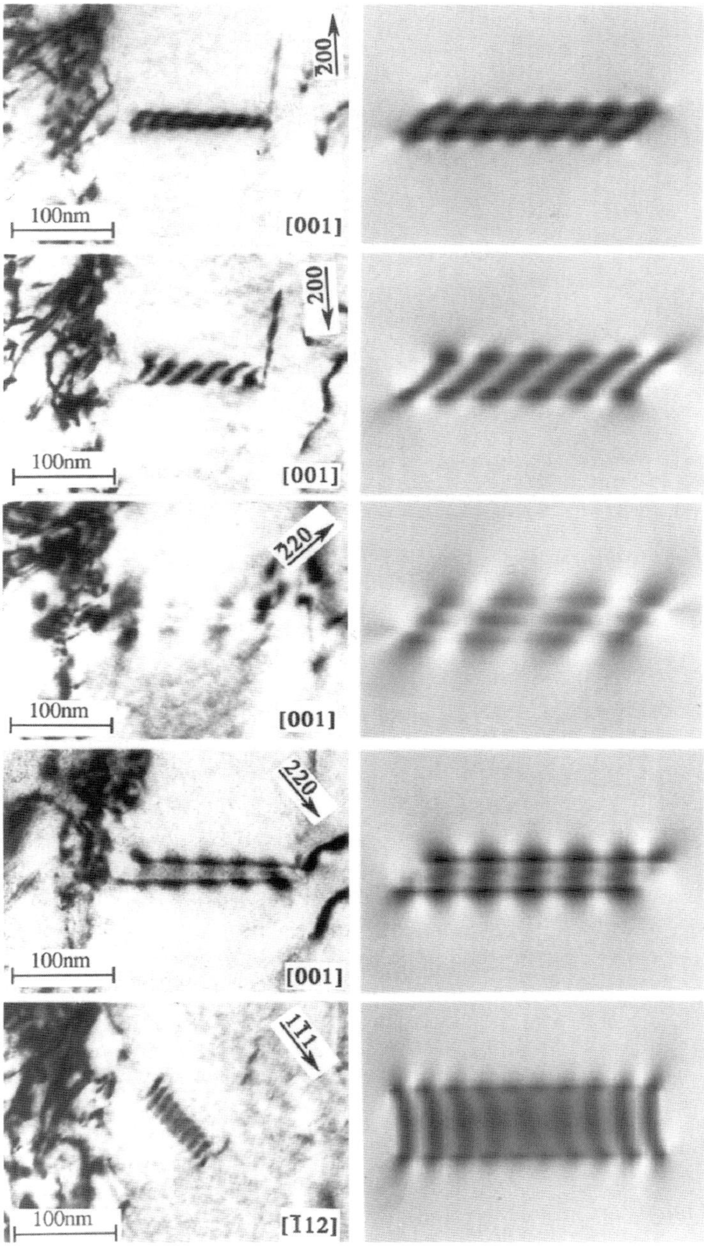

Fig. 44. TEM micrographs of René N5 creep specimen tested to 0.5% creep strain at 982°C (30 hours, ~ 40% of expected life) along with corresponding simulated images showing $\frac{1}{3}[112]$ dislocation dipoles (line direction = [011]) and associated stacking fault on the [11$\bar{1}$] plane within the γ' (line direction = [011]) (longitudinal cross-section)].

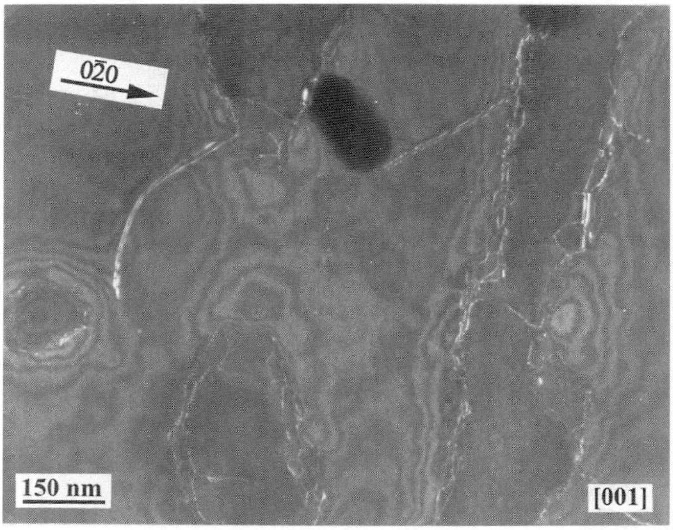

Fig. 45. TEM micrograph (weak beam) of René N5 creep specimen tested to 0.9% creep strain at 982°C (150 hours, ~40% of expected life) showing paired $\frac{1}{2}\langle 110\rangle$ dislocations within the γ' (longitudinal cross-section).

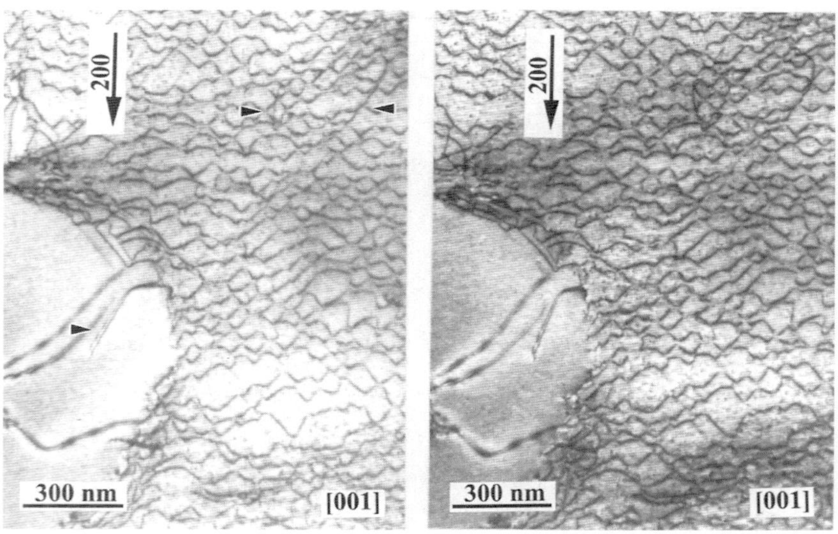

Fig. 46. TEM micrographs (stereo pair) of a René N5 creep specimen which was pre-rafted at 1093°C and subsequently tested at 982°C for 235 h. Note $\frac{1}{2}\langle 110\rangle$ dislocations extending out of the interfacial networks into the γ matrix (upper arrows) and paired $\frac{1}{2}\langle 110\rangle$ dislocations extending into the γ' (lower arrow) (transverse cross-section).

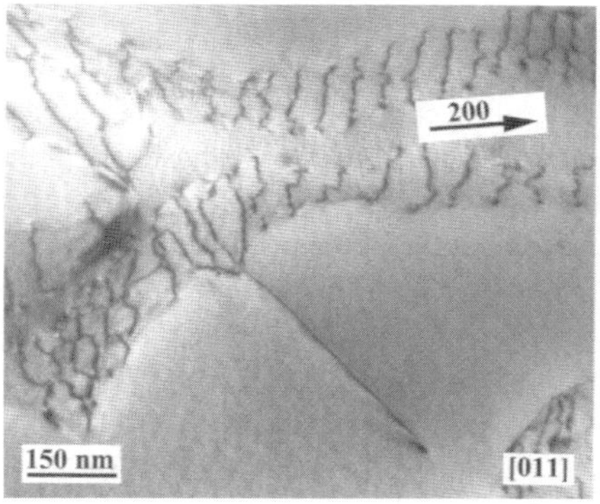

(a)

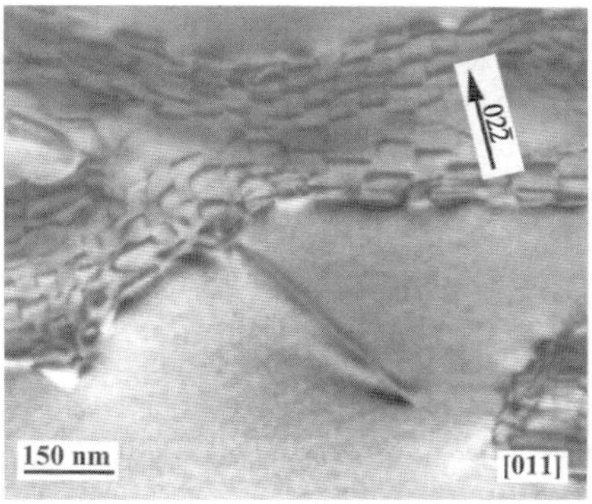

(b)

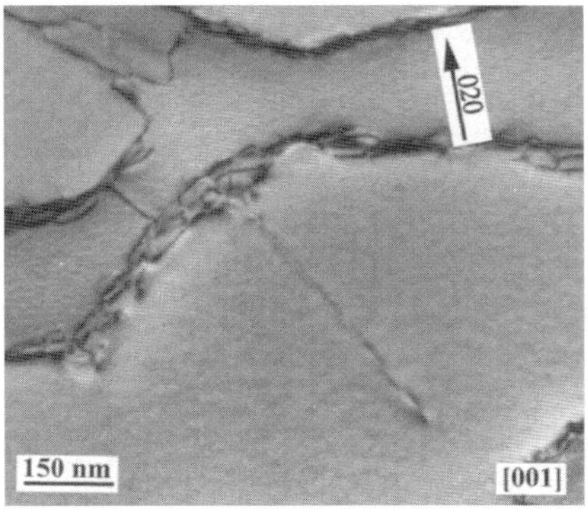

(c)

Fig. 47. TEM micrograph of René N5 creep specimen tested to 0.2% creep strain at 1093°C (40 hours, ~10% of expected life) (20B) showing the reaction of two $\boldsymbol{b} = \frac{1}{2}\langle 101 \rangle$ dislocations in the interfacial net to form a $\boldsymbol{b} = [100]$ dislocation within the γ' (longitudinal cross-section).

($\sim$ 10% expected life). In this case, the reaction $\frac{1}{2}[101] + \frac{1}{2}[10\bar{1}] \rightarrow [100]$ has occurred, with the $\boldsymbol{b} = [100]$ dislocation extending into the γ' phase.[1] Observations such as these demonstrate the knitting and unknitting processes by which dislocations continue to move through the specimen in fully rafted structures. These processes, particularly the unknitting of dislocations out of the networks and into the γ' phase, are believed to be rate controlling in these specimens. These processes probably are similar to those described above for network formation. However, the exact mechanisms and, more importantly, factors which influence their kinetics, are not well understood. As will be discussed in the following section, these mechanisms, and how they are influenced by alloy composition, represent an important and relatively unexplored aspect of high-temperature creep behavior in this class of alloys.

In the latter stages of life, as the creep rate increases, more dislocation activity can be seen within the γ' phase. Figure 48 shows evidence of this activity, resulting in much more dislocation interaction to form low energy structures, including $\langle 100 \rangle$ Burgers vectors, in a René N5 specimen tested to failure at 982°C. Low-energy dislocation structures within the γ' phase of crept specimens have also been observed by other investigators [164].

5.3.3. Influence of rafting on properties

The effects of the lattice mismatch and rafting on creep properties of Ni-base superalloys have long been the subject of debate. Before the advent of single-crystal alloys, numerous investigations of the effects of lattice mismatch on creep strength were conducted [165–172].

In the rafted microstructure, there is no continuous path for dislocations through the matrix, forcing cutting of the γ'. This can be considered as a strengthening mechanism, beneficial to creep resistance. The interfacial dislocation network can also be considered as a barrier to dislocation motion, since dislocations must pass through the networks in order to cut the γ' [89]. Alternatively, these networks can act as a source of dislocations for continued deformation as well as reduce coherency strains which resist dislocation motion, thus increasing the creep rate [117]. Heat treatments designed to enhance γ' alignments have been found to result in finer rafts, with correspondingly better high-temperature creep properties [20]. A similar effect has been noted for higher mismatch alloys [89]. Additionally, rafting has been reported to reduce the anisotropy of creep as a function of crystal orientation [173]. Thus, rafting has been reported to be both beneficial [20,89,95, 143,144] and detrimental [117,146,174] to creep properties. It has further been concluded that rafting can improve properties under conditions of high temperature and low stress, while degrading properties otherwise [175]. However, to date, these conclusions have been drawn based on the study of a single alloy, and often over a relatively narrow range of temperature and stress history. Here we report new results on a systematic study of the properties of "pre-rafted" structures over a relatively wide range of stress. Furthermore,

[1] Recent investigations of specimens tested in shear creep have elaborated on the role of $\boldsymbol{b} = \langle 100 \rangle$ dislocations in γ' cutting (e.g., [162]). These dislocations have been found to be composed of closely spaced $\boldsymbol{b} = \frac{1}{2}\langle 110 \rangle$ dislocations which move by a cooperative slip/climb mechanism within the γ' phase [163]. Although $\boldsymbol{b} = \langle 100 \rangle$ dislocations might be expected to contribute to creep deformation for other orientations, there is no resolved shear stress for this Burgers vector in the [001] orientation; therefore, this slip system will not be considered further here.

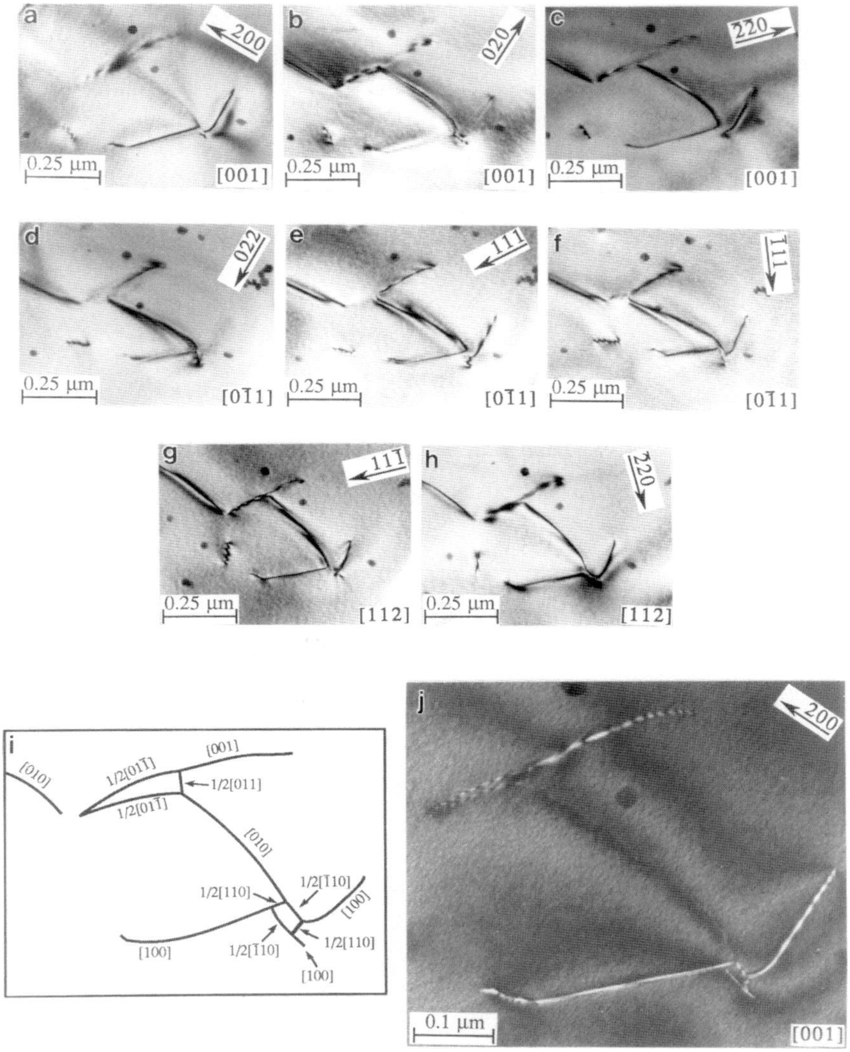

Fig. 48. TEM micrographs and schematic showing dislocation reactions within a γ' precipitate in an René N5 specimens tested to failure at 982°C.

the properties of rafted microstructures have been studied with changes in precipitate composition.

The study of the effects of pre-rafting on creep at 982°C has been conducted for two alloys: René N4 and René N5. One of the major differences in these two single-crystal compositions is the 3 wt% Re in René N5. Also, René N4 contains higher levels of γ' solid solution strengtheners, particularly Ti (see table 1). For these experiments, a pre-rafting treatment consisting of 1093°C for 10 and 20 hours for N4 and N5, respectively

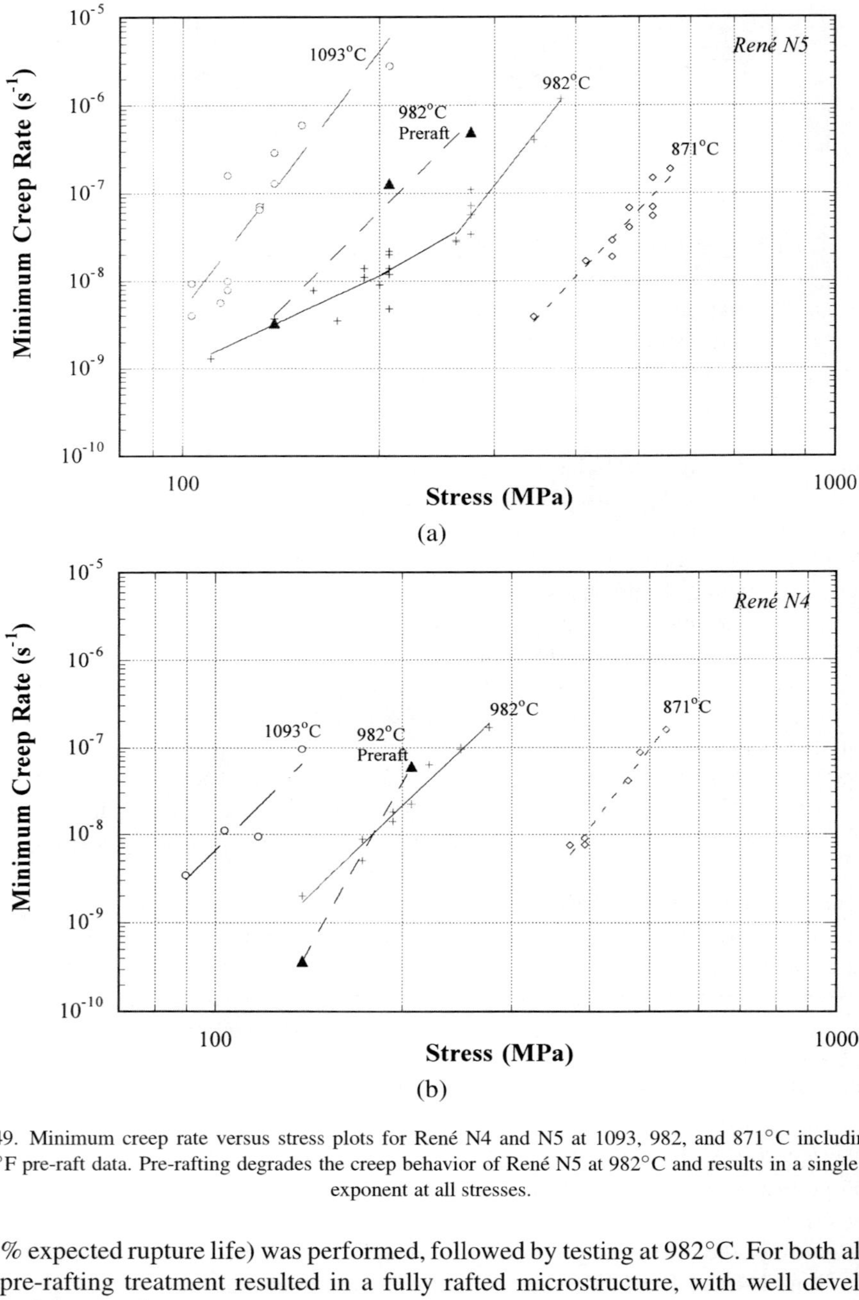

Fig. 49. Minimum creep rate versus stress plots for René N4 and N5 at 1093, 982, and 871°C including the 1800°F pre-raft data. Pre-rafting degrades the creep behavior of René N5 at 982°C and results in a single stress exponent at all stresses.

(~ 5% expected rupture life) was performed, followed by testing at 982°C. For both alloys, this pre-rafting treatment resulted in a fully rafted microstructure, with well developed interfacial networks.

In fig. 49, minimum creep rate data from the pre-rafted specimens has been plotted along with data for the as-heat treated specimens. For the as-heat treated specimens, René

Table 9
Compositions of modified René N5 alloys (wt%)

Alloy	Co	Cr	Mo	W	Ta	Al	Re	Ti	Hf
N5	7.5	7.0	1.5	5.0	6.5	6.2	3.0	0.0	0.15
N5-2	7.3	6.8	1.5	4.8	9.4	5.6	3.0	0.0	0.15
N5-4	7.4	7.0	1.5	4.9	6.6	5.6	3.0	0.8	0.15

N5 displays two regimes in the data at 982°C, with a high stress exponent at high stresses and a lower stress exponent at low stresses. The effect of pre-rafting in N5 is to increase the minimum creep rate and decrease the rupture life, particularly at high stresses. The pre-raft data can be fitted to a straight line which approaches the as-heat treated data at low stresses. N4 is much less affected by the pre-rafting treatment, and at very low stresses the creep rate of the pre-rafted specimens is lower than that of the fully heat treated material. This is very interesting, as the width of the horizontal channels increases as the rafts form, reducing the Orowan stress for glide through these channels. Thus, it is clear that the precipitates more strongly influence the behavior of the material as it evolves to the rafted form.

Several observations can be made from these data. First, the rafted morphology is detrimental to creep behavior of N5 at 982°C. Second, since the pre-rafted data fall on a straight line with a single stress exponent and approach the as-heat treated data at low stresses, where rafting occurs early in life for N5, the change in mechanism, implied by the stress exponent change with stress in the as-heat treated data, appears to be associated with the change to a rafted morphology during creep. Finally, the creep stress exponent in the lower stress (rafted) range in the René N5 is in the same range as the creep stress exponents of single phase Ni_3Al alloys, section 3.2.

A series of alloys with modifications of the N5 composition were made by substituting γ' strengtheners (Ti and Ta) for Al. A list of the compositions is given in table 9. The results from this series of alloys are summarized in fig. 50. Due to the limited number of data points for each alloy, the temptation to over-analyze the data must be resisted. This is particularly true when considering such quantities as the stress exponent with only three data points. However, several observations may be made from the data. The value of the stress exponent in the low-stress region appears to have increased and the low-stress values have been particularly improved. This is accompanied by approximately a $2\times$ increase in rupture life. The degradation of minimum creep rate caused by pre-rafting was also considerably reduced for these alloys.

The effects of rafting on creep properties of René N4 and N5 at 982°C are believed to be associated with the balance of properties for the matrix and precipitate phases in this alloy. When the γ' maintains a cuboidal morphology, as is the case in low-temperature creep and high-stress 982°C testing of N5, the dislocations are restricted to the thin γ channels between the γ' precipitates. Under these circumstances, the high solid solution strengthener content of the γ matrix in N5 (particularly Re which partitions almost exclusively to the γ, as shown in table 10) affords considerable creep resistance. As the γ' rafts, dislocations are forced to either climb around the γ' precipitates or pass through them. If the latter occurs, the strength of the γ' phase plays a much larger role in the overall creep strength of the alloy. These results indicate that the balance of properties between the

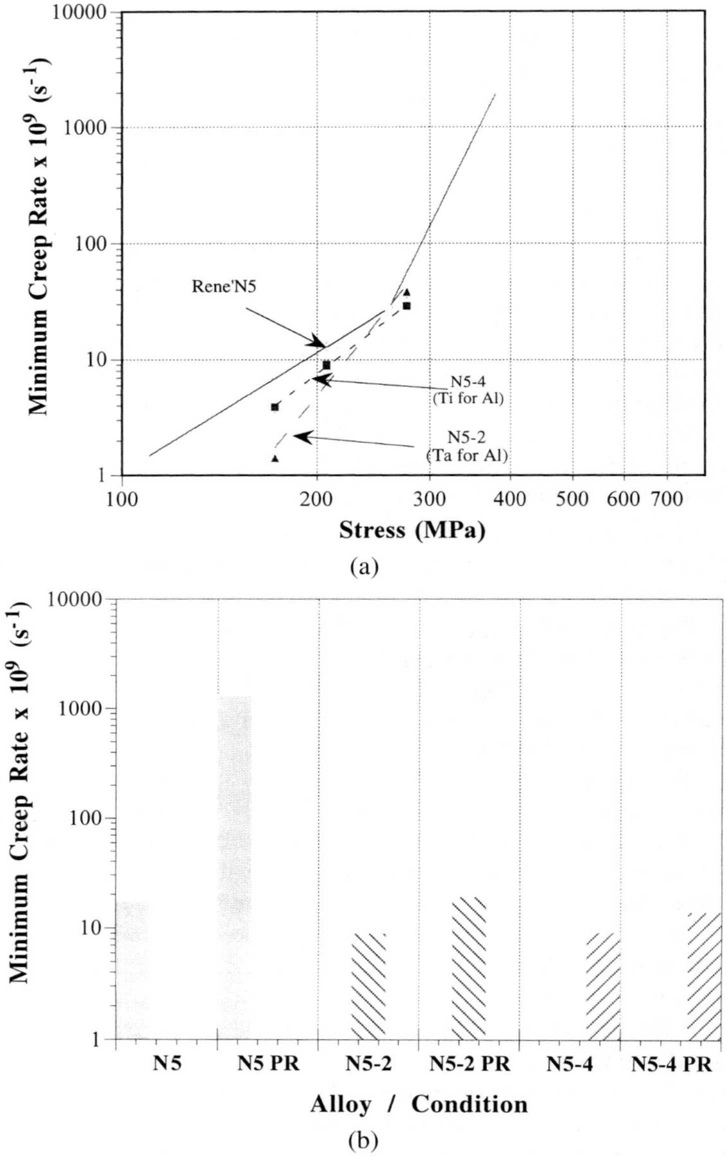

Fig. 50. Minimum creep rate data for René N5 composition modifications: (a) minimum creep rate versus stress plot; (b) bar graph showing the effect of pre-rafting treatment.

γ and γ' phases should be an important consideration in the design of advanced Ni-base superalloys.

In order to cut the γ' phase, dislocations must pass through the interfacial nets. Indeed, the nets are expected to act as a source for dislocations in the rafted structure, and the

Table 10
Phase compositions in René N4 and N5 (wt%)

Phase	Ni	Co	Cr	W	Al	Ta	Re	Hf	Mo	Ti
N4-γ'	72.7	4.87	2.30	3.42	5.86	4.47	na	0.3	0.42	5.07
N4-γ	51.4	11.50	21.50	9.05	1.56	0.85	na	nd	3.18	0.71
N5-γ'	72.8	4.52	2.42	4.0	7.47	7.26	0.53	0.2	0.78	na
N5-γ	54.6	12.1	15.0	5.27	2.34	0.39	7.62	nd	2.73	na

na – not applicable, nd – not detected.

reduction in creep strength of N5 after pre-rafting may be partially understood in terms of the increase in available dislocations for deformation. Differences in the ease of γ' cutting might be expected to be manifested in the structure of these nets for N4 and N5. Exhaustive studies of these nets in pre-rafted and subsequently tested specimens of both alloys revealed no discernible differences. It appears that the operant mechanism by which the dislocations enter and exit the interfacial networks occurs rapidly and, therefore, is not observed in the interrupted creep specimens.

At 1093°C both N4 and N5 raft early in life. Based on the argument presented above, N4 would be expected to display some advantage over N5 at this temperature. This is not the case, however, as N5 is significantly stronger than N4 at 1093°C. This may be due to the fact that diffusion processes, such as dislocation climb, are more dominant at this temperature. The higher rhenium content of N5 provides solid solution strengthening as well as reduced diffusion kinetics, slowing dislocation climb.

The results from the modified René N5 composition alloys are generally consistent with the "balanced γ–γ' properties" hypothesis. The additions of Ta and Ti increase the low-stress rupture strength at 982°C, where rafting occurs during testing, and ameliorates the degradation caused by pre-rafting. Thus, the strengthening of the γ' phase by the addition of either Ti or Ta appears to have a pronounced positive effect on the creep and rupture properties. Considerably more study of the effects of alloying on dislocation kinetics within the interfacial networks and dislocation cutting through them will be required to fully understand these effects.

6. *Concluding remarks*

Nickel-base superalloys remain unsurpassed by any other system for their ability to tolerate substantial stresses at temperatures up to 90% of melting. This is possible due to the high volume fraction of ordered precipitates that are highly resistant to shearing. A wide variety of deformation phenomena occur in these constrained two-phase systems, and many aspects of the deformation are unique to the two-phase materials, and not directly related to deformation mechanisms in their single-phase ordered or disordered counterparts. Detailed studies by many investigators on the influence of the compositions of the γ and γ' phases, precipitate shape and misfit and mechanisms of precipitate shearing have permitted continued improvements in the properties of superalloy single crystals. However, a number of aspects of monotonic as well as cyclic deformation in these materials are not yet well understood in terms of dislocation mechanisms. Since future systems that replace superalloys in applications such as the aircraft engine will undoubtedly possess many of the same deformation characteristics of these two-phase systems, further study of mechanisms of deformation in single crystal superalloys is likely to have broad impact.

Acknowledgments

The authors would like to acknowledge previous collaborators who have made important contributions to this work, and in particular W.H. Murphy, A.S. Argon, M. Fahrmann, E. Fahrmann, W.S. Walston, K.S. O'Hara, E.W. Ross. The authors are grateful for the support of the General Electric Company and one of the authors (TMP) acknowledges the support of the National Science Foundation, Grant DMR925287.

References

[1] G.I. Taylor, Proc. Roy. Soc. London A 145 (1934) 362.
[2] E. Orowan, Z. Phys. 89 (1934) 605.
[3] R.F. Decker and C.T. Sims, in: The Superalloys, eds C.T. Sims and W.C. Hagel (Wiley, New York, 1972) p. 33.
[4] E.W. Ross and C.T. Sims, in: Superalloys II, eds C.T. Sims, N.S. Stoloff and W.C. Hagel (Wiley, New York, 1987) p. 97.
[5] E.W. Ross and K.S. O'Hara, in: Superalloys 1996, eds R.D. Kissinger et al. (TMS, Warrendale, PA, 1996) p. 19.
[6] M. Gell, D.N. Duhl and A.F. Giamei, in: Superalloys 1980, eds J.K. Tien et al. (ASM, Metals Park, OH, 1980) p. 205.
[7] K. Harris, G.L. Erickson and R.E. Schwer, in: High Temperature Alloys for Gas Turbines and Other Applications, eds W. Betz et al. (D. Reidel, Dordrecht, 1986) p. 709.
[8] K. Harris, G.L. Erickson and R.E. Schwer, in: Superalloys 1984, eds M. Gell et al. (TMS, Warrendale, PA, 1984) p. 227.
[9] T.E. Strangman, G.S. Hoppin, C.M. Phillips, K. Harris and R.W. Schwer, in: Superalloys 1980, eds J.K. Tien et al. (ASM, Metals Park, OH, 1980) p. 215.
[10] T. Khan and P. Caron, in: Advances in High Temperature Structural Materials and Protective Coatings, eds A.K. Koul et al. (NRC, Ottawa, Canada, 1994).
[11] D.A. Ford and R.P. Arthey, in: Superalloys 1984, eds M. Gell et al. (TMS, Warrendale, PA, 1984) p. 115.
[12] T. Khan and P. Caron, High Temperature Materials for Power Engineering 1990 (Kluwer Acad. Publ., Dordrecht, 1990) p. 1261.
[13] M. Doner and J. Heckler, Presentation at Aero. Tech. Conf., Long Beach, CA (1985).
[14] C.S. Wukusick and L. Buchakjian, U.K. Patent GB2235697 (1991).
[15] A.D. Cetel and D.N. Duhl, in: Superalloys 1988, eds D.N. Duhl et al. (TMS, Warrendale, PA, 1988) p. 235.
[16] K. Harris, G.L. Erickson, S.L. Sikkenga, W.D. Brentnall, J.M. Aurrecoechea and K.G. Kubarych, in: Superalloys 1992, eds S.D. Antolovich et al. (TMS, Warrendale, PA, 1992) p. 297.
[17] W.S. Walston, K.S. O'Hara, E.W. Ross, T.M. Pollock and W.H. Murphy, in: Superalloys 1996, eds R.D. Kissinger et al. (TMS, Warrendale, PA, 1996) p. 27.
[18] G.L. Erickson, in: Superalloys 1996, eds R.D. Kissinger et al. (TMS, Warrendale, PA, 1996) p. 35.
[19] T.M. Pollock and W.H. Murphy, Metall. Mater. Trans. A 27 (1996) 1081.
[20] P. Caron and T. Khan, Mater. Sci. Eng. A 61 (1983) 173.
[21] M.V. Nathal and L.J. Ebert, Metall. Trans. A 16 (1985) 1849.
[22] T.M. Pollock, Mater. Sci. Eng. B 32 (1995) 255.
[23] J. Weertman and P. Shahinian, Trans. Metall. Soc. AIME 206 (1956) 1223.

[24] P.R. Landon, J.L. Lytton, L.A. Shephard and J.E. Dorn, Trans. Am. Soc. Metals 51 (1959) 900.
[25] C.M. Sellars and A.G. Quarrell, J. Inst. Metals 90 (1961–1962) 329.
[26] V.M. Rozenberg, Fiz. Metal. Metalloved 14 (1) (1962) 114.
[27] P. Dennison, R.J. Llewellyn and B. Wilshire, J. Inst. Metals 94 (1966) 130.
[28] S. Karashima, H. Oikawa and T. Motomiya, Trans. Jpn. Inst. Metals 28 (1964) 253.
[29] E.G. Norman and S.A. Duran, Acta Metall. 18 (1970) 723.
[30] A.M. Brown and M.F. Ashby, Scripta Metall. 14 (1980) 1297.
[31] A.S. Argon and A.K. Bhattacharaya, Acta Metall. 35 (1987) 1499.
[32] A.K. Mukherjee, J.E. Bird and J.E. Dorn, Trans. ASM 62 (1969) 155.
[33] H. Frost and M.F. Ashby, Deformation Mechanism Maps (Pergamon, 1980).
[34] R.E. Hoffman, F.W. Pikus and R.A. Ward, Trans. AIME 483 (1956).
[35] S. Takeuchi and A.S. Argon, J. Mater. Sci. 11 (1976) 1542.
[36] F. Prinz and A.S. Argon, Acta Metall. 32 (1984) 7.
[37] W.D. Nix, J.C. Gibeling and D.A. Hughes, Metall. Trans. A 16 (1985) 2215.
[38] W.R. Johnson, C.R. Barrett and W.D. Nix, Metall. Trans. A 3 (1972) 963.
[39] K. Monma, H. Suto and H. Oikawa, J. Jpn. Inst. Metals 28 (1964) 253.
[40] R.A. Swalin and A. Martin, Trans. AIME 206 (1956) 567.
[41] R.A. Swalin, A. Martin and R. Oison, Trans. AIME 207 (1957) 936.
[42] R.C. Ruder and C.E. Birchenall, Trans. AIME 191 (1951) 142.
[43] K. Monma, H. Suto and H. Oikawa, J. Jpn. Inst. Metals 28 (1964) 188.
[44] A.S. Argon and W.C. Moffatt, Acta Metall. 29 (1981) 293.
[45] F. Prinz, A.S. Argon and W.C. Moffatt, Acta Metall. 30 (1982) 821.
[46] P.S. Kotval and O.H. Nestor, Trans. AIME 245 (1969) 1275.
[47] B.E. Beeston and L.K. France, J. Inst. Metals 96 (1968) 105.
[48] N. Tierny and N.J. Grant, Metall. Trans. A 13 (1982) 1827.
[49] M.S.A. Karunaratne, D.C. Cox, P. Carter and R.C. Reed, in: Superalloys 2000, eds T.M. Pollock et al. (TMS, Warrendale, PA, 2000) p. 263.
[50] M.V. Nathal, J.O. Diaz and R.V. Miner, in: High Temperature Ordered Intermetallics III, Mat. Res. Soc. Symp. 133, 269, eds C.T. Liu et al. (MRS, Pittsburgh, PA, 1987) p. 269.
[51] D.P. Pope and S.S. Ezz, Intl. Metals Rev. 29 (1984) 136.
[52] C.T. Liu and D.P. Pope, in: Intermetallic Compounds: Principles and Practice, eds J.H. Westbrook and R.L. Fleischer (Wiley, New York, 1995) p. 2, 17.
[53] F.R.N. Nabarro and H.L. de Villiers, The Physics of Creep (Taylor and Francis, London, 1995) p. 103.
[54] F.R.N. Nabarro (ed.), Dislocations in Solids, Vol. 10 (Elsevier, Amsterdam, 1997).
[55] K.J. Hemker, M.J. Mills and W.D. Nix, Acta Metall. 39 (1991) 1901.
[56] D.L. Anton, D.D. Peason and D.B. Snow, in: High Temp. Ordered Intermetallics, MRS Symp. Proc, eds N.S. Stoloff et al. (MRS, Pittsburgh, PA, 1987) p. 81, 287.
[57] B.H. Kear and H.G.F. Wilsdorf, Trans. Met. Soc. AIME 224 (1962) 382.
[58] P.M. Hazzledine and J.H. Schneibel, Scripta Metall. 23 (1989) 1887 .
[59] Y.Q. Sun, P.M. Hazzledine, M.A. Crimp, A. Couret, Phil. Mag. A 64 (1991) 311.
[60] D.M. Shah, Scripta Metall. 17 (1983) 997.
[61] K.J. Hemker and W.D. Nix, Metall. Trans. A 24 (1993) 335.
[62] C.P. Flinn, Trans. Met. Soc. AIME 218 (1960) 145.
[63] R.K. Ham, R.H. Cook, G.R. Purdy and G. Willoughby, Metal Sci. J. 6 (1972) 205.
[64] M.V. Nathal and L.J. Ebert, Metall. Trans. A 16 (1985) 427.
[65] G.R. Leverant and B.H. Kear, Metall. Trans. A 1 (1970) 491.
[66] C. Carry and J.L. Strudel, Acta Metall. 26 (1978) 859.
[67] K. Yu, PhD thesis (MIT, 1985).
[68] M.V. Nathal and L.J. Ebert, Scripta Metall. 17 (1985) 1151.
[69] T.M. Pollock and A.S. Argon, Acta Metall. 40 (1992) 1.
[70] W. Schneider, J. Hammer and H. Mughrabi, in: Superalloys 1992, eds S.D. Antolovich et al. (TMS, Warrendale, PA, 1992) p. 589.
[71] A.J. Ardell, R.B. Nicholson and J.D. Eshelby, Acta Metall. 14 (1966) 1295.

[72] D.J. Chellman and A.J. Ardell, Acta Metall. 22 (1974) 577.
[73] C.K. Davies, P. Nash and R.N. Stevens, J. Mater. Sci. 15 (1980) 1521.
[74] A.F. Giamei and D.L. Anton, Metall. Trans. A 16 (1985) 1997.
[75] R.A. Ricks, A.J. Porter and R.C. Ecob, Acta Metall. 31 (1983) 43.
[76] M. Fahrmann, P. Fratzl, O. Paris, E. Fahrmann and W.C. Johnson, Acta Metall. Mater. 43 (1995) 1007.
[77] D.A. Grose and G.S. Ansell, Metall. Trans. A 12 (1981) 1631.
[78] M.V. Nathal, R.A. MacKay and R.G. Garlick, Mater. Sci. Eng. A 75 (1985) 195.
[79] H.-A. Kuhn, H. Biermann, T. Ungar and H. Mughrabi, Acta. Metall. 39 (1991) 2783.
[80] H. Mughrabi, H. Biermann and T. Ungar, in: Superalloys 1992, eds S.D. Antolovich et al. (TMS, Warrendale, PA, 1992).
[81] H. Biermann, M. Strehler and H. Mughrabi, Metall. Trans. A 27 (1996) 1003.
[82] D.F. Lahrman, R.D. Field, R. Darolia and H.L. Fraser, Acta Metall. 36 (1988) 1309.
[83] M. Fahrmann, J.G. Wolf and T.M. Pollock, Mater. Sci. Eng. A 210 (1996) 8.
[84] U. Glatzel, Script. Metall. Mater. 31 (1994) 291.
[85] A. Royer, P. Bastie and M. Veron, Acta Mater. 46 (1998) 5357.
[86] R.C. Ecob, M.P. Shaw, A.J. Porter and B. Ralph, Phil. Mag. 44 (1981) 1117.
[87] A.J. Porter, M.P. Shaw, R.C. Ecob and B. Ralph, Phil. Mag. 44 (1981) 1135.
[88] R.C. Ecob, R.A. Ricks and A.J. Porter, Scripta Metall. 16 (1982) 1085.
[89] T.P. Gabb, S.L. Draper, D.R. Hull, R.A. MacKay and M.V. Nathal, Mater. Sci. Eng. A 118 (1989) 59.
[90] R.R. Keller, H.J. Maier and H. Mughrabi, Scripta Metall. 28 (1993) 23.
[91] T.M. Pollock and A.S. Argon, in: Superalloys 1988, eds D.N. Duhl et al. (TMS, Warrendale, PA, 1988) p. 285.
[92] T.M. Pollock and A.S. Argon, Acta Metall. Mater. 42 (1994) 1859.
[93] U. Glatzel and M. Feller-Kniepmeier, Scripta Metall. 23 (1989) 1839.
[94] J.F. Ganghoffer, A. Hazotte, S. Denis and A. Simon, Scripta Metall. Mater. 25 (1991) 2491.
[95] L. Muller, U. Glatzel and M. Feller-Kniepmeier, Acta Metall. Mater. 40 (1992) 1331.
[96] H. Feng, H. Biermann and H. Mughrabi, Mater. Sci. Eng. A 214 (1996) 1.
[97] B.H. Kear, G.R. Leverant and J.M. Oblak, Trans. ASM 62 (1969) 639.
[98] B.H. Kear, A.F. Giamei, G.R. Leverant and J.M. Oblak, Scripta Metall. 3 (1969) 123.
[99] B.H. Kear, A.F. Giamei, G.R. Leverant and J.M. Oblak, Scripta Metall. 3 (1969) 455.
[100] G.R. Leverant, B.H. Kear and J.M. Oblak, Metall. Trans. 2 (1971) 2305.
[101] G.R. Leverant and D.N. Duhl, Metall. Trans. 2 (1971) 907.
[102] A.J. Huis in't Veld, G. Boom, P.M. Bronsveld and J.Th.M. De Hosson, Scripta Metall. 19 (1985) 1123.
[103] M. Condat and B. Decamps, Scripta Metall. 21 (1987) 607.
[104] P. Caron, T. Kahn and P. Veyssière, Phil. Mag. A 57 (1988) 859.
[105] J. Courbon, F. Louchet, M. Ignat, J. Pelissier and P. Debrenne, Phil. Mag. Lett. 63 (1991) 73.
[106] D.L. Lin and M. Wen, Acta Metall. 37 (1989) 3099.
[107] P.R. Bhowal, E.F. Wright and E.L. Raymond, Metall. Trans. 21 (1990) 1709.
[108] T. Link, Scripta Metall. Mater. 31 (1994) 671.
[109] Y.Q. Sun, M.A. Crimp and P.M. Hazzledine, Phil. Mag. A 64 (1991) 223.
[110] Y.Q. Sun and P.M. Hazzledine, Phil. Mag. A 58 (1988) 603.
[111] R.A. MacKay, R.L. Dreshfield and R.D. Maier, in: Superalloys 1980, eds J.K. Tien et al. (ASM, Metals Park, OH, 1980) p. 385.
[112] R.A. MacKay and R.D. Maier, Metall. Trans. A 13 (1982) 1747.
[113] P. Caron, Y. Ohta, Y.G. Nakagawa and T. Khan, in: Superalloys 1988, eds D.N. Duhl et al. (TMS, Warrendale, PA, 1988) p. 215.
[114] K. Kakehi, Mat. Sci. Eng. A 278 (2000) 135.
[115] S.S.K. Gunturi, D.W. MacLachlan and D.M. Knowles, Mat. Sci. Eng. A 289 (2000) 289.
[116] P. Caron and T. Khan, ICSMA 7 (1985).
[117] M.V. Nathal, R.A. MacKay and R.V. Miner, Metall. Trans. A 20 (1989) 133.
[118] H. Mughrabi, W. Schneider, V. Sass and C. Lang, in: ICSMA 10, eds H. Oikawa et al. (Japan Institute of Metals, 1994) p. 705.
[119] T. Khan, P. Caron and C. Duret, in: Superalloys 1984, eds M. Gell et al. (TMS, Warrendale, PA, 1984) p. 145.

[120] G.S. Hillier, C.M.F. Rae and H.K.D.H. Bhadeshia, Acta Metall. 36 (1988) 95.
[121] A.F. Giamei, J.M. Oblak, B.H. Kear and W.H. Rand, in: Proc. 29th Annual Meeting EMSA (Claitor's Pub., Baton Rouge, 1971) p. 112.
[122] C.M.F. Rae, N. Matan, D.C. Cox, M.A. Rist and R.C. Reed, Metall. Trans. A 31 (2000) 2219.
[123] C.M.F. Rae, N. Matan and R.C. Reed, Mater. Sci. Eng. A 300 (2001) 125.
[124] M.V. Nathal and L.J. Ebert, Metall. Trans. A 16 (1985) 1863.
[125] G.A. Webster and B.J. Piearcy, Metal. Sci. J. 1 (1967) 97.
[126] C. Carry and J.L. Strudel, Acta Metall. 25 (1977) 767.
[127] C. Carry, S. Dermarkar, J.L. Strudel and B.C. Wonsiewicz, Metall. Trans. A 10 (1979) 855.
[128] H. Mughrabi, H. Feng and H. Biermann, in: Micromechanics of Plasticity and Damage of Multiphase Materials, eds A. Pineau and A. Zaoui (Kluwer Acad. Publ., 1996) p. 115.
[129] M.V. Nathal, Metall. Trans. A 18 (1987) 1961.
[130] T. Khan, in: High Temperature Alloys for Gas Turbines and Other Applications, eds W. Betz et al. (Riedel, Dordrecht, 1986) p. 21.
[131] M. Feller-Kniepmeier and T. Link, Metall. Trans. A 20 (1989) 1233.
[132] V. Sass, U. Glatzel and M. Feller-Kniepmeier, in: Superalloys 1996, eds R.D. Kissinger et al. (TMS, Warrendale, PA, 1996) p. 283.
[133] M. Feller-Kniepmeier and T. Link, Mater. Sci. Eng. A 113 (1989) 191.
[134] G.R. Leverant, B.H. Kear and J.M. Oblak, Metall. Trans. 4 (1973) 355.
[135] E. Nembach and G. Neite, Prog. Mater. Sci. 29 (1985) 177.
[136] W. Schroeder and D.A. Webster, J. Appl. Mech. 16 (1949) 289.
[137] A. Lasalmonie and J.L. Strudel, Phil. Mag. 32 (1975) 937.
[138] R.D. Field, T.M. Pollock and W.H. Murphy, in: Superalloys 1992, eds S.D. Antolovich et al. (TMS, Warrendale, PA, 1992) p. 557.
[139] R.R. Keller, H.J. Maier and H. Mughrabi, Scripta Metall. 28 (1993) 23.
[140] A.J. Ardell, R.B. Nicholson and J.D. Eshelby, Acta Metall. 14 (1966) 1295.
[141] J.K. Tien and S.M. Copley, Metall. Trans. 2 (1971) 215.
[142] J.K. Tien and S.M. Copley, Metall. Trans. 2 (1971) 543.
[143] D.D. Pearson, F.D. Lemkey and B.H. Kear, in: Superalloys 1980, eds J.K. Tien et al., eds (ASM, Metals Park, OH, 1980) p. 513.
[144] D.D. Pearson, B.H. Kear and F.D. Lemkey, in: Creep and Fracture of Engineering Materials and Structures, eds B. Wilshire and D.R.J. Owen (Pineridge Press, Swansea, U.K., 1981) p. 213.
[145] M.V. Nathal and L.J. Ebert, Scripta Metall. 17 (1983) 1151.
[146] R.A. MacKay and L.J. Ebert, Scripta Metall. 17 (1983) 1217.
[147] A.K. Singh, N. Louat and K. Sadananda, Metall. Trans. A 19 (1988) 2965.
[148] M. Kolbe, A. Dlouhy and G. Eggeler, Mater. Sci. Eng. A 246 (1998) 133.
[149] C. Mayr, G. Eggeler and A. Dloughy, Mater. Sci. Eng. A 207 (1996) 51.
[150] M. Miura, Y. Kondo and N. Ohi, in: Superalloys 2000, eds T.M. Pollock et al. (TMS, Warrendale, PA, 2000) p. 377.
[151] R.R. Keller, H.J. Maier, H. Renner and H. Mughrabi, Scripta Metall. 27 (1992) 1167.
[152] H. Mughrabi, H. Biermann and T. Ungar, in: Superalloys 1992, eds S.D. Antolovich et al. (TMS, Warrendale, PA, 1992) p. 599.
[153] H. Biermann, M. Strehler and H. Mughrabi, Script. Metall. Mater. 32 (1995) 1405.
[154] M. Véron, Y. Brechet and F. Louchet, Acta Mater. 44 (1996) 3633.
[155] M. Fahrmann, E. Fahrmann, O. Paris, P. Fratzl and T.M. Pollock, in: Superalloys 1996, eds R.D. Kissinger et al. (TMS, Warrendale, PA, 1996) p. 191.
[156] O. Paris, M. Fahrmann, E. Fahrmann, T.M. Pollock and P. Fratzl, Acta Mater. 45 (1997) 1085.
[157] S. Socrate and D.M. Parks, Acta Metall. 41 (1993) 2185.
[158] H. Feng, H. Biermann and H. Mughrabi, Metall. Mater. Trans. A 31 (2000) 585.
[159] J. Svodboda and P. Lukas, Acta Mater. 44 (1996) 2557.
[160] A.K. Head, P. Humble, P. Clarebrough, L.M. Morton, and C.T. Forwood, in: Computed Electron Micrographs and Defect Identification (North-Holland, New York, 1976).
[161] W.W. Milligan and S.D. Antolovich, Metall. Trans. A 20 (1989) 1888.

[162] G. Eggeler and A. Dlouhy, Acta Mater. 45 (1997) 4251.
[163] R. Srinivasan, G.F. Eggeler and M.J. Mills, Acta Mater. 48 (2000) 4867.
[164] F. Louchat and M. Ignat, Acta Metall. 34 (1986) 1681.
[165] R. Nordheim and N.J. Grant, J. Metals 6 (1954) 211.
[166] J.R. Mihalisin and R.F. Decker, Trans. Met. Soc. AIME 218 (1960) 507.
[167] P.K. Pitler and G.S. Ansell, Trans. ASM 57 (1964) 220.
[168] C.M. Hammond and G.S. Ansell, Trans. ASM 57 (1964) 727.
[169] W.C. Bigelow, J.A. Amy, C.L. Corey and J.W. Freeman, ASTM STP 245 (1958) 73.
[170] G.N. Maniar, J.E. Bridge, Jr., H.M. James and G.B. Heydt, Metall. Trans. 1 (1970) 31.
[171] G.N. Maniar and J.E. Bridge, Jr., Metall. Trans. 2 (1971) 95.
[172] D.A. Grose and G.S. Ansell, Metall. Trans. A 12 (1981) 1631.
[173] V. Sass and M. Feller-Kniepmeier, Mater. Sci. Eng. A 245 (1998) 19.
[174] W. Schneider, J. Hammer and H. Mughrabi, in: Superalloys 1992, eds S.D. Antolovich et al. (TMS, Warrendale, PA, 1992) p. 589.
[175] H. Mughrabi, in: Proc. J. Weertman Symp., eds R.J. Arsenault et al. (TMS, Warrendale, PA, 1996).

Author Index

Roman numbers refer to pages on which the author (or his/her work) is mentioned. Italic numbers refer to reference pages. Numbers between brackets are the reference numbers. No distinction is made between first and co-author(s).